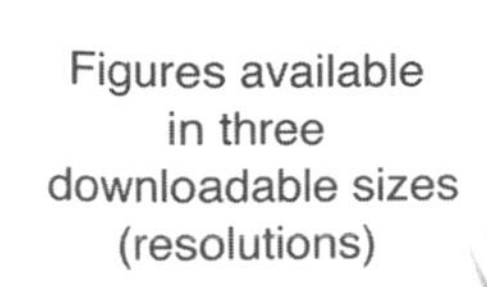

STRUCTURE AND MECHANISM OF Na,K-ATPASE: Functional Sites and Their Inter...

Figure 4 (*A*) Ball-and-stick representation of Na^+-binding sites I and II; homology model of Na,K-pump side chains on the backbone of the high-resolution structure of SR Ca-ATPase (1EUL) (16). (*B*) Proposed mechanism of Na^+ translocation. The entrance of the third Na^+ from the cytoplasmic site and the release of the first ion on the extracellular side are electrogenic owing to a bucket brigade mechanism. Binding of the third Na^+ will displace the second Na^+ to site I. The E_1-P-E_2-P transition displaces the third Na^+ ion from the entry site to site I and expels the second Na^+ ion (K.O. Håkansson & P.L. Jorgensen, submitted). (*C*) Probing for channels through the model of Na,K-ATPase structure with the HOLE software package (K.O. Håkansson & P.L. Jorgensen, submitted).

View larger version (54K)

Figure 5 A homology model of Na,K-ATPase with transmembrane helices seen from the extracellular side. Note residues highlighted in extracellular loops, particularly L9/10.

View larger version (80K)

The transmembrane regions of P-type ATPases differ from those of known ion channels by the absence of an obvious transport path in the form of an open water-filled channel. Presumably, this reflects the difference between passive and active transport mechanisms and the necessity of occluding the cations in the ion pumps. Probing the structure with the Hole software (61) in Figure 4c shows that a channel with dimensions suitable for passage of Na^+

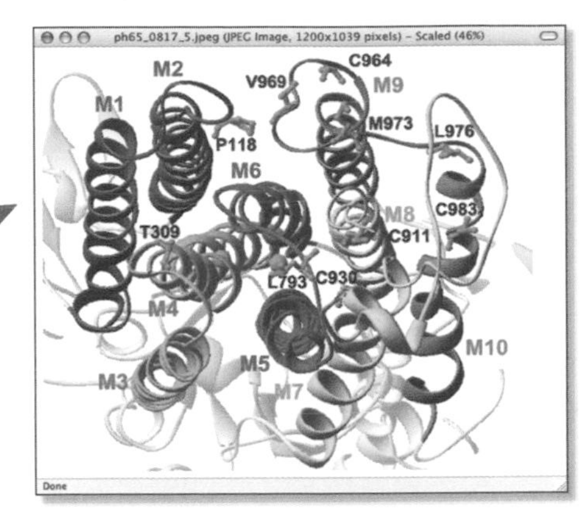

65(1):817 -

61. Smart OS, Neduvelil JG, Wang X, Wallace BA, Sansom MS. 1996. HOLE: a program for the analysis of the pore dimensions of ion channel structural models. *J. Mol. Graph.* 14:354–60 [CrossRef] [Medline]

Citations in text link to references in bibliography

References in Annual Reviews chapter bibliography link out to sources of cited articles online

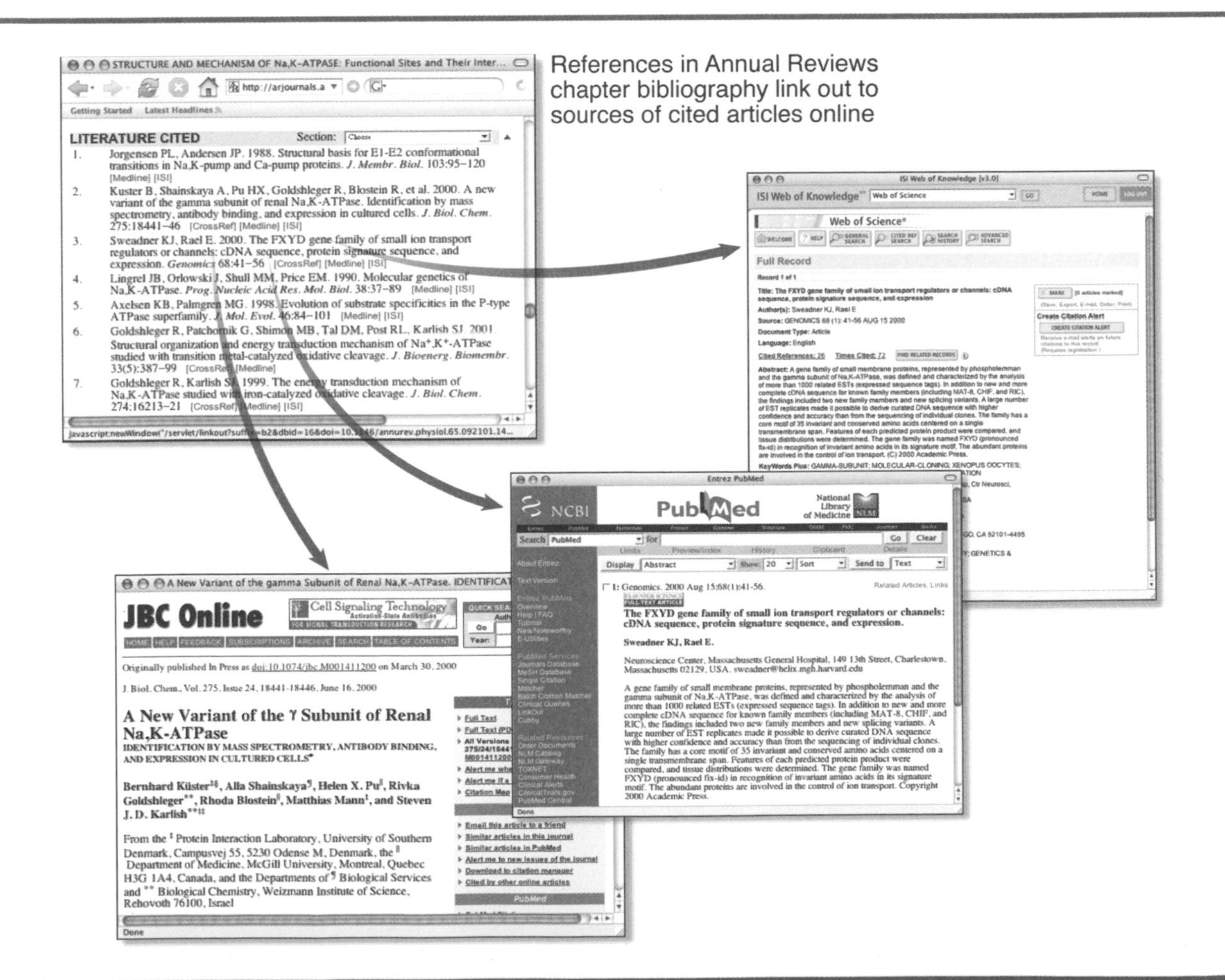

Annual Review of
Physiology

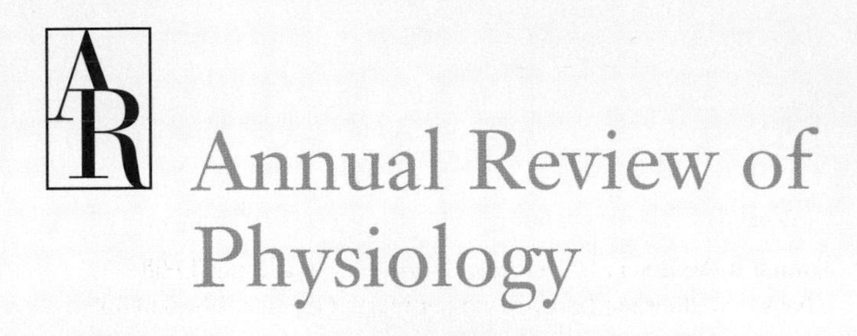

Annual Review of Physiology

Volume 69, 2007

David Julius, *Editor*
University of California, San Francisco

David L. Garbers, *Editor (2005-2006)*
In dedication for his years of service

David E. Clapham, *Associate Editor*
Harvard Medical School

www.annualreviews.org • science@annualreviews.org • 650-493-4400

Annual Reviews
4139 El Camino Way • P.O. Box 10139 • Palo Alto, California 94303-0139

Annual Reviews
Palo Alto, California, USA

International Standard Serial Number: 0066-4278
International Standard Book Number: 978-0-8243-0369-3
Library of Congress Catalog Card Number: 39-15404

∞ The paper used in this publication meets the minimum requirements of American National Standards for Information Sciences—Permanence of Paper for Printed Library Materials, ANSI Z39.48-1992.

TYPESET BY TECHBOOKS, FALLS CHURCH, VIRGINIA
PRINTED AND BOUND BY SHERIDAN BOOKS, INC., CHELSEA, MICHIGAN

Preface

This year's volume brings to you another collection of uniquely interesting and eclectic chapters, covering topics spanning the gamut from cellular to comparative aspects of physiology.

We are especially pleased that this year's prefatory chapter, *Life Among the Axons*, has been contributed by Clay Armstrong, Professor of Physiology at the University of Pennsylvania and an internationally recognized figure in the world of neurophysiology. Dr. Armstrong's work has helped to elucidate many of the basic principles underlying ion channel gating, permeation, selectivity, and inactivation, thereby establishing a biochemical and molecular framework for understanding electrical signaling by excitable cells. His chapter takes us through a personal journey connecting the electrical signaling concepts of Hodgkin and Huxley with today's atomic-level view of the ion channel proteins that account for these classical transmembrane conductances.

We are also very pleased to feature in this volume a Special Topic Section on the arrestins, regulators of G protein–coupled receptor (GPCR) function. This section is organized by Robert Lefkowitz, Investigator of the Howard Hughes Medical Institute and James B. Duke Professor of Medicine and Biochemistry at Duke University Medical Center. Dr. Lefkowitz is among the foremost contributors to this field, and his work has helped to shape the way we think about mechanisms of neurotransmitter and hormone action, particularly in relationship to molecular processes governing receptor–G protein coupling, and receptor desensitization and recycling. Undoubtedly, these concepts have great relevance to drug design and therapeutics, as some 50% of prescribed drugs target GPCRs affecting physiological processes in every major organ system in the body. Although initially discovered as regulators of GPCR function, arrestins turn out to be surprisingly more versatile than previously appreciated and, as described in these chapters, influence the activity and output of numerous other cellular signaling pathways.

The credit for organizing this exceptional volume goes to David Garbers, our respected and beloved friend and Editor, whose sudden and untimely death in September of 2006 came as a shock to all of us. David was a world leader in reproductive biology, especially in regard to characterizing biochemical steps governing sperm-egg interaction and fertilization. Behind David's innovative, unique, and groundbreaking research was an extraordinarily broad knowledge of cellular and systems physiology and a tremendous intuition and talent for biochemistry and cell biology. We are extremely grateful to Drs. Joel Hardman (David's postdoctoral mentor at Vanderbilt) and Alfred Gilman (Chair of David's department at UT Southwestern), who have written a piece for this volume that describes David's many contributions to science while giving the reader a sense of what made Dave and his research so special.

In 2005, I joined the *Annual Review of Physiology* team at the request of Dave Garbers, who was a friend and something of an "unofficial" mentor to me from the time I was a postdoctoral fellow. In Dave's absence, I shall now serve as Editor. My goal is simply to appreciate the many fascinating and important aspects of cellular, molecular, systems, and organ physiology and to convince the very best scientists to contribute to the *ARP* series. This job came naturally to Dave, given his broad scope and reputation for excellence, and I know that it will be challenging to follow in his footsteps. Fortunately, it is my good luck to have David Clapham assume the post of Associate Editor, thereby maintaining Dave Garber's brand of scientific excellence, scope, and good humor. We will also be joined by my UCSF colleague, Roger Nicoll, a world-renowned neuroscientist who will take over as Editor for the section on neurophysiology. Together with the other members of the Editorial Committee and Shirley Park, our superb Production Editor, we shall continue to produce a series that Dave Garbers would have loved to read.

David Julius for the *ARP* Editorial Committee

David L. Garbers

David L. Garbers (1944–2006)

We lost a good friend and colleague and the field of biomedical science lost one of its best when David Garbers, Editor of the *Annual Review of Physiology*, died of a heart attack on September 5, 2006. He was 62 years old.

David was born in La Crosse, Wisconsin and grew up on his family's farm. After receiving his PhD degree in 1972 from the University of Wisconsin, he went to Vanderbilt University for postdoctoral work. He joined the faculty there in 1974 and rose quickly through the faculty ranks, becoming Professor of Pharmacology and Molecular Physiology and Biophysics. Appointed to the Howard Hughes Medical Institute on July 1, 1976 (along with Robert Lefkowitz), David and Bob were co-record holders for longevity as HHMI investigators at the time of David's death. In 1990 David moved to the University of Texas Southwestern Medical Center at Dallas. There he held a Professorship in Pharmacology and the Cecil H. and Ida M. Green Distinguished Chair in Reproductive Biology and was director of the Green Center for Reproductive Biology Sciences.

David was a private, modest, and easy-going person who never sought the scientific spotlight, although it found him many times. He was the recipient of many honors and awards for his research including memberships in the National Academy of Sciences and the American Academy of Arts and Science. His long-standing interest in reproductive biology was first expressed in highly regarded research when he was a graduate student. His pioneering research on mechanisms of fertilization and communication between eggs and spermatozoa gave him insights into other biological systems, and his research broadened over time to include such diverse topics as vision, olfaction, and hypertension.

David blazed a beautiful trail from fertilization to hypertension, and the story is emblematic of his approach to science. Trained as a biochemist, David turned to the study of fertilization in sea urchins because of the availability of large numbers of gametes from urchins of both sexes. Sea urchin eggs secrete factors that activate and attract spermatozoa of the same species, presumably facilitating fertilization in the ocean. David was the first to identify an animal sperm chemoattractant—a small peptide secreted by the sea urchin egg. Later, he and colleagues demonstrated that different species of urchins secrete different peptides, which accounts for the species specificity of the observed effects. Given the diversity of peptide structures but similarity of physiological responses of sperm cells among different species of urchins, David reasoned that the intracellular signaling pathway would be conserved among the species but that the detector region of a cell-surface receptor would vary. If this were true, then hybridization with a sea urchin cDNA probe based on the conserved region would permit detection of related receptors in other phyla. David identified the cell-surface receptors for two of the urchin egg peptides (which he termed speract and resact). In a brilliant technical study, he showed that not only was there a highly conserved intracellular region, but that the egg peptide receptors were also enzymes: guanylyl cyclases that convert GTP into the second messenger cyclic GMP. This provided the first demonstration that a cell-surface receptor could *itself* generate a low-molecular-weight second messenger. The cloning of the receptor also demonstrated its unique structure—an extracellular ligand binding site, a single transmembrane span, and intracellular protein kinase–like and cyclase domains. The kinase-like domain is necessary for transmembrane signaling and imparts a requirement for ATP for hormonal stimulation of cyclic GMP synthesis. This is fascinatingly reminiscent of and complementary to the requirement for GTP to promote cyclic AMP synthesis.

David subsequently cloned and expressed membrane-bound guanylyl cyclases from mammalian sources. One of these is a receptor for atrial natriuretic peptide; another, found in the gut, is the receptor for heat-stable enterotoxins—small peptides produced by pathogenic bacteria that cause an acute secretory diarrhea; still others are present in the retina and olfactory neuroepithelium. David thus became the best friend of cardiologists, gastroenterologists, and ophthalmologists, and a plethora of fascinating findings followed.

Despite these marvelous distractions, David's interest in fertilization never waned. Recent work focused on the discovery of sperm-specific proteins required for such critical phenomena as motility and recognition and fertilization of the egg. He sought and found conditions necessary for the culture, transfection, and differentiation of rat spermatogonial stem cells, with a particular eye on general methods for manipulating the rat genome. David was often annoyed with mice; "too small for physiology" was a frequent complaint.

David's research was driven by his intense curiosity. His ability to see and pursue broader implications of his work in one biological system to others required both imagination and wide knowledge of the current state of science in many areas, traits that served him well as Editor of this publication. He had a knack for asking the right questions and never hesitated to adapt or develop even the most formidable techniques to get answers to them. He had a powerful work ethic and a keen sense for quality in science and could be a stern critic of his own research. He never jumped to conclusions. Exciting though some of his preliminary research results could be, he would keep them to himself until he was convinced of their reproducibility and significance. His research was known and highly respected internationally, and his laboratories were magnets for graduate students, postdoctoral fellows, and visiting scientists from many countries.

Joel G. Hardman
Alfred G. Gilman
October 2006

Introduction to Special Section on β-Arrestins

The G protein–coupled receptors, also known as seven-transmembrane receptors (7TMRs), represent by far the largest family of plasma membrane receptors and mediate a dizzying array of physiological functions. There is no branch of medicine or field of physiology that is not centrally involved with 7TMR function. Moreover, 7TMRs represent the most common target of therapeutic drugs used in current medical practice, accounting for a substantial fraction of all prescription drug sales worldwide.

Three families of proteins mediate the function of the receptors: the heterotrimeric G proteins, the G protein–coupled receptor kinases (GRKs), and the β-arrestins. The G proteins mediate the signaling function of the receptors, classically by activating second messenger–generating enzymes. The GRKs and β-arrestins were originally discovered as molecules that desensitize G protein–mediated signaling. However, over the past decade the formulation of GRKs and β-arrestins as simply a desensitization system has proven inadequate to explain many cellular phenomena. In fact, β-arrestins have emerged as remarkably versatile adaptor molecules that regulate receptor endocytosis and also serve as signal transducers in their own right. None of these functions were imagined when β-arrestins and GRKs were initially discovered in the specific context of receptor desensitization.

The following interrelated chapters discuss in detail these new functions of β-arrestins. Benovic and colleagues review how the β-arrestin/GRK system mediates the receptors' interaction with the cell's endocytic machinery. DeWire et al. comprehensively review β-arrestins as signaling molecules, including recent information that indicates that these functions are in no way limited to 7TMRs but rather operate for an increasingly wide array of other types of plasma membrane receptors. Premont & Gainetdinov review in vivo studies in genetically altered mice; such studies demonstrate the important physiological impact of β-arrestin/GRK-mediated mechanisms. Although most of this work has focused on the desensitization functions of the system, it also covers very recent information regarding the novel signaling roles of the system in vivo. DeFea analyzes in detail the role of β-arrestin-mediated signaling in the important cellular process of chemotaxis. As she discusses, the receptors' regulation of the actin cytoskeleton may be one of the most important consequences of β-arrestin-mediated signaling. Finally, Hupfeld & Olefsky present a detailed review of how β-arrestins regulate signaling by receptor tyrosine kinases (RTKs) as well as cross talk between the TMR and RTK systems. Taken together, this group of reviews presents a comprehensive and up-to-the-minute review of the rapidly expanding network of cellular signaling roles mediated by the β-arrestins and their GRK partners.

Robert Lefkowitz
Duke University Medical Center

Annual Review of Physiology

Volume 69, 2007

Contents

Indexes

Errata

An online log of corrections to *Annual Review of Physiology* chapters (if any, 1997 to the present) may be found at http://physiol.annualreviews.org/errata.shtml

Other Reviews of Interest to Physiologists

From the ***Annual Review of Biochemistry***, Volume 75 (2006)

Signaling Pathways in Skeletal Muscle Remodeling
Rhonda Bassel-Duby and Eric N. Olson

Energy Transduction: Proton Transfer Through the Respiratory Complexes
Jonathan P. Hosler, Shelagh Ferguson-Miller, and Denise A. Mills

Obesity-Related Derangements in Metabolic Regulation
Deborah M. Muoio and Christopher B. Newgard

Axonal Transport and Alzheimer's Disease
Gorazd B. Stokin and Lawrence S.B. Goldstein

G Protein–Coupled Receptor Rhodopsin
Krzysztof Palczewski

From the ***Annual Review of Biomedical Engineering***, Volume 8 (2006)

Mechanism and Dynamics of Cadherin Adhesion
Deborah Leckband and Anil Prakasam

Running Interference: Prospects and Obstacles to Using Small Interfering RNAs as Small Molecule Drugs
Derek M. Dykxhoorn and Judy Lieberman

From the ***Annual Review of Cell and Developmental Biology***, Volume 22 (2006)

How Does Voltage Open an Ion Channel?
Francesco Tombola, Medha M. Pathak, and Ehud Y. Isacoff

Cholesterol Sensing, Trafficking, and Esterification
Ta-Yuan Chang, Catherine C.Y. Chang, Nobutaka Ohgami, and Yoshio Yamauchi

Intrinsic Regulators of Pancreatic β-Cell Proliferation
Jeremy J. Heit, Satyajit K. Karnik, and Seung K. Kim

The Formation of TGN-to-Plasma-Membrane Transport Carriers
Frédéric Bard and Vivek Malhotra

Iron-Sulfur Protein Biogenesis in Eukaryotes: Components and Mechanisms
Roland Lill and Ulrich Mühlenhoff

The Cellular Basis of Kidney Development
Gregory R. Dressler

Caspase-Dependent Cell Death in *Drosophila*
Bruce A. Hay and Ming Guo

From the ***Annual Review of Neuroscience***, Volume 29 (2006)

Adaptive Roles of Programmed Cell Death During Nervous System Development
Robert R. Buss, Woong Sun, and Ronald W. Oppenheim

Locomotor Circuits in the Mammalian Spinal Cord
Ole Kiehn

Neural Mechanisms of Addiction: The Role of Reward-Related Learning and Memory
Steven E. Hyman, Robert C. Malenka, and Eric J. Nestler

From the ***Annual Review of Pharmacology and Toxicology***, Volume 47 (2007)

Allosteric Modulation of G Protein–Coupled Receptors
Lauren T. May, Katie Leach, Patrick M. Sexton, and Arthur Christopoulos

Pharmacogenomic and Structural Analysis of Constitutive G Protein–Coupled Receptor Activity
Martine J. Smit, Henry F. Vischer, Remko A. Bakker, Aldo Jongejan, Henk Timmerman, Leonardo Pardo, and Rob Leurs

Mitochondrial Oxidative Stress: Implications for Cell Death
Sten Orrenius, Vladimir Gogvadze, and Boris Zhivotovsky

The Functional Impact of SLC6 Transporter Genetic Variation
Maureen K. Hahn and Randy D. Blakely

The Obesity Epidemic: Current and Future Pharmacologic Treatments
Karl G. Hofbauer, Janet R. Nicholson, and Olivier Boss

Targeting Antioxidants to Mitochondria by Conjugation to Lipophilic Cations
Michael P. Murphy and Robin A.J. Smith

Acute Effects of Estrogen on Neuronal Physiology
Catherine S. Woolley

Nicotinic Acetylcholine Receptors and Nicotinic Cholinergic Mechanisms of the Central Nervous System
John A. Dani and Daniel Bertrand

Annual Reviews is a nonprofit scientific publisher established to promote the advancement of the sciences. Beginning in 1932 with the *Annual Review of Biochemistry*, the Company has pursued as its principal function the publication of high-quality, reasonably priced *Annual Review* volumes. The volumes are organized by Editors and Editorial Committees who invite qualified authors to contribute critical articles reviewing significant developments within each major discipline. The Editor-in-Chief invites those interested in serving as future Editorial Committee members to communicate directly with him. Annual Reviews is administered by a Board of Directors, whose members serve without compensation.

2007 Board of Directors, Annual Reviews

Management of Annual Reviews

Annual Reviews of

Clay M. Armstrong

Life Among the Axons

Clay M. Armstrong

Department of Physiology, University of Pennsylvania, Philadelphia, Pennsylvania 19104; email: carmstro@mail.med.upenn.edu

Annu. Rev. Physiol. 2007. 69:1–18

First published online as a Review in Advance on October 19, 2006

The *Annual Review of Physiology* is online at http://physiol.annualreviews.org

This article's doi: 10.1146/annurev.physiol.69.120205.124448

0066-4278/07/0315-0001$20.00

Key Words

ion channel, ion selectivity, voltage gating, inactivation

Abstract

A blink in history's eye has brought us an understanding of electricity, and with it a revolution in human life. From the frog leg twitch experiments of Galvani and the batteries of Volta, we have progressed to telegraphs, motors, telephones, computers, and the Internet. In the same period, the ubiquitous role of electricity in animal and plant life has become clear. A great milestone in this journey was the elucidation of electrical signaling by Hodgkin & Huxley in 1952. This chapter gives a personal account of a small part of this story, the transformation of the rather abstract electrical conductances of Hodgkin & Huxley into the more tangible gated ion channel.

INTRODUCTION: MY LIFE AS AN ASPIRING SCIENTIST

My early years were in some respects quite tragic. I grew up a male in Texas too small to play football. Perhaps this fact more than any other destined me to be a scientist. To compensate, I struggled for a time on the tennis team, without notable success. Helpful in taking my mind off these failures were my high school English teachers, both spinsters, one shrunken but vital and witty, the other romantic and in thrall to Beowulf. Together they gave me a permanent love for literature. Math, chemistry, and physics scarcely existed in the curriculum, leaving me with deficits I am still struggling to overcome. All in all, it was not a promising start for a scientific career.

Things changed when I got to Rice, where my heart belongs, educationally speaking. At the time Rice was tough enough that a third of the freshman class did not return the second year. The idea was to whip those Texas country boys into shape (football players were exempt), and for the survivors it seemed to work quite well. My love of literature persisted and grew; in literature I found a continual awakening. And at Rice I got an excellent grounding in chemistry and physics but demonstrated no great talent.

What does a middle-class boy with a good memory and no particular aptitudes do with himself? He goes to medical school, in my case Washington University. It is and was a fine school, but after the first year I was not a fine student. Fatefully, in physiology class I was exposed to the Hodgkin & Huxley (1) formulation and was permanently imprinted by its mysteries. I was also fascinated by brain electrophysiology, and under the supervision of Bill Landau, I worked in the lab of the legendary George H. Bishop, using his handmade equipment. It was in his darkened, electrically shielded room that I discovered just how remote and specialized science is and learned many things that I could never explain to my mother. All the big questions about my place in the universe were at least temporarily shunted aside. In compensation, I found it exciting to walk mentally in obscure places where there were few footprints. One of my favorite pieces of work was done in Bishop's lab, deciphering the electrical waves elicited in the visual cortex by lateral geniculate stimulation. The pieces of this puzzle all fell together one memorable day as I walked in Forest Park. Alas, only single-cell recordings were considered significant in those days, and my technically unfashionable work sank without a trace (2, 3).

Life, which is a series of chances, changed direction when the threat of the draft pushed me from Washington University to the NIH, where I worked in the lab of the famous K.S. (Kacy) Cole. Kacy's lab had a mind-hand duality, and I signed on as a pair of unskilled hands. I learned an enormous amount there, studying, as my aunt Roberta called it, little bags of salt (cells). The focus in Kacy's lab was, of course, nerve conduction, and as I saw it when I got my license to think a bit, there were three questions remaining after Hodgkin & Huxley's (1) great work. (*a*) How do ions get through membranes? (*b*) How does the membrane distinguish between Na^+ and K^+? (*c*) How does the membrane change its electrical resistance and ion selectivity in a fraction of a millisecond? The dominant view was that carriers, one carrier for Na^+ and one for K^+ ions, ferry ions across membranes. Channels were a less popular idea because selectivity seemed harder to explain for a channel. Some guesses were pretty wild. For a time there was widespread intoxication with lipids and phase transitions and even a proposal that the conducting path in a membrane is formed of lipid, with protein as the insulator. One prominent biochemically oriented lab was certain that axon propagation was mediated by acetylcholine. Others, certain that ions were bound in cytoplasm, denied the importance of membranes. In Kacy's physically oriented lab I was safe from many of these turbulent currents and had excellent exposure to electricity, feedback, mathematics, and physics.

In closing this short personal section, I have many debts to mentors, postdocs, and students. None is greater than my debt to A.F. Huxley, in whose lab I struggled unsuccessfully, after my NIH years, to become a worthy student of muscle. My failure in this regard in no way dampens my admiration and respect for this great and imaginative scientist.

TEA^+ AND THE GROSS ARCHITECTURE OF THE K CHANNEL

My first (and perhaps only) good research idea came as a result of a short visit to A.F. Huxley's lab, where a talented student, Denis Noble, was modifying the Hodgkin-Huxley equations to fit cardiac action potentials (4). These equations describe an action potential that is sharply depolarized to a peak by an increase of Na conductance and then repolarized by an increase of K conductance. One of Noble's tasks was to explain why the peak of a cardiac action potential is followed by a long plateau rather than repolarizing promptly, as in an axon. For this purpose, he invoked the "anomalous rectifier," recently discovered by Katz (5) in muscle fibers: anomalous in the sense that K permeability decreases with voltage, rather than increasing like the normal K permeability in axon membrane. On returning to the NIH I found that Tasaki & Hagiwara (6) had recently published a paper showing that internal TEA^+ causes a cardiac-like plateau action potential when injected into squid axons. Their data suggested that total conductance during the plateau was lower than at rest, as though TEA^+ somehow produced anomalous rectification. This was good enough to get started, so Leonard Binstock and I spent a summer at Woods Hole injecting squid axons with TEA^+ and examining the membrane currents under voltage clamp. Neither of us was very good with axons, but luck eventually smiled on us, and we saw that with ~40-mM TEA^+ inside, the action potential indeed had a long plateau and that the outward I_K measured in voltage clamp was completely suppressed. In contrast, external TEA^+ had almost no effect. Most excitingly, when we raised the extracellular $[K^+]$, we found that inward K^+ current was near normal even though there was no outward current! With intracellular TEA^+, K conductance became a K-anomalous rectifier. Activation of the rectifier on depolarization (in present terms, the gating) seemed to occur normally; i.e., the conducting structures, whatever they were, activated normally but allowed only influx of K^+ ions and no efflux (7).

Further work on the very giant axons of Chilean squid showed that with low concentration, when only half of I_K was blocked in the steady state, the time course of block was slow enough to be directly measurable (8). This made it possible to estimate the rate of K^+ movement through the conductor simply by multiplying the TEA^+ entry rate by the ratio of $[K^+]$ to $[TEA^+]$, yielding an intriguing number, approximately 600 K^+ ions ms^{-1}.

How does TEA^+ work? I studied TEA^+ with molecular models and found it to be roughly tetrahedral in shape, with little flexibility. Moreover, it is approximately the size of a K^+ ion with one hydration shell (~8 Å). The simplicity of the TEA^+ molecule seemed to require an equally simple model, and the idea grew irresistibly in my mind that TEA^+ was blocking a channel through the membrane that had a wide inner vestibule and a narrower outer section too small for TEA^+ to pass through. Furthermore, it seemed that a gate at the inner end of the channel prevented TEA^+ entry when it was closed (at rest). To investigate the features of TEA^+ that affected its entry rate into the channels, I had Eastman Kodak synthesize derivatives of TEA^+ with one of its ethyl arms replaced by a progressively longer hydrocarbon chain. The experiments were performed in Chile, where the ocean was inspiring, the facilities inspiringly primitive, and the companionship wonderful. Much to my surprise, the blocking potency increased with chain length. Kinetically, the main effect was not on the entry but on the exit rate, roughly consistent with the idea that each

Anomalous or inward rectifier: prechannel names for K_{IR} channels

Rest: resting V_m. Loosely, V_m < −60 mV

I_K or I_{Na}: current through K or Na channels, respectively. Inward current is plotted down. In prechannel days, one spoke of Na conductance or permeability rather than Na channels

Depolarization: driving V_m positive relative to rest

additional CH_2 group added ~600 cal mol^{-1} to the binding energy. When TEA^+ was applied at low concentration there was a dramatic "inactivation" of the current, shown in **Figure 1*a*** for C_9^+ (nonyltriethylammonium ion) (9). In **Figure 1*a***, the top trace is control I_K, and below this are traces of two families of I_K taken at low and higher internal C_9^+ concentrations. Each family was elicited by a series of voltage steps to −30 through +90 mV, with 2-s recovery intervals between steps. The inactivation rate is faster with high C_9^+ and changes little or not at all as the voltage is driven more positive. This implies that C_9^+ simply diffuses into the channel at a rate proportional to its concentration and that voltage has little effect on the entry rate.

The cartoons in **Figure 1** depict what is occurring during C_9^+ block of I_K. At rest a gate at the inner end of the channel is closed, and the vestibule just above it is empty. After depolarization, several steps lead to the all-or-none opening of the gate, and internal K^+ ions move into the vestibule, with their hydration shell in place. To pass through the channel's narrow selectivity filter (a term coined

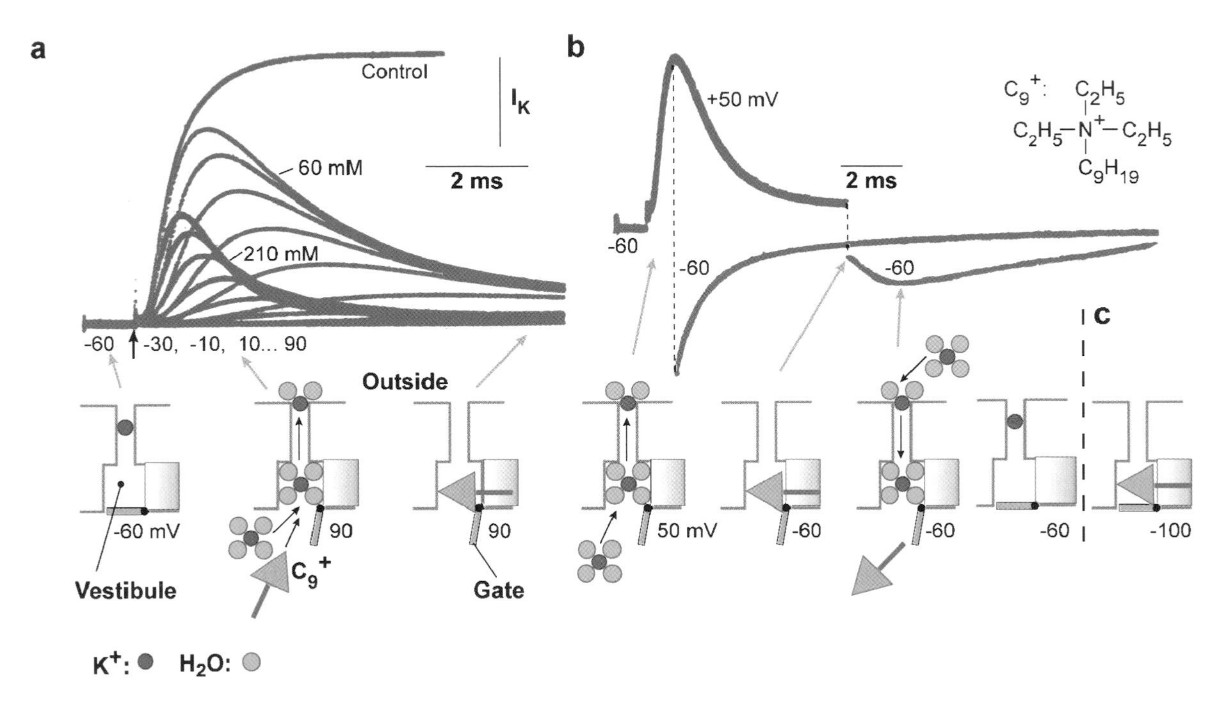

Figure 1

C_9^+ block of I_K resembles inactivation. The cartoons underneath parts *a* and *b* depict what is occurring at corresponding stages. (*a*) I_K (in squid axon) normally increases, with a lag after a step depolarization (*arrow*), to a fixed level (trace labeled *control*). With 60 μM C_9^+ in the axon, I_K increases normally and then "inactivates." At rest (−60 mV) a gate at the inner end of the channel protects a relatively wide vestibule. When the gate is opened by depolarization, K^+ ions enter the vestibule, dehydrate as they go through a narrow tunnel (filter), and rehydrate at the outer end of the tunnel. C_9^+, present at low concentration, takes time to find the channel but then enters the vestibule and gets stuck, with its hydrophobic nonyl chain bound to a hydrophobic region in the vestibule wall. "Inactivation" is faster at higher C_9^+ concentrations. A family of depolarizations is shown at each concentration for steps to −30 though +90 mV. (*b*) At the end of a short depolarization, I_K is inward (external [K^+] is high) but decays rapidly as the channels deactivate. Deactivation is much slower after a long step because the gate is unlikely to close when the vestibule contains C_9^+. In this case, I_K initially increases as C_9^+ is driven from the vestibule by inward-moving K^+, and the gate then closes. (*c*) The gate can be slammed shut owing to repolarization to −100 mV, trapping C_9^+ in the vestibule. Figure modified from Reference 9.

by Bert Hille), they must dehydrate and then rehydrate as they leave the filter. After a time that depends on concentration, a C_9^+ ion diffuses into the vestibule and gets stuck because it cannot move through the filter. Its hydrocarbon nonyl chain increases affinity by binding to a hydrophobic portion of the vestibule's lining.

On repolarization C_9 comes out of the channels, and there are several points of interest, illustrated in **Figure 1*b***. The traces show I_K during a depolarizing step, with repolarization to −60 mV, for a C_9^+-containing axon bathed in high external K^+. When the step is short, there is little block, and the channel gates close relatively quickly, with a simple monotonic time course. After the longer step, when block is complete, I_K on repolarization is at first quite small but increases with time as external K^+ ions move inward through the channel and push out the C_9^+ ions. In low external K^+ (not shown in **Figure 1**), this knock-off effect does not occur, and recovery is much slower. Even with high K^+, closing of a blocked channel is greatly prolonged: C_9 has a "foot-in-the-door" effect (an apt phrase from Hodgkin) that hinders the gate from closing. However, if voltage is made very negative it is possible to close the gate and trap C_9^+ in some of the channels for many seconds (**Figure 1*c***).

All these phenomena were so easy to explain with a channel model that I accepted it and never looked back. At approximately the same time, Hille (10) was finding compelling reasons to believe that Na^+ also traveled through channels. The main features of K channel architecture seemed as follows. (*a*) A gate at the inner end was, following Hodgkin and Huxley, fully open or fully closed and for this reason is portrayed as an open-or-closed trap door in **Figure 1**. How voltage operated this door remained to be determined. (*b*) The gate opened into a vestibule large enough for TEA^+ or a hydrated K^+ ion. Part of the lining was hydrophobic. (*c*) Above the vestibule was a narrow tunnel, permeable to a dehydrated or partially dehydrated K^+ ion but not to TEA^+, with its covalently bound side chains.

TEA^+ sites are at both ends of K channels in frog myelinated axons. The inner site is similar to the one just described for squid channels, but the outer one is quite different in character and not sensitive to C_9^+ (11).

Repolarization: restoring V_m toward, to, or negative to rest

Pancho: Francisco Bezanilla

SELECTIVITY AND CONDUCTION

With a mental picture of the channel's architecture, could we imagine how the channel distinguishes between Na^+ and K^+ ions? According to the gross architectural picture, K^+ apparently dehydrates at least partially while going through the filter. Both Na^+ and K^+ have a charge of 1 e (electronic charge) and are simple hard-sphere cations (12), so why cannot the smaller Na^+ (which has a Pauling radius of 0.95 Å) pass through a filter that accepts K^+ (1.33 Å)? This problem became acute when Pancho and I discovered that Na^+ in the internal medium partially blocked K channels at high membrane voltage (13). Aided in our thinking by the close-fit hypothesis used by Lorin Mullins (14), we proposed the scheme given in **Figure 2*a***. The top row shows K^+ and Na^+ ions in water, with the oxygens of water molecules packed closely around both, binding more tightly to the smaller Na^+ than to K^+ (which has a lower hydration energy). The filter is lined with hydrophilic groups, e.g., carbonyl oxygens, and its diameter makes it a good fit for K^+ (**Figure 2*a***, lower left). Thus, K^+ is bound about equally well in water or in the channel's filter. For Na^+, however, binding in the filter is not favorable because the filter's walls are rigidly fixed and prevent close approach of the carbonyl oxygens to the ion (**Figure 2*a***, lower right). The result is that Na^+ has much higher energy in the filter than in water; it is therefore unlikely to leave water and enter the filter. This is precisely what is needed for selective transport through a channel: Selectivity is conferred by having a low entry

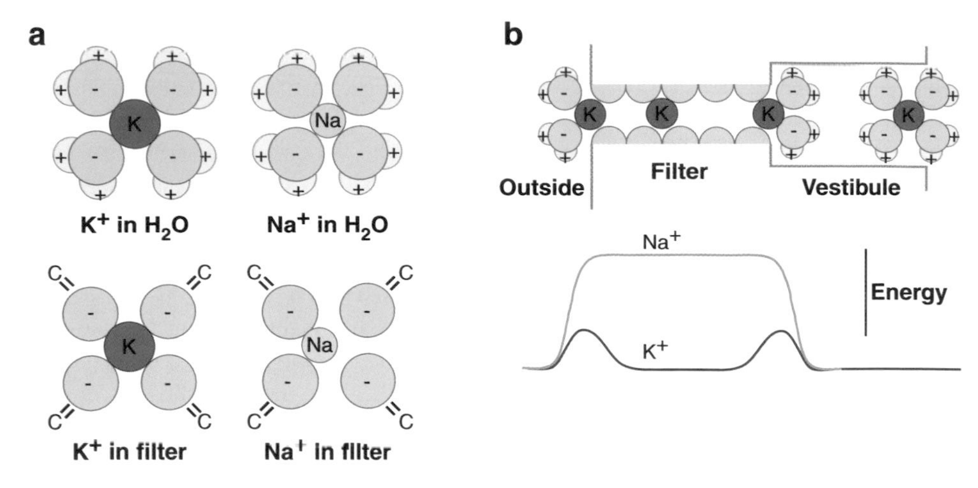

Figure 2

(*a*) A theory for K^+ selectivity. K^+ and Na^+ ions are shown in water and in a carbonyl-lined filter that is a good fit for K^+. The energy of K^+ in water and in the filter is approximately the same. The energy of Na^+ in the filter is much higher than in water because the filter walls are too rigid to allow the carbonyls to close around this small ion. Through the use of an elementary calculation, the situation drawn would provide a selectivity that is orders of magnitude greater than observed, implying that the walls are not completely rigid. Modified from Reference 13. (*b*) K^+ is hydrated in the vestibule, dehydrates to enter the filter, and rehydrates at the outer end. Modified from Reference 16. The energy diagram (*lower diagram*) shows that K^+ faces no large barriers in transit through the channel. Na^+ is excluded by a high barrier (cf. the appendix to Reference 13).

barrier for the selected ion rather than an energy well or, put another way, by selective exclusion rather than selective binding (see appendix in Reference 13).

It is useful to think of the energy of a K^+ ion in five positions as it passes through the filter (**Figure 2*b***, right to left): (*1*) hydrated in the water of the vestibule, (*2*) in the transition state as it dehydrates to enter the filter, (*3*) in the filter, (*4*) as it leaves the filter and rehydrates, and (*5*) rehydrated in the external solution. Because the transport rate of K^+ through the channel is fast, K^+ must encounter no large energy barriers as it passes through. Dehydration at both ends of the filter must be performed quite efficiently to keep the barrier low. Additionally, the energy of K^+ in the filter must be similar to the energy of K^+ in H_2O, where the ion is bound tightly (hydration energy 79.3 kCal mol^{-1}; see Reference 15). If so, K^+ binding in the selectivity filter must be approximately as tight as in H_2O. A clever experiment verifies this: K^+ remains bound in the filter for minutes when K^+ is removed from internal and external solutions while the channel gate is closed (17). Ion replacement, on the other hand, is rapid because the energy barrier for an entering, replacement ion is low (**Figure 2*b***).

GATING

To me the most fascinating aspect of channels has always been gating: how the channels respond to voltage and how their gates open and close in fractions of milliseconds. Any student of gating willing to study the papers of Hodgkin & Huxley (1) has a big head start. Here I greatly simplify their work. They showed that axons conduct action potentials in much the same way that submarine telephone cables (the first trans-Atlantic telephone cable was laid in 1956) conduct telephone messages. The objective is to get high-frequency transmission along an axon or cable that consists of a conductor surrounded by an insulator, both immersed in salt water. The signal current flowing along an axon/telephone cable continually weakens as current leaks out of the conductor (axoplasm/copper) through the resistance and capacitance of the insulator (membrane/insulator). These losses are particularly severe in axons, in which the signal attenuates in millimeters, versus miles for a telephone cable. In both cases the solution is to put in boosters, which detect the weakened signal coming in from upstream and amplify it back to full strength. The boosters in axons are Na channels, and in telephone cables they are electronic amplifiers. Axons also have K channels to get them repolarized and ready for the next action potential.

Hodgkin & Huxley (1) put all this together in a masterful summary that deserves to live forever. They quantitatively described separate Na and K conductances (now known to be mediated by channels). Both conductances can be activated over approximately the same voltage range; activation begins near −50 mV and is complete near 0 mV. The activation process for both is similar, but Na conductance is faster, and the action potential is well established by Na^+ influx before the K conductance turns on. Activation of the K conductance allows an efflux of K^+ that pulls membrane voltage back toward the resting level. This is made easier by a second gate on the Na channel, termed the inactivation gate, which shuts down the Na conductance, facilitating repolarization. Hodgkin & Huxley did not and could not know the mechanisms involved in selectivity, activation, and inactivation. They seemed to favor the idea that Na^+ and K^+ each had a selective carrier that ferried (conducted) ions across the membrane when the carrier was activated by control particles sensitive to membrane voltage. K^+ carriers, e.g., were switched on or off by the movement of four hypothetical particles termed n, each with a charge of 1–2 e. The particles could be either positive and move outward to activate the carrier, or negative and move inward; Hodgkin & Huxley had no way to know. All four particles had to be activated to switch on the carrier/conductance. The Na^+ carrier was similar but with three control particles (m) rather than four. Movement of a single control particle (h) with a charge of ~2 e "inactivated" the Na carrier.

The Hodgkin & Huxley scheme accounted nicely for the kinetics of activation and deactivation of the carriers/conductances. At rest all the control particles were deactivated on one side of the membrane, and they activated one by one following depolarization, moving to the other side. Because all had to be activated to switch on the carrier, there was a pronounced lag in activation, making both Na and K conductance increase sigmoidally. In contrast, on repolarization a conducting carrier was switched off by deactivation of a single one of its control particles; the time course was exponential (for sufficiently negative voltage) and had no initial lag. The requirement for multiple control particles to explain the kinetics gave a premonition of the multisubunit (or multidomain) nature of the molecules underlying the conductances.

DESTROYING Na INACTIVATION

The most accessible of the gating functions turned out to be Na inactivation. The finding that K channels could be given a semblance of inactivation by C_9^+ raised the question of whether inactivation of Na channels had a similar inactivation mechanism, a terrifying departure from the formulation of Hodgkin & Huxley. The question received additional impetus from the finding that Na inactivation could be removed without much effect on activation by perfusing an axon internally with pronase, a proteolytic enzyme mix first used in axons to facilitate internal perfusion. An overdose of pronase caused changes in the voltage clamp currents that were first interpreted as an effect on K channels (18). I thought the effect looked more like removal of Na inactivation, so Pancho, Eduardo Rojas, and I set out to see, and got, the result shown in **Figure 3** (19).

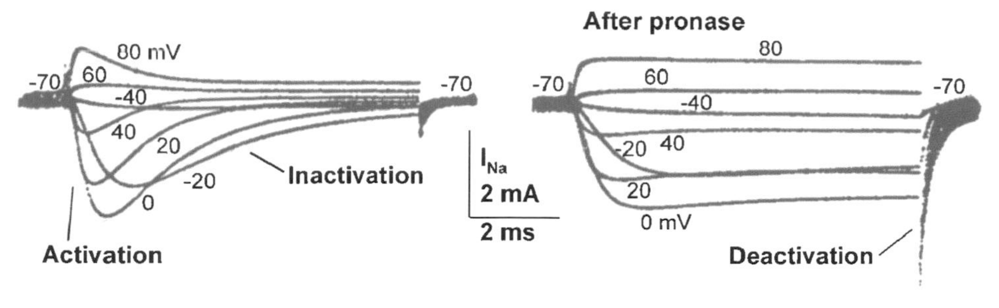

Figure 3

Pronase destroys inactivation when applied internally. Families of superimposed I_{Na} traces are shown before and after pronase application, in an axon with I_K blocked by TEA^+. (*Left*) Before pronase application I_{Na} increases after depolarization as the channels activate, then decreases as inactivation occurs. On repolarization there is a small inward tail of I_{Na} through the small fraction of channels that did not inactivate. (*Right*) After pronase application, inactivation does not occur. On repolarization there is a large inward current through the still-active channels. This current decays rapidly as the channels deactivate. Each trace shown represents a depolarization from −70 mV to the indicated voltage, with a 2-s interval between each depolarization. I_{Na} is outward at 60 and 80 mV where the voltage drives Na^+ outward through the channels. Modified from Reference 19.

After blocking of the K channels of an axon with TEA^+, the Na channels fail to inactivate after pronase treatment, even though activation and deactivation occur normally. Pronase removes inactivation from a channel in a single clip: The inactivation rate of those channels not yet touched by pronase is completely normal (not shown in **Figure 3**). The picture of a ball-and-chain mechanism for Na inactivation was almost irresistible: A single internal C_9^+-like particle (ball) diffuses in and blocks a Na channel after its internally located activation gate is partially or fully activated. Pronase cuts a peptide chain that links the ball to the inner end of a channel, leaving the activation gate untouched. This seemed nice but was distinctly unlike the model of Hodgkin & Huxley, who had never been wrong! Further elucidation had to wait on other evidence.

Activation gate: gate at the inner end of a K_V or Na_V channel, which can be open (activated) or closed (deactivated). S4 segment charge moves outward as the gate opens, inward as it closes

ACTIVATION

The experiments described above provided evidence that ions go through channels that have gates at their cytoplasmic end. It was a necessity from the physicist's point of view that the gate be controlled by something analogous to the charged particles (m, n) postulated by Hodgkin & Huxley (see above section, "Gating"). A magnetic sensor, for example, would be sensitive only to current or to dV_m/dt and could be excluded. Whatever the identity of the charges, they inevitably produce a current as they move through the membrane. The control, or gating, current would presumably be quite small and of short duration because the charges would at most move from one side of the membrane to the other. It would be outward as the gate opens and inward as it closes. But could it be detected? With stimulus from Rojas, Pancho and I set out to find it. We suppressed all the ionic current, subtracted out the linear portion of the capacitive current, and turned up the amplification.

After intense effort and much sorting through artifacts, we were rewarded by the recordings in **Figure 4** (20). Trace i was in very low external Na^+ and shows inward I_{Na} preceded by a transient outward current. After adding tetrodotoxin (TTX) to block ion movement through Na channels, we got trace ii. The peak of the current is the same as in the absence of TTX, and the current's decay toward zero is visible thanks to the absence of I_{Na}. Trace ii has the time course expected for gating current: It starts early and dies away as the channels open. The current is outward on depolarization and inward on repolarization (not shown in **Figure 4**), as expected for the passive movement of charge in the electric field of the membrane. There is no way to tell from the current itself whether the gating charge is positive and moving outward on depolarization, or whether it is negative and moving inward.

From trace ii (**Figure 4**) it seems that the gate opens normally in the presence of TTX. Hille (21) had postulated that TTX binds to a receptor near the outside end of the Na channel. The absence of an effect of external TTX on gating current is consistent with an internal activation gate that is unaffected by TTX on the other side of the membrane. This further supports the idea that both the activation and the inactivation gates of Na channels are inside. Above all, we were happy that gating current, which had to exist, not only existed but was detectable.

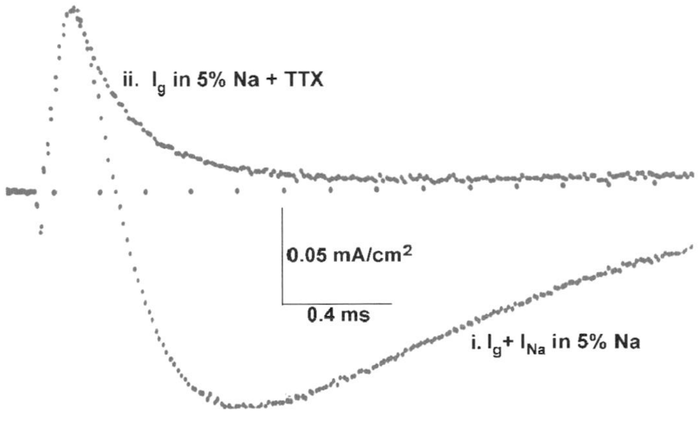

Figure 4

Gating current (I_g) and sodium current (I_{Na}). I_g is generated by the voltage-driven movement within the membrane of charged "particles," now known to be arginine and lysine residues. This movement forces the conformational changes that open and close the activation gate. (*Trace i*) I_K was eliminated by removing all K^+, and I_{Na} was reduced by lowering $[Na^+]$. Capacitive current was removed by subtraction. I_{Na} is preceded by an outward gating current that starts very quickly after depolarization. (*Trace ii*) After adding TTX to block I_{Na} movement, I_g is seen in isolation. The same result can be achieved by complete removal of Na^+. Modified from Reference 20.

Gating current is small and sometimes frustrating to work with. One must continually remember that it is obtained by subtracting out the best estimate of the linear capacitive current and that this estimate can never be perfect. Furthermore, I_g is a combined signal generated by all the steps required to open or close a gate, and the components are temporally overlapped and often inseparable. Nonetheless, there were two immediate lessons, one about activation and one about inactivation (the latter is described below). Before inactivation occurs, the (inward) gating current on repolarization decays at approximately the same rate as I_{Na} as the activation gate closes. The Hodgkin & Huxley model of three independent gating particles predicted that it should be three times faster. This made it clear that information on activation/deactivation would have to be assembled empirically and that no preexisting theory could help much (22, 23). If the channel were made of protein, as seemed likely from the pronase results, one could think of the gating transitions not in terms of gating particles but as transitions among conformational states of a channel protein.

A lover of speculation, I proposed the model channel shown in **Figure 5** (24). The hypothetical protein had (*a*) four subunits surrounding a pore and (*b*) a gate that could open only when all four subunits were activated. Activation and deactivation for each subunit were governed by a zipper-like arrangement of charges (**Figure 5*b***). Making the inside of the membrane positive caused relative motion between the two halves of the zipper, driving a conformational change that, by an unknown mechanism, activated the subunit. This made it possible to activate the next subunit and so on until all subunits were activated and the gate could open. When Na (25) and K channels (26, 27) were cloned some years later, definitively proving at last that the channels were proteins, I was pleased to see some echoes of these speculations evident in the amino-acid sequence.

I_g: gating current, generated by outward or inward S4 movement

Activation/deactivation: (*a*) Driving the activation gate toward open/closed. (*b*) Driving S4 segments outward/inward

K_V or Na_V channel: voltage-dependent K^+- or Na^+-selective channel made of four identical subunit peptides (K) or one peptide with four similar domains (Na). Each subunit peptide (K) or domain (Na) has six transmembrane segments, S1–S6

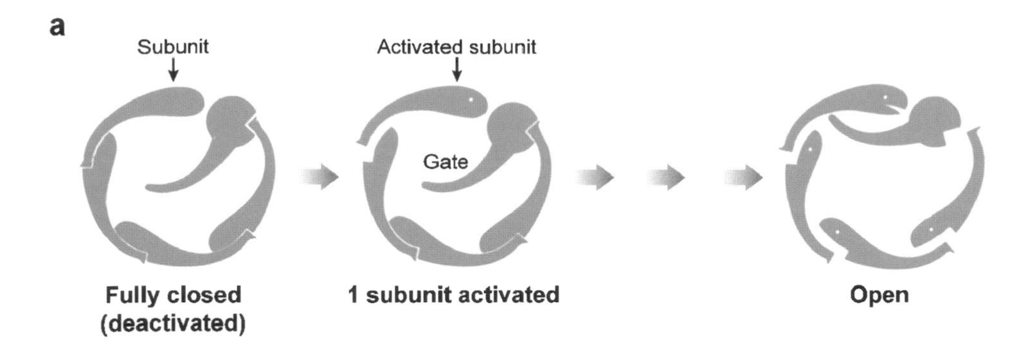

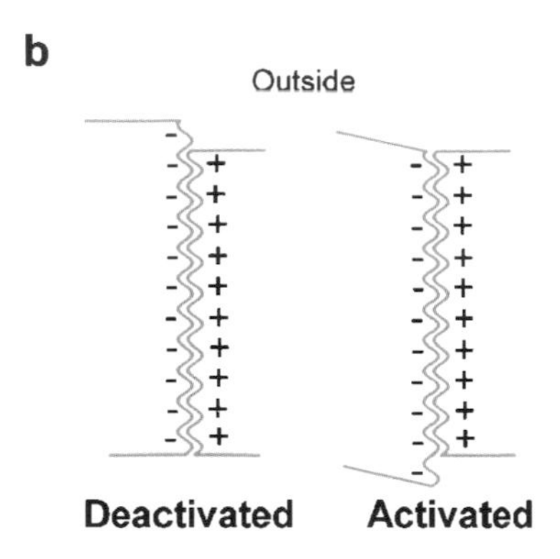

Figure 5

An early speculation about channel gating. The model channel had four subunits and a gate. In each subunit, or in the junction between, was a zipper-like arrangement of charge (*b*). The small relative motion shown (deactivated < > activated) transferred one full electronic charge across the membrane because all the charges moved in the membrane field. This motion drove a conformational change (details unknown but easy to imagine) (*a*) that activated the zipper's subunit, with activation pictured diagramatically as outward rotation. This made it possible for the next subunit to be activated by its zipper and so on, until the gate was free to open. Modified from Reference 24.

THE SEQUENCING OF Na AND K CHANNELS

K_{IR} channel: a K channel that acts like a one-way valve, blocking outward current. K_{IR} channels are opened by K^+ influx, caused by negative V_m and/or high extracellular K^+

S4: the fourth transmembrane segment, containing four to eight positively charged amino acids

Once channels were sequenced they became much more tangible. Abandoning chronology for simplicity, let us begin with the two families of K channels, K_V and K_{IR}. Members of both families are made of four identical subunits (peptides); K_{IR} peptides are approximately one-fourth as large as those of the K_V family. The N and C termini are inside for both. For each K_{IR} subunit there are 2 transmembrane crossings (T1 and T2), yielding a total of 8 crossings for a K_{IR} channel, whereas a K_V subunit has 6 (S1–S6), and a K_V channel, thus, 24. T1 and T2 are analogous to S5 and S6 and are now known to form the ion conduit of the channel. Segments S1–S4 are found in voltage-sensitive channels. The very obvious voltage sensor is the S4 segment, which contains seven positively charged amino acids in some K_V peptides, much like the positive side of the zipper. The complementary negative charges, which are an energetic necessity, are still something of a mystery, as discussed below.

Na channels seem to have come later in evolution. They involve mutations that stitched the subunit peptides together to form a very large peptide containing four domains. As in the K_V family, the Na channel apparently has an ion conduit made by apposition of S5 and S6 from the four domains, forming a central core and four subdomains (S1–S4) surrounding the core (details are still uncertain). After the stitching, the domains differentiated, presumably to confer Na selectivity, faster gating, and inactivation.

THE DISCOVERY OF THE PORE REGION

A prescient Na channel model from Guy & Seetharamulu (28) aided the effort to associate regions of the sequence with function. These researchers identified the region between S5 and S6 (in present terminology) as the likely pore and selectivity filter. A search for the binding site of the potent K channel blocker charybdotoxin (29) led within a few years to the identification of the pore region in K channels (30, 31) and of a signature sequence (32) that made it possible to identify potential K channels in the protein database. Among these was a bacterial K_{IR} family channel, KcsA (33).

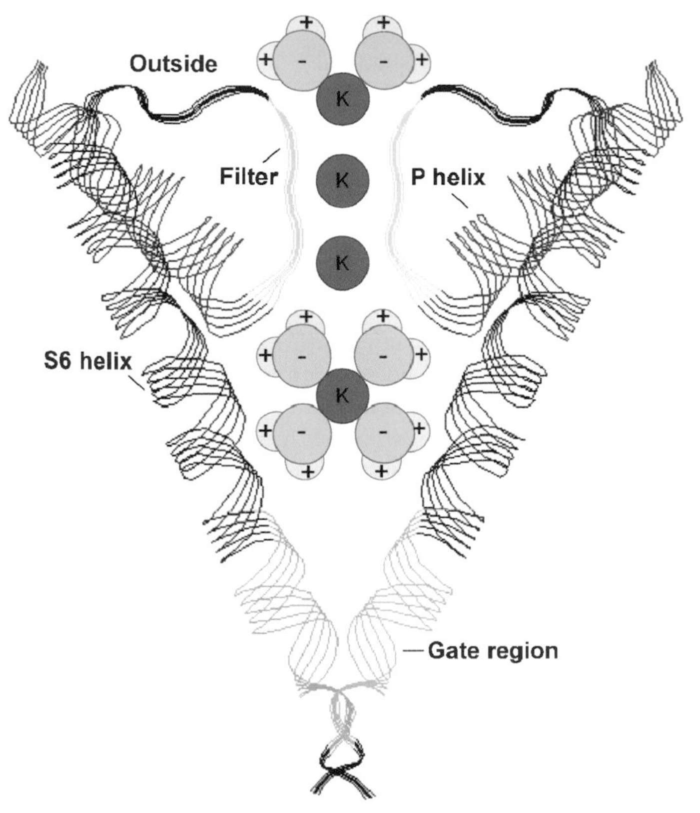

Figure 6

The core of a bacterial K channel in cutaway showing, from bottom to top, the gate region (closed), a hydrated K^+ ion in the cavity, two dehydrated K^+ ions in the filter, and a partially hydrated K^+ ion at the outer mouth of the filter. After Reference 34.

PORE, FILTER, AND GATE: THE KcsA CRYSTAL STRUCTURE

MacKinnon and colleagues (34) then began the arduous task of crystallizing the KcsA channel and subjecting it to X-ray analysis. The following is a brief summary of their work, but the original papers are a pleasure that should not be missed. In each of the four subunits that comprise the channel, there are two membrane-crossing helices, T1 and T2, that correspond to S5 and S6 in the K_V family and are so labeled here. S5 and S6 are joined by a long linker that dips into the membrane to make a pore helix that supports the selectivity filter. **Figure 6** shows the KcsA structure with S5 omitted. The S6 helices form a cone that contains the pore helix and the filter, form the walls of a cavity, and line the inner part of the pore. They converge toward the inner side of the membrane; the gate region is at the convergence. The gate is closed in KcsA: There are three points in the convergence zone at which methylene groups crowd together too closely for even a dehydrated K^+ ion to pass. Because these groups are hydrophobic, each of the three seals makes a very high energy barrier to ion movement, a very effective gate. The K_V family also has a gate in this part of the S6, although the structure of this family's S6, based on the sequence, is probably quite different, as Yellen (35) has pointed out. The K_V gate closes tightly enough to exclude small cations (36). The open state of KcsA has not been crystallized, but a reasonable model is provided by another bacterial channel, MthK (37), in which the convergence zone is widely spread apart and more than wide enough to admit a hydrated K^+ ion.

A K^+ ion going from axoplasm to external fluid passes through the gate region, enters the water-filled cavity, and approaches the inner end of the selectivity filter. To enter the filter, the ion must dehydrate and the carbonyl groups at both ends of the filter must be carefully arranged to lower the energy barrier for dehydration/rehydration. Once in the filter, the dehydrated ion is surrounded by carbonyls of the filter wall. Starting inside, there are four binding sites—4, 3, 2, and 1 (38)—and probably only two are simultaneously occupied, 4 and 2, or 3 and 1. An ion approaching from the cavity would first repel the occupying ions from positions 4 and 2 to positions 3 and 1. When the entering ion occupies site 4, the ion in 3 would move to 2, and the ion in 1 would rehydrate as it left the filter. The occupying ions at this point are again in 4 and 2, and the cycle can repeat. The figures of Reference 39 show dehydration in progress.

INACTIVATION OF Na CHANNELS

After the gating current measurements (22, 23), it was reasonable to believe that Na inactivation resembled C_9^+ block in that the voltage dependence of block/inactivation arose from coupling to activation: Both C_9^+ and the inactivation particle simply diffused into their blocking site when the activation process made the site available. However, there were clear differences. The Na channel can close readily when inactivated, useful functionally in preventing leak during recovery from inactivation. Related to this, approximately one-third of gating charge remains mobile after Na inactivation. This is unlike in TEA^+ block, which immobilizes all gating charge in K channels. The not-immobilized charge moves on repolarization to close the channel promptly (no foot in the door), averting leak during recovery. Also, Na channels need not open fully before inactivating (40), although most inactivation in an action potential occurs from the open state. There is much recent evidence on Na inactivation, which will be the subject of a future communication.

INACTIVATION OF K CHANNELS

Some K channels, notably those termed Shaker B, inactivate rapidly. Yellen and colleagues (40a) found that K inactivation resembles C_9^+ block in that it occurs only when the channel is open. Furthermore, like C_9^+, the inactivation particle is subject to knock off by K^+ ions moving inward through the filter after repolarization, and it acts like a foot in the door, preventing closing of the activation gate. Hoshi et al. (41) identified the inactivation particle as a "ball" at the N terminus of the Shaker B peptide. Four balls are attached by flexible chains to the inner surface of the channel and diffuse into the channel mouth when the gate is open. Mutations that shorten the chain reduce the number of possible conformations of the chain and thus increase the inactivation rate. Decreasing the number of balls by mutation (42) or enzymatically (43) reduces the rate of inactivation, proving that there are four inactivation particles rather than one, as in the Na channel (19). MacKinnon and colleagues (44) have X-ray pictures of the ball within the cavity.

Aldrich and colleagues (45) described a second type of inactivation, referring to it as C inactivation to distinguish it from N-type (or ball-and-chain) inactivation. The mechanism for inactivation is not completely clear, but the experiments of Yellen and colleagues (46) strongly suggest that it involves a constriction near the outer mouth of the selectivity filter.

S4 MOTION DURING ACTIVATION

As a voltage-dependent channel activates, gating charge must move; as noted above, there is no physical alternative. Common sense and some evidence tie this to motion of the S4 segments (47, 48). The original estimate of Hodgkin and Huxley was that approximately six electronic charges move to activate a channel, but current estimates for the Shaker K channel are more than twice that (49, 50). Precisely how the motion occurs is not yet clear. Simplest conceptually are the zipper model (**Figure 5**), its helical screw variant (51), and the paddle model (52). All three models have problems. The negative counterions for the first two models have never been identified in sufficient number. The absence of counterions in the paddle model raises, in my mind, the insurmountable energetic problem of putting multiple arginines (charged and hydrophilic) into lipid, at a cost of approximately 15 kCal mol^{-1} for each arginine (53).

A more complex model postulates crevices that penetrate around the S4 segments and go part way through the membrane from both sides (54, 55). These crevices admit anions from the internal and external solution to serve as counterions. A new crystal structure of Kv1.2 is helpful but not definitive because it shows the S4 only in one state, probably open (56). There are no obvious crevices. How the S4 moves during gating remains a challenging question.

HOLDING THE ACTIVATION GATE OPEN

Yellen and colleagues (57) found a thought-provoking way of holding open the gate of a K_V channel even when V_m dictates it should be closed. They introduced a cysteine residue (replacing a valine) in the gating region and added cadmium. The cadmium binds to the introduced cysteine and forms a bridge to a native histidine 10 residues further down and in a different subunit. Remarkably, the gate is tied open, unable to close on repolarization.

HOLDING THE GATE CLOSED WITH 4-AMINO PYRIDINE

$4AP^+$ is a K channel blocker with an interesting mode of action (58, 59). It enters and leaves the vestibule from inside when the activation gate is open, as do TEA^+ and $C_9{}^+$. Its blocking affinity is very low when the gate is open ($K_D \sim 55$ mM) but high when it is closed ($K_D \sim 30$ μM). Presumably, $4AP^+$ fits loosely in the vestibule in the open state but the vestibule contracts around it in the closed state. $4AP^+$ then serves as an adhesive to keep the vestibule contracted. The gate can open, but slowly and with low probability. In short, $4AP^+$ enters through the open gate and pulls the gate closed behind it. This is the opposite of TEA^+, which tends to hold the gate open, like a foot in the door. Interestingly, when $4AP^+$ is trapped in the vestibule by the closed gate, the S4 segments of all subunits can move in and out almost normally in response to voltage changes (58, 60). Total charge movement, however, is reduced by 5% or 10% because the final opening step and its associated charge movement are unlikely to occur. Also, the Q-V curve, which relates gating charge movement and V_m, is shifted to the right by 5–10 mV, reflecting the small amount of extra work that is required to move the S4s outward with $4AP^+$ trapped in the vestibule. This fact tells a good deal about the coupling of S4 to the gate.

V_m: voltage at the inside of a membrane, with outside defined as 0 mV

COUPLING OF S4 TO THE GATE

Open probability P_o can vary from $\sim 10^{-9}$ at -80 mV (49, 50) to ~ 0.7 at $+20$ mV. Do the S4s pull the gate open when they are activated? Or do they lock the gate shut when they are deactivated? Or both? TEA^+ and $4AP^+$ interact with the gate, and studying these interactions is quite helpful in deciding such questions. First, consider deactivation of the channel. TEA^+ in the vestibule holds the gate open and freezes S4 motion (61). **Figure 7*a*** shows a mechanical explanation: Flanges on the S6 segments prevent the S4s from moving downward as long as the gate is held open by TEA^+. (K^+ and Rb^+ in the vestibule also impede closing, suggesting that the vestibule contains only water when the gate closes normally.) Apparently there is rigid coupling between S4 deactivation and gate closing: The S4 cannot move inward while the gate is open.

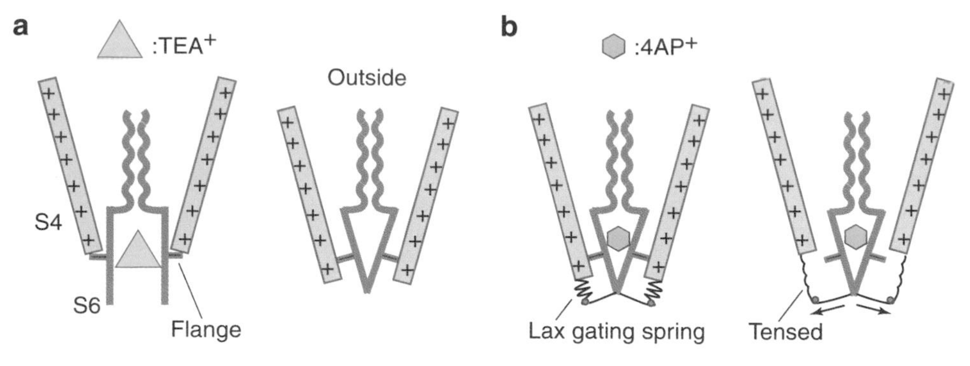

Figure 7

Thinking about the coupling of S4 and the activation gate. (*a*) When TEA^+ is in the vestibule, the channel gate is held open, and the S4s are immobilized in activated position. Projections (flanges) on the S6 prevent the S4s from moving inward as long as the gate is open. When TEA^+ leaves the vestibule, the gate can close, and negative internal voltage drives the S4s inward to the deactivated position. The deactivated S4s lock the gate shut. (*b*) $4AP^+$ in the vestibule stabilizes the closed state by an energy roughly equivalent to that of a hydrogen bond, thus making the gate unlikely to open. Experimentally, with $4AP^+$ in the vestibule the S4s can move almost normally, from deactivated (*left*) to activated (*right*). Thus, an elastic connection such as the gating spring diagrammed must be very compliant. Arrows show the forces exerted on the gate by the gating springs (via imaginary pulleys) when they have been tensed by outward movement of the S4s. The forces are quite small. Modified from Reference 63.

Does S4 motion pull the S6 open by means of some so-far-unknown gating spring (62)? $4AP^+$ holds the gate closed but impedes S4 motion only slightly (see above). This suggests that any link between the S4 and the gate is very weak and easy to stretch (63). In **Figure 7*b***, a hypothetical S4-S6 connection is drawn as an S4-linked gating spring attached via a pulley (to get the force in the right direction) and an inelastic cord to S6. If the spring is very stiff, the S4s cannot move when $4AP^+$ holds the gate shut. At the other extreme, if the spring is completely compliant, no extra work will be required to move the S4s, and the center point of the Q-V curve will be unaffected. As noted above, the experimental curve shows only a small displacement, signifying enough energy to change P_o by a factor of less than 10, many orders of magnitude less than the experimental range of P_o. This strongly suggests that the main function of the S4 is to lock the gate closed and that S4 makes only a minor contribution to pulling it open. This idea gains some support from the crystal structure of Kv1.2 (64), which shows that the S4-S5 linker is in a position that would appear to lock the gate shut when S4 is deactivated. Mutational alterations in this region of the S6 and the facing region of the S4-S5 linker profoundly alter gating (65, 66).

WHY IS THE ACTIVATION GATE ALL OR NONE (USUALLY)?

The gate-opening step is all or none and must involve coordinated movement in the S6 segment of all four subunits. This is clear from single-channel records, in which the opening is a single jump that cannot be resolved into smaller jumps. One factor almost surely is that the gate region is either large enough to admit a hydrated K^+ ion, or it is not: Even partially dehydrating a K^+ in a hydrophobic environment like the gate region would create a large energy barrier and make ion flux too small to detect. A related but more difficult question is why all four S4s must be activated to open the gate. A possible answer is that the gate region's hydrophobic lining must be pulled in four directions to effectively break the hydrophobic bonds that almost certainly help to hold it closed. An analogy would be the need to pull in at least two directions to open a plastic sack whose sides are clinging together electrostatically (63).

SUMMARY AND CONCLUSIONS

Ion channels have gone from concept to reality in the past 50 years. They have been probed, sequenced, and crystallized, and much is now understood about their general properties, selectivity, and mechanisms of gating. They are essential in virtually all physiological regulatory mechanisms, perception, memory, and thought. Their genetic defects are involved in an ever-growing number of diseases, and they offer enormous possibilities for therapeutic intervention. Ah, to be able to start over again!

LITERATURE CITED

1. **Hodgkin AL, Huxley AF. 1952. A quantitative description of membrane current and its application to conduction and excitation in nerve. *J. Physiol. (Lond.)* 117:500–44**
2. Armstrong CM. 1968. Monosynaptic activation of pyramidal cells in area 18 by optic radiation fibers. *Exp. Neurol.* 21:413–28
3. Armstrong CM. 1968. The inhibitory path from the lateral geniculate body to the optic cortex in the cat. *Exp. Neurol.* 21:429–41
4. Noble D. 1962. A modification of the Hodgkin-Huxley equations applicable to Purkinje fiber action and pace-maker potentials. *J. Physiol. (Lond.)* 160:317–52
5. Katz B. 1949. Les constants électriques de la membrane du muscle. *Arch. Sci. Physiol.* 2:285–99
6. Tasaki I, Hagiwara S. 1957. Demonstration of two stable potential states in the squid giant axon under tetraethylammonium chloride. *J. Gen. Physiol.* 40:859–85
7. Armstrong CM, Binstock L. 1965. Anomalous rectification in the squid giant axon injected with tetraethylammonium chloride. *J. Gen. Physiol.* 48:859–72
8. Armstrong CM. 1968. Time course of TEA^+-induced anomalous rectification in squid giant axons. *J. Gen. Physiol.* 21:413–503
9. **Armstrong CM. 1971. Interaction of tetraethylammonium ion derivatives with the potassium channels of giant axons. *J. Gen. Physiol.* 59:413–37**
10. **Hille B. 2001. *Ion Channels of Excitable Membranes*. Sunderland, MA: Sinauer. 722 pp. 3rd ed.**
11. Armstrong CM, Hille B. 1972. The inner quaternary ammonium ion receptor in potassium channels of the node of Ranvier. *J. Gen. Physiol.* 59:388–400
12. Cotton FA, Wilkinson G, Murillo CA, Bochmann M. 1999. *Advanced Inorganic Chemistry*. New York: John Wiley
13. **Bezanilla F, Armstrong CM. 1972. Negative conductance caused by the entry of sodium and cesium ions into the potassium channels of squid axons. *J. Gen. Physiol.* 60:588–608**
14. Mullins LJ. 1960. An analysis of pore size in excitable membranes. *J. Gen. Physiol.* 43:105–17
15. Robinson RA, Stokes RH. 1965. *Electrolyte Solutions*. London: Buttersworth
16. Armstrong CM. 1975. Ionic pores, gates and gating currents. *Quart. Rev. Biophys.* 7:179–210
17. Gomez-Lagunas F. 1997. Shaker B K^+ conductance in Na^+ solutions lacking K^+ ions: a remarkably stable nonconducting state produced by membrane depolarizations. *J. Physiol. (Lond.)* 499:3–15

1. The foundation of ion conduction. Difficult to read because of older conventions but rewarding.

9. Elucidates the gross architecture of K channels. Among my papers, this is my favorite.

10. Unrivaled as an instructional and reference book on channels.

13. An awful title, but the selectivity theory is good.

18. Rojas E, Atwater I. 1967. Blocking of potassium currents by pronase in perfused giant axons. *Nature* 215:850–52
19. Armstrong CM, Bezanilla FM, Rojas F. 1973. Destruction of sodium conductance inactivation in squid axons perfused with pronase. *J. Gen. Physiol.* 62:375–91
20. Armstrong CM, Bezanilla F. 1973. Currents related to movement of the gating particles of the sodium channels. *Nature* 242:459–61
21. Hille B. 1975. The receptor for tetrodotoxin and saxitoxin. A structural hypothesis. *Biophys. J.* 15:615–19
22. **Bezanilla F, Armstrong CM. 1977. Inactivation of the sodium channel. I. Sodium current experiments. *J. Gen. Physiol.* 70:549–66**
23. **Armstrong CM, Bezanilla F. 1977. Inactivation of the sodium channel. II. Gating current experiments. *J. Gen. Physiol.* 70:567–90**

22, 23. These papers provide a still-useful theory of coupled activation-inactivation.

24. Armstrong CM. 1981. Sodium channels and gating currents. *Physiol. Rev.* 61:645–83
25. **Noda M, Shimuzu S, Tanabe T, Takai T, Kayano T, et al. 1984. Primary structure of *Electrophorus electricus* sodium channel deduced from cDNA sequence. *Nature* 312:121–17**

25. The Holy Grail, part one: the amino-acid sequence of a Na channel.

26. Tempel BL, Papazian DM, Schwarz TL, Jan YN, Jan LY. 1987. Sequence of a probable potassium channel component encoded at Shaker locus of *Drosophila*. *Science* 237:770–75
27. Pongs O, Kecskemethy N, Muller R, Krah-Jentgens I, Baumann A, et al. 1988. Shaker encodes a family of putative potassium channel proteins in the nervous system of *Drosophila*. *EMBO J.* 7:1087–96
28. Guy HR, Seetharamulu P. 1986. Molecular model of the action potential sodium channel. *Proc. Natl. Acad. Sci. USA* 83:508–12
29. Miller C, Moczydlowski E, LaTorre R, Phillips M. 1985. Charybdotoxin, a protein inhibitor of single Ca^{2+}-activated K^+ channels from mammalian skeletal muscle. *Nature* 313:316–18
30. Yellen G, Jurman ME, Abramson T, MacKinnon R. 1991. Mutations affecting internal TEA blockade identify the probable pore-forming region of a K^+ channel. *Science* 251:939–42
31. Hartmann HA, Kirsch GE, Drewe JA, Tagliatella M, Joho RH, Brown AM. 1991. Exchange of conduction pathways between two related K^+ channels. *Science* 251:942–44
32. Heginbotham L, Lu Z, Abramson T, MacKinnon R. 1994. Mutations in the K^+ channel signature sequence. *Biophys. J.* 66:1061–67
33. Schrempf H, Schmidt O, Kummerlen R, Hinnah S, Muller D, et al. 1995. A prokaryotic potassium ion channel with two predicted transmembrane segments from *Streptomyces lividans*. *EMBO J.* 14:5170–78
34. **Doyle DA, Carbal JM, Pfuetzner RA, Kuo A, Gullbis JM, et al. 1998. The structure of the potassium channel: molecular basis of K^+ conduction and selectivity. *Science* 280:69–77**

34. The Holy Grail, part two: the structure of a simple K channel.

35. Yellen G. 2002. The voltage-gated potassium channels and their relatives. *Nature* 419:35–42
36. del Camino D, Yellen G. 2001. Tight steric closure at the intracellular activation gate of a voltage-gated K^+ channel. *Neuron* 32:649–56
37. Jiang Y, Lee A, Chen J, Cadene M, Chait BT, MacKinnon R. 2002. The open pore conformation of potassium channels. *Nature* 417:523–26
38. Morais-Cabral JH, Zhou Y, MacKinnon R. 2001. Energetic optimization of ion conduction rate by the K^+ selectivity filter. *Nature* 414:37–42

39. **Zhou Y, Morais-Cabral JH, Kaufman A, MacKinnon R. 2001. Chemistry of ion coordination and hydration revealed by a K^+ channel-Fab complex at 2.0 Å resolution. *Nature* 414:43-48**
40. Bean BP. 1980. Sodium channel inactivation in the crayfish giant axon. Must channels open before inactivating? *Biophys. J.* 35:595–614

40a. Demo SD, Yellen G. 1991. The inactivation gate of the Shaker K^+ channel behaves like an open-channel blocker. *Neuron* 7:743–53

41. Hoshi T, Zagotta WN, Aldrich RW. 1990. Biophysical and molecular mechanisms of *Shaker* potassium channel inactivation. *Science* 250:533–38
42. MacKinnon R, Aldrich RW, Lee AW. 1993. Functional stoichiometry of Shaker potassium channel inactivation. *Science* 262:757–59
43. Gomez-Lagunas F, Armstrong CM. 1995. Inactivation in *Shaker* K^+ channels: a test for the number of inactivating particles on each channel. *Biophys. J.* 68:89–95
44. Zhou M, Morais-Cabral JH, Mann S, MacKinnon R. 2001. Potassium channel receptor site for the inactivation gate and quaternary amine inhibitors. *Nature* 411:657–61
45. Hoshi T, Zagotta WN, Aldrich RW. 1991. Two types of inactivation in Shaker K^+ channels: effects of alterations in the carboxy-terminal region. *Neuron* 7:547–56
46. Yellen G, Sodickson D, Chen TY, Jurman ME. 1994. An engineered cysteine in the external mouth of a K^+ channel allows inactivation to be modulated by metal binding. *Biophys. J.* 66:1068–75
47. Stühmer W, Conti F, Suzuki H, Wang XD, Noda M, et al. 1989. Structural parts involved in activation and inactivation of the sodium channel. *Nature* 339:597–603
48. Yang N, Horn R. 1995. Evidence for voltage-dependent S4 movement in sodium channels. *Neuron* 15:213–18
49. Hirschberg B, Rovner A, Lieberman M, Patlak J. 1995. Transfer of twelve charges is needed to open skeletal muscle Na^+ channels. *J. Gen. Physiol.* 106:1053–68
50. Islas L, Sigworth F. 1999. Voltage sensitivity and gating charge in Shaker and Shab family potassium channels. *J. Gen. Physiol.* 114:723–41
51. Catterall WA. 1995. Structure and function of voltage-gated ion channels. *Annu. Rev. Biochem.* 64:493–531
52. Jiang Y, Ruta V, Chen J, Lee A, MacKinnon R. 2003. The principle of gating charge movement in a voltage-dependent K^+ channel. *Nature* 423:42–48
53. Radzicka A, Wolfenden R. 1988. Comparing the polarities of the amino-acids: side-chain distribution coefficients between the vapor phase, cyclohexaon, 1-octanol, and neutral aqueous solution. *Biochemistry* 27:1670–77
54. Ahern CA, Horn R. 2004. Specificity of charge-carrying residues in the voltage sensor of potassium channels. *J. Gen. Physiol.* 123:205–16
55. Chanda B, Asamoah OK, Blunck R, Roux B, Bezanilla F. 2005. Gating charge displacement in voltage-gated ion channels involves limited transmembrane movement. *Nature* 436:852–56
56. Long SB, Campbell EB, MacKinnon R. 2005. Crystal structure of a mammalian voltage-dependent Shaker family K^+ channel. *Science* 309:897–903
57. Holmgren M, Shin KS, Yellen G. 1998. The activation gate of a voltage-gated K^+ channel can be trapped in the open state by an intersubunit metal bridge. *Neuron* 21:617–21
58. Armstrong CM, Loboda A. 2001. A model for 4-aminopyridine action on K channels. Similarities to TEA^+ action. *Biophys. J.* 81:895–904
59. Del Camino D, Kanevsky M, Yellen G. 2005. Status of the intracellular gate in the activated-not-open state of shaker K^+ channels. *J. Gen. Physiol.* 126:419–28

39. Unbelievable! You can see K^+ ions disrobing.

60. Loboda A, Armstrong CM. 2001. Resolving the gating charge movement associated with late transitions in K channel activation. *Biophys. J.* 81:905–16
61. Perozo E, Papazian DM, Stefani E, Bezanilla F. 1992. Gating currents in Shaker K^+ channels. Implications for activation and inactivation models. *Biophys. J.* 62:160–68
62. Howard J, Hudspeth AJ. 1987. Mechanical relaxation of the hair bundle mediates adaptation in mechanoelectrical transduction by the bullfrog's saccular hair cell. *Proc. Natl. Acad. Sci. USA* 84:3064–68
63. Armstrong CM. 2003. Voltage-gated K channels. *Sci. STKE* 2003(188):RE10
64. Long B, Campbell EB, MacKinnon R. 2005. Voltage sensor of Kv1.2: structural basis of electromechanical coupling. *Science* 309:903–8
65. Lu Z, Klem AM, Ramu Y. 2002. Mechanism of rectification in inward-rectifier K^+ channels. *J. Gen. Physiol.* 120:663–76
66. Hackos DH, Chang TH, Swartz KJ. 2002. Scanning the intracellular S6 activation gate in the shaker K^+ channel. *J. Gen. Physiol.* 119:521–32

Mitochondrial Ion Channels

Brian O'Rourke

Institute of Molecular Cardiobiology, Division of Cardiology, Department of Medicine, The Johns Hopkins University, Baltimore, Maryland 21205; email: bor@jhmi.edu

Annu. Rev. Physiol. 2007. 69:19–49

First published online as a Review in Advance on October 20, 2006

The *Annual Review of Physiology* is online at http://physiol.annualreviews.org

This article's doi: 10.1146/annurev.physiol.69.031905.163804

0066-4278/07/0315-0019$20.00

Key Words

bioenergetics, ischemia, preconditioning, mitochondria, oxidative phosphorylation, ion transport, energy metabolism, ATP synthesis

Abstract

In work spanning more than a century, mitochondria have been recognized for their multifunctional roles in metabolism, energy transduction, ion transport, inheritance, signaling, and cell death. Foremost among these tasks is the continuous production of ATP through oxidative phosphorylation, which requires a large electrochemical driving force for protons across the mitochondrial inner membrane. This process requires a membrane with relatively low permeability to ions to minimize energy dissipation. However, a wealth of evidence now indicates that both selective and nonselective ion channels are present in the mitochondrial inner membrane, along with several known channels on the outer membrane. Some of these channels are active under physiological conditions, and others may be activated under pathophysiological conditions to act as the major determinants of cell life and death. This review summarizes research on mitochondrial ion channels and efforts to identify their molecular correlates. Except in a few cases, our understanding of the structure of mitochondrial ion channels is limited, indicating the need for focused discovery in this area.

1. INTRODUCTION

1.1. Historical Perspectives

Interest in mitochondria has risen and fallen over the past century in the quest to describe the fundamental processes of the cell. In many cases, the significance of claims about the role of mitochondria is realized only after a period of intense criticism, gradual accumulation of evidence, and later recognition that part or all of the original conjecture was correct. This began in 1890 with Altmann's (1) description of mitochondria as the fundamental living elements of the protoplasm. He noted the similar morphology and staining characteristics of bacteria and the grains and filaments of cells (mitochondria), which he called bioblasts (bio, "life"; blasts, "germs"), and proposed that these bioblasts were the centers of biological function. Although he was roundly criticized at the time for his enthusiastic view of mitochondria, it seems that the next 100 years largely proved his point.

Novel functional roles for mitochondria often survive in modified form after the initial hypothesis has been overstated. The suggestion that mitochondria may be responsible for genetic inheritance came prematurely and was supplanted by recognition that the nucleus contained the bulk of the genetic material, only to be resurrected as our understanding of the mitochondrial genome and maternal inheritance emerged. Based on the observation that mitochondria can take up enormous amounts Ca^{2+}, researchers initially proposed that mitochondria could be the main intracellular storage organelles involved in Ca^{2+} cycling. This suggestion was superceded by the elucidation of the specialized role of the endo- (or sarco)plasmic reticulum. However, recent experiments have rejuvenated interest in the role of mitochondria in fast Ca^{2+} responses (2).

Other roles for mitochondria have only grown in importance since they were originally suggested. Mitochondria's participation in oxidation-reduction reactions and cellular respiration was proposed early in the twentieth century, and the mechanisms underlying mitochondrial bioenergetics have been elucidated through a steady series of discoveries and conceptual realignments, including the description of the Krebs cycle, the concept of oxidative phosphorylation, and the chemiosmotic hypothesis (3, 4).

The vectorial movement of ions is central to the chemiosmotic theory, and the four main postulates are as follows: (*a*) H^+ translocation down its electrochemical gradient across the mitochondrial inner membrane is reversibly coupled to ATP phosphorylation through the ATP synthase (F_1F_O ATPase). (*b*) The flow of electrons down the respiratory chain is coupled to H^+ pumping from the matrix to the intermembrane space to establish the protonmotive force. (*c*) Exchange-diffusion carrier proteins are present on the inner membrane to transport metabolites and selected inorganic ions into and out of the matrix. (*d*) The mitochondrial inner membrane is generally impermeable to ions other than H^+.

The last point applies only to mitochondria with tight coupling between oxygen consumption and the phosphorylation of ADP. The extent of coupling varies with membrane leakiness. An increase in the ion permeability of the mitochondrial inner membrane can be induced artificially with protonophoric chemical uncouplers, by a change in metabolic demand [through the stimulation of proton flux through the ATP synthase by ADP and phosphate (P_i)], or in response to increased flux through mitochondrial ion channels (e.g., by Ca^{2+} influx via the uniporter). The energy dissipated by ion flux has a depolarizing influence on mitochondrial membrane potential ($\Delta\Psi_m$), which stimulates NADH oxidation, proton pumping, and respiration. An increase in NADH production is required to compensate for higher rates of respiration and to avoid a mismatch in energy supply and demand.

To understand mitochondrial ion movements and their consequences, mitochondria must be considered as a subsystem of the cell's integrated processes, which include ion transport across the plasmalemma, intracellular Ca^{2+} cycling, and the mechanical or synthetic work of a given cell type. Only then can one appreciate the central role of mitochondria in maintaining the pseudohomeostasis known as normal cell physiology and understand how the failure of mitochondrial function can lead to cell death through a catastrophic necrotic event, autophagy, or the orchestrated process of apoptosis. Recognition of the role of mitochondria in cell life and death in the production of reactive oxygen species (ROS) and nitrogen species and in aging and disease has fueled a resurgence of mitochondrial interest in recent years. Mitochondrial ion channels, although still a nascent subject of investigation, appear to play a fundamental role in most of these functions.

1.2. Ion Transport Across the Inner Membrane: Background

Early observational studies of mitochondria in living cells suggested that mitochondria were sensitive to changes in osmolarity, suggestive of ion and water transport across the mitochondrial compartment. In 1915, Lewis & Lewis (5) reported that mitochondria in cultured embryonic chick cells could reversibly swell and contract in response to changes in the pH or osmotic pressure of the medium. These researchers also observed presciently that the morphology of mitochondria was extremely plastic (undergoing fission, fusion, elongation, etc.) and that mitochondrial structure degenerates in the presence of noxious agents, including hydrogen peroxide.

Other studies showed that a tissue could accumulate potassium against its concentration gradient in a metabolism-dependent manner, which supported the concept of active transport of ions. This in turn led investigators to study how isolated mitochondria actively accumulate ions. Lehninger (6) reviewed early work correlating mitochondrial electrolyte and water movement with oxidative phosphorylation and also discussed the behavior of mitochondria as "osmometers." The technique of measuring light-scattering changes as an indicator of mitochondrial volume has since been extensively employed to assay the activity of mitochondrial ion channels.

The binding (or uptake) of Ca^{2+} to mitochondria was first recognized in the 1950s. Work by Chance & Saris reported that energized mitochondria took up Ca^{2+} to uncouple mitochondria transiently (reviewed in Reference 7). Mitochondria eventually irreversibly uncoupled and discharged Ca^{2+} when a certain threshold level of Ca^{2+} was exceeded [now understood as the mitochondrial permeability transition (MPT)]. The ability of mitochondria to take up large quantities of Ca^{2+} in a manner that depended on the presence of the respiratory substrates ATP, P_i, and Mg^{2+} was reported soon after. An elegant study by Chance (8) in 1965 demonstrated that Ca^{2+} (*a*) induces a decrease in pH in the suspending medium, (*b*) stimulates respiration, (*c*) causes a cycle of swelling and contraction of the mitochondria, (*d*) oxidizes NADH, and (*e*) alters the redox state of respiratory chain carriers with millisecond response times (t1/2 ~70 msec). Additionally, Chance noted a crossover point in the redox chain between cytochromes *b* and *c*. This effect on the respiratory chain was relieved by the addition of 1.8 mM P_i, which also enhanced the maximal Ca^{2+}-stimulated rates of respiration and proton accumulation. Small additions of Ca^{2+} yielded cycles of oxidation and reduction. However, when the total amount of Ca^{2+} added exceeded 300 μM, irreversible, large-amplitude swelling occurred and was accompanied by sustained oxidation of the pyridine nucleotide pool and cytochrome *b* (8). These early studies have inspired abundant interest in the role of mitochondrial Ca^{2+} uptake in cell physiology.

Although the relative impermeability of the mitochondrial inner membrane to ions is a basic tenet of the chemiosmotic hypothesis, Mitchell recognized that several modes of ion transport are present, and he referred to them as symporters, antiporters, and uniporters (discussed in Reference 4). Symporters cotransport multiple ions (or an ion and a metabolite) in the same direction across the membrane (e.g., the mitochondrial P_i/H^+ carrier or the plasma membrane Na^+/glucose transporter), often utilizing the asymmetric electrochemical ion gradient to drive the transport in a thermodynamically favorable direction. Antiporters exchange ions on different sides of the membrane and can be either electroneutral (e.g., the Na^+/H^+ antiporter of the mitochondrial or plasma membrane) or electrogenic (e.g., the Na^+/Ca^{2+} exchanger of the plasma membrane). For electrogenic transporters, ion flux is driven by both the electrochemical gradients of the transported ions and the membrane potential. In a uniporter, ions flow electrophoretically down their electrochemical gradient, with transport rates in the range of 10^4–10^6 ions s^{-1} (e.g., the ion channels of the plasma membrane, the mitochondrial Ca^{2+} uniporter, etc.). With advancements in the ability to record ion channel activity in membranes and bilayers, the concept of the uniporter has evolved into the study of mitochondrial ion channels.

The large electrical driving force for ion movement (~180 mV) across the mitochondrial inner membrane strongly favors ion flow through any open ion channel, unless there is an equal and opposite chemical gradient for the ion (which is normally never the case for mitochondria). The theoretically enormous accumulation of cations in the mitochondrial matrix is limited by the restrictions imposed by the mitochondrial membrane–delimited compartment: Charge neutrality must be maintained, and changes in the osmolarity of the matrix have limits. For example, the entry of a cation through an open channel depolarizes $\Delta\Psi_m$ and stimulates respiration and proton pumping. This increases the pH gradient component of the protonmotive force and also balances the positive charge entering with the ejection of a proton. If the cation accumulation in the matrix is accompanied by either the movement of a permeant weak acid (such as acetate) or the electroneutral uptake of P_i driven by the proton gradient, the pH change is prevented, and there is a net increase in matrix osmolarity and an increase in mitochondrial volume. Mitochondrial swelling can be counteracted by the concomitant stimulation of cation/H^+ antiporters. A futile cycle of K^+ influx [through a mitochondrial ATP–sensitive K^+ channel (mitoK_{ATP})] and efflux (via the K^+/H^+ antiporter) may regulate physiological mitochondrial volume (9) (see Section 3.4.1). Conversely, the efflux of anions through an inner membrane anion channel [e.g., inner membrane anion channel (IMAC); see Section 3.5.1], coupled with cation efflux, may mediate mitochondrial contraction (10).

If the mitochondrial permeability transition pore (PTP) is not activated, mitochondria can take up large amounts of Ca^{2+}. This capacity as a Ca^{2+} sink is facilitated by the fact that P_i accumulates in parallel and can reversibly precipitate with Ca^{2+}, effectively lowering the free Ca^{2+} level to the μM range and minimizing osmotic effects. Ca^{2+} uptake is indirectly coupled to proton fluxes through direct H^+/Ca^{2+} exchange in some tissues, through the simultaneous uptake of P_i with H^+, and through Na^+/H^+ exchange secondary to the action of the Na^+/Ca^{2+} exchange Ca^{2+} efflux pathway. Mitochondrial Ca^{2+} overload can also result in rapid Ca^{2+} efflux from the matrix through PTP opening or PTP-independent pathways. In the case of PTP activation, matrix constituents with mass <~1.5 kDa exit the mitochondrion, and the ion permeability barrier of the inner membrane is lost, effectively short-circuiting proton-coupled energy transduction and dissipating $\Delta\Psi_m$.

The other principal rapid cation uptake pathway that has been studied extensively is the K^+ uniporter. K^+ conductance can be substantial in energized mitochondria, but recent work has indicated that energy depletion (e.g., ischemia) or treatment with K^+ channel openers activates fast K^+ uptake. K^+-selective ion channels may account for this response and are currently the subject of intense investigation because they confer protection against ischemia- or oxidative stress–mediated cell injury and apoptotic cell death (Section 3.4).

Researchers have also recognized for some time the transport of anions across the inner membrane in concert with cation movement. Mitchell & Moyle (11) reported that anions, including P_i, succinate, and malonate, accelerated the rate of decay of the pH gradient induced by a pulse of oxygen. This suggested the presence of anion transport systems coupled to proton movement, leading to the identification of the anion/metabolite-coupled cotransporter family. Inner membrane anion uniporters have been less well studied, but in the 1980s, an IMAC was postulated to account for anion-selective mitochondrial swelling responses (12). Subsequently, several anion channels were found in single-channel patch-clamp studies of mitoplasts (Section 3.5). Furthermore, investigators have identified certain mitochondrial membrane proteins that display anion channel activity under some conditions, including the mitochondrial uncoupling protein (UCP) (Section 3.1).

The best-studied mitochondrial ion channel in terms of structure and function is the voltage-dependent anion channel (VDAC), which is abundant in the mitochondrial outer membrane. VDAC is the primary route of entry and exit of metabolites and ions across the outer membrane and is a component of the mitochondrial contact site between the outer and inner membrane. Recent studies have focused on the regulation of outer membrane permeability as a physiological or pathophysiological control mechanism (see Section 2.1).

Table 1 summarizes known or postulated mitochondrial ion channels. The properties of mitochondrial ion channels are described below.

2. MITOCHONDRIAL ION CHANNELS OF THE OUTER MEMBRANE

2.1. The Voltage-Dependent Anion Channel

VDAC was originally purified and reconstituted in phospholipid bilayers in 1975 (13). A highly conserved protein with homology to bacterial porins, it forms an outer membrane pore with a diameter of 2.5–3 nm in the full conductance state (approximately 3 nS in 1 M NaCl). The channel is thought to be a polypeptide monomer that forms a beta barrel with 13 β-strands and one α-helix (14). It is partially anion selective in the open state, allowing the passage of metabolites such as ATP, ADP, and P_i, but it also permits the free diffusion of cations, including Ca^{2+}, K^+, and Na^+ (14). Nonelectrolytes (<3 nm diameter; <6 kDa) can also pass through the open channel. When the channel is in the closed state, the pore is still conductive but constricts to ~1.8 nm, at which point it favors cation permeability. The conductance to K^+ decreases by ~60%, and Ca^{2+} ions still permeate, but ATP flux is blocked, owing to a shift in the electrostatic profile and selectivity of the pore (15).

VDAC gating is highly voltage dependent, with peak currents near 0 mV and a dramatic reduction in channel open probability at positive or negative voltages in the range of 20–40 mV (15). This gating property may be physiologically important; recent evidence suggests that there is a significant membrane potential across the outer membrane (16). Although the large conductance of VDAC would tend to dissipate any membrane potential, the partial anion selectivity and the fact that fixed anionic charges on immobile macromolecules are asymmetrically disposed across the outer membrane would result in differential mobility of ions and the establishment of a Donnan potential. Assuming that protons rapidly equilibrate across the outer membrane, investigators hypothesized that if a pH gradient

Table 1 Mitochondrial ion channels. A summary of mitochondrial ion channel types identified either in isolated mitochondria, proteolipid bilayers, or patch-clamp experiments. Detected single-channel conductances have been tentatively assigned to a given class of channel, but these assignments have not been unequivocally proven. See text for details and abbreviations

Location	Type		Conductance (~150 mM salt)	Modulators/inhibitors	Putative role(s)	Selected refs.
Outer membrane	VDAC (porin)		0.5–4 nS	Bax/Bak/Bcl-xL/Bcl-2, TOM20, Ca^{2+}, pH, ΔV, NADH, VDAC modulator	Metabolite transport, cytochrome *c* release/apoptosis, PTP complex	16
	TOM40 (PSC)		0.5–1 nS	Signal peptides	Protein transport	54
	MAC (BH proteins)		2.5 nS	Bax/Bak	Cytochrome *c* release/apoptosis	37
	Miscellaneous		10–307 pS	ΔV (for >100 pS)	-	31
Inner membrane	Ca^{2+} uniporter		6 pS	Divalents, nucleotides, RuRed, ryanodine	Ca^{2+} uptake	94
	PTP	MCC	0.03–1.5 nS	Ca^{2+}, ΔV, signal peptides, CsA	Protein transport	174
		MMC	0.3–1.3 nS	CsA, pH, Ca^{2+}, thiols, Bax, ANT inhibitors	Necrosis, apoptosis	178
	UCP		75 pS	Fatty acids	Thermogenesis	88
	K_{Ca}		295 pS	Ca^{2+}, ΔV, ChTx, IbTx	Volume regulation	137, 138
	K_{ATP}		9.7 pS	ATP, GTP, palmitoyl CoA, Mg^{2+}, Ca^{2+}	Volume regulation, protection against apoptosis/ischemic injury	116, 119
	$K_V1.3$		17 pS	Margatoxin	Cell death	147
	IMACs		45, 450 pS	ATP	Volume regulation (in yeast)	155
			15 pS (LCC) 107 pS (centum pS)	Mg^{2+}, pH, P_i, thiols, DIDS,cationic amphiphiles	Volume regulation	149, 150, 153

were detected, it could be justified only if a membrane potential existed. In fact, Porcelli et al. (17) detected a pH gradient by using pH-sensitive fluorescent proteins targeted to the intermembrane space. They reported a pH gradient of ~0.7 between the cytosol (pH 7.59) and the intermembrane space (pH 6.88) and calculated that the outer membrane potential necessary to support such a gradient would have to be 43 mV. If this is true, then VDAC would tend to be in the closed (low anion permeability) state in intact cells, raising the question of whether VDAC regulates mitochondrial metabolism by limiting metabolite flux in vivo.

Notably, several cytosolic factors decrease the open probability of VDAC channels reconstituted in bilayers. For example, the reduced pyridine nucleotides NADH and Mg^{2+}-NAD(P)H decrease ADP flux through VDAC, with K_is in the μM range (18). Moreover, endogenous inhibitors of VDAC, including an intermembrane space protein (19) and tbid (a cleaved form of Bid) (20), have been reported. Using a novel microfluidic and surface-plasmon resonance method, Roman and coworkers (21) recently surveyed a vast number of epitopes expressed from a liver cDNA library for these epitopes' ability to bind to reconstituted VDAC. Roman et al. were able to identify (*a*) positive interactions with ~40% of known VDAC binding partners and (*b*) 55 novel interactions, some of which inhibited VDAC in swelling assays.

VDAC is phosphorylated by protein kinase A (PKA) (22), protein kinase Cε (PKCε) (23), and tyrosine kinases (24). The phosphorylation of mitochondrial proteins by PKCε was correlated with the inhibition of PTP opening and may be related to protection against ischemic injury (23). VDAC is also a binding site for hexokinase II (HKII), and the dissociation of HKII from mitochondria has been correlated with induction of the MPT and apoptosis. The binding of HKII to VDAC appears to prevent VDAC closure (25), whereas hexokinase I may have the opposite effect (26).

Whether VDAC modulation of metabolite flux is a physiological control mechanism remains to be determined, and little is known about VDAC conductance in intact cells. However, recent reports indicate that a large outer membrane channel can be activated in intact neurons by a train of synaptic action potentials (27) or by a proteolytically cleaved form of Bcl-xL and that NADH or hypoxia can inhibit these channels (28). Additionally, ethanol metabolism, by shifting the cytosolic $NADH/NAD^+$ redox potential to a more reduced state, may inhibit ADP flux across the outer membrane (29). Also, one study has reported that the molecular cut-off size and conductance of VDAC are Ca^{2+} dependent at submicromolar Ca^{2+} concentrations (30). Ca^{2+} pretreatment of partially permeabilized cells enhanced subsequent ATP uptake, implying VDAC-mediated regulation of metabolism. Notably, the large VDAC conductances seen in bilayers are not typically observed in patch-clamp experiments of intact mitochondria; rather, a variety of conductances in the range of 10–307 pS have been reported (31).

The implications of changes in outer membrane permeability on metabolism and cell death have been recognized in the context of various pathological states (29, 32). Apoptosis is initiated by the release of proapoptotic factors from the mitochondria, including cytochrome *c*, Smac/Diablo, HtrA2/Omi, AIF and Endo G, and others (reviewed in Reference 33). For some noxious stimuli [e.g., ceramide (34)], mitochondrial outer membrane permeability increases prior to, or in the absence of, an immediate change in inner membrane permeability (35, 36). At least three different mechanisms to explain the loss of intermembrane factors such as cytochrome *c* have been hypothesized: (*a*) physical rupture of the outer membrane as a result of mitochondrial swelling (usually linked to PTP opening), (*b*) a modification of VDAC structure, perhaps induced by proapoptotic proteins, such that the pore size increases enough to allow cytochrome *c* release (36), and (*c*) the formation of a new pore as a consequence of oligomerization and membrane insertion of proapoptotic proteins (37).

An oft-repeated model of the structure of the PTP (Section 3.6) includes, at a minimum, the adenine nucleotide translocase (ANT), VDAC, cyclophilin, and the F_1F_O ATPase (38). VDAC has been implicated both because it coimmunoprecipitates with the other proteins present at mitochondrial contact sites and because it is the main pathway for molecular transport across the outer membrane under normal conditions. Moreover, reconstitution of (at least) VDAC, ANT, and cyclophilin in phospholipid bilayers can form cyclosporin-sensitive channels (38). In addition, several common factors—for example, NADH and HKII binding—inhibit both the reconstituted VDAC channel and the PTP. The precise requirement for VDAC and the three-dimensional orientation of the components in the complex that form the pore are presently unknown, but possibly either a long pore spanning the two mitochondrial membranes forms from the association of VDAC with ANT or mitochondrial swelling simply ruptures the outer membrane (the role of VDAC is unclear in this case). In evaluating whether the current model of the PTP is correct (see Section 3.6), recent data demonstrating that the permeability transition can still be observed in ANT knockout mice (39) must be taken into account (40). Furthermore, large multiconductance channels in the inner membrane can still be observed in mitochondria devoid of VDAC (41), and Ca^{2+}-induced MPTs with properties indistinguishable from controls occurred in mitochondria from VDAC1$^{-/-}$ mice (42).

An ongoing debate is whether apoptosis is favored by VDAC opening or closure (14). Capano & Crompton (43) showed that the proapoptotic BH3-homology protein Bax coimmunoprecipitates with VDAC. In reconstitution studies in liposomes and phospholipid bilayers, Shimizu and coworkers (36) have reported that Bax can induce a novel high-conductance state of VDAC that permits cytochrome *c* to escape from the intermembrane space. Similarly, a recent report demonstrated that Bax and tBid increased the conductance of VDAC in planar bilayers and enhanced cytochrome *c* translocation from the *cis* to the *trans* chamber (44). Interestingly, VDAC phosphorylation by PKA inhibited the effects of Bax and tBid (45). Hexokinase II dissociation from VDAC and enhanced VDAC permeability as a trigger for cell death have been reviewed recently (32, 46).

In contrast, others have observed no effect of monomeric or oligomeric Bax on VDAC gating or conductance (20), although the antiapoptotic protein Bcl-x_L promoted VDAC opening and helped to maintain the rate of ATP/ADP exchange across the outer membrane (47). In addition, the proapoptotic tBid was found to induce VDAC closure (20), and other studies have also linked the induction of apoptosis with a decrease in outer membrane permeability to metabolites (48).

2.2. The Mitochondrial Apoptosis–Induced Channel

Although debate is ongoing about whether VDAC constitutes a route for cytochrome *c* release, another candidate has emerged as a possible outer membrane permeation pathway. A recent study found that a novel high-conductance channel was frequently observed in proteoliposomes prepared from mitochondrial outer membranes that were isolated from cells undergoing apoptosis induced by growth factor withdrawal (49). This mitochondrial apoptosis–induced channel (MAC) had a fully open conductance of ~2.5 nS (~5 nm diameter) and multiple substates (37). Unlike VDAC, it was partially cation selective ($P_{K+}:P_{Cl-} = 3$) and was voltage independent. Because both the antiapoptotic Bcl-2 (50) and proapoptotic Bax (51) proteins can form channels when reconstituted in lipid bilayers, their role as mediators/regulators of MAC was investigated. MAC was not detected in proteoliposomes from cells overexpressing the antiapoptotic Bcl-2 protein. In contrast, the expression of human Bax in a VDAC-less yeast strain resulted in the appearance of channels identical to MAC (49). Moreover, the increased probability of observing MAC correlated with the translocation of Bax to the outer membrane and cytochrome *c* loss from the intermembrane space, and interventions that prevented Bax activation (e.g., Bcl-2 overexpression) reduced MAC formation. In agreement with MAC constituting a cytochrome *c* efflux pathway, cytochrome *c* may modify the conductance of reconstituted MAC. Bax and Bak appear to be functionally redundant with respect to the formation of MAC because MAC can still be observed in Bax knockout cell lines but not in Bax + Bak double knockouts.

The role of other outer membrane proteins in the formation of MAC and the precise time course of MAC activation in intact cells remain to be elucidated. With respect to the latter, Guihard et al. (52) detected MAC in mitochondria isolated from apoptotic rat liver only during the later stages of apoptosis, but Dejean et al. (53) argued that this was because of the particular apoptotic stimulus used (which activated the extrinsic pathway of apoptosis).

2.3. Translocase of the Outer Membrane

The insertion and translocation of mitochondrial preproteins encoded by the nucleus require specialized molecular machinery consisting of complexes of proteins on the outer and inner membranes. The constituent proteins of the translocase of the outer membrane of mitochondria (TOM) and its partner on the inner membrane, TIM, can form large conductance channels (54). Early electrophysiological studies of reconstituted mitochondrial membranes showed channels that were blocked by small basic peptides derived from mitochondrial presequences (e.g., the first 12 amino acids of cytochrome oxidase preprotein), and this peptide-sensitive channel (PSC) was proposed to be involved in mitochondrial protein import (55). Subsequent studies have shown that immunodepletion of the preparations with antibodies against TOM40p correlates with loss of channel activity, whereas control antibodies have no effect (56). Furthermore, antibodies against the carboxy terminus of TOM40p altered channel activity (56).

The properties of the channels formed by TIM or TOM proteins are almost identical (54). Both show a full conductance open state of 1 nS and a half-open state of 500 pS, and they have partial cation selectivity ($P_{K+}:P_{Cl-} = 5$) (54). TOM channels show an asymmetric voltage dependence, with closure favored at positive voltages between +20 and +60 mV, but little inactivation at negative voltages. Estimates of the pore diameter based on conductance are in the range of 2.4–2.7 nm, whereas polymer exclusion methods give a molecular weight cutoff of ~1000 and a dimension of ~2 nm (54). The slightly larger estimates of pore size based on conductance are complicated by the finding that the channel may consist of a double-barrel structure in which there may be cooperative gating of the two pores.

2.4. Peripheral (Mitochondrial) Benzodiazepine Receptor

Mitochondria contain binding sites for benzodiazepine receptor (BR) ligands that are pharmacologically distinguishable from the central-type receptor. Because these sites were also found in tissues outside the nervous system, they were labeled peripheral benzodiazepine receptors (PBR) (57). The PBR was preferentially enriched in tissues with high rates of oxidative phosphorylation, and membrane subfractionation studies demonstrated that the receptor was mitochondrial (58). The influence of the mitochondrial benzodiazepine receptor (mBzR) on mitochondrial physiology is incompletely understood, but we know that it plays a vital role in cholesterol transport and steroid biogenesis, accounting for the high density of this receptor in the adrenal gland and testes (59). The mBzR resides in the outer membrane as a complex of proteins; the primary component is the 18-kDa isoquinoline carboxamide–binding protein (IBP). As its name implies, IBP has a high affinity for isoquinoline carboxamide ligands like PK11195 (60), and it is found in close association with VDAC, ANT, and other proteins of the mitochondrial contact site. The binding of isoquinoline carboxamides and benzodiazepines is mutually competitive for the mBzR, and it was originally proposed that the binding site for benzodiazepines, such as the prototypical Ro 5–4864 (4′ Cl-diazepam), included the IBP, VDAC, and ANT (61). However, studies of yeast transformed with the IBP, but lacking either the VDAC or ANT protein, show high-affinity binding sites for isoquinoline carboxamides and benzodiazepines that are no different from binding sites of native receptors (62). More recently, other proteins have been reported to associate with the IBP; these include the PBR-associated protein 1 [PRAX-1 (63)], the PBR-associated protein 7 [PAP7, or PKA with regulatory subunit 1α (64)], and a 10-kDa protein [pk10 (65)]. Several endogenous ligands (endozepines) interact directly with the mBzR, including an 86-amino-acid polypeptide known as the diazepam binding inhibitor (DBI; homologous to liver acyl CoA–binding proteins) (66, 67) and its truncated derivative triakontatetraneuropeptide (TTN; DBI fragment 17–50) (68) as well as porphyrins (69), which bind to the receptor with nM affinities.

The first step of steroid synthesis and metabolism is the conversion of cholesterol to pregnenolone. This reaction is catalyzed by the cytochrome P-450 side-chain cleavage enzyme (P-450scc) located on the mitochondrial inner membrane. The overall reaction rate is limited by the transport of cholesterol across the outer membrane (70). PBR ligands facilitate this cholesterol transport (71) and enhance steroid production (72), whereas knocking out the IBP by homologous recombination in a cell line decreases steroid production (73). Molecular modeling indicated that the mBzR may form a barrel-like outer membrane cholesterol-permeable transport pathway composed of five membrane-inserted α-helices (74). Transmission electron and atomic force microscopic studies showed that anti-PBR antibodies are organized into complexes of 4–6 proteins (75).

If this structure forms a membrane-spanning pore in the outer membrane, it will need to accommodate the ~6.4-Å diameter of cholesterol, so ions (with Pauling radii <2 Å) should also be permeable. It is well known that benzodiazepine binding to $GABA_A$ receptors activates Cl^- conductance in the central nervous system. However, ion flux through the outer membrane via mBzR has not been demonstrated, although the close association between the IBP and VDAC begs the question. Several early studies showed that anions competitively inhibit ^{3}[H]Ro 5–4864 binding in kidney membranes with a profile correlating perfectly with their permeability through Cl^- channels, and inhibitors of anion transport similarly inhibited Ro 5–4864, but not PK11195, binding (76). Other modulators of mBzR ligand binding include lipids (77), phospholipase A2 (78), König's polyanion, and cyclosporin A (CsA) (79), but their effects on mBzR-associated mitochondrial ion flux have not been investigated. The relationship between mBzR ligands, mitochondrial inner membrane ion channels, and function is discussed below (see Section 3.5.1).

3. MITOCHONDRIAL ION CHANNELS OF THE INNER MEMBRANE

3.1. Proton Leak and Uncoupling Protein

Proton transport across the mitochondrial inner membrane is at the heart of chemiosmotic theory. However, in addition to the passage of protons down the electrochemical gradient via the mitochondrial ATP synthase, other routes of "proton leak" are present. Proton leak can contribute significantly to the control of respiration in mitochondria in state 4 (limited by the availability of ADP) and, to a lesser extent, in mitochondria in state 3 (in the presence of substrate and ADP) (80). The proton conductance of the inner membrane displays an ohmic voltage dependence at low to intermediate $\Delta\Psi_m$ and a steep increase in conductance (nonohmic) at large $\Delta\Psi_m$ (>200 mV). The mechanisms responsible for the background conductances are incompletely understood, but several possible explanations include "slips" in the redox-driven proton pump stoichiometries and leaks of protons across the lipid bilayer.

In the context of this review, the contribution of proton-permeable channels must be considered. The pioneering work of Nicholls showed that in brown fat mitochondria, which have high rates of respiration dissipated as heat during nonshivering thermogenesis, GDP and albumin (which binds fatty acids) could regulate the proton leak (81). Later, this proton leak was attributed to a 32-kDa UCP (82). Three isoforms (UCP1, UCP2, and UCP3) are expressed in a variety of tissues (83). Besides the role of UCP1 in heat production, the function of UCP in physiology remains obscure; however, recent interest has focused on the modulation of ROS production, particularly in the context of diseases such as diabetes and obesity. Mild uncoupling of mitochondria, perhaps by altering the rate of production of free radicals, may be protective (84). Alternatively, partial uncoupling by UCP2 and/or UCP3 may optimize the efficiency of energy metabolism (85). Theoretical work suggests that the optimal ATP flow is achieved when the conductance of oxidative phosphorylation is matched to the conductance of the workload, most efficiently at coupling ratios slightly less than one (86).

It is not clear whether the proton and Cl^- conductance (87) mediated by UCP involves a carrier or ion channel–type mechanism. A 75pS anion channel was recorded in giant liposomes reconstituted with UCP1 (88), and reconstitution of the individual transmembrane domains of human UCP2 showed that all six transmembrane peptides form helical conformations in lipid model membranes. Only the second transmembrane peptide exhibited voltage-dependent anion channel behavior (89). Other evidence indicates that the anion and proton conductances are separable, depending on the presence or absence of coenzyme Q (90). Alternatively, UCP-mediated uncoupling may involve fatty acid cycling accompanied by protonation/deprotonation to cause net H^+ transport (85).

As a cautionary note, a leftward shift in the respiration-versus-$\Delta\Psi_m$ curve (i.e., higher oxygen consumption at any given $\Delta\Psi_m$) is often referred to as an increase in "H^+ leak." However, the ion selectivity of the leak pathway is usually not examined, so a sizable increase in the conductance of an ion channel with an equilibrium potential different from $\Delta\Psi_m$ (essentially any ion channel), be it selective or nonselective, would affect the curve similarly, by partially depolarizing $\Delta\Psi_m$. For example, mitochondria isolated from postischemic hearts have an increased proton (ion) leak rate that can be reversed by agents that block the PTP, a large nonselective channel (91). In mitochondria from hearts subjected to an ischemic preconditioning (IPC) protocol, GDP reversed the increase in proton leak, implicating UCP as a leak pathway potentially responsible for the cardioprotective effect of preconditioning. Unfortunately, there are no selective pharmacological tools available to block UCP in intact cells or tissues, so GDP and/or albumin sensitivity of isolated mitochondria is the only way to test for the contribution of UCP to a given process.

3.2. The Mitochondrial Ca^{2+} Uniporter

As mentioned above, mitochondrial Ca^{2+} uptake has been recognized for more than 50 years, yet the protein mediating this essential physiological transport process is still unidentified. The mitochondrial Ca^{2+} uniporter (MCU) is present at low density (0.001 nmol mg^{-1} (92) on the mitochondrial inner membrane, but it has a high V_{max} of >1000 nmol min^{-1} mg^{-1}, corresponding to a flux of ~20,000 Ca^{2+} s^{-1} (93), in the range of a fast gated pore.

Other evidence also supports a model of the Ca^{2+} uniporter as an ion channel. For example, the dependence of Ca^{2+} uptake on $\Delta\Psi_m$ is consistent with electrodiffusion through a pore according to the Goldman constant field equation. Moreover, in a recent patch-clamp study of intact mitoplasts (94), an inwardly rectifying, highly Ca^{2+} selective (affinity 2 nM), voltage-dependent Ca^{2+} channel (MiCa) with properties that match the Ca^{2+} uniporter was recorded. Extramitochondrial concentrations of Ca^{2+} in the range of 20–100 μM yielded whole mitoplast currents of 20–30 pA, for a current density of ~55 pA pF^{-1} at −160 mV. Remarkably, the Ca^{2+} flux did not saturate until [Ca^{2+}] exceeded 100 mM, with a $K_{1/2}$ of 19 mM. Single-channel recordings in 105 mM Ca^{2+} showed a single-channel amplitude of 0.5–1 pA and a channel density of 10–40 channels μm^{-2} (channel conductance ~6 pS). MiCa had a high open probability at negative potentials and displayed rapid, but partial, Ca^{2+}-independent inactivation. It was argued that the much higher fluxes (5×10^6 Ca^{2+} s^{-1}) and $K_{1/2}$ recorded for MiCa in comparison to earlier studies of the Ca^{2+} uniporter were due to the inability to clamp the membrane potential in isolated mitochondria with such large currents flowing (94). Divalent ion selectivity for MiCa was also very similar to that reported for the MCU ($Ca^{2+} \approx Sr^{2+} \gg Mn^{2+} \approx Ba^{2+}$; blocked by Mg^{2+}), and the channel was sensitive to ruthenium red and Ru360 at low nM concentrations.

Despite a number of attempts to identify the MCU protein over the years, its molecular structure is still unknown. A soluble mitochondrial Ca^{2+}-binding glycoprotein, whose Ca^{2+} binding could be inhibited by La^{3+} or ruthenium red, was isolated in 1972 (95), and antibodies against it inhibited Ca^{2+} uptake (96). Later, Mironova and coworkers (97) also isolated a 40-kDa Ca^{2+}-binding glycoprotein from beef heart mitochondria and found that a 2-kDa peptide fragment in the preparation could reconstitute a ruthenium red–sensitive Ca^{2+} selective conductance in black-lipid membranes. Antibodies against this peptide inhibited Ca^{2+} uniporter activity in mitoplasts (98). The 2-kDa protein was further purified and formed 20-pS ruthenium red–sensitive channels upon reconstitution (99).

In other studies, antibodies were raised against a mitochondrial Ca^{2+} transporting extract; an antibody that recognized a 20-kDa protein was most effective in inhibiting Ca^{2+} transport (100). An 18-kDa protein fraction with high specificity for ^{103}Ru360 was further purified, and antibodies against it inhibited Ca^{2+} uptake in cytochrome oxidase–containing vesicles (101). None of the putative Ca^{2+} transporter proteins have been sequenced thus far.

Gunter and coworkers (102) have reported a second rapid mode of mitochondrial Ca^{2+} uptake (RaM). In liver and heart mitochondria, RaM transports Ca^{2+} very rapidly at the onset of a Ca^{2+} pulse and inactivates during a pulse. As compared with liver mitochondria, heart RaM requires a longer time to reset between pulses when a series of pulses is applied. The rapid uptake is less sensitive to activation by spermine, ATP, and GTP but is inhibited by AMP (103). It is presently unknown whether RaM is mediated by a separate transport protein or if it is a second kinetic mode of the Ca^{2+} uniporter. The proteins involved in RaM-mediated Ca^{2+} uptake have not been identified.

3.3. Ryanodine Receptors

Mitochondrial Ca^{2+} uptake and endoplasmic reticular Ca^{2+} release are inhibited by common pharmacological inhibitors, including ruthenium red and ryanodine. Recent studies have reported that type I (skeletal type) ryanodine receptors (RyR1) are present on the cardiac mitochondrial inner membrane and may participate in either Ca^{2+} influx or efflux (104). Intriguingly, RyR1 immunoreactivity could not be detected in mitochondria from RyR1 knockout mice, providing strong support for the specificity of the antibodies used in these immunoprecipitation studies (104).

Any consideration of the ryanodine receptor as a viable candidate for the protein responsible for Ca^{2+} uniport activity in mitochondria must reconcile the discrepancy between the extremely high Ca^{2+} selectivity of the uniporter (94), which allows for mitochondrial Ca^{2+} uptake at submillimolar concentrations without permitting energy dissipation by monovalent ion flux, and the nonselectivity of ion permeation through ryanodine receptors. In addition, ion channels resembling RyR1 have not yet been reported in intact mitoplast patch-clamp recordings. Hence, further studies will be required to elucidate the role of RyR1 in mitochondrial Ca^{2+} handling.

Depending on the orientation of the Ca^{2+}-triggering site of the RyR1, i.e., toward the matrix or toward the intermembrane space, RyR1 may be a candidate either for rapid uptake of Ca^{2+} by mitochondria or for Ca^{2+}-triggered Ca^{2+} release from the mitochondrion through a non-PTP mechanism. Ruthenium red–sensitive Ca^{2+} efflux from Ca^{2+}-loaded mitochondria, attributed to reversal of the Ca^{2+} uniporter, has been reported (105).

3.4. K^+ Channels and Protection Against Ischemic Injury

More than 20 years ago, Lamping & Gross (106) demonstrated that pharmacological agents capable of opening K^+ channels protect hearts against ischemia-reperfusion injury. K^+ channel openers have come to be viewed as chemical preconditioners, i.e., compounds that can mimic the protective effects of brief cycles of ischemia and reperfusion. Moreover, the finding that K^+ channel inhibitors such as glybenclamide and 5-hydroxydecanoate could block the protective effects of either IPC or K^+ channel openers (107) suggested that K^+ channels were a native trigger and/or effector of preconditioning. Although earlier studies naturally presumed that the target of the K^+ channel openers was the sarcolemmal ATP–sensitive K^+ channel (sarcK_{ATP}), the focus has shifted recently to the mitochondria as the primary target of these compounds.

Plasma membrane K_{ATP} channels have been extensively studied at the molecular level, and their physiological roles in insulin release and in the modulation of vascular tone are indisputable. However, the presence of high densities of sarcK_{ATP} in muscle cells has not been adequately explained. The recent availability of gene knockout mouse models in which components of the sarcK_{ATP} channel have been ablated provides a new opportunity to answer this question. In the mouse, the primary physiological role of sarcK_{ATP} apparently is to help the animal cope with metabolic stress. Mice that lack the pore-forming subunit of the cardiac sarcK_{ATP} (Kir6.2) have a severely compromised ability to tolerate ischemia—even short periods of ischemia lead to rapid ischemic contracture of the heart (108, 109). Similarly, when the K_{ATP} channel is pharmacologically inhibited in mice with HMR1098, a sarcK_{ATP} blocker, ischemic dysfunction is accentuated. Neither ischemic nor chemical preconditioning can mitigate this effect. Function is also compromised with exercise, a more physiological form of metabolic stress (110). These studies indicate that the mouse is highly dependent on sarcK_{ATP} channels for survival under conditions of high energy demand.

In contrast, in larger animal species (e.g., rabbits) (111) and in humans (112), sarcK_{ATP} appears to play a minor role in protecting the heart during ischemia. Selective pharmacological inhibition of sarcK_{ATP} has little or no effect on infarct size after ischemia and reperfusion or on the cardiac preconditioning response (113). Rather, sarcK_{ATP} contributes to postischemic electrical dysfunction by increasing the dispersion of repolarization and the heterogeneity of electrical excitability (114, 115). Moreover, the action potential–shortening effects of sarcK_{ATP} activation during ischemia are not correlated with the extent of protection afforded by K^+ channel openers (107). Thus, other targets of these compounds, including the mitochondria, have been investigated in the context of protection against metabolic stress.

3.4.1. Mitochondrial K_{ATP}. In 1991, Inoue et al. (116) reported the presence of ATP-sensitive K^+ channels in the liver mitochondrial inner membrane, using the direct mitoplast patch-clamp method. The mitochondrial K_{ATP} channel (mitoK_{ATP}) had properties similar to the K_{ATP} channel observed in the sarcolemma of cardiac cells, albeit with a lower conductance. Thus, a link was established between the effects of K^+ channel openers on mitochondrial function and a specific ion channel target. The effects of K^+ channel openers on mitoK_{ATP} have been correlated with protection against ischemic injury in intact hearts (117) and in isolated myocytes (118). The general hypothesis that an increase in mitochondrial inner membrane permeability to K^+ improves cellular tolerance to ischemia-reperfusion injury has found widespread support in various tissues, including the liver, gut, brain, kidney, and heart.

Although studies of mitoK_{ATP} in mitoplasts by patch-clamp have been scarce (116, 119), functional evidence supporting mitoK_{ATP} has been obtained by several methods, including (*a*) electrophysiological recordings of channels in proteoliposomes reconstituted with mitochondrial membrane proteins, (*b*) measurements of K^+ uptake into mitochondria or reconstituted liposomes, (*c*) mitochondrial swelling assays, (*d*) changes in mitochondrial redox potential, and (*e*) alterations in mitochondrial energetic parameters such as respiration and $\Delta\Psi_m$. Several groups have also reported the reconstitution of diazoxide-activated, 5-hydroxydecanoate-inhibited ion channels in planar lipid bilayers (for recent reviews of the evidence supporting a specific mitoK_{ATP} channel in the inner membrane, see References 120–122).

As for most of the mitochondrial channels, the lack of a specific molecular entity associated with mitoK_{ATP} has largely restricted putative identification to the use of pharmacological agents with known properties. A strong caveat is that some of these agents have substantial nonspecific effects on mitochondria that can alter the response to ischemia-reperfusion injury. A variety of K^+ channel openers activate mitoK_{ATP}; these include diazoxide, nicorandil, BMS191095, cromakalim, levcromakalim, EMD60480, EMD57970, pinacidil, RP66471, minoxidil sulfate, and KRN2391. Only the first three show significant selectivity toward the mitochondrial versus the sarcolemmal isoform of the K_{ATP} channel in cardiac myocytes. Another drawback is that there is only one widely available K^+ inhibitor that selectively inhibits mitoK_{ATP} without blocking sarcK_{ATP} (123), 5-hydroxydecanoate. The classical K_{ATP} channel inhibitor, glybenclamide, is a sulfonylurea that blocks both the sarcK_{ATP} and mitoK_{ATP} isoforms, whereas HMR1098 is usually selective for the sarcolemmal channel. In general, the pharmacological profile of the protective effect is consistent with mitoK_{ATP}, rather than sarcK_{ATP}, as the prime target (111).

Recent reports suggest that a number of other compounds—including sildenafil (124), levosimedan (125), YM934 (126), and MCC-134 (127)—modulate mitoK_{ATP}. MCC-134 could inhibit mitoK_{ATP} while activating sarcK_{ATP}. Importantly, MCC-134 prevented diazoxide-mediated protection against simulated ischemia, again supporting the argument that mitoK_{ATP} and not the sarcK_{ATP} channels mediated protection (127).

Another emerging area of interest is how signal transduction pathways either activate or are activated by the mitoK_{ATP} channel. Signaling pathways linked to phosphoinositide hydrolysis, protein kinase C (PKC) activation, or tyrosine kinases are mediators and/or effectors of cellular protection (128). In many cases, the downstream effects of receptor activation can be blocked not only by inhibitors of the kinases but by inhibition of the mitoK_{ATP} channel (129). This begs the question of whether the channel lies upstream or downstream of the posttranslational modifications mediated by either PKC or other kinases. One plausible link between the activation of mitoK_{ATP} and signaling would be a change in redox-sensitive pathways as a result of an increase in mitochondrially derived ROS (130). This could be due to an increase in respiratory rate (and consequent leak of electrons to superoxide) induced by the opening of the K^+ channel. A common effector, glycogen synthase kinase 3β (GSK-3β), has been proposed as the integrator of various preconditioning stimuli, including the actions of K^+ channel openers (131). Activation of GSK-3β blunts the effects of laser-induced oxidative stress on PTP activation in isolated cardiomyocytes. The mitoK_{ATP} channel is likely to be both a target and an effector in these pathways. For example, nitric oxide donors (132) and PKC activators (123) can enhance the activation of mitoK_{ATP} in isolated cardiac cells. In some studies, the mitoK_{ATP} inhibitor 5-hydroxydecanoate not only prevented IPC when applied during the preconditioning phase but also blocked the protection against infarction when given before the long index ischemia: this was the case for both early and delayed preconditioning protocols (133). More recently, the nitric oxide/cyclic GMP/G kinase signal cascade has been implicated in the activation of mitoK_{ATP} (134–136).

POSSIBLE STRUCTURE OF mitoK_{ATP}

Because the pharmacology of mitoK_{ATP} parallels certain combinations of known plasma membrane sulfonylurea receptor (SUR) and inward rectifier K^+ channel (Kir) subunits (191, 192), these proteins have been obvious molecular candidates for the mitochondrial channel. Antibodies against either SUR or Kir6.x (193–200) have been reported to give a strong immunoreactive band in purified mitochondrial membranes. However, there are several criteria to meet when considering these results: (*a*) The mitochondrial inner membrane preparation must be shown to be completely free of other contaminating components, (*b*) the subunit of interest is co-enriched with mitochondrial markers, and (*c*) independent methods must confirm that the protein identified by the antibody is actually the original target (i.e., proteomic analysis should return an SUR or a Kir sequence in the preparation). Although some studies have satisfied the first two criteria, the last has not been achieved, leaving open the question of whether or not mitoK_{ATP} is a homolog of surface membrane K_{ATP} channels.

3.4.2. Mitochondrial K_{Ca}. The second selective K^+ channel to be identified in the mitochondrial inner membrane by direct patch-clamp of mitoplasts was the Ca^{2+}-activated K^+ channel (mitoK_{Ca}), detected in mitochondria from human glioma cells in 1999 (137) and in cardiac mitochondria by our laboratory in 2002 (138). In mitoplast patch-clamp experiments, this channel had a conductance of ~300 pS and was inhibited by the K^+ channel toxins charybdotoxin or iberiotoxin at nM concentrations, thus resembling the properties of Ca^{2+}-activated K^+ channels found in the plasma membranes of some cells. In partially permeabilized adult cardiomyocytes, these toxins also blunted K^+ uptake into mitochondria (138). A K_{Ca} opener, NS-1619, accelerated mitochondrial K^+ uptake and decreased infarct size in rabbit hearts subjected to 30 min of global ischemia and 2 h of reperfusion (138). We suggested that mitoK_{Ca} may be activated under pathophysiological conditions that increase mitochondrial Ca^{2+} uptake as a safeguard against excessive mitochondrial Ca^{2+} accumulation and may also play a physiological role to fine-tune mitochondrial volume and/or Ca^{2+} accumulation under conditions of increased cardiac workload. Ca^{2+} activation of this channel would be expected to cause a partial depolarization of $\Delta\Psi_m$, which would decrease the driving force for Ca^{2+} entry under conditions of positive inotropic stimulation or ischemia.

Subsequent reports have confirmed that K_{Ca} channel openers protect hearts against ischemic injury (139, 140). Moreover, similar to the effect of mitoK_{ATP} activation, mitoK_{Ca}

opening has been implicated in early and delayed preconditioning (141) and may participate in the cardioprotection triggered by ischemia or receptor activation (142, 143). NS-1619-mediated preconditioning can be prevented by blocking K_{Ca} channels with paxilline or by scavenging ROS during the exposure to the opener (140). The latter effect is reminiscent of the role of mitoK_{ATP} in preconditioning and implies that mitoK_{Ca} may be both a trigger and an effector of protection. A recent report (144) showed modulation of mitoK_{Ca} by PKA pathway but not by the PKC pathway, in contradistinction to the regulation of mitoK_{ATP}. Although both mitochondrial K^+ channels appear to have similar effects on mitochondrial function, each has a distinct and nonoverlapping pharmacology (138, 144). This strongly supports the idea that enhanced mitochondrial K^+ uptake is the common factor associated with resistance to cell injury.

Antibodies against the BK type K_{Ca} channel cross-react with purified cardiac mitochondrial membranes (138), and a recent immuno-electron microscopy study revealed BK immunoreactivity in brain mitochondria (145). In addition, the β subunit of the BK channel is present in the mitochondria (146). However, it is still not clear whether the mitochondrial protein is identical, or just antigenically similar, to the surface membrane channel.

3.4.3. Kv1.3. Kv1.3, a Shaker-type K^+ channel, was reported to be present in the mitochondrial inner membrane of Jurkat T lymphocytes (147). In symmetric (134 mM) KCl, channels with a conductance of ~17 pS were detected in a mitoplast preparation and inhibited by margatoxin, a Kv1.3-selective toxin. The mitochondrial channels had properties that were identical to those of plasma membrane Kv1.3 channels, which are also present in this cell line. The channels were not present in mitoplasts from Kv1.3-null cells (CTLL-2) and were restored when Kv1.3 was overexpressed in a mouse cytotoxic T lymphocyte cell line (CTLL-2). Interpretation of the results is complicated by the concomitant expression of the Kv1.3 channel in the plasma membrane, possibly contributing surface membrane contamination to the preparation. However, the authors provide several lines of evidence supporting a mitochondrial inner membrane localization, including observations of the small conductance channel in patches that also contained the 107-pS channel (see Section 3.5.1) and the PTP (see Section 3.6) and the finding that margatoxin can hyperpolarize $\Delta\Psi_m$.

3.5. Anion Channels

3.5.1. IMAC. Mitochondrial swelling induced by cation and anion movements into the matrix compartment has been extensively employed to define ion permeabilities across the mitochondrial inner membrane (4). Following early studies of mitochondrial anion flux from the laboratories of Azzone, Selwyn, and Brierly (reviewed in Reference 12), a series of papers published in the 1980s provided experimental evidence that an inner membrane anion channel (IMAC) was active under certain conditions (e.g., Mg^{2+} depletion, alkalinization) (10, 12, 148). IMAC is permeable to a variety of inorganic (e.g., $SCN^- > NO_3^- > Cl^- > P_i$) and organic (e.g., oxaloacetate^{2-} > citrate^{3-} > malate^{2-} > ATP^{4-}) anions (10). It is inhibited by protons and Mg^{2+} and by many different cationic amphiphiles, including amiodarone ($IC_{50} \sim 1\ \mu M$), amitriptyline ($IC_{50} \sim 10\ \mu M$), dibucaine ($IC_{50} \sim 20\ \mu M$), propranolol ($IC_{50} \sim 25\ \mu M$), and Ro 5–4864 (4′-chlorodiazepam; $IC_{50} \sim 34\ \mu M$), among others (10). IMAC is modulated by thiol crosslinkers, including mersalyl and *N*-ethylmaleamide (10). IMAC activity is thought to be kept in check by endogenous inhibitors and by its strong voltage dependence under physiological conditions. However, the inhibition of IMAC by protons and Mg^{2+} is very temperature sensitive, suggesting that IMAC is poised to respond to small changes in pH or Mg^{2+} under normal conditions (148), perhaps as a mechanism for modulating mitochondrial volume.

Direct single-channel patch-clamp methods revealed a number of partially anion-selective conductances in the inner membrane; the most prominent is the so-called centum pS (or mCtS) channel, a strongly voltage-dependent outwardly rectifying current (149–151). In the first patch-clamp study of the mitochondrial inner membrane, a 107-pS channel was recorded with a P_{Cl^-}:P_{K^+} permeability ratio of 4.5; this channel comprises the main background conductance of the inner membrane (149). In this initial study, no pH dependence was observed, so the authors concluded that the 107 pS was not likely to be the single-channel equivalent of IMAC. However, subsequent investigations revealed properties of the ~100-pS channel in brown fat mitochondria that closely match the IMAC of swelling assays, including inhibition by propranolol, dihydropyridines, and the nucleotide analog cibacron Blue (10, 150). More recently, Schonfeld et al. (152) showed that IMAC was activated by long-chain fatty acids in both intact rat liver mitochondria and in patch-clamp studies of the centum pS channel. Alternatively, a 15-pS anion-selective channel displaying many of the same properties as IMAC (i.e., both are activated by matrix alkalization and blocked by amiodarone, propranolol, and tributyltin) has been described (153).

Of particular interest is the inhibition of IMAC by benzodiazepines (10, 154), considering that the mBzR (see Section 2.4) is present on mitochondrial membranes (albeit on the mitochondrial outer membrane). In studies comparing the effects of Ro 5–4894 and PK11195 on the mCtS and the large multiple-conductance channel (MCC) in mitoplasts, both channel types were blocked by mBzR ligands at concentrations of less than 1 μM (151), whereas central benzodiazepine ligands had no effect at concentrations up to 20 μM. The MCC was also sensitive to inhibition by CsA, but the mCtS channel was not. Another feature discriminating the two channel types was their sensitivity to protoporphyrin IX (PPIX), which is a known ligand of the mBzR (69). The MCC was inhibited by PPIX at low concentrations ($IC_{50} \sim$ 24 nM) but was activated by PPIX at high concentrations (EC_{50} = 244 nM). In contrast, the mCtS was inhibited by PPIX at all concentrations ($IC_{50} \sim$ 35 nM) (151). It is still unclear how the modulation of a presumed outer membrane receptor, mBzR, regulates inner membrane conductances such as the MCC and the mCtS. Researchers have reported other anion-selective channels, including ATP-sensitive anion channels found in yeast inner membranes (155). However, nothing is known about their physiological role and distribution in other species.

3.5.2. IMAC and mitochondrial criticality. IMAC and the mBzR play a role in postischemic electrical and contractile dysfunction in the heart (115). Inhibitors of IMAC and/or the mBzR (e.g., 4′-chlorodiazepam or PK11195) prevent or reverse oscillatory (PTP-independent) mitochondrial depolarizations induced by substrate deprivation (156) or oxidative stress (157) in adult cardiomyocytes. The underlying mechanism of the mitochondrial oscillator, explored in both experimental and computational studies (157–159), incorporates the concept of mitochondrial ROS–induced ROS release [originally coined to describe direct laser-induced PTP-mediated depolarization of $\Delta\Psi_m$ (160)], which triggers IMAC activation in a positive feedback loop (158). Remarkably, the local depolarization of just a few mitochondria in a cardiac cell can lead to cell-wide self-sustaining oscillations of $\Delta\Psi_m$ (157). Hence, the term mitochondrial criticality was forwarded to refer to the cellular conditions leading up to a breakpoint between stable and unstable $\Delta\Psi_m$ in the mitochondrial network (159). The approach to the critical state depends on whether mitochondrial ROS production by the respiratory chain exceeds a threshold level in a significant fraction of mitochondria (~60%) in the network. At the critical point, the weakly coupled fluctuations of individual mitochondria transition into an emergent spatiotemporal pattern of synchronized limit-cycle oscillations. This mitochondrial network phenomenon is tightly coupled to the cardiomyocyte's electrophysiological response through energy-sensitive sarcK_{ATP} channels on the surface membrane, leading to scaling of the organelle-level dysfunction to the whole organ during ischemia and reperfusion (115).

3.5.3. CLC, mCLIC. Given the lack of information on the molecular structure of the anion-permeable mitochondrial channels described above, it is worthwhile to examine potential candidate channels known to be localized to mitochondrial membranes. Two classes of chloride channels, the ClC type and the CLIC type, have been reported to be present in mitochondria. The voltage-dependent chloride channel (ClC) family comprises a large class of structurally related membrane proteins with putative chloride channel activities. Although some members of the ClC family (ClC-3, ClC-4, ClC-5) may be targeted to intracellular locations, there is no evidence for mitochondrial localization. However, ClC-Nt, an anion channel isolated from tobacco plants, is enriched in mitochondrial membranes and has been proposed as a candidate for IMAC (161).

CLICs (chloride intracellular channels) represent a new class of intracellular anion channels that have been identified by their homology to the p64 protein. A mitochondrial homolog, mtCLIC, has been identified from differential display analysis of differentiating mouse keratinocytes from $p53^{+/+}$ and $p53^{-/-}$ mice. MtCLIC colocalized with cytochrome oxidase in keratinocyte mitochondria but also was detected in the cytoplasmic compartment. This p53-regulated putative channel has been associated with apoptosis (162) and may exist as either a soluble or transmembrane form that may translocate to the nucleus in response to cell stress (163).

3.6. Permeability Transition Pore

Permeabilization of the mitochondrial inner membrane, first detected as a large-amplitude swelling response in response to a variety of effectors (e.g., thyroxine, Ca^{2+}, P_i), has been known for more than 50 years (8, 164). In the 1970s, the MPT was characterized in greater detail (165). It was proposed to involve the activation of a PTP (reviewed in References 40, 166, and 167). The finding that the immunosuppressant drug CsA could inhibit the MPT provided a crucial tool for subsequent investigation (168, 169) and spurred the identification of its mitochondrial binding protein, cyclophilin D (170, 171). Definitive evidence that the PTP was truly an ion-permeable channel was obtained upon the application of the technique of patch-clamping mitoplasts and the detection of MCC (153, 172–175), or mitochondrial megachannels (176–179), which displayed conductances up to ~1.3 nS that were inhibited by CsA. The PTP allows the passage of ions and metabolites up to ~1500 molecular weight, for an apparent pore diameter of ~3 nm. In addition to the most commonly employed trigger, Ca^{2+} overload of the mitochondria, PTP opening is promoted by $\Delta\Psi_m$ depolarization, P_i, ROS, and thiol modification, and it can be inhibited by adenine nucleotides, Mg^{2+}, or matrix protons. Evidence that the ANT is a component of the PTP or even the pore itself was provided by the findings that bongkrekic acid, an inhibitor that stabilizes the ANT in the "m" conformation, inhibits the PTP and that atractylosides, which stabilize ANT in the "c" conformation, activate the PTP (180). Moreover, purified ANT preparations formed large Ca^{2+}-activated ion channels upon reconstitution in lipid bilayers (181). Similar channels of 600-pS conductance were obtained when *Neurospora crassa* ANT was expressed in *Escherichia coli* and reconstituted, a preparation that is devoid of possible contamination with other mitochondrial membrane components found in heart mitochondrial preparations (182). These channels were inhibited by

ADP and bongkrekate, but not by carboxyatractylate, and were modulated by cyclophilin, whereas CsA abolished the cyclophilin effect.

Although the structure of the PTP is widely portrayed as consisting of a multiprotein complex prominently featuring the ANT, VDAC, cyclophilin, the F_1F_O ATPase, hexokinase, and other modulatory proteins, recent observations have challenged this model. The two proteins central to prior models of PTP structure, ANT (39) and cyclophilin D (183–186), have been ablated in recent gene knockout mouse studies. The results support a modulatory (i.e., involving an increase in the threshold for Ca^{2+} activation) rather than an obligatory role for these proteins in the MPT. As discussed above (Section 2.1), data from mitoplast patch-clamp studies and gene-targeted mice also indicate that VDAC is not an obligatory member of the PTP protein complex.

Regardless of the structural details of the PTP, multiple lines of evidence support the idea that in tissues such as the heart, PTP opening occurs only during reperfusion after ischemia (115, 187) and is a major checkpoint on the route toward cell injury and death, although classic stimuli can readily induce apoptotic cell death in MPT-resistant mouse strains (186). However, mitochondrial depolarization in response to stress is not always the result of PTP opening (157). Thus, multiple assays must be used to determine whether a MPT has occurred; these include sensitivity to PTP inhibitors (e.g., CsA or sanglifehrin) and the direct demonstration that small molecules can permeate the mitochondrial inner membrane.

3.7. Connexin 43

Rodriguez-Sinovas et al. (188) recently demonstrated that connexin 43 (Cx43), the protein responsible for forming gap-junctional channels between cells, is present in mitochondrial membrane fractions and is increased in mitochondria from hearts subjected to an IPC protocol. Moreover, the protection afforded by IPC in wild-type mice is absent in heterozygous Cx43-deficient ($Cx43^{+/-}$) mice (189). Diazoxide-mediated protection is also lost in $Cx43^{+/-}$ mice. This effect was tentatively attributed to a decrease in cardiomyocyte ROS production during chemical preconditioning, as measured using a probe that loads (depending on $\Delta\Psi_m$) into the mitochondrial matrix, where it is then oxidized (190). Cx43 was specifically enriched in mitochondrial inner membrane fractions (188), and coimmunoprecipitation studies demonstrated an interaction of Cx43 with Tom20, part of the protein import machinery (see Section 2.5). However, it is presently unclear whether Cx43 forms a channel in the inner membrane. Several factors make it unlikely that it forms a pore: (*a*) Cx43 hemichannels generally do not conduct until the connexon is formed when the hemichannels meet at the junctional membrane, (*b*) there is no evidence that mitochondrial respiration or $\Delta\Psi_m$ is influenced by the mitochondrial Cx43 content, and (*c*) the Cx43 appeared to be in the phosphorylated form, which can induce channel closure. However, the activity of Cx43 may be altered by ischemia and/or reperfusion. Inhibiting the Hsp-90-dependent protein import pathway with geldanamycin prevents the IPC-mediated Cx43 translocation to the mitochondria. But the role of mitochondrial Cx43 in cardioprotection is unclear because geldamycin did not alter the protective effect (188). Furthermore, although geldanamycin abolished diazoxide-mediated protection, there was no correlation between this protection and Cx43 levels (188), nor was there a correlation between the extent of injury and the level of Cx43 in the absence of the chemical preconditioning stimulus.

4. CONCLUDING REMARKS

Ion channels can transport millions of ions per second, and the electrochemical driving forces for ion movement across the inner membrane are enormous. Thus, the knowledge gap regarding mitochondrial ion channel structure is perhaps understandable—these proteins must be present in extremely low abundance or have a very low open probability to maintain the low permeability to ions required to exploit the protonmotive force for ATP generation. Nevertheless, although mitochondrial ion channels may be highly controlled and open only briefly, the significance of their effects cannot be overstated. Mitochondrial ion channels are crucial to the mechanism of energy supply and demand matching and are the decisive factor in determining whether a cell lives or dies.

Mitochondrial ion channels for Ca^{2+}, K^+, or anions have been functionally and pharmacologically characterized at many levels, spanning from the single channel to the intact cell and to whole-organ function. The challenge ahead lies in defining the molecular structures responsible for forming the ion-selective mitochondrial pores. Achieving this goal will undoubtedly spur the development of novel and specific therapeutic agents targeted to the mitochondria. As the organelles responsible for integrating and responding to environmental challenges, mitochondria are the hub of all cellular functions and play a central role as determinants of cell life and death in a variety of pathologies, including acute coronary syndrome, neurodegeneration, cancer, and aging. Perhaps the key to managing these health problems will come from the next phase in the history of mitochondria to arise from mitochondrial ion channel discovery.

LITERATURE CITED

1. Altmann R. 1890. *Die Elementarorganismen und ihre Beziehungen zu den Zellen*. Leipzig: Veit & Co. 145 pp.
2. Rizzuto R, Duchen MR, Pozzan T. 2004. Flirting in little space: the ER/mitochondria Ca^{2+} liaison. *Sci. STKE* 2004:re1
3. Mitchell P. 1961. Coupling of phosphorylation to electron and hydrogen transfer by a chemi-osmotic type of mechanism. *Nature* 191:144–48
4. Nicholls DG, Ferguson SJ. 2002. *Bioenergetics3*. London: Academic. 297 pp.
5. Lewis MR, Lewis WH. 1915. Mitochondria (and other cytoplasmic structures) in tissue cultures. *Amer. J. Anat.* 17:339–401
6. Lehninger AL. 1962. Water uptake and extrusion by mitochondria in relation to oxidative phosphorylation. *Physiol. Rev.* 42:467–517
7. Saris NE, Carafoli E. 2005. A historical review of cellular calcium handling, with emphasis on mitochondria. *Biochemistry* 70:187–94
8. Chance B. 1965. The energy-linked reaction of calcium with mitochondria. *J. Biol. Chem.* 240:2729–48
9. Garlid KD. 1996. Cation transport in mitochondria—the potassium cycle. *Biochim. Biophys. Acta* 1275:123–26
10. Beavis AD. 1992. Properties of the inner membrane anion channel in intact mitochondria. *J. Bioenerg. Biomembr.* 24:77–90
11. Mitchell P, Moyle J. 1967. Respiration-driven proton translocation in rat liver mitochondria. *Biochem. J.* 105:1147–62
12. Garlid KD, Beavis AD. 1986. Evidence for the existence of an inner membrane anion channel in mitochondria. *Biochim. Biophys. Acta* 853:187–204
13. Schein SJ, Colombini M, Finkelstein A. 1976. Reconstitution in planar lipid bilayers of a voltage-dependent anion-selective channel obtained from paramecium mitochondria. *J. Membr. Biol.* 30:99–120

14. Rostovtseva TK, Tan W, Colombini M. 2005. On the role of VDAC in apoptosis: fact and fiction. *J. Bioenerg. Biomembr.* 37:129–42
15. Levadny V, Colombini M, Li XX, Aguilella VM. 2002. Electrostatics explains the shift in VDAC gating with salt activity gradient. *Biophys. J.* 82:1773–83
16. Colombini M. 2004. VDAC: the channel at the interface between mitochondria and the cytosol. *Mol. Cell Biochem.* 256–257:107–15
17. Porcelli AM, Ghelli A, Zanna C, Pinton P, Rizzuto R, Rugolo M. 2005. pH difference across the outer mitochondrial membrane measured with a green fluorescent protein mutant. *Biochem. Biophys. Res. Commun.* 326:799–804
18. Zizi M, Forte M, Blachly-Dyson E, Colombini M. 1994. NADH regulates the gating of VDAC, the mitochondrial outer membrane channel. *J. Biol. Chem.* 269:1614–16
19. Holden MJ, Colombini M. 1993. The outer mitochondrial membrane channel, VDAC, is modulated by a protein localized in the intermembrane space. *Biochim. Biophys. Acta* 1144:396–402
20. Rostovtseva TK, Antonsson B, Suzuki M, Youle RJ, Colombini M, Bezrukov SM. 2004. Bid, but not Bax, regulates VDAC channels. *J. Biol. Chem.* 279:13575–83
21. Roman I, Figys J, Steurs G, Zizi M. 2006. Hunting interactomes of a membrane protein: obtaining the largest set of voltage-dependent anion channel-interacting protein epitopes. *Mol. Cell Proteom.* 5:1667–80
22. Blachly-Dyson E, Peng S, Colombini M, Forte M. 1990. Selectivity changes in site-directed mutants of the VDAC ion channel: structural implications. *Science* 247:1233–36
23. Baines CP, Song CX, Zheng YT, Wang GW, Zhang J, et al. 2003. Protein kinase Cε interacts with and inhibits the permeability transition pore in cardiac mitochondria. *Circ. Res.* 92:873–80
24. Liberatori S, Canas B, Tani C, Bini L, Buonocore G, et al. 2004. Proteomic approach to the identification of voltage-dependent anion channel protein isoforms in guinea pig brain synaptosomes. *Proteomics* 4:1335–40
25. Vander Heiden MG, Chandel NS, Li XX, Schumacker PT, Colombini M, Thompson CB. 2000. Outer mitochondrial membrane permeability can regulate coupled respiration and cell survival. *Proc. Natl. Acad. Sci. USA* 97:4666–71
26. Azoulay-Zohar H, Israelson A, Abu-Hamad S, Shoshan-Barmatz V. 2004. In self-defense: Hexokinase promotes voltage-dependent anion channel closure and prevents mitochondria-mediated apoptotic cell death. *Biochem. J.* 377:347–55
27. Jonas EA, Buchanan J, Kaczmarek LK. 1999. Prolonged activation of mitochondrial conductances during synaptic transmission. *Science* 286:1347–50
28. Jonas EA, Hickman JA, Chachar M, Polster BM, Brandt TA, et al. 2004. Proapoptotic N-truncated BCL-xL protein activates endogenous mitochondrial channels in living synaptic terminals. *Proc. Natl. Acad. Sci. USA* 101:13590–95
29. Lemasters JJ, Holmuhamedov E. 2006. Voltage-dependent anion channel (VDAC) as mitochondrial governator—thinking outside the box. *Biochim. Biophys. Acta* 1762:181–90
30. Bathori G, Csordas G, Garcia-Perez C, Davies E, Hajnoczky G. 2006. Ca^{2+}-dependent control of the permeability properties of the mitochondrial outer membrane and voltage-dependent anion-selective channel (VDAC). *J. Biol. Chem.* 281:17347–58
31. Moran O, Sciancalepore M, Sandri G, Panfili E, Bassi R, et al. 1992. Ionic permeability of the mitochondrial outer membrane. *Eur. Biophys. J.* 20:311–9
32. Pastorino JG, Hoek JB. 2003. Hexokinase II: the integration of energy metabolism and control of apoptosis. *Curr. Med. Chem.* 10:1535–51
33. Sharpe JC, Arnoult D, Youle RJ. 2004. Control of mitochondrial permeability by Bcl-2 family members. *Biochim. Biophys. Acta* 1644:107–13

34. Siskind LJ, Kolesnick RN, Colombini M. 2006. Ceramide forms channels in mitochondrial outer membranes at physiologically relevant concentrations. *Mitochondrion* 6:118–25
35. Madesh M, Hajnoczky G. 2001. VDAC-dependent permeabilization of the outer mitochondrial membrane by superoxide induces rapid and massive cytochrome *c* release. *J. Cell Biol.* 155:1003–15
36. Shimizu S, Tsujimoto Y. 2000. Proapoptotic BH3-only Bcl-2 family members induce cytochrome *c* release, but not mitochondrial membrane potential loss, and do not directly modulate voltage-dependent anion channel activity. *Proc. Natl. Acad. Sci. USA* 97:577–82
37. Dejean LM, Martinez-Caballero S, Kinnally KW. 2006. Is MAC the knife that cuts cytochrome *c* from mitochondria during apoptosis? *Cell Death Differ.* 13:1387–95
38. Crompton M, Barksby E, Johnson N, Capano M. 2002. Mitochondrial intermembrane junctional complexes and their involvement in cell death. *Biochimie* 84:143–52
39. Kokoszka JE, Waymire KG, Levy SE, Sligh JE, Cai J, et al. 2004. The ADP/ATP translocator is not essential for the mitochondrial permeability transition pore. *Nature* 427:461–65
40. Di Lisa F, Bernardi P. 2006. Mitochondria and ischemia-reperfusion injury of the heart: Fixing a hole. *Cardiovasc. Res.* 70:191–99
41. Lohret TA, Kinnally KW. 1995. Multiple conductance channel activity of wild-type and voltage-dependent anion-selective channel (VDAC)-less yeast mitochondria. *Biophys. J.* 68:2299–309
42. Krauskopf A, Eriksson O, Craigen WJ, Forte MA, Bernardi P. 2006. Properties of the permeability transition in VDAC1$^{-/-}$ mitochondria. *Biochim. Biophys. Acta* 1757:590–95
43. Capano M, Crompton M. 2002. Biphasic translocation of Bax to mitochondria. *Biochem. J.* 367:169–78
44. Banerjee J, Ghosh S. 2004. Bax increases the pore size of rat brain mitochondrial voltage-dependent anion channel in the presence of tBid. *Biochem. Biophys. Res. Commun.* 323:310–14
45. Banerjee J, Ghosh S. 2006. Phosphorylation of rat brain mitochondrial voltage-dependent anion as a potential tool to control leakage of cytochrome *c*. *J. Neurochem.* 98:670–76
46. Shoshan-Barmatz V, Israelson A, Brdiczka D, Sheu SS. 2006. The voltage-dependent anion channel (VDAC): function in intracellular signaling, cell life and cell death. *Curr. Pharm. Des.* 12:2249–70
47. Vander Heiden MG, Li XX, Gottleib E, Hill RB, Thompson CB, Colombini M. 2001. Bcl-xL promotes the open configuration of the voltage-dependent anion channel and metabolite passage through the outer mitochondrial membrane. *J. Biol. Chem.* 276:19414–19
48. Duan S, Hajek P, Lin C, Shin SK, Attardi G, Chomyn A. 2003. Mitochondrial outer membrane permeability change and hypersensitivity to digitonin early in staurosporine-induced apoptosis. *J. Biol. Chem.* 278:1346–53
49. Pavlov EV, Priault M, Pietkiewicz D, Cheng EH, Antonsson B, et al. 2001. A novel, high conductance channel of mitochondria linked to apoptosis in mammalian cells and Bax expression in yeast. *J. Cell Biol.* 155:725–31
50. Schendel SL, Xie Z, Montal MO, Matsuyama S, Montal M, Reed JC. 1997. Channel formation by antiapoptotic protein Bcl-2. *Proc. Natl. Acad. Sci. USA* 94:5113–18
51. Antonsson B, Conti F, Ciavatta A, Montessuit S, Lewis S, et al. 1997. Inhibition of Bax channel-forming activity by Bcl-2. *Science* 277:370–72
52. Guihard G, Bellot G, Moreau C, Pradal G, Ferry N, et al. 2004. The mitochondrial apoptosis-induced channel (MAC) corresponds to a late apoptotic event. *J. Biol. Chem.* 279:46542–50

53. Dejean LM, Martinez-Caballero S, Manon S, Kinnally KW. 2006. Regulation of the mitochondrial apoptosis-induced channel, MAC, by BCL-2 family proteins. *Biochim. Biophys. Acta* 1762:191–201
54. Grigoriev SM, Muro C, Dejean LM, Campo ML, Martinez-Caballero S, Kinnally KW. 2004. Electrophysiological approaches to the study of protein translocation in mitochondria. *Int. Rev. Cytol.* 238:227–74
55. Henry JP, Chich JF, Goldschmidt D, Thieffry M. 1989. Ionic mitochondrial channels: characteristics and possible role in protein translocation. *Biochimie* 71:963–68
56. Juin P, Thieffry M, Henry JP, Vallette FM. 1997. Relationship between the peptide-sensitive channel and the mitochondrial outer membrane protein translocation machinery. *J. Biol. Chem.* 272:6044–50
57. Veenman L, Gavish M. 2006. The peripheral-type benzodiazepine receptor and the cardiovascular system. Implications for drug development. *Pharmacol. Ther.* 110:503–24
58. Anholt RR, Pedersen PL, De Souza EB, Snyder SH. 1986. The peripheral-type benzodiazepine receptor. Localization to the mitochondrial outer membrane. *J. Biol. Chem.* 261:576–83
59. Papadopoulos V. 2004. In search of the function of the peripheral-type benzodiazepine receptor. *Endocr. Res.* 30:677–84
60. Desjardins P, Bandeira P, Raghavendra Rao VL, Ledoux S, Butterworth RF. 1997. Increased expression of the peripheral-type benzodiazepine receptor-isoquinoline carboxamide binding protein mRNA in brain following portacaval anastomosis. *Brain Res.* 758:255–58
61. McEnery MW, Snowman AM, Trifiletti RR, Snyder SH. 1992. Isolation of the mitochondrial benzodiazepine receptor: association with the voltage-dependent anion channel and the adenine nucleotide carrier. *Proc. Natl. Acad. Sci. USA* 89:3170–74
62. Joseph-Liauzun E, Farges R, Delmas P, Ferrara P, Loison G. 1997. The M_r 18,000 subunit of the peripheral-type benzodiazepine receptor exhibits both benzodiazepine and isoquinoline carboxamide binding sites in the absence of the voltage-dependent anion channel or of the adenine nucleotide carrier. *J. Biol. Chem.* 272:28102–6
63. Galiègue S, Jbilo O, Combes T, Bribes E, Carayon P, et al. 1999. Cloning and characterization of PRAX-1. A new protein that specifically interacts with the peripheral benzodiazepine receptor. *J. Biol. Chem.* 274:2938–52
64. Steinberg RA, Symcox MM, Sollid S, Ogreid D. 1996. Arginine 210 is not a critical residue for the allosteric interactions mediated by binding of cyclic AMP to site A of regulatory (RIα) subunit of cyclic AMP-dependent protein kinase. *J. Biol. Chem.* 271:27630–36
65. Blahos J, Whalin ME, Krueger KE. 1995. Identification and purification of a 10-kDa protein associated with mitochondrial benzodiazepine receptors. *J. Biol. Chem.* 270:20285–91
66. Guidotti A, Forchetti CM, Corda MG, Konkel D, Bennett CD, Costa E. 1983. Isolation, characterization, and purification to homogeneity of an endogenous polypeptide with agonistic action on benzodiazepine receptors. *Proc. Natl. Acad. Sci. USA* 80:3531–35
67. Costa E, Guidotti A. 1991. Diazepam binding inhibitor (DBI): a peptide with multiple biological actions. *Life Sci.* 49:325–44
68. Duparc C, Lefebvre H, Tonon MC, Vaudry H, Kuhn JM. 2003. Characterization of endozepines in the human testicular tissue: effect of triakontatetraneuropeptide on testosterone secretion. *J. Clin. Endocrinol. Metab.* 88:5521–28
69. Verma A, Nye JS, Snyder SH. 1987. Porphyrins are endogenous ligands for the mitochondrial (peripheral-type) benzodiazepine receptor. *Proc. Natl. Acad. Sci. USA* 84:2256–60
70. Simpson ER, Waterman MR. 1988. Regulation of the synthesis of steroidogenic enzymes in adrenal cortical cells by ACTH. *Annu. Rev. Physiol.* 50:427–40

71. Krueger KE. 1991. Peripheral-type benzodiazepine receptors: a second site of action for benzodiazepines. *Neuropsychopharmacology* 4:237–44
72. Mukhin AG, Papadopoulos V, Costa E, Krueger KE. 1989. Mitochondrial benzodiazepine receptors regulate steroid biosynthesis. *Proc. Natl. Acad. Sci. USA* 86:9813–16
73. Papadopoulos V, Amri H, Li H, Boujrad N, Vidic B, Garnier M. 1997. Targeted disruption of the peripheral-type benzodiazepine receptor gene inhibits steroidogenesis in the R2C Leydig tumor cell line. *J. Biol. Chem.* 272:32129–35
74. Bernassau JM, Reversat JL, Ferrara P, Caput D, Lefur G. 1993. A 3D model of the peripheral benzodiazepine receptor and its implication in intra mitochondrial cholesterol transport. *J. Mol. Graph.* 11:236–44
75. Papadopoulos V, Boujrad N, Ikonomovic MD, Ferrara P, Vidic B. 1994. Topography of the Leydig cell mitochondrial peripheral-type benzodiazepine receptor. *Mol. Cell Endocrinol.* 104:R5–9
76. Lueddens HW, Skolnick P. 1987. 'Peripheral type' benzodiazepine receptors in the kidney: regulation of radioligand binding by anions and DIDS. *Eur. J. Pharmacol.* 133:205–14
77. Beaumont K, Skowronski R, Vaughn DA, Fanestil DD. 1988. Interactions of lipids with peripheral-type benzodiazepine receptors. *Biochem. Pharmacol.* 37:1009–14
78. Havoundjian H, Cohen RM, Paul SM, Skolnick P. 1986. Differential sensitivity of "central" and "peripheral" type benzodiazepine receptors to phospholipase A2. *J. Neurochem.* 46:804–11
79. Hirsch JD, Beyer CF, Malkowitz L, Loullis CC, Blume AJ. 1989. Characterization of ligand binding to mitochondrial benzodiazepine receptors. *Mol. Pharmacol.* 35:164–72
80. Brand MD, Chien LF, Ainscow EK, Rolfe DF, Porter RK. 1994. The causes and functions of mitochondrial proton leak. *Biochim. Biophys. Acta* 1187:132–39
81. Nicholls DG, Grav HJ, Lindberg O. 1972. Mitochondrial from hamster brown-adipose tissue. Regulation of respiration in vitro by variations in volume of the matrix compartment. *Eur. J. Biochem.* 31:526–33
82. Heaton GM, Wagenvoord RJ, Kemp AJ, Nicholls DG. 1978. Brown-adipose-tissue mitochondria: photoaffinity labeling of the regulatory site of energy dissipation. *Eur. J. Biochem.* 82:515–21
83. Ricquier D, Bouillaud F. 2000. Mitochondrial uncoupling proteins: from mitochondria to the regulation of energy balance. *J. Physiol.* 529(Pt. 1):3–10
84. Starkov AA. 1997. "Mild" uncoupling of mitochondria. *Biosci. Rep.* 17:273–79
85. Jezek P. 1999. Fatty acid interaction with mitochondrial uncoupling proteins. *J. Bioenerg. Biomembr.* 31:457–66
86. Stucki JW. 1980. The optimal efficiency and the economic degrees of coupling of oxidative phosphorylation. *Eur. J. Biochem.* 109:269–83
87. Nicholls DG. 1974. Hamster brown-adipose-tissue mitochondria. The chloride permeability of the inner membrane under respiring conditions, the influence of purine nucleotides. *Eur. J. Biochem.* 49:585–93
88. Huang SG, Klingenberg M. 1996. Chloride channel properties of the uncoupling protein from brown adipose tissue mitochondria: a patch-clamp study. *Biochemistry* 35:16806–14
89. Yamaguchi H, Jelokhani-Niaraki M, Kodama H. 2004. Second transmembrane domain of human uncoupling protein 2 is essential for its anion channel formation. *FEBS Lett.* 577:299–304
90. Klingenberg M, Winkler E, Echtay K. 2001. Uncoupling protein, H^+ transport and regulation. *Biochem. Soc. Trans.* 29:806–11

91. Nadtochiy SM, Tompkins AJ, Brookes PS. 2006. Different mechanisms of mitochondrial proton leak in ischaemia/reperfusion injury and preconditioning: implications for pathology and cardioprotection. *Biochem. J.* 395:611–18
92. Reed KC, Bygrave FL. 1974. The inhibition of mitochondrial calcium transport by lanthanides and ruthenium red. *Biochem. J.* 140:143–55
93. Gunter TE, Buntinas L, Sparagna G, Eliseev R, Gunter K. 2000. Mitochondrial calcium transport: mechanisms and functions. *Cell Calcium* 28:285–96
94. Kirichok Y, Krapivinsky G, Clapham DE. 2004. The mitochondrial calcium uniporter is a highly selective ion channel. *Nature* 427:360–64
95. Sottocasa G, Sandri G, Panfili E, De Bernard B, Gazzotti P, et al. 1972. Isolation of a soluble Ca^{2+} binding glycoprotein from ox liver mitochondria. *Biochem. Biophys. Res. Commun.* 47:808–13
96. Panfili E, Sandri G, Sottocasa GL, Lunazzi G, Liut G, Graziosi G. 1976. Specific inhibition of mitochondrial Ca^{2+} transport by antibodies directed to the Ca^{2+}-binding glycoprotein. *Nature* 264:185–86
97. Mironova GD, Sirota TV, Pronevich LA, Trofimenko NV, Mironov GP, et al. 1982. Isolation and properties of Ca^{2+}-transporting glycoprotein and peptide from beef heart mitochondria. *J. Bioenerg. Biomembr.* 14:213–25
98. Saris NE, Sirota TV, Virtanen I, Niva K, Penttila T, et al. 1993. Inhibition of the mitochondrial calcium uniporter by antibodies against a 40-kDa glycoproteinT. *J. Bioenerg. Biomembr.* 25:307–12
99. Mironova GD, Baumann M, Kolomytkin O, Krasichkova Z, Berdimuratov A, et al. 1994. Purification of the channel component of the mitochondrial calcium uniporter and its reconstitution into planar lipid bilayers. *J. Bioenerg. Biomembr.* 26:231–38
100. Zazueta C, Masso F, Paez A, Bravo C, Vega A, et al. 1994. Identification of a 20-kDa protein with calcium uptake transport activity. Reconstitution in a membrane model. *J. Bioenerg. Biomembr.* 26:555–62
101. Zazueta C, Zafra G, Vera G, Sanchez C, Chavez E. 1998. Advances in the purification of the mitochondrial Ca^{2+} uniporter using the labeled inhibitor ^{103}Ru360. *J. Bioenerg. Biomembr.* 30:489–98
102. Sparagna GC, Gunter KK, Sheu SS, Gunter TE. 1995. Mitochondrial calcium uptake from physiological-type pulses of calcium. A description of the rapid uptake mode. *J. Biol. Chem.* 270:27510–15
103. Buntinas L, Gunter KK, Sparagna GC, Gunter TE. 2001. The rapid mode of calcium uptake into heart mitochondria (RaM): comparison to RaM in liver mitochondria. *Biochim. Biophys. Acta* 1504:248–61
104. Beutner G, Sharma VK, Lin L, Ryu SY, Dirksen RT, Sheu SS. 2005. Type 1 ryanodine receptor in cardiac mitochondria: transducer of excitation-metabolism coupling. *Biochim. Biophys. Acta* 1717:1–10
105. Bernardi P, Paradisi V, Pozzan T, Azzone GF. 1984. Pathway for uncoupler-induced calcium efflux in rat liver mitochondria: inhibition by ruthenium red. *Biochemistry* 23:1645–51
106. Lamping KA, Gross GJ. 1985. Improved recovery of myocardial segment function following a short coronary occlusion in dogs by nicorandil, a potential new antianginal agent, and nifedipine. *J. Cardiovasc. Pharmacol.* 7:158–66
107. Grover GJ, Garlid KD. 2000. ATP-sensitive potassium channels: a review of their cardioprotective pharmacology. *J. Mol. Cell Cardiol.* 32:677–95

108. Suzuki M, Sasaki N, Miki T, Sakamoto N, Ohmoto-Sekine Y, et al. 2002. Role of sarcolemmal K_{ATP} channels in cardioprotection against ischemia/reperfusion injury in mice. *J. Clin. Invest.* 109:509–16
109. Suzuki M, Saito T, Sato T, Tamagawa M, Miki T, et al. 2003. Cardioprotective effect of diazoxide is mediated by activation of sarcolemmal but not mitochondrial ATP-sensitive potassium channels in mice. *Circulation* 107:682–85
110. Zingman LV, Hodgson DM, Bast PH, Kane GC, Perez-Terzic C, et al. 2002. Kir6.2 is required for adaptation to stress. *Proc. Natl. Acad. Sci. USA* 99:13278–83
111. Sato T, Sasaki N, Seharaseyon J, O'Rourke B, Marban E. 2000. Selective pharmacological agents implicate mitochondrial but not sarcolemmal K_{ATP} channels in ischemic cardioprotection. *Circulation* 101:2418–23
112. Ghosh S, Standen NB, Galinanes M. 2000. Evidence for mitochondrial K_{ATP} channels as effectors of human myocardial preconditioning. *Cardiovasc. Res.* 45:934–40
113. Tanno M, Miura T, Tsuchida A, Miki T, Nishino Y, et al. 2001. Contribution of both the sarcolemmal K_{ATP} and mitochondrial K_{ATP} channels to infarct size limitation by K_{ATP} channel openers: differences from preconditioning in the role of sarcolemmal K_{ATP} channels. *Naunyn-Schmiedebergs Arch. Pharmacol.* 364:226–32
114. Billman GE, Englert HC, Scholkens BA. 1998. HMR 1883, a novel cardioselective inhibitor of the ATP-sensitive potassium channel. Part II: effects on susceptibility to ventricular fibrillation induced by myocardial ischemia in conscious dogs. *J. Pharmacol. Exp. Ther.* 286:1465–73
115. Akar FG, Aon MA, Tomaselli GF, O'Rourke B. 2005. The mitochondrial origin of postischemic arrhythmias. *J. Clin. Invest.* 115:3527–35
116. Inoue I, Nagase H, Kishi K, Higuti T. 1991. ATP-sensitive K^+ channel in the mitochondrial inner membrane. *Nature* 352:244–47
117. Garlid KD, Paucek P, Yarov-Yarovoy V, Murray HN, Darbenzio RB, et al. 1997. Cardioprotective effect of diazoxide and its interaction with mitochondrial ATP-sensitive K^+ channels. Possible mechanism of cardioprotection. *Circ. Res.* 81:1072–82
118. Liu Y, Sato T, O'Rourke B, Marban E. 1998. Mitochondrial ATP-dependent potassium channels: novel effectors of cardioprotection? *Circulation* 97:2463–69
119. Dahlem YA, Horn TF, Buntinas L, Gonoi T, Wolf G, Siemen D. 2004. The human mitochondrial K_{ATP} channel is modulated by calcium and nitric oxide: a patch-clamp approach. *Biochim. Biophys. Acta* 1656:46–56
120. O'Rourke B. 2004. Evidence for mitochondrial K^+ channels and their role in cardioprotection. *Circ. Res.* 94:420–32
121. Garlid KD, Dos Santos P, Xie ZJ, Costa AD, Paucek P. 2003. Mitochondrial potassium transport: the role of the mitochondrial ATP-sensitive K^+ channel in cardiac function and cardioprotection. *Biochim. Biophys. Acta* 1606:1–21
122. Facundo HT, Fornazari M, Kowaltowski AJ. 2006. Tissue protection mediated by mitochondrial K^+ channels. *Biochim. Biophys. Acta* 1762:202–12
123. Sato T, O'Rourke B, Marban E. 1998. Modulation of mitochondrial ATP-dependent K^+ channels by protein kinase C. *Circ. Res.* 83:110–14
124. Ockaili R, Salloum F, Hawkins J, Kukreja RC. 2002. Sildenafil (Viagra) induces powerful cardioprotective effect via opening of mitochondrial K_{ATP} channels in rabbits. *Am. J. Physiol. Heart Circ. Physiol.* 283:H1263–69
125. Kopustinskiene DM, Pollesello P, Saris NE. 2001. Levosimendan is a mitochondrial K_{ATP} channel opener. *Eur. J. Pharmacol.* 428:311–14

126. Tanonaka K, Taguchi T, Koshimizu M, Ando T, Morinaka T, et al. 1999. Role of an ATP-sensitive potassium channel opener, YM934, in mitochondrial energy production in ischemic/reperfused heart. *J. Pharmacol. Exp. Ther.* 291:710–16
127. Sasaki N, Murata M, Guo Y, Jo SH, Ohler A, et al. 2003. MCC-134, a single pharmacophore, opens surface ATP-sensitive potassium channels, blocks mitochondrial ATP-sensitive potassium channels, and suppresses preconditioning. *Circulation* 107:1183–88
128. Cohen MV, Baines CP, Downey JM. 2000. Ischemic preconditioning: from adenosine receptor to K_{ATP} channel. *Annu. Rev. Physiol.* 62:79–109
129. Uchiyama Y, Otani H, Okada T, Uchiyama T, Ninomiya H, et al. 2003. Integrated pharmacological preconditioning in combination with adenosine, a mitochondrial K_{ATP} channel opener and a nitric oxide donor. *J. Thorac. Cardiovasc. Surg.* 126:148–59
130. Pain T, Yang XM, Critz SD, Yue Y, Nakano A, et al. 2000. Opening of mitochondrial K_{ATP} channels triggers the preconditioned state by generating free radicals. *Circ. Res.* 87:460–66
131. Juhaszova M, Zorov DB, Kim SH, Pepe S, Fu Q, et al. 2004. Glycogen synthase kinase-3β mediates convergence of protection signaling to inhibit the mitochondrial permeability transition pore. *J. Clin. Invest.* 113:1535–49
132. Sasaki N, Sato T, Ohler A, O'Rourke B, Marban E. 2000. Activation of mitochondrial ATP-dependent potassium channels by nitric oxide. *Circulation* 101:439–45
133. Ockaili R, Emani VR, Okubo S, Brown M, Krottapalli K, Kukreja RC. 1999. Opening of mitochondrial K_{ATP} channel induces early and delayed cardioprotective effect: role of nitric oxide. *Am. J. Physiol.* 277:H2425–34
134. Xu Z, Ji X, Boysen PG. 2004. Exogenous nitric oxide generates ROS and induces cardioprotection: involvement of PKG, mitochondrial K_{ATP} channels, and ERK. *Am. J. Physiol. Heart Circ. Physiol.* 286:H1433–40
135. Costa AD, Garlid KD, West IC, Lincoln TM, Downey JM, et al. 2005. Protein kinase G transmits the cardioprotective signal from cytosol to mitochondria. *Circ. Res.* 97:329–36
136. Dang VC, Kim N, Youm JB, Joo H, Warda M, et al. 2006. Nitric oxide–cGMP–protein kinase G signaling pathway induces anoxic preconditioning through activation of ATP-sensitive K^+ channels in rat hearts. *Am. J. Physiol. Heart Circ. Physiol.* 290:H1808–17
137. Siemen D, Loupatatzis C, Borecky J, Gulbins E, Lang F. 1999. Ca^{2+}-activated K channel of the BK-type in the inner mitochondrial membrane of a human glioma cell line. *Biochem. Biophys. Res. Commun.* 257:549–54
138. Xu W, Liu Y, Wang S, McDonald T, Van Eyk JE, et al. 2002. Cytoprotective role of Ca^{2+}-activated K^+ channels in the cardiac inner mitochondrial membrane. *Science* 298:1029–33
139. Shintani Y, Node K, Asanuma H, Sanada S, Takashima S, et al. 2004. Opening of Ca^{2+}-activated K^+ channels is involved in ischemic preconditioning in canine hearts. *J. Mol. Cell Cardiol.* 37:1213–18
140. Stowe DF, Aldakkak M, Camara AK, Riess ML, Heinen A, et al. 2006. Cardiac mitochondrial preconditioning by Big Ca^{2+}-sensitive K^+ channel opening requires superoxide radical generation. *Am. J. Physiol. Heart Circ. Physiol.* 290:H434–40
141. Wang X, Yin C, Xi L, Kukreja RC. 2004. Opening of Ca^{2+}-activated K^+ channels triggers early and delayed preconditioning against I/R injury independent of NOS in mice. *Am. J. Physiol. Heart Circ. Physiol.* 287:H2070–77
142. Cao CM, Xia Q, Gao Q, Chen M, Wong TM. 2005. Calcium-activated potassium channel triggers cardioprotection of ischemic preconditioning. *J. Pharmacol. Exp. Ther.* 312:644–50

143. Gao Q, Zhang SZ, Cao CM, Bruce IC, Xia Q. 2005. The mitochondrial permeability transition pore and the Ca^{2+}-activated K^+ channel contribute to the cardioprotection conferred by tumor necrosis factor-α. *Cytokine* 32:199–205
144. Sato T, Saito T, SaegUSA N, Nakaya H. 2005. Mitochondrial Ca^{2+}-activated K^+ channels in cardiac myocytes: a mechanism of the cardioprotective effect and modulation by protein kinase A. *Circulation* 111:198–203
145. Douglas RM, Lai JC, Bian S, Cummins L, Moczydlowski E, Haddad GG. 2006. The calcium-sensitive large-conductance potassium channel (BK/MAXI K) is present in the inner mitochondrial membrane of rat brain. *Neuroscience* 139:1249–61
146. Ohya S, Kuwata Y, Sakamoto K, Muraki K, Imaizumi Y. 2005. Cardioprotective effects of estradiol include the activation of large-conductance Ca^{2+}-activated K^+ channels in cardiac mitochondria. *Am. J. Physiol. Heart Circ. Physiol.* 289:H1635–42
147. Szabo I, Bock J, Jekle A, Soddemann M, Adams C, et al. 2005. A novel potassium channel in lymphocyte mitochondria. *J. Biol. Chem.* 280:12790–98
148. Beavis AD, Powers M. 2004. Temperature dependence of the mitochondrial inner membrane anion channel: the relationship between temperature and inhibition by magnesium. *J. Biol. Chem.* 279:4045–50
149. Sorgato MC, Keller BU, Stuhmer W. 1987. Patch-clamping of the inner mitochondrial membrane reveals a voltage-dependent ion channel. *Nature* 330:498–500
150. Borecky J, Jezek P, Siemen D. 1997. 108-pS channel in brown fat mitochondria might be identical to the inner membrane anion channel. *J. Biol. Chem.* 272:19282–89
151. Kinnally KW, Zorov DB, Antonenko YN, Snyder SH, McEnery MW, Tedeschi H. 1993. Mitochondrial benzodiazepine receptor linked to inner membrane ion channels by nanomolar actions of ligands. *Proc. Natl. Acad. Sci. USA* 90:1374–78
152. Schonfeld P, Sayeed I, Bohnensack R, Siemen D. 2004. Fatty acids induce chloride permeation in rat liver mitochondria by activation of the inner membrane anion channel (IMAC). *J. Bioenerg. Biomembr.* 36:241–48
153. Antonenko YN, Kinnally KW, Tedeschi H. 1991. Identification of anion and cation pathways in the inner mitochondrial membrane by patch clamping of mouse liver mitoplasts. *J. Membr. Biol.* 124:151–58
154. Beavis AD. 1989. On the inhibition of the mitochondrial inner membrane anion uniporter by cationic amphiphiles and other drugs. *J. Biol. Chem.* 264:1508–15
155. Ballarin C, Sorgato MC. 1995. An electrophysiological study of yeast mitochondria. Evidence for two inner membrane anion channels sensitive to ATP. *J. Biol. Chem.* 270:19262–68
156. O'Rourke B. 2000. Pathophysiological and protective roles of mitochondrial ion channels. *J. Physiol.* 529(Pt. 1):23–36
157. Aon MA, Cortassa S, Marban E, O'Rourke B. 2003. Synchronized whole cell oscillations in mitochondrial metabolism triggered by a local release of reactive oxygen species in cardiac myocytes. *J. Biol. Chem.* 278:44735–44
158. Cortassa S, Aon MA, Winslow RL, O'Rourke B. 2004. A mitochondrial oscillator dependent on reactive oxygen species. *Biophys. J.* 87:2060–73
159. Aon MA, Cortassa S, O'Rourke B. 2004. Percolation and criticality in a mitochondrial network. *Proc. Natl. Acad. Sci. USA* 101:4447–52
160. Zorov DB, Filburn CR, Klotz LO, Zweier JL, Sollott SJ. 2000. Reactive oxygen species (ROS)-induced ROS release: a new phenomenon accompanying induction of the mitochondrial permeability transition in cardiac myocytes. *J. Exp. Med.* 192:1001–14

161. Lurin C, Guclu J, Cheniclet C, Carde JP, Barbier-Brygoo H, Maurel C. 2000. CLC-Nt1, a putative chloride channel protein of tobacco, colocalizes with mitochondrial membrane markers. *Biochem. J.* 348(Pt. 2):291–95
162. Fernandez-Salas E, Suh KS, Speransky VV, Bowers WL, Levy JM, et al. 2002. mtCLIC/CLIC4, an organellular chloride channel protein, is increased by DNA damage and participates in the apoptotic response to p53. *Mol. Cell Biol.* 22:3610–20
163. Suh KS, Mutoh M, Nagashima K, Fernandez-Salas E, Edwards LE, et al. 2004. The organellular chloride channel protein CLIC4/mtCLIC translocates to the nucleus in response to cellular stress and accelerates apoptosis. *J. Biol. Chem.* 279:4632–41
164. Cooper C, Tapley DF. 1956. The effect of thyroxine and related compounds on oxidative phosphorylation. *J. Biol. Chem.* 222:341–49
165. Hunter DR, Haworth RA, Southard JH. 1976. Relationship between configuration, function, and permeability in calcium-treated mitochondria. *J. Biol. Chem.* 251:5069–77
166. Bernardi P, Krauskopf A, Basso E, Petronilli V, Blachly-Dyson E, et al. 2006. The mitochondrial permeability transition from in vitro artifact to disease target. *FEBS J.* 273:2077–99
167. Halestrap AP. 2006. Calcium, mitochondria and reperfusion injury: a pore way to die. *Biochem. Soc. Trans.* 34:232–37
168. Fournier N, Ducet G, Crevat A. 1987. Action of cyclosporine on mitochondrial calcium fluxes. *J. Bioenerg. Biomembr.* 19:297–303
169. Crompton M, Ellinger H, Costi A. 1988. Inhibition by cyclosporin A of a Ca^{2+}-dependent pore in heart mitochondria activated by inorganic phosphate and oxidative stress. *Biochem. J.* 255:357–60
170. Connern CP, Halestrap AP. 1992. Purification and N-terminal sequencing of peptidyl-prolyl *cis-trans*-isomerase from rat liver mitochondrial matrix reveals the existence of a distinct mitochondrial cyclophilin. *Biochem. J.* 284(Pt. 2):381–85
171. Woodfield KY, Price NT, Halestrap AP. 1997. cDNA cloning of rat mitochondrial cyclophilin. *Biochim. Biophys. Acta* 1351:27–30
172. Kinnally KW, Campo ML, Tedeschi H. 1989. Mitochondrial channel activity studied by patch-clamping mitoplasts. *J. Bioenerg. Biomembr.* 21:497–506
173. Kinnally KW, Antonenko YN, Zorov DB. 1992. Modulation of inner mitochondrial membrane channel activity. *J. Bioenerg. Biomembr.* 24:99–110
174. Zorov DB, Kinnally KW, Perini S, Tedeschi H. 1992. Multiple conductance levels in rat heart inner mitochondrial membranes studied by patch clamping. *Biochim. Biophys. Acta* 1105:263–70
175. Zorov DB, Kinnally KW, Tedeschi H. 1992. Voltage activation of heart inner mitochondrial membrane channels. *J. Bioenerg. Biomembr.* 24:119–24
176. Petronilli V, Szabo I, Zoratti M. 1989. The inner mitochondrial membrane contains ion-conducting channels similar to those found in bacteria. *FEBS Lett.* 259:137–43
177. Szabo I, Zoratti M. 1991. The giant channel of the inner mitochondrial membrane is inhibited by cyclosporin A. *J. Biol. Chem.* 266:3376–79
178. Szabo I, Zoratti M. 1992. The mitochondrial megachannel is the permeability transition pore. *J. Bioenerg. Biomembr.* 24:111–17
179. Zoratti M, Szabo I. 1994. Electrophysiology of the inner mitochondrial membrane. *J. Bioenerg. Biomembr.* 26:543–53
180. Schultheiss HP, Klingenberg M. 1984. Immunochemical characterization of the adenine nucleotide translocator. Organ specificity and conformation specificity. *Eur. J. Biochem.* 143:599–605

181. Brustovetsky N, Klingenberg M. 1996. Mitochondrial ADP/ATP carrier can be reversibly converted into a large channel by Ca^{2+}. *Biochemistry* 35:8483–88
182. Brustovetsky N, Tropschug M, Heimpel S, Heidkamper D, Klingenberg M. 2002. A large Ca^{2+}-dependent channel formed by recombinant ADP/ATP carrier from *Neurospora crassa* resembles the mitochondrial permeability transition pore. *Biochemistry* 41:11804–11
183. Nakagawa T, Shimizu S, Watanabe T, Yamaguchi O, Otsu K, et al. 2005. Cyclophilin D-dependent mitochondrial permeability transition regulates some necrotic but not apoptotic cell death. *Nature* 434:652–58
184. Baines CP, Kaiser RA, Purcell NH, Blair NS, Osinska H, et al. 2005. Loss of cyclophilin D reveals a critical role for mitochondrial permeability transition in cell death. *Nature* 434:658–62
185. Basso E, Fante L, Fowlkes J, Petronilli V, Forte MA, Bernardi P. 2005. Properties of the permeability transition pore in mitochondria devoid of cyclophilin D. *J. Biol. Chem.* 280:18558–61
186. Schinzel AC, Takeuchi O, Huang Z, Fisher JK, Zhou Z, et al. 2005. Cyclophilin D is a component of mitochondrial permeability transition and mediates neuronal cell death after focal cerebral ischemia. *Proc. Natl. Acad. Sci. USA* 102:12005–10
187. Halestrap AP, Clarke SJ, Javadov SA. 2004. Mitochondrial permeability transition pore opening during myocardial reperfusion—a target for cardioprotection. *Cardiovasc. Res.* 61:372–85
188. Rodriguez-Sinovas A, Boengler K, Cabestrero A, Gres P, Morente M, et al. 2006. Translocation of connexin 43 to the inner mitochondrial membrane of cardiomyocytes through the heat shock protein 90-dependent TOM pathway and its importance for cardioprotection. *Circ. Res.* 99:93–101
189. Schwanke U, Konietzka I, Duschin A, Li X, Schulz R, Heusch G. 2002. No ischemic preconditioning in heterozygous connexin43-deficient mice. *Am. J. Physiol. Heart Circ. Physiol.* 283:H1740–42
190. Heinzel FR, Luo Y, Li X, Boengler K, Buechert A, et al. 2005. Impairment of diazoxide-induced formation of reactive oxygen species and loss of cardioprotection in connexin 43 deficient mice. *Circ. Res.* 97:583–86
191. Liu Y, Ren G, O'Rourke B, Marban E, Seharaseyon J. 2001. Pharmacological comparison of native mitochondrial K_{ATP} channels with molecularly defined surface K_{ATP} channels. *Mol. Pharmacol.* 59:225–30
192. Mironova GD, Negoda AE, Marinov BS, Paucek P, Costa AD, et al. 2004. Functional distinctions between the mitochondrial ATP-dependent K channel (mitoKATP) and its inward rectifier subunit (mitoKIR). *J. Biol. Chem.* 279:32562–68
193. Jiang MT, Ljubkovic M, Nakae Y, Shi Y, Kwok WM, et al. 2006. Characterization of human cardiac mitochondrial ATP-sensitive potassium channel and its regulation by phorbol ester in vitro. *Am. J. Physiol. Heart Circ. Physiol.* 290:H1770–76
194. Suzuki M, Kotake K, Fujikura K, Inagaki N, Suzuki T, et al. 1997. Kir6.1: a possible subunit of ATP-sensitive K channels in mitochondria. *Biochem. Biophys. Res. Commun.* 241:693–97
195. Zhou M, Tanaka O, Sekiguchi M, Sakabe K, Anzai M, et al. 1999. Localization of the ATP-sensitive potassium channel subunit (Kir6. 1uK_{ATP}-1) in rat brain. *Brain Res. Mol. Brain Res.* 74:15–25
196. Lacza Z, Snipes JA, Miller AW, Szabo C, Grover G, Busija DW. 2003. Heart mitochondria contain functional ATP-dependent K channels. *J. Mol. Cell Cardiol.* 35:1339–47

197. Lacza Z, Snipes JA, Kis B, Szabo C, Grover G, Busija DW. 2003. Investigation of the subunit composition and the pharmacology of the mitochondrial ATP-dependent K channel in the brain. *Brain Res.* 994:27–36
198. Singh H, Hudman D, Lawrence CL, Rainbow RD, Lodwick D, Norman RI. 2003. Distribution of Kir6.0 and SUR2 ATP-sensitive potassium channel subunits in isolated ventricular myocytes. *J. Mol. Cell Cardiol.* 35:445–59
199. Cuong DV, Kim N, Joo H, Youm JB, Chung JY, et al. 2005. Subunit composition of ATP-sensitive potassium channels in mitochondria of rat hearts. *Mitochondrion* 5:121–33
200. Zhou M, Tanaka O, Sekiguchi M, He HJ, Yasuoka Y, et al. 2005. ATP-sensitive K-channel subunits on the mitochondria and endoplasmic reticulum of rat cardiomyocytes. *J. Histochem. Cytochem.* 53:1491–500

Preconditioning: The Mitochondrial Connection

Elizabeth Murphy[1,*] and Charles Steenbergen[2]

[1]National Institute of Environmental Health Sciences, National Institutes of Health, Research Triangle Park, North Carolina 27709; email: murphy1@niehs.nih.gov

[2]Department of Pathology, Johns Hopkins University, Baltimore, Maryland 21205; email: csteenb1@jhmi.edu

Annu. Rev. Physiol. 2007. 69:51–67

First published online as a Review in Advance on September 28, 2006

The *Annual Review of Physiology* is online at http://physiol.annualreviews.org

This article's doi: 10.1146/annurev.physiol.69.031905.163645

0066-4278/07/0315-0051$20.00

Key Words

mitochondrial permeability transition, glycogen synthase kinase, protein kinase C, voltage-dependent anion channel

Abstract

Over the past decade there has been considerable progress in elucidating the signaling pathways involved in cardioprotection. Considerable recent data suggest that many of these signaling pathways converge on the mitochondria, where such pathways alter the activity of key mitochondrial proteins, leading to reduced apoptosis and necrosis. Inhibition of the mitochondrial permeability transition pore is emerging as a central mechanism in cardioprotection. This review focuses on mechanisms by which cardioprotection alters mitochondrial proteins and channels that regulate cell death and survival.

ROS: reactive oxygen species

Preconditioning (PC): brief intermittent periods of ischemia and reperfusion that reduce ischemia-reperfusion injury during a subsequent sustained period of ischemia

PKC: protein kinase C

GSK-3β: glycogen synthase kinase-3β

NOS: nitric oxide synthase

INTRODUCTION

Mitochondria, the site of electron transport and ATP production, are particularly important for ATP generation in the heart. Mitochondria comprise ~30–40% of the myocyte volume and generate more than 90% of the ATP. Mitochondria are the major source of reactive oxygen species (ROS) and can also avidly accumulate calcium (Ca^{2+}). Although excessively high levels of ROS and Ca^{2+} can be toxic to the cell, ROS and Ca^{2+} can be important in signal transduction. Thus, it is not surprising that prolonged mitochondrial dysfunction is associated with cell death. However, it is somewhat surprising that the release of mitochondrial proteins, such as cytochrome *c*, is key in the regulation of apoptosis. Given the important role of mitochondria in the regulation of cell death, energetics, ROS, and Ca^{2+}, it is not surprising that mitochondria are also an important target of cardioprotective signals.

A number of drugs and agonists reduce ischemia-reperfusion injury; these are referred to as cardioprotective. Brief intermittent periods of ischemia and reperfusion, termed preconditioning (PC), are also cardioprotective. PC and many of these cardioprotective agonists activate a signaling cascade that involves PI3-kinase, protein kinase C epsilon (PKCε), nitric oxide synthase (NOS), glycogen synthase kinase (GSK)-3β, and ERK. Recent data suggest that the mitochondria are a major target of these cardioprotective signaling cascades. The precise mitochondrial targets and mechanisms that lead to reduced myocyte death, which are just beginning to be elucidated, are the focus of this review.

CARDIOPROTECTION

Although researchers have discovered a number of drugs that reduce ischemia-reperfusion injury, the study of cardioprotection was energized by the observation that PC reduces subsequent lethal ischemia-reperfusion injury (1, 2). Investigators have made considerable progress in elucidating the signaling pathways that are activated and required for PC; these data are summarized in **Figure 1**. PC leads to release of agonists such as adenosine, bradykinin, and opioids. These agonists bind to G protein–coupled receptors and activate a signaling cascade that results in cardioprotection. Such binding to G protein–coupled receptors results in PI3-kinase activation (3). This in turn leads to altered activity (usually activation, although GSK-3β is inactivated) of a number of downstream signaling molecules, such as AKT (3), protein kinase C (PKC) (3–5), GSK-3ß (6–8), NOS (3, 9, 10), extracellular regulated kinase (ERK) (11), protein kinase G (12), and the mitochondrial K_{ATP} channel (13, 14). Additionally, activation of PI3-kinase plays a key role in endosomal receptor

Figure 1

Signaling pathways activated by preconditioning (PC). (*a*) PC leads to the release of agonists such as adenosine, bradykinin, and opioids, which bind to G protein–coupled receptors (GPCRs) and activate a signaling cascade that results in cardioprotection. The activation of PI3-kinase (PI3K) leads to altered activity [usually activation, although glycogen synthase kinase-3β (GSK-3β) is inactivated] of a number of downstream signaling molecules, such as AKT, protein kinase C (PKC), GSK-3β, nitric oxide synthase (NOS), extracellular regulated kinase (ERK), protein kinase G (PKG) (12), guanylyl cyclase (GC), ribosomal S6 kinase ($p70^{s6k}$), and the mitochondrial K_{ATP} channel. (*b*) These signals converge on the mitochondria. Mitochondrial ERK, PKCε, GSK-3β, connexin 43 (Cx43), Bcl-2, and BAD either translocate to the mitochondria or have altered mitochondrial activity in response to cardioprotection. PKCε and GSK-3β have been reported to phosphorylate VDAC, which may be a component of the mitochondrial permeability transition (MPT). Activation of mitoK_{ATP} may inhibit the MPT. ERK has also been reported to phosphorylate BAD; phosphorylated BAD dissociates from Bcl-2, enhancing the antiapoptotic effects of Bcl-2.

signaling, which is important for cardioprotection (15). Endosomal signaling may be important in targeting signaling molecules to the correct location, as many of these signaling proteins and kinases localize to the mitochondria (7, 16–18), which is important in cardioprotection (19–21). Many signaling molecules may target the mitochondria, where they improve energetics and reduce apoptosis and necrosis (16, 19, 22). Activation of the mitochondrial K_{ATP} channel may play an important role in cardioprotection, although the

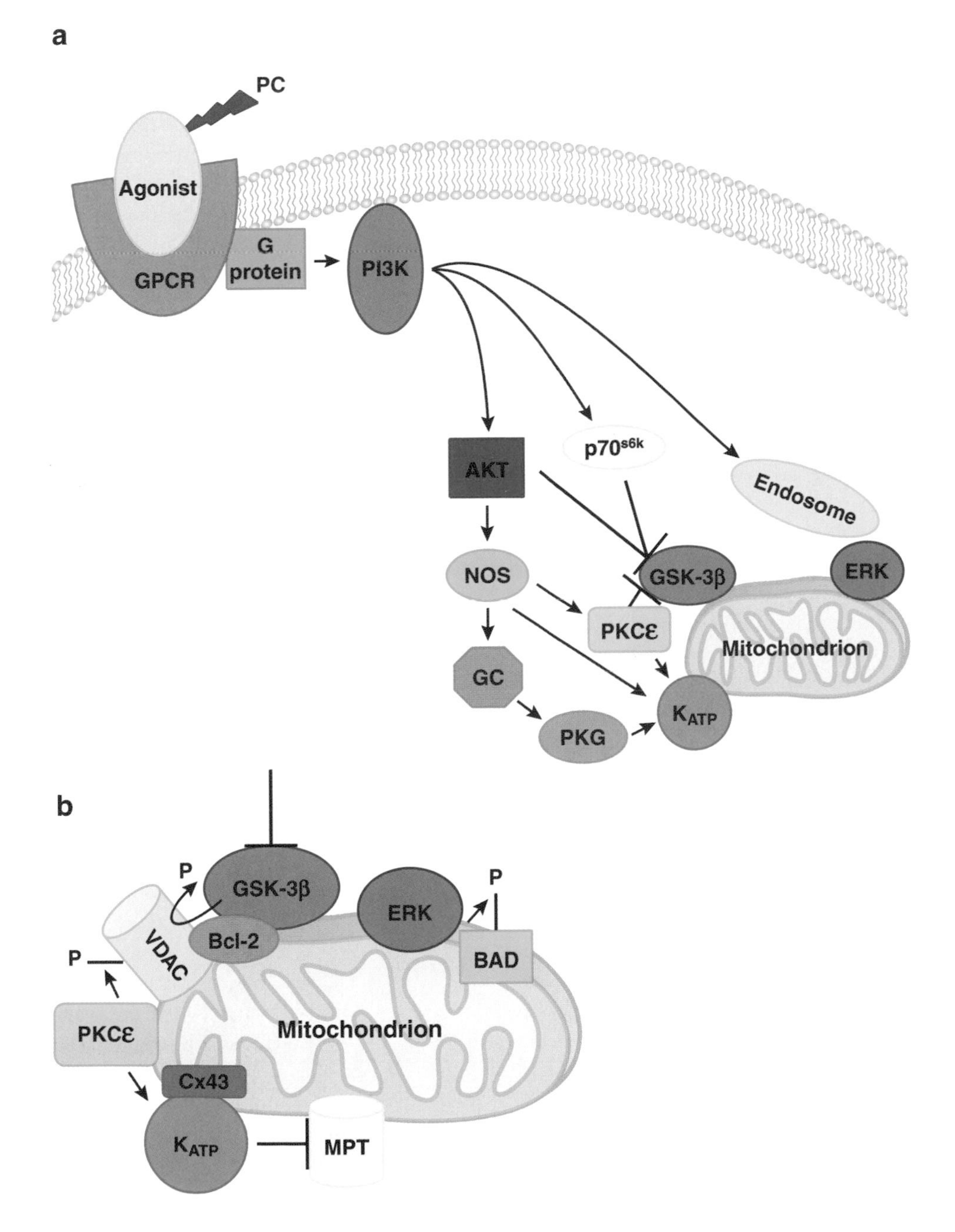

$\Delta\psi$: membrane potential across the inner mitochondrial membrane

ΔpH: pH gradient across the inner mitochondrial membrane

Δp: mitochondrial protomotive force; the sum of $\Delta\psi$ and ΔpH

F_1F_0ATPase: mitochondrial enzyme that uses the energy in the proton gradient across the inner mitochondrial membrane to generate ATP; can also consume ATP to expel protons and thereby generate $\Delta\psi$

Adenine nucleotide translocator (ANT): transports adenine nucleotides across the inner mitochondrial membrane

CK: creatine kinase

Voltage-dependent anion channel (VDAC): transports metabolites, anions, and cations across the outer mitochondrial membrane

precise mechanism is still debated (19, 21, 23). Addition of diazoxide, an activator of the putative mitochondrial K_{ATP} channel, reduces apoptosis and necrosis following ischemia and reperfusion (24). One exciting new development in cardioprotection, which has important clinical implications, is the observation that cardioprotection can be initiated by the activation of signaling pathways at the start of reperfusion (25, 26).

MITOCHONDRIA

Electron Transport

Mitochondria are the major site of ATP synthesis via electron transport and oxidative phosphorylation. Electrons donated by NADH and/or FADH2 are transferred through a series of redox reactions, with oxygen as the final electron acceptor. The energy is coupled to extrusion of protons from the mitochondria matrix at three coupling sites. There are several inhibitors of electron transport, which inhibit at different sites. Mitochondrial electron transport is one of the primary sources of cellular ROS, and different inhibitors of electron transport can enhance or reduce ROS generation, depending on the inhibition site (27, 28). The extrusion of protons from the mitochondrial matrix results in the generation of a mitochondrial membrane potential ($\Delta\psi$) and pH gradient (ΔpH), which together comprise the protomotive force (Δp). The extruded protons reenter the mitochondria via the F_1F_0ATPase, which couples the transfer of protons down their electrochemical gradient to generate ATP. Protonophores such as dinitrophenol and FCCP are classified as uncouplers because they dissipate Δp and stimulate electron transport and oxygen consumption independently of ADP and thus uncouple electron transport and oxygen consumption from the conversion of ADP to ATP. Transport of Ca^{2+} into the matrix via the Ca^{2+} uniporter also stimulates oxygen consumption in the absence of ADP and lowers Δp in proportion to Ca^{2+} influx. Similarly, the electrogenic transport of other cations such as K^+ into the matrix increases oxygen consumption and lowers Δp in proportion to the rate and amount of K^+ influx. Researchers have also described mitochondrial uncoupling proteins, which mediate proton leak across the mitochondrial inner membrane (29).

The ATP synthesized by the F_1F_0ATPase must be transported back to the cytosol, to be used to fuel cellular processes. Furthermore, the ADP generated in the cytosol must be transported into the mitochondrial matrix to regenerate ATP. Adenine nucleotide transport is accomplished by the adenine nucleotide translocator (ANT) in the inner mitochondrial membrane and the voltage-dependent anion channel (VDAC) in the outer mitochondrial membrane. ANT and VDAC likely are positioned at contact sites between the inner and outer mitochondrial membranes (30). Intermediate filaments such as desmin may play a role in the regulation of ATP and ADP transport across the mitochondria, perhaps by the regulation of VDAC (31, 32). Cardiac muscle also contains creatine kinase (CK), which facilitates high-energy phosphate transfer from the mitochondria to the cytosol. Mitochondrial CK is localized to the intermembrane space. CK uses the ATP synthesized by the F_1F_0ATPase and transported by ANT to form phosphocreatine and ADP; ADP is returned to the matrix via ANT as a substrate for the F_1F_0ATPase. The CK system effectively lowers the mitochondrial K_m for ADP (33). Below we discuss VDAC in more detail because of its role in cell death and cardioprotection.

VDAC

VDAC, a 30–32-kDa protein, is the major protein of the outer mitochondrial membrane. It facilitates the transport of metabolites, anions, and cations across the outer mitochondrial membrane (34–38). Prior to the identification of VDAC, researchers assumed that the outer mitochondrial membrane was permeable to small metabolites and solutes. However, we now know that this apparent permeability is due to the presence of large amounts of VDAC. In lipid bilayers, VDAC can be gated by voltage, as implied by its name. Voltage can alter the permeability of the channel and its selectivity for anions versus cations (34–38). The open state, which occurs at low voltage (~10 mV, either positive or negative), is a high-conductance state that allows transport of metabolites such as ATP and has weak selectivity for anions over cations. The closed state, which predominates at higher voltages (more than 30 mV, either positive or negative), is a low-conductance state for anions, with very low transport of adenine nucleotides and high permeability for cations. These selectivity studies have all been done in vitro, and the presence of a voltage across the outer mitochondrial membrane is debated (see Reference 39). In addition to regulating transport of adenine nucleotides, VDAC also regulates Ca^{2+} uptake into the mitochondria (40, 41). Overexpression of VDAC increases mitochondrial Ca^{2+} uptake following agonist stimulation. Rapizzi et al. (40) suggest that, in the agonist-mediated transfer of Ca^{2+} released by the endoplasmic reticulum to mitochondria, the permeability of the outer mitochondrial membrane is a bottleneck and that increasing VDAC enhances Ca^{2+} entry into the inner mitochondrial membrane, where it can reach concentrations high enough to approach the K_m of the Ca^{2+} uniporter.

VDAC appears to serve as a scaffolding protein for several proteins, particularly kinases that use ATP. For example, CK, hexokinase (HK), and glucokinase bind to VDAC (42, 43). As discussed above for CK, binding to VDAC gives these kinases preferential access to ATP supplied by ANT. Such binding may also regulate VDAC function. Bcl-2 also binds to VDAC and regulates its opening (see below). Phosphorylation of VDAC by kinases such as GSK may regulate its binding to proteins such as HK (44). It has been difficult to study the role of VDAC in intact cells or organs because of the lack of suitable inhibitors. DIDS has been used as a nonselective inhibitor of VDAC. Lai et al. (45) recently suggested that G3139, an 18-mer phosphorothioate oligonucleotide with sequence antisense to bcl-2, is a specific inhibitor of VDAC.

HK: hexokinase

$MitoK_{ATP}$ channel: an ATP-sensitive K channel in the mitochondrial inner membrane that allows entry of K into the mitochondrial matrix, resulting in an increase in matrix volume

Mitochondrial Volume Regulation and the $MitoK_{ATP}$ Channel

The mitochondrial inner membrane is highly invaginated, with a large surface area relative to matrix volume, and there is normally very little space between the inner and outer membranes. The matrix volume, and thus the space between the inner and outer mitochondrial membranes, is regulated by a K/H antiporter and a mitochondrial K^+ channel ($mitoK_{ATP}$ channel) (46). Because of the high negative mitochondrial membrane potential, there is a large driving force for K^+ entry into the mitochondrial matrix. K^+ entry into the matrix is followed by entry of osmotically driven water, which results in mitochondrial matrix swelling. The mitochondrial K/H antiporter, which is activated by swelling, results in K^+ and thus water efflux, thereby restoring matrix volume. However, proton influx counters K^+ efflux, resulting in a decrease in Δp (owing to a decrease in ΔpH). Furthermore, when mitochondrial matrix volume contracts, as occurs during high ATP synthesis rates, the $mitoK_{ATP}$ channel is activated, which increases matrix volume and may result in cardioprotection (24). A number of pharmacological agents such as diazoxide activate the $mitoK_{ATP}$ channel (47). Diazoxide also reduces acidification during ischemia as well as ischemia-reperfusion injury (48). Researchers have extensively studied the mechanism by which opening of this $mitoK_{ATP}$ channel results in protection but have not yet established the precise mechanism. Murata et al. (49) have suggested that activation of the $mitoK_{ATP}$ channel slightly reduces the mitochondrial membrane potential, which in turn decreases mitochondrial Ca^{2+} uptake during ischemia. Dos Santos et al. (50) have suggested

MPT: mitochondrial permeability transition

that the activation of the mitoK$_{ATP}$ channel and the resultant mitochondrial matrix swelling alter the K_m for ADP entry into the mitochondria, perhaps by altering the conformation of VDAC or by altering contact sites between VDAC and ANT. Other research groups have proposed that a change in mitochondrial matrix volume alters metabolism (51–53). According to yet another theory, activation of the mitoK$_{ATP}$ channel results in generation of ROS, which activate signaling pathways, resulting in cardioprotection (48, 54). These mechanisms are not mutually exclusive and may all play a role.

Mitochondrial Permeability Transition

The mitochondrial permeability transition (MPT) is a large-conductance mitochondrial megachannel that allows the passage of solutes with molecular weights up to ~1500 Da. Therefore, opening of this channel results in mitochondrial swelling and dissipation of the mitochondrial membrane potential. $\Delta\psi$ is required for ATP generation. Opening of the MPT is promoted by high matrix Ca^{2+}, inorganic phosphate, ROS, fatty acids, and a reduction in $\Delta\psi$, which all occur during ischemia-reperfusion. Mg^{2+}, adenine nucleotides, and low pH inhibit MPT opening. Although the MPT was described more than 40 years ago (55, 56), the molecular composition of the channel is still poorly understood. The MPT may be a multiprotein channel composed of the ANT in the inner mitochondrial membrane; VDAC in the outer mitochondrial membrane; and cyclophilin A, which mediates the channel's Ca^{2+} sensitivity, in the matrix. The peripheral benzodiazepine receptor also may be involved in MPT. However, with the exception of cyclophilin, confirmation of an obligatory role of these proteins in MPT is lacking. In fact, the MPT was still present in mitochondria from livers of mice lacking all three isoforms of ANT (57). Although mitochondria lacking all three isoforms of VDAC have not been tested, mice lacking VDAC1 still undergo MPT (58).

Based on a large library screen, Cesura et al. (59) reported that Ro 68-3400 blocks MPT. They also reported that 3H-labeled Ro 68-3400 bound to a 32-kDa protein that they initially identified as VDAC1. However, in a recent subsequent study of mice lacking VDAC1, Krauskopf et al. (58) found that MPT was still blocked by Ro 68-3400. Krauskopf et al. were able to separate all three isoforms of VDAC from the 32-kDa protein that binds to labeled Ro 68-3400. This study of VDAC1-null mitochondria does not rule out a role for other VDAC isoforms in the MPT, but it does suggest that VDAC is not required for inhibition of MPT by Ro 68-3400 (58). The identity of the 32-kDa protein that binds Ro 68-3400 has not been established. If ANT (and possibly VDAC) is not required for the MPT, what is the MPT? He & Lemasters (60) and Kowaltowski et al. (61) have suggested that the MPT is formed by mitochondrial proteins damaged by oxidative stress. In this model, if one protein is deleted, others damaged by oxidative stress can interact to form the MPT.

As discussed more fully below, several lines of data suggest that the MPT is a focal point for cell death and cardioprotection. Data show that loss of cyclophilin significantly reduces the amount of necrosis in a model of ischemia-reperfusion injury, suggesting a crucial role for MPT in ischemia-reperfusion injury and necrosis (62, 63). Similarly, the inhibition of MPT with cyclosporin reduces ischemia-reperfusion injury (64). The inhibition of MPT on reperfusion also reduces ischemia-reperfusion-induced cell death, suggesting that the activation of MPT at the start of reperfusion contributes to cell death (65, 66). Furthermore, Yellon and coworkers (67) reported that inhibition of MPT during preconditioning blocks the protection afforded by PC. Flickering of the MPT may be required to initiate PC (67). Clearly a better understanding of the molecular components of the MPT is a major focus of future research on cell death and cardioprotection.

Bcl-2 Family Members

Bcl-2 is the founding member of a family of proteins that regulate apoptosis. Bcl-2 localizes to the mitochondria, where it reduces apoptotic cell death. Inhibitors of apoptosis, such as Bcl-2, may reduce what appears to be necrotic or oncotic cell death (68, 69). Imahashi et al. (69) have shown that cardiac-specific overexpression of Bcl-2 significantly reduces ischemia-induced necrosis as measured by TTC staining after two hours of reperfusion. Necrosis was 38 ± 5% in wild-type hearts and was significantly lower at 18 ± 3% in hearts with overexpression of Bcl-2. Apoptosis, as assessed by TUNEL staining, was also reduced in hearts with Bcl-2 overexpression (1.2 ± 0.9%) compared with wild-type hearts (4.3 ± 1.7%). However, only a small proportion of total cell death was due to apoptosis, as measured by TUNEL. For example, in wild-type hearts 38% of the heart was dead as measured by TTC, yet only 4.3% of this death could be attributed to apoptosis. With overexpression of Bcl-2, 18% of the heart was dead, as measured by TTC, but this reduction in cell death cannot be accounted for solely by a reduction in apoptosis because apoptosis was only 4% in wild-type hearts. These data suggest that Bcl-2 also reduces what is commonly classified as necrotic or oncotic cell death. These results are consistent with the findings of Chen et al. (70), although in that study it is difficult to compare the proportion of total death (TTC staining) to apoptotic death because apoptosis was measured after 30 min of ischemia, whereas TTC measurements were made after 50 min of ischemia. Consistent with a role for Bcl-2 in inhibition of necrosis, recent data suggest that oncotic cell death can be regulated (68).

In spite of intense efforts, the precise mechanism by which Bcl-2 inhibits cell death is not well understood. BAX, a proapoptotic Bcl-2 family member, may be involved in forming channels (either alone or with other mitochondrial proteins such as VDAC) that result in the release of cytochrome *c* and the induction of apoptosis. Binding and sequestering of BAX by Bcl-2 can mediate the anti-apoptotic effect of Bcl-2. Bcl-2 also can bind to VDAC and thereby inhibit apoptosis (71–73). However, there are conflicting data as to whether the binding of Bcl-2 to VDAC results in VDAC opening or closing (71–73). The reasons for this discrepancy are unclear. Under conditions of growth factor deprivation at normal oxygen, Bcl-2 binding maintains VDAC in an open conformation, allowing ATP generated by the mitochondria to enter the cytosol (71). In contrast, Shimizu et al. (72), using reconstituted liposomes and radiolabeled sucrose uptake as a measure of VDAC activity, reported that BAX enhances VDAC opening, whereas Bcl-2 promotes VDAC closure. Additional studies by this group showed that BAX-induced apoptosis was blocked by injection of antibodies that inhibited VDAC opening (73). Perhaps the effect of Bcl-2 on VDAC depends on matrix Ca^{2+} or oxygen levels or some other factor (perhaps ROS). Imahashi et al. (69) found that Bcl-2 overexpression, in addition to reducing ischemia-reperfusion injury, also reduces ischemic acidification and the rate of decline in ATP during ischemia; these data are consistent with Bcl-2-mediated inhibition of consumption of glycolytically generated ATP. Fifty to eighty percent of the ATP consumed during ischemia is via reverse mode of the F_1F_0ATPase (74, 75). Consistent with this hypothesis, the reduction in acidification and the rate of ATP decline during ischemia depend on the activity of the mitochondrial F_1F_0-ATPase. Bcl-2-mediated inhibition of glycolytically generated ATP may be accomplished by limiting ATP entry into the mitochondria through VDAC or ANT or by direct inhibition of the F_1F_0-ATPase. Consistent with these possibilities, Bcl-2 overexpression reduces the rate of mitochondrial ATP consumption when ATP hydrolysis is stimulated. Thus, Bcl-2, in addition to reducing cytochrome *c* release, has other, significant effects on mitochondrial function.

SIGNALING PATHWAYS AND MITOCHONDRIA

Signaling Pathways and Cardioprotection

Several cardioprotective signaling molecules are localized to the mitochondria, where they can alter mitochondrial proteins involved in energetics, metabolism, and cell death. Although much progress has been made in elucidating the signaling pathways involved in preconditioning (see **Figure 1**), the mechanism(s) by which these pathways reduce cell death is still poorly understood. Because mitochondria are key organelles in regulating apoptotic and necrotic cell death, there has been considerable interest regarding how these signaling pathways interact with the mitochondria to improve energetics and reduce apoptosis and necrosis. Conceptually, it is useful to consider how PC and cardioprotection can reduce cell death. During and following ischemia-reperfusion, cell death is a mixture of apoptosis and necrosis. Interestingly, the same mitochondrial proteins, such as VDAC and the MPT, seem involved in regulating both apoptosis and necrosis. Furthermore, antiapoptotic proteins such as Bcl-2 reduce necrosis as well as apoptosis. Let us consider how PC might decrease necrosis. Necrosis appears to involve rupture of the plasma membrane. This is thought to occur at least in part because of a decrease in ATP to levels so low that the cell cannot maintain vital functions, including the maintenance of cell volume and cytoskeletal integrity, thereby predisposing the cell to plasma membrane rupture. Activation of the MPT appears to be very important in necrosis (62, 63). In fact, recent data suggest that mice lacking cyclophilin A, a component of MPT, have reduced ischemia-reperfusion injury but do not have reduced BAX-induced apoptosis. Activation of the MPT would result in a very rapid loss of ATP and release of Ca^{2+} into the cytosol, which would trigger hypercontracture. Sustained activation of the MPT, and the resultant ATP loss, would result in necrosis.

Cardioprotection and Reduced ATP Consumption During Ischemia

One of the early findings regarding PC was that it resulted in a reduced rate of ATP consumption. Interestingly, diazoxide and Bcl-2 overexpression, which are both cardioprotective, reduce the rate of ATP consumption during ischemia (69). PC may reduce the rate of ATP decline during ischemia by reducing ATP entry into the mitochondria, which may occur by inhibition of VDAC, ANT, or the F_1F_0ATPase. Inhibition of mitochondrial consumption of ATP may result in a faster rate of decline in $\Delta\psi$ during ischemia (76, 77). Previous studies have shown that glycolytically generated ATP is used during ischemia to maintain the $\Delta\psi$ and that oligomycin, an inhibitor of the F_1F_0-ATPase, results in a more rapid decline in $\Delta\psi$ and a slower rate of decline in ATP (76, 77). The reduced rate of decline in ATP observed with overexpression of Bcl-2 is similar to the rate of decline in ischemic ATP observed with oligomycin, and the effect is not additive, suggesting that oligomycin and Bcl-2 act on the same pathway. A faster or more complete decline in $\Delta\psi$ may inhibit mitochondrial Ca^{2+} uptake during ischemia and early reperfusion, and this may delay MPT opening. Consistent with this hypothesis, Belisle & Kowaltowski (78) report that addition of diazoxide to isolated mitochondria in anoxic media reduces the rate of ATP hydrolysis. Diazoxide decreases mitochondrial Ca^{2+} uptake during ischemia-reperfusion; this may be mediated by a decrease in $\Delta\psi$, which is secondary to inhibition of consumption of glycolytic ATP. This hypothesis is consistent with Garlid et al.'s (23) data showing that under normoxic conditions, diazoxide does not uncouple mitochondria. Also consistent with this hypothesis, Imahashi et al. (69) find that addition of Bcl-2 to isolated mitochondria reduces the uncoupler-stimulated consumption of added ATP.

How Does Preconditioning Inhibit VDAC, ANT, and the F_1F_0ATPase?

As shown in **Figure 1**, mitochondrial ERK, PKCε, GSK-3β, Cx43, Bcl-2, and nitric oxide either translocate to the mitochondria or have altered mitochondrial activity in response to cardioprotection (7, 16–18, 69). In addition, large numbers of kinases and phosphatases reside in the mitochondria (79); their activity may be altered by cardioprotective signaling. These signaling pathways may lead to the posttranslational modification of mitochondrial proteins such as VDAC, ANT, or the F_1F_0ATPase. A number of mitochondrial targets may contribute to cardioprotection. ERK phosphorylates BAD (17); 14-3-3 sequesters away phosphorylated BAD from Bcl-2, allowing Bcl-2 to oppose cell death. PKCε phosphorylates VDAC (16). VDAC is thought to be a component of the MPT, although the functional consequences of VDAC phosphorylation have not been established. Phosphorylation and inactivation of mitochondrial GSK-3β are associated with inhibition of the MPT (7), although the precise mechanism by which inhibition of GSK alters MPT is unclear. In noncardiac tissue, inhibition of GSK-3β results in decreased phosphorylation of VDAC, which increases binding of HK (44). The binding of HK to VDAC inhibits apoptosis. Nitric oxide activates the mitoK_{ATP} channel, which is cardioprotective (80, 81). PC also leads to an increase in *S*-nitrosylation of mitochondria complex I (82). Additionally, PC results in translocation of Cx43 to the mitochondrial inner membrane (18). Clearly, a number of signaling kinases and signaling molecules localize to the mitochondria with cardioprotection, and several systematic proteomic studies are underway to characterize better the changes in proteins and protein modifications associated with cardioprotection.

CONCLUSION AND WORKING HYPOTHESIS

PC activates a number of signaling pathways, some of which act on cytosolic, nuclear (these are likely important in the second window of PC), and mitochondrial targets. PC results in activation of mitochondrial ERK and PKC, inactivation of GSK, and alterations in nitric oxide and ROS generation. Through phosphorylation and *S*-nitrosylation (and other nitric oxide signaling mechanisms), these PC signaling pathways may induce posttranslational modification of mitochondrial proteins, which alters mitochondrial function, leading to reduced necrosis and apoptosis (see **Figures 1** and **2**).

Activation of the mitoK_{ATP} reduces Ca^{2+} uptake into the mitochondria (49). Opening of mitoK_{ATP} channels may uncouple mitochondria and thereby reduce $\Delta\psi$(83). However, this view has been challenged by Garlid et al. (23), who report that opening of mitoK_{ATP} channels does not significantly reduce $\Delta\psi$. Interestingly, Dos Santos et al. (50) suggest that opening of the mitoK_{ATP} channel by increasing mitochondrial volume would reduce ATP/ADP entry into the mitochondria. This is consistent with the results from Belisle & Kowaltowski (78) showing that diazoxide reduces the rate of ATP consumption under de-energized conditions. This decrease in ATP consumption would indirectly reduce $\Delta\psi$ during ischemia, when reverse-mode operation of the F_1F_0ATPase is used to generate $\Delta\psi$. The mechanism by which PC and diazoxide reduce ATP consumption by the mitochondria under de-energized conditions is still uncertain. Inhibition of ATP entry into the mitochondria via VDAC or ANT would reduce the consumption of ATP during ischemia, as would inhibition of the F_1F_0ATPase. Imahashi et al. (69) found that Bcl-2, which also inhibits mitochondrial consumption of ATP under de-energized

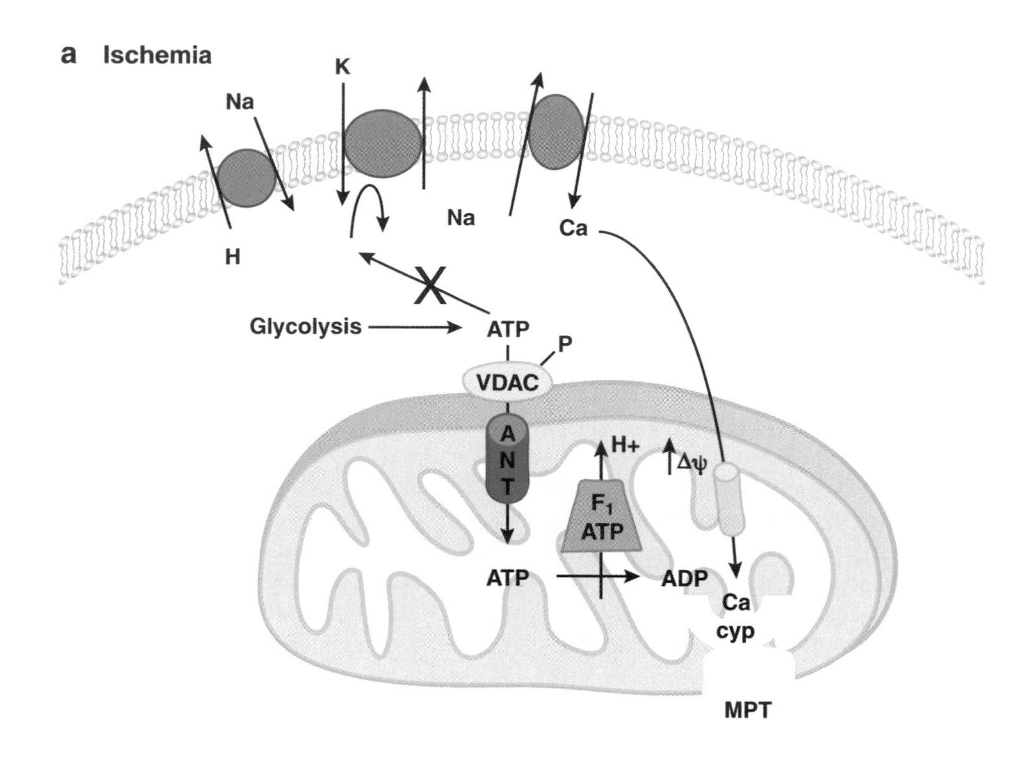
a Ischemia
Na
K
H
Na
Ca
Glycolysis
ATP
P
VDAC
A
N
T
H+
Δψ
F1
ATP
ATP
ADP
Ca
cyp
MPT

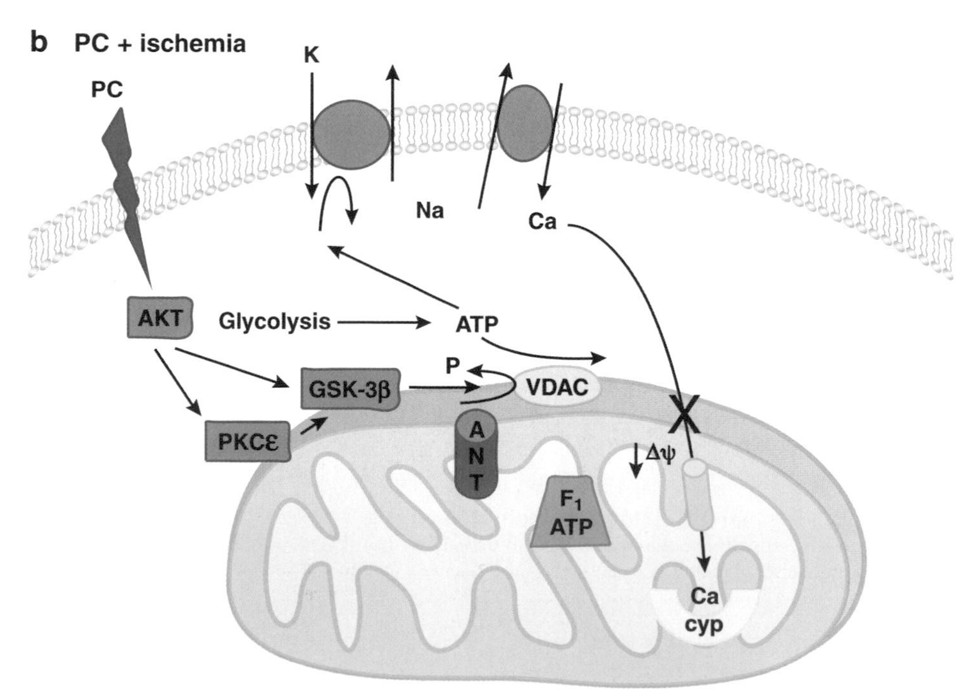
b PC + ischemia
PC
K
Na
Ca
AKT
Glycolysis
ATP
P
GSK-3β
VDAC
PKCε
A
N
T
Δψ
F1
ATP
Ca
cyp

conditions, associates with VDAC. These data support a hypothesis in which the cardioprotective effect of Bcl-2 is related to the increased association of Bcl-2 with mitochondrial proteins such as VDAC, decreased ATP consumption during ischemia, and a more rapid decline in $\Delta\psi$, leading to reduced mitochondrial Ca^{2+} uptake and reduced cell death. PKCε (16) and GSK-3β (44) increase phosphorylation of VDAC. It will be important to determine the sites of altered phosphorylation as well as the functional consequences of altered VDAC phosphorylation. Pastorino et al. (44) report that the decreased phosphorylation of VDAC, which occurs with inhibition of GSK, results in the increased association of VDAC with HK, as HK does not bind to VDAC when it is phosphorylated by GSK.

Others suggested that PC and diazoxide may reduce the rate of ATP consumption during ischemia. This may occur via inhibition of the F_1F_0ATPase by enhanced binding of an inhibitor factor (84). Inhibition of the F_1F_0ATPase may conserve glycolytically generated ATP during ischemia and not allow the ATP to be used to maintain $\Delta\psi$, thus resulting in a larger more rapid decline in $\Delta\psi$ and reducing the driving force for Ca^{2+} uptake into the mitochondria. However, others (85, 86) have reported that preconditioning does not result in inhibition of the F_1F_0ATPase. Ala-Rami et al. (84) attribute their finding that PC inhibits the F_1F_0ATPase to their rapid sample preparation in an alkaline medium and suggest that pH influences binding of the inhibitor factor. Additional studies will be necessary to determine whether PC and/or diazoxide inhibit the F_1F_0ATPase.

PC increases *S*-nitrosylation of mitochondrial complex I (82), which also may contribute to inhibition of MPT (87). Other means of reducing cell Ca^{2+}, such as inhibition of the plasma membrane Na-H exchanger, also reduce the Ca^{2+} available to be taken up by the mitochondria (88). Thus, the emerging hypothesis (see **Figure 2**) proposes that a reduction in MPT plays an essential role in cardioprotection. A reduction in MPT can occur due to (*a*) a decrease in cytosolic Ca^{2+}, (*b*) a decrease in $\Delta\psi$, which decreases the driving force for Ca^{2+} uptake into the mitochondria, or (*c*) loss of inhibition of cyclophilin D, which binds Ca^{2+} and leads to activation of MPT. Also of interest is that inhibition of MPT during the first few minutes of reperfusion appears to reduce cell death. Thus, PC activates signaling cascades that alter mitochondrial function to reduce necrosis and apoptosis. Future studies will be needed to determine the precise mitochondrial targets.

Figure 2

Proposed mechanism by which preconditioning leads to inhibition of mitochondrial permeability transition (MPT) and of necrosis and apoptosis. In ischemia (without PC), glycolytically generated ATP is used to maintain $\Delta\psi$, and $\Delta\psi$ provides the energy for Ca^{2+} uptake into the mitochondria. Ca^{2+} binds to cyclophilin D (cyp), resulting in activation of the MPT. The consumption of glycolytic ATP also decreases cytosolic ATP, leading to increased Ca^{2+} entry into the cell via Na-H and Na-Ca exchange. Preconditioning (PC) signaling pathways induce posttranslational modification of mitochondrial proteins. This decreases mitochondrial consumption of glycolytically generated ATP, which is used to maintain mitochondrial membrane potential ($\Delta\psi$), thus resulting in a decline in $\Delta\psi$. The decline in $\Delta\psi$ reduces the driving force for mitochondrial Ca^{2+} uptake and thereby reduces Ca^{2+} activation of MPT. Reduced activation of MPT conserves ATP and reduces necrosis. The posttranslational modification of mitochondrial proteins is likely also to block release of cytochrome *c* and inhibit apoptosis. Abbreviations used: AKT, a serine-threonine homology of the viral oncogene *v-akt*; ANT, adenine nucleotide translocator; cyp, cyclophilin D; GSK, glycogen synthase kinase; PKC, protein kinase C; VDAC, voltage-dependent anion channel.

SUMMARY POINTS

1. Cardioprotection activates a signaling cascade involving PI3-kinase, AKT, ERK, PKCε, GSK-3β, mitoK_{ATP}, VDAC, and mitochondrial permeability transition (MPT).
2. Cardioprotection leads to alterations (location or activity) in mitochondrial signaling molecules.
3. Mitochondrial F_1F_0ATPase hydrolysis is responsible for >50% of ATP consumed during ischemia.
4. Cardioprotection decreases ATP consumption during ischemia by inhibiting mitochondrial consumption of glycolytically generated ATP.
5. Glycolytically generated ATP is used to maintain mitochondrial membrane potential during ischemia.
6. Therefore, a preconditioning-mediated inhibition of consumption of glycolytic ATP enhances the decline in mitochondrial membrane potential ($\Delta\psi$) during ischemia. This enhanced reduction in $\Delta\psi$ reduces the driving force for mitochondrial Ca^{2+} uptake and thereby reduces Ca^{2+} activation of MPT.

FUTURE ISSUES

1. What is the mechanism by which PC reduces mitochondrial consumption of ATP and reduces $\Delta\psi$ during the subsequent sustained period of ischemia?
2. What is the purpose of the MPT? Does the MPT exist only to initiate cell death, or does it have some other purpose? The MPT may be involved in Ca^{2+}-induced Ca^{2+} release from the mitochondria, which may amplify Ca^{2+} signals from the endoplasmic reticulum (89). The ryanodine receptor, the Ca^{2+} release channel in the sarcoplasmic reticulum that is responsible for Ca^{2+}-induced Ca^{2+} release from the sarcoplasmic reticulum, is also associated with a peptidyl-prolyl *cis-trans* isomerase, FKBP. Cyclophilin is also a peptidyl-prolyl *cis-trans* isomerase. The ryanodine receptor can also localize to the mitochondria (90). Some researchers have speculated that the MPT exists to release high levels of Ca^{2+} or other ions from the matrix. Consistent with this proposal, Altschuld (91) reported that cyclosporin inhibits mitochondrial Ca^{2+} efflux. Others have suggested that the MPT may open to dissipate $\Delta\psi$ under conditions of high $\Delta\psi$. High $\Delta\psi$ would increase ROS (27, 28, 92), and opening of MPT may be a safety valve. It is interesting that MPT is regulated by Ca^{2+} via activation of cyclophilin, as Ca^{2+} regulates both mitochondrial dehydrogenases and F_1F_0ATPase activity (93, 94). Perhaps a high $\Delta\psi$ in the presence of a high matrix Ca^{2+}, which would be a driving force to increase electron transport further, results in opening of MPT to reduce $\Delta\psi$ and prevent excess generation of ROS.
3. Does Bcl-2 have a role in regulating mitochondrial function and/or energetics in addition to its role in inhibiting apoptosis? The precise mechanism by which Bcl-2 inhibits apoptosis is not well understood, despite intense research over the past 15 years.

LITERATURE CITED

1. Murry CE, Jennings RB, Reimer KA. 1986. Preconditioning with ischemia: a delay of lethal cell injury in ischemic myocardium. *Circulation* 74:1124–36
2. Murry CE, Richard VJ, Reimer KA, Jennings RB. 1990. Ischemic preconditioning slows energy metabolism and delays ultrastructural damage during a sustained ischemic episode. *Circ. Res.* 66:913–31
3. Tong H, Chen W, Steenbergen C, Murphy E. 2000. Ischemic preconditioning activates phosphatidylinositol-3-kinase upstream of protein kinase C. *Circ. Res.* 87:309–15
4. Ytrehus K, Liu Y, Downey JM. 1994. Preconditioning protects ischemic rabbit heart by protein kinase C activation. *Am. J. Physiol.* 266:H1145–52
5. Ping P, Zhang J, Qiu Y, Tang XL, Manchikalapudi S, et al. 1997. Ischemic preconditioning induces selective translocation of protein kinase C isoforms epsilon and eta in the heart of conscious rabbits without subcellular redistribution of total protein kinase C activity. *Circ. Res.* 81:404–14
6. Tong H, Imahashi K, Steenbergen C, Murphy E. 2002. Phosphorylation of glycogen synthase kinase-3β during preconditioning through a phosphatidylinositol-3-kinase-dependent pathway is cardioprotective. *Circ. Res.* 90:377–9
7. Juhaszova M, Zorov DB, Kim SH, Pepe S, Fu Q, et al. 2004. Glycogen synthase kinase-3β mediates convergence of protection signaling to inhibit the mitochondrial permeability transition pore. *J. Clin. Invest.* 113:1535–49
8. Gross ER, Hsu AK, Gross GJ. 2004. Opioid-induced cardioprotection occurs via glycogen synthase kinase beta inhibition during reperfusion in intact rat hearts. *Circ. Res.* 94:960–66
9. Ping P, Takano H, Zhang J, Tang XL, Qiu Y, et al. 1999. Isoform-selective activation of protein kinase C by nitric oxide in the heart of conscious rabbits: a signaling mechanism for both nitric oxide-induced and ischemia-induced preconditioning. *Circ. Res.* 84:587–604
10. Bolli R, Dawn B, Tang XL, Qiu Y, Ping P, et al. 1998. The nitric oxide hypothesis of late preconditioning. *Basic Res. Cardiol.* 93:325–38
11. Ping P, Zhang J, Huang S, Cao X, Tang XL, et al. 1999. PKC-dependent activation of p46/p54 JNKs during ischemic preconditioning in conscious rabbits. *Am. J. Physiol.* 277:H1771–85
12. Costa AD, Garlid KD, West IC, Lincoln TM, Downey JM, et al. 2005. Protein kinase G transmits the cardioprotective signal from cytosol to mitochondria. *Circ. Res.* 97:329–36
13. Garlid KD, Paucek P, Yarov-Yarovoy V, Murray HN, Darbenzio RB, et al. 1997. Cardioprotective effect of diazoxide and its interaction with mitochondrial ATP-sensitive K^+ channels. Possible mechanism of cardioprotection. *Circ. Res.* 81:1072–82
14. Liu Y, Sato T, O'Rourke B, Marban E. 1998. Mitochondrial ATP-dependent potassium channels: novel effectors of cardioprotection? *Circulation* 97:2463–69
15. Tong H, Rockman HA, Koch WJ, Steenbergen C, Murphy E. 2004. G protein-coupled receptor internalization signaling is required for cardioprotection in ischemic preconditioning. *Circ. Res.* 94:1133–41
16. Baines CP, Song CX, Zheng YT, Wang GW, Zhang J, et al. 2003. Protein kinase Cε interacts with and inhibits the permeability transition pore in cardiac mitochondria. *Circ. Res.* 92:873–80
17. Baines CP, Zhang J, Wang GW, Zheng YT, Xiu JX, et al. 2002. Mitochondrial PKCε and MAPK form signaling modules in the murine heart: enhanced mitochondrial PKCε-MAPK interactions and differential MAPK activation in PKCε-induced cardioprotection. *Circ. Res.* 90:390–97

18. Boengler K, Dodoni G, Rodriguez-Sinovas A, Cabestrero A, Ruiz-Meana M, et al. 2005. Connexin 43 in cardiomyocyte mitochondria and its increase by ischemic preconditioning. *Cardiovasc. Res.* 67:234–44
19. Rourke B. 2004. Evidence for mitochondrial K^+ channels and their role in cardioprotection. *Circ. Res.* 94:420–32
20. Murphy E. 2004. Primary and secondary signaling pathways in early preconditioning that converge on the mitochondria to produce cardioprotection. *Circ. Res.* 94:7–16
21. Lim KH, Javadov SA, Das M, Clarke SJ, Suleiman MS, Halestrap AP. 2002. The effects of ischemic preconditioning, diazoxide and 5-hydroxydecanoate on rat heart mitochondrial volume and respiration. *J. Physiol.* 545:961–74
22. Weiss JN, Korge P, Honda HM, Ping P. 2003. Role of the mitochondrial permeability transition in myocardial disease. *Circ. Res.* 93:292–301
23. Garlid KD, Dos Santos P, Xie ZJ, Costa AD, Paucek P. 2003. Mitochondrial potassium transport: the role of the mitochondrial ATP-sensitive K^+ channel in cardiac function and cardioprotection. *Biochim. Biophys. Acta* 1606:1–21
24. O'Rourke B. 2000. Myocardial K_{ATP} channels in preconditioning. *Circ. Res.* 87:845–55
25. Zhao ZQ, Corvera JS, Halkos ME, Kerendi F, Wang NP, et al. 2003. Inhibition of myocardial injury by ischemic postconditioning during reperfusion: comparison with ischemic preconditioning. *Am. J. Physiol. Heart Circ. Physiol.* 285:H579–88
26. Hausenloy DJ, Tsang A, Mocanu MM, Yellon DM. 2005. Ischemic preconditioning protects by activating prosurvival kinases at reperfusion. *Am. J. Physiol. Heart. Circ. Physiol.* 288:H971–76
27. Turrens JF. 2003. Mitochondrial formation of reactive oxygen species. *J. Physiol.* 552:335–44
28. Balaban RS, Nemoto S, Finkel T. 2005. Mitochondria, oxidants, and aging. *Cell* 120:483–95
29. Hoerter J, Gonzalez-Barroso MD, Couplan E, Mateo P, Gelly C, et al. 2004. Mitochondrial uncoupling protein 1 expressed in the heart of transgenic mice protects against ischemic-reperfusion damage. *Circulation* 110:528–33
30. van der Klei IJ, Veenhuis M, Neupert W. 1994. A morphological view on mitochondrial protein targeting. *Microsc. Res. Tech.* 27:284–93
31. Appaix F, Kuznetsov AV, Usson Y, Kay L, Andrienko T, et al. 2003. Possible role of cytoskeleton in intracellular arrangement and regulation of mitochondria. *Exp. Physiol.* 88:175–90
32. Milner DJ, Mavroidis M, Weisleder N, Capetanaki Y. 2000. Desmin cytoskeleton linked to muscle mitochondrial distribution and respiratory function. *J. Cell Biol.* 150:1283–98
33. Dzeja PP, Terzic A. 2003. Phosphotransfer networks and cellular energetics. *J. Exp. Biol.* 206:2039–47
34. Pavlov E, Grigoriev SM, Dejean LM, Zweihorn CL, Mannella CA, Kinnally KW. 2005. The mitochondrial channel VDAC has a cation-selective open state. *Biochim. Biophys. Acta* 1710:96–102
35. Blachly-Dyson E, Forte M. 2001. VDAC channels. *IUBMB Life* 52:113–18
36. Hodge T, Colombini M. 1997. Regulation of metabolite flux through voltage-gating of VDAC channels. *J. Membr. Biol.* 157:271–79
37. Zizi M, Byrd C, Boxus R, Colombini M. 1998. The voltage-gating process of the voltage-dependent anion channel is sensitive to ion flow. *Biophys. J.* 75:704–13
38. Komarov AG, Deng D, Craigen WJ, Colombini M. 2005. New insights into the mechanism of permeation through large channels. *Biophys. J.* 89:3950–59

39. Lemeshko VV, Lemeshko SV. 2004. The voltage-dependent anion channel as a biological transistor: theoretical considerations. *Eur. Biophys. J.* 33:352–59
40. Rapizzi E, Pinton P, Szabadkai G, Wieckowski MR, Vandecasteele G, et al. 2002. Recombinant expression of the voltage-dependent anion channel enhances the transfer of Ca^{2+} microdomains to mitochondria. *J. Cell Biol.* 159:613–24
41. Gincel D, Zaid H, Shoshan-Barmatz V. 2001. Calcium binding and translocation by the voltage-dependent anion channel: a possible regulatory mechanism in mitochondrial function. *Biochem. J.* 358:147–55
42. Nakashima RA, Mangan PS, Colombini M, Pedersen PL. 1986. Hexokinase receptor complex in hepatoma mitochondria: evidence from *N,N′*-dicyclohexylcarbodiimide-labeling studies for the involvement of the pore-forming protein VDAC. *Biochemistry* 25:1015–21
43. Schlattner U, Dolder M, Wallimann T, Tokarska-Schlattner M. 2001. Mitochondrial creatine kinase and mitochondrial outer membrane porin show a direct interaction that is modulated by calcium. *J. Biol. Chem.* 276:48027–30
44. Pastorino JG, Hoek JB, Shulga N. 2005. Activation of glycogen synthase kinase 3β disrupts the binding of hexokinase II to mitochondria by phosphorylating voltage-dependent anion channel and potentiates chemotherapy-induced cytotoxicity. *Cancer Res.* 65:10545–54
45. Lai JC, Tan W, Benimetskaya L, Miller P, Colombini M, Stein CA. 2006. A pharmacologic target of G3139 in melanoma cells may be the mitochondrial VDAC. *Proc. Natl. Acad. Sci. USA* 103:7494–99
46. Garlid KD. 1996. Cation transport in mitochondria—the potassium cycle. *Biochim. Biophys. Acta* 1275:123–26
47. Paucek P, Mironova G, Mahdi F, Beavis AD, Woldegiorgis G, Garlid KD. 1992. Reconstitution and partial purification of the glibenclamide-sensitive, ATP-dependent K^+ channel from rat liver and beef heart mitochondria. *J. Biol. Chem.* 267:26062–69
48. Forbes RA, Steenbergen C, Murphy E. 2001. Diazoxide-induced cardioprotection requires signaling through a redox-sensitive mechanism. *Circ. Res.* 88:802–9
49. Murata M, Akao M, O'Rourke B, Marban E. 2001. Mitochondrial ATP-sensitive potassium channels attenuate matrix Ca^{2+} overload during simulated ischemia and reperfusion: possible mechanism of cardioprotection. *Circ. Res.* 89:891–98
50. Dos Santos P, Kowaltowski AJ, Laclau MN, Seetharaman S, Paucek P, et al. 2002. Mechanisms by which opening the mitochondrial ATP-sensitive K^+ channel protects the ischemic heart. *Am. J. Physiol. Heart. Circ. Physiol.* 283:H284–95
51. Halestrap AP. 1994. Regulation of mitochondrial metabolism through changes in matrix volume. *Biochem. Soc. Trans.* 22:522–29
52. Territo PR, French SA, Dunleavy MC, Evans FJ, Balaban RS. 2001. Calcium activation of heart mitochondrial oxidative phosphorylation: rapid kinetics of mVO2, NADH, and light scattering. *J. Biol. Chem.* 276:2586–99
53. Korge P, Honda HM, Weiss JN. 2005. K^+-dependent regulation of matrix volume improves mitochondrial function under conditions mimicking ischemia-reperfusion. *Am. J. Physiol. Heart Circ. Physiol.* 289:H66–77
54. Pain T, Yang XM, Critz SD, Yue Y, Nakano A, et al. 2000. Opening of mitochondrial K_{ATP} channels triggers the preconditioned state by generating free radicals. *Circ. Res.* 87:460–66
55. Lehninger AL, Remmert LF. 1959. An endogenous uncoupling and swelling agent in liver mitochondria and its enzymic formation. *J. Biol. Chem.* 234:2459–64
56. Azzi A, Azzone GF. 1965. Swelling and shrinkage phenomena in liver mitochondria. I. Large amplitude swelling induced by inorganic phosphate and by ATP. *Biochim. Biophys. Acta* 105:253–64

57. Kokoszka JE, Waymire KG, Levy SE, Sligh JE, Cai J, et al. 2004. The ADP/ATP translocator is not essential for the mitochondrial permeability transition pore. *Nature* 427:461–65
58. Krauskopf A, Eriksson O, Craigen WJ, Forte MA, Bernardi P. 2006. Properties of the permeability transition in VDAC1$^{-/-}$ mitochondria. *Biochim. Biophys. Acta* 1752:590–95
59. Cesura AM, Pinard E, Schubenel R, Goetschy V, Friedlein A, et al. 2003. The voltage-dependent anion channel is the target for a new class of inhibitors of the mitochondrial permeability transition pore. *J. Biol. Chem.* 278:49812–18
60. He L, Lemasters JJ. 2002. Regulated and unregulated mitochondrial permeability transition pores: a new paradigm of pore structure and function? *FEBS Lett.* 512:1–7
61. Kowaltowski AJ, Castilho RF, Vercesi AE. 2001. Mitochondrial permeability transition and oxidative stress. *FEBS Lett.* 495:12–15
62. Baines CP, Kaiser RA, Purcell NH, Blair NS, Osinska H, et al. 2005. Loss of cyclophilin D reveals a critical role for mitochondrial permeability transition in cell death. *Nature* 434:658–62
63. Nakagawa T, Shimizu S, Watanabe T, Yamaguchi O, Otsu K, et al. 2005. Cyclophilin D-dependent mitochondrial permeability transition regulates some necrotic but not apoptotic cell death. *Nature* 434:652–58
64. Griffiths EJ, Halestrap AP. 1993. Protection by Cyclosporin A of ischemia/reperfusion-induced damage in isolated rat hearts. *J. Mol. Cell Cardiol.* 25:1461–69
65. Javadov SA, Clarke S, Das M, Griffiths EJ, Lim KH, Halestrap AP. 2003. Ischaemic preconditioning inhibits opening of mitochondrial permeability transition pores in the reperfused rat heart. *J. Physiol.* 549:513–24
66. Halestrap AP, Clarke SJ, Javadov SA. 2004. Mitochondrial permeability transition pore opening during myocardial reperfusion—a target for cardioprotection. *Cardiovasc. Res.* 61:372–85
67. Hausenloy D, Wynne A, Duchen M, Yellon D. 2004. Transient mitochondrial permeability transition pore opening mediates preconditioning-induced protection. *Circulation* 109:1714–17
68. Zong WX, Thompson CB. 2006. Necrotic death as a cell fate. *Genes Dev.* 20:1–15
69. Imahashi K, Schneider MD, Steenbergen C, Murphy E. 2004. Transgenic expression of Bcl-2 modulates energy metabolism, prevents cytosolic acidification during ischemia, and reduces ischemia/reperfusion injury. *Circ. Res.* 95:734–41
70. Chen Z, Chua CC, Ho YS, Hamdy RC, Chua BH. 2001. Overexpression of Bcl-2 attenuates apoptosis and protects against myocardial I/R injury in transgenic mice. *Am. J. Physiol. Heart Circ. Physiol.* 280:H2313–20
71. Vander Heiden MG, Chandel NS, Li XX, Schumacker PT, Colombini M, Thompson CB. 2000. Outer mitochondrial membrane permeability can regulate coupled respiration and cell survival. *Proc. Natl. Acad. Sci. USA* 97:4666–71
72. Shimizu S, Narita M, Tsujimoto Y. 1999. Bcl-2 family proteins regulate the release of apoptogenic cytochrome *c* by the mitochondrial channel VDAC. *Nature* 399:483–87
73. Shimizu S, Ide T, Yanagida T, Tsujimoto Y. 2000. Electrophysiological study of a novel large pore formed by Bax and the voltage-dependent anion channel that is permeable to cytochrome *c*. *J. Biol. Chem.* 275:12321–25
74. Rouslin W, Erickson JL, Solaro RJ. 1986. Effects of oligomycin and acidosis on rates of ATP depletion in ischemic heart muscle. *Am. J. Physiol.* 250:H503–8
75. Rouslin W, Broge CW, Grupp IL. 1990. ATP depletion and mitochondrial functional loss during ischemia in slow and fast heart-rate hearts. *Am. J. Physiol.* 259:H1759–66
76. Di Lisa F, Menabo R, Canton M, Petronilli V. 1998. The role of mitochondria in the salvage and the injury of the ischemic myocardium. *Biochim. Biophys. Acta* 1366:69–78

77. Leyssens A, Nowicky AV, Patterson L, Crompton M, Duchen MR. 1996. The relationship between mitochondrial state, ATP hydrolysis, $[Mg^{2+}]_i$ and $[Ca^{2+}]_i$ studied in isolated rat cardiomyocytes. *J. Physiol.* 496(Pt. 1):111–28
78. Belisle E, Kowaltowski AJ. 2002. Opening of mitochondrial K^+ channels increases ischemic ATP levels by preventing hydrolysis. *J. Bioenerg. Biomembr.* 34:285–98
79. Hopper RK, Carroll S, Aponte AM, Johnson DT, French S, et al. 2006. Mitochondrial matrix phosphoproteome: effect of extra mitochondrial calcium. *Biochemistry* 45:2524–36
80. Sasaki N, Sato T, Ohler A, O'Rourke B, Marban E. 2000. Activation of mitochondrial ATP-dependent potassium channels by nitric oxide. *Circulation* 101:439–45
81. Sato T, O'Rourke B, Marban E. 1998. Modulation of mitochondrial ATP-dependent K^+ channels by protein kinase C. *Circ. Res.* 83:110–14
82. Burwell LS, Nadtochiy SM, Tompkins AJ, Young S, Brookes PS. 2006. Direct evidence for S-nitrosation of mitochondrial complex I. *Biochem. J.* 394:627–34
83. Holmuhamedov EL, Jahangir A, Oberlin A, Komarov A, Colombini M, Terzic A. 2004. Potassium channel openers are uncoupling protonophores: implication in cardioprotection. *FEBS Lett.* 568:167–70
84. Ala-Rami A, Ylitalo KV, Hassinen IE. 2003. Ischaemic preconditioning and a mitochondrial K_{ATP} channel opener both produce cardioprotection accompanied by F_1F_0-ATPase inhibition in early ischemia. *Basic Res. Cardiol.* 98:250–58
85. Green DW, Murray HN, Sleph PG, Wang FL, Baird AJ, et al. 1998. Preconditioning in rat hearts is independent of mitochondrial F_1F_0 ATPase inhibition. *Am. J. Physiol.* 274:H90–97
86. Vander Heide RS, Hill ML, Reimer KA, Jennings RB. 1996. Effect of reversible ischemia on the activity of the mitochondrial ATPase: relationship to ischemic preconditioning. *J. Mol. Cell Cardiol.* 28:103–12
87. Bernardi P, Krauskopf A, Basso E, Petronilli V, Blalchy-Dyson E, et al. 2006. The mitochondrial permeability transition from in vitro artifact to disease target. *FEBS J.* 273:2077–99
88. Murphy E, Perlman M, London RE, Steenbergen C. 1991. Amiloride delays the ischemia-induced rise in cytosolic free calcium. *Circ. Res.* 68:1250–58
89. Ichas F, Mazat JP. 1998. From calcium signaling to cell death: two conformations for the mitochondrial permeability transition pore. Switching from low- to high-conductance state. *Biochim. Biophys. Acta* 1366:33–50
90. Beutner G, Sharma VK, Lin L, Ryu SY, Dirksen RT, Sheu SS. 2005. Type 1 ryanodine receptor in cardiac mitochondria: transducer of excitation-metabolism coupling. *Biochim. Biophys. Acta* 1717:1–10
91. Altschuld RA, Hohl CM, Castillo LC, Garleb AA, Starling RC, Brierley GP. 1992. Cyclosporin inhibits mitochondrial calcium efflux in isolated adult rat ventricular cardiomyocytes. *Am. J. Physiol.* 262:H1699–704
92. Aon MA, Cortassa S, Marban E, O'Rourke B. 2003. Synchronized whole cell oscillations in mitochondrial metabolism triggered by a local release of reactive oxygen species in cardiac myocytes. *J. Biol. Chem.* 278:44735–44
93. Territo PR, French SA, Balaban RS. 2001. Simulation of cardiac work transitions, in vitro: effects of simultaneous Ca^{2+} and ATPase additions on isolated porcine heart mitochondria. *Cell Calcium* 30:19–27
94. Denton RM, McCormack JG. 1980. The role of calcium in the regulation of mitochondrial metabolism. *Biochem. Soc. Trans.* 8:266–68

Iron Homeostasis

Nancy C. Andrews and Paul J. Schmidt

Children's Hospital Boston and Harvard Medical School, Karp Family Research Laboratories, Boston, Massachusetts 02115; email: nancy_andrews@hms.harvard.edu

Annu. Rev. Physiol. 2007. 69:69–85

First published online as a Review in Advance on October 2, 2006

The *Annual Review of Physiology* is online at http://physiol.annualreviews.org

This article's doi: 10.1146/annurev.physiol.69.031905.164337

0066-4278/07/0315-0069$20.00

Key Words

hemochromatosis, transferrin, anemia, transport

Abstract

Iron is needed by all mammalian cells but is toxic in excess. Specialized transport mechanisms conduct iron across cellular membranes. These are regulated to ensure homeostasis both systemically in living organisms and within individual cells. Over the past decade, major advances have been made in identifying and characterizing the proteins involved in the transport, handling, and homeostatic regulation of iron. Molecular understanding of these processes has provided important insights into the pathophysiology of human iron disorders.

Enterocyte: absorptive cell of the intestinal epithelium

DCYTB: duodenal cytochrome *b*, an intestinal ferrireductase

Divalent metal transporter 1 (DMT1): the major transmembrane transporter of divalent metal cations into cells

INTRODUCTION

Iron is a functional component of oxygen-carrying globin proteins, cytochromes, and enzymes that transfer electrons. It is abundant in the environment but insoluble in aqueous solutions at physiological pH. The challenge for most organisms is to acquire adequate amounts of iron for critical biological processes yet avoid the toxicity associated with free iron. Meticulously regulated mechanisms have evolved to move iron across biological membranes and orchestrate its distribution in multicellular organisms. Over the past ten years our understanding of mammalian iron transport and homeostasis has advanced dramatically.

INTESTINAL IRON ABSORPTION

Normally, mammals obtain iron exclusively from the diet. Inorganic, nonheme iron is inefficiently absorbed but present in a wide variety of foodstuffs. Heme iron is more efficiently absorbed, primarily from animal sources in the form of hemoglobin or myoglobin. Iron absorption takes place in the proximal small intestine, near the gastroduodenal junction. There, the epithelium is organized in finger-like villous structures, increasing the surface area and aiding absorption. The absorptive cells (enterocytes) are polarized, with an apical microvillus brush border separated from the basolateral surface by tight junctions. Because there is little or no paracellular iron transport under normal circumstances, iron must traverse both the apical and basolateral membranes to gain access to the circulation. Each transmembrane transport step requires specific transporter proteins and accessory enzymes that change the oxidation state of iron.

Dietary nonheme iron is reduced by brush border ferrireductases to Fe^{2+}, the substrate for divalent metal transporters. Duodenal cytochrome *b* (DCYTB) was the first intestinal ferrireductase to be described (1). On the basis of its homology to cytochrome *b*561 reductases, DCYTB is presumed to be a heme protein that uses ascorbate to facilitate ferrireduction. The expression of DCYTB increases in response to iron deficiency and hypoxia, consistent with an important role in apical iron uptake (1). However, mice lacking DCYTB (*Cybrd1−/−*) do not develop an iron deficit when maintained on a normal lab diet (2), suggesting that other mechanisms for iron reduction can substitute, at least in mice. Other candidates for brush border reductases include members of the STEAP protein family (3). No mutations in human DCYTB or other putative intestinal ferrireductases have been reported to date.

Fe^{2+}, like several other divalent transition metals, is transported across the cellular membrane by divalent metal transporter 1 (DMT1, also known as SLC11A2, NRAMP2, and DCT1), a member of the Nramp family of 12-transmembrane-segment proteins (4–6). DMT1 is a proton symporter, requiring low pH for efficient metal transport (6). A unique spontaneous mutation in *DMT1* (G185R), identified in microcytic anemia (*mk*) mice and Belgrade (*b*) rats, has provided some insight into the in vivo functions of DMT1 and the transport mechanism (4, 7). Both mutant strains have significant defects in intestinal iron absorption and assimilation of iron by erythroid precursor cells, indicating that the protein has lost iron transport function in those tissues. Electrophysiological studies of G185R DMT1 showed that the mutation also effectively converts DMT1 into a calcium channel (8). This is intriguing because L-type calcium channels permeate iron, suggesting a continuum of function between high-flux divalent ion channels and low-flux metal ion transporters.

Targeted mutation of the murine *DMT1* gene (*Slc11a2−/−* mice) confirmed roles for DMT1 in intestinal iron absorption and erythroid iron uptake (5). Those experiments suggested, however, that DMT1 is not essential for placental iron transfer or iron acquisition by other tissues, notably the liver and brain. DMT1 is highly expressed in discrete areas of the brain (6), suggesting that it has subtle roles there that were not detected by initial phenotypical examination. Human mutations have also been identified in patients with congenital anemia (9–11). The phenotype resulting from loss-of-function mutations in human *DMT1* differs from that seen in rodents; human patients develop hepatic iron overload. This suggests either slight differences in DMT1 activity between the species or, perhaps more likely, better intestinal iron absorption in human subjects due to a greater proportion of heme iron in their diet.

Heme iron appears to be transported intact from the gut lumen into enterocytes, through the activity of heme carrier protein 1 (12). The intracellular metabolism of heme is not fully understood. It is likely that, after it is disassembled by heme oxygenase, the liberated iron enters the same storage and transport pathways taken by inorganic iron. Some of the iron extracted from the diet is stored in ferritin within the enterocyte, and some is exported across the enterocyte's basolateral membrane. The iron retained in endothelial ferritin is lost after two to three days owing to the sloughing of enterocytes into the gut lumen.

A single protein, ferroportin (FPN; also known as SLC40A1, IREG1, and MTP1), is important for basolateral transfer of iron to the circulation. FPN has 10 to 12 predicted transmembrane segments but bears no homology to DMT1 or other mammalian proteins (13–15). Whereas DMT1 has homologs in many species, including yeast and bacteria, FPN homologs are not found in unicellular organisms. The iron transport function of FPN has not been studied as thoroughly as that of DMT1, primarily because systems for analyzing iron export are less robust than systems available to study iron import. Nonetheless, the normal function can be deduced from gene targeting in mice (*Slc40a1−/−* mice) (16). FPN is critically important for materno-embryonic iron transfer, for basolateral transport of iron out of enterocytes, and for the export of iron from tissue macrophages. It appears to play a lesser role in the export of iron from hepatocytes.

Similar to DMT1, ferroportin likely conducts Fe^{2+} ions. Cellular iron export requires an associated ferroxidase activity. In the intestine, ceruloplasmin, a circulating multicopper oxidase, and hephaestin, a homolog of ceruloplasmin, appear to supply this activity (17–19). Hepatocytes and macrophages probably use ceruloplasmin exclusively. The precise site of ferroxidase activity remains uncertain, but Fe^{2+} must be oxidized to Fe^{3+} to circulate bound to plasma proteins.

Absorbed iron is rapidly bound to transferrin (TF), an abundant, high-affinity iron-binding protein. Under normal circumstances, TF carries nearly all serum iron, which solubilizes iron and dampens its reactivity. Very small amounts of iron may be loosely associated with albumin or small molecules. In normal human subjects, iron occupies approximately 30% of the iron-binding sites on plasma TF. In mice the TF saturation is higher, typically ranging from 60% to 80%. The saturation of TF by iron varies on a diurnal cycle (20) and rapidly responds to local circumstances. It is likely to be higher in the portal circulation, where recently absorbed iron from the intestine enters the circulation and passes through the liver. This first-pass effect may explain the periportal hepatocyte distribution of iron observed in iron overload disorders associated with inappropriately increased intestinal iron absorption. Conversely, TF saturation is likely lower than average in plasma leaving the erythroid bone marrow, where most of the iron present is extracted for use by erythroid precursor cells.

Ferritin: the major iron storage protein

Ferroportin (FPN): the major transmembrane transporter transferring iron out of cells

Transferrin (TF): the primary plasma iron carrier protein

IRON UTILIZATION

The erythroid bone marrow is the largest consumer of iron. Normally, two-thirds of the body iron endowment is found in developing erythroid precursors and mature red blood cells. Approximately a billion iron atoms are

Transferrin receptor (TFR): the primary receptor for transferrin

used each day to form hemoglobin in new erythrocytes. Erythroid precursors express cell-surface transferrin receptors (TFRs) that take up Fe-TF by receptor-mediated endocytosis (reviewed in Reference 21). Although TFRs are widely expressed, most other cells apparently can use non-TFR mechanisms to assimilate iron. Targeted disruption of the murine *TFR* gene (*Trfr−/−* mice) causes embryonic lethality, attributable to severe anemia (22). At the time of death, approximately embryonic day 11.5 (E11.5), most nonhematopoietic tissues appear normal, suggesting that TFR is not required for their early development. However, in the absence of TFR, an increased fraction of neuroepithelial precursor cells undergoes apoptosis, and the neural tube is kinked. No human mutations in *TFR* have been identified, but antibodies against TFR have been demonstrated in a patient with severe, acquired anemia (24).

Spontaneous mutations disrupting the *TF* gene demonstrate the importance of TF in animals (25, 26) and humans (27, 28). Consistent with the *TFR* knockout results, profound deficiency of TF results in severe anemia and, possibly, abnormalities of the central nervous system (29). However, iron deficiency occurs only in hematopoietic cells; other tissues develop massive iron overload (30). This underscores the importance of the TF-TFR endocytic cycle in erythropoiesis. It further demonstrates that nonhematopoietic cells have alternative mechanisms to assimilate iron and highlights changes in overall iron homeostasis that result from iron-deficient erythropoiesis. Intestinal iron absorption is markedly increased, apparently to try to compensate for the lack of iron available to erythroid precursors. This is compelling evidence for a signaling mechanism that allows the erythroid bone marrow to communicate its iron needs to the intestine's absorptive epithelium (discussed in Reference 31).

Most cells can assimilate iron without the TF cycle. However, it is not clear why hematopoietic cells remain so dependent upon the TF cycle. Presumably, receptor-mediated endocytosis of Fe-TF allows iron to be extracted from TF and concentrated in a low-pH compartment to facilitate rapid uptake through DMT1. After endocytosis, the TF endosome is acidified to pH 5.5 through the action of a proton pump. TF and TFR undergo conformational changes leading to iron release. Fe^{3+} released from TF must be reduced to Fe^{2+} for transport by DMT1. Elegant genetic experiments suggest that STEAP3 serves this function (32). A mouse mutant lacking STEAP3 develops iron-deficiency anemia attributable to a defect in endosomal ferrireductase function. It is not yet known whether STEAP3 interacts directly with either TF or DMT1 to transfer iron from carrier to transporter.

Once iron leaves the endosome, it must move to the mitochondrion for incorporation into protoporphyrin IX by ferrochelatase to form heme. Heme biosynthesis begins and ends in the mitochondrion, but intermediate steps occur in the cytoplasm (reviewed in Reference 33). Mitoferrin (also known as SLC25A37) carries out mitochondrial iron import (34). Zebrafish and mice lacking mitoferrin fail to incorporate erythroid iron into heme. Most heme in erythroid precursors is used for hemoglobin production. However, a recent report suggests that a protein capable of cellular heme export, FLVCR, plays a key role in erythropoiesis (35). The role(s) of FLVCR in other tissues has not yet been fully defined.

Although muscle cells also require large amounts of iron to produce myoglobin, far less is known about how they assimilate iron. The phenotypes of mouse embryos lacking TFR and newborn mice lacking DMT1 suggest that neither protein is essential for myogenesis. It is not clear whether mitoferrin is also involved in mitochondrial iron uptake in muscle and other nonhematopoietic tissues. This is likely to be an area of active investigation over the next few years.

TFR plays an important, but poorly understood, role in lymphopoiesis (23). It was recognized as a prominent marker of lymphocyte activation more than two decades ago (36, 37), but its function remains uncertain. It is not clear whether TFR is needed for accelerated iron uptake by rapidly proliferating cells or for a different, unrelated function. TFR is recruited to the immunological synapse in response to T cell receptor engagement (38) and associates with the T cell receptor zeta chain (39). Accordingly, Woodward et al. (40) have proposed anti-TFR antibodies for use as immunosuppressants.

Iron is an essential growth factor in early kidney development. In vitro, it can be supplied by lipocalin 2 (also known as Ngal and 24p3) as a protein-siderophore complex (41, 42). However, animals lacking lipocalin 2 develop normally, suggesting that it is not essential. Such animals show a defect in innate immunity when challenged by bacterial infection, suggesting that lipocalin 2 serves to sequester iron-containing siderophores from microorganisms that depend upon them for survival (43).

Iron deposition is a characteristic feature of several neurodegenerative diseases. These include common disorders such as Parkinson and Alzheimer diseases and much rarer disorders such as aceruloplasminemia, Hallervorden-Spatz disease (neurodegeneration with brain iron accumulation), Friedreich ataxia, and neuroferritinopathy (reviewed in Reference 44). Iron has been implicated in such diseases for its capacity to increase oxidative stress, and chelators ameliorate Parkinson-like disease in animal models (45). However, its precise role in these disorders remains unknown. It is still not clear whether abnormal iron metabolism leads to neurodegeneration or simply occurs as a consequence of damage to neurons.

IRON STORAGE AND RECYCLING

Intestinal absorption accounts for only a fraction of iron circulating bound to TF. Because there is no known regulated mechanism for iron excretion, and the amount of iron entering the body each day represents less than 0.1% of the total body iron endowment, most circulating iron must be derived from the recycling of iron already within the system. Quantitatively, recovery of iron from senescent erythrocytes contributes the most. Old and damaged erythrocytes are phagocytosed by tissue macrophages, particularly in the spleen. The cells are lysed, and hemoglobin is catabolized, presumably by heme oxygenase, to liberate iron. Some iron remains in storage in macrophages, although some is exported to plasma TF. FPN is critical for macrophage iron export (16) and, as discussed below, can be regulated to change the ratio between stored and released iron.

Hepatocytes also serve as depots for iron storage, but they acquire their iron load in a different fashion. Although the TF cycle may be involved in hepatocyte iron acquisition to some extent, non–transferrin-bound iron (NTBI) uptake pathways become particularly important when serum iron levels exceed TF binding capacity (46–51). The identity of the hepatocyte NTBI uptake system is not known, but DMT1 is unlikely to be involved, because hepatocytes can accumulate iron in the absence of DMT1 (5) and because DMT1 functions poorly at neutral pH (52, 53). Candidates for the NTBI transporter include L-type calcium channels (54), transient receptor potential canonical protein TRPC6 (47), and transporters already identified for their ability to transport other metal ions.

Hepatocytes have a large capacity to store excess iron. Most storage iron is probably in the form of ferritin, which can be mobilized when needed elsewhere in the body. Eventually, however, massive iron overload results in hepatotoxicity; hepatitis leads to fibrosis and cirrhosis. The liver is probably exposed to more NTBI than are other tissues because of the first-pass effect of the portal circulation. However, other tissues have NTBI uptake activities and load iron when NTBI is present in the plasma. The heart and endocrine tissues are particularly susceptible; cardiomyopathy and endocrinopathies are the

NTBI: non–transferrin-bound iron

predominant nonhepatic complications of iron overload.

Iron regulatory protein (IRP): binds to IREs

Iron regulatory element (IRE): an RNA stem-loop structure found in noncoding sequences

REGULATION OF CELLULAR IRON BALANCE

Individual cells must maintain internal iron homeostasis to ensure that there is adequate iron for basal functions but no free iron that could promote formation of reactive oxygen species. Cells accomplish this by at least two mechanisms. First, all mammalian cells produce ferritin, an iron storage protein. Ferritin is a 24-subunit polymer composed of varying ratios of two similar polypeptide chains, L-ferritin (light or liver ferritin) and H-ferritin (heavy or heart ferritin) (55). The two ferritin polypeptides are related, but H-ferritin contains an enzymatic oxidase activity. Ferritin polymers are cage-like structures with a central cavity to store hydrated iron oxides. Each ferritin polymer can accommodate up to 4500 iron atoms. Ferritin acts as a depot, accepting excess iron and allowing for the mobilization of iron when needed. A distinct intronless gene encoding a mitochondrial ferritin protein has been identified (56), and its product has been detected in sideroblastic anemia patients with erythroid mitochondrial iron accumulation (57). However, the function of mitochondrial ferritin remains uncertain.

The second protective mechanism involves iron regulatory proteins (IRPs), which, when iron is limiting, bind to RNA stem-loop iron regulatory elements (IREs) found in the untranslated regions of mRNAs involved in iron transport and storage. IRP1 can incorporate an iron-sulfur cluster, which acts as an iron sensor (58). When the 4Fe-4S cluster is complete, IRP1 cannot bind RNA but assumes an enzymatic function as a cytoplasmic aconitase (59). IRP2 does not contain an iron-sulfur cluster, but it is rapidly ubiquinated and degraded in the presence of excess iron (60).

The binding of IRPs to IREs serves either of two purposes, depending upon the location of the IREs (reviewed in Reference 61). IRP binding to IREs found in the 5′ untranslated regions of mRNAs encoding ferritin, ferroportin, and the heme biosynthetic enzyme aminolevulinate synthase sterically blocks the initiation of translation by interfering with ribosome assembly at the start codon. In the case of ferritin, this interrupts the production of both ferritin subunits when iron storage would be counterproductive. IREs found in the 3′ untranslated region of *TFR1* mRNA serve a very different function. There, five tandem IREs bind IRP to stabilize the molecule by inhibiting nuclease digestion. This allows more TFR1 to be produced when iron is limiting. A single 3′ UTR presumably serving a similar function is found in two of four possible isoforms of *DMT1* mRNA.

The roles of IREs and IRPs in vivo remain uncertain. Mice lacking IRP1 (*Ireb1−/−*) have a very subtle phenotype in kidneys and brown fat, suggesting that in most tissues IRP2 plays the major role in iron-dependent translational regulation (62). Two different laboratories have developed mice lacking IRP2 (*Ireb2−/−*). One group reported deregulation of iron homeostasis in the intestinal mucosa, microcytic anemia, and neurodegeneration with a long latency (63). However, the other group reported only systemic abnormalities in iron homeostasis, with very subtle neurological abnormalities (63a). The difference in phenotypes may be attributable to off-target effects of the first group's gene targeting or to strain differences. Recently, Meyron-Holtz et al. (64) reported a functional difference between IRP1 and IRP2. IRP2 appears to predominate under low-oxygen conditions similar to tissue oxygenation in vivo (64).

Beaumont et al. (65) have reported mutations in the L-ferritin IRE in human patients with hyperferritinemia-cataract syndrome. Although these patients come to clinical attention because of either elevated serum ferritin or early cataracts, they have no apparent defects in iron homeostasis. In contrast, a mutation in the H-ferritin IRE has been associated with genetic iron overload in a Japanese family (66). No human mutations have been identified in either of the two IRPs.

REGULATION OF SYSTEMIC IRON BALANCE

Iron balance must be meticulously regulated to provide iron as needed but avoid the toxicity associated with iron excess. This starts at the level of intestinal absorption but also involves macrophage iron recycling and hepatocyte iron mobilization. The stimuli known to modulate the iron homeostatic mechanism are erythroid iron needs, hypoxia, iron deficiency, iron overload, and inflammation. We now know that hepcidin, a circulating peptide hormone, controls much of this regulation (67–69).

Functional hepcidin is a 25-amino-acid protein that is primarily produced by hepatocytes and derived from an 84-amino-acid precursor. It has eight cysteine residues, similar to antimicrobial defensin proteins, and measurable antimicrobial activity. Although an unusual hairpin structure has been determined for hepcidin by NMR (70), recent reports suggest that cysteine residues in the native protein may hold a copper atom (71). Hepcidin is primarily produced by the liver but can also be produced by the heart, pancreas, and hematopoietic cells (72, 73). There are two hepcidin genes in mice, likely owing to a recent gene duplication event, but only mouse hepcidin 1 appears to have biological activity (74).

Two observations initially established the importance of this peptide in iron homeostasis. First, Pigeon et al. (69) demonstrated that, although hepcidin is normally expressed at high levels in the liver, its expression increases in iron-overloaded animals. Soon after, Nicolas et al. (67) showed that fortuitous inactivation of the hepcidin gene led to an iron overload phenotype. Subsequent studies have shown that forced expression of a hepcidin transgene causes anemia by interrupting intestinal iron absorption and inducing the retention of iron within recycling macrophages (75). Targeted disruption of the hepcidin gene itself (*Hamp1−/−* mice) causes severe iron overload (76). Hepcidin expression is altered in response to each of the stimuli known to affect iron homeostasis: It is increased in response to increased serum iron, iron overload, and inflammation and diminished in response to increased erythroid drive, hypoxia, and iron deficiency (69, 77, 78).

Hepcidin: a circulating peptide hormone that regulates ferroportin activity; also known as hepcidin antimicrobial peptide (HAMP)

BMP: bone morphogenetic protein

The molecular activity of hepcidin is now understood in some detail. Early observations provided compelling evidence that hepcidin interrupts cellular iron export in at least two sites: the intestinal epithelium and tissue macrophages. This suggested a functional link to ferroportin, the only known iron exporter. Studies in cultured cells subsequently established that hepcidin binds directly to ferroportin, triggering its internalization and degradation within lysosomes (79). Through this elegant mechanism, hepcidin directly regulates cellular iron release. It is not yet known whether hepcidin regulates ferroportin activity in other cell types. In particular, no definitive role has been established for either ferroportin or hepcidin in iron mobilization from hepatocyte stores.

Much of the regulation of hepcidin expression appears to occur at the level of transcription. An early study implicated cEBP-alpha in the transcriptional regulation of hepcidin, but its significance has not been fully established (80). Two other signal transduction pathways modulate binding of transcription factors to the hepcidin promoter. Under basal conditions, hepcidin expression depends upon signaling through a bone morphogenetic protein (BMP)/SMAD pathway (81). Liver-specific inactivation of the co-SMAD protein SMAD4 causes a failure of hepcidin production and an iron overload phenotype similar to that observed in hepcidin knockout mice (82). Hepcidin production is increased by treatment with BMPs (81, 82) and inhibited by expression of a dominant-negative BMP receptor or a dominant-negative regulatory SMAD protein (81). The ubiquitous BMP signaling apparatus is coopted for the regulation

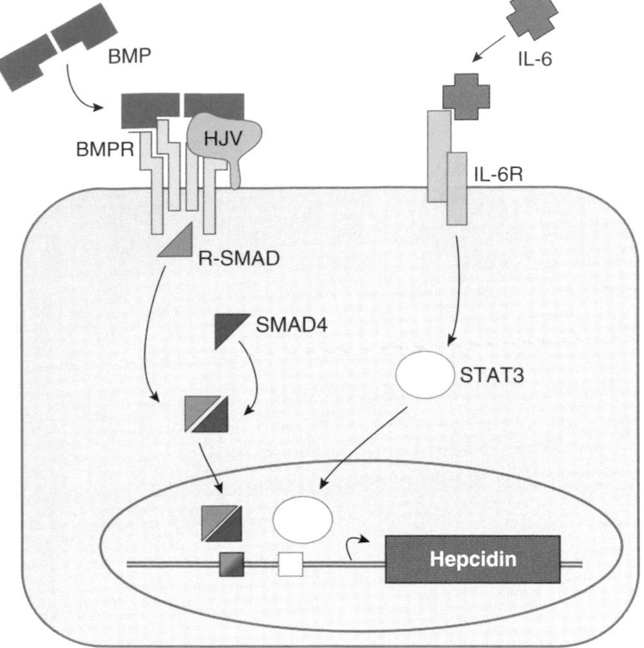

Figure 1

Transcriptional regulation of hepcidin expression. Two pathways regulating hepcidin transcription are well characterized. Basal expression depends upon signaling through bone morphogenetic protein receptors (BMPR) and downstream SMAD factors. Hemojuvelin (HJV) acts as a BMP coreceptor. Hepcidin induction in inflammation results, at least in part, from the signaling of interleukin-6 (IL-6) through its receptor and STAT3 (signal transducer and activator of transcription 3).

of hepcidin expression through the interaction of hemojuvelin (HJV) with BMP and BMP receptors (81). HJV is a homolog of repulsive guidance molecule proteins important in neurodevelopment (83). It is mutated in patients with severe, early-onset juvenile hemochromatosis.

HJV: hemojuvelin

Hemochromatosis: genetic iron overload resulting from mutations in any of five genes involved in iron homeostasis

***HFE*:** the gene mutated in classical hemochromatosis

Transferrin receptor 2 (TFR2): a homolog of TFR1

A second type of transcriptional regulation occurs in inflammatory conditions. Interleukin-6 (IL-6) and possibly other inflammatory cytokines induce transcription of the hepcidin gene in hepatocytes in vitro and in vivo (84). This induction involves the activation of STAT3 (signal transducer and activator of transcription 3) and binding of STAT3 to a regulatory element in the hepcidin promoter (85). These two mechanisms for hepcidin regulation are depicted in **Figure 1**.

Spontaneous mutations in human patients and experimental animals have provided a great deal of insight into systemic iron homeostasis and genetic iron disorders. Mice (4, 18, 32), rats (7), and zebrafish (14, 34, 86) with easily observed anemia phenotypes facilitated the identification of DMT1, hephaestin, STEAP3, ferroportin, and other proteins through positional cloning approaches (reviewed in Reference 87). Although iron-deficiency anemia in human patients rarely results from genetic causes, hereditary iron overload (hemochromatosis) disorders have provided important insights into genes that regulate iron homeostasis. Similar clinical phenotypes of varying severity result from homozygosity for mutations in the *HFE*, transferrin receptor 2 (*TFR2*), *HJV*, and hepcidin (*HAMP*) genes (88–91). A related, autosomal-dominant disorder is caused by mutations in ferroportin (92, 93) that disrupt its regulation by hepcidin (94–96). Affected patients with genetic hemochromatosis have parenchymal iron deposition in the liver, heart, and endocrine tissues but a paucity of iron in intestinal epithelial cells and tissue macrophages (97). In severe cases, tissue iron leads to cirrhosis, cardiomyopathy, diabetes, and other endocrinopathies. Patients with *HFE*, *TFR2*, and ferroportin mutations typically present in midlife. Most patients with HFE hemochromatosis never develop clinical disease, and the proportion who do remains controversial (98–101). In contrast, patients with *HJV* and *HAMP* mutations are severely affected early in life, typically dying of cardiomyopathy before the fourth decade if not treated. Aggressive phlebotomy (bloodletting) is a safe and effective treatment for all these disorders.

All forms of genetic hemochromatosis are associated with decreased hepcidin production or activity (90, 91, 102, 103). Considering this common feature, the diseases can be organized into three functional classes (**Figure 2**). First, mutations in the hepcidin gene prevent

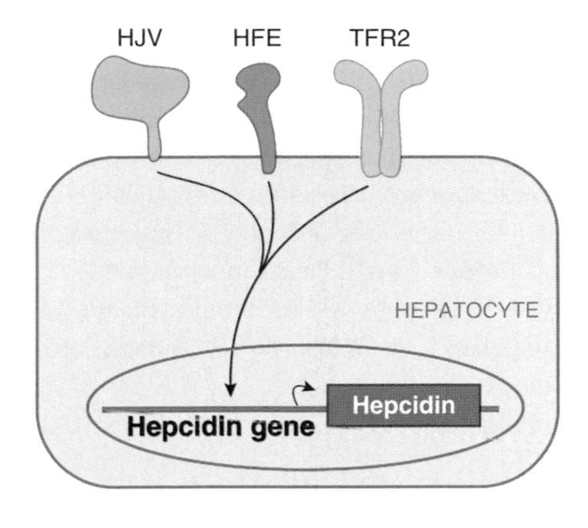

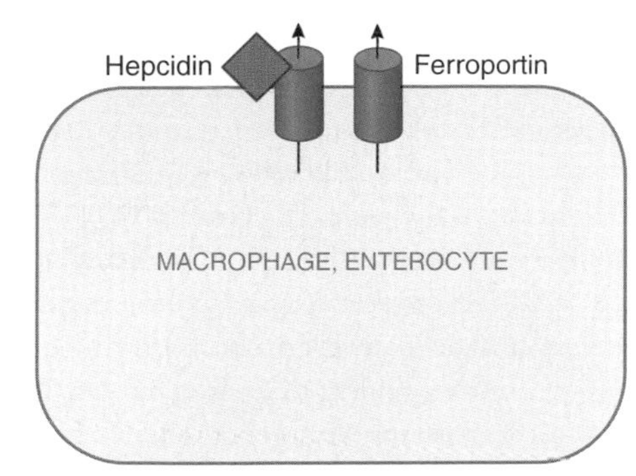

Figure 2

Three classes of hemochromatosis disorders. Hemochromatosis disorders result from mutations that affect the hepcidin/ferroportin axis on any of three levels. They may perturb the hepcidin gene, the regulation of ferroportin by hepcidin, or the activity of proteins (HJV, HFE, or TFR2) necessary for normal hepcidin expression. Abbreviations used: HJV, hemojuvelin; TFR2, transferrin receptor 2.

the production of active hepcidin protein, leading to unregulated ferroportin activity. Second, hemochromatosis-associated mutations in ferroportin prevent hepcidin binding and/or consequent ferroportin internalization and degradation, similarly leading to unregulated ferroportin activity (95, 96). The third class of mutations is in other genes—*HFE*, *TFR2*, and *HJV*—that can be inferred to be necessary for appropriate hepcidin gene regulation.

HFE, the gene most commonly mutated in patients with hemochromatosis, encodes an atypical major histocompatibility class I protein that complexes with β-2 microglobulin but cannot bind small peptides (104). The prevalent C282Y mutation in *HFE* apparently occurred in a single founder individual more than two centuries ago, as demonstrated by a small, shared haplotype block found in all patients (105). C282Y disrupts HFE conformation but does not entirely prevent its function (106). A common polymorphism in HFE, H63D, is sometimes associated with iron overload, particularly when it is found in compound heterozygotes carrying C282Y. The discovery that HFE forms a protein-protein complex with TFR led to several hypotheses for its role in iron homeostasis (107). However, recent results suggest that the HFE/TFR complex itself does not have a direct iron-regulatory activity; rather, TFR appears to sequester HFE to prevent it from acting elsewhere (N. Andrews & P. Schmidt, unpublished data).

TFR2 encodes a homolog of TFR (108). It can bind and internalize Fe-TF but probably does not serve a primary role in cellular iron uptake. Rather, it likely acts as an iron sensor; TFR2 protein levels increase in response to increased ambient Fe-TF (109, 110). TFR2 is highly expressed in hepatocytes and in developing erythroid precursor cells. In contrast to *HFE*, various *TFR2* mutations have been described, including a truncation mutation that is expected to inactivate the protein.

HFE, TFR2, and hemojuvelin are all expressed on the surface of hepatocytes. As discussed above, HJV acts as a BMP coreceptor. To date, the roles of HFE and TFR2 have not been reported, but it is reasonable to speculate that they, too, may participate in BMP signaling in some way.

Iron overload can also be a secondary feature in patients with congenital anemias. Patients with thalassemia due to mutations in the human globin genes have ineffective erythropoiesis. As a compensatory response to attempt to increase red blood cell production, these patients' hepcidin levels fall (111, 112), and intestinal iron absorption is markedly increased. In severe thalassemia syndromes, the anemia must be treated with chronic red blood cell transfusions, adding to the iron burden. In contrast to patients with genetic hemochromatosis, who have normal erythropoiesis and can be treated by phlebotomy, treatment with iron chelator agents is the only option to remove the excess iron. Until recently, infusion of the natural product chelator deferoxamine was the standard of care, but new oral agents (deferiprone and deferasirox) are becoming important therapeutic options, particularly in patients who are poorly compliant with a daily infusion regimen.

Iron overload also occurs in patients lacking ceruloplasmin, the serum multicopper oxidase that oxidizes iron in concert with export from cells (113). Familial aceruloplasminemia, due to mutations in the ceruloplasmin gene, can be confused with hemochromatosis because it also results in marked liver iron overload. However, it is important to make the correct diagnosis for two reasons. First, aceruloplasminemia results in brain iron accumulation and neurodegenerative disease. Genetic hemochromatosis has no nervous system manifestations. Second, phlebotomy therapy is contraindicated in aceruloplasminemia because most patients have mild anemia attributable to a defect in macrophage iron export.

CONCLUDING REMARKS

Over the past decade, substantial progress has been made in understanding mammalian iron transport and its regulation. The key players involved in maintaining iron homeostasis have been identified, and their molecular functions have been partially elucidated. In the next few years, this information is likely to lead to a better understanding of the molecular pathophysiology of iron disorders. The larger challenge will be to integrate our understanding of iron metabolism with knowledge of other physiological processes in health and disease.

SUMMARY POINTS

1. Iron is necessary for the basal functioning of all mammalian cells but is particularly important in cells producing hemoglobin and myoglobin.
2. Excess iron promotes the formation of toxic oxygen radicals, leading to cellular damage.
3. Specialized transmembrane transporters are required to move iron across biological membranes.
4. Ferrireductases and ferroxidases aid in iron transport and storage by changing the oxidation state of inorganic iron.
5. Transferrin carries iron in the plasma and facilitates its delivery to cells bearing transferrin receptors.
6. Intracellular iron balance is achieved through posttranscriptional regulation of key mRNA molecules by iron regulatory proteins (IRPs).

7. Systemic iron homeostasis is maintained by hepcidin, a circulating peptide hormone that binds to the iron exporter ferroportin and causes it to be degraded.
8. Genetic iron overload (hemochromatosis) results from mutations in molecules that regulate hepcidin production or activity, including hepcidin itself, HFE, transferrin receptor 2 (TFR2), hemojuvelin (HJV), and ferroportin.

LITERATURE CITED

1. McKie AT, Barrow D, Latunde-Dada GO, Rolfs A, Sager G, et al. 2001. An iron-regulated ferric reductase associated with the absorption of dietary iron. *Science* 291:1755–59
2. Gunshin H, Starr CN, Direnzo C, Fleming MD, Jin J, et al. 2005. Cybrd1 (duodenal cytochrome *b*) is not necessary for dietary iron absorption in mice. *Blood* 106(8):2879–83.
3. Ohgami RS, Campagna DR, McDonald A, Fleming MD. 2006. The Steap proteins are metalloreductases. *Blood* 108:1388–94
4. Fleming MD, Trenor CC, Su MA, Foernzler D, Beier DR, et al. 1997. Microcytic anaemia mice have a mutation in *Nramp2*, a candidate iron transporter gene. *Nat. Genet.* 16:383–86
5. **Gunshin H, Fujiwara Y, Custodio AO, Direnzo C, Robine S, Andrews NC. 2005. Slc11a2 is required for intestinal iron absorption and erythropoiesis but dispensable in placenta and liver. *J. Clin. Invest.* 115:1258–66**
6. Gunshin H, Mackenzie B, Berger UV, Gunshin Y, Romero MF, et al. 1997. Cloning and characterization of a mammalian proton-coupled metal-ion transporter. *Nature* 388:482–88
7. Fleming MD, Romano MA, Su MA, Garrick LM, Garrick MD, Andrews NC. 1998. *Nramp2* is mutated in the anemic Belgrade (b) rat: evidence of a role for Nramp2 in endosomal iron transport. *Proc. Natl. Acad. Sci. USA* 95:1148–53
8. Xu H, Jin J, DeFelice LJ, Andrews NC, Clapham DE. 2004. A spontaneous, recurrent mutation in divalent metal transporter-1 exposes a calcium entry pathway. *PLoS Biol.* 2:E50
9. Beaumont C, Delaunay J, Hetet G, Grandchamp B, de Montalembert M, Tchernia G. 2006. Two new human DMT1 gene mutations in a patient with microcytic anemia, low ferritinemia, and liver iron overload. *Blood* 107:4168–70
10. Iolascon A, d'Apolito M, Servedio V, Cimmino F, Piga A, Camaschella C. 2006. Microcytic anemia and hepatic iron overload in a child with compound heterozygous mutations in DMT1 (SCL11A2). *Blood* 107:349–54
11. Mims MP, Guan Y, Pospisilova D, Priwitzerova M, Indrak K, et al. 2005. Identification of a human mutation of DMT1 in a patient with microcytic anemia and iron overload. *Blood* 105:1337–42
12. **Shayeghi M, Latunde-Dada GO, Oakhill JS, Laftah AH, Takeuchi K, et al. 2005. Identification of an intestinal heme transporter. *Cell* 122:789–801**
13. Abboud S, Haile DJ. 2000. A novel mammalian iron-regulated protein involved in intracellular iron metabolism. *J. Biol. Chem.* 275:19906–12
14. Donovan A, Brownlie A, Zhou Y, Shepard J, Pratt SJ, et al. 2000. Positional cloning of zebrafish ferroportin1 identifies a conserved vertebrate iron exporter. *Nature* 403:776–81

5. Targeted disruption of the murine DMT1 gene demonstrated its importance in intestinal iron absorption and erythropoiesis.

12. Heme carrier protein 1 appears to be the transporter responsible for the uptake of dietary heme iron.

15. McKie AT, Marciani P, Rolfs A, Brennan K, Wehr K, et al. 2000. A novel duodenal iron-regulated transporter, IREG1, implicated in the basolateral transfer of iron to the circulation. *Mol. Cell* 5:299–309

16. Donovan A, Lima CA, Pinkus JL, Pinkus GS, Zon LI, et al. 2005. The iron exporter ferroportin/Slc40a1 is essential for iron homeostasis. *Cell Metab.* 1:191–200

17. Targeted disruption of murine ferroportin demonstrates its roles in intestinal iron absorption and macrophage iron release.

17. Harris ZL, Durley AP, Man TK, Gitlin JD. 1999. Targeted gene disruption reveals an essential role for ceruloplasmin in cellular iron efflux. *Proc. Natl. Acad. Sci. USA* 96:10812–17

18. Vulpe CD, Kuo YM, Murphy TL, Cowley L, Askwith C, et al. 1999. Hephaestin, a ceruloplasmin homologue implicated in intestinal iron transport, is defective in the *sla* mouse. *Nat. Genet.* 21:195–99

19. Cherukuri S, Potla R, Sarkar J, Nurko S, Harris ZL, Fox PL. 2005. Unexpected role of ceruloplasmin in intestinal iron absorption. *Cell Metab.* 2:309–19

20. Uchida T, Akitsuki T, Kimura H, Tanaka T, Matsuda S, Kariyone S. 1983. Relationship among plasma iron, plasma iron turnover, and reticuloendothelial iron release. *Blood* 61:799–802

21. Hentze MW, Muckenthaler MU, Andrews NC. 2004. Balancing acts: molecular control of mammalian iron metabolism. *Cell* 117:285–97

22. Levy JE, Jin O, Fujiwara Y, Kuo F, Andrews NC. 1999. Transferrin receptor is necessary for development of erythrocytes and the nervous system. *Nat. Genet.* 21:396–99

23. Ned RM, Swat W, Andrews NC. 2003. Transferrin receptor 1 is differentially required in lymphocyte development. *Blood* 102:3711–18

24. Larrick J, Hyman E. 1984. Acquired iron-deficiency anemia caused by an antibody against the transferrin receptor. *N. Engl. J. Med.* 311:214–18

25. Bernstein SE. 1987. Hereditary hypotransferrinemia with hemosiderosis, a murine disorder resembling human atransferrinemia. *J. Lab. Clin. Med.* 110:690–705

26. Trenor CC, Campagna DR, Sellers VM, Andrews NC, Fleming MD. 2000. The molecular defect in hypotransferrinemic mice. *Blood* 96:1113–18

27. Asada-Senju M, Maeda T, Sakata T, Hayashi A, Suzuki T. 2002. Molecular analysis of the transferrin gene in a patient with hereditary hypotransferrinemia. *J. Hum. Genet.* 47:355–59

28. Beutler E, Gelbart T, Lee P, Trevino R, Fernandez MA, Fairbanks VF. 2000. Molecular characterization of a case of atransferrinemia. *Blood* 96:4071–74

29. Dickinson T, Connor JR. 1994. Histological analysis of selected brain regions of hypotransferrinemic mice. *Brain Res.* 635:169–78

30. Craven CM, Alexander J, Eldridge M, Kushner JP, Bernstein S, Kaplan J. 1987. Tissue distribution and clearance kinetics of nontransferrin-bound iron in the hypotransferrinemic mouse: a rodent model for hemochromatosis. *Proc. Natl. Acad. Sci. USA* 84:3457–61

31. Finch C. 1994. Regulators of iron balance in humans. *Blood* 84:1697–702

32. Ohgami RS, Campagna DR, Greer EL, Antiochos B, McDonald A, et al. 2005. Identification of a ferrireductase required for efficient transferrin-dependent iron uptake in erythroid cells. *Nat. Genet.* 37:1264–69

33. Ponka P. 1997. Tissue-specific regulation of iron metabolism and heme synthesis: distinct control mechanisms in erythroid cells. *Blood* 89:1–25

34. Shaw GC, Cope JJ, Li L, Corson K, Hersey C, et al. 2006. Mitoferrin is essential for erythroid iron assimilation. *Nature* 440:96–100

35. Quigley JG, Yang Z, Worthington MT, Phillips JD, Sabo KM, et al. 2004. Identification of a human heme exporter that is essential for erythropoiesis. *Cell* 118:757–66

36. Neckers LM, Cossman J. 1983. Transferrin receptor induction in mitogen-stimulated human T lymphocytes is required for DNA synthesis and cell division and is regulated by interleukin 2. *Proc. Natl. Acad. Sci. USA* 80:3494–98
37. Neckers LM, Yenokida G, James SP. 1984. The role of the transferrin receptor in human B lymphocyte activation. *J. Immunol.* 133:2437–41
38. Batista A, Millan J, Mittelbrunn M, Sanchez-Madrid F, Alonso MA. 2004. Recruitment of transferrin receptor to immunological synapse in response to TCR engagement. *J. Immunol.* 172:6709–14
39. Salmeron A, Borroto A, Fresno M, Crumpton MJ, Ley SC, Alarcon B. 1995. Transferrin receptor induces tyrosine phosphorylation in T cells and is physically associated with the TCR zeta-chain. *J. Immunol.* 154:1675–83
40. Woodward JE, Bayer AL, Chavin KD, Boleza KA, Baliga P. 1998. Anti-transferrin receptor monoclonal antibody: a novel immunosuppressant. *Transplantation* 65:6–9
41. Yang J, Goetz D, Li JY, Wang W, Mori K, et al. 2002. An iron delivery pathway mediated by a lipocalin. *Mol. Cell* 10:1045–56
42. Goetz DH, Holmes MA, Borregaard N, Bluhm ME, Raymond KN, Strong RK. 2002. The neutrophil lipocalin NGAL is a bacteriostatic agent that interferes with siderophore-mediated iron acquisition. *Mol. Cell* 10:1033–43
43. Flo TH, Smith KD, Sato S, Rodriguez DJ, Holmes MA, et al. 2004. Lipocalin 2 mediates an innate immune response to bacterial infection by sequestrating iron. *Nature* 432:917–21
44. Zecca L, Youdim MB, Riederer P, Connor JR, Crichton RR. 2004. Iron, brain ageing and neurodegenerative disorders. *Nat. Rev. Neurosci.* 5:863–73
45. Kaur D, Yantiri F, Rajagopalan S, Kumar J, Mo JQ, et al. 2003. Genetic or pharmacological iron chelation prevents MPTP-induced neurotoxicity in vivo: a novel therapy for Parkinson's disease. *Neuron* 37:899–909
46. Kielmanowicz MG, Laham N, Coligan JE, Lemonnier F, Ehrlich R. 2005. Mouse HFE inhibits Tf-uptake and iron accumulation but induces nontransferrin bound iron (NTBI)-uptake in transformed mouse fibroblasts. *J. Cell Physiol.* 202:105–14
47. Mwanjewe J, Grover AK. 2004. Role of transient receptor potential canonical 6 (TRPC6) in nontransferrin-bound iron uptake in neuronal phenotype PC12 cells. *Biochem. J.* 378:975–82
48. Mwanjewe J, Martinez R, Agrawal P, Samson SE, Coughlin MD, et al. 2000. On the Ca^{2+} dependence of nontransferrin-bound iron uptake in PC12 cells. *J. Biol. Chem.* 275:33512–15
49. Barisani D, Berg CL, Wessling-Resnick M, Gollan JL. 1995. Evidence for a low Km transporter for nontransferrin-bound iron in isolated rat hepatocytes. *Am. J. Physiol.* 269:G570–76
50. Randell EW, Parkes JG, Olivieri NF, Templeton DM. 1994. Uptake of nontransferrin bound iron by both reductive and nonreductive processes is modulated by intracellular iron. *J. Biolog. Chem.* 269:16046–53
51. Inman RS, Coughlan MM, Wessling-Resnick M. 1994. Extracellular ferrireductase activity of K562 cells is coupled to transferrin-independent iron transport. *Biochemistry* 33:11850–57
52. Worthington MT, Browne L, Battle EH, Luo RQ. 2000. Functional properties of transfected human DMT1 iron transporter. *Am. J. Physiol. Gastrointest. Liver Physiol.* 279:G1265–73

53. Lam-Yuk-Tseung S, Govoni G, Forbes J, Gros P. 2003. Iron transport by Nramp2/DMT1: pH regulation of transport by 2 histidines in transmembrane domain 6. *Blood* 101:3699–707
54. Oudit GY, Sun H, Trivieri MG, Koch SE, Dawood F, et al. 2003. L-type Ca^{2+} channels provide a major pathway for iron entry into cardiomyocytes in iron-overload cardiomyopathy. *Nat. Med.* 9:1187–94
55. Theil EC. 2003. Ferritin: at the crossroads of iron and oxygen metabolism. *J. Nutr.* 133:S1549–53
56. Levi S, Corsi B, Bosisio M, Invernizzi R, Volz A, et al. 2001. A human mitochondrial ferritin encoded by an intronless gene. *J. Biol. Chem.* 276:24437–40
57. Cazzola M, Invernizzi R, Bergamaschi G, Levi S, Corsi B, et al. 2003. Mitochondrial ferritin expression in erythroid cells from patients with sideroblastic anemia. *Blood* 101:1996–2000
58. Rouault TA, Stout CD, Kaptain S, Harford JB, Klausner RD. 1991. Structural relationship between an iron-regulated RNA-binding protein (IRE-BP) and aconitase: functional implications. *Cell* 64:881–83
59. Kaptain S, Downey WE, Tang C, Philpott C, Haile D, et al. 1991. A regulated RNA binding protein also possesses aconitase activity. *Proc. Natl. Acad. Sci. USA* 88:10109–13
60. Iwai K, Drake SK, Wehr NB, Weissman AM, LaVaute T, et al. 1998. Iron-dependent oxidation, ubiquitination, and degradation of iron regulatory protein 2: implications for degradation of oxidized proteins. *Proc. Natl. Acad. Sci. USA* 95:4924–28
61. Pantopoulos K. 2004. Iron metabolism and the IRE/IRP regulatory system: an update. *Ann. N.Y. Acad. Sci.* 1012:1–13
62. Meyron-Holtz EG, Ghosh MC, Iwai K, LaVaute T, Brazzolotto X, et al. 2004. Genetic ablations of iron regulatory proteins 1 and 2 reveal why iron regulatory protein 2 dominates iron homeostasis. *EMBO J.* 23:386–95
63. LaVaute T, Smith S, Cooperman S, Iwai K, Land W, et al. 2001. Targeted deletion of the gene encoding iron regulatory protein-2 causes misregulation of iron metabolism and neurodegenerative disease in mice. *Nat. Genet.* 27:209–14
63a. Galy B, Holter SM, Klopstock T, Ferring D, Becker L, et al. 2006. Iron homeostasis in the brain: complete iron regulatory protein 2 deficiency without symtpmatic neurodegeneration in the mouse. *Nat. Genet.* 38:967–69
64. Meyron-Holtz EG, Ghosh MC, Rouault TA. 2004. Mammalian tissue oxygen levels modulate iron-regulatory protein activities in vivo. *Science* 306:2087–90
65. Beaumont C, Leneuve P, Devaux I, Scoazec JY, Berthier M, et al. 1995. Mutation in the iron responsive element of the L ferritin mRNA in a family with dominant hyperferritinemia and cataract. *Nat. Genet.* 11:444–46
66. Kato J, Fujikawa K, Kanda M, Fukuda N, Sasaki K, et al. 2001. A mutation, in the iron-responsive element of H ferritin mRNA, causing autosomal dominant iron overload. *Am J. Hum. Genet.* 69:191–97
67. **Nicolas G, Bennoun M, Devaux I, Beaumont C, Grandchamp B, et al. 2001. Lack of hepcidin gene expression and severe tissue iron overload in upstream stimulatory factor 2 (USF2) knockout mice. *Proc. Natl. Acad. Sci. USA* 98:8780–85**
68. Park CH, Valore EV, Waring AJ, Ganz T. 2001. Hepcidin, a urinary antimicrobial peptide synthesized in the liver. *J. Biol. Chem.* 276:7806–10
69. **Pigeon C, Ilyin G, Courselaud B, Leroyer P, Turlin B, et al. 2001. A new mouse liver-specific gene, encoding a protein homologous to human antimicrobial peptide hepcidin, is overexpressed during iron overload. *J. Biol. Chem.* 276:7811–19**

67. Initial evidence of the importance of hepcidin in mammalian iron homeostasis.

69. The first report of hepcidin as a gene induced in iron overload.

70. Hunter HN, Fulton DB, Ganz T, Vogel HJ. 2002. The solution structure of human hepcidin, a peptide hormone with antimicrobial activity that is involved in iron uptake and hereditary hemochromatosis. *J. Biol. Chem.* 277:37597–603
71. Melino S, Garlando L, Patamia M, Paci M, Petruzzelli R. 2005. A metal-binding site is present in the amino terminal region of the bioactive iron regulator hepcidin-25. *J. Pept. Res.* 66:65–71
72. Ilyin G, Courselaud B, Troadec MB, Pigeon C, Alizadeh M, et al. 2003. Comparative analysis of mouse hepcidin 1 and 2 genes: evidence for different patterns of expression and coinducibility during iron overload. *FEBS Lett.* 542:22–26
73. Peyssonnaux C, Zinkernagel AS, Datta V, Lauth X, Johnson RS, Nizet V. 2006. TLR4-dependent hepcidin expression by myeloid cells in response to bacterial pathogens. *Blood* 107:3727–32
74. Lou DQ, Nicolas G, Lesbordes JC, Viatte L, Grimber G, et al. 2004. Functional differences between hepcidin 1 and 2 in transgenic mice. *Blood* 103:2816–21
75. Nicolas G, Bennoun M, Porteu A, Mativet S, Beaumont C, et al. 2002. Severe iron deficiency anemia in transgenic mice expressing liver hepcidin. *Proc. Natl. Acad. Sci. USA* 99:4596–601
76. Viatte L, Lesbordes-Brion JC, Lou DQ, Bennoun M, Nicolas G, et al. 2005. Deregulation of proteins involved in iron metabolism in hepcidin-deficient mice. *Blood* 105:4861–64
77. Nicolas G, Chauvet C, Viatte L, Danan JL, Bigard X, et al. 2002. The gene encoding the iron regulatory peptide hepcidin is regulated by anemia, hypoxia, and inflammation. *J. Clin. Invest.* 110:1037–44
78. Weinstein DA, Roy CN, Fleming MD, Loda MF, Wolfsdorf JI, Andrews NC. 2002. Inappropriate expression of hepcidin is associated with iron refractory anemia: implications for the anemia of chronic disease. *Blood* 100:3776–81
79. **Nemeth E, Tuttle MS, Powelson J, Vaughn MB, Donovan A, et al. 2004. Hepcidin regulates cellular iron efflux by binding to ferroportin and inducing its internalization. *Science* 306:2090–93**
80. Courselaud B, Pigeon C, Inoue Y, Inoue J, Gonzalez FJ, et al. 2002. C/EBPα regulates hepatic transcription of hepcidin, an antimicrobial peptide and regulator of iron metabolism. Cross-talk between C/EBP pathway and iron metabolism. *J. Biol. Chem.* 277:41163–70
81. **Babitt JL, Huang FW, Wrighting DM, Xia Y, Sidis Y, et al. 2006. Bone morphogenetic protein signaling by hemojuvelin regulates hepcidin expression. *Nat. Genet.* 38:531–39**
82. **Wang RH, Li C, Xu X, Zheng Y, Xiao C, et al. 2005. A role of SMAD4 in iron metabolism through the positive regulation of hepcidin expression. *Cell Metab.* 2:399–409**
83. Niederkofler V, Salie R, Sigrist M, Arber S. 2004. Repulsive guidance molecule (RGM) gene function is required for neural tube closure but not retinal topography in the mouse visual system. *J. Neurosci.* 24:808–18
84. Nemeth E, Rivera S, Gabayan V, Keller C, Taudorf S, et al. 2004. IL-6 mediates hypoferremia of inflammation by inducing the synthesis of the iron regulatory hormone hepcidin. *J. Clin. Invest.* 113:1271–76
85. Wrighting DM, Andrews NC. 2006. Interleukin-6 induces hepcidin expression through STAT3. *Blood.* In press
86. Donovan A, Brownlie A, Dorschner MO, Zhou Y, Pratt SJ, et al. 2002. The zebrafish mutant gene *chardonnay* (*cdy*) encodes divalent metal transporter 1 (DMT1). *Blood* 100:4655–59

79. The first report of a mechanism of action for hepcidin and its role in regulating ferroportin activity.

81. Demonstration that the hemochromatosis-associated protein hemojuvelin acts as a bone morphogenetic protein coreceptor to regulate hepcidin production.

82. Evidence that hepcidin expression is regulated through a BMP-SMAD signaling pathway.

87. Andrews NC. 2000. Iron homeostasis: insights from genetics and animal models. *Nat. Rev. Genet.* 1:208–17
88. Camaschella C, Roetto A, Cali A, De Gobbi M, Garozzo G, et al. 2000. The gene *TFR2* is mutated in a new type of hemochromatosis mapping to 7q22. *Nat. Genet.* 25:14–15
89. Feder JN, Gnirke A, Thomas W, Tsuchihashi Z, Ruddy DA, et al. 1996. A novel MHC class I-like gene is mutated in patients with hereditary hemochromatosis. *Nat. Genet.* 13:399–408
90. Papanikolaou G, Samuels ME, Ludwig EH, MacDonald ML, Franchini PL, et al. 2004. Mutations in HFE2 cause iron overload in chromosome 1q-linked juvenile hemochromatosis. *Nat. Genet.* 36:77–82
91. Roetto A, Papanikolaou G, Politou M, Alberti F, Girelli D, et al. 2003. Mutant antimicrobial peptide hepcidin is associated with severe juvenile hemochromatosis. *Nat. Genet.* 33:21–22
92. Montosi G, Donovan A, Totaro A, Garuti C, Pignatti E, et al. 2001. Autosomal-dominant hemochromatosis is associated with a mutation in the ferroportin (*SLC11A3*) gene. *J. Clin. Invest.* 108:619–23
93. Njajou OT, Vaessen N, Joosse M, Berghuis B, van Dongen JW, et al. 2001. A mutation in *SLC11A3* is associated with autosomal dominant hemochromatosis. *Nat. Genet.* 28:213–14
94. De Domenico I, Ward DM, Musci G, Kaplan J. 2006. Iron overload due to mutations in ferroportin. *Haematologica* 91:92–95
95. Drakesmith H, Schimanski LM, Ormerod E, Merryweather-Clarke AT, Viprakasit V, et al. 2005. Resistance to hepcidin is conferred by hemochromatosis-associated mutations of ferroportin. *Blood* 106:1092–97
96. De Domenico I, Ward DM, Nemeth E, Vaughn MB, Musci G, et al. 2005. The molecular basis of ferroportin-linked hemochromatosis. *Proc. Natl. Acad. Sci. USA* 102:8955–60
97. Fillet G, Beguin Y, Baldelli L. 1989. Model of reticuloendothelial iron metabolism in humans: abnormal behavior in idiopathic hemochromatosis and in inflammation. *Blood* 74:844–51
98. Ajioka RS, Kushner JP. 2003. Clinical consequences of iron overload in hemochromatosis homozygotes. *Blood* 101:3351–53; discussion 4–8
99. Ajioka RS, Kushner JP. 2003. Rebuttal to Beutler. *Blood* 101:3358
100. Beutler E. 2003. Penetrance of hemochromatosis. *Gut* 52:610–11
101. Beutler E. 2003. Rebuttal to Ajioka and Kushner. *Blood* 101:3354–57
102. Bridle KR, Frazer DM, Wilkins SJ, Dixon JL, Purdie DM, et al. 2003. Disrupted hepcidin regulation in HFE-associated hemochromatosis and the liver as a regulator of body iron homoeostasis. *Lancet* 361:669–73
103. Nemeth E, Roetto A, Garozzo G, Ganz T, Camaschella C. 2005. Hepcidin is decreased in TFR2 hemochromatosis. *Blood* 105:1803–6
104. Lebron JA, Bennett MJ, Vaughn DE, Chirino AJ, Snow PM, et al. 1998. Crystal structure of the hemochromatosis protein HFE and characterization of its interaction with transferrin receptor. *Cell* 93:111–23
105. Ajioka RS, Jorde LB, Gruen JR, Dimitrova D, Barrow J, et al. 1997. Haplotype analysis of hemochromatosis: evaluation of different linkage-disequilibrium approaches and evolution of disease chromosomes. *Am J. Hum. Genet.* 60:1439–47
106. Levy JE, Montross LK, Cohen DE, Fleming MD, Andrews NC. 1999. The C282Y mutation causing hereditary hemochromatosis does not produce a null allele. *Blood* 94:9–11

107. Roy CN, Enns CA. 2000. Iron homeostasis: new tales from the crypt. *Blood* 96:4020–27
108. Kawabata H, Yang R, Hirama T, Vuong PT, Kawano S, et al. 1999. Molecular cloning of transferrin receptor 2. A new member of the transferrin receptor-like family. *J. Biol. Chem.* 274:20826–32
109. Robb A, Wessling-Resnick M. 2004. Regulation of transferrin receptor 2 protein levels by transferrin. *Blood* 104:4294–99
110. Johnson MB, Enns CA. 2004. Regulation of transferrin receptor 2 by transferrin: Diferric transferrin regulates transferrin receptor 2 protein stability. *Blood* 104:4287–93
111. Kearney SL, Nemeth E, Neufeld EJ, Thapa D, Ganz T, et al. 2005. Urinary hepcidin in congenital chronic anemias. *Pediatr. Blood Cancer*
112. Papanikolaou G, Tzilianos M, Christakis JI, Bogdanos D, Tsimirika K, et al. 2005. Hepcidin in iron overload disorders. *Blood* 105:4103–5
113. Harris ZL, Klomp LW, Gitlin JD. 1998. Aceruloplasminemia: an inherited neurodegenerative disease with impairment of iron homeostasis. *Am. J. Clin. Nutr.* 67:S972–77

RELATED RESOURCES

Donavan A, Roy CN, Andrews NC. 2006. The ins and outs of iron homeostasis. *Physiology* 21:115–23

Transporters as Channels

Louis J. DeFelice[1] and Tapasree Goswami[2]

[1]Department of Pharmacology and Molecular Neuroscience, Vanderbilt University Medical Center, Nashville, Tennessee 37232; email: lou.defelice@vanderbilt.edu

[2]Department of Cell Biology, Harvard Medical School, Boston, Massachusetts 02115; email: tapasree_goswami@hms.harvard.edu

Annu. Rev. Physiol. 2007. 69:87–112

First published online as a Review in Advance on October 23, 2006

The *Annual Review of Physiology* is online at http://physiol.annualreviews.org

This article's doi: 10.1146/annurev.physiol.69.031905.164816

0066-4278/07/0315-0087$20.00

Key Words

membrane, synapse, biophysics, neuroscience

Abstract

This review investigates some key aspects of transport mechanisms and recent advances in our understanding of this ubiquitous cellular process. The prevailing model of cotransport is the alternating access model, which suggests that large conformational changes in the transporter protein accompany cotransport. This model rests on decades of research and has received substantial support because many transporter characteristics are explained using its premises. New experiments, however, have revealed the existence of channels in transporters, an idea that is in conflict with traditional models. The alternating access model is the subject of previous detailed reviews. Here we concentrate on the relatively recent data that document primarily the channel properties of transporters. In some cases, namely, the observation of single-transporter currents, the evidence is direct. In other cases the evidence—for example, from fluctuation analysis or transporter currents too large to be described as anything other than channel-like—is indirect. Although the existence of channels in transporters is not in doubt, we are far from understanding the significance of this property. In the online Supplemental Material, we review some pertinent aspects of ion channel theory and cotransport physiology to provide background for the channels and transporters presented here. We discuss the existence of channels in transporters, and we speculate on the biological significance of this newly unveiled property of transport proteins.

INTRODUCTION

Cotransporters employ the electrochemical gradient of certain ions to concentrate specific substrates. Since the early 1990s, new data from fluorescence microscopy and electrophysiology have suggested that conformational changes predicted by the alternating access model (1) of cotransport (see also **Supplemental Figure 3**) may be smaller than predicted and more similar to channel gating (1–5). One of the most astonishing experiments of the new era of channel and transporter characterization is a study of pre- and postsynaptic signaling at an intact synapse (6). **Figure 1** shows a salient feature of this experiment.

Figure 1 shows the pre- and postsynaptic responses to the release of serotonin [5-hydroxytryptamine (5-HT)] in a leech synapse under voltage clamp. Before the 5-HT-gated, postsynaptic receptor current occurs, a larger and faster presynaptic current, associated entirely with the serotonin transporter (SERT), occurs. This latter current is associated with transporting 5-HT back into the presynaptic terminal (uptake). Thus, as early as 1993, we had a strong indication that transporter currents can be as large as channel currents. Although this does not in itself define channel-like activity within a transporter, traditional models for cotransport predict less than 1 pA of uptake current for a million transporters operating at once. Thus, some activity extraneous to conventional explanations must be occurring. One outstanding question might be to what extent the uptake current is actually carried by the substrate (in this case, 5-HT^+) in addition to other ions. We discuss the details surrounding this question in the **Supplemental Text**, Section B3 (follow the Supplemental Material link from the Annual Reviews home page at **http://www.annualreviews.org**).

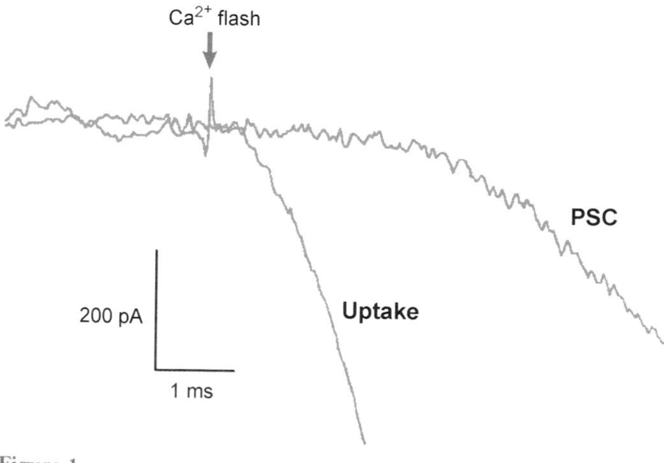

Figure 1

Pre- and postsynaptic responses in a serotonergic synapse. Only a submillisecond delay occurs between a presynaptic Ca^{2+} flash (*arrow*) and presynaptic serotonin uptake current, which is faster and larger than the serotonin-induced postsynaptic current (PSC). Pre- and postneurons were voltage clamped at −70 mV and −60 mV, respectively. From Reference 6.

The point to note from **Figure 1** is that in real synapses large currents are associated with neurotransmitter transporters. In this review, we consider carefully what purpose these presynaptic currents serve and whether they can be explained by traditional models with fixed stoichiometry and tight coupling or whether they require new models with channel-like properties.

CHANNELS IN TRANSPORTERS

Na^+/K^+ ATPase

Transport pumps are bonafide enzymes, and the prime example is Na^+/K^+ ATPase (the Na^+ pump). For supporting details, please see the **Supplemental Text** (Section B2) and also References 7 and 8. Even the archetypical Na^+ pump, however, appears to become an ion channel under certain conditions. Palytoxin (PTX), a marine toxin, is found in reef corals that are armed with a stinging apparatus (9). PTX depolarizes cells via nonselective cation channels of 10 pS (10). Na^+/K^+ ATPase is known to be the target because pump-specific oubaine antagonizes PTX (11). PTX, which forms ion channels in the pump (**Figure 2**), presumably binds the Na^+ pump, forming a pore coincident with the Na^+/K^+ pathway, and activates a conductance with an affinity for $Na^+ > K^+$. Oubaine accelerates

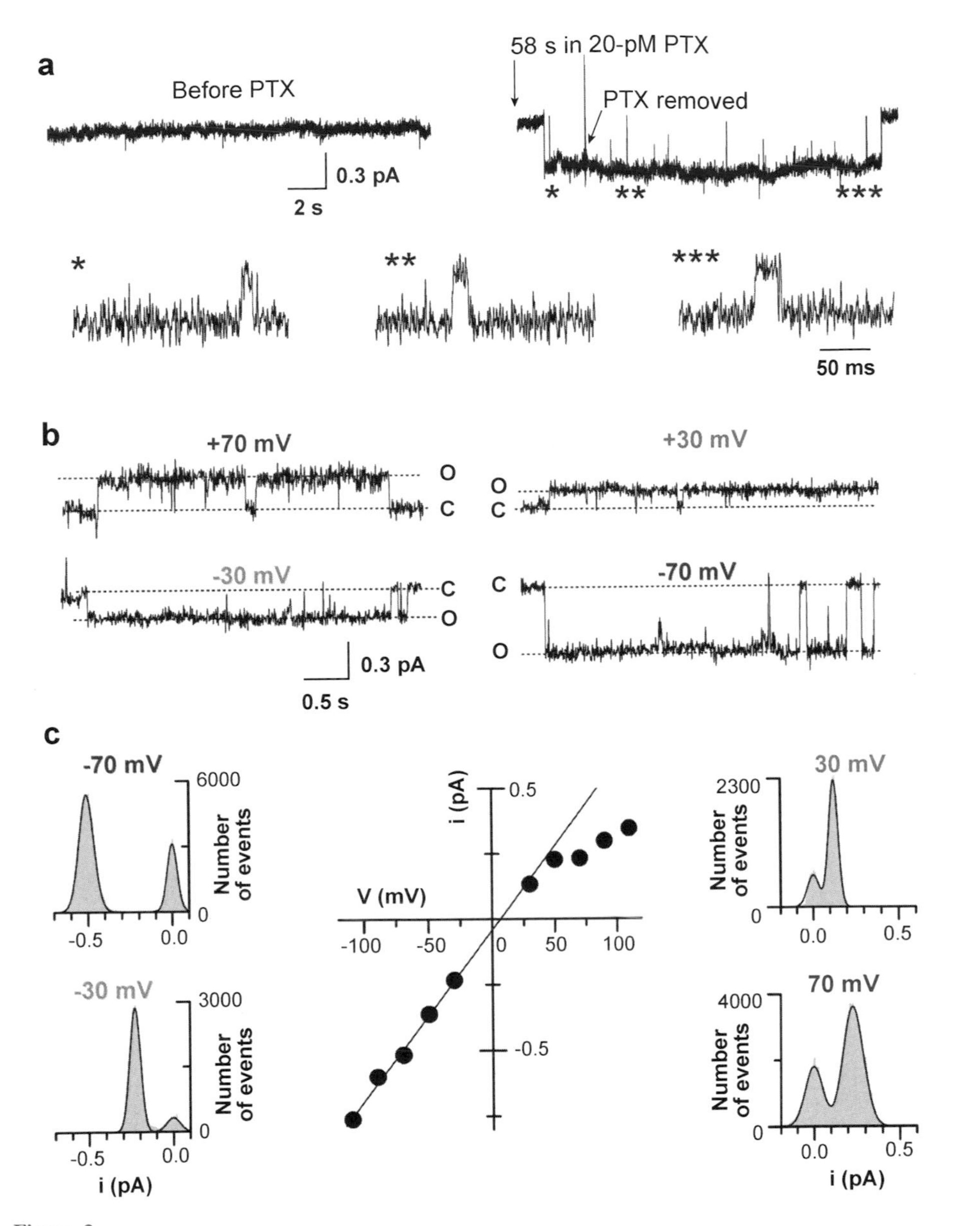

Figure 2

A channel-like component of Na^+/K^+ ATPase mediates the palytoxin (PTX)-induced current. (*a*) (*Top Left*) Absence of channel activity in an outside-out, ventricular-myocyte patch held at 40 mV, with Na^+ on both sides of the membrane and 5-mM internal MgATP, before PTX application. (*Right*) A single PTX-induced channel opens 1 min after 20-pM PTX is applied, characterized by long-open-time bursts with brief closures (*asterisks*). (*Bottom*) Examples of brief closures on an expanded timescale. (*b*) Channel currents at different voltages. Closed (c) and open (o) current levels are marked. (*c*) (*Left* and *right*) Histograms of baseline-corrected records 20 s long, fitted with sums of two Gaussians. (*Center*) Single-channel current, I (difference between peaks), plotted against V gave channel conductance of 7 pS. From Artigas & Gadsby (12, 13) and Hilgemann and colleagues (14–16).

PTX washing-out effects, and preincubating the ATPase with the steroid slows subsequent activation of the PTX-induced conductance. Thus, PTX and ouabain likely occupy the Na^+ pump simultaneously, each destabilizing the other. PTX-induced channels are also permeable to large organic cations, including NMDG (*N*-methyl-D-glucamine), suggesting that the narrowest section of the pore must be at least 7.5 Å wide.

Na^+/Ca^{2+} Exchangers

The Na^+/Ca^{2+} exchangers (NCXs) are a ubiquitously expressed group of transporters. They transport Ca^{2+} across membranes against the ion's electrochemical gradient by using the electrochemical gradient of Na^+. The cotransport is therefore bidirectional and is controlled by membrane potential and by both Na^+ and Ca^{2+} (substrate) gradients. In cardiac muscle, exchangers extrude intracellular Ca^{2+} during the excitation-contraction cycle. The stoichiometry for the exchange is thought to be 1 Ca^{2+}:3 Na^+; hence, the exchanger is electrogenic, and membrane current is a quantitative readout of NCX activity (17). Intracellular Ca^{2+} concentration also assesses NCX function (18). In cardiac myocytes, NCX is additionally crucial for Ca^{2+} homeostasis and muscle relaxation after contraction and can play an important role in excitation-contraction coupling when running in reverse. During cardiac ischemia, malfunction of the exchanger causes Ca^{2+} overload and cardiomyocyte dysfunction. Mammalian NCXs comprise a multigene family of homologous proteins with three isoforms: NCX1–3. NCX1 is expressed predominantly in the heart, kidney, and brain, whereas NCX2 and 3 are restricted primarily to skeletal muscle and brain. The three isoforms bear 70% identity to each other and have similar predicted membrane topologies, consisting of nine transmembrane domains (TMDs) and a large central cytoplasmic loop containing the exchanger inhibitory peptide (XIP) region, Ca^{2+}-binding sites, and phosphorylation sites. Another exchanger, NCKX2, is a $Na^+/Ca^{2+}/K^+$ exchanger.

As mentioned above, Ca^{2+} import/export depends partly on the Na^+:Ca^{2+} exchange ratio or transport stoichiometry. Whereas Ca^{2+} flux equilibrium experiments indicate Na^+:Ca^{2+} ratios of 3:1, reversal potential data suggest 4:1 ratios (16). Using an ion-selective electrode to quantify ion fluxes in giant patches, Hilgemann (19) showed that ion flux ratios during maximal transport in either direction are 3:2. Because Na^+ and Ca^{2+} are present on both sides of the membrane, the net current and Ca^{2+} flux additionally are dependent upon and can be reversed at different membrane potentials. Hilgemann proposes that transport of substrates by NCX1 is not restricted to a ratio of 3:1 but that other transport stoichiometries are indeed possible. These include 1:1 at a lower rate, a Na^+-conducting mode that exports 1 Ca^{2+}, and an electroneutral Ca^{2+} influx mode that exports 3 Na^+ (**Figure 3**). The two minor transport modes may potentially contribute to and determine resting concentrations of free Ca^{2+} and background inward current in heart muscle (17, 20). This release from strict substrate coupling ratios in NCX1 therefore shifts the definition of the mechanism involved in this molecule's function away from one that can be described wholly by the alternating access model and suggests characteristics more akin to those of channel proteins.

Noise analysis also suggests that NCX currents have a channel-like component (16). NCX1 has a relatively high turnover rate at physiological membrane potentials, which, in terms of a unitary current, is on the order of 1 fA (6000 e sec^{-1}). The channels, although not seen directly, are thought to fluctuate between open and closed states, similar to ion channels. The large transporter-associated currents observed therefore are somewhat expected, and net NCX1 currents in heart are comparable to the ion channel currents that comprise the cardiac action potential (**Figure 4**).

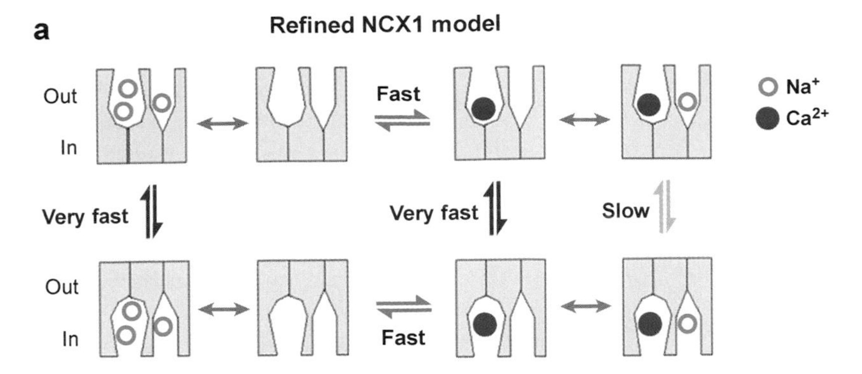

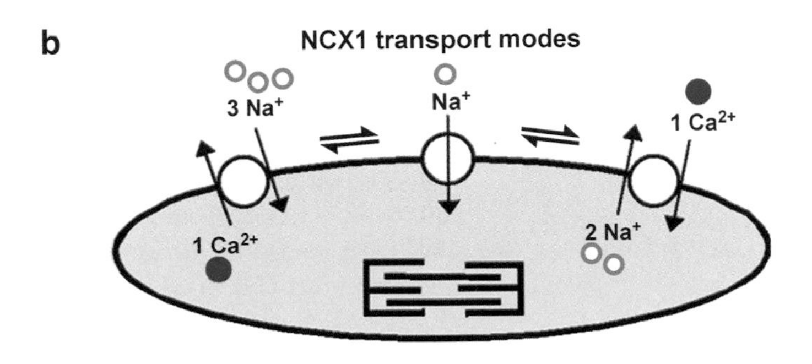

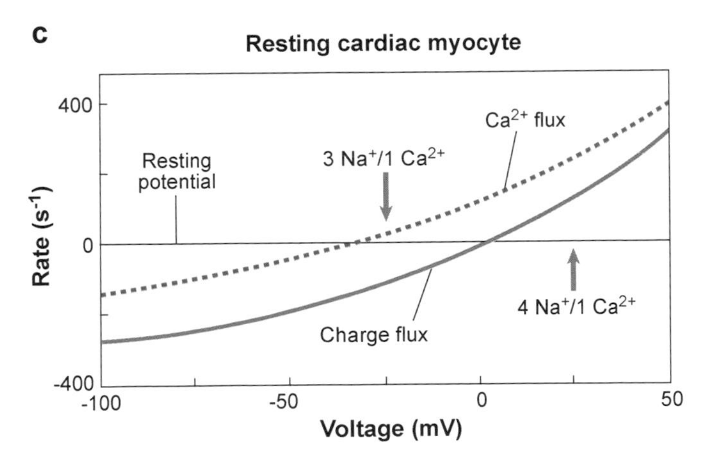

Figure 3

Na^+/Ca^{2+} exchange model. (*a*) At one site, 2 Na^+ or 1 Ca^{2+} bind, whereas only 1 Na^+ binds at the second site. When sites are occupied by 3 Na^+ or 1 Ca^{2+}, ion translocation is fast (30,000 sec^{-1}). Translocation of 1 Na^+ together with 1 Ca^{2+} occurs ten times slower (2500 sec^{-1}), and then Ca^{2+} release rates are intermediate (10,000 sec^{-1}). Three exchange modes are postulated. (*b*) NCX1 function in a resting myocyte. Ca^{2+} import by the 2 Na^+/1 Ca^{2+} mode is balanced by Ca^{2+} export in the 3 Na^+/ 1 Ca^{2+} mode. Simultaneously, the Na^+-conducting mode generates a background inward current. (*c*) Predicted voltage dependence of single-exchanger Ca^{2+} flux and I(V) in the resting state of a cardiac myocyte. Positive y-axis values give inward Ca^{2+} flux and outward cation flux rates. Relevant to cardiac physiology, the Na^+/Ca^{2+} transport ratio is approximately 4 in the negative potential range when free internal Ca^{2+} rises to 2 mM. From Reference 17.

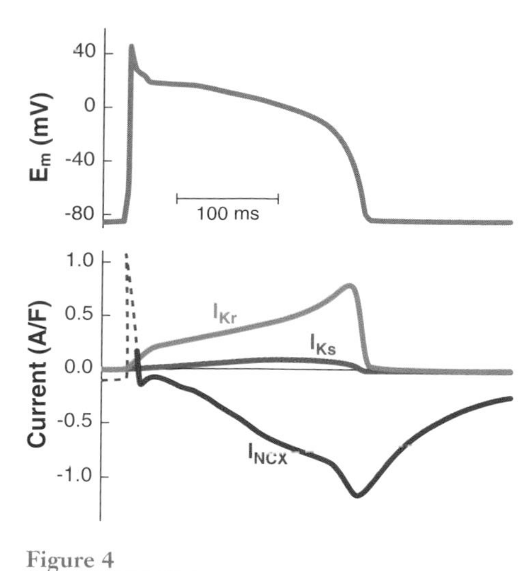

Figure 4

Ionic current through the Na^+/Ca^{2+} exchanger (NCX) during a cardiac action potential. Model results based on data in References 21 and 22. Figure is courtesy of Donald Bers. See also Reference 22.

Glutamate Transporters

Glutamate transporters belong to a class of membrane transporters known as excitatory amino acid transporters (EAATs) and are expressed in neurons and glial cells, where they limit extracellular glutamate concentrations. Climbing fiber Purkinje cell synapses in the cerebellum release glutamate that is rapidly bound by neuronal EAATs located postsynaptically. Photolysis of caged glutamate was used to characterize the current in Purkinje cells resistant to glutamate receptor antagonists and inhibited by the high-affinity EAAT antagonist TBOA (DL-threo-benzyloxyaspartic acid). Through the subtraction of this residual non-EAAT current from the response recorded in the absence of glutamate receptor antagonists, estimates of postsynaptic uptake were obtained. Analyses of such synaptic EAAT currents suggest that, on average, postsynaptic EAATs take up 1,300,000 glutamate molecules in response to a single climbing fiber action potential. Isolation of the synaptic EAAT current from the contaminating current suggests that, near physiological temperatures, only a small fraction of glutamate (~1%) is removed postsynaptically (23–26).

Additionally, EAATs exhibit an anion channel activity, which, although uncoupled from the glutamate transport mechanisms, is nevertheless gated by glutamate. The EAAT channel is Cl^- selective, and activation of the EAAT Cl^- channel alters synaptic currents more rapidly than Na^+-coupled glutamate transport turnover (100 msec) could. Thus, charge movements in cotransporters, including those in this example of the glutamate transporter, are not always stoichiometrically coupled to neurotransmitter flux (see **Supplemental Text**, Sections B3 and B4). In voltage-clamped *Xenopus laevis* oocytes, in which it is possible to measure ion current and substrate flux simultaneously, charge-to-flux ratios may vary from 1–2 to more than 100. Early studies of current fluctuations in photoreceptors (27) first hinted at the existence of the Cl^- channel activity in the glutamate transporter. Heterologous expression of neurotransmitter transporters (28) made it possible to study their transport properties under controlled conditions. Characterization of their electrical properties eventually confirmed the presence of Cl^- channels in glutamate transporters and showed that the transporter's channel kinetics were closely related to its glutamate transport kinetics (28, 29). Amara, Kavanaugh, and colleagues (30–33) have extensively studied EAATs and their dual channel-like and transporter-like properties.

The ion channel in the Na^+-dependent glutamate transporter selects for Cl^-, but glutamate transport is independent of Cl^-. This is in marked contrast to the Na^+/Cl^--dependent monoamine transporters, in which function depends on both ions. The kinetics of the Cl^- channel and the glutamate transporter are related. However, it is unknown whether mechanistically there are different pathways for the movement of glutamate and Cl^- ions, or simply a single pore that can switch between two, sometimes separate, modes. Glutamate presumably does not permeate the

Cl^- channel activated during transport, and the coupling of glutamate to Na^+, K^+, and H^+ gradients is extremely tight (24).

GLAST is the predominant glutamate transporter expressed in Bergmann glia (34). In particular, the application of L-glutamate to outside-out patches pulled from cerebellar Bergmann glial cells activates EAAT transporter currents in less than 1 ms (35) (**Figure 5**). Glutamate released from climbing fibers likely escapes the synaptic cleft and reaches glial membranes shortly after its release.

The intrinsic kinetics of transporters from glial cells indicate that the glutamate concentration at glial cell membranes peaks at a lower level than the 1–3 mM in the synaptic cleft and likely persists extrasynaptically for up to 10 ms following release. Glutamate transporters expressed in glial cells have rapid kinetics and are capable of binding glutamate on a submillisecond timescale, similar to ionotropic glutamate receptors (37). Rapid binding of glutamate provides an efficient mechanism for capturing the glutamate that escapes the cleft, reducing glutamate concentration and hence preventing elevated glutamate levels in the extracellular space. Glutamate transport, on the other hand, is much slower, on the order of 10 msec per transport cycle for the glutamate transporter (GLT-1) (30–32). The exact mechanistic steps in the transport cycle that permit the channel mode and flow of anions are unknown.

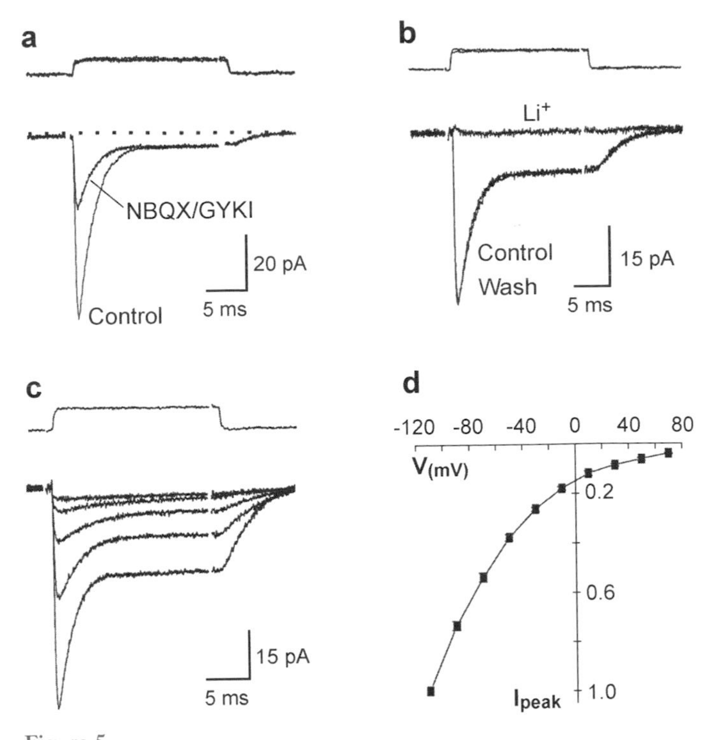

Figure 5

Glutamate transporter currents and alpha-amino-3-hydroxy-5-methyl-4-isoxazolepropionic acid (AMPA) receptor currents evoked in outside-out patches from Bergmann glial cells. (*a*) L-Glutamate (10 mM) activates a transient current composed of AMPA receptor current blocked by NBQX and GYKI-52466 and a smaller biphasic current associated with the glutamate transporter. $V_m = 90$ mV. (*b*) Substituting extracellular Li^+ for Na^+ blocks the transporter current evoked by 10-mM L-glutamate. $V_m = 90$ mV. (*c*) Currents at membrane potentials 10, 30, 50, 70, and 90 mV. (*d*) The I(V) curve for the transporter current is inwardly rectifying and does not reverse. From Reference 36.

Glucose Transporter

A new study by Naftalin and colleagues (38) uses a template model of the facilitated glucose transporter GLUT1 to reveal nine substrate-binding clusters spanning a hydrophilic channel through the transporter. D-Glucose binds to five sites at the external opening, with increasing affinity as substrates approach the pore. The substrates then appear to pass through a narrow channel into an internal vestibule with four low-affinity sites. This channel view of molecular flux presents a very different picture of glucose transport than the alternating access model predicts. Contrast, for example, GLUT1 with the Na^+-coupled glucose transporter (SGLT1), which uses the transmembrane Na^+ gradient to concentrate glucose (39–42). The Na^+-coupled glucose transporter itself, however, exhibits a proton channel behavior under some conditions (41). Furthermore, electrophysiological characterization of another member of the same gene family, the human choline transporter (43), reveals large nonstoichiometric currents (44) strongly suggestive of channel-like properties.

Monoamine Transporters

An in-depth characterization of Na^+/Cl^--coupled neurotransmitter transporters has occurred since their early discovery by Axelrod and colleagues (45). New research has revealed currents that obey electrodiffusion laws (please see **Supplemental Text**, Section A1). Single-transporter events have been observed in γGABA (γ-aminobutyric acid) (46, 47), NE (norepinephrine) (48), 5-HT (49), and DA (dopamine) (50) transporters. Larger-than-expected currents, as observed for the $Na^+/H^+/K^+$-coupled glutamate transporters discussed above, are also encountered with the monoamine transporters. Single-channel or -transporter events, however, have not yet been observed in glutamate transporters. We therefore repeatedly observe that transporters can exhibit large currents and, in some cases, form observable channel-like pores that conduct both ions and substrates. These substrates may include the relatively large neurotransmitter molecules, which appear to permeate channel-like pores, at least for the monoamine transporters. For further theoretical discussion on this topic, please see **Supplemental Text**, Sections B3 and B4.

Investigators have thought for some time that a channel mode may exist for the Na^+/Cl^--coupled transporters in the GAT/NET gene family (51–54). In one mechanistic model, 5-HT and Na^+ permeate a narrow pore in SERT, and the experimentally measured coupling results from queuing of the copermeating species (55, 56). Single-file diffusion of ions and substrate may account for transport against an electrochemical gradient and amplification of ion gradients into substrate gradients by raising the gradient ratio to a power (45, 56–58; see also **Supplemental Text**, Sections C3 and C4). This model of cotransport was inspired by Hodgkin & Keynes (59), who used a hypothetical single-file, multi-ion pore model to explain their data. Numerous subsequent experiments also support the model (60). Hodgkin's autobiography (61) and the **Supplemental Text**, Sections A2 and B4, elaborate further on the single-file electrodiffusion model and its relevance to the present discussion of channel-like activities of transporters.

Serotonin transporters. SERT belongs to the GAT/NET family of cotransporters. SERT modulates serotonergic signaling in the central nervous system and is implicated in human mood and appetite behavior (62). Serotonin-selective reuptake inhibitors block human SERT (hSERT) as well as 5-HT uptake and are used to treat depression and panic disorders (63, 64). hSERT also functions as a receptor for methamphetamine (MDMA, or ecstasy), amphetamine (AMPH), and cocaine (65). Like other Na^+-coupled transporters, hSERT concentrates 5-HT against its gradient by using the energy stored in coupled ion gradients (4, 66). SERTs expressed heterologously produce currents of magnitudes greatly in excess of those predicted by the alternating access model of cotransport. Interestingly, the observed currents are far more amenable to explanation by a channel model of transport (67, 68). It is less clear, however, if the SERT-associated channel-type currents exist in native cells and have physiological significance. The data of Bruns et al. (6) (**Figure 1**) for the serotonin transporter compare well with the results presented in the section on glutamate transporters. Presynaptic uptake and the postsynaptic response to Ca^{2+} uncaging both decay as single exponentials, with a strong correlation between the pre- and postsynaptic time constant. The response to a flash of Ca^{2+} uncaging light demonstrated the faster activation of the presynaptic uptake current compared with the postsynaptic current. A delay of less than 1 ms was observed between flash and onset of the presynaptic uptake current. The presynaptic uptake current showed an initial fast decay followed by a slower decline that was paralleled by the decay of the postsynaptic current. Both components of the presynaptic current were strongly reduced in 5-mM Na^+.

In *X. laevis* oocytes expressing rat serotonin transporters, serotonin activates macroscopic currents that are at least 10 times larger than the flux of (^{3}H) serotonin. Furthermore, large serotonin-independent currents that were blocked by selective transporter blockers were also observed (69). Lin et al. (49) identified a point mutation in the serotonin transporter that increases the unitary conductance. Petersen & DeFelice (68) presented evidence that *Drosophila* SERT (dSERT) conducts not only Na^+ but also 5-HT^+ itself in a channel mode of conduction.

Quick (70) has shown that the SNARE protein syntaxin 1A (STX1A) binds the N-terminal domain of SERT and that this binding regulates two of the conducting states. STX1A binding abolishes the 5-HT-induced and leak currents. These two currents occur naturally in neurons, and they can cause endogenous depolarization of these cells. Hence, molecules that disrupt the SERT-STX1A interaction effectively influence neuronal excitability by regulating SERT activity (70).

Norepinephrine transporters. Galli et al. (48, 71, 72) have performed several key experiments investigating the transport mechanisms of the norepinephrine transporters (NETs) (**Figures 6** and **7**). In these studies, Galli et al. recorded currents in patches from NET-expressing cells while simultaneously monitoring NE flux. These authors observed that the neurotransmitter permeated the patches in discrete bursts that correlated with single-channel openings. Inorganic ions permeate the channel in addition to the neurotransmitter itself. One question these experiments raised was how NETs are able to concentrate NE against its electrochemical gradient if NE can permeate an open channel. The initial, large inward gradient of neurotransmitter transiently present in synapses just after NE release may be the driving force for clearance. When the extracellular concentration declines and the electrochemical gradient diminishes, another mechanism may take over. Additionally, single-file diffusion through the NE channel may help concentrate NE through the use of the Na^+ gradient (57, 58).

Galli et al. (48, 71, 72) therefore demonstrate authentic channels in NETs. They also show bursts of NE release associated with discrete channel-opening events (**Figure 7**).

Dopamine transporters. Dopamine transporters (DATs) play a crucial role in the reuptake and release of neurotransmitters. Their mechanism of transport has classically been described using an alternating access model. Binding of DA and Na^+ and Cl^- ions to an extracellularly oriented transporter is thought to induce a conformational rearrangement to an intracellularly oriented transporter from which the cargo is released into the cytosol, completing the transport process (73–75). DA transport by DAT generates an electrical current because of the net movement of positive charge into the cell. DA efflux induced by the psychostimulant AMPH is believed to result from the ability of AMPH to reverse this inward transport process. Hence, the inward transport of AMPH by DAT increases the number of inward-facing transporter-binding sites and thereby increases the rate of outward DA transport through an exchange process. Curiously, however, there are modulators and mutations that differentially affect AMPH-induced DA efflux and uptake. For example, Khoshbouei et al. (74, 75) recently reported that N-terminal phosphorylation of human DAT (hDAT) plays a critical role in AMPH-induced DA efflux but not in DA uptake. These and other data suggest that the mechanism for the AMPH-induced DA efflux mediated by the reversal of DAT flux direction is more complex than a simple exchange process.

Cocaine, AMPH, and heroin elevate extracellular DA in the brain. Substrates like AMPH induce nonvesicular release of DA mediated by DAT. Outside-out patches from heterologous DAT-expressing cells or from dopaminergic neurons show that AMPH

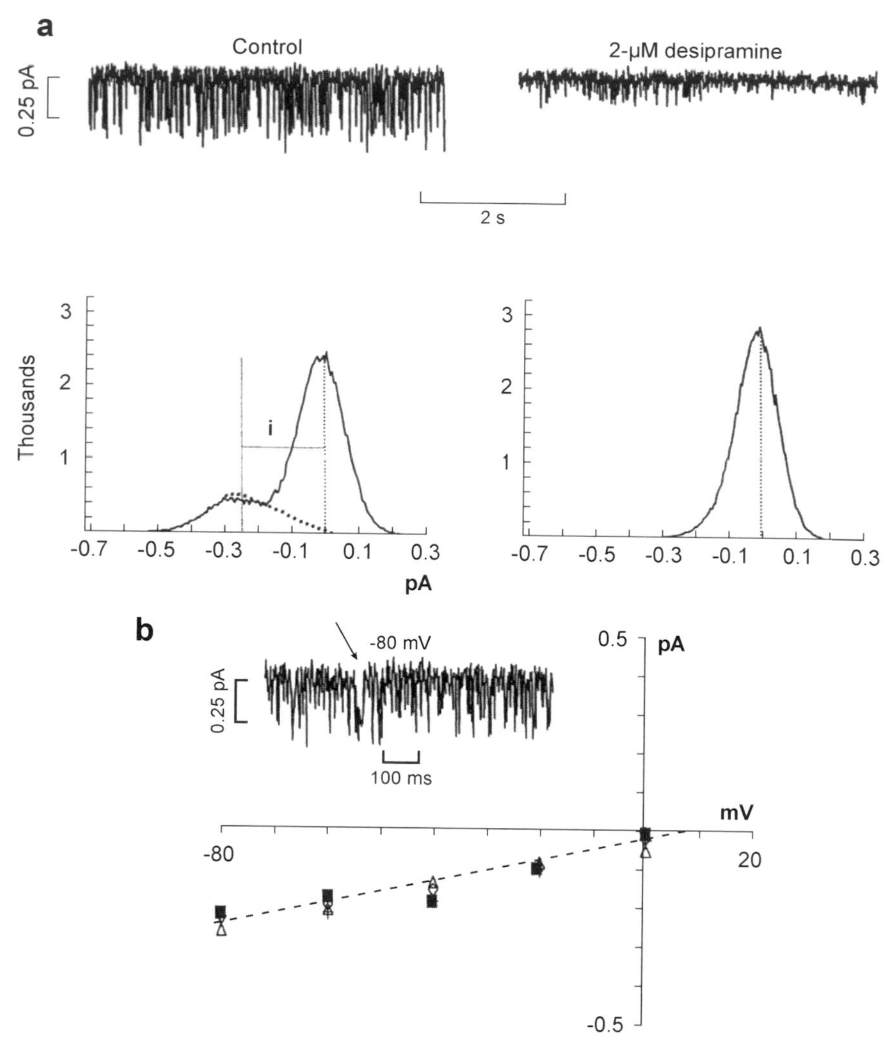

Figure 6

Inside-out patches. Norepinephrine or the substrate quanethidine induces channel activity in inside-out patches. (*a*) (*Left*) Raw data and an amplitude histogram from an inside-out patch containing 30-μM GU (quanethidine) (−80 mV). At −80 mV, distinct inward current spikes occur. The amplitude of the current at each voltage is defined as the center of the revealed inward component with respect to zero. This difference is shown as vertical dashed lines: in this example i = −0.25 pA. (*Right*) The patch 1 min after adding 2-μM DS [desipramine, a norepinephrine transporter (NET) blocker] to the bath. (*b*) i(V) curve constructed by the method outlined in the left panel of *a*. The inset in the i(V) curve shows an example of rare, long events (*arrow*). The NET single-channel conductance is linear in the range −20 to −80 mV, with the value 2.95 + 0.17 pS ($n = 4$).

causes DAT-mediated DA efflux by a slow process consistent with an exchanger and a fast process of millisecond bursts of DA efflux through a channel in DAT. A single burst of DA is on par with the amount of DA released from a vesicle (76). Such channel-like behavior associated with neurotransmitter efflux is also observed in the NETs.

Caenorhabditis elegans DA transporters (DAT-1) generate single-channel currents

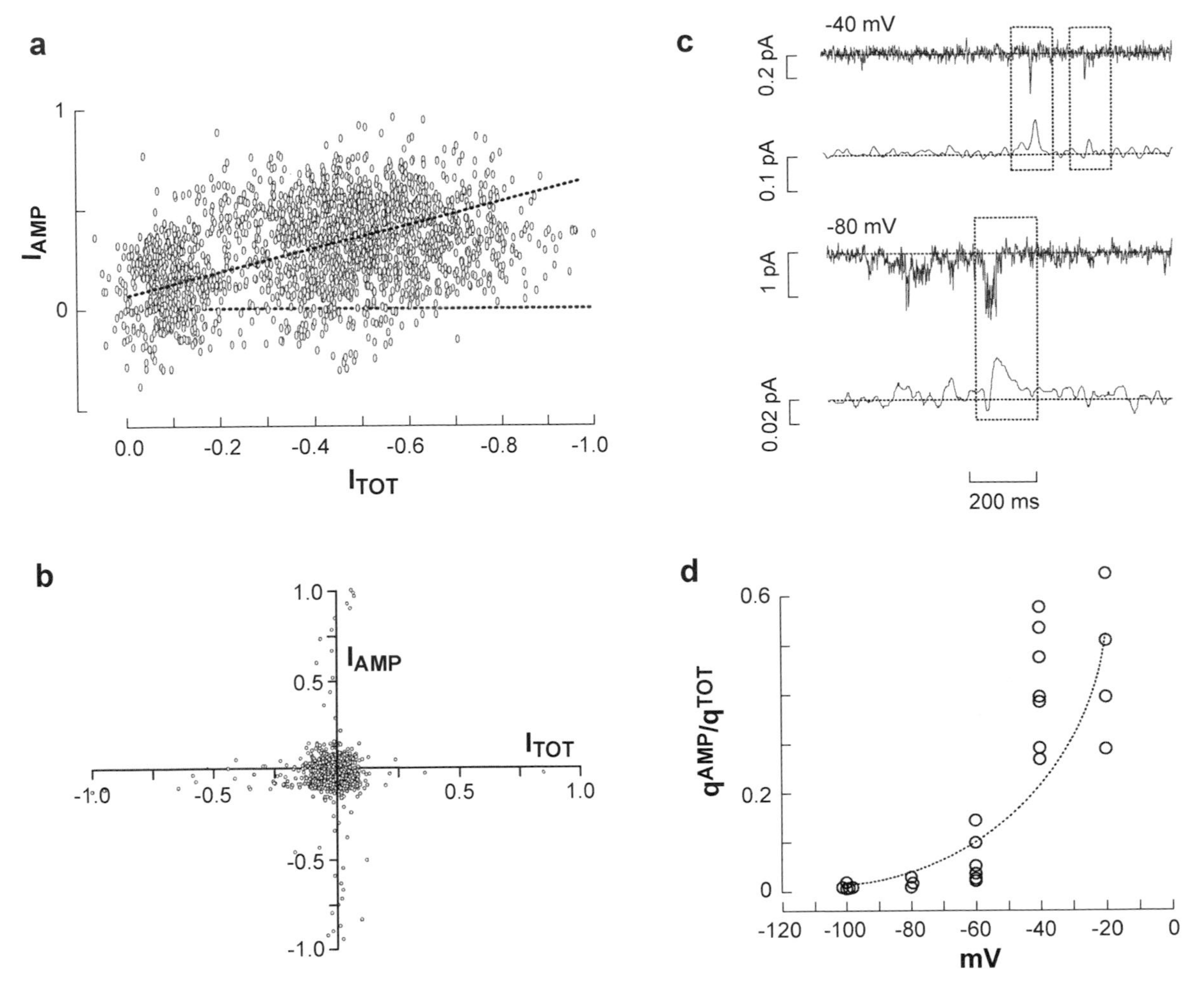

Figure 7

(*a*) Correlation between norepinephrine (NE)-induced current through the patch and the NE released from the same patch (4-mM NE in the pipette). Holding potential stepped from 0 to 100 mV, the plot is normalized time-varying NE-induced current (I_{TOT}) against normalized time-varying NE uptake (I_{AMP}). The flat line indicates zero correlation. (*b*) Time relation between patch and amperometric currents from a parental cell patch with 4-mM NE in the pipette. (*c*) Isolated NE-induced, DS (desipramine)-sensitive current events at 40 and 80 mV and simultaneous recordings from the amperometric electrode. The dotted rectangles are pairs selected for analysis. (*d*) Ratio of amperometric charge, q^{AMP}, to the corresponding patch-current charge, q^{TOT} (from seven patches), as a function of patch voltage. Figure from Reference 71.

that are selective to Cl^- ions (50). In 1-μM DA, small (−0.8 ± 0.1 pA) single-channel events were recorded (n = 7) at −120 mV (**Figure 8**). These channels were abolished in 10-μM IMP (imipramine, a DAT-selective blocker) with 1-μM DA present. When external Na^+ was replaced by $NMDG^+$, no IMP-sensitive events were recorded, and DA-induced, single-channel events were absent in patches pulled from *dat-1*-deleted dopaminergic neurons. Inhibitors that block DA receptors did not eliminate DA-induced channel events, whereas IMP blocked these currents. Furthermore, DAT-mediated macroscopic

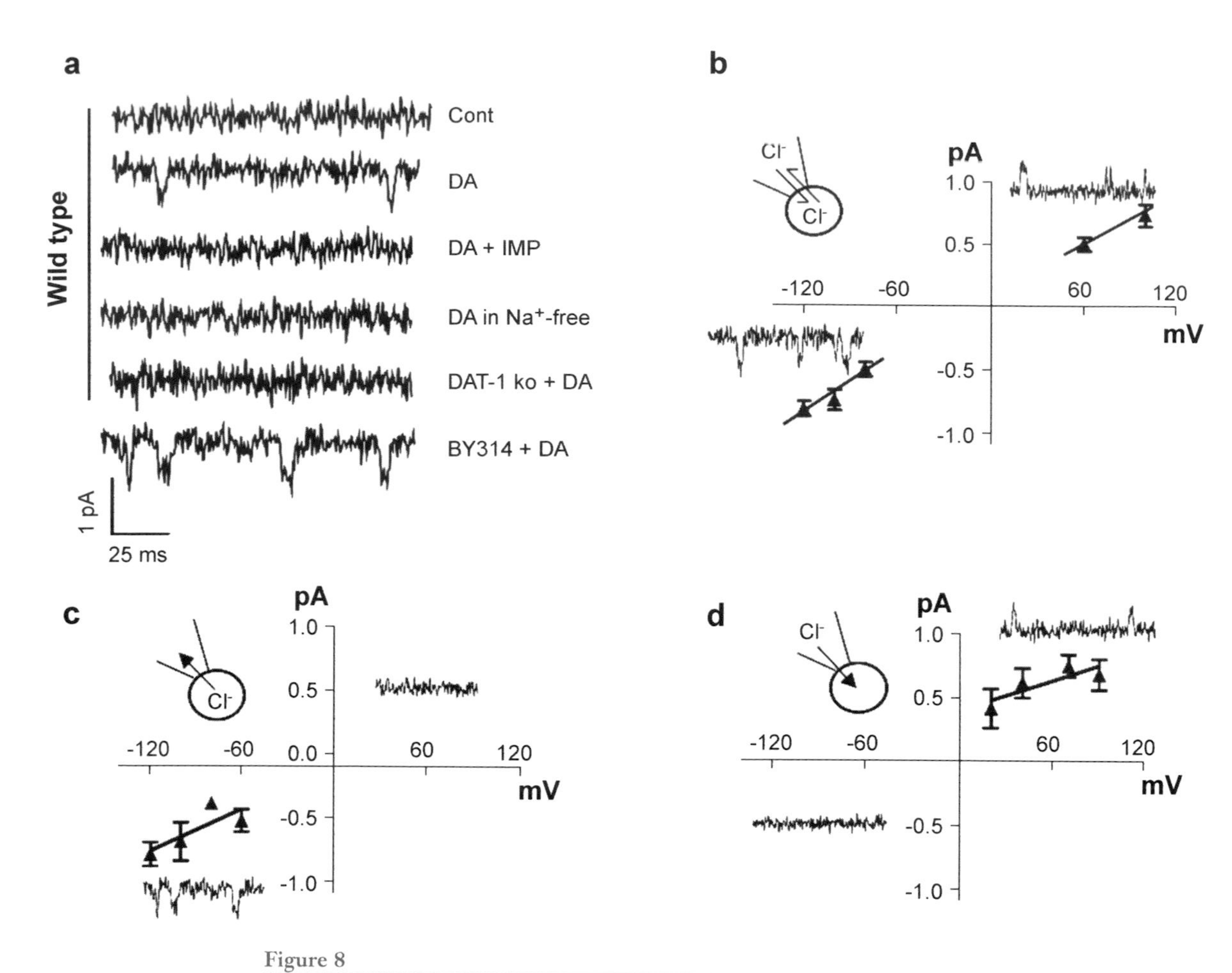

Figure 8

Transporter channels in *C. elegans* dopaminergic neurons. (*a*) Channel events from wild-type and BY314 *C. elegans* dopaminergic neurons in outside-out patches at −120 mV. Abbreviations used: Cont, control; DA, dopamine; DAT-1 ko, dopamine transporter knockout; IMP, imipramine (a dopamine inhibitor). (*b–d*) Single-channel amplitude recorded in cell-attached configuration and plotted against the voltage, when the extracellular Cl^- concentration on both sides of the patch was equal to (*b*) or was lower (*c*) or higher (*d*) than the intracellular Cl^- concentration. From Reference 50.

and microscopic currents had a specific requirement for Na^+ and Cl^-, a characteristic of the SLC6A family to which DAT-belongs.

DA and AMPH increase the firing activity of rat dopaminergic neurons in culture, elicit inward currents primarily comprised of anions, and result in an excitatory response. In addition to clearing extracellular DA, currents associated with DAT therefore appear to modulate excitability and regulate neurotransmitter release from the dopaminergic midbrain neurons (77).

Metal Ion Transporters

Exact cellular concentrations of metal ions are essential for the proper function of life processes. Disruption of normal, homeostatically controlled cellular concentrations of metal ions leads to cell disease and death either by starving the cell of a vital metabolic element or by causing irreversible cell damage through the generation of toxic metabolites. Metal ion homeostasis is therefore highly regulated and dependent on specific transporters, channels, and pore proteins that balance uptake, storage, and secretion across cell membranes. Predominant disorders associated with metal ion overload or deficiencies include microcytic anemia, hereditary hemochromatosis, Menke's disease, Wilson's disease, and neurodegenerative diseases such as Alzheimer's and Parkinson's disease, Friedreich's ataxia, and pica.

Transition metal ions are essential for oxidative phosphorylation, DNA replication, neuronal signaling, oxygen carriage, muscle contraction, endocrine and exocrine secretion, and sensory transduction. The functional cloning and characterization of divalent cation transporter 1 [DCT1, also known as divalent metal ion transporter 1 (DMT1)], the rat homolog of mouse Nramp2 (natural resistance–associated macrophage protein 2) and human NRAMP2, were the first demonstration of an active cellular transport mechanism for divalent metal cations (e.g., Fe^{2+}) in mammalian cells (78). DMT1 was isolated from an iron-deficient rat duodenal cDNA library, using a *X. laevis* oocyte expression cloning system. The library was screened by radiolabeled iron ($^{55}Fe^{2+}$) uptake activity in *Xenopus* ooctyes. Expression of DMT1 in oocytes stimulated greater-than-200-fold increases in $^{55}Fe^{2+}$ uptake compared with control water-injected oocytes. Electrophysiological characterization using two-microelectrode voltage-clamp analysis of injected ooctyes showed that the activity of the divalent cation transporter was electrogenic with reversible currents up to 1000 nA induced by 100-mM Fe^{2+} at pH 5.5. The characterization also demonstrated that DMT1 transports a wide range of divalent cations such as Fe^{2+}, Zn^{2+}, Mn^{2+}, Cu^{2+}, Co^{2+}, Cd^{2+}, and Pb^{2+}. The large Fe^{2+}-induced, DMT1-mediated currents were H^+ dependent; the size of the inward current was significantly enhanced when cells were shifted from an extracellular pH of 7.5 to 5.5. Steady-state Fe^{2+}-evoked currents were saturable with Michaelis-Menten kinetics and affinity constants of 2 mM and 1–2 mM for Fe^{2+} and H^+, respectively, and showed a curvilinear voltage dependence approaching a zero-current asymptote. Concomitant voltage-clamp and intracellular pH measurements showed that Fe^{2+} uptake was accompanied by intracellular acidification, indicative of a symport mechanism of cotransport. This and subsequent studies therefore identified DMT1 as a bonafide transporter responsible for the H^+-dependent cotransport of divalent metal cations across the plasma membrane of cells, including intestinal epithelial cells, and from the lumen of the early endosomal compartment into the cytoplasm.

Interestingly, however, in response to a decrease in extracellular pH, DMT1-injected oocytes demonstrated substantial inward currents compared with control ooctyes, even in the absence of divalent cation substrate. In addition to operating as an H^+/divalent cation cotransporter, therefore, DMT1 appeared to exhibit an additional H^+ uniporter function, constituting the H^+ leak, as it has been termed, as this H^+ current is independent of the divalent cation transport pathway. A subsequent study by Chen et al. (79) further investigated this H^+ leak current (see **Figure 9**). Chen et al. simultaneously performed voltage-clamp analysis and radioisotope uptake measurements on DMT1-injected oocytes in the presence of 10-mM $^{55}Fe^{2+}$ and at a constant pH of 5.5. Calculation of the ratio of the total Fe^{2+}-evoked current to the $^{55}Fe^{2+}$ uptake showed that this ratio varied with membrane potential. At +10 mV, the charge-to-uptake ratio was close to 3:1, indicative of a cotransport stoichiometry of 1:1. At increasingly negative membrane potentials, however, the value of this ratio became unexpectedly high (20:1 at −80 mV), inconsistent with a 1:1 ratio for Fe^{2+} and H^+ cotransport and favorable of a stoichiometry-free, uncoupled H^+-mediated current. The charge-to-uptake ratios were apparently unaffected by the removal of extracellular Na^{2+}, Ca^{2+}, or Mg^{2+} or the depletion of intracellular Cl^-. The linear, nonsaturable kinetics of this uncoupled H^+ current suggested its conduction via a channel or pore-type mechanism

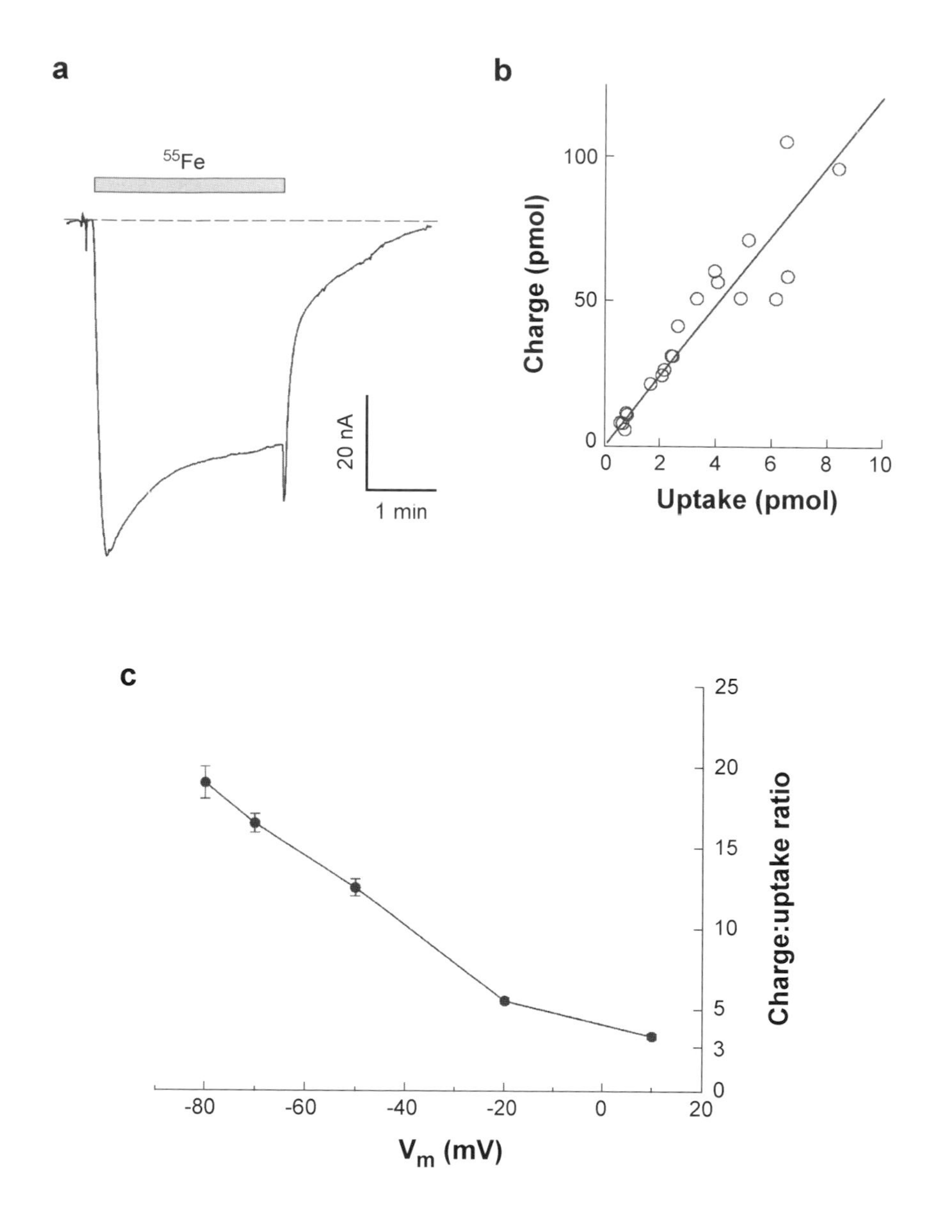

Figure 9

Voltage dependence of the charge-to-Fe^{2+}-uptake ratio of DMT1 mediated Fe^{2+} transport. Fe^{2+}-evoked currents and Fe^{2+} uptake were measured simultaneously under voltage-clamp conditions. (*a*) Example of currents generated by 10-μM $^{55}Fe^{2+}$ at 50 mV and pH 5.7. The charge moved was calculated by integrating the Fe^{2+}-evoked current over the uptake period. (*b*) The charge moved at 50 mV was converted to pmol and plotted against Fe^{2+} uptake. The slope of the linear fit, which is equal to the mean charge-to-uptake ratio, is 12.5 ± 0.5. (*c*) Dependence of charge to Fe^{2+} uptake on the membrane voltage. Figure from Reference 79; see also Reference 80.

and that the cotransport stoichiometry of H^+ and X^{2+} (where X is a metal ion) is most likely variable and determined by $[H^+]_o$:$[H^+]_i$, as well as $[X^{2+}]_o$:$[X^{2+}]_i$, and more dependent on the first ratio. These findings add complexity to our picture of DMT1 as a transporter molecule, imparting to it a significant channel character.

The importance of iron as a biomolecule has led to extensive and multidisciplinary investigations of this interesting metal ion transporter. A recent study provides sufficient evidence for DMT1 as an example of transporters upon which this review concentrates (81). In an attempt to reconfirm the transport characterization of an important disease-associated mutation, $DMT1^{G185R}$, that had previously been shown to abrogate the Fe^{2+} transport function of DMT1, Xu et al. (81) unveiled that this single-nucleotide point mutation increased Ca^{2+} permeability mediated by DMT1 in a channel-type pathway.

Previously, the initial characterization using *Xenopus* oocytes had shown that Ca^{2+} was not a substrate for DMT1 transport. Ca^{2+} did not mimic the effect of Fe^{2+} in the evocation of inward currents. In fact, Ca^{2+} present at 10-mM concentration produced only a small outward current at pH 5.5 and only partially inhibited Fe^{2+}-evoked currents. Similarly, when cells expressing wild-type DMT1 were subjected to electrophysiological patch-clamp analyses, no significant current was detected in the presence of 10-mM Ca^{2+} and 140-mM Na^+ at pH 7.4. However, in stark contrast, under the same conditions, cells expressing $DMT1^{G185R}$ exhibited large, stable inward currents. These currents were inhibited by substitution of Ca^{2+} and Na^+ in the extracellular medium with $NMDG^+$. Unlike the X^{2+} currents observed with wild-type DMT1, Ca^{2+} currents mediated by $DMT1^{G185R}$ were inhibited by a reduction of extracellular pH to 5.8. Transient kinetics of the Ca^{2+} currents of the G185R mutant protein were distinct from those of the X^{2+} currents mediated by wild-type DMT1, including fast voltage-dependent inactivation that was enhanced with increasing $[Ca^{2+}]_o$. Cationic selectivity studies using $DMT1^{G185R}$ showed that the permeability sequence of this channel ($Ca^{2+} > Sr^{2+} > Ba^{2+} > Li^+ > Na^+ > K^+ > Cs^+$) was similar to L-type Ca^{2+} channels. $DMT1^{G185R}$ was, however, less selective for Ca^{2+} than are other known voltage-gated channels. Additionally, its I(V) curve, kinetics, and sensitivity to pharmacological agents differed significantly in terms of mechanism from other known Ca^{2+} channels, including the voltage-gated Ca^{2+} channels, transient receptor proteins (TRPs), and Ca^{2+} release–activated Ca^{2+} channels. A biological purpose for the G185R gain-of-function mutation may be that elevated intracellular Ca^{2+} concentrations increase uptake of non-transferrin-bound iron (81, 82). Indeed, studies using reticulocytes from mice inherently carrying the G185R mutation (mk/mk) have shown that iron uptake in these cells represents 45% of that seen in wild-type cells, a value higher than that expected for a severe loss-of-function mutation. It is difficult at present, however, to speculate on the biological reason behind this phenomenon.

Another interesting discovery is associated with the electrophysiological characterization of the suppressor mitochondrial import function (SMF) proteins, SMF1, 2, and 3, which are yeast homologs of DMT1. These proteins bear 33–36% identity to DMT1 and 51–54% identity to each other. SMF1 expression is localized to the yeast plasma membrane, whereas SMF2 and SMF3 are mitochondrial membrane proteins. Complementation studies have demonstrated that SMF1 is a general metal ion transporter that transports a wide range of divalent metal cations, including Mn^{2+}, Zn^{2+}, Cu^{2+}, Fe^{2+}, Cd^{2+}, Ni^{2+}, and Co^{2+}. SMF2 and 3 may be broad-range metal ion transporters that have specificity distinct from SMF1. Chen et al. (79) showed that SMF1 mediates Fe^{2+} uptake in *Xenopus* oocytes in a H^+- and voltage-dependent manner. Metal ion transport in SMF1 was electrogenic and saturable with Michaelis-Menten kinetics similar to DMT1. However, in contrast to DMT1, Na^+ inhibited divalent metal ion uptake by SMF1. Furthermore, in the absence of metal ions, Na^+ evoked large SMF1-specific inward currents at -50 mV. Both Fe^{2+} and increased H^+ concentration inhibited this Na^+ current. The Na^+ currents were voltage dependent but exhibited neither saturation by hyperpolarization nor saturable Michaelis-Menten kinetics. Moreover, unlike the H^+/Fe^{2+} cotransport observed in SMF1, these Na^+ currents were susceptible to partial inhibition by Ca^{2+}. Therefore, the Na^+ currents observed in SMF1 seem to be uncoupled to the H^+/Fe^{2+} cotransport. The Na^+ conduction pathway was nonselective to other monovalent ions, including Li^+, K^+, and Rb^+, and mediated flux of Ca^{2+} ions. Hence, metal ion transport in SMF1 (and possibly also SMF2) appears to have at least two main components: a proton-coupled divalent cation cotransport mechanism and an uncoupled channel-like pathway that mediates the permeation of Na^+, other monovalent cations, and Ca^{2+}. This double identity may be a signature of not only DMT1 and the SMF proteins but also other members of this family, such as the natural resistance–associated macrophage protein 1 (Nramp1). Indeed, in the characterization of Nramp1 as a H^+-coupled divalent cation antiporter, the large inward currents evoked by Fe^{2+} were greater in amplitude than expected from $^{55}Fe^{2+}$ radioisotope uptake measurements (83).

Perhaps channel-like fluxes are integral to the mechanism of molecules identified essentially as transporters. Examples with relation to transition metal ion transport include members of both families of membrane proteins responsible for mammalian copper homeostasis: the P-type ATPase copper pumps and the CTR transporters. Active ClC-4, an extensively studied intracellular chloride channel, physically associates with and specifically modulates copper transport via the P-type ATPase copper pump ATP7B in hepatocytes (84). Also, determination of the 6-Å projection structure of the copper uptake transporter CTR1 revealed the presence of a copper-permeable pore along the center threefold axis of the trimeric transporter complex, thus placing the structural design of this transporter in close proximity to that normally observed in channel proteins (85).

A Cl^- Channel Is Really a Cl^-/H^+ Exchanger

Both structural and electrophysiological data are known for only one specific transporter, and the results are surprising. Accardi, Miller, and colleagues provide startling evidence that ClC-ec1, a membrane protein predicted by homology to be a Cl^- channel, is actually a Cl^-/H^+ countertransporter (86–88). Their paper and associated commentary by Gadsby (89) challenge us to ask how transporters differ from ion channels and vice versa. Channels dissipate ion gradients by passive electrodiffusion. However, although ions predominantly move down their electrochemical gradient, they may by chance move uphill. Thus, moment to moment, molecules may disobey thermodynamics, just as Boltzmann had described (90), and this will apply to both channel and transporter mechanisms (see **Supplemental Text**, Section A2, and **Supplemental Figure 1**). Accardi & Miller bring out the salient features of the transporter-channel question in a powerful illustration that is strengthened greatly by the dual presence of structural and functional data (91, 92).

If the signature of channels is allowing the downhill flux of ions (with the chance occurence of uphill flow), the signature of cotransporters is using this downhill energy to build other gradients (see **Supplemental Text**, Sections C1 and C2). These seemingly opposite characteristics led quite early to the failure of electrodiffusion as an adequate explanation for the mechanism for metabolite concentration (93). Subsequently, the transport of molecules against their electrochemical gradients was more carefully explained mechanistically using an enzymatic theory with carrier kinetics developed by Jardetzky (94) in 1966 (see **Supplemental Text**, Section C3). Classically, cotransporters couple the electrochemical gradients of Na^+ or protons to the movement of other substrates; thus, they may be said to have both channel properties (ions flowing downhill) and transporter properties (substrates flowing uphill) (95).

In the example of ClC-ec1, evidence for transporter properties comes from the observation that Cl^- flux by this molecule in one direction is positively coupled with H^+ flux in the opposite direction (that is, antiport, exchange, or countertransport). Surely the ability of one ion to drive another in the opposite direction lies wholly within the domain of transporters. To examine this, we would need to know more about what countertransport means at both molecular and structural levels. Unfortunately, there is a dearth of knowledge concerning structure in a dynamic sense. The postulated conformational changes for *lac* permease are far too massive (3), and the subtle changes for the leucine transporter (96) and the H^+/Cl^- exchanger (97) are speculative. More evidence for the countertransporter behavior of ClC-ec1 is provided by the observation that when pH is varied symmetrically in asymmetric Cl^-, it does not affect the reversal potential, whereas when pH is varied asymmetrically in symmetric Cl^-, it does. To explain these data, one assumes that the reversal potential for the exchanger was

$$V_{rev} = 1/(1+r)(E_{Cl} + rE_H),$$

where $r = m/n$ is the fixed H^+/Cl^- stoichiometry of countertransport, and E represents the Nernst potentials for Cl^- or H^+. This equation follows from the reaction

$$n\mathrm{Cl}_{ex} + m\mathrm{H}_{in} => n\mathrm{Cl}_{in} + m\mathrm{H}_{ex}.$$

This formula predicts the measured reversal potentials better than the Goldman equation (98, 99), which is written for a channel that is copermeable to Cl^- and H^+ (see **Supplemental Text**, Section C4, and **Supplemental Figure 4**). However, Hodgkin & Horowicz in 1959 (100) wrote an expression for V_{rev} that is identical to the above formula; the Hodgkin & Horowicz equation is, however, the reversal potential for a parallel-branch circuit used to model a membrane conductance (101). In the transporter interpretation, r is a fixed ratio independent of voltage and ion concentration; in the channel interpretation, r is a conductance ratio and in general depends on voltage and concentrations. Thus, the channel interpretation of the Accardi & Miller data requires the H^+/Cl^- ratio to be fixed. The possibility remains that single-file conduction in a non-voltage-gated channel renders r approximately constant, and then calling the H^+/Cl^- exchanger an ion channel seems semantic.

Accardi & Miller show that a point mutation removes the proton dependence of the substrate flux, and causes ClC-ec1 to become purely Cl^- selective. Curiously, however, this change creates a uniporter, not a channel. At a detailed molecular level, subtle molecular movements are likely involved. Thus, the difference between (*a*) a uniporter mediating the flux of one ion species and (*b*) an ion-selective channel is unclear. Examples like ClC-ec1 demonstrate that it may be exceedingly difficult to distinguish channels from transporters. For ion swapping at rates of 100 times per millisecond, as in ClC-ec1, there may not be much difference, aside from nomenclature, between the two mechanisms employed for ion/substrate flux.

A Proton Transporter Inside a K^+ Channel

The voltage sensitivity of K^+ channels arises from the presence of charged residues in the fourth TMD (S4) of each subunit in the tetrameric channel. When histidine (His) replaces all S4 charges except for the most extracellular arginine (Arg) (R362), S4 acts as a proton transporter (102). Replacing R362 with His creates a proton pore when the channel is hyperpolarized. When His replaces the fourth S4 charge (R371H), both proton transport and conduction occur. However, proton conduction is small and occurs only at depolarized potentials (**Figure 10**). The protein-bilayer interface is broken by aqueous crevices to allow proton access to R362H from both sides of the membrane (**Figure 10*e***). Operation of this proton pore revealed a unique structural feature of the voltage-sensing region, which has implications for naturally occurring voltage-dependent proton channels. In K^+ channels, S1–S4 functions as a self-contained voltage-sensing domain (VSD) that opens and closes the S5–S6 pore domain. A mammalian VSD protein, Hv1, which lacks a pore domain, exhibits voltage-sensitive, proton-selective ion channel activity (103). Mutagenesis of Hv1 identified three Arg residues in S4 that regulate gating. Hv1 is widely expressed in the immune system and belongs to a family of mammalian proton-conducting channels that straddles the border between channels and transporter proteins.

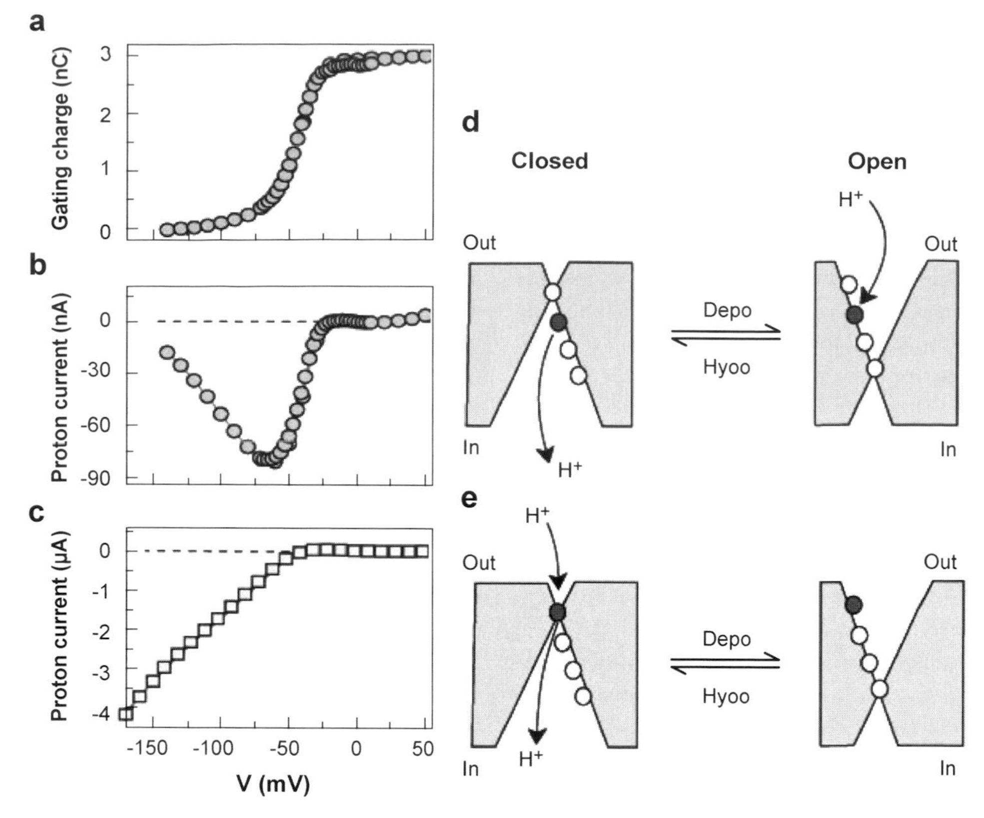

Figure 10

Proton transport and conduction: models of the conformational changes of the voltage sensor. (*a*) Voltage dependence of the gating-charge movement of the Shaker R365H K^+ channel. (*b*) Voltage dependence of proton-transport current of the R365H K^+ channel. (*c*) Voltage dependence of proton channel current of the R362H K^+ channel. (*d*) Conformational changes during proton transport. In a proton gradient (inward), the His at 365 (*filled circle*) transports one proton each time the voltage sensor goes from the open to the closed conformation. (*e*) Representation of the R362H K^+ channel proton pore. Internal and external protons have simultaneous access to the His (*filled circle*) only in the closed position, which results in a continuous proton current when the sensor is closed. The protein core is gray. Depo and hyper indicate a more positive and a more negative membrane potential, respectively. Figure from Reference 104; see also Reference 105.

Transporter Structure

Protein structure enters strongly into the Accardi-Miller argument because ClC-ec1 does not have an obvious channel motif, but then neither do homologous members of the CLC family, which are all thought to be authentic Cl^- channels. The structures of channels and transporters, as far as they are known, are highly varied. The structures of enzymatic transporters, such as the Ca^{2+} pump, are completely different from that of the proton-coupled lactose cotransporter (3, 106). In turn, *lac* permease is structurally distinct from the Na^+-coupled glutamate transporter (107) and the leucine transporter (96). Furthermore, individual studies presenting structures for these transport proteins each evoke radically different mechanisms to achieve compliance with the traditional alternating access model. *Lac* permease ostensibly undergoes enormous conformational changes between outward- and inward-facing forms, whereas glutamate transport is envisaged as a complex of coordinated gates, like trap doors at the base of an immobile, membrane-imbedded vestibule. Similarly, the structure of the leucine transporter gave rise to a hypothetical mechanism again involving a set of gates at the inner and outer faces, in line with previous suggestions for transporter mechanisms (51, 89). *Lac* permease is a monomer, whereas the glutamate transporter is a trimer. A trimeric structure is also indicated for the ammonia channel (108), a molecule previously thought to be a proton-coupled transporter.

IMPLICATIONS

Synaptic Transmission

Half a century ago, Alan Hodgkin discovered the ionic basis of excitability, and Bernard Katz discovered the fundamentals of synaptic transmission (61). The ability to generate action potentials, which in turn effect synaptic transmission, requires that Na^+ ions move across membranes down their electrochemical gradients. This drives action potentials to initiate neurotransmitter release and terminate the action of transmitters via Na^+-coupled reuptake mechanisms. Even the enzymatically degraded transmitter acetylcholine (ACh) relies on Na^+-coupled choline uptake for its resynthesis (109). Transporters located on presynaptic terminals, varicosities, and glial cells in the central nervous system rapidly and efficiently return neurotransmitters for repackaging and re-release. By this mechanism, transporters "time" the action of transmitters, localize transmitters to specific synapses, and recycle transmitters for reuse. Within the presynaptic terminal, proton-coupled transporters pack transmitters such as ACh, serotonin, and glutamate into synaptic vesicles.

Na^+-coupled glutamate transporters representing one gene family, serotonin transporters representing another family, and choline transporters in yet another family have been cloned and characterized using multidisciplinary techniques. In addition, crystal structures of some transporters are becoming known, and these will contribute immensely to our understanding of transport mechanisms (110). The main challenge remains to correlate molecular conformations with actual transport mechanisms. Despite new data, transmitter coupling to ion gradients looks roughly as it did 50 years ago (please see **Supplemental Figure 2**).

In the case of glutamate and monoamine transmitters, the evidence is overwhelming that large presynaptic currents accompany reuptake. Transporters thus return transmitter to the presynaptic cell and shape its physiology. It is difficult to accommodate these large currents, which in several cases are measured in actual neurons, with traditional models. For glutamate transporters, a syncretic model is proposed: Tight-coupling, fixed stoichiometry occurs in parallel with a Cl^- channel imbedded in the transporter but not involved in glutamate transport. In the monoamine transporters, this does not appear to be the case. Serotonin, DA, and NE are themselves conducted through the transporter in its channel mode. Furthermore, the stoichiometry is not fixed but varies with voltage and both ionic and transmitter concentration. Moreover, in the monoamine transporters, authentic single-channel activity is observed in connection with transport. The conclusion seems inevitable that the molecular events underlying the large transporter currents are bonafide channel events.

If monoamine transporters are channels, how can they concentrate substrates? One proposed explanation is that presynaptic transporters never drive monoamines uphill, because an efficient packaging mechanism exists in synaptic vesicles. This begs to be questioned, however, because transmitter concentration inside vesicles itself requires cotransport. One may counter that no channel properties of vesicle transporters have been observed and thus there is no contradiction. However, whether or not the presynaptic membrane concentrates monoamines intracellularly, SERTs, NETs, and DATs *can* do the job under controlled situations.

We propose the following scenario. After transmitter is released, transporters near the release sites sense a huge gradient. Transmitter and Na^+ gate the transporter to open, and initially Na^+ rushes in, accounting for the initial spike in controlled conditions. Because of its greater affinity, the less-concentrated transmitter enters the pore and initially flows downhill. Stoichiometry thus changes with time and gradually adjusts from hundreds to several ions per transmitter but is never fixed. Under conditions in which the transporter does concentrate transmitter, single-file electrodiffusion comes into play, which allows coupled transport in a channel mode (**Supplemental Text**, Sections A2 and C4, and **Supplemental Figure 3**).

A certain role of large inward transporter currents is membrane depolarization, which affects the excitability of the releasing neuron. Antidepressants and other drugs that block reuptake, whose actions are generally regarded as postsynaptic, may in addition have early and forceful presynaptic effects owing to blocked currents.

Cardiac Excitability

The case for channels in transporters is solid for neurotransmitter transporters, and in a few cases possible functions have been identified. This is less true for exchangers, such as the NCX, the anion exchanger (Band III), the Cl^-/H^+ exchanger, and the $Na^+/K^+/Cl^-$ transporters. Nevertheless, in these cases, too, the observed currents are larger than one might expect from a purely carrier perspective and are in fact clearly comparable to channels. Unfortunately, however, no transporter channels in these exchangers have been directly observed.

Metal Ion Metabolism

The uncoupled flux of molecules may introduce into a transport system some degree of elasticity and hence adaptability. In this case, the overall processes to which the transport events contribute and the cells that express these transport systems may adjust constantly in relation to the environment. The uncoupled flux of molecules involved in cotransport processes has been referred to as a molecular slip mechanism. However, in the examples above as well as in other studies, experiments have shown that such uncoupled molecular flux exhibits channel-like properties in terms of ion conductance. Despite the growing evidence for the existence of uncoupled flux mechanisms or channels in transporters, the biological function of these currents remains unclear. This type of mechanism may provide a kind of buffering action or built-in autoregulatory role distinct from other modes of regulation. It may help maintain stability and contribute to homeostasis in cells, despite sometimes abrupt changes in the concentration of substrates or of other influencing conditions in the environment. According to Nelson and colleagues (111), uncoupled flux mechanisms in transporters may serve a general protective role for the cell.

CONCLUSIONS

Regardless of whether uncoupled currents have a true biological purpose, the existence of channels in transporters has implications for the therapeutic modulations of these molecules. Channel mechanisms generally are better understood than the detailed workings of transporter function, a fact brought to light by the studies discussed in this review. The modulation of channel activities in transporter molecules may thus provide an alternative therapeutic route to alleviate transporter-associated disease conditions. We can perhaps look forward to a new genre of classification in which transporter molecules are no longer dichotomized into the rigid and classically opposing houses of channels and transporters but are treated with a more relaxed and unifying nomenclature that will reflect more closely and comprehensively the true functional properties of these extraordinary multifunctional molecules.

LITERATURE CITED

1. Posson DJ, Ge PH, Miller C, Bezanilla F, Selvin PR. 2005. Small vertical movement of a K^+ channel voltage sensor measured with luminescence energy transfer. *Nature* 436:848–51
2. Jiang YX, Lee A, Chen JY, Ruta V, Cadene M, et al. 2003. X-ray structure of a voltage-dependent K^+ channel. *Nature* 423:33–41
3. DeFelice LJ. 2004. Transporter structure and mechanism. *Trends Neurosci.* 27:352–59
4. Gu H, Wall SC, Rudnick G. 1994. Stable expression of biogenic amine transporters reveals differences in inhibitor sensitivity, kinetics, and ion dependence. *J. Biol. Chem.* 269:7124–30
5. Gu HH, Wall S, Rudnick G. 1996. Ion coupling stoichiometry for the norepinephrine transporter in membrane vesicles from stably transfected cells. *J. Biol. Chem.* 271:6911–16
6. Bruns D, Engert F, Lux HD. 1993. A fast activating presynaptic reuptake current during serotonergic transmission in identified neurons of Hirudo. *Neuron* 10:559–72
7. Läuger P. 1991. *Electronic Ion Pumps*. Sunderland, MA: Sinauer
8. DeFelice LJ. 1992. Ion pumps as stochastic machines. *Cell* 70:9–10
9. Moore RE, Scheuer PJ. 1971. Palytoxin: a new marine toxin from a coelenterate. *Science* 172:495–98
10. Ikeda M, Mitan K, Ito K. 1988. Palytoxin induces a nonselective cation channel in single ventricular cells of rat. *Naunyn-Schmiedeberg's Arch. Pharmacol.* 337:591–93
11. Habermann E. 1989. Palytoxin acts through Na^+,K^+-ATPase. *Toxicon* 27:1171–87
12. Artigas P, Gadsby DC. 2003. Na/K-pump ligands modulate gating of palytoxin-induced ion channels. *Proc. Natl. Acad. Sci. USA* 100:501–5
13. Artigas P, Gadsby DC. 2004. Large diameter of palytoxin-induced Na/K pump channels and modulation of palytoxin interaction by Na/K pump ligands. *J. Gen. Physiol.* 123:357–76
14. Hilgemann DW. 1994. Channel-like function of the Na,K pump probed at microsecond resolution in giant membrane patches. *Science* 263:1429–32
15. Lu CC, Kabakov A, Markin VS, Mager S, Frazier GA, Hilgemann DW. 1995. Membrane transport mechanisms probed by capacitance measurements with megahertz voltage clamp. *Proc. Natl. Acad. Sci. USA* 92:11220–24
16. Hilgemann DW. 1996. Unitary cardiac Na^+, Ca^{2+} exchange current magnitudes determined from channel-like noise and charge movements of ion transport. *Biophys. J.* 71:759–68
17. Kang TM, Hilgemann DW. 2004. Multiple transport modes of the cardiac Na^+/Ca^{2+} exchanger. *Nature* 427:544–48
18. Hilgemann DW. 1990. Regulation and deregulation of cardiac Na^+-Ca^{2+} exchange in giant excised sarcolemmal membrane patches. *Nature* 344:242–45
19. Hilgemann DW, Lu CC. 1998. Giant membrane patches: improvements and applications. *Methods Enzymol.* 293:267–80
20. Hilgemann DW, Nicoll DA, Philipson KD. 1991. Charge movement during Na^+ translocation by native and cloned cardiac Na^+/Ca^{2+} exchanger. *Nature* 352:715–18

21. Weber CR, Piacentino V III, Ginsburg KS, Houser SR, Bers DM. 2002. Na^+-Ca^{2+} exchange current and submembrane [Ca^{2+}] during the cardiac action potential. *Circ. Res.* 90:182–89
22. Shannon TR, Wang F, Puglisi J, Weber C, Bers DM. 2004. A mathematical treatment of integrated Ca dynamics within the ventricular myocyte. *Biophys. J.* 87:3351–71
23. Brasnjo G, Otis TS. 2001. Neuronal glutamate transporters control activation of postsynaptic metabotropic glutamate receptors and influence cerebellar long-term depression. *Neuron* 31:607–16
24. Dzubay JA, Otis TS. 2002. Climbing fiber activation of metabotropic glutamate receptors on cerebellar Purkinje neurons. *Neuron* 36:1159–67
25. Otis TS, Kavanaugh MP. 2000. Isolation of current components and partial reaction cycles in the glial glutamate transporter EAAT2. *J. Neurosci.* 20:2749–57
26. Otis TS, Kavanaugh MP, Jahr CE. 1997. Postsynaptic glutamate transport at the climbing fiber-Purkinje cell synapse. *Science* 277:1515–18
27. Larsson HP, Picaud SA, Werblin FS, Lecar H. 1996. Noise analysis of the glutamate-activated current in photoreceptors. *Biophys. J.* 70:733–42
28. Blakely RD, Robinson MB, Amara SG. 1988. Expression of neurotransmitter transport from rat brain mRNA in *Xenopus laevis* oocytes. *Proc. Natl. Acad. Sci. USA* 85:9846–50
29. Fairman WA, Vandenberg RJ, Arriza JL, Kavanaugh MP, Amara SG. 1995. An excitatory amino-acid transporter with properties of a ligand-gated chloride channel. *Nature* 375:599–603
30. Wadiche JI, Amara SG, Kavanaugh MP. 1995. Ion fluxes associated with excitatory amino acid transport. *Neuron* 15:721–28
31. Wadiche JI, Arriza JL, Amara SG, Kavanaugh MP. 1995. Kinetics of a human glutamate transporter. *Neuron* 14:1019–27
32. Wadiche JI, Kavanaugh MP. 1998. Macroscopic and microscopic properties of a cloned glutamate transporter/chloride channel. *J. Neurosci.* 18:7650–61
33. Arriza JL, Eliasof S, Kavanaugh MP, Amara SG. 1997. Excitatory amino acid transporter 5, a retinal glutamate transporter coupled to a chloride conductance. *Proc. Natl. Acad. Sci. USA* 94:4155–60
34. Rothstein JD, Kuncl RW. 1995. Neuroprotective strategies in a model of chronic glutamate-mediated motor neuron toxicity. *J. Neurochem.* 65:643–51
35. Bergles DE, Tzingounis AV, Jahr CE. 2002. Comparison of coupled and uncoupled currents during glutamate uptake by GLT-1 transporters. *J. Neurosci.* 22:10153–62
36. Bergles DE, Dzubay JA, Jahr CE. 1997. Glutamate transporter currents in Bergmann glial cells follow the time course of extrasynaptic glutamate. *Proc. Natl. Acad. Sci. USA* 94:14821–25
37. Wong AY, Fay AM, Bowie D. 2006. External ions are coactivators of kainate receptors. *J. Neurosci.* 26:5750–55
38. Cunningham P, Afzal-Ahmed I, Naftalin RJ. 2006. Docking studies show that D-glucose and quercetin slide through the transporter GLUT1. *J. Biol. Chem.* 281:5797–803
39. Hediger MA, Ikeda T, Coady M, Gundersen CB, Wright EM. 1987. Expression of size-selected mRNA encoding the intestinal Na/glucose cotransporter in *Xenopus laevis* oocytes. *Proc. Natl. Acad. Sci. USA* 84:2634–37
40. Loo DD, Hirayama BA, Gallardo EM, Lam JT, Turk E, Wright EM. 1998. Conformational changes couple Na^+ and glucose transport. *Proc. Natl. Acad. Sci. USA* 95:7789–94
41. Quick M, Loo DD, Wright EM. 2001. Neutralization of a conserved amino acid residue in the human Na^+/glucose transporter (hSGLT1) generates a glucose-gated H^+ channel. *J. Biol. Chem.* 276:1728–34

42. Quick M, Tomasevic J, Wright EM. 2003. Functional asymmetry of the human Na^+/glucose transporter (hSGLT1) in bacterial membrane vesicles. *Biochemistry* 42:9147–52
43. Apparsundaram S, Ferguson SM, George AL Jr, Blakely RD. 2000. Molecular cloning of a human, hemicholinium-3-sensitive choline transporter. *Biochem. Biophys. Res. Commun.* 276:862–67
44. Iwamoto H, Blakely RD, DeFelice LJ. 2006. Na^+, Cl^-, and pH dependence of the human choline transporter in *Xenopus* oocytes: the proton inactivation hypothesis of hCHT in synaptic vesicles. *J. Neurosci.* 26:9851–59
45. Axelrod J, Kopin IJ. 1969. The uptake, storage, release and metabolism of noradrenaline in sympathetic nerves. *Prog. Brain Res.* 31:21–32
46. Cammack JN, Rakhilin SV, Schwartz EA. 1994. A GABA transporter operates asymmetrically and with variable stoichiometry. *Neuron* 13:949–60
47. Cammack JN, Schwartz EA. 1996. Channel behavior in a γ-aminobutyrate transporter. *Proc. Natl. Acad. Sci. USA* 93:723–27
48. Galli A, Blakely RD, DeFelice LJ. 1996. Norepinephrine transporters have channel modes of conduction. *Proc. Natl. Acad. Sci. USA* 93:8671–76
49. Lin F, Lester HA, Mager S. 1996. Single-channel currents produced by the serotonin transporter and analysis of a mutation affecting ion permeation. *Biophys. J.* 71:3126–35
50. Carvelli L, McDonald PW, Blakely RD, DeFelice LJ. 2004. Dopamine transporters depolarize neurons by a channel mechanism. *Proc. Natl. Acad. Sci. USA* 101:16046–51
51. Lester HA, Mager S, Quick MW, Corey JL. 1994. Permeation properties of neurotransmitter transporters. *Annu. Rev. Pharmacol. Toxicol.* 34:219–49
52. Lester HA, Cao Y, Mager S. 1996. Listening to neurotransmitter transporters. *Neuron* 17:807–10
53. DeFelice LJ, Blakely RD. 1996. Pore models for transporters? *Biophys. J.* 70:579–80
54. Sonders MS, Amara SG. 1996. Channels in transporters. *Curr. Opin. Neurobiol.* 6:294–302
55. Su A, Mager S, Mayo SL, Lester HA. 1996. A multi-substrate single-file model for ion-coupled transporters. *Biophys. J.* 70:762–77
56. DeFelice LJ, Adams SV, Ypey DL. 2001. Single-file diffusion and neurotransmitter transporters: Hodgkin and Keynes model revisited. *Biosystems* 62:57–66
57. Adams SV, DeFelice LJ. 2002. Flux coupling in the human serotonin transporter. *Biophys. J.* 83:3268–82
58. Adams SV, DeFelice LJ. 2003. Ionic currents in the human serotonin transporter reveal inconsistencies in the alternating access hypothesis. *Biophys. J.* 85:1548–59
59. Hodgkin AL, Keynes RD. 1955. The potassium permeability of a giant nerve fibre. *J. Physiol.* 128:61–88
60. Doyle DA, Morais Cabral J, Pfuetzner RA, Kuo A, Gulbis JM, et al. 1998. The structure of the potassium channel: Molecular basis of K^+ conduction and selectivity. *Science* 280:69–77
61. Hodgkin A. 1994. *Chance and Design: Reminiscences of Science in Peace and War*. Cambridge, UK: Cambridge Univ. Press
62. Blakely RD, Berson HE, Fremeau RT Jr, Caron MG, Peek MM, et al. 1991. Cloning and expression of a functional serotonin transporter from rat brain. *Nature* 354:66–70
63. Ramamoorthy S, Cool DR, Mahesh VB, Leibach FH, Melikian HE, et al. 1993. Regulation of the human serotonin transporter. Cholera toxin-induced stimulation of serotonin uptake in human placental choriocarcinoma cells is accompanied by increased serotonin transporter mRNA levels and serotonin transporter-specific ligand binding. *J. Biol. Chem.* 268:21626–31

64. Blakely RD, Ramamoorthy S, Schroeter S, Qian Y, Apparsundaram S, et al. 1998. Regulated phosphorylation and trafficking of antidepressant-sensitive serotonin transporter proteins. *Biol. Psychiatry* 44:169–78
65. Rudnick G, Wall SC. 1992. The molecular mechanism of "ecstasy" [3,4-methylenedioxymethamphetamine (MDMA)]: Serotonin transporters are targets for MDMA-induced serotonin release. *Proc. Natl. Acad. Sci. USA* 89:1817–21
66. Keyes SR, Rudnick G. 1982. Coupling of transmembrane proton gradients to platelet serotonin transport. *J. Biol. Chem.* 257:1172–76
67. Galli A, Petersen CI, deBlaquiere M, Blakely RD, DeFelice LJ. 1997. *Drosophila* serotonin transporters have voltage-dependent uptake coupled to a serotonin-gated ion channel. *J. Neurosci.* 17:3401–11
68. Petersen CI, DeFelice LJ. 1999. Ionic interactions in the *Drosophila* serotonin transporter identify it as a serotonin channel. *Nat. Neurosci.* 2:605–10
69. Mager S, Min C, Henry DJ, Chavkin C, Hoffman BJ, et al. 1994. Conducting states of a mammalian serotonin transporter. *Neuron* 12:845–59
70. Quick MW. 2003. Regulating the conducting states of a mammalian serotonin transporter. *Neuron* 40:537–49
71. Galli A, Blakely RD, DeFelice LJ. 1998. Patch-clamp and amperometric recordings from norepinephrine transporters: channel activity and voltage-dependent uptake. *Proc. Natl. Acad. Sci.USA* 95:13260–65
72. Galli A, DeFelice LJ, Duke BJ, Moore KR, Blakely RD. 1995. Sodium-dependent norepinephrine-induced currents in norepinephrine-transporter-transfected HEK-293 cells blocked by cocaine and antidepressants. *J. Exp. Biol.* 198(Pt. 10):2197–212
73. Kahlig KM, Javitch JA, Galli A. 2003. Amphetamine regulation of dopamine transport: Combined measurements of transporter currents and transporter imaging support the endocytosis of an active carrier. *J. Biol. Chem.* 279:8966–75
74. Khoshbouei H, Sen N, Guptaroy B, Johnson L, Lund D, et al. 2004. N-terminal phosphorylation of the dopamine transporter is required for amphetamine-induced efflux. *PLoS Biol.* 2:E78
75. Khoshbouei H, Wang H, Lechleiter JD, Javitch JA, Galli A. 2003. Amphetamine-induced dopamine efflux. A voltage-sensitive and intracellular Na^+-dependent mechanism. *J. Biol. Chem.* 278:12070–77
76. Kahlig KM, Binda F, Khoshbouei H, Blakely RD, McMahon DG, et al. 2005. Amphetamine induces dopamine efflux through a dopamine transporter channel. *Proc. Natl. Acad. Sci. USA* 102:3495–500
77. Ingram SL, Prasad BM, Amara SG. 2002. Dopamine transporter-mediated conductances increase excitability of midbrain dopamine neurons. *Nat. Neurosci.* 5:971–78
78. Gunshin H, Mackenzie B, Berger UV, Gunshin Y, Romero MF, et al. 1997. Cloning and characterization of a mammalian proton-coupled metal-ion transporter. *Nature* 388:482–88
79. Chen XZ, Peng JB, Cohen A, Nelson H, Nelson N, Hediger MA. 1999. Yeast SMF1 mediates H^+-coupled iron uptake with concomitant uncoupled cation currents. *J. Biol. Chem.* 274:35089–94
80. Peng JB, Chen XZ, Berger UV, Vassilev PM, Tsukaguchi H, et al. 1999. Molecular cloning and characterization of a channel-like transporter mediating intestinal calcium absorption. *J. Biol. Chem.* 274:22739–46
81. Xu H, Jin J, DeFelice LJ, Andrews NC, Clapham DE. 2004. A spontaneous, recurrent mutation in divalent metal transporter-1 exposes a calcium entry pathway. *PLoS Biol.* 2:E50

82. Kaplan J, Jordan I, Sturrock A. 1991. Regulation of the transferrin-independent iron transport system in cultured cells. *J. Biol. Chem.* 266:2997–3004
83. Goswami T, Bhattacharjee A, Babal P, Searle S, Moore E, et al. 2001. Natural-resistance-associated macrophage protein 1 is an H^+/bivalent cation antiporter. *Biochem. J.* 354:511–19
84. Wang T, Weinman SA. 2004. Involvement of chloride channels in hepatic copper metabolism: ClC-4 promotes copper incorporation into ceruloplasmin. *Gastroenterology* 126:1157–66
85. Aller SG, Unger VM. 2006. Projection structure of the human copper transporter CTR1 at 6-Å resolution reveals a compact trimer with a novel channel-like architecture. *Proc. Natl. Acad. Sci. USA* 103:3627–32
86. Accardi A, Kolmakova-Partensky L, Williams C, Miller C. 2004. Ionic currents mediated by a prokaryotic homologue of CLC Cl^- channels. *J. Gen. Physiol.* 123:109–19
87. Accardi A, Miller C. 2004. Secondary active transport mediated by a prokaryotic homologue of ClC Cl^- channels. *Nature* 427:803–7
88. Iyer R, Iverson TM, Accardi A, Miller C. 2002. A biological role for prokaryotic ClC chloride channels. *Nature* 419:715–18
89. Gadsby DC. 2004. Ion transport: Spot the difference. *Nature* 427:795–97
90. Lindley D. 2001. *Boltzmann's Atom: The Great Debate that Launched a Revolution in Physics*. New York: Free Press
91. Dutzler R, Campbell EB, Cadene M, Chait BT, MacKinnon R. 2002. X-ray structure of a ClC chloride channel at 3.0 Å reveals the molecular basis of anion selectivity. *Nature* 415:287–94
92. Dutzler R, Campbell EB, MacKinnon R. 2003. Gating the selectivity filter in ClC chloride channels. *Science* 300:108–12
93. Widdas WF. 1952. Inability of diffusion to account for placental glucose transfer in the sheep and consideration of the kinetics of a possible carrier transfer. *J. Physiol.* 118:23–39
94. Jardetzky O. 1966. Simple allosteric model for membrane pumps. *Nature* 211:969–70
95. DeFelice LJ. 2004. Going against the flow. *Nature* 432:279
96. Yamashita A, Singh SK, Kawate T, Jin Y, Gouaux E. 2005. Crystal structure of a bacterial homologue of Na^+Cl^- dependent neurotransmitter transporters. *Nature* 437:215–23
97. Miller C. 2006. ClC chloride channels viewed through a transporter lens. *Nature* 440:484–89
98. Hille B. 2001. *Ion Channels of Excitable Membranes*. Sunderland, MA: Sinauer. 3rd ed.
99. DeFelice LJ. 1981. *Introduction to Membrane Noise*. New York: Plenum
100. Hodgkin AL, Horowicz P. 1959. The influence of potassium and chloride ions on the membrane potential of single muscle fibres. *J. Physiol.* 148:127–60
101. Hodgkin AL, Horowicz P. 1959. Movements of Na and K in single muscle fibres. *J. Physiol.* 145:405–32
102. Starace DM, Bezanilla F. 2004. A proton pore in a potassium channel voltage sensor reveals a focused electric field. *Nature* 427:548–52
103. Ramsey IS, Moran MM, Chong JA, Clapham DE. 2006. A voltage-gated proton-selective channel lacking the pore domain. *Nature* 440:1213–16
104. Starace DM, Bezanilla F. 2001. Histidine scanning mutagenesis of basic residues of the S4 segment of the shaker K^+ channel. *J. Gen. Physiol.* 117:469–90
105. Starace DM, Bezanilla F. 2004. A proton pore in a potassium channel voltage sensor reveals a focused electric field. *Nature* 427:548–53

106. Abramson J, Smirnova I, Kasho V, Verner G, Iwata S, Kaback HR. 2003. The lactose permease of *Escherichia coli*: overall structure, the sugar-binding site and the alternating access model for transport. *FEBS Lett.* 555:96–101
107. Yernool D, Boudker O, Jin Y, Gouaux E. 2004. Structure of a glutamate transporter homologue from Pyrococcus horikoshii. *Nature* 431:811–18
108. Khademi S, O'Connell J 3rd, Remis J, Robles-Colmenares Y, Miercke LJ, Stroud RM. 2004. Mechanism of ammonia transport by Amt/MEP/Rh: structure of AmtB at 1.35 Å. *Science* 305:1587–94
109. Ferguson SM, Blakely RD. 2004. The choline transporter resurfaces: new roles for synaptic vesicles? *Mol. Interv.* 4:22–37
110. Henry LK, DeFelice LJ, Blakely RD. 2006. Getting the message across: A recent transporter structure shows the way. *Neuron* 49:791–96
111. Nelson N, Sacher A, Nelson H. 2002. The significance of molecular slips in transport systems. *Nat. Rev. Mol. Cell Biol.* 3:876–81

Hypoxia Tolerance in Mammals and Birds: From the Wilderness to the Clinic

Jan-Marino Ramirez,[1] Lars P. Folkow,[2] and Arnoldus S. Blix[2]

[1]Department of Organismal Biology & Anatomy, University of Chicago, Chicago, Illinois 60637; email: jramire@uchicago.edu

[2]Department of Arctic Biology, University of Tromso, Tromso, N-9037 Norway; email: larsf@fagmed.uit.no, asblix@fagmed.uit.no

Annu. Rev. Physiol. 2007. 69:113–43

First published online as a Review in Advance on October 12, 2006

The *Annual Review of Physiology* is online at http://physiol.annualreviews.org

This article's doi:
10.1146/annurev.physiol.69.031905.163111

0066-4278/07/0315-0113$20.00

Key Words

diving, hibernation, high altitude, sleep apnea, ischemia

Abstract

All mammals and birds must develop effective strategies to cope with reduced oxygen availability. These animals achieve tolerance to acute and chronic hypoxia by (*a*) reductions in metabolism, (*b*) the prevention of cellular injury, and (*c*) the maintenance of functional integrity. Failure to meet any one of these tasks is detrimental. Birds and mammals accomplish this triple task through a highly coordinated, systems-level reconfiguration involving the partial shutdown of some but not all organs. This reconfiguration is achieved through a similarly complex reconfiguration at the cellular and molecular levels. Reconfiguration at these various levels depends on numerous factors that include the environment, the degree of hypoxic stress, and developmental, behavioral, and ecological conditions. Although common molecular strategies exist, the cellular and molecular changes in any given cell are very diverse. Some cells remain metabolically active, whereas others shut down or rely on anaerobic metabolism. This cellular shutdown is temporarily regulated, and during hypoxic exposure, active cellular networks must continue to control vital functions. The challenge for future research is to explore the cellular mechanisms and conditions that transform an organ or a cellular network into a hypometabolic state, without loss of functional integrity. Much can be learned in this respect from nature: Diving, burrowing, and hibernating animals living in diverse environments are masters of adaptation and can teach us how to deal with hypoxia, an issue of great clinical significance.

INTRODUCTION

Oxygen is among the elements whose reduction provides the largest free energy release per electron transfer. Not surprisingly, most multicellular organisms reduce O_2 (during aerobic metabolism) to meet metabolic needs (1). The capacity for aerobiosis has expanded dramatically during evolution from ecto- to endothermy. Resting and maximal rates of O_2 consumption are nearly an order of magnitude greater in endotherms than in ectotherms of equal size and body temperature (2). As a result, endotherms have much greater stamina and range of aerobically supported work output. This increased efficiency enables endothermic vertebrates to hunt over long distances, whereas reptiles fatigue at velocities of only 0.5–1 $km \cdot h^{-1}$. Most reptiles are limited to burst-like hunting behavior that is largely supported through anaerobic metabolism (3). However, the enormous metabolic costs to maintain endothermy turned mammals and birds into slaves of aerobic metabolism. As a consequence, coping with hypoxia became a much more daunting task for endothermic than for ectothermic animals. Whereas some turtles and fish can survive anoxic conditions for several weeks, as discussed in the accompanying review in this volume (4), mammals maximally tolerate 2 h of apnea (cessation of breathing), and birds, even less. In humans, severe hypoxia leads to the loss of consciousness within seconds (5).

As this review discusses, the best strategy to tolerate hypoxia involves the orchestration of a complex hypometabolic state that retains functional integrity and also prevents many of the detrimental effects associated with energy depletion. We focus on the organismal and cellular adaptations of hypoxia-tolerant birds and mammals and arrive at conclusions that have important biological as well as clinical implications. Every year, several hundred thousands of humans are irreversibly affected by hypoxia due to, e.g., cerebrovascular accidents, myocardial ischemia, sleep apneas, chronic bronchitis, emphysema, recurrent apneas in premature babies, and sudden infant death syndrome.

BIRDS AND MAMMALS IN HYPOXIC ENVIRONMENTS

Several species of birds and mammals have developed remarkable hypoxia tolerance to conquer environments that expose them, either acutely or chronically, to severe limitations in O_2 supply. Diving birds and mammals, as well as some hibernating mammals, routinely experience acute hypoxia, whereas birds and mammals that live at high altitude or in burrows are exposed to chronic hypoxia. An existence at 5000 m altitude or in subterranean burrows holding O_2 partial pressure (P_{O2}) of <50 mm Hg would render any nonacclimatized mammal hypoxic but does not necessarily cause problems in adapted animals. For example, arterial O_2 saturation (S_{aO2}) in the llama (*Lama glama*) is $>92\%$ at an arterial O_2 tension (P_{aO2}) of only ~50 mm Hg (6), a P_{aO2} that would represent a hypoxic challenge to nonacclimatized sea-level residents.

Organismal Versus Environmental Hypoxia

Hypoxia is a state of deficiency in O_2 availability to, or utilization by, body tissues. The states of hypoxia can be further specified. Asphyxia is characterized by restricted gas exchange with the environment, resulting in hypoxia and hypercapnia. Environmental hypoxia concerns inadequate P_{O2}, whereas stagnant hypoxia involves inadequate local circulation. Ischemia occurs when a tissue receives insufficient blood to meet its metabolic demands, resulting in hypoxia and hypoglycemia. Hypoxia is often a state of the organism, not of the environment. Hypoxic conditions may occur during diving, in burrows, at high altitude, in hibernation, and in clinical conditions such as stroke, emphysema, and asthma. Responses to hypoxia can be acute (lasting seconds to minutes) or chronic (lasting hours to days).

The distinction between organismal and environmental hypoxia is not always straightforward. Organismal hypoxia may occur in a normoxic environment, for example, under clinical conditions or in hibernation. Moreover, adapted organisms living in high-altitude and subterrain burrows that are low in O_2 can create an internal environment that is not hypoxic.

Specific Versus General Adaptations to Hypoxia

Hypoxia-tolerant animals have significantly extended their capabilities to cope with acute and chronic hypoxia through adaptations that are largely absent in other animals. For example, most hypoxia-tolerant animals exhibit increased blood O_2 carrying capacity, which is caused by increased red blood cell (RBC) mass and blood volume. The high body mass of most diving mammals can also be considered an adaptation because an increased body mass is associated with a decreased metabolic rate. The mass-specific basal metabolic rate of a whale is almost ten times lower than a human's specific metabolic rate (7). However, all mammals must obey the same physiological rules. Thus, the distinction between adaptations specific to hypoxia-tolerant animals and those in all mammals is often blurred. Indeed, many physiological adaptations—such as differential vasoconstriction, bradycardia, and hypometabolism of hypoperfused tissue—are highly conserved in all mammals.

The overarching theme of this review is that tolerance to acute and chronic hypoxia is achieved through the complex coordination, reorganization, and partial shutdown of different organ systems, which is accomplished by an equally complex reconfiguration at the cellular level. This massive reorganization is driven by the need to reduce O_2 requirements, prevent cellular injury, and maintain functional integrity. The specifics of this reorganization depend on numerous factors that include the hypoxic environment and the degree of hypoxic stress as well as many species-, development-, and behavior-specific aspects.

Oxygen Sensing

To respond adequately to changes in oxygenation status, animals require internal O_2-sensing mechanisms. The efferent response pattern produced by chemoreceptors is characteristic: The primary effects on circulation are bradycardia, a diminution of cardiac output, and peripheral vasoconstriction. However, this primary response is greatly modified by the reflex hyperpnea simultaneously induced in the spontaneously breathing animal. This secondary response to chemoreceptor stimulation, when breathing is allowed, is characterized by increased cardiac output and heart rate and by changes in peripheral flow pattern. These changes depend on the balance between constrictor fiber distribution and discharge and the local dilator effect of hypoxia itself on vascular muscular tone, as modified by blood-borne catecholamines. Cerebral and coronary blood flow increases, whereas gastrointestinal and renal blood flows usually decrease (9). Below we further discuss the cellular mechanisms that underlie the process of O_2 sensing.

ACUTE HYPOXIA: DIVING ANIMALS

Most terrestrial mammals, such as humans, have relatively little tolerance to acute changes in O_2 levels. Humans suffocate if breathing is stopped for more than 3–4 min (10). Within 15–20 s of brain anoxia, humans lose consciousness and purposeful behavior, electroencephalographic (EEG) activity, and evoked EEG potentials (5, 11).

By contrast, sperm whales (*Physeter catodon*) (12) and southern elephant seals (*Mirounga leonina*) (13), which normally dive to 300–600 m, may dive to more than 1000 m and occasionally remain submerged for 2 h. Hooded seals (*Cystophora cristata*) also normally dive to 300–600 m, with dive durations

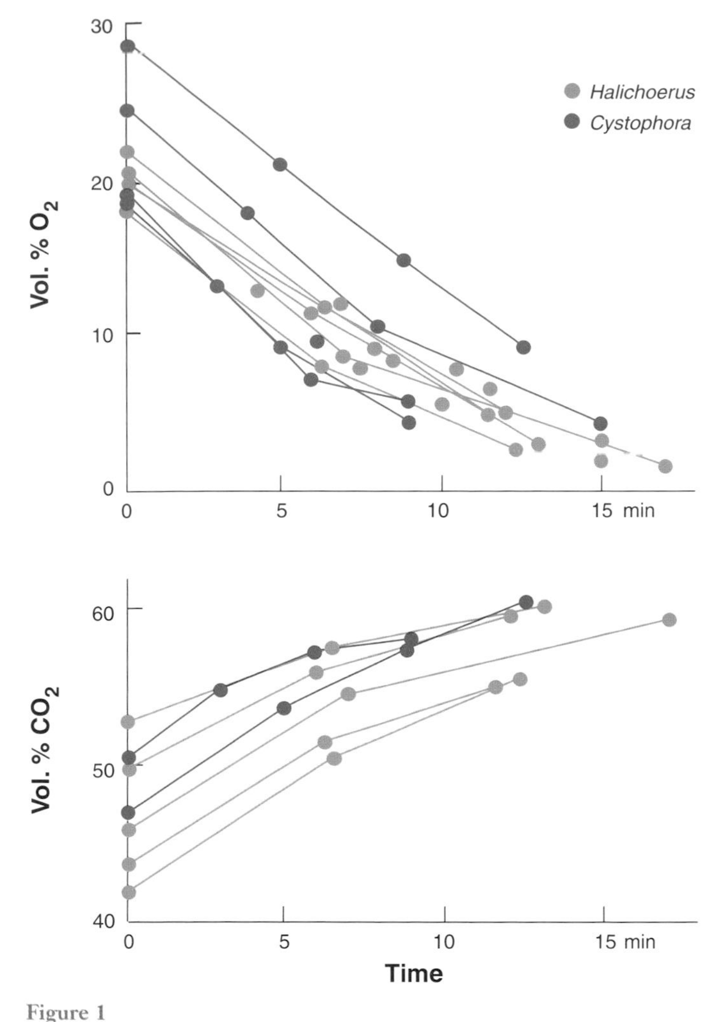

Figure 1

Arterial blood during diving. The basic problem facing air-breathing animals when submerged is a constantly decreasing arterial content of O_2 (*upper graph*) and constantly increasing arterial content of CO_2 (*lower graph*). Data are from gray seals (*Halichoerus grypus*) and hooded seals (*Cystophora cristata*). From Scholander (16), with permission from the Norwegian Academy of Science and Letters.

of 5–25 min, but some individuals specialize in repetitive deep diving to more than 1000 m, with durations of up to 1 h (14). Even birds like the emperor penguin (*Aptenodytes forsteri*) dive to depths of 500 m, with durations of more than 15 min (15).

When mammals dive, ventilation must stop immediately to avoid drowning. Because their tissues and cells continue to metabolize, arterial O_2 content continues to decrease and arterial CO_2 continues to increase (16; **Figure 1**). How can air-breathing animals like whales, seals, penguins, and ducks survive such conditions, which would kill any human?

IMPROVED OXYGEN STORAGE AND SUPPLY

All diving mammals and birds have an increased O_2 storing capacity. O_2 is stored in blood, muscle, and lungs.

Hemoglobin in the Blood

Hematocrit (Hct) and blood hemoglobin concentration ($[Hb]_{blood}$) of deep-diving phocid seals may exceed 60% and 25 $g \cdot dL^{-1}$, respectively (16, 17). Hct and $[Hb]_{blood}$ levels in shallow-diving otariid seals and small cetaceans appear to be somewhat lower (17, 18). Hct and $[Hb]_{blood}$ values for ducks and penguins range between 45–53% and 11–20 $g \cdot dL^{-1}$, respectively (19, 20). The RBC mass of divers correlates positively with diving capacity (21), but a high Hct is not maintained without increased blood viscosity. Seals partially overcome this problem by sequestering RBCs in the spleen when the animals are not diving (22). These RBCs are then released into circulation when diving commences. Cabanac et al. (23) modeled the structure and blood-storing function of the spleen in the deep-diving hooded seal, a species with a particularly large spleen. The polycythemia of breath-hold divers is accompanied by substantially elevated plasma volume, resulting in a blood volume of up to 100–200 $ml \cdot kg^{-1}$ (17, 20). Thus, the blood O_2 stores of the divers are three to four times greater than the average for terrestrial mammals (24).

Hb-O_2 affinities of diving birds and mammals are not particularly high (24, 25), which makes sense because diving animals do not experience low O_2 tensions while breathing. Deep-diving animals may instead benefit from efficient Hb O_2 unloading during asphyxic hypoxia, owing to a low Hb temperature coefficient and, in particular, a high Bohr coefficient (25), which facilitate O_2 unloading as acidosis develops during diving.

Myoglobin in Muscles

The monomeric myoglobin (Mb) molecule has an extraordinary high affinity for O_2 ($P_{50} = 2.5$ torr) and therefore serves as an O_2 store (26). But Mb also facilitates O_2 transport from the cell membrane to the mitochondria (26) and provides antioxidant defenses (27). Diving mammals and birds have very high Mb concentrations ([Mb]) in their skeletal muscles (50–80 $mg \cdot g^{-1}$; 17, 24). The highest [Mb] yet recorded (94 $mg \cdot g^{-1}$) is in the swimming muscles of hooded seals (28). Shallow-diving otariid seals, small cetaceans, sea otters, sirenians, and diving rodents have lower [Mb] than do deep-diving animals (17, 24, 29) but higher [Mb] than do most terrestrial species (24). This is also the case in diving birds, which have [Mb] of 4–64 $mg \cdot g^{-1}$ (19, 20, 30). Mb is also found in high concentrations (28 $mg \cdot g^{-1}$) in the hearts of diving animals (31).

Oxygen in the Lung

Ventilation in diving mammals is characterized by exchange of high tidal volumes (32, 33) during brief surfacing periods, enabling the animals to dispose excess CO_2 and reload O_2 rapidly. Lung volumes of deep-diving mammals conform to allometric relationships for terrestrial species (17, 34, 35). However, such animals normally exhale before diving (16, 33), reflecting both lack of reliance on lung O_2 stores and the need to avoid decompression sickness and reduce buoyancy during diving. By contrast, shallow-diving species, such as sea otters and some otariids, have relatively larger lung volumes (17, 34) and thus rely heavily on lung O_2 stores during diving (17).

Tissue Capillary Density

The brain tissue of the harbor seal (*Phoca vitulina*) has higher capillary densities than does that of terrestrial mammals (36), implying increased O_2 conductance to neural tissue in this species. Muscle capillary density in harbor seals, on the other hand, is a factor of 1.5–3 less than that of dogs, possibly reflecting an increased reliance on intramuscular (Mb-based) O_2 stores (37).

OXYGEN ECONOMY (HYPOMETABOLISM)

Increased O_2 storage alone is not sufficient to allow the long-duration dives of whales, seals, and penguins. This adaptation must be supported by additional strategies to sustain acute hypoxia. Perhaps the most important defense strategy to cope with hypoxia is hypometabolism, a strategy first conceptualized by Scholander (16) and later extended by Hochachka and others (21). The concept is that an animal assumes a hypometabolic state to reduce its O_2 requirements and to avoid the detrimental effects of hypoxia, including necrosis and apoptosis. This hypometabolic state involves probably all organ systems and thus requires very tight coordination of various concurrently occurring processes.

A Complex Decision-Making Process

The transition into a hypometabolic state can be abrupt (38, 39). As mentioned above, metabolism continues during a dive, and thus blood O_2 levels decrease continuously (**Figure 1**). One would therefore assume that most functions are affected gradually, but this is not necessarily the case. Diving mammals anticipate many of these changes in a feedforward manner. Whereas some organs are rendered hypoxic early, others are maintained at near-normal O_2 levels for a prolonged time. To accomplish this, animals must coordinate different organ systems and different metabolic processes occurring concurrently. This requires a complex multilevel decision-making process. Throughout the hypoxic response, the organism must continuously decide whether a given function must be maintained, altered, or halted. The decision-making process must be regulated dynamically and temporarily, and a hierarchy of events must govern this process. Studies of whales and seals have shown that this process is not purely involuntary but includes conscious decisions involving higher-brain areas.

Forced (Experimental) and Long-Duration Natural Dives

Experiments in which a diving mammal, such as the seal, was experimentally forced under water provided the first insights into diving physiology (38). Investigators have known for more than a century that habitually diving animals respond to forced submergence with a profound and, in the case of seals, abrupt bradycardia (38). Scholander (16), stimulated by the discoveries of Irving (40), first put this dramatic event into perspective. In an elegant series of experiments, Scholander demonstrated that bradycardia is developed in concert with a widespread selective peripheral arterial constriction that ensures that the much-reduced cardiac output is almost exclusively distributed to the most O_2-sensitive tissues (16) (**Figure 2*a***).

Of all the tissues the brain is unquestionably the most favored for the supply of blood during diving. But even in the brain there is an initial 50% reduction in blood flow, which by the end of the dive gradually increases to well above predive values for most parts of the brain. The perfusion of different parts of the brain is very different and variable over time (41) (**Figure 2*b***).

Myocardial blood flow decreases almost instantaneously in experimental and long duration natural dives to an average of only 10% of predive values in seals (**Figure 2*c***). The much-reduced coronary blood flow is oscillating and frequently stops entirely for periods as long as 45 s (42). There is increasing myocardial lactate and H^+ production throughout the dive, and after surfacing there is an immediate return to myocardial uptake of lactate (43). The seal heart manages this in part because of its large stores of glycogen (44) and its hypoxia-tolerant enzyme systems (21, 45). Seal hearts are significantly more hypoxia-tolerant than those of terrestrial mammals (46, 47). Moreover, even during prolonged dives there is no evidence of ischemic dilation of the left ventricle or S-T segment elevation in the electrocardiogram (43). Thus, the seal appears to maintain myocardial function during dives with a reduction of coronary blood flow comparable to that observed in the infarcted dog myocardium (48). Further studies of myocardial function in diving seals may have relevance for therapeutic approaches to reduce myocardial ischemic injury, particularly with regard to preconditioning effects in humans.

Figure 2

(*a*) Angiograms of peripheral (abdominal) arteries of a harbor seal (*Phoca vitulina*). (*1*) During breathing in air at surface position, well-filled arteries of flanks (*upper arrow*) and hind flippers (*lower arrow*) are seen. (*2*) During experimental diving, the same arteries in the same animal are profoundly constricted and consequently poorly filled with contrast medium. bl, urinary bladder. From Reference 156. (*b*) Tissue blood flow of *P. vitulina* at four different brain regions was determined by the use of radioactively labeled microspheres before diving and at 2, 5, and 10 min of experimental submersion as well as after 40 s of recovery after a 10-min dive. Tissue blood flow appears in percent of predive (control) values. The number of samples is in parentheses above the columns. From Reference 41. (*c*) Left circumflex coronary artery (LCCA) blood flow and blood velocity in a 3-min experimental dive of a harbor seal *P. vitulina*. Flow abruptly diminishes immediately after the beginning of the dive and is transiently restored at 30- to 45-s intervals. The response suggests rhythmic, neurogenic, spasm-like coronary vasoconstrictions modulated by myocardial metabolic demand. From Reference 42.

Kidney perfusion is completely shut down in experimental (49) and prolonged voluntary dives in seals (50, 53). Moreover, Halasz et al. (51) demonstrated that seal kidneys can endure 1 h of warm (32–34°C) ischemia and still show prompt recovery of urine production upon reperfusion.

Skeletal muscle blood flow is apparently also turned off during prolonged and experimental dives (16). However, this does not

a **Abdominal arteries**

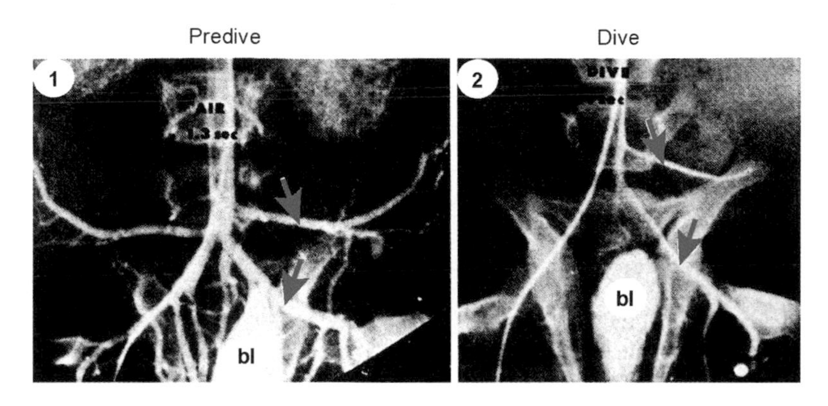

b **Brain**

Pons/medulla
Thalamus/hypothalamus
Cerebellum
Neocortex
Tissue blood flow
300
200
100
2 3
2 3
3
3
6
3
3
4
9
Predive (control) level
2-min dive
5-min dive
10-min dive
40-s recovery

c **Circumflex coronary artery**

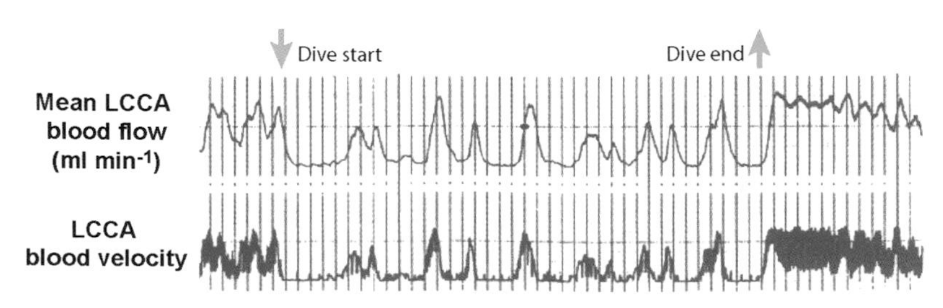

indicate that all skeletal muscles are metabolizing anaerobically, as was first assumed. Most likely, the large resting muscle mass, such as that involved in respiration, may rely on their endogenous stores of glycogen and oxy-myoglobin or simply shut down. The muscles actively engaged in swimming, however, must ultimately endure anaerobic metabolism after exhaustion of the endogenous oxy-myoglobin stores. The muscles that are rich in hypoxia-tolerant enzyme systems (21, 45) are therefore most likely the main sources of the postdive rise in arterial lactate (16). This lactate accumulation is the price paid for extended diving capacity and must be eliminated by the liver before the animal can dive again. Thus, Kooyman and associates (33) showed that in the Weddell seal (*Leptonychotes weddellii*) recovery time at the surface increases exponentially with the duration of the previous dive. This has important implications. If the animal indulges in very long dives, then, paradoxically, the total time the animal can spend submerged during a day is lower than if the animal makes a series of short dives within its aerobic dive capacity and thus does not accumulate lactate, as addressed below.

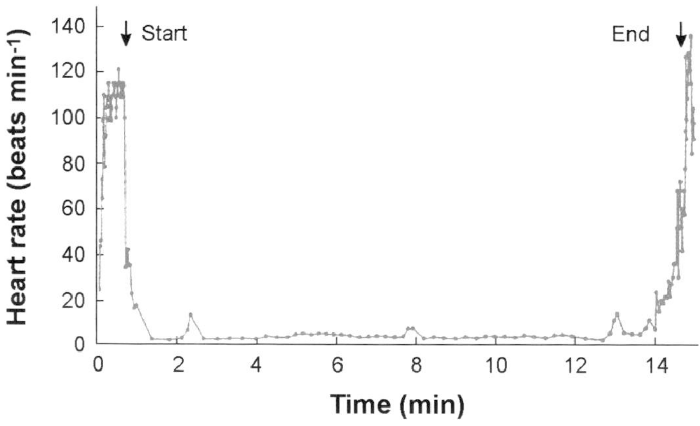

Figure 3

Heart rate during a voluntary dive in gray seal (*Halichoerus grypus*) during diving at sea, obtained by radiotelemetry. Arrows mark the start and end of the dive. During this particular dive, heart rate averaged 6.5 beats min^{-1} and was below 4 beats min^{-1} for 90% of the time. From Reference 55.

Cortical Influences on Cardiovascular Function

With the introduction of biotelemetry in the 1970s, it soon became clear that most seals and whales usually do not take advantage of their full diving capacity. Instead, they perform series of short dives, during which the dramatic cardiovascular adjustments described above may be much reduced or not be expressed at all (39). This was confusing at the time (52). However, it is now generally accepted that expert divers have cortical (suprabulbar) control over their cardiovascular system and are able to determine their diving strategy in accordance with the challenge of each individual dive (38). Murdaugh et al. (53) reported prompt bradycardia and heart rates as low as 7 beats min^{-1} in freely diving seals and tachycardia in anticipation of surfacing. Ridgway et al. (54) showed that in sea lions the bradycardia response could be conditioned both when they were in air and submersed. Finally, Tompson & Fedak (55) demonstrated conclusively that freely diving gray seals (*Halichoerus grypus*), some of which habitually perform long-duration dives, display spectacular bradycardia, whereas other gray seals, which habitually perform series of short dives, do not (**Figure 3**).

These results are consistent with the notion that seals can actively decide how to deal with hypoxia. If an animal cannot predict the duration of submersion, its nervous system will immediately activate the full O_2-conserving responses that involve powerful, modulating descending pathways. If the animal can anticipate that the voluntary dive is brief, higher central nervous system (CNS) centers suppress such responses.

Short, Aerobic, Natural Dives

What happens when animals decide to perform a series of dives that are short enough not to exceed the animals' aerobic capacity?

In the much-studied dabbling ducks, which usually dip their heads under water only for a few seconds, the answer is nothing, except for cessation of breathing. This is because the "dive" is too short to activate the peripheral chemoreceptors, which, when activated, result in spectacular cardiovascular responses during forced dives in ducks (39, 56).

In freely diving seals, the activation of the cardiovascular responses does not depend on the peripheral chemoreceptors (57) and is instead variable and dependent on factors such as the anticipated duration of the dive and the extent of swimming activity. In very short dives, the autonomic responses are again usually not expressed. In contrast, in dives approaching the animal's aerobic capacity, a somewhat reduced heart rate reflects some degree of peripheral vasoconstriction (58). Because no lactate is produced during this latter category of dives, the animal is operating (33) fully aerobically. This raises the questions of where and why this minor vasoconstriction and bradycardia occurs and how it can occur without resulting in lactate production. We must assume that the brain and heart are adequately supplied with blood during such dives, because even the kidneys are adequately supplied (50). Ignoring the guts and the liver for simplicity, that leaves us with the skeletal muscles, which are very rich in Mb.

Because Mb has a much higher O_2 affinity than does Hb, the blood and muscle O_2 stores must remain separate during prolonged diving. Without such a separation, Mb would act as an O_2 sink by depleting the O_2 stores from the blood. In short aerobic dives, this is not an issue, and the seal can differentially control blood supply to the various muscles. In this case, the best O_2 economy will be achieved by perfusing the active swimming muscles: Because of the short duration of the dive, their Mb will remain fully saturated and these muscles will compete with other perfused tissues for the blood O_2 stores. By contrast, supply to the inactive muscles, like those involved in respiration, should be shut off from circulation. These inactive muscles will then be able to sustain themselves on their vast amounts of endogenous oxy-myoglobin (59) and/or enter into a state of shutdown.

Using this strategy, the seal returns to the surface to breathe, with the blood O_2 stores fairly depleted and the Mb of the swimming muscles fully oxygenated. Although the Mb of the inactive muscles may be low in O_2, the animal has avoided a lactate load. Consequently, the animal may continue to perform such short dives all day long, separated only by short stops at the surface for gas exchange. This strategy allows the animal to stay submerged for approximately 80% of the time (13, 14) but with little time for lengthy underwater activities.

Cooling and Hypometabolism

Investigators have calculated the aerobic dive limit for several species of seals from estimates of blood and muscle O_2 stores and the animal's assumed diving metabolic rate. The values thus obtained often deviated from the actual diving times of many species of seals (13, 14) and penguins (15). These discrepancies may, in part, be due to erroneous calculations but also to regional cellular shut-down, cooling, and peripheral vasoconstriction during diving, which thus lead to unaccounted-for hypometabolism. Scholander et al. (60), in an often-forgotten paper, reported a 2°C decrease in brain and rectal temperature during experimental dives in a seal (species not named). Moreover, Hill et al. (58) found that the aortic temperature of freely diving Weddell seals dropped, sometimes approximately 2°C, during diving and that the cooling often started prior to bouts of repetitive diving (**Figure 4*b***). Seals (61) and ducks (62) can reduce their brain temperature by approximately 3°C during experimental diving (**Figure 4*a***). This ability, which in seals is assumed to be achieved by a physiologically controlled perfusion of the skin of primarily the foreflippers (63), brings cold blood into circulation and cools those tissues that are perfused. This reduces these tissues' metabolic rate and, assuming an average Q_{10} of 2.5, also reduces

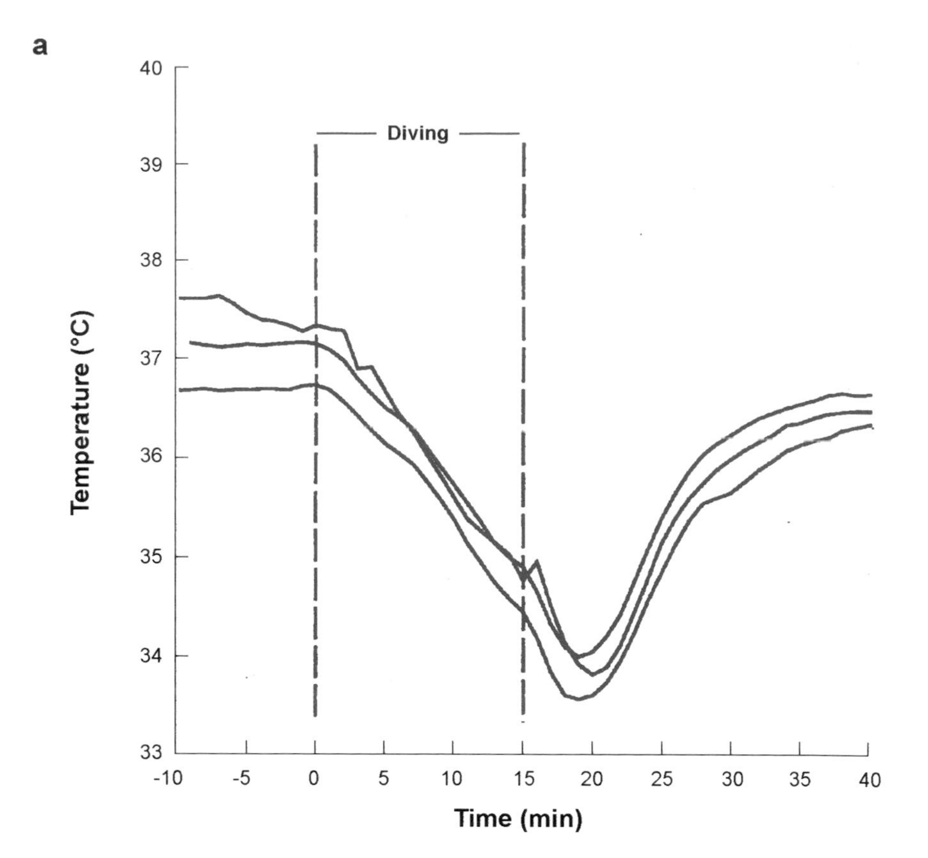

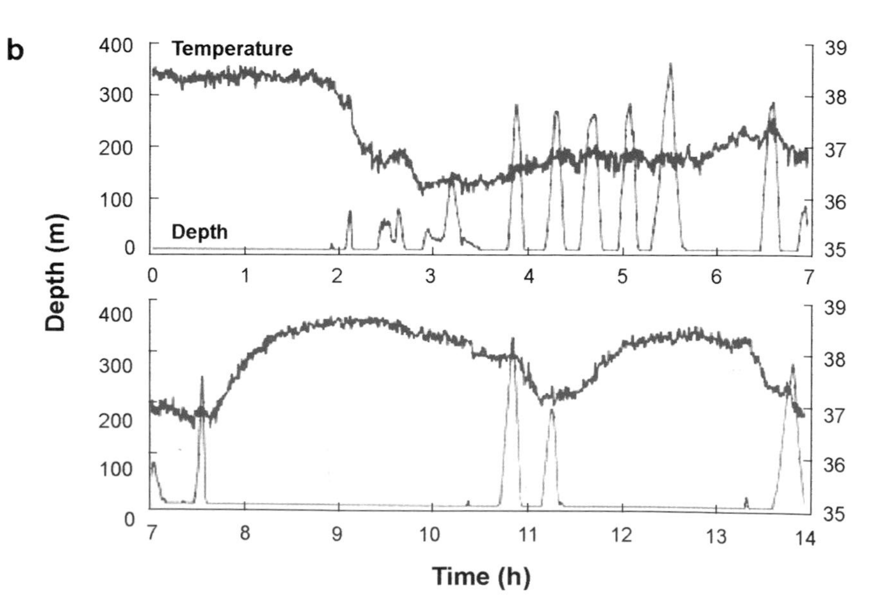

Figure 4

(*a*) Changes in brain temperature in a hooded seal during three experimental dives each lasting 15 min. From Reference 61. (*b*) Selected period of 14 h (from a total of 118 h) of continuous recording of temperature in aortic blood and diving behavior (depth) of voluntary diving Weddell seal (*Leptonychotes weddellii*). From Reference 58.

their O_2 consumption by approximately 20%. This in turn significantly reduces O_2 stress and confers cell protection.

In mammals, however, any cooling of the body, and in particular cooling of the brain, normally elicits vigorous shivering, which would compromise the cooling. What happens to the shivering response in hypothermic seals? Recently, Kvadsheim et al. (64) have shown that hypothermic hooded seals respond to (experimental) diving with an almost instantaneous inhibition of shivering. This thermogenic occlusion allows cooling of the brain and other perfused tissues, whereby O_2 consumption is decreased and diving in aerobic mode is prolonged. This is particularly advantageous in prolonged dives when the brain is the main consumer of O_2.

CELLULAR MECHANISMS IN ACUTE HYPOXIA

As described above, hypoxic tolerance is achieved through a complex coordination, reorganization, and partial shutdown of different organ systems to reduce metabolism while simultaneously maintaining organ system function. Here we show that a similarly complex coordination, reorganization, and partial shutdown occurs at the cellular level. Our discussion emphasizes the nervous system, for which functional reconfiguration is a well-established principle (64a). But many of these insights may apply to other organs as well.

Coordinating Hypometabolism, Neuroprotection, and Functional Integrity

To tolerate and survive hypoxia, the mammalian nervous system must (*a*) reduce metabolism, (*b*) prevent cell death and injury, and (*c*) maintain functional integrity. This review centers around one key hypothesis: In the mammalian nervous system, these three aspects—hypometabolism, neuroprotection, and functional integrity—operate as an inseparable unit.

This conclusion has important consequences: Although increasing evidence suggests that hypoxia leads to substantial downregulation of ion channels within neuronal systems (4, 10, 65–68), the hypoxic response of individual neurons is very diverse. As we elaborate below, this heterogeneity is best explained by the necessity to maintain functional integrity. Ion channel arrest cannot lead to the shutdown of entire networks, nor can networks operate by all cellular components entering a general hypometabolic-neuroprotective state. In biological terms, a seal would not benefit from a general shutdown of *N*-methyl-D-aspartate (NMDA)- or Ca^{2+}-dependent processes. Such a shutdown would severely compromise memory and other higher-brain functions.

Unfortunately, hypoxia research increasingly focuses on the understanding of hypometabolism and neuroprotection in the absence of a functional context. Although it is advantageous to study Ca^{2+} homeostasis or aerobic and anaerobic metabolism in isolated cells, our understanding of strategies that maintain network functions in hypoxia is lagging. This disconnect may explain why the prevention of hypoxic brain injury remains a major medical challenge. We know that the activation of NMDA receptors or intracellular Ca^{2+} may lead to necrosis and apoptosis, but drugs aimed at inhibiting these mechanisms are not very useful because they also compromise functional integrity. As Bickler (66) notes, "the list of failed stroke treatments involving blocking calcium increases is particularly long."

Managing Metabolic Costs While Maintaining Functional Integrity

Most neuronal networks maintain persistent activity at all times; they cannot simply shut down to save metabolic energy. This is especially the case for networks that control autonomic functions such as heart rate or blood pressure. But for most mammals and in particular for diving mammals, shutdown of networks that control higher-brain functions is not advantageous.

Yet the maintenance of persistent activity is metabolically very costly, as it is associated with continuous changes in ionic gradients (69). Indeed, 50% of a neuron's energy is used to maintain ionic gradients and fluxes (69). This expenditure depends on ATP production, which normally relies on aerobic and anaerobic metabolism (69). Acute hypoxia inhibits oxidative phosphorylation, which leads to a drop in ATP levels, unless anaerobic ATP supply pathways make up the energy deficits (70). The capacity of central neurons for anaerobic glycolysis depends on various factors, including glucose supply, intracellular conditions such as pH_i, and the activity, expression level, and types of glycolytic enzymes (71). The increased glycolytic capacity of marine mammals such as Weddell seals (72) enhances their neurons' ability to rely on anaerobic metabolism during severe hypoxia. Weddell seals' glucose consumption increases by 40% during dives (72). Moreover, enzymes that operate more efficiently during anaerobic glycolysis, including the M-type lactate dehydrogenase (45), are abundant in seals.

The increased glycogen content of marine mammals (44) may be due to the enhanced capacity of glia to support glycolysis in hypoxic states. In dolphins and whales, astrocytes are abundant (10), which may be protective, because almost all of the brain's glycogen is in these cells (73). In terrestrial mammals, elevated glucose concentrations also protect against hypoxic neuronal (74) and glial damage (75). Some hypoxia-induced changes in synaptic transmission can be prevented by conditions favoring anaerobic metabolism (74). An increased efficiency for anaerobic metabolism may also be responsible for the increased hypoxia tolerance of neonatal mammals (10, 67, 69, 76, 77).

Although an efficient anaerobic metabolism may convey enhanced hypoxia tolerance, it also leads to increased H^+ concentrations and decreased pHi (16, 78). H^+ concentration surges are detrimental for synaptic function (79). Thus, despite an enhanced glycolytic capacity of diving mammals, it remains unknown how these mammals maintain persistent neocortical activity during severe hypoxia.

Diversity of Neuronal Responses to Acute Hypoxia

Even within the same brain region, the hypoxic response of individual neurons can be surprisingly heterogeneous. Neurons in the hippocampus and neocortex can initially depolarize then hyperpolarize, or hyperpolarize then depolarize, or continuously depolarize (11, 67, 80). Some neurons in the locus coeruleus hyperpolarize, whereas others depolarize (81). Midbrain dopaminergic neurons hyperpolarize (82), whereas neurons of the striatum (83) and thalamus (84) depolarize.

The diversity of cellular responses in any given network may be the key for understanding hypoxia-induced hypometabolism and neuroprotection. To maintain functional integrity, not all neurons can afford to shut down, especially not at the same time. Some neurons will shut down to save metabolic energy and avoid excitotoxicity and cell death, whereas others remain active to preserve neuronal function. The relationship of active versus inactive neurons may reflect the importance of a given neuronal network. Some networks may become silent without detrimental effects for the organism, whereas other networks must maintain a high percentage of active neurons because the organism cannot allow the complete shutdown of this network.

The mammalian respiratory network is a good example of a network that during hypoxia maintains activity resulting from a selective shutdown of certain neuronal elements. In the normoxic state (**Figure 5**, upper panel), this network contains neurons that can burst intrinsically (so-called pacemaker neurons) and neurons that normally do not burst (nonpacemaker neurons, which in **Figure 5** are denoted by I). Pacemaker neurons are of two distinct types: (*a*) neurons that burst based on Ca^{2+} mechanisms involving presumably the CAN current (denoted by Ca in **Figure 5**) and (*b*) pacemakers that depend on the persistent Na^+ current (denoted by Na in **Figure 5**) (86). In the hypoxic state, nonpacemakers hyperpolarize, and Ca^{2+}-dependent pacemakers cease to burst, whereas Na^+-dependent pacemakers continue to burst (85–87) (**Figure 5**, lower panel). The shutdown of Ca^{2+}-dependent bursting may be neuroprotective, as the activation of Ca^{2+} mechanisms may promote necrosis and apoptosis. The shutdown of nonpacemakers and Ca^{2+}-dependent pacemakers is hypometabolic because these neurons constitute the majority of respiratory neurons (**Figure 5**, lower panel). This differential hypoxic response results in a reconfiguration that may transform the respiratory output from normal respiration to gasping in hypoxia (86, 89) (**Figure 5**).

The diverse neuronal responses are produced by similarly diverse changes in specific ionic conductances. To describe all known hypoxia-induced subcellular changes is beyond the scope of this review, and this topic is reviewed well elsewhere (10, 65, 66). A few examples are given here. Hippocampal neurons

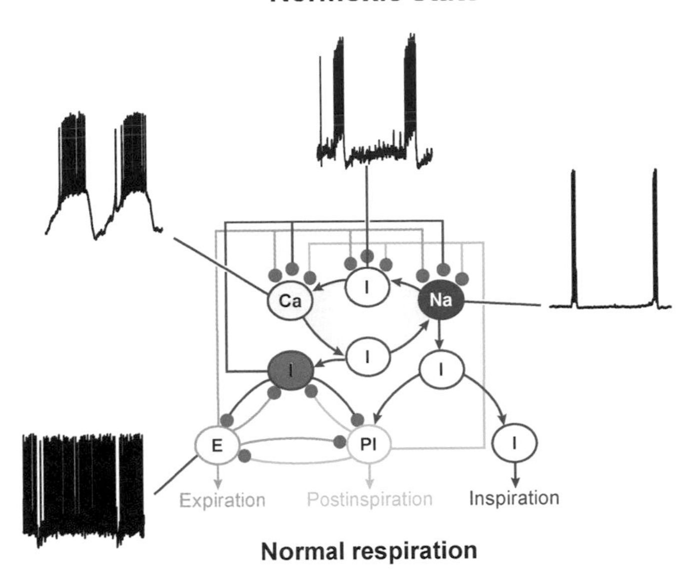

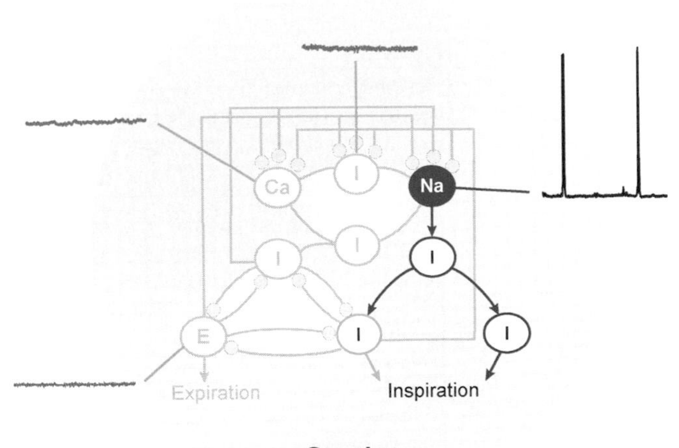

Figure 5

Hypothetical model illustrating the reconfiguration of the respiratory network during the transition from normal respiration (*upper panel*) to gasping (*lower panel*). In the normoxic state, the network controls different phases of respiration: expiration, postinspiration, and inspiration. The network depends on the activity of nonpacemaker neurons (I), expiratory neurons (E), postinspiratory neurons (PI), Ca^{2+}-dependent pacemaker neurons (Ca), and Na^+-dependent pacemaker neurons as well as inhibitory (*red circles*) and excitatory (*dark-blue arrows*) synaptic interactions. Example traces of four representative respiratory neurons [E, Ca, I, and Na^+-dependent pacemaker neurons (Na)] are shown. In the hypoxic state, most neurons cease to discharge, and the respiratory network is driven by Na^+-dependent pacemaker neurons (Na). The network output consists of only inspiration, which characterizes gasping activity. Traces of the same four representative respiratory neurons are shown in the hypoxic state. Data taken from References 86, 88, and 89.

cease action potential generation either because of a pronounced intrinsic hyperpolarization or because of the inactivation of transient Na^+ channels (11). The hyperpolarization can be caused by the drop in ATP that activates ATP-sensitive K^+ currents (90) or by the rise in Ca^{2+} that activates Ca^{2+}-activated K^+ channels (11, 80). Hypoxia may also have opposite effects on K^+ channels and depolarize neurons by inhibiting background two-pore (2P)-domain K^+ channels (91). Similarly complex is the modulation of Na^+ and Ca^{2+} channels, as Reference 67 describes in more detail. Hypoxia can also activate (84) or depress hyperpolarization activated currents (92). Moreover, the decline in intracellular ATP reduces the activity of Na^+/K^+ ATPase. This in turn results in a massive depolarization (77, 93), which may be responsible for the flat EEG seen in humans seconds after the onset of anoxia (5).

This diversity of ion channel modulation may be understandable only in a functional context. Again, the mammalian respiratory network serves as a good example (**Figure 6**). In this network, hypoxia inhibits N-type Ca^{2+} channels (94), which could be considered neuroprotective, as this inhibition reduces Ca^{2+} influx. However, the reduced Ca^{2+} influx results in the closing of Ca^{2+}-dependent K^+ channels, which in turn decrease the inhibition that typically follows an inspiratory burst. This decreased inhibition results in enhanced frequency of respiratory activity (95) (**Figure 6**, inset). The early inhibition of the N-type Ca^{2+} channels may underlie the hypoxia-induced augmentation, a critical component of the ventilatory response to hypoxia. During prolonged hypoxia, the opening of K_{ATP} channels hyperpolarizes respiratory neurons but, as described above, only a subset of these neurons (**Figure 6**). Some neurons remain active and are thought to generate gasping (86; A.K. Tryba, J.C. Viemari, & J.M. Ramirez, unpublished observations). If K_{ATP} channels inhibited all respiratory neurons, respiration would cease in hypoxia. Thus, the differential modulation of ion channels in different types of neurons is adaptive in the context of network responses.

Reconfiguration of the Nervous System: A General Principle of Hypoxic Defense

Based on the above considerations, we conclude that attempts to generalize strategies that convey hypoxia tolerance and neuroprotection may be fruitless unless they consider the functional context of neurons. Ion channels need to be altered in a very diverse

Figure 6

Hypothetical model illustrating how different ion channels contribute to the reconfiguration of respiratory neurons during the transition from the normoxic to the hypoxic state. The activity of respiratory neurons is determined by the opening and closing of various ion channels, including leak K^+ currents (TASK), ATP-dependent K^+ (K_{ATP}), Ca-dependent K^+ channels (K_{Ca}), L-type (Ca_L) and N-type (Ca_N) Ca^{2+} channels, persistent Na^+ current (Na_p), and the nonspecific cation current (CAN). The activity is also determined by synaptic mechanisms, including the activation of NMDA receptors (NMDA). The differential activation of these ion channels determines the specific membrane properties of the three representative respiratory neurons shown. In the normoxic state (*upper panel*), Ca^{2+}-dependent pacemaker activity is largely determined by CAN current activation, Na^+-dependent pacemaker activity is primarily determined by the persistent Na^+ current, and nonpacemaker neurons exhibit no pacemaker properties owing to a prominent activation of the TASK channels. In the hypoxic state (*lower panel*), Ca^{2+}-dependent pacemakers and nonpacemakers cease to discharge owing to the activation of the K_{ATP} channels, and Na^+-dependent pacemakers continue to burst owing to the pronounced activation of the persistent Na^+ current. (*Inset, lower right*) Pharmacological blockade of the N-type Ca^{2+} channel leads to the activation of respiratory activity, mimicking the hypoxic augmentation. Data from Reference 95.

Normoxic state

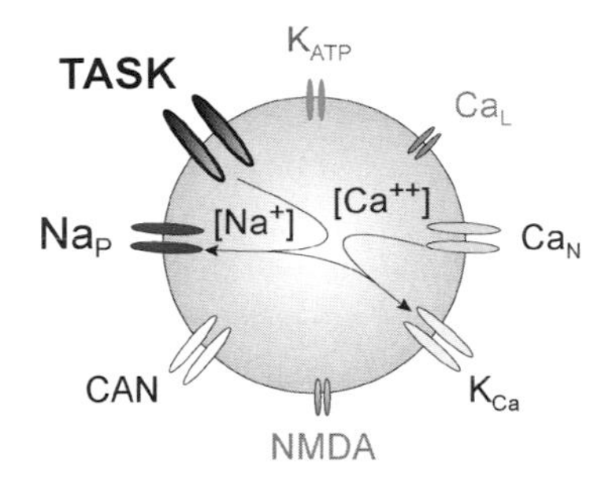

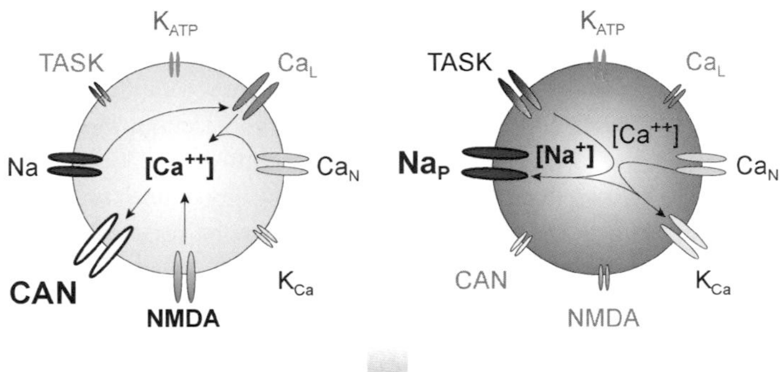

Hypoxic state

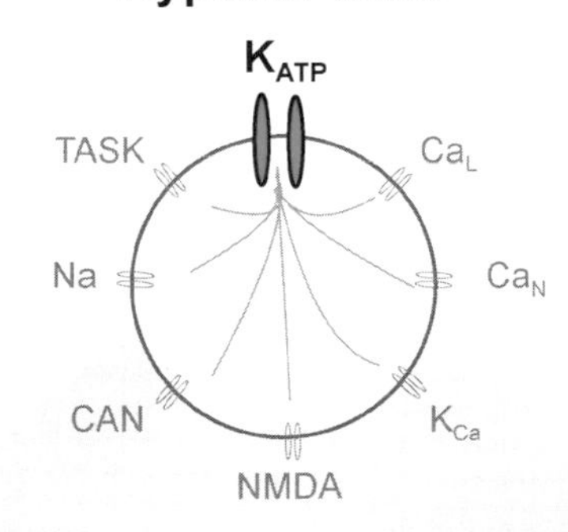

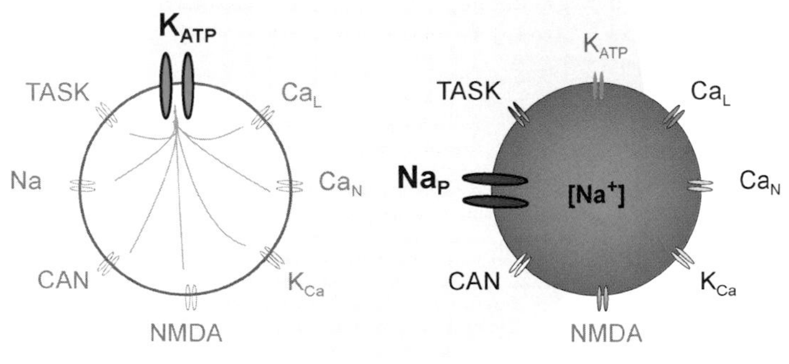

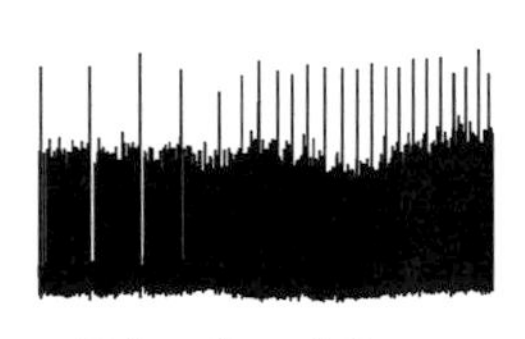

Block of Ca_N

manner to accomplish massive network reconfigurations (**Figures 5, 6**). The same is true for the differential modulation of synaptic transmission, an aspect of equal importance for the hypoxic response, as further discussed in Reference 67.

The reconfiguration of neuronal networks occurs in a behavior-, development-, and metabolism-specific manner to enable the nervous system to function under hypoxic conditions. Neuronal networks that control autonomic functions such as digestion, breathing, and cardiovascular control must be differentially reconfigured and metabolically altered to produce the changes in cardiac output, ventilation, and digestion that confer hypoxia tolerance. Reconfigurations affect all neuronal networks, including those that control higher-brain functions. A diving animal cannot afford to compromise higher-brain functions if it wants to catch prey 1000 m below the surface during the Arctic night.

In summary, we propose that the mammalian nervous system adapts to hypoxia through network reconfiguration, which is aimed at meeting the triple task of hypoxia tolerance: hypometabolism, neuroprotection, and functional integrity. Our hypothesis departs from conventional hypoxia research that is aimed at understanding hypometabolism and neuroprotection in the absence of functional network considerations.

OXYGEN SENSING: CELLULAR MECHANISMS

To respond adequately to changes in oxygenation status, animals require internal O_2-sensing mechanisms (10). O_2 sensing occurs through neurons or other cells that increase their activity during hypoxia.

The arterial O_2–sensing chemoreceptor in the carotid body represents a peripheral chemoreceptor that contributes fundamentally to the regulation of respiratory drive and PaO_2 (96). In hypoxia, glomus (type I) cells release transmitters causing depolarization of nearby afferent nerve endings. The sensory information is subsequently relayed to brain stem neurons regulating breathing (96). The transduction process is not yet fully understood but appears to involve O_2 sensing by heme-containing mitochondrial (e.g., cytochrome oxidases) or nonmitochondrial (e.g., NADPH oxidases) enzymes and the inhibition of K^+ currents through a variety of O_2-sensitive K^+ [$K^+_{(O2)}$] channels in the glomus cells. This leads to the depolarization of these channels via influx of Ca^{2+} through L-type voltage-gated Ca^{2+} channels (97).

The small-resistance pulmonary artery smooth-muscle cell (PASMC) is another peripheral chemoreceptor. It responds to hypoxia with contraction, thereby producing a hypoxic pulmonary vasoconstriction response that diverts blood flow to better-ventilated alveoli. It may be responsible for pulmonary hypertension at high altitude (98).

Denervating peripheral chemoreceptors abolishes the hypoxic response as well as the response to changes in H^+ blood concentration in the pH range of 7.3–7.5. The response to arterial CO_2 is only slightly affected. Such lesioning experiments suggest that O_2 is measured in the periphery, whereas CO_2 is measured within the CNS. However, this interpretation may be oversimplified and possibly wrong. Peripheral chemosensory neurons are continuously active under control conditions, and their lesioning will decrease chemosensory drive, which essentially signals the existence of hyperoxia to the CNS. Thus, following peripheral chemodeafferentiation the CNS receives mixed messages: peripheral hyperoxia versus hypoxia measured by central chemosensitive neurons. The hypoxic response may be blunted as a result of this mismatch and not because peripheral chemoreceptors provide the sole O_2 information. Indeed, evidence is increasing that multiple sites within the CNS are sensitive to O_2, and these chemosensitive responses may activate local networks. Central O_2 sensors may monitor brain O_2 levels and, when activated, coordinate critical functions necessary for the survival of the whole organism

(99). A good example is the hypoxic activation of pacemaker neurons that may be critical to generate the hypoxic response of the respiratory network (85) (**Figure 5**).

O_2 sensors that are a part of a distributed and general network (as opposed to a few peripheral sensors that act as master controllers) may be adaptive for the same reasons as described above: The hypoxic response of different neuronal networks is heterogeneous. Heterogeneity characterizes even the response of networks that contribute to the same overall behavior. For example, the pre-Bötzinger complex (PBC), hypoglossal (XII) nucleus, and phrenic nucleus are all activated during inspiration. All three networks contribute to an initial augmentation of respiratory drive that leads to enhanced ventilation in response to hypoxia. However, such augmentation has different time courses and characteristics for each of the three networks (100). In the XII nucleus, the amplitude of inspiratory bursts is enhanced and maintained throughout hypoxia. The frequency, but not the amplitude, of inspiratory bursts is enhanced in PBC neurons. XII and PBC neurons tonically depolarize in response to hypoxia. However, the tonic depolarization of XII is much more pronounced than that of PBC neurons, and the time courses of the depolarizations are different in PBC and XII, suggesting that the hypoxic response is generated independently in both nuclei (100). Thus, is the hypoxic reconfiguration of neuronal networks generated by local chemosensitive neurons that constitute integral parts of these circuits? Future research will need to unravel the different roles of these central versus peripheral chemoreceptors.

Subcellular Oxygen-Sensing Mechanisms

Hypoxia-inducible factor (HIF), a transcription factor protein complex activated by hypoxia, has a key molecular role in O_2 sensing (10, 96, 97, 101). The dimeric HIF complex is composed of an α and a β subunit, of which several versions exist. Whereas the β subunit (e.g., HIF-1β) is a constitutive, O_2-independent subunit, the α subunit (e.g., HIF-1α or its paralogs HIF-2α and HIF-3α) is hypoxia regulated. In normoxia, the HIF-α subunit is bound by the von Hippel–Landau tumor suppressor protein (pVHL), which makes HIF-α susceptible to proteosomal degradation. In hypoxia, the binding of pVHL to HIF-α is inhibited and the unit is stabilized, allowing the dimerization of HIF-α with HIF-β and subsequent binding of this dimer to DNA. Combinatorial interactions of HIF with other identified transcription factors are required for the hypoxic induction of physiologically important genes. The consequences of HIF activation vary among cell types. Transcriptional targets of HIF include hundreds of genes that are involved in, e.g., erythropoiesis, angiogenesis, vascular reactivity, glucose and energy metabolism (glucose uptake, expression of glycolytic enzymes), cell proliferation and survival, and apoptosis (10, 101). The role of the ancient and conserved HIF in hypoxia-tolerant birds and mammals is only beginning to emerge, as data on its amino acid and DNA sequences in such species are being generated (102).

CELLULAR PROTECTION UPON RECOVERY FROM HYPOXIA

The recovery from hypoxia/ischemia can be as detrimental as the exposure to acute hypoxia itself. Sudden cardiac death accounts for more than half of all cardiovascular deaths in the United States. These deaths are largely attributed to cardiac arrest that occurs after an ischemic insult, i.e., following the return to apparently normal conditions (103). Upon recovery from hypoxia, intracellular Ca^{2+} concentration ($[Ca^{2+}]_i$) rises to levels that are detrimental for the heart and ultimately the organism (104). The rise in $[Ca^{2+}]_i$ is preceded by a rise in $[Na^+]_i$, which is caused by several mechanisms, including an enhanced persistent Na^+ current and Na-HCO_3

symporter activity. The intracellular acidification contributes also to an enhanced activation of the Na^+-H^+ exchanger, which increases $[Na^+]_i$ but slows or reverses the Na^+-Ca^{2+} exchanger, resulting in a rise in $[Ca^{2+}]_i$ (104).

A sudden increase in oxidative stress is associated with the rise in $[Ca^{2+}]_i$ upon recovery from hypoxia. Opposing oxidative stress is an antioxidant network that includes enzymes such as catalase, superoxide dismutases, glutathione peroxidase, and low-molecular-weight scavengers such as melatonin; the water-soluble glutathione, urate, and ascorbate; and lipid-soluble scavengers such as α-tocopherol (vitamin E). Reperfusion after ischemia provides O_2 as a substrate for numerous enzyme oxidation reactions that produce free radicals to such an extent that antioxidant systems may be overwhelmed. This leads to oxidative damage such as lipid peroxidation, protein oxidation, and DNA damage. In humans, reactive oxygen species are central to the genesis of cardiac ischemic and reperfusion injury, contributing to infarction, apoptosis, and arrhythmias (105). Diving mammals generally have constitutively higher antioxidant capacity than do nondiving mammals (106).

Detrimental and Beneficial Effects of Intermittent Hypoxia

The daily routine of diving mammals involves frequent and abrupt changes in O_2 levels. Thus, these animals must deal with continuous de- and reoxygenation. This hypoxic pattern, which is observed clinically, can also be characterized as intermittent hypoxia. In obstructive sleep apnea (OSA), episodic airway obstructions can occur more than 60 times per hour, causing significant desaturations of Hb levels. These events are associated not only with hypoxia but also hypercapnia. OSA patients develop neurocognitive deficits associated with regional gray-matter loss, increased circulating markers of oxidative stress, and inflammation (67, 107). Chronic intermittent hypoxia has been linked to pulmonary vascular remodeling and pulmonary hypertension (108), and it evokes sustained hyperexcitability of hippocampal neurons that involves the activation of L-type Ca^{2+} channels and NMDA receptors (109). How seals avoid these detrimental effects is still a mystery, and lessons learned from seals may have important clinical implications.

Protocols of intermittent hypoxia also have beneficial effects. Intermittent hypoxia has antiarrhythmic effects in acute myocardial ischemia (110) and produces long-lasting increases in respiratory activity (111, 112). Brief hypoxic exposures (hypoxic preconditioning) can convey hypoxia tolerance to various organs, including the nervous system, liver, lung, kidney, and skeletal muscles. Numerous studies have investigated the cellular and molecular mechanisms underlying preconditioning (e.g., 10, 67, 113). Indeed, the beneficial effects of ischemic preconditioning are now clinically used in cardiac surgery (114). Diving animals routinely precondition their hearts and other organs. Thus, preconditioning mechanisms may contribute much to the hypoxic tolerance of diving animals.

HIBERNATION: THE GENERATION OF EXTREME HYPOMETABOLISM

A photoperiodically determined hypometabolic state enables hibernators to survive food shortages. A hallmark of hibernation is the highly regulated drop in body core temperature (T_b), which is subsequently defended though autonomic thermoregulation (115–118). For reasons still not understood, hibernating rodents intermittently undergo rapid, brief, and energetically costly arousals back to a euthermic metabolic rate and T_b between bouts of deep hibernation (115, 117, 119) (**Figure 7**).

Small hibernators commonly enter hibernation while in a normoxic state (119). The dramatic drop in T_b is preceded by abrupt decreases in metabolic rate, cardiac output,

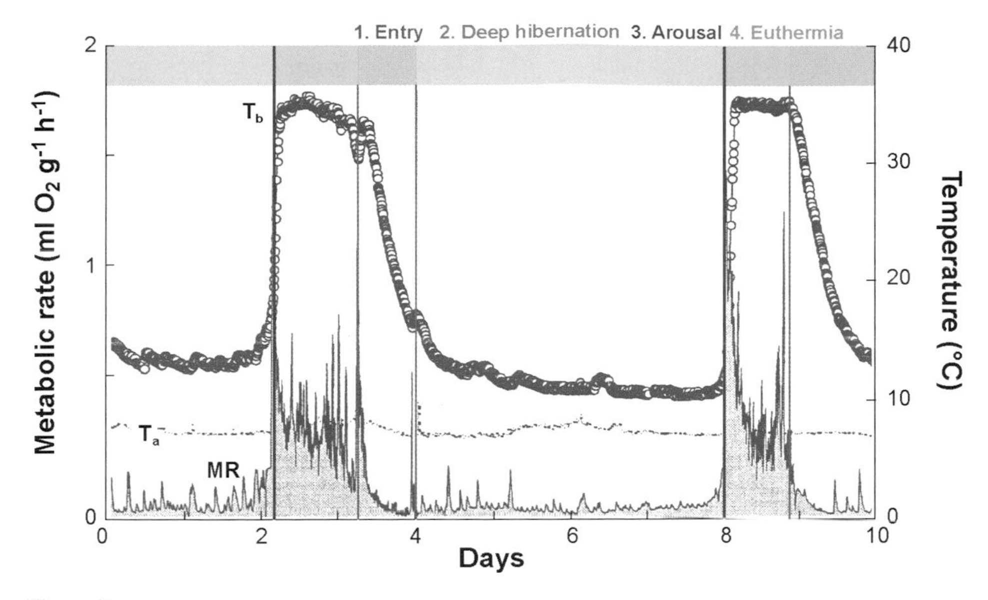

Figure 7

Hibernation bout in Alpine marmot (*Marmota marmota*). Continuous record of metabolic rate and body temperature reveals the development of hypometabolism and hypothermia during entrance into hibernation (*1*), maintenance of hypometabolism during deep hibernation (*2*), rapid rewarming during arousal (*3*), and a constant high level of metabolic rate and body temperature in the euthermic state (*4*). MR, metabolic rate; T_a, ambient temperature; T_b, body temperature. Reprinted from Reference 119 with permission from Elsevier.

and ventilation (115, 118, 120, 121), whereas peripheral vasoconstriction reduces blood flow to most tissues (118). At the same time, increased Hb-O_2 affinity promotes blood O_2 binding at low P_{O2}. The high Hb-O_2 affinity is due partly to reduced RBC concentrations of 2,3-DPG and partly to reduced T_b (122). Conversely, O_2 affinity falls as T_b increases during arousal. In some, but not all, species, blood concentrations of Hb and skeletal muscle Mb also increase in preparation for hibernation to levels that exceed those of nonhibernating animals. These adaptations enhance blood O_2 carrying capacity and possibly improve myocyte O_2 uptake during arousal (122, 123).

Owing to the combined effects of a hypometabolic state, reduced muscular and cardiac activity, and thermodynamic (Q_{10}) effects, O_2 consumption declines to less than 5% of euthermic resting levels (117, 119). Thus, despite periods of apnea and a P_{aO2} that temporarily drops to close to 10 mm Hg, most hibernators probably experience only mild or no hypoxia (121). Indeed, hibernating Arctic ground squirrels (*Sphermophilus parryii*) exhibit high average P_{aO2} (80–130 mm Hg), low plasma lactate, and low brain levels of HIF-1α and inducible nitric oxide synthase (iNOS), a marker of cellular stress (124).

Are Arousals Necessary to Reset Homeostatic Mechanisms During Hibernation?

The above section indicates that hibernating mammals avoid hypoxic conditions by using similar strategies as diving mammals. Yet hibernating mammals undergo a much more extreme hypometabolic state than any other mammals: They become immobile, and CNS activity is largely reduced. Achieving an extreme hypometabolic brain state is not a trivial task, however.

Continuous neocortical activity is characteristic of all normal brain states, such as wakefulness and sleep. None of these brain states is characterized by hypometabolism. Cerebral O_2 metabolism during REM sleep is as high as during wakefulness (125), and during the transition from wakefulness to sleep cerebral metabolic rate may even increase (126). Thus, diving mammals cannot achieve a metabolic advantage by going to sleep. Sleeping with only half the brain is a fascinating phenomenon in marine mammals (127). Yet this sleep behavior does not convey an obvious metabolic advantage. Direct measurements in seals indicate that O_2 consumption during slow-wave sleep is not significantly different than during wakefulness (128).

Thus, to achieve hypometabolism, hibernating mammals had to develop a brain state that is very different from any mammalian wake or sleep state. During hibernation, the EEG appears isoelectric, and cerebral blood flow is only 10% of normal, yet the animal shows no neuropathologies (129). The brain is not completely shut down. The hippocampus, septum, and hypothalamus retain periodic EEGs at temperatures below which EEGs in other structures become isoelectric (130).

Many issues associated with this hypometabolic brain state are unresolved. In nonhibernating mammals, activity-dependent homeostatic mechanisms continuously adjust synaptic and intrinsic membrane properties in the neocortex and hippocampus and presumably throughout the brain (131). An activity decrease strengthens glutamatergic synaptic transmission through postsynaptic insertions of AMPA receptors (131). The silencing of neuronal networks alters not only synaptic connectivity but also intrinsic membrane properties; silencing a brain area may evoke homeostatic changes that result in epilepsy (132). This raises fascinating questions: How can hibernating animals afford to silence neocortical activity without activating homeostatic responses that trigger seizures? Is low brain temperature sufficient to suspend these homeostatic mechanisms? Are the frequent arousals of small hibernating mammals perhaps necessary to reset homeostatic regulation periodically? It has been hypothesized that arousals are necessary for sleep homeostasis, but this hypothesis has not found experimental support (117). Future research will be necessary to understand these very important questions and to understand better the consequences of body cooling (121).

Hypermetabolism and the Need for Maximal Cellular Protection

Although why hibernating mammals periodically arouse remains unknown, these arousals (**Figure 7**) come with tremendous costs. During these intense arousals, a hibernating mammal encounters hypoxia and oxidative stress (121, 124). The 10–100-fold increase in metabolic rate is initially caused by massive sympathetic activation of nonshivering thermogenesis based on abundant thoracic and interscapular stores of brown adipose tissue (BAT). Metabolic rate subsequently increases further through muscular shivering, which may raise T_b to euthermic levels in only 2–3 h (115, 119). The increase in metabolic rate is accompanied by a rapid rise in ventilation and heart rate, which is followed by a progressive vasodilation (115, 118, 124). During peak arousal, tissue O_2 requirements, particularly in BAT and the shivering skeletal muscles, increase almost exponentially.

At peak arousal, the hibernating mammal loses some of its hypothermic protection. Despite the accompanying increases in ventilation and cardiac output, a mismatch between O_2 supply and demand becomes unavoidable. P_{aO2} values sometimes drop below 10 mm Hg in Arctic ground squirrels, leaving some tissues hypoxic (121, 124). Brain levels of HIF-1α increase (124), as do plasma levels of urate (120), a marker for the production of reactive oxygen species. Unexpectedly, however, the extent of cellular stress appears modest. Plasma lactate and brain iNOS levels remain unchanged from hibernation levels

(115, 121, 124). The lack of a lactate response suggests a reduced reliance on anaerobic metabolism, which may be linked to the shift from carbohydrate-based metabolism to the lipid-based metabolism typical of hibernators (115, 121).

Hibernators show various other adaptive mechanisms. These animals exhibit membrane-related protective mechanisms, including ion channel arrest (133). Hippocampal slices from hibernating hamsters display less NMDA-induced Ca^{2+} influx than do active animals (134). As proposed by Dave et al. (133), hibernators may also benefit from preconditioning effects caused by low P_{aO2}/high P_{aCO2} in hibernation and the early stages of arousal.

DEALING WITH CHRONIC HYPOXIA

The situation of animals living with chronic hypoxia is very different from animals adapted to acute yet transient hypoxia. The former animals cannot expect that their environment will ever become normoxic. Consequently, developing large O_2 stores, assuming extreme hypometabolic states, and compromising behavioral functions are not long-term options. The major mammalian strategy to cope with chronic hypoxia involves improving O_2 affinity and supply properties, which creates an internal environment that is well oxygenated. Yet environmental and metabolic changes, e.g., during heavy exercise, can bring these mammals into acute hypoxic states that are not too different from those faced by other hypoxia-tolerant animals. Thus, mammals adapted to chronic and acute hypoxic conditions share many of the same strategies.

BURROWING ANIMALS

Some mammals, such as the blind subterranean mole rat (superspecies *Spalax ehrenbergi*), live in subterranean burrows in which P_{O2} may drop to 50 mmHg or less (135). In fact, mole rats remain active at P_{O2} of 40 mm Hg (which is below P_{O2} levels at the Mount Everest summit) and P_{CO2} of 100 mm Hg (136). Experimentally exposed to P_{O2} of 20 mm Hg for up to 11 h, mole rats show no deleterious effects, whereas white rats (*Rattus norvegicus*) succumb after 2.5 h (137).

The chronic hypoxic conditions force mole rats to extract as much O_2 from their environment as possible by relying on a wide repertoire of adaptations. These animals achieve enhanced pulmonary O_2 diffusing capacity by a larger alveolar surface and higher lung capillarization, which improves lung O_2 uptake (138). Low levels of RBC 2,3-DPG, which drop even further with hypoxia exposure (139), increase Hb-O_2 affinity. As a consequence, arterial blood is saturated with O_2 at much lower P_{O2} than in rats (139). Hct levels and blood Hb contents of mole rats are in the upper range for terrestrial mammals (140). Arieli & Nevo (141) have demonstrated a positive correlation between Hct levels and the ability of mole rats to handle hypoxic stress.

Mole rats display high capillary densities in skeletal and cardiac muscles and threefold higher skeletal muscle Mb levels (138). Levels of HIF-1α (142) are constitutively high in mole rat muscles and remain unchanged after exposure to acute hypoxia. Thus, mole rat muscles may maintain adequate O_2 supply even during acute hypoxia (137, 142).

Mole rats are hypometabolic, even in normoxia, and remain so at O_2 concentrations as low as 4.6%. Yet at low P_{O2}, maximum O_2 uptake of exercising mole rats is higher than in white rats (138). The heart rate of mole rats under resting normoxic conditions is lower than expected on the basis of body size, but in hypoxia the heart rate increases to almost twice that of white rats (139). Thus, mole rats maintain a chronic hypometabolic state throughout much of their life. Yet despite low P_{O2} they can increase aerobic metabolism substantially in connection with physical activity.

BIRDS AND MAMMALS AT HIGH ALTITUDE

Because P_{O2} falls with increasing altitude, exposure to high altitude triggers a hypoxic response in nonacclimatized humans. Lung O_2 uptake is increased through carotid body–mediated hyperventilation (96, 97). O_2 delivery to tissues is enhanced owing to increased cardiac output (9) and subsequent, gradual increases in RBC and Hb content (143), which are induced by HIF-1-mediated increase in erythropoietin (144). O_2 delivery is further improved owing to a reduction in Hb-O_2 affinity caused by the increased formation of 2,3-DPG (145, 146). Although these hypoxic responses may appear appropriate, they are partly maladaptive: The large hyperventilatory response produces a state of hypocapnia and disturbance of acid-base balance (145). Hypoxia triggers pulmonary hypertension through the constriction of small pulmonary arteries (98, 145), whereas polycythemia increases blood viscosity, thereby placing additional burdens on the heart (145).

Not surprisingly, these acute, phenotypic responses differ from the largely genotypic adaptations displayed by several species of birds and mammals that permanently live at high altitude with few signs of hypoxic stress (143, 147). At any level of hypoxic challenge, the hyperventilation response of such animals is generally attenuated compared with that of lowland residents (6, 145, 146, 148). The hypoxia-induced pulmonary vasoconstrictor response is also blunted in high-altitude mammals (98, 145), whose lungs instead display a high O_2 diffusing capacity (145, 149). They have generally high, not low, Hb-O_2 affinity (6, 145, 150, 151) because of low RBC 2,3-DPG levels (145, 146), reduced sensitivity to 2,3-DPG and other organic phosphates, and specific mutations in Hb globin structure (145). They also maintain low to moderate $[Hb]_{blood}$ (13–15 $g \cdot dL^{-1}$) and Hct levels (35–47%) (145, 148, 151), and hypoxia-induced erythropoiesis appears blunted (6, 143, 148, 151). This strategy balances the benefits of a high O_2 carrying capacity with the disadvantages of high blood viscosity levels, at optimal Hb concentrations (145). Some birds and mammals living at high altitude may also have increased muscle capillarization (145, 150) and elevated skeletal muscle Mb concentrations (151), thus further promoting efficient cellular O_2 uptake.

Metabolically, high-altitude-residing humans display a lower maximal aerobic and glycolytic work capacities than do sea-level-residing humans, but work capacity is much less affected by acute hypoxia in highland than in lowland residents. Thus, highland-residing humans accumulate less lactate during exercise and display improved endurance under submaximal work conditions as compared with lowland-residing humans. With regard to the CNS, some high-altitude residents may employ a mild hypometabolic strategy as a hypoxia defense strategy (152).

The impressing tolerance to high-altitude exposure of some bird species deserves particular attention. For example, both bar-headed geese (*Anser indicus*) and Ruppell's griffon (*Gyps rueppellii*) can fly at 9,000–11,000 m altitude (150), levels at which their P_{aO2} could not possibly exceed 30–35 mm Hg and which would cause unconsciousness and death in unacclimatized humans. This unusual hypoxia tolerance is due in part to various adaptations, as listed above. But it also lies in the parabronchial cross-current circulation of the avian lung, which has a higher efficiency of O_2 exchange than does the alveolar mammalian lung (150, 153). Above all, the particularly high avian tolerance to hypocapnia explains these birds' high hypoxia tolerance. During hypoxia and hypocapnia, cerebral blood flow decreases in mammals but increases in birds manyfold (150, 153).

Effects of Chronic Hypoxia on the Central Nervous System and Carotid Body

Chronic hypoxia occurs not only in high-altitude or subterranean burrows but also in normoxic environments, e.g., owing to insufficient blood flow (cerebrovascular hemorrhage, brain tumor, vascular occlusion, cardiac arrest, or bypass surgery) or respiratory dysfunction (airway obstruction, asthma, emphysema, lung dysfunction, or neural control failure). Thus, there is great clinical interest in understanding the effects of chronic as opposed to acute hypoxia. Chronic hypoxic conditions typically result in protein expression (154) and gene expression that appear to be controlled by the activation of HIF-1. Chronic hypoxia upregulates glycolytic and related enzymes such as lactate dehydrogenase, pyruvate kinase, and hexokinase but not nonglycolytic enzymes such as fatty acyl-CoA synthetase, cytoplasmic malate dehydrogenase, and glucose-6-phosphate dehydrogenase (67). Chronic hypoxia also alters the expression of various Ca^{2+}, K^+, and Na^+ channels (67).

In the carotid body, chronic hypoxia induces similar changes that lead to, e.g., HIF-1α upregulation. These changes begin within minutes of exposure but progress such that chronic exposure results in morphological and biochemical changes in the carotid body, such as enlarged cells, increased catecholamine levels, and altered cellular appearance. The chronically adapted carotid body shows enhanced responsiveness to changes in O_2 pressure (155).

CONCLUSIONS

Hypoxia tolerance is achieved through the integration of (*a*) a reduction in metabolism, (*b*) protection against hypoxic cell death and injury, and (*c*) the maintenance of functional integrity. Much research has been devoted to understanding the molecular mechanisms that underlie hypometabolism and cell protection in the absence of functional considerations. This rather limited approach may not suffice to develop novel strategies for the prevention of hypoxic damage in the clinic. This review of the cellular- and systems-level adaptations to hypoxia illustrates that hypometabolism and cell protection are not achieved through any one molecular mechanism. Instead, hypometabolism and cellular protection occur through the reconfiguration of all organ systems and networks as well as their underlying cellular and subcellular components.

This reconfiguration leads to a hypometabolic and protective state that is very different from the organismal state seen in normoxia and that is characterized by very diverse cellular and subcellular changes. Some but not all cellular components may exhibit ion channel arrest. Some cells are hypometabolic, whereas others maintain a high metabolic state to maintain functional integrity. This conclusion has important clinical implications. To convey hypoxia tolerance and prevent hypoxic damage, it will be important to mimic a natural hypometabolic systems-level state. This is a very different aim than trying to find a specific molecular target that will uniformly affect all cells in an organism.

Thus, the challenge for future hypoxia research is to explore the cellular mechanisms and conditions that reconfigure an organ or cellular network into a hypometabolic state. Habitually hypoxia-exposed animals, such as diving, burrowing, and hibernating animals, know exactly how to attain this hypometabolic state, and our task should be to follow their lead.

ACKNOWLEDGMENTS

Studies associated with this review were supported by grants from the National Institutes of Health to J.M.R. and the Norwegian Research Council to L.P. Folkow and A.S. Blix.

LITERATURE CITED

1. Catling DC, Glein CR, Zahnle KJ, McKay CP. 2005. Why O_2 is required by complex life on habitable planets and the concept of planetary "oxygenation time." *Astrobiology* 5:415–38
2. Bennett AF, Ruben JA. 1979. Endothermy and activity in vertebrates. *Science* 206:649–54
3. Bennett A. 1980. The metabolic foundations of vertebrate behavior. *Bioscience* 30:452–56
4. Bickler PE, Buck LT. 2007. Hypoxia tolerance in reptiles, amphibians, and fishes: life with variable oxygen availability. *Annu. Rev. Physiol.* 69:145–70
5. Rossen R, Kabat H, Anderson JP. 1943. Acute arrest of cerebral circulation in man. *Arch. Neurol. Psychiatry* 50:510–28
6. Banchero N, Grover RF, Will JA. 1971. Oxygen transport in the llama (*Lama glama*). *Respir. Physiol.* 13:102–15
7. Singer D, Bach F, Bretschneider HJ, Kuhn HJ. 1993. Metabolic size allometry and the limits to beneficial metabolic reduction: hypothesis of a uniform specific minimal metabolic rate. See Ref. 8, pp. 447–58
8. Hochachka PW, Lutz PI, Sick T, Rosenthal M, Van den Thillart G, eds. 1993. *Surviving Hypoxia: Mechanisms of Control and Adaptation*. London/Tokyo/Boca Raton/Ann Arbor: CRC Press
9. Folkow B, Neil E. 1971. *Circulation*. London: Oxford Univ. Press. 593 pp
10. Lutz PL, Nilsson GE, Prentice HM. 2003. *The Brain Without Oxygen: Causes of Failure—Physiological and Molecular Mechanisms for Survivial*. Dordrecht/Boston/London: Kluwer Acad. Publ. 3rd ed. 252 pp
11. Leblond J, Krnjevic K. 1989. Hypoxic changes in hippocampal neurons. *J. Neurophysiol.* 62:1–14
12. Watkins WA, Moore KE, Tyack P. 1985. Investigations of sperm whale acoustic behaviours in the southeast Carribean. *Cetology* 49:1–15
13. Hindell MA, Slip DJ, Burton HR. 1991. The diving behaviour of adult male and female southern elephant seals, *Mirounga leonina* (Pinnipedia: Phocidae). *Aust. J. Zool.* 39:595–619
14. Folkow LP, Blix AS. 1999. Diving behaviour of hooded seals (*Cystophora cristata*) in the Greenland and Norwegian Seas. *Polar Biol.* 22:61–74
15. Kooyman GL, Kooyman TG. 1995. Diving behavior of emperor penguins nurturing chicks at Coulman Island, Antarctica. *Condor* 97:536–49
16. Scholander PF. 1940. Experimental investigations on the respiratory function in diving mammals and birds. *Hvalrådets Skr.* 22:1–131
17. Lenfant C, Johansen K, Torrance JD. 1970. Gas transport and oxygen storage capacity in some pinnipeds and the sea otter. *Respir. Physiol.* 9:277–86
18. Ridgway SH, Johnston DG. 1966. Blood oxygen and ecology of porpoises of three genera. *Science* 151:456–58
19. Ponganis PJ, Starke LN, Horning M, Kooyman GL. 1999. Development of diving capacity in emperor penguins. *J. Exp. Biol.* 202:781–86
20. Stephenson R, Turner DL, Butler PJ. 1989. The relationship between diving activity and oxygen storage capacity in the tufted duck (*Aythya fuligula*). *J. Exp. Biol.* 141:265–75
21. Hochachka PW. 2000. Pinniped diving response mechanism and evolution: a window on the paradigm of comparative biochemistry and physiology. *Comp. Biochem. Physiol.* 126A:435–58
22. Elsner R, Meiselman HJ. 1995. Splenic oxygen storage and blood viscosity in seals. *Mar. Mamm. Sci.* 11:93–96

23. Cabanac AJ, Messelt EB, Folkow LP, Blix AS. 1999. The structure and blood-storing function of the spleen of the hooded seal (*Cystophora cristata*). *J. Zool. London* 248:75–81
24. Snyder GK. 1983. Respiratory adaptations in diving mammals. *Respir. Physiol.* 54:269–94
25. Willford DC, Gray AT, Hempleman SC, Davis RW, Hill EP. 1990. Temperature and the oxygen-hemoglobin dissociation curve of the harbor seal, *Phoca vitulina*. *Respir. Physiol.* 79:137–44
26. Wittenberg BA, Wittenberg JB. 1989. Transport of oxygen in muscle. *Annu. Rev. Physiol.* 51:857–78
27. Flögel U, Godecke A, Klotz LO, Schrader J. 2004. Role of myoglobin in the antioxidant defense of the heart. *FASEB J.* 18:1156–58
28. Burns JM, Blix AS, Folkow LP. 2000. Physiological constraint and diving ability: a test in hooded seals, *Cystophora cristata*. *FASEB J.* 14:A440, abstract no. 317.8
29. Polasek LK, Davis RW. 2001. Heterogeneity of myoglobin distribution in the locomotory muscles of five cetacean species. *J. Exp. Biol.* 204:209–15
30. Weber RE, Hemmingsen EA, Johansen K. 1974. Functional and biochemical studies of penguin myoglobin. *Comp. Biochem. Physiol.* 49B:197–214
31. O'Brien PJ, Shen H, McCutcheon LJ, O'Grady M, Byrne P, et al. 1992. Rapid, simple and sensitive microassay for skeletal and cardiac muscle myoglobin and hemoglobin: use in various animals indicates functional role of myohemoproteins. *Mol. Cell Biochem.* 112:42–52
32. Olsen CR, Elsner R, Hale FC, Kenney DW. 1969. "Blow" of the pilot whale. *Science* 163:953–55
33. Kooyman GL, Ponganis PJ. 1998. The physiological basis of diving to depth: birds and mammals. *Annu. Rev. Physiol.* 60:19–32
34. Tenney SM, Remmers JE. 1963. Comparative quantitative morphology of the mammalian lung: diffusing area. *Nature* 197:54–56
35. Folkow LP, Blix AS. 1992. Metabolic rates of minke whales (*Balaenoptera acutorostrata*) in cold water. *Acta Physiol. Scand.* 146:141–50
36. Kerem D, Elsner R. 1973. Cerebral tolerance to asphyxial hypoxia in the harbor seal. *Respir. Physiol.* 19:188–200
37. Kanatous SB, Elsner R, Mathieu-Costello O. 2001. Muscle capillary supply in harbor seals. *J. Appl. Physiol.* 90:1919–26
38. Blix AS, Folkow B. 1983. Cardiovascular adjustments to diving in mammals and birds. In *Handbook of Physiology. The Cardiovascular System III. Peripheral Circulation and Organ Blood Flow*, ed. JT Shepherd, FM Abboud, pp. 917–45. Bethesda: Am. Physiol. Soc.
39. Butler PJ, Jones DR. 1997. Physiology of diving of birds and mammals. *Physiol. Rev.* 77:837–99
40. Irving L. 1938. Changes in the blood flow through the brain and muscles during the arrest of breathing. *Am. J. Physiol.* 122: 207–14
41. Blix AS, Elsner R, Kjekshus JK. 1983. Cardiac output and its distribution through A-V shunts and capillaries during and after diving in seals. *Acta Physiol. Scand.* 118:109–16
42. Elsner R, Millard RW, Kjekshus JK, White F, Blix AS, Kemper WS. 1985. Coronary blood flow and myocardial segment dimensions during simulated dives in seals. *Am. J. Physiol.* 249:H1119–26
43. Kjekshus JK, Blix AS, Hol R, Elsner R, Amundsen E. 1982. Myocardial blood flow and metabolism in the diving seal. *Am. J. Physiol.* 242:R97–104
44. Kerem D, Hammond DD, Elsner R. 1973. Tissue glycogen levels in the Weddell seal (*Leptonychotes weddellii*): a possible adaptation to asphyxial hypoxia. *Comp. Biochem. Physiol. A* 45:731–36

45. Blix AS, From SH. 1971. Lactate dehydrogenase in diving animals—a comparative study with special reference to the eider (*Somateria mollissima*). *Comp. Biochem. Physiol.* 40B:579–84
46. White FC, Elsner R, Willford D, Hill E, Merhoff E. 1990. Responses of harbor seal and pig heart to progressive and acute hypoxia. *Am. J. Physiol.* 259:R849–56
47. Henden T, Aasum E, Folkow L, Mjos OD, Lathrop DA, Larsen TS. 2004. Endogenous glycogen prevents Ca^{2+} overload and hypercontracture in harp seal myocardial cells during simulated ischemia. *J. Mol. Cell Cardiol.* 37:43–50
48. Kjekshus JK, Maroko PK, Sobel BE. 1972. Distribution of myocardial injury and its relation to epicardial ST-segment changes after coronary artery occlusion in the dog. *Cardiovasc. Res.* 6:490–99
49. Elsner R, Franklin DL, Van Citters RL, Kenney DW. 1966. Cardiovascular defense against asphyxia. *Science* 153:941–49
50. Davis RW, Castellini MA, Kooyman GL, Maue R. 1983. Renal glomerular filtration rate and hepatic blood flow during voluntary diving in Weddell seals. *Am. J. Physiol.* 245:R743–48
51. Halasz NA, Elsner R, Garvie RS, Grotke GT. 1974. Renal recovery from ischemia: a comparative study of harbor seal and dog kidneys. *Am. J. Physiol.* 227:1331–35
52. Blix AS. 1987. Diving responses: fact or fiction. *NIPS* 2:64–66
53. Murdaugh HV Jr, Schmidt-Nielsen B, Wood JW, Mitchell WL. 1961. Cessation of renal function during diving in the trained seal (*Phoca vitulina*). *J. Cell Comp. Physiol.* 58:261–65
54. Ridgway SH, Carder DA, Clark W. 1975. Conditioned bradycardia in the sea lion *Zalophus californianus*. *Nature* 256:37–38
55. Thompson D, Fedak MA. 1993. Cardiac responses of gray seals during diving at sea. *J. Exp. Biol.* 174:139–64
56. Blix AS, Berg T. 1974. Arterial hypoxia and the diving responses of ducks. *Acta Physiol. Scand.* 92:566–68
57. Daly MdeB, Elsner R, Angell-James JE. 1977. Cardiorespiratory control by the carotid chemoreceptors during experimental dives in the seal. *Am. J. Physiol.* 232:H508–16
58. Hill RD, Schneider RC, Liggins GC, Schuette AH, Elliott RL, et al. 1987. Heart rate and body temperature during free diving of Weddell seals. *Am. J. Physiol.* 253:R344–51
59. Guyton GP, Stanek KS, Schneider RC, Hochachka PW, Hurford WE, et al. 1995. Myoglobin saturation in free-diving Weddell seals. *J. Appl. Physiol.* 79:1148–55
60. Scholander PF, Irving L, Grinnell SW. 1942. On the temperature and metabolism of the seal during diving. *J. Cell Comp. Physiol.* 19:67–78
61. Odden Å, Folkow LP, Caputa M, Hotvedt R, Blix AS. 1999. Brain cooling in diving seals. *Acta Physiol. Scand.* 166:77–78
62. Caputa M, Folkow L, Blix AS. 1998. Rapid brain cooling in diving ducks. *Am. J. Physiol.* 275:R363–71
63. Blix AS, Folkow LP, Walloe L. 2002. How seals may cool their brains during prolonged diving. *J. Physiol.* 543P:7P(Abstr.)
64. Kvadsheim PH, Folkow LP, Blix AS. 2005. Inhibition of shivering in hypothermic seals during diving. *Am. J. Physiol.* 289:R326–31
64a. Marder E, Bucher D. 2007. Understanding circuit dynamics using the stomatogastric nervous system of lobsters and crabs. *Annu. Rev. Physiol.* 69:in press
65. Bickler PE, Donohoe PH, Buck LT. 2000. Hypoxia-induced silencing of NMDA receptors in turtle neurons. *J. Neurosci.* 20:3522–28
66. Bickler PE. 2004. Clinical perspectives: neuroprotection lessons from hypoxia-tolerant organisms. *J. Exp. Biol.* 207:3243–49

67. Peña F, Ramirez JM. 2005. Hypoxia-induced changes in neuronal network properties. *Mol. Neurobiol.* 32:251–83
68. Boutilier RG. 2001. Mechanisms of cell survival in hypoxia and hypothermia. *J. Exp. Biol.* 204:3171–81
69. Hansen AJ. 1985. Effect of anoxia on ion distribution in the brain. *Physiol. Rev.* 65:101–48
70. Hong SS, Gibney GT, Esquilin M, Yu J, Xia Y. 2004. Effect of protein kinases on lactate dehydrogenase activity in cortical neurons during hypoxia. *Brain Res.* 1009:195–202
71. Robin ED, Murdaugh HV Jr, Pyron W, Weiss E, Soteres P. 1963. Adaptations to diving in the harbor seal: gas exchange and ventilatory responses to CO_2. *Am. J. Physiol.* 205:1175–77
72. Murphy B, Zapol WM, Hochachka PW. 1980. Metabolic activities of heart, lung, and brain during diving and recovery in the Weddell seal. *J. Appl. Physiol.* 48:596–605
73. Swanson RA, Choi DW. 1993. Glial glycogen stores affect neuronal survival during glucose deprivation in vitro. *J. Cereb. Blood Flow Metab.* 13:162–69
74. Tian GF, Baker AJ. 2000. Glycolysis prevents anoxia-induced synaptic transmission damage in rat hippocampal slices. *J. Neurophysiol.* 83:1830–39
75. Callahan DJ, Engle MJ, Volpe JJ. 1990. Hypoxic injury to developing glial cells: protective effect of high glucose. *Pediatr. Res.* 27:186–90
76. Jiang C, Agulian S, Haddad GG. 1991. O_2 tension in adult and neonatal brain slices under several experimental conditions. *Brain Res.* 568:159–64
77. Haddad GG, Jiang C. 1993. O_2 deprivation in the central nervous system: on mechanisms of neuronal response, differential sensitivity and injury. *Prog. Neurobiol.* 40:277–318
78. Siesjo BK, Katsura K, Kristian T. 1996. Acidosis-related damage. *Adv. Neurol.* 71:209–33
79. Giffard RG, Monyer H, Christine CW, Choi DW. 1990. Acidosis reduces NMDA receptor activation, glutamate neurotoxicity, and oxygen-glucose deprivation neuronal injury in cortical cultures. *Brain Res.* 506:339–42
80. Erdemli G, Xu YZ, Krnjevic K. 1998. Potassium conductance causing hyperpolarization of CA1 hippocampal neurons during hypoxia. *J. Neurophysiol.* 80:2378–90
81. Yang JJ, Chou YC, Lin MT, Chiu TH. 1997. Hypoxia-induced differential electrophysiological changes in rat locus coeruleus neurons. *Life Sci.* 61:1763–73
82. Mercuri NB, Bonci A, Johnson SW, Stratta F, Calabresi P, Bernardi G. 1994. Effects of anoxia on rat midbrain dopamine neurons. *J. Neurophysiol.* 71:1165–73
83. Calabresi P, Pisani A, Mercuri NB, Bernardi G. 1995. Hypoxia-induced electrical changes in striatal neurons. *J. Cereb. Blood Flow Metab.* 15:1141–45
84. Erdemli G, Crunelli V. 1998. Response of thalamocortical neurons to hypoxia: a whole-cell patch-clamp study. *J. Neurosci.* 18:5212–24
85. Thoby-Brisson M, Ramirez JM. 2000. Role of inspiratory pacemaker neurons in mediating the hypoxic response of the respiratory network in vitro. *J. Neurosci.* 20:5858–66
86. Peña F, Parkis MA, Tryba AK, Ramirez JM. 2004. Differential contribution of pacemaker properties to the generation of respiratory rhythms during normoxia and hypoxia. *Neuron* 43:105–17
87. Ramirez JM, Tryba AK, Peña F. 2004. Pacemaker neurons and neuronal networks: an integrative view. *Curr. Opin. Neurobiol.* 14:665–74
88. Thoby-Brisson M, Ramirez JM. 2001. Identification of two types of inspiratory pacemaker neurons in the isolated respiratory neural network of mice. *J. Neurophysiol.* 86:104–12
89. Lieske SP, Thoby-Brisson M, Telgkamp P, Ramirez JM. 2000. Reconfiguration of the neural network controlling multiple breathing patterns: eupnea, sighs and gasps. *Nat. Neurosci.* 3:600–7

90. Campanucci VA, Fearon IM, Nurse CA. 2003. A novel O_2-sensing mechanism in rat glossopharyngeal neurones mediated by a halothane-inhibitable background K^+ conductance. *J. Physiol.* 548:731–43
91. Yamada K, Ji JJ, Yuan H, Miki T, Sato S, et al. 2001. Protective role of ATP-sensitive potassium channels in hypoxia-induced generalized seizure. *Science* 292:1543–46
92. Mironov SL, Langohr K, Richter DW. 2000. Hyperpolarization-activated current, I_h, in inspiratory brainstem neurons and its inhibition by hypoxia. *Eur. J. Neurosci.* 12:520–26
93. Luhmann HJ. 1996. Ischemia and lesion induced imbalances in cortical function. *Prog. Neurobiol.* 48:131–66
94. Elsen FP, Ramirez JM. 1997. Hypoxia reduces the amplitude of voltage dependent calcium currents in the isolated respiratory system of mice. *Soc. Neurosci. Abstr.* 27:495.11
95. Lieske SP, Ramirez JM. 2006. Pattern-specific synaptic mechanisms in a multifunctional network. I. Effects of alterations in synapse strength. *J. Neurophysiol.* 95:1323–33
96. Prabhakar NR. 2000. Oxygen sensing by the carotid body chemoreceptors. *J. Appl. Physiol.* 88:2287–95
97. Prabhakar NR. 2006. O_2 sensing at the mammalian carotid body: why multiple O_2 sensors and multiple transmitters? *Exp. Physiol.* 91:17–23
98. Groves BM, Droma T, Sutton JR, McCullough RG, McCullough RE, et al. 1993. Minimal hypoxic pulmonary hypertension in normal Tibetans at 3,658 m. *J. Appl. Physiol.* 74:312–18
99. Neubauer JA, Sunderram J. 2004. Oxygen-sensing neurons in the central nervous system. *J. Appl. Physiol.* 96:367–74
100. Telgkamp P, Ramirez JM. 1999. Differential responses of respiratory nuclei to anoxia in rhythmic brain stem slices of mice. *J. Neurophysiol.* 82:2163–70
101. Maxwell PH. 2005. Hypoxia-inducible factor as a physiological regulator. *Exp. Physiol.* 90:791–97
102. Johnson P, Elsner R, Zenteno-Savín T. 2004. Hypoxia-inducible factor in ringed seal (*Phoca hispida*) tissue. *Free Radic. Res.* 38:847–54
103. Cascio WE, Yang H, Muller-Borer BJ, Johnson TA. 2005. Ischemia-induced arrhythmia: the role of connexins, gap junctions, and attendant changes in impulse propagation. *J. Electrocardiol.* 38:55–59
104. Saint DA. 2006. The role of the persistent Na^+ current during cardiac ischemia and hypoxia. *J. Cardiovasc. Electrophysiol.* 17:S96–103
105. Kevin LG, Novalija E, Stowe DF. 2005. Reactive oxygen species as mediators of cardiac injury and protection: the relevance to anesthesia practice. *Anesth. Analg.* 101:1275–87
106. Zenteno-Savín T, Clayton-Hernández E, Elsner R. 2002. Diving seals: are they a model for coping with oxidative stress? *Comp. Biochem. Physiol.* 133C:527–36
107. Beebe DW, Gozal D. 2002. Obstructive sleep apnea and the prefrontal cortex: towards a comprehensive model linking nocturnal upper airway obstruction to daytime cognitive and behavioral deficits. *J. Sleep Res.* 11:1–16
108. Nattie EE, Doble EA. 1984. Threshold of intermittent hypoxia-induced right ventricular hypertrophy in the rat. *Respir. Physiol.* 56:253–59
109. Godukhin O, Savin A, Kalemenev S, Levin S. 2002. Neuronal hyperexcitability induced by repeated brief episodes of hypoxia in rat hippocampal slices: involvement of ionotropic glutamate receptors and L-type Ca^{2+} channels. *Neuropharmacology* 42:459–66
110. Meerson FZ, Ustinova EE, Orlova EH. 1987. Prevention and elimination of heart arrhythmias by adaptation to intermittent high altitude hypoxia. *Clin. Cardiol.* 10:783–89
111. Feldman JL, Mitchell GS, Nattie EE. 2003. Breathing: rhythmicity, plasticity, chemosensitivity. *Annu. Rev. Neurosci.* 26:239–66

112. Blitz DM, Ramirez JM. 2002. Long-term modulation of respiratory network activity following anoxia in vitro. *J. Neurophysiol.* 87:2964–71
113. Gidday JM. 2006. Cerebral preconditioning and ischaemic tolerance. *Nat. Rev. Neurosci.* 7:437–48
114. Cheung MM, Kharbanda RK, Konstantinov IE, Shimizu M, Frndova H, et al. 2006. Randomized controlled trial of the effects of remote ischemic preconditioning on children undergoing cardiac surgery: first clinical application in humans. *J. Am. Coll. Cardiol.* 47:2277–82
115. Lyman CP, Willis JS, Malan A, Wang LCH. 1982. *Hibernation and Torpor in Mammals and Birds*. New York: Academic Press. 265 pp.
116. Heller HC. 1979. Hibernation: neural aspects. *Annu. Rev. Physiol.* 41:305–21
117. Carey HV, Andrews MT, Martin SL. 2003. Mammalian hibernation : Cellular and molecular responses to depressed metabolism and low temperature. *Physiol. Rev.* 83:1153–81
118. Milsom WK, Zimmer MB, Harris MB. 1999. Regulation of cardiac rhythm in hibernating mammals. *Comp. Biochem. Physiol.* 124A:383–91
119. Heldmaier G, Ortmann S, Elvert R. 2004. Natural hypometabolism during hibernation and daily torpor in mammals. *Resp. Physiol. Neurobiol.* 141:317–29
120. Tøien Ø, Drew KL, Chao ML, Rice ME. 2001. Ascorbate dynamics and oxygen consumption during arousal from hibernation in Arctic ground squirrels. *Am. J. Physiol.* 281:R572–83
121. Drew KL, Harris MB, LaManna JC, Smith MA, Zhu XW, Ma YL. 2004. Hypoxia tolerance in mammalian heterotherms. *J. Exp. Biol.* 207:3155–62
122. Maginniss LA, Milsom WK. 1994. Effects of hibernation on blood oxygen transport in the golden-mantled ground squirrel. *Respir. Physiol.* 95:195–208
123. Postnikova GB, Tselikova SV, Kolaeva SG, Solomonov NG. 1999. Myoglobin content in skeletal muscle of hibernating ground squirrels rises in autumn and winter. *Comp. Biochem. Physiol.* 124A:35–37
124. Ma YL, Zhu X, Rivera PM, Tøien Ø, Barnes BM, et al. 2005. Absence of cellular stress in brain after hypoxia induced by arousal from hibernation in Arctic ground squirrels. *Am. J. Physiol.* 289:R1297–1306
125. Silvani A, Asti V, Berteotti C, Ferrari V, Franzini C, et al. 2006. Sleep-dependent changes in cerebral oxygen consumption in newborn lambs. *J. Sleep Res.* 15:206–11
126. Hoshi Y, Mizukami S, Tamura M. 1994. Dynamic features of hemodynamic and metabolic changes in the human brain during all-night sleep as revealed by near-infrared spectroscopy. *Brain Res.* 652:257–62
127. Lyamin OI, Mukhametov LM, Siegel JM, Nazarenko EA, Polyakova IG, Shpak OV. 2002. Unihemispheric slow wave sleep and the state of the eyes in a white whale. *Behav. Brain Res.* 129:125–29
128. Skinner LA, Milsom WK. 2004. Respiratory chemosensitivity during wake and sleep in harbour seal pups (*Phoca vitulina richardsii*). *Physiol. Biochem. Zool.* 77:847–63
129. Frerichs KU, Kennedy C, Sokoloff L, Hallenbeck JM. 1994. Local cerebral blood flow during hibernation, a model of natural tolerance to "cerebral ischemia." *J. Cereb. Blood Flow Metab.* 14:193–205
130. Pakhotin PI, Pakhotina ID, Belousov AB. 1993. The study of brain slices from hibernating mammals in vitro and some approaches to the analysis of hibernation problems in vivo. *Prog. Neurobiol.* 40:123–61
131. Turrigiano GG, Nelson SB. 2004. Homeostatic plasticity in the developing nervous system. *Nat. Rev. Neurosci.* 5:97–107

132. Houweling AR, Bazhenov M, Timofeev I, Steriade M, Sejnowski TJ. 2005. Homeostatic synaptic plasticity can explain post-traumatic epileptogenesis in chronically isolated neocortex. *Cereb. Cortex* 15:834–45
133. Dave KR, Prado R, Raval AP, Drew KL, Perez-Pinzon MA. 2006. The Arctic ground squirrel brain is resistant to injury from cardiac arrest during euthermia. *Stroke* 37:1261–65
134. Igelmund P, Spangenberger H, Nikmanesh FG, Gabriel A, Lütke K, et al. 1996. Hibernation and hippocampal synaptic transmission. In *Adaptation to the Cold: Tenth International Hibernation Symposium*, ed. F Geiser, AJ Hubert, SC Nicol, pp. 159–65. Armidale, Australia: Univ. New Es. Press
135. Shams I, Avivi A, Nevo E. 2005. Oxygen and carbon dioxide fluctuations in burrows of subterranean blind mole rats indicate tolerance to hypoxic-hypercapnic stresses. *Comp. Biochem. Physiol.* 142A:376–82
136. Arieli R, Ar A, Shkolnik A. 1977. Metabolic responses of a fossorial rodent (*Spalax ehrenbergi*) to simulated burrow conditions. *Physiol. Zool.* 50:61–75
137. Avivi A, Resnick MB, Nevo E, Joel A, Levy AP. 1999. Adaptive hypoxic tolerance in the subterranean mole rat *Spalax ehrenbergi*: the role of vascular endothelial growth factor. *FEBS Lett.* 452:133–40
138. Widmer HR, Hoppeler H, Nevo E, Taylor CR, Weibel E. 1997. Working underground: respiratory adaptations in the blind mole rat. *Proc. Natl. Acad. Sci. USA* 94:2062–67
139. Nevo E. 1979. Adaptive convergence and divergence of subterranean mammals. *Annu. Rev. Ecol. Syst.* 10:269–308
140. Arieli R, Heth G, Nevo E, Hoch D. 1986. Hematocrit and hemoglobin concentration in four chromosomal species and some isolated populations of actively speciating subterranean mole rats in Israel. *Experentia* 42:441–43
141. Arieli R, Nevo E. 1991. Hypoxic survival differs between two mole rat species (*Spalax ehrenbergi*) of humid and arid habitats. *Comp. Biochem. Physiol.* 100A:543–45
142. Avivi A, Shams I, Joel A, Lache O, Levy AP, Nevo E. 2005. Increased blood vessel density provides the mole rat physiological tolerance to its hypoxic subterranean habitat. *FASEB J.* 19:1314–16
143. Schmidt W, Spielvogel H, Eckardt KU, Quintela A, Penaloza R. 1993. Effects of chronic hypoxia and exercise on plasma erythropoietin in high-altitude residents. *J. Appl. Physiol.* 74:1874–78
144. Wang GL, Jiang B-H, Rue EA, Semenza GL. 1995. Hypoxia-inducible factor 1 is a basic-helix-loop-helix-PAS heterodimer regulated by cellular O_2 tension. *Proc. Natl. Acad. Sci. USA* 92:5510–14
145. Monge C, Léon-Velarde F. 1991. Physiological adaptations to high altitude: oxygen transport in mammals and birds. *Physiol. Rev.* 71:1135–72
146. Samaja M, Mariani C, Prestini A, Cerretelli P. 1997. Acid-base balance and O_2 transport at high altitude. *Acta Physiol. Scand.* 159:249–56
147. Hochachka PW, Rupert JL. 2003. Fine tuning the HIF-1 'global' O_2 sensor for hypobaric hypoxia in Andean high-altitude natives. *BioEssays* 25:515–19
148. Black CP, Tenney SM. 1980. Oxygen transport during progressive hypoxia in high-altitude and sea-level waterfowl. *Respir. Physiol.* 39:217–39
149. Lindsted SL, Hokanson JF, Wells DJ, Swain SD, Hoppeler H, Navarro V. 1991. Running energetics in the pronghorn antelope. *Nature* 353:748–50
150. Faraci FM. 1991. Adaptations to hypoxia in birds: how to fly high. *Annu. Rev. Physiol.* 53:59–70

151. Reynafarje C, Faura J, Villavicenchio D, Curaca A, Reynafarje B, et al. 1975. Oxygen transport of hemoglobin in high-altitude animals (Camelidae). *J. Appl. Physiol.* 38:806–10
152. Hochachka PW, Gunga HC, Kirsch K. 1998. Our ancestral physiological phenotype: an adaptation for hypoxia tolerance and for endurance performance? *Proc. Natl. Acad. Sci. USA* 95:1915–20
153. Piiper J, Scheid P. 1975. Gas transport efficiency of gills, lungs and skin: theory and experimental data. *Respir. Physiol.* 13:292–304
154. Peers C, Kemp PJ. 2004. Ion channel regulation by chronic hypoxia in models of acute oxygen sensing. *Cell Calcium* 36:341–48
155. Lahiri S, Roy A, Baby SM, Hoshi T, Semenza GL, Prabhakar NR. 2006. Oxygen sensing in the body. *Prog. Biophys. Mol. Biol.* 91:249–86
156. Bron KM, Murdaugh HV Jr, Millen JE, Lenthall R, Rzaskin P, Robin ED. 1966. Arterial constrictor response in a diving mammal. *Science* 152:540–43

Hypoxia Tolerance in Reptiles, Amphibians, and Fishes: Life with Variable Oxygen Availability

Philip E. Bickler[1,*] and Leslie T. Buck[2]

[1]Department of Anesthesia, University of California, San Francisco, California 94143; email: bicklerp@anesthesia.ucsf.edu

[2]Department of Zoology, University of Toronto, Toronto, Ontario, Canada M5S 3G5; email: buckl@zoo.utoronto.ca

Annu. Rev. Physiol. 2007. 69:145–70

First published online as a Review in Advance on October 12, 2006

The *Annual Review of Physiology* is online at http://physiol.annualreviews.org

This article's doi: 10.1146/annurev.physiol.69.031905.162529

0066-4278/07/0315-0145$20.00

*Corresponding author

Key words

metabolic arrest, dormancy, reoxygenation, lower vertebrates

Abstract

The ability of fishes, amphibians, and reptiles to survive extremes of oxygen availability derives from a core triad of adaptations: profound metabolic suppression, tolerance of ionic and pH disturbances, and mechanisms for avoiding free-radical injury during reoxygenation. For long-term anoxic survival, enhanced storage of glycogen in critical tissues is also necessary. The diversity of body morphologies and habitats and the utilization of dormancy have resulted in a broad array of adaptations to hypoxia in lower vertebrates. For example, the most anoxia-tolerant vertebrates, painted turtles and crucian carp, meet the challenge of variable oxygen in fundamentally different ways: Turtles undergo near-suspended animation, whereas carp remain active and responsive in the absence of oxygen. Although the mechanisms of survival in both of these cases include large stores of glycogen and drastically decreased metabolism, other mechanisms, such as regulation of ion channels in excitable membranes, are apparently divergent. Common themes in the regulatory adjustments to hypoxia involve control of metabolism and ion channel conductance by protein phosphorylation. Tolerance of decreased energy charge and accumulating anaerobic end products as well as enhanced antioxidant defenses and regenerative capacities are also key to hypoxia survival in lower vertebrates.

INTRODUCTION

Dormancy: sustained period of behavioral inactivity often associated with profound reductions in metabolism and hypothermia

Q_{10}: the factor by which the rate of a process such as metabolism changes per ten-degree change in temperature

Compared with most birds and mammals, lower vertebrates (fishes, amphibians, and reptiles) are tolerant of variable oxygen availability. In fact, several species of fishes and turtles are the most anoxia-tolerant vertebrates, surviving complete lack of oxygen for prolonged periods. Perhaps the most remarkable example is the Western painted turtle (*Chrysemys picta*). This North American species survives without oxygen for up to four months during winter dormancy (1, 2). After accounting for the reduced metabolic rate of dormant turtles, this is the metabolic equivalent of enhancing the hypoxic survival of a typical mammal or bird by a factor of 1000 to 10,000. This survival is even more remarkable when one considers that the metabolic rate of the turtle's brain, corrected for temperature, is similar to that of mammals (3). As we illustrate below, studies of this and other reptiles, fishes, and amphibians reveal some of the mechanisms required for cellular survival during severely reduced oxygen availability.

The central thesis of this review is that long-term tolerance of variable oxygen levels requires three basic capabilities: (*a*) a capacity to decrease overall metabolic rate during hypoxia (metabolic depression), (*b*) a tolerance of increased levels of metabolic by-products (particularly protons), and (*c*) the capacity to avoid and/or repair cellular injury following reoxygenation. Lower vertebrates have realized this core triad of capabilites in diverse ways, which enables lower vertebrates to survive tissue anoxia far better than can mammals and birds on average [see accompanying review by Ramirez and colleagues (3a)]. Because of these capacities, studies on lower vertebrates are identifying novel targets for the therapy of human diseases involving tissue hypoxia, such as cardiopulmonary disease, vascular disease, and stroke.

Figure 1 depicts the challenges that cells face in dealing with variable oxygen levels. This figure shows that avoidance of oxidative injury during high oxygen availability and reoxygenation must complement the tolerance of low oxygen. Thus, to achieve hypoxia tolerance, an organism must be tolerant of cycles of highly variable oxygen availability.

Adaptations to Variable Oxygen in Lower Vertebrates Are Highly Diverse

The variety of body structures, lifestyles, and environments common to fishes, amphibians, and reptiles is equaled by the diversity of adaptations to variable oxygen. Aquatic animals, more so than air breathers, are likely to experience variation in ambient oxygen. For example, ocean tide pools, especially in the tropics, range from hyperoxia during daylight hours to profound hypoxia at night (4). Waters in the Amazon basin decline in oxygen for entire seasons when stagnant water and decaying organic matter combine. Similar fluctuations in dissolved oxygen occur in coastal and pelagic marine waters as well. Winter ice and organic decay create severely hypoxic water in northern freshwater lakes. During environmental hypoxia or anoxia, gills and lungs become much less useful for gas exchange, and cellular respiration can become severely compromised. Although terrestrial amphibians and reptiles may encounter hypoxic burrow environments, the degree of hypoxia is probably not as severe as that experienced by aquatic species.

Life in the Cold: Is Hypothermia Tolerance Related to Hypoxia Tolerance?

Lower vertebrates' tolerance of variable oxygen is frequently associated with inactivity and hypometabolism (dormancy) induced by low temperature. Cold temperatures reduce metabolic demand (often with a Q_{10} exceeding 2.5), which the decreased energy production during hypoxia can supply. The energy savings and substrate conservation required for surviving cold dormancy and hypoxia probably involve universal mechanisms:

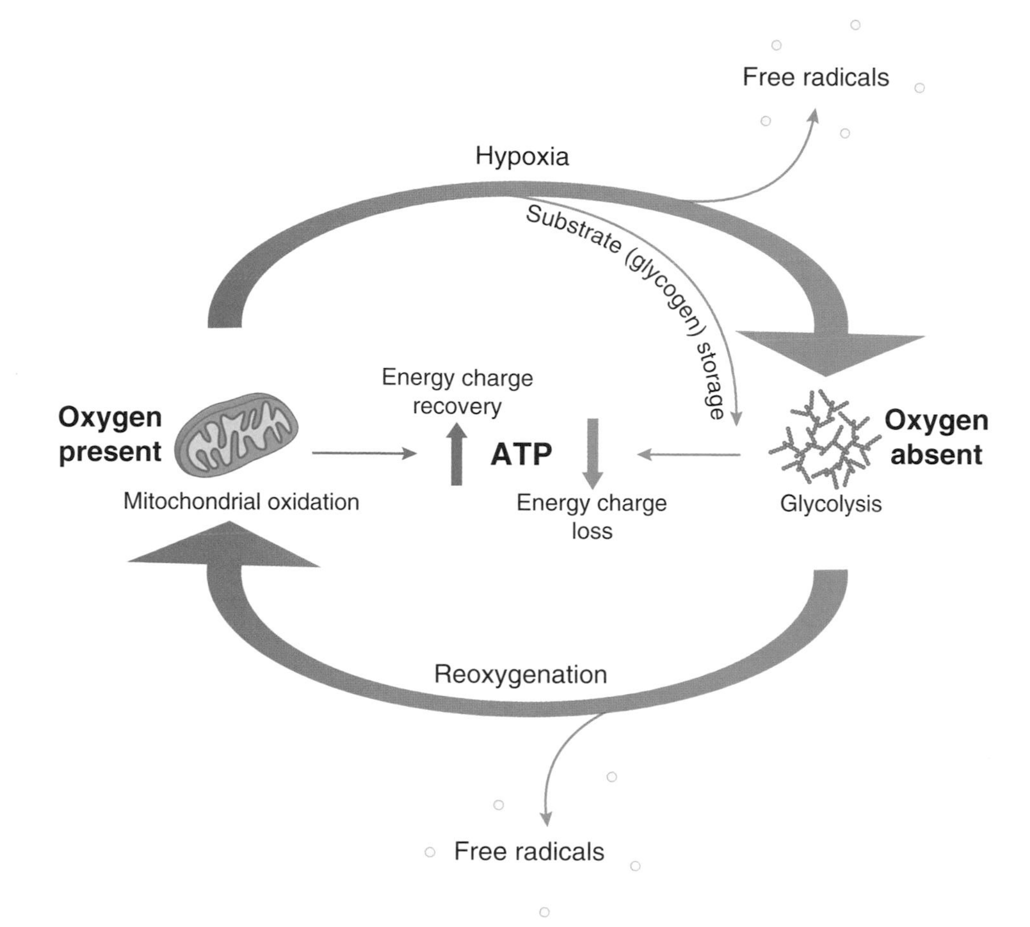

Figure 1

Fundamentals of oxidative stress in tissues experiencing cycles of oxygen availability, illustrating the cyclic nature of free-radical generation during oxygen/hypoxia transitions. Alterations in oxygen availability also lead to cycles of metabolic rate variation, with low metabolism during anaerobiosis and higher metabolism when oxygen is available. When oxygen is absent, glycogen is the major fuel. With prolonged anoxia, both rates of formation and the concentration of ATP and other high-energy intermediates will decrease. The generation of free radicals causes free-radical signaling, defense, and injury.

downregulation of energy production (supply-side bioenergetics) and of energy consumption (demand-side bioenergetics) (5). We discuss this in more detail below.

Twenty years ago, Hochachka (6) hypothesized that because cold and hypoxia decrease metabolic demands and energy production, tolerance of hypothermia and hypoxia may involve similar mechanisms. This hypothesis posits that energy conservation (metabolic arrest and ion channel arrest) must accompany decreased energy production to maintain energy charge, substrate stores, and ionic homeostasis during hypoxia or hypothermia. However, the precise mechanistic link between cold and hypoxia tolerance remains elusive. With respect to ion homeostasis (i.e., preventing the collapse of transcellular ion gradients in excitable cells), reduced body temperature generally decreases the activity of ion pumps more than ion leaks. Ion channels responsible for ionic leaks (e.g., potassium channels) are relatively insensitive to temperature changes ($Q_{10} < 2$), whereas ion-motive pumps such as

Metabolic arrest: the near-cessation of metabolism exhibited by many dormant organisms; often used to describe the state of metabolic suppression in anoxia-tolerant organisms during dormancy

the Na^+/K^+ pump are more strongly inhibited by temperature ($Q_{10} > 3$ in some cases). The disparity between ion pumping and ion channel leak rates leads to a redistribution of ions, depolarization of transcellular electrical potentials, and water accumulation in cold-intolerant cells (7).

Ion channel arrest: a substantial reduction in net transcellular ionic movements; a theoretical construct often not specific to the types of ion channels involved

Evolution of Tolerance of Variable Oxygen in Fish, Amphibians, and Reptiles

The appearance of molecular oxygen in the atmosphere approximately 2.2 billion years ago created both the stress of oxygen-related free-radical biotoxicity and the biochemical opportunity for complex metabolic and bioenergetic transformations (8). During their long evolutionary history, lower vertebrates have experienced wide variations in atmospheric oxygen. Isotopic and chemical analysis of ice cores and rock strata reveal that atmospheric oxygen levels at the time of the Permian-Triassic transition plummeted from approximately 30% oxygen to approximately 10% oxygen—less than half of present-day levels (9). Therefore, ancestors of modern fishes, amphibians, and some reptiles (turtles, but probably not lizards and snakes) underwent dramatic selection for hypoxia tolerance and experienced a limitation of altitudinal distributions (10). In contrast to this severe oxygen limitation faced by the ancestors of modern lower vertebrates, the adaptive radiation of modern mammals occurred when oxygen levels were steadily increasing from this minimum. The evolution of the mammalian placenta may be related to this increasing level of oxygen availability (10, 11).

DIVERSITY OF ANOXIA TOLERANCE STRATEGIES IN LOWER VERTEBRATES

Fish

The 20,000 extant species of fishes vary greatly in hypoxia tolerance. Species that rely extensively on aerobic metabolism for rapid and sustained swimming, such as salmon and tuna, are moderately to extremely sensitive to anoxia (12–14), whereas carp, eels, and hagfish can perform well at low oxygen levels (14–17). This impressive diversity prompts several questions that may be particularly well addressed in fishes: In fish, is hypoxia tolerance basal or a derived adaptation, and why are high aerobic capabilities, as seen in fish such as trout or tuna, associated with intolerance of hypoxia?

At the greatest extreme of hypoxia tolerance is the crucian carp (*Carassius carassius*), a species that endures months of hypoxia at low temperature [survival data reviewed by Nilsson & Renshaw (18)], a capacity approaching that of some North American freshwater turtles (see below). Their cousins, common goldfish (*Carassius auratus*), have half-lethal times of 45 h under anoxia at 5°C and 22 h at 20°C. Strong metabolic depression (to approximately 30% of normal) during anoxia is key for survival of these *Carassius* species because it is essential for the conservation of glycogen stores. Recent publications detail the survival mechanisms of *Carassius* (18–21).

Nilsson and his group in Norway have made substantial advances in understanding the diversity and mechanisms of anoxia tolerance in fishes (20). Fishes such as the crucian carp and the goldfish differ from turtles in that they remain active during anoxia (19). Both *Carassius* rely on large muscle and liver glycogen reserves, reduced metabolism, and avoidance of lactic acidosis by converting lactate to ethanol and CO_2. These end products can be excreted through the gills, helping the animals to avoid the severe acid-base disturbances that dormant turtles endure (22). This is discussed further below (see Mechanisms of Hypoxia Tolerance).

Sustained locomotor activity during anoxia requires that the myocardium retain contractility, a capacity that carp and goldfish have (21). Information is increasing on the molecular adaptations of the myocardium to anoxia in these species. The activation of ATP-sensitive potassium (K_{ATP}) channels contributes to the anoxic survival of these cells (23, 24), and thus these channels have a similar role as in preconditioning mechanisms in mammalian heart tissue. K_{ATP} channels on the mitochondrial membrane may play critical roles in adaptation to hypoxia, but this hypothesis is presently controversial (25).

The sustained activity of anoxic goldfish also correlates with a maintenance, rather than a suppression, of Na^+-K^+ ATPase activity in hepatocytes (26), which is perhaps critical to the conversion of lactate to ethanol (see below).

Diverse hypoxia-tolerant fishes inhabit the Amazon basin (reviewed in Reference 27). Some tolerate six hours of anoxia at 28°C (27–29). Larger body size is correlated with tolerance of prolonged anoxia in these fishes. Almeida-Val et al. (27) have suggested that this is related to the lower mass-specific metabolism in larger species and to a greater absolute mass of fermentable substrate (glycogen) in larger animals (30).

Diverse coral reef fishes also tolerate highly variable levels of oxygen (4). Fishes in this environment will no doubt be a resource for identifying adaptations to variable oxygen.

Estivating fishes, such as African lungfish, enter a dormant, suspended animation state during drought, which is possibly associated with hypoxia. Reduced metabolism and a dormancy-like state of prolonged inactivity in estivating fishes are more similar to patterns for winter-dormant turtles than for overwintering carp or anoxic goldfish.

Variable environmental oxygen may be an important developmental signal in fishes. In Zebrafish embryos, low oxygen exposure is associated with developmental arrest. Low oxygen causes cell division arrest in the G and S portions of the cell cycle (31), which may be important in the developmental program of other vertebrates as well.

Amphibians

Frogs' anoxia tolerance is in between the anoxia sensitivity of mammals and anoxia tolerance of turtles and carp. For example, ranids survive a few days of anoxia at low temperatures (32–35) and approximately 3 h of anoxia at room temperature (36, 37). Larvae tolerate hypoxia better than adult frogs of the same species (38, 39). This is consistent with the general pattern that embryonic and neonatal vertebrates are more hypoxia tolerant than adults (40).

Na^+-K^+ ATPase: sodium-potassium adenosine triphosphatase pump

Because amphibians are apparently intermediate to turtles/carp and mammals in anoxia tolerance, they are useful for investigating the nature of the ultimate failure of tolerance with prolonged anoxia. The slow loss of ATP and dissipation of ion gradients observed in amphibians must also occur in the more tolerant turtles and carp but over a much longer timescale.

Some amphibians overwinter dormant in hypoxic water. Leopard frogs (*Rana pipiens*) hibernate under ice in cold water, where they can withstand anoxia for several days (41). Metabolic rate depression during cold hypoxic submergence is advantageous for adult hibernating *Rana temporaria*. *R. temporaria* brain maintains ATP levels for two days at 3°C in anoxia before declining (42). Frogs can maintain brain ATP levels in hypoxia (water P_{O2} 30–60 mm Hg) for up to 16 weeks, although briefer anoxia at 25°C is associated with ATP loss (36). The ability to depress metabolic rate so that low levels of oxidative phosphorylation meet ATP demands is probably key to the frogs' survival.

During anoxia at room temperature, brain ATP levels in frogs decrease slowly. At approximately 35% of normoxic ATP levels, ionic homeostasis fails, and extracellular K^+ concentration ($[K^+]_o$) gradually increases (36). With continued anoxia, this slow increase of $[K^+]_o$ is followed by a larger increase in $[K^+]_o$ and release of the excitatory neurotransmitters GABA and glutamate. Thus, the slow death of the anoxic frog brain exhibits all the features of acute energy failure seen in mammals but on a longer timescale (20, 36, 37, 43).

Amphibians may also show tolerance of ionic disturbances during hypoxia. Hedrick at al. (44) found that, compared with mammalian neurons, which experience increased $[Ca^{2+}]_i$ and die rapidly following even brief (<10 min) hypoxia, tadpole forebrain cells survive prolonged hypoxia and tolerate substantial (>1 μM) increases in $[Ca^{2+}]_i$ while doing so. This observation is important because researchers had assumed that increased $[Ca^{2+}]_i$ during anoxia is a central (and universal) cause of cell death during anoxia (45). That increased $[Ca^{2+}]_i$ may be associated with survival mechanisms rather than cell death is an example of the insights that are possible only by studying anoxia-tolerant cells.

Estivation may involve energetic and oxidative stress and also involve low ambient oxygen. During dry seasons, burrows and cocoons of the Australian frog *Cyclorana platycephala* or the American spadefoot toad *Scaphiopus couchii* may become moderately hypoxic. Spadefoot toads decrease metabolism substantially during dormancy, with a hypometabolic state regulated by phosphorylation of glycolytic pathway components (46).

Reptiles: Turtles, Snakes, Lizards

Although turtles may be the most anoxia-tolerant group of reptiles, not all reptiles and not even all turtles are similarly tolerant. For example, soft-shelled turtles (*Apalone*) are relatively intolerant of hypoxia, with mortality of approximately 50% after 14 days of submergence in anoxic water at 3°C (47). Other temperate species are more tolerant; for example, snapping turtles (*Chelydra serpentina*) tolerate 100 days of submergence in anoxic water (48). Sea turtles are probably relatively tolerant of hypoxia, based on their diving performance, but published studies are lacking.

In turtles, the heart, liver, and brain are adapted to hypoxia. The brain is of special interest because its metabolic rate is comparable with that of mammals at the same temperature (3). The regulation of metabolism and ion channel regulation in anoxic turtle brain is discussed at length below.

Hypoxia tolerance has been investigated in only a few species of snakes. Garter snakes (*Thamnophis sirtalis parietalis*) are quite tolerant of hypoxia, tolerating several days without oxygen at 5°C (49, 50). These animals are cold tolerant and have substantial free-radical antioxidant defenses as compared with nonhypoxia-tolerant reptiles (51). Sea snakes encounter some hypoxia with dives, but they do not dive long enough to become severely hypoxic (52).

Relatively little information is available to generalize regarding hypoxia sensitivity or tolerance in lizards. Although many lizards burrow, adaptation to low ambient oxygen, other than that associated with pulmonary ventilation adjustments, has received little attention.

MECHANISMS OF HYPOXIA TOLERANCE: GENERAL ISSUES FACING ALL TISSUES AND ORGANISMS

Evidence obtained over the past several decades shows that the key adaptation to long-term hypoxia is a simultaneous reduction in metabolic rate and metabolic demands, i.e., a reduction or near-suspension of many bioenergetic processes. Although it remains unclear whether there is some minimal survivable cellular energy charge (i.e., ATP or ATP plus phosphocreatine) concentration, anoxic survival is associated with reductions in both ATP production and ATP utilization. To match energy supply with energy demand and maintain ATP levels above some minimum, an organism has two options: increase supplies, using anaerobic pathways, or reduce demand (6). Increasing ATP supply via anaerobiosis wastes substrate and serves only to shorten survival time; thus, reducing ATP demand is the only viable long-term strategy for a vertebrate to survive without oxygen. Thus, organisms capable of long-term anoxic survival do not increase glycolytic ATP production to maintain aerobic respiratory rates; that is, they lack a Pasteur effect. A Pasteur effect, aside from rapidly depleting glycogen stores, would increase lactic acid concentration. By contrast, a reduction in energetic demand inevitably leads to an overall reduction in ATP turnover and thus metabolic rate or, as termed by Hochachka (6), metabolic arrest. The general concepts of the Hochachka framework have withstood the test of time and can accommodate new findings related to free-radical defenses, mechanisms of metabolic and ion channel regulation, and oxygen-sensing mechanisms, as shown in **Figure 2**.

Hypoxia

Signaling

Metabolic suppression

- Hypothermia
- Substrate conservation
- Arrest of protein synthesis

Ion channel suppression

- Selected tissues remain active

Effects

- Slow alterations in H, lactate, Ca^{2+}
- Slow ATP loss
- Delayed apoptosis

Reoxygenation

- Free radical defenses

Repair and recovery

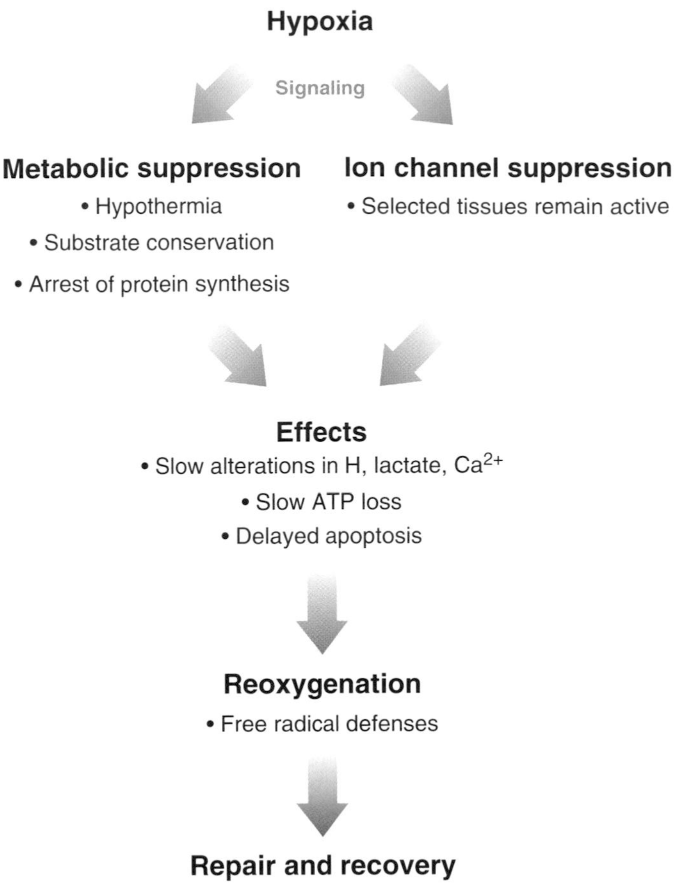

Figure 2

Triad of essential adaptations for surviving the hypoxia/reoxygenation cycles: metabolic suppression, ion channel suppression, and recovery and repair. Lower vertebrates that survive long-term hypoxia all use metabolic suppression to conserve fermentable substrate (glycogen) during anoxia. Depending on the species and tissues, ion channel suppression complements metabolic suppression as a necessary way to avoid ion translocations during decreased energy production. A major exception is crucian carp, in which the heart and at least some parts of the brain maintain some degree of ion channel activity. The consequences of the first two phases of the triad include, e.g., Ca^{2+}, lactate, protons, delayed ATP loss, and avoidance of apoptosis. The repair-and-recovery phase involves constitutive or induced free-radical defenses and the capacity for regeneration of damaged cells and tissues.

Organisms tolerating variable oxygen levels for any significant time must accomplish the following goals:

1. Metabolic re-engineering. This entails a switch from aerobic metabolism to a reduced level of anaerobic metabolism (i.e., without a significant Pasteur effect). Cellular sensors of oxygen and/or energy charge are required to coordinate this response. Prolonged survival requires large stores of fermentable substrate (glycogen). Energy conservation is necessary for substrate conservation and for energy balance during decreased energy production. This involves decreases in ionic conductance/ion pumping and in protein synthesis.
2. Tolerance of metabolic derangements associated with prolonged anoxia, in severe cases including ion translocations, decreases in ATP levels, and profound accumulation of metabolic end products such as lactate, protons, and ethanol. Increases in $[Ca^{2+}]_i$ must be tolerated and potentially used as a signaling mechanism.
3. Recovery and repair involves a sensor for emergence from dormancy and defense against the free radicals produced with reoxygenation. Such a strategy may involve the deferral of repairs associated with long-term anoxic dormancy or the regeneration of damaged tissues, such as in neurogenesis in the central nervous system.

These goals are diverse, and their accomplishment requires multiple mechanisms in multiple organ systems. This tolerance is clearly the result of changes in many genes over long periods.

$\Delta\Psi_m$: mitochondrial membrane potential

PKC: protein kinase C

Metabolic Depression, Mechanisms

Metabolic depression associated with hypoxia tolerance is well documented in virtually all major animal phyla. Guppy & Withers (53), Storey (46), and Guppy (5) provide good and phylogenetically broad summaries of metabolic depression. Hypothermia commonly accompanies dormancy and low-oxygen conditions, and hypothermia per se is a strategy to conserve endogenous fuel reserves.

Metabolic suppression: regulation of supply, demand, or both? The mechanisms by which hypoxia-tolerant organisms reduce their metabolic rate to minimal levels during oxygen deprivation are not understood. Specifically, it is uncertain whether the regulation of pathways of ATP supply (glycolysis, oxidative phosphorylation), ATP demand (protein synthesis, ion pumping), or both drives metabolic suppression. Although metabolic arrest refers to lowered energy or ATP turnover in the hypometabolic state, it is also loosely used to refer to the ATP-supply pathways of glycolysis and oxidative phosphorylation. Ion pumping by the Na^+-K^+ ATPase and protein synthesis utilize a similar 20–30% fraction of the total ATP-coupled oxygen consumption. The remaining oxygen consumption is due to mitochondrial proton leak (20%), the Ca^{2+} ATPase (8%), actinomyosin (10%), gluconeogenesis (10%), and ureagenesis (3%). Protein synthesis is a large ATP consumer, although it rapidly decreases in hypoxic or anoxic tissues of the turtle and goldfish but not in hypoxia-intolerant species (54). A sensitive detector of cellular energy status and a regulator of protein synthesis, AMP kinase (AMPK), may regulate the hypoxia-mediated decreases in protein synthesis (55). AMPK is activated by increases in cellular [AMP] and inhibited by high [ATP] and is regulated by differential phosphorylation. Although cellular adenylate levels do not change appreciably in anoxia-tolerant organisms, regional heterogeneity may occur. The key to AMPK regulation may again lie in phosphorylation of the enzyme. Regardless, this enzyme has not been investigated in anoxia-tolerant species.

The important glycolytic control enzymes phosphofructokinase and pyruvate kinase consistently show phosphorylation and decreased activities of this pair during hibernation, estivation, and overwintering in squirrel, snail, and turtle tissues (55a). There are also coordinate changes in the enzymes likely responsible for phosphorylation, protein kinase C (PKC) and protein kinase A (PKA) (55a). In fact, PKA and PKC activity and cAMP concentration increase within the first hour of anoxia in turtle liver and decreases thereafter, implicating a glycolytic regulatory role during the transition to anoxia. Furthermore, protein phosphatase 1 activity decreases, consistent with a general increase in cellular protein phosphorylation levels. The concept of regulation from the supply side garners some support from experiments on mitochondrial proton leak. Proton leak may account for 15% to 30% of the standard metabolic rate in both endo- and ectothermic species and therefore result in significant energy inefficiencies during hypometabolic periods (56b).

The hypothesis that proton leak is actively regulated in estivating and hibernating species in a channel arrest fashion and leads to increased energy efficiency proved to be incorrect. Hulbert et al. (55b) showed that proton leak was not actively regulated. Rather, it decreased passively as a result of a reduction in mitochondrial membrane potential ($\Delta\Psi_m$). These researchers divided energy metabolism into three components: (*a*) substrate oxidation (glycolysis, Kreb's cycle, and electron transport), (*b*) proton leak (across the mitochondrial inner membrane), and (*c*) ATP turnover (ATP synthase and ATP consumers such as ion pumps). The investigators determined the contribution of each component to metabolic rate in control and estivating snail populations. Using hepatopancreas cells from these populations and the mitochondrial ATP synthase inhibitor ouabain to block proton movement into the matrix, the researchers estimated proton leak. They also tested response of the substrate oxidation component by titration with FCCP (carbonyl cyanide *p*-trifluoromethoxyphenylhydrazone), a proton ionophore that dissipates the proton gradient. This methodology demonstrated that mitochondria from estivating snails have a reduced ability to increase respiration in response to the same concentration of FCCP as compared with control animals. ATP-consuming reactions (or ATP turnover) differed little in estivating and control animals. Approximately 75% of the total response of mitochondrial respiration to estivation was due to changes in substrate oxidation, and only 25% was due to reduction in ATP turnover and demand (56). Proton leak experiments on hibernating arctic ground squirrels and frogs yield similar conclusions (57, 58).

One difficulty in interpreting such data is that the respiratory data are normalized to milligram cell protein or cell number. Both measurements can obscure decreases in the concentrations of the enzymes of substrate oxidation. Furthermore, the proton leak studies are of animals after several weeks of hibernation or estivation when metabolism has reached a new hypometabolic steady state, making mechanistic inferences difficult. Indeed, hepatocytes and hepatopancreas cells retain their lower metabolic rates even after isolation from hypometabolic animals. Although this property makes them excellent model systems, the lower rates are likely due to lower titers of oxidative enzymes. For example, levels of citrate synthase and cytochrome oxidase decrease by approximately 20% and 40% in mitochondria isolated from hepatopancreas of estivating snails and control snails, respectively (58). Moreover, the response is tissue- and species-specific. Skeletal muscle in vitro from estivating frogs exhibits lowered respiration rates, but intestine, liver, skin, and fat do not; arctic ground squirrel hepatocytes and snail hepatopancreas cells do show a respiratory depression (58). Reduced substrate oxidation rates in cells from hibernating and estivating animals seem the result and not the cause of metabolic arrest. To determine the ultimate mechanisms and regulators of metabolic depression, studies performed during the transition to a depressed metabolic state are required. A reduction in ATP demand, either before or in concert with reductions in ATP supply, still remains the most logical occurrence for entry into a hypometabolic state. As we outline next, ion channel arrest seems the most important first step in reducing ATP demand.

Energy conservation: regulation of ionic conductance and pumps. How unnecessary energy demands are decreased during hypoxia is well understood. Hochachka's (6) ion channel arrest hypothesis is a key statement of the underlying issue. This hypothesis makes two predictions: (*a*) that the plasma membrane of anoxia-tolerant organisms has an inherently low permeability (either due to low channel densities or low channel activities) and (*b*) that membrane permeability decreases further in low-oxygen conditions (further channel arrest by suppression of either channel densities or channel activities). The first prediction appears correct. Comparisons of membrane permeability to Na^+ and K^+ in mammals and reptiles of similar size and body temperature indicate that reptile membranes are approximately fivefold less leaky than mammalian membranes (59). The documentation of channel arrest has been more difficult than that of metabolic arrest, but there are data supporting the channel arrest hypothesis. The oxygen-sensing mechanisms and second-messenger pathways that must be part of such a cell-based mechanism are even more elusive and have been recently reviewed (60).

Indirect indications of ion channel arrest include the maintenance of membrane potential while Na^+-K^+ ATPase activity decreases by 75% in anoxic turtle hepatocytes (61), an anoxia-mediated 42% decrease in voltage-gated Na^+ channel density in turtle cerebellum (62), and a hypoxia-mediated 50%

NMDAR: *N*-methyl-D-aspartate receptor

PP1 and PP2A: protein phosphatases types 1 and 2A

NR1: NMDAR subunit 1

mK_{ATP} channels: mitochondrial ATP-sensitive potassium channels

decrease in Na^+-K^+ ATPase activity coupled with decreased Na^+ permeability and maintenance of membrane potential in frog skeletal muscle (42). The only direct measures of ion channel arrest come from our laboratories and focus on anoxia-mediated reductions in NMDA receptor (NMDAR) open probability and whole-cell currents in turtle neurons (63, 64) and, more recently, in goldfish (L.T. Buck, M. Wilkie, & M.E. Pamenter, unpublished observations). The NMDAR is an important target for an effective anoxia-tolerance strategy in the turtle brain because activating the receptor in anoxic brain sheets results in a high degree of cell death, similar to that observed in anoxic mammalian brain (63). In turtle brain cerebral cortical sheets, a 62% decrease in NMDAR single-channel open time (P_{open}) occurs within 15 min of the onset of anoxia (65). This decrease may be reproduced by adenosine or during normoxia with the A_1 receptor agonist cyclopentyladenosine (CPA). Thus far, the ability to regulate NMDAR activity acutely in response to anoxia has been documented only in the freshwater turtle (*C. picta*) but may generally be a neuronal adaptation of anoxia-tolerant species.

The NMDAR is a high-flux ligand-gated cation channel that is highly permeable to Ca^{2+} and is required for fast excitatory neurotransmission in the central nervous system. Anoxic regulation of this receptor/ion channel is likely critical because ionized calcium levels in cerebral spinal fluid increase by approximately 6.5-fold during prolonged anoxia (66). The intracellular domain of the NMDAR contains multiple phosphorylation sites. In general, phosphorylation of NMDARs increases Ca^{2+} currents, and dephosphorylation decreases Ca^{2+} currents. Single-channel patch-clamp and whole-cell patch-clamp recording methods have shown that the channel is phosphorylated by protein kinases such as protein kinase A, protein kinase C, CaM kinase II, and tyrosine kinase. The channel is dephosphorylated by the associated protein phosphatases, Ca^{2+}/calmodulin-dependent protein phosphatase 2B, protein-tyrosine phosphatase, and protein phosphatases types 1 and 2A (PP1 and PP2A, respectively) (67, 68).

Protein phosphatases and Ca^{2+}/calmodulin independently modulate NMDAR activity in the anoxic turtle brain. Our data suggest that both mechanisms work in concert to modulate NMDAR activity during anoxia. When PP1 and PP2A were inhibited during anoxia, the anoxia-mediated decrease in NMDAR activity was completely abolished. If Ca^{2+}/calmodulin modulated NMDAR activity independently of PP1 and PP2A, then a partial decrease in NMDAR current would be observed during anoxia; this, however, was not the case. Furthermore, when Ca^{2+} was chelated and calmodulin inhibited, the anoxic response was also completely abolished. These results suggest that PP1, PP2A, Ca^{2+}, and calmodulin act in concert to decrease NMDAR activity in the anoxic turtle brain. In 1996, Ehlers et al. (69) reported two calmodulin-binding sites (CBS1 and CBS2) on the C terminus of the NMDAR subunit 1 (NR1) subunit.

Based on these findings, we propose mechanisms involving PP1, PP2A, Ca^{2+}, and calmodulin-mediated attenuation of NMDAR activity in the turtle cortex (**Figure 3**). During anoxia, PP1 and PP2A dephosphorylate the serine residue on the C terminus of the NR1 subunit. This enables Ca^{2+}/calmodulin to disrupt the NMDAR from the cytoskeleton by outcompeting binding with α-actinin-2. Thus, the reduction in NMDAR activity in turtle cortical neurons during anoxia is mediated by its dephosphorylation via PP1/PP2A. This attenuation is Ca^{2+} and calmodulin dependent but does not involve protein phosphatase type 2B (PP2B) after 20 min of anoxia.

The Mitochondrion and Ion Channel Arrest: mK_{ATP} Channels

The mitochondrion is the source of signaling events that modulate ion channel activity in anoxia-tolerant cells. The most compelling evidence comes from nontolerant cells

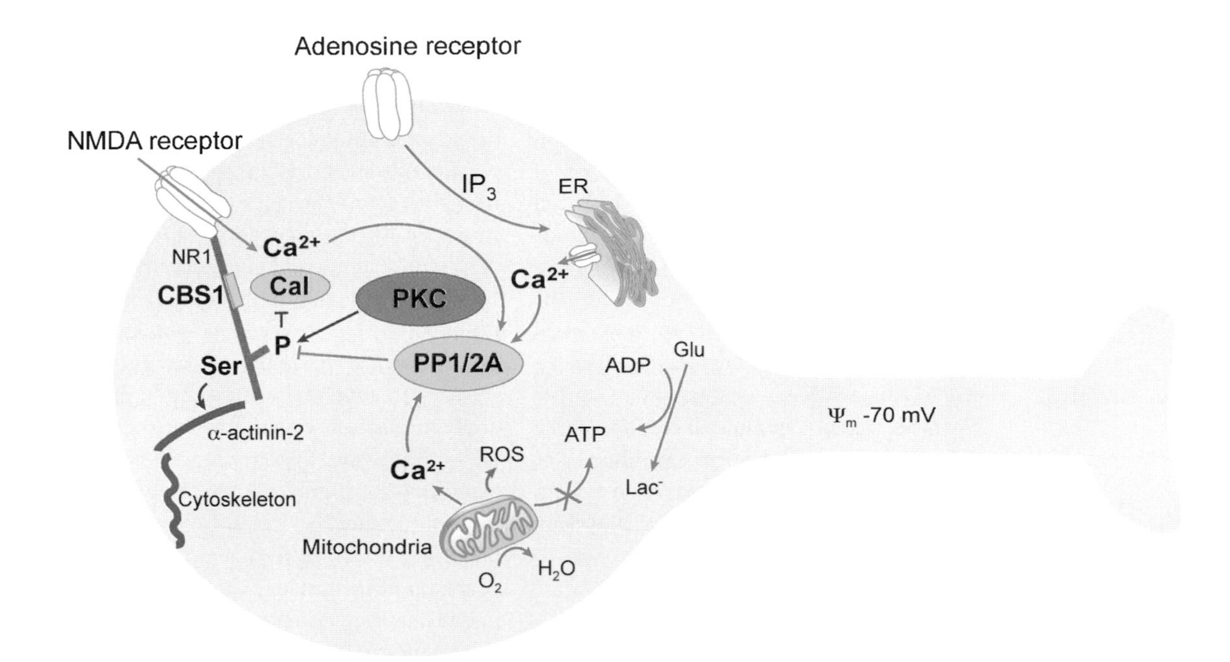

Figure 3

NMDA receptor regulation in anoxic turtle neurons. In the presence of oxygen, PKC phosphorylates (*red pathway*) a serine residue (Ser) on the NR1 subunit C terminus, blocking binding of calmodulin (Cal) to its binding site (CBS1) and promoting interaction with α-actinin-2 and the cytoskeleton. With anoxia and the loss of oxidative phosphorylation, there is a perturbation in ATP concentrations such that a rapid rise in adenosine levels activates adenosine receptors. Adenosine receptor activation may lead to an elevation in cytosolic Ca^{2+} (*blue pathway*) by an inositol-3-phosphate (IP_3)-mediated pathway. Depolarization of mitochondria and/or opening of mK_{ATP} channels may also trigger a rise in cytosolic Ca^{2+}. A role for reactive oxygen species (ROS) has not been investigated. The rise in Ca^{2+} activates protein phosphatases (PP1/2A), which dephosphorylate the serine residue and permit binding of calmodulin to the CBS1 site. This promotes the dissociation of α-actinin-2 from the cytoskeleton and the NR1 C-terminal domain, decreasing NMDA receptor activity (122, 123). Other abbreviations used: Ψ_m, membrane potential; ER, endoplasmic reticulum; Glu, glucose; Lac^-, lactate.

during hypoxic preconditioning, a brief sublethal hypoxia/ischemic bout that confers protection against a subsequent potentially injurious hypoxia/ischemic bout. Mitochondria confer preconditioning protection through mitochondrial K_{ATP} channels (mK_{ATP} channels) [for a review of the role of mK_{ATP} channels in ischemic preconditioning (IPC), see References 70–72]. This mechanism may be a link to channel arrest in anoxia-tolerant species as well.

mK_{ATP} channels are located on the mitochondrial inner membrane and thought to be structurally similar to plasmalemmal K_{ATP} channels (72a). L.T. Buck & M.E. Pamenter (unpublished observations) found that attenuation of NMDAR activity in the anoxic turtle brain may involve the mK_{ATP} channels in a fashion similar to a mechanism proposed by Holmuhamedov et al. (73): mK_{ATP} channel activation in heart causes K^+ influx into the mitochondria, resulting in depolarization of the mitochondrial membrane. As a consequence, Ca^{2+} is released from the mitochondria, elevating $[Ca^{2+}]_i$. In anoxic turtle cerebrocortical neurons, $[Ca^{2+}]_i$ increased by 35%, and inactivation of NMDARs scaled in a dose-dependent manner with changes

ROS: reactive oxygen species

in $[Ca^{2+}]_i$ (63). The elevation of $[Ca^{2+}]_i$ may originate from mK_{ATP} channel–induced Ca^{2+} release from the mitochondria. Our laboratory's studies on turtle cortical neuron NMDAR regulation show that activation of mK_{ATP} channels reduces NMDAR whole-cell currents by approximately 40%. Additionally, the anoxic reduction in NMDAR currents is abolished by perfusion of either the mK_{ATP}-specific blocker 5HD or the general K_{ATP} channel blocker glibenclamide. $[Ca^{2+}]_i$ may mediate mK_{ATP} attenuation of anoxic NMDAR activity because inclusion of the Ca^{2+} chelator BAPTA in the recording electrode abolishes both the anoxia- and mK_{ATP} channel–induced reductions in NMDAR currents.

Free Radicals Coordinate Responses to Variable Oxygen

Mitochondria are a major source of reactive oxygen species (ROS) under normal physiological conditions. ROS production is directly linked to the rate of oxidative phosphorylation, which is regulated partially by mitochondrial Ca^{2+} concentration ($[Ca^{2+}]_m$) and partially by $\Delta\Psi_m$. As this potential is reduced owing to ischemic opening of ion channels, ROS production is altered. ROS are potent cellular messengers, and redox signaling has been implicated in the regulation of many cellular processes. The primary radical produced in the mitochondrion is the superoxide anion (O_2^-). This radical is produced by complexes I and III of the electron transport chain. The specific site of superoxide production within complex I is not known; however, superoxide formation here occurs when electron transport is reversed owing to a high proton motive force. Experiments with the complex I inhibiter rotenone abolish the majority of superoxide formation, suggesting that superoxide production is due primarily to reverse electron transport from succinate to NAD^+ and not to complex III activity (74). Similarly, experiments by St.-Pierre et al. (75) implicated complex I as the primary production site of H_2O_2 via reverse electron flow. The same study found that complex III produces H_2O_2 only in the presence of inhibitors and not under normal physiological conditions.

The mitochondrial proton gradient regulates the rate of the reverse electron flow responsible for the generation of free radicals in the mitochondria. Therefore, a partial uncoupling of $\Delta\Psi_m$ alters the rate of free-radical production. The opening of mitochondrial ion channels such as the mK_{ATP} and mK_{Ca} channels uncouples the mitochondrial proton gradient and subsequently alters the rate of ROS production. Superoxide anions in large quantities are highly deleterious to the cell. However, mitochondria are constantly producing these anions, and alterations in the rate of radical formation may act as a redox signaling mechanism, potentially regulating downstream messengers such as PKC (76).

Oxygen-Sensing Mechanisms and Hypoxia Tolerance

There is considerable evidence that signaling based on ROS plays a role in tolerance of hypoxia or ischemia. Vanden Hoek et al. (77) reported an early increase in ROS production during hypoxic preconditioning. The blockade of mK_{ATP} channels with 5HD abolished both the protective effects of IPC as well as the hypoxia-induced generation of ROS. Similarly, Pain et al. (78) found that opening mK_{ATP} channels with diazoxide reduced infarct size in ischemic rabbit hearts and that this protection was abolished by inclusion of free-radical scavengers, suggesting that diazoxide triggers cardioprotection via the generation of free radicals. In addition, ROS may form a positive-enforcing feedback loop on mK_{ATP} channels. Mitochondria-derived nitric oxide (NO) can activate mK_{ATP} channels, and this messenger has also been implicated in IPC (for a review, see Reference 79).

A decrease in available oxygen such as that which occurs during ischemia or anoxia may decrease the rate of oxidative phosphorylation, changing mitochondrial ROS production. This altered redox signal may then feed-forward to activate mitochondrial ion channels such as the mK_{ATP} channel. Opening of these channels would partially uncouple the mitochondria, potentially altering cellular Ca^{2+} dynamics and activating other second messengers to downregulate energy-expensive ion channels, pumps, and cellular processes.

Hypoxia-inducible factor (HIF)-1 is involved in the expression of hypoxia response genes required for metabolic adjustments to hypoxia (80). The role of HIF-1 in the anoxia tolerance of turtles and carp remains to be studied. HIF-1α is at very low levels in rainbow trout, a species with poor ability to withstand prolonged hypoxia (81), although mild hypoxia (5% oxygen) induced HIF expression in this species. Contrary to expectations, a recent study with a late-stage brain HIF-1α knockout mouse demonstrated that the knockout mice were more tolerant to acute hypoxia/ischemia than were the control wild-type mice (82). This suggests that HIF-1α is of limited value during severe hypoxia. Moreover, an important role for HIF-2 is also possible. At present, whether HIF is an important organizer of the response to anoxia in hypoxia-tolerant lower vertebrates is unresolved.

Adenosine as a Low-Oxygen Signal

Adenosine is a good candidate for a metabolic signal that is produced locally and rapidly from the breakdown of ATP in response to decreasing oxygen availability. Cellular ATP concentrations range from 2 mM to approximately 3 mM, whereas cellular and extracellular adenosine concentrations are approximately 1 μM to 3 μM. This represents a 1000-fold concentration gradient similar in magnitude to the 10,000-fold difference in extra- and intracellular calcium concentrations. A 1% decrease in ATP concentration, which would be difficult to detect, corresponds to a 900% increase in adenosine concentration (from 3 μM to 30 μM). Because various stresses, such as intense metabolic activity and severe hypoxia, can decrease cellular ATP levels, the resulting rapid increase in adenosine concentration can serve as a sensitive signal of metabolic stress. Acting through its specific receptors (A_1, A_{2A}, A_{2B}, and A_3), adenosine has numerous physiological and metabolic effects that are consistent with this role. Indeed, adenosine has even been referred to as a "retaliatory metabolite" (83). Effects of adenosine include (*a*) vasodilation, to increase blood flow and therefore substrate delivery; (*b*) stimulation of glycogenolysis, providing a substrate for anaerobic glycolysis; (*c*) stimulation of anaerobic glycolysis, increasing ATP production to meet utilization; and (*d*) decreased neuronal excitability (postsynaptic inhibition) as well as neurotransmitter release (presynaptic inhibition), effectively reducing neuronal energy requirements [for a review, see Collis & Hourani (84)].

HIF-1α: hypoxia-inducible factor-1α

Studies on the effects of adenosine on various cellular and tissue functions in brain and liver offer six lines of evidence that adenosine plays such a regulatory role in the transition from normoxia to anoxia in anoxia-tolerant vertebrates. (*1*) Extracellular adenosine in the turtle intracerebral space accumulates during 2 h of anoxia to levels tenfold greater than during normoxia (from approximately 2 μM to 20 μM) (85). (*2*) Antagonizing the A_1 receptor (A_1R) during anoxia in isolated turtle cerebellum leads to a release of intracellular K^+ (86); retention of K^+ is characteristic of anoxia tolerance in the turtle brain (87). (*3*) Blood flow increases and decreases in the turtle cerebral vasculature in concert with the appearance and disappearance of adenosine during anoxia, responses that are inhibited by A_1R antagonists (88). (*4*) The application of adenosine to turtle brain sheets reduces NMDAR open probability and whole-cell conductance (G_w) (65, 89, 90). (*5*) In the anoxia-tolerant goldfish (*C. auratus*), adenosine has a powerful depressant effect on protein synthesis and Na^+-K^+ ATPase activity (91). (*6*) In anoxic crucian carp, adenosine receptor blockade results in a threefold increase in ethanol release

(92). Interestingly, although adenosine depresses neither hepatocyte oxygen consumption nor anaerobic lactate production in goldfish, it does decrease both of these processes in the anoxia-sensitive trout hepatocyte model (91). Unpublished results (R. Centritto & L.T. Buck) demonstrate that, as in the goldfish model, acute adenosine application does depress oxygen consumption or anaerobic lactate production in turtle hepatocytes. The role that adenosine plays in anoxic turtle liver function has not been investigated; specifically, its effect on turtle hepatocyte Na^+-K^+ ATPase activity is unknown.

Not only may adenosine function as an acute signaling molecule initiating a metabolic suppression, but it may also be involved in the long-term maintenance of a suppressed state. Microdialysis measurements of adenosine concentrations in the striatum of anoxic *Trachemys scripta* show that extracellular adenosine concentration cycles from a control normoxic level of approximately 1 μM to 5 μM during anoxia, with a mean period of 151 min (93). Changes in extracellular adenosine concentration also correspond inversely with changes in intracellular ATP concentration, strongly suggesting that ATP breakdown is the source of the adenosine increase. These data suggest that adenosine is part of a negative feedback oscillatory mechanism. When the system initially encounters an anoxic episode and glycolysis alone cannot provide sufficient ATP to meet cellular needs, a net dephosphorylation of ATP occurs, increasing adenosine levels. Adenosine levels increase extracellularly and interact with adenosine receptors, which decreases metabolic rate via an as-yet-unknown second-messenger pathway. Once metabolic rate decreases, glycolysis alone can meet the cellular ATP demand, and adenosine levels decrease, resulting in the removal of adenosine's suppressive effect; the cycle repeats until the organism again encounters oxygen.

Cerebral blood flow (CBF) in the turtle also seems to be cyclic and influenced by adenosine concentration. Epi-illumination microscopy reveals a 1.7-fold increase in CBF in *T. scripta* following adenosine administration during hypoxia. In contrast, adenosine application during normoxia increased CBF 3.8-fold (88). The adenosine receptor blocker, aminophylline, blocked both the anoxia- and adenosine-mediated changes in CBF (88). CBF increased and returned to basal after approximately 100 min of anoxia. Interestingly, this time frame is similar to the period of adenosine cycling discussed above and thus supports a causal relationship between extracellular adenosine concentration and CBF. However, long-term measurements of anoxic CBF cycling are required to support the oscillator model proposed by Lutz & Kabler (93).

Adenosine and adenosine receptor blockers have surprising impacts on CBF in several other vertebrates. In the crucian carp (*C. carassius*), CBF increases 2.2-fold during anoxia, and superfusing the brain with aminophylline abolishes this increase (94). Furthermore, this result can be replicated by the normoxic administration of adenosine and blocked by aminophylline. In both the turtle brain and the carp brain, adenosine seems to be an important anoxic signal, mediating vasodilation and increased CBF.

Similar studies in two other vertebrates—an anoxia-tolerant frog (*R. pipiens*) (95) and a hypoxia-tolerant epaulette shark (*Hemiscyllium ocellatum*) (96)—reveal an adenosine-independent anoxia-mediated increase in CBF. In the frog, topical application of adenosine during normoxic perfusion increased CBF, and aminophylline blocked CBF, but this was not observed during anoxia. Interestingly, in the hypoxia-tolerant epaulette shark, severe hypoxia did not increase CBF, nor did aminophylline affect hypoxic blood flow. However, normoxic application of adenosine can increase CBF, and it can be blocked by aminophylline. As in *R. pipiens*, in *H. ocellatum* adenosine does not appear to be involved in regulating CBF during anoxia or hypoxia.

TOLERANCE OF ACIDIC METABOLIC END PRODUCTS, ALTERED IONIC DISTRIBUTIONS, AND DECREASED ATP

Available fermentable substrate must ultimately limit tolerance of anoxia; as substrate is depleted, ATP levels will inevitably fall, even if metabolism is at a nadir. A gradual fall in [ATP] indeed occurs in hypoxic frogs, correlating with a slow leak of potassium from cells, in a pattern that Milton, Lutz, and colleagues (69a) termed slow death. A similar phenomenon may occur in dormant turtles and crucian carp after a much longer span of time.

While dormant, species such as turtles are essentially self-contained units, exchanging little with the environment. The work of Donald Jackson has been central to revealing that buffering by shell-derived calcium carbonate in turtles is critical to reducing the changes in pH that may occur during prolonged anaerobiosis (97–102). The binding of calcium released by the shell during the buffering process with lactate (Ca-Lactate) is similarly crucial to avoiding excessive increases in ionized calcium in the extracellular and intracellular fluids. These relationships are illustrated in **Figure 4**. Jackson has pointed out that, whereas the affinity of lactate for calcium ions is low in comparison to other calcium-binding compounds such as albumin, the very large increases in plasma lactate concentration during dives in turtles (>150 mmol l^{-1}) make the association of lactate and calcium quantitatively significant. The capacity of the shell to buffer pH changes may be critical to turtles' tolerance of anoxia for months. Furthermore, the survival success of yearling turtles critically depends on the amount of calcium in the shell at the initiation of dormancy.

Most interesting is the use of ethanol as an anaerobic end product in anoxic goldfish (**Figure 5**). In anoxic goldfish hepatocytes, ion gradients, and therefore probably metabolic functions such as the conversion of lactate to ethanol, are maintained during anoxia (26). Ethanol can be eliminated by diffusion across the gills, avoiding the gradual but eventually substantial load of protons associated with the lactic acid accumulation in prolonged dormancy that turtles experience. This strategy is obviously valuable to goldfish and carp, as the loss of ethanol is energetically costly.

RECOVERY FROM HYPOXIA: ANTIOXIDANT DEFENSES AND REPAIR MECHANISMS

Recovery from anoxia involves significant oxidative stress, and in the most anoxia-tolerant organisms, reoxygenation induces more molecular mediators of the stress response than does anoxia itself.

Antioxidant Defenses

A most significant issue in the tolerance of low oxygen is the avoidance of free-radical-mediated injury during reoxygenation (**Figure 1**). Ramaglia & Buck (103) found that heat shock protein (HSP) expression in anoxic painted turtles always increased upon reoxygenation from an anoxic bout, whereas more than 18 h of anoxia was required to induce a significant increase in HSP 72, 73, or 90 expression. This indicates that reoxygenation and a likely increase in free-radical stress following an anoxic bout may be more stressful than the anoxic period. Studies of free-radical stress in hypoxia-tolerant lower vertebrates reveal two general adaptive patterns: (*a*) Many anoxia-tolerant species constitutively express high levels of antioxidant defenses, such as that evinced by turtles, and (*b*) some species can generate free-radical defenses during hypoxia in apparent preparation for the ensuing oxidative stress of reoxygenation. In some species, the constitutive expression of amounts of defense molecules is not remarkable; it is only during hypoxia/reoxygenation that the increase occurs. Goldfish, red-sided garter snakes, and leopard frogs use this latter strategy (51).

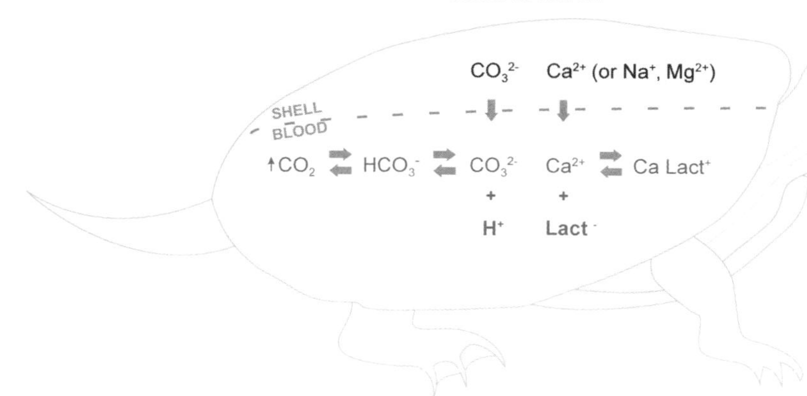

Blood to shell

$CO_2 \rightleftharpoons H^+ + CO_3^{2-}$ $Lact^- + Ca^{2+} \rightleftharpoons Ca\ Lact^+$

SHELL / BLOOD

$\uparrow CO_2$ H^+ $Lact^-$

Effects of anoxia adaptations

- Increased glycogen stores
- Enhanced blood buffering
- Metabolic supression
- Decreased activity
- Anaerobic metabolism

Figure 4

The role of the shell in buffering protons during prolonged anoxic dormancy in painted turtles, based on Donald Jackson's (97–102) studies. During prolonged anoxia, turtles develop high plasma concentrations of both lactate and calcium. Because of the very high concentrations of these ions, calcium lactate becomes a quantitatively significant species. This complex significantly reduces the free concentrations of both lactate and calcium. In addition, lactate is taken up by the shell and skeleton to an extent that strongly indicates that calcium lactate formation occurs in these organs as well. The binding of calcium to lactate thus contributes to the efflux of lactic acid from the anoxic cells and to the utilization of the considerable buffering capacity of the shell and skeleton. The buffering of protons produced by anaerobic metabolism, combined with a decreased metabolic rate and enhanced glycogen stores, is the cornerstone of anoxic survival in this species. For a review, see Jackson (97).

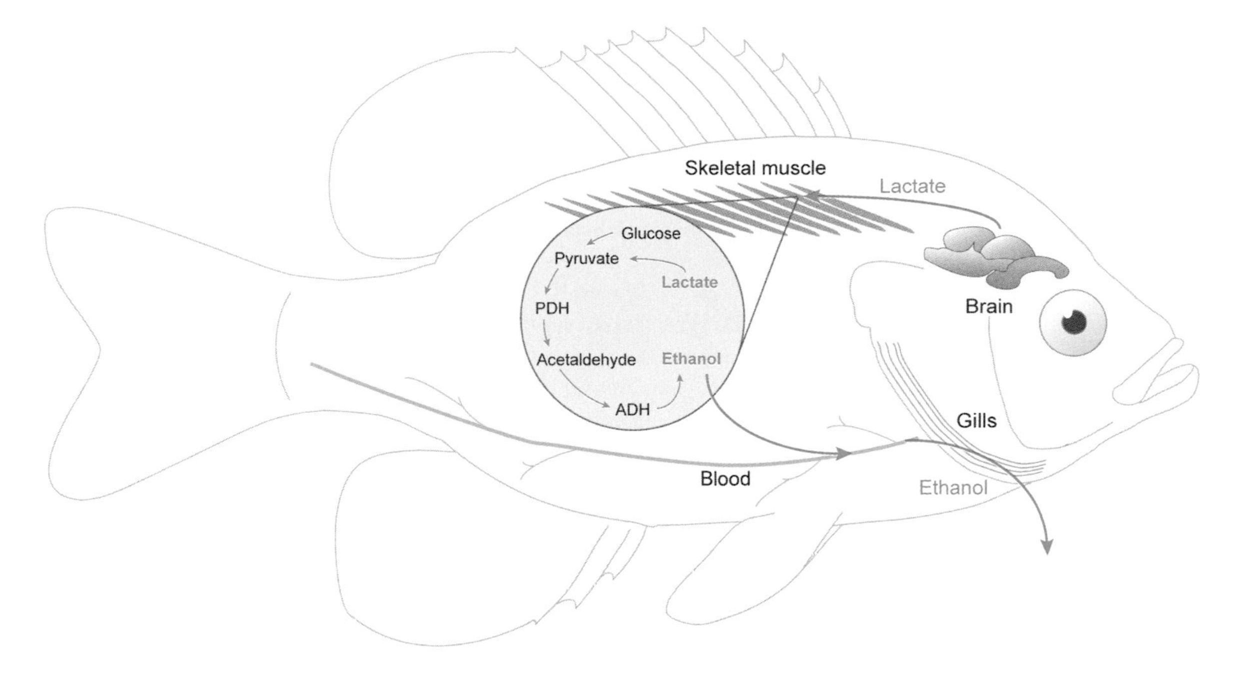

Effects of anoxia adaptations

- Acid-base balance maintained
- Regional brain activity maintained
- Activity level reduced, vigilance maintained
- Enhanced glycogen stores
- Decreased energy consumption

Figure 5

Metabolic adaptations to prolonged anoxia in crucian carp and goldfish, based on the work of Goran Nilsson and his group in Norway (reviewed in Reference 3). Lactate produced by the metabolism of glycogen in a range of tissues, including the brain, is converted to ethanol in skeletal muscle by lactate dehydrogenase (*not shown*), pyruvate dehydrogenase (PDH), and alcohol dehydrogenase (ADH). Because it is highly soluble, ethanol diffuses readily from the gills, avoiding the build up of lactic acid that would otherwise be associated with prolonged anaerobiosis.

In both leopard frogs and goldfish, antioxidant enzymes increase in activity during anoxia, whereas most other enzyme activities do not. Specific examples include goldfish brain Se-GPX (glutathione peroxidase) (a 79% increase in activity), goldfish liver catalase (a 38% increase), frog heart and muscle catalase (47% and 53% increases, respectively), frog heart Se-GPX (a 75% increase), and frog brain GST (a 66% increase). Glucose-6-phosphate dehydrogenase (G6PDH) activity, which is relevant for the production of NADPH and maintenance of the GSH cycle, increased by 26% in anoxic goldfish brain. Total GSH levels (GSH-eqsGSHq2GSSG) remained constant in frog liver, skeletal muscle, and heart during anoxia but decreased by 32% in anoxic brain. GSH-eq levels were also maintained during anoxia-reoxygenation in goldfish organs, with the exception of a small decrease in anoxic kidney GSH-eq (104). For goldfish, lipid peroxidation, determined as conjugated dienes, increased 114% in liver after 1 h reoxygenation and 75% in brain after 14 h recovery. Because the animals survive, manageable and

NADPH: reduced nicotinamide dinucleotide phosphate

physiological oxidative stress occurs in these two organs following reperfusion.

Liver, brain, and gill superoxide dismutase (SOD) activity increase in carp, *Cyprinus carpio*, after several hours of extreme hypoxia (105). Elevated SOD activity is a relevant adaptive mechanism against posthypoxic ROS insult.

Turtles constitutively express high levels of antioxidant enzymes and GSH-eq levels compared with other nonmammalian vertebrates. Activities of antioxidant enzymes and GSH-eq levels are similar to those in mammals despite the much lower aerobic metabolic rate of turtles (51, 106). Unlike garter snakes, leopard frogs, goldfish, and carp (see above), hypoxic turtles exhibit little anticipatory adjustment in antioxidant defenses; they apparently rely on constitutively high levels of defense compounds (51).

Cellular Repair Mechanisms Following Hypoxia

Winter dormancy in hypoxic water undoubtedly kills numerous turtles, particularly immature individuals (107, 108). Other individuals, no doubt, are stressed and near death. Extended periods of hypoxia probably involve gradual changes in homeostasis, resulting in injury that, if survivable, must be repaired at the end of hypoxia.

Lower vertebrates show considerable repair and regeneration potential, similar to the stem cells present in many adult tissues (109). Do stem cells play an important role in recovery from long-term hypoxic dormancy in organisms such as turtles? Reptiles may be the only amniotic vertebrate capable of extensive spontaneous repair of the central nervous system. The regeneration can occur in the cortex (110) and involves the restoration of the glial framework, new neurons, and newly formed glia. Radial glial cells produce neurons in the ventricular zone and also have a role in the guidance of migration of the newly formed neurons (111). In the brains of adult mammals, neurogenesis is spatially restricted to several regions, whereas in the brains of adult fish (112–114), amphibians (115), and reptiles, neurogenesis is widespread (116–118). Adult neurons arise in these groups of animals in the periventricular regions, where the adult radial glia-like cells (ependymo-glia) divide (118, 119). In fish, amphibians, and reptiles, radial glia persist into adulthood, generating neurons throughout life. In adult and juvenile turtles, injections of BrdU initially label radial ependymo-glial cell nuclei, but several months later BrdU labeling is evident in telencephalon (111, 120). Radial glial cells also contribute to central nervous system regeneration, following injury in the spinal cord of amphibians (121) and in the cortex of reptiles.

CONCLUSIONS

Long-term anoxia tolerance in lower vertebrates (with turtles and carp as primary examples) involve large reductions in metabolism, storage and efficient use of substrates, tolerance of metabolic products formed during anaerobiosis, and strategies for avoiding injury from free radicals during re-encounters with oxygen. Although these broad outlines of the adaptations for anoxic survival are now relatively clear, much work remains in the realm of mechanistic understandings of the signals involved in these interrelated processes.

SUMMARY POINTS

1. The adaptations of fishes, amphibians, and reptiles to brief or prolonged hypoxia and reoxygenation are highly diverse, reflecting the variety of body structure, habitats, and environmental stresses experienced by these groups of organisms.

2. Whereas lower vertebrates are generally more tolerant of hypoxia/reoxygenation cycles than are birds and mammals, animals such as Western painted turtles and crucian carp are extremely tolerant of prolonged hypoxia, surviving months with no oxygen. Hypothermia is essential for long-term survival without oxygen in these species.
3. Hypoxia tolerance involves a core triad of adaptations: metabolic suppression, tolerance of metabolite accumulation, and free-radical defenses during reoxygenation. The storage of large amounts of glycogen is also key for surviving long-term oxygen limitation.

FUTURE ISSUES

1. The processes that link reductions in metabolic rate to reductions in energy consumption during hypoxia should be identified. Studies may focus on mitochondrial-based signaling processes and their link to ion channel activity and other energy consuming processes.
2. The necessary repair-and-recovery processes following prolonged hypoxic dormancy need exploration.
3. Studies elucidating the link between hypoxia tolerance and hypothermia tolerance in cells from lower vertebrates are needed.

ACKNOWLEDGMENTS

We thank Dan Warren and Paul Donohoe for helpful comments. We dedicate this review to the memory of three great comparative physiologists who contributed greatly to both the theoretical basis of and experimental studies material covered by this review: Peter Hochachka, Peter Lutz, and Bob Boutilier.

LITERATURE CITED

1. Jackson DC, Herbert CV, Ultsch GR. 1984. The comparative physiology of diving in North American freshwater turtles. II. Plasma ion balance during prolonged anoxia. *Physiol. Zool.* 57:632–40
2. Ultsch GR. 1985. The viability of nearctic freshwater turtles submerged in anoxia and normoxia at 3 and 10°C. *Comp. Biochem. Physiol. A* 81:607–11
3. Lutz PL, Nilsson GE, Prentice HM. 2003. *The Brain Without Oxygen: Causes of Failure-Physiological and Molecular Mechanisms for Survival.* Dordrecht, Boston, London: Kluwer Acad. 252 pp.

3a. Ramirez J-M, Folkow LP, Blix AS. 2007. Hypoxia tolerance in mammals and birds: from the wilderness to the clinic. *Annu. Rev. Physiol.* 69:113–43

4. Nilsson GE, Ostlund-Nilsson S. 2004. Hypoxia in paradise: widespread hypoxia tolerance in coral reef fishes. *Proc. Biol. Sci.* 271(Suppl. 3):S30–33
5. Guppy M. 2004. The biochemistry of metabolic depression: a history of perceptions. *Comp. Biochem. Physiol. B* 139:435–42
6. **Hochachka P. 1986. Defense strategies against hypoxia and hypothermia. *Science* 231:234–41**

6. First clear outline of the hypothesis that simultaneous ion channel arrest and metabolic arrest are required for both anoxia and hypothermia tolerance.

7. Boutilier RG. 2001. Mechanisms of cell survival in hypoxia and hypothermia. *J. Exp. Biol.* 204:3171–81
8. Raymond J, Segre D. 2006. The effect of oxygen on biochemical networks and the evolution of complex life. *Science* 311:1764–67
9. Berner RA. 1999. Atmospheric oxygen over Phanerozoic time. *Proc. Natl. Acad. Sci. USA* 96:10955–57
10. Huey RB, Ward PD. 2005. Hypoxia, global warming, and terrestrial late Permian extinctions. *Science* 308:398–401
11. Falkowski PG, Katz ME, Milligan AJ, Fennel K, Cramer BS, et al. 2005. The rise of oxygen over the past 205 million years and the evolution of large placental mammals. *Science* 309:2202–4
12. Bushnell PG, Brill RW, Bourke RE. 1990. Cardiorespiratory responses of skipjack tuna (*Katsuwonus pelamis*), yellowfin tuna (*Thunnus albacares*), and bigeye tuna (*Thunnus obesus*) to acute reductions of ambient oxygen. *Can. J. Zool.* 68:1857–65
13. Gamperl AK, Todgham AE, Parkhouse WS, Dill R, Farrell AP. 2001. Recovery of trout myocardial function following anoxia: preconditioning in a nonmammalian model. *Am. J. Physiol.* 281:R1755–63
14. Gesser H. 1977. The effects of hypoxia and reoxygenation on force development in myocardia of carp and rainbow trout: protective effects of CO/HCO^-. *J. Exp. Biol.* 69:199–206
15. Axelsson M, Farrell AP, Nilsson S. 1990. Effects of hypoxia and drugs on the cardiovascular dynamics of the Atlantic hagfish *Myxine glutinosa*. *J. Exp. Biol.* 151:297–316
16. Bailey JR, Val AL, Almeida-Val VMF, Driedzic WR. 1999. Anoxic cardiac performance in Amazonian and north-temperate-zone teleosts. *Can. J. Zool.* 77:683–89
17. Faust HA, Gamperl AK, Rodnick KJ. 2004. All rainbow trout (*Oncorhynchus mykiss*) are not created equal: intraspecific variation in cardiac hypoxia tolerance. *J. Exp. Biol.* 207:1005–15
18. Nilsson GE, Renshaw GM. 2004. Hypoxic survival strategies in two fishes: extreme anoxia tolerance in the North European crucian carp and natural hypoxic preconditioning in a coral-reef shark. *J. Exp. Biol.* 207:3131–39
19. Nilsson GE. 2001. Surviving anoxia with the brain turned on. *News Physiol. Sci.* 16:217–21
20. Lutz PL, Nilsson GE. 2004. Vertebrate brains at the pilot light. *Resp. Physiol. Neurobiol.* 141:285–96
21. Stecyk JA, Stenslokken KO, Farrell AP, Nilsson GE. 2004. Maintained cardiac pumping in anoxic crucian carp. *Science* 306:77
22. **Shoubridge EA, Hochachka PW. 1980. Ethanol: novel end product of vertebrate anaerobic metabolism. *Science* 209:308–9**

22. Identification of ethanol as a unique metabolic end product in carp.

23. Cameron JS, Hoffmann KE, Zia C, Hemmett HM, Kronsteiner A, Lee CM. 2003. A role for nitric oxide in hypoxia-induced activation of cardiac KATP channels in goldfish (*Carassius auratus*). *J. Exp. Biol.* 206:4057–65
24. Chen J, Zhu JX, Wilson I, Cameron JS. 2005. Cardioprotective effects of K ATP channel activation during hypoxia in goldfish *Carassius auratus*. *J. Exp. Biol.* 208:2765–72
25. Hanley PJ, Daut J. 2005. K(ATP) channels and preconditioning: a re-examination of the role of mitochondrial K(ATP) channels and an overview of alternative mechanisms. *J. Mol. Cell Cardiol.* 39:17–50
26. Krumschnabel G, Schwarzbaum PJ, Lisch J, Biasi C, Wieser W. 2000. Oxygen-dependent energetics of anoxia-tolerant and anoxia-intolerant hepatocytes. *J. Exp. Biol.* 203:951–59

27. Almeida-Val VM, Val AL, Duncan WP, Souza FC, Paula-Silva MN, Land S. 2000. Scaling effects on hypoxia tolerance in the Amazon fish *Astronotus ocellatus* (Perciformes: Cichlidae): contribution of tissue enzyme levels. *Comp. Biochem. Physiol. B* 125:219–26
28. Chippari-Gomes AR, Gomes LC, Lopes NP, Val AL, Almeida-Val VM. 2005. Metabolic adjustments in two Amazonian cichlids exposed to hypoxia and anoxia. *Comp. Biochem. Physiol. B* 141:347–55
29. Mesquita-Saad LS, Leitao MA, Paula-Silva MN, Chippari-Gomes AR, Almeida-Val VM. 2002. Specialized metabolism and biochemical suppression during aestivation of the extant South American lungfish—*Lepidosiren paradoxa*. *Braz. J. Biol.* 62:495–501
30. Sloman KA, Wood CM, Scott GR, Wood S, Kajimura M, et al. 2006. Tribute to R.G. Boutilier: the effect of size on the physiological and behavioural responses of oscar, *Astronotus ocellatus*, to hypoxia. *J. Exp. Biol.* 209:1197–205
31. Padilla PA, Roth MB. 2001. Oxygen deprivation causes suspended animation in the zebrafish embryo. *Proc. Natl. Acad. Sci. USA* 98:7331–35
32. Christiansen J, Penney D. 1973. Anaerobic glycolysis and lactic acid accumulation in cold-submerged *Rana pipiens*. *J. Comp. Physiol.* 87:237–45
33. Hutchison VH, Dady MJ. 1964. The viability of *Rana pipiens* and *Bufo terrestris* submerged at different temperatures. *Herpetologica* 20:149–62
34. Lillo RS. 1980. Heart rate and blood pressure in bullfrogs during prolonged maintenance in water at low temperature. *Comp. Biochem. Physiol.* 65A:251–53
35. Stewart ER, Reese SA, Ultsch GR. 2004. The physiology of hibernation in Canadian leopard frogs (*Rana pipiens*) and bullfrogs (*Rana catesbeiana*). *Physiol. Biochem. Zool.* 77:65–73
36. Knickerbocker DL, Lutz PL. 2001. Slow ATP loss and the defense of ion homeostasis in the anoxic frog brain. *J. Exp. Biol.* 204:3547–51
37. Lutz P, Nilsson G. 1997. Contrasting strategies for anoxic brain survival—glycolysis up or down. *J. Exp. Biol.* 200:411–19
38. Bradford DF. 1983. Winterkill, oxygen relations, and energy metabolism of a submerged dormant amphibian, *Rana muscosa*. *Ecology* 64:1171–83
39. Crowder WC, Nie M, Ultsch GR. 1998. Oxygen uptake in bullfrog tadpoles (*Rana catesbeiana*). *J. Exp. Zool.* 280:121–34
40. Duffy TE, Kohle SJ, Vannucci RC. 1975. Carbohydrate and energy metabolism in perinatal rat brain: relation to survival in anoxia. *J. Neurochem.* 24:271–76
41. Pinder AW, Storey KB, Ultsch GR. 1992. Estivation and hibernation. In *Environmental Biology of the Amphibia*, ed. ME Feder, WW Burggren, pp. 250–74. Chicago: Univ. Chicago Press
42. Donohoe PH, West TG, Boutilier RG. 2000. Factors affecting membrane permeability and ionic homeostasis in the cold-submerged frog. *J. Exp. Biol.* 203:405–14
43. Wegner G, Krause U. 1993. Environmental and exercise anaerobiosis in frogs. In *Surviving Hypoxia*, ed. PW Hochachka, PL Lutz, TJ Sick, M Rosenthal, G van den Thillart, pp. 217–36. Boca Raton, FL: CRC Press
44. Hedrick M, Fahlman CS, Bickler P. 2005. Intracellular calcium and survival of tadpole forebrain cells in anoxia. *J. Exp. Biol.* 208:618–86
45. Kristian T, Siesjo BK. 1998. Calcium in ischemic cell death. *Stroke* 29:705–18
46. Storey KB. 2002. Life in the slow lane: molecular mechanisms of estivation. *Comp. Biochem. Physiol.* 133:733–54
47. Reese SA, Jackson DC, Ultsch GR. 2003. Hibernation in freshwater turtles: softshell turtles (*Apalone spinifera*) are the most intolerant of anoxia among North American species. *J. Comp. Physiol. B* 173:263–68

51. Comprehensive review of antioxidant defense strategies in hypoxia-tolerant animals.

53. An excellent and encyclopedic review of metabolic depression in many species.

48. Reese SA, Jackson DC, Ultsch GR. 2002. The physiology of overwintering in a turtle that occupies multiple habitats, the common snapping turtle (*Chelydra serpentina*). *Physiol. Biochem. Zool.* 75:432–38
49. Hermes-Lima M, Storey KB. 1993. Antioxidant defenses in the tolerance of freezing and anoxia by garter snakes. *Am. J. Physiol.* 265:R646–52
50. Rice ME, Lee EJ, Choy Y. 1995. High levels of ascorbic acid, not glutathione, in the CNS of anoxia-tolerant reptiles contrasted with levels in anoxia-intolerant species. *J. Neurochem.* 64:1790–99
51. **Hermes-Lima M, Zenteno-Savin T. 2002. Animal response to drastic changes in oxygen availability and physiological oxidative stress. *Comp. Biochem. Physiol. C* 133:537–56**
52. Seymour RS, Webster ME. 1975. Gas transport and blood acid-base balance in diving sea snakes. *Exp. Zool.* 191:169–81
53. **Guppy M, Withers PC. 1999. Metabolic depression in animals: physiological perspectives and biochemical generalizations. *Biol. Rev.* 74:1–40**
54. Hochachka P, Buck L, Doll C, Land S. 1996. Unifying theory of hypoxia tolerance: Molecular/metabolic defense and rescue mechanisms for surviving oxygen lack. *Proc. Natl. Acad. Sci. USA* 93:9493–98
55. Lindsley JE, Rutter J. 2004. Nutrient sensing and metabolic decisions. *Comp. Biochem. Physiol. B* 139:543–59
55a. Storey KB. 1998. Survival under stress: molecular mechanisms of metabolic rate depression in animals. *S. Afr. J. Zool.* 33:55–64
55b. Hulbert AJ, Else PL, Manolis SC, Brand MD. 2002. Proton leak in hepatocytes and liver mitochondria from archosaurs (crocodiles) and the allometric relationships for ectotherms. *J. Comp. Physiol. B* 172:387–97
56. Bishop T, St-Pierre J, Brand MD. 2002. Primary causes of decreased mitochondrial oxygen consumption during metabolic depression in snail cells. *Am. J. Physiol. Regul. Integr. Comp. Physiol.* 282:R372–82
57. Barger JL, Brand MD, Barnes BM, Boyer BB. 2003. Tissue-specific depression of mitochondrial proton leak and substrate oxidation in hibernating arctic ground squirrels. *Am. J. Physiol. Regul. Integr. Comp. Physiol.* 284:R1306–13
58. Boutilier RG, St-Pierre J. 2002. Adaptive plasticity of skeletal muscle energetics in hibernating frogs: mitochondrial proton leak during metabolic depression. *J. Exp. Biol.* 205:2287–96
59. Else PL, Hulbert AJ. 1987. Evolution of mammalian endothermic metabolism: "leaky" membranes as a source of heat. *Am. J. Physiol.* 253:R1–7
60. Bickler PE, Donohoe PH, Buck LT. 2002. Molecular adaptations for survival during anoxia: lessons from lower vertebrates. *Neuroscientist* 8:234–42
61. Buck LT, Hochachka PW. 1993. Anoxic suppression of Na^+-K^+-ATPase and constant membrane potential in hepatocytes: support for channel arrest. *Am. J. Physiol.* 265:R1020–25
62. Perez-Pinzon M, Rosenthal M, Sick T, Lutz P, Pablo J, Mash D. 1992. Downregulation of sodium channels during anoxia: a putative survival strategy of turtle brain. *Am. J. Physiol.* 262:R712–15
63. Bickler PE, Donohoe PH, Buck LT. 2000. Hypoxia-induced silencing of NMDA receptors in turtle neurons. *J. Neurosci.* 20:3522–28
64. Shin DS, Buck LT. 2003. Effect of anoxia and pharmacological anoxia on whole-cell NMDA receptor currents in cortical neurons from the western painted turtle. *Physiol. Biochem. Zool.* 76:41–51

65. **Buck LT, Bickler PE. 1998. Adenosine and anoxia reduce *N*-methyl-D-aspartate receptor open probability in turtle cerebrocortex. *J. Exp. Biol.* 210:289–97**
66. Cserr H, DePasquale M, Jackson D. 1988. Brain and cerebrospinal fluid composition after long-term anoxia in diving turtles. *Am. J. Physiol.* 255:R338–43
67. Buck LT. 2004. Adenosine as a signal for ion channel arrest in anoxia-tolerant organisms. *Comp. Biochem. Physiol. B* 139:401–14
68. Shin DS, Wilkie MP, Pamenter ME, Buck LT. 2005. Calcium and protein phosphatase 1/2 A attenuate *N*-methyl-D-aspartate receptor activity in the anoxic turtle cortex. *Comp. Biochem. Physiol. A* 142:50–57
69. Ehlers MD, Zhang S, Berhardt JP, Huganir RL. 1996. Inactivation of NMDA receptors by direct interaction of calmodulin with the NR1 subunit. *Cell* 84:745–55
69a. Milton SL, Manuel L, Lutz PL. 2003. Slow death in the leopard frog *Rana pipiens*: neurotransmitters and anoxia tolerance. *J. Exp. Biol.* 206:4021–28
70. Oldenburg O, Cohen MV, Downey JM. 2003. Mitochondrial K_{ATP} channels in preconditioning. *J. Mol. Cell. Cardiol.* 35:569–75
71. Kis B, Nagy K, Snipes JA, Rajapakse NC, Horiguchi T, et al. 2004. The mitochondrial KATP channel opener BMS-191095 induces neuronal preconditioning. *Neuroreport* 15:345–49
72. Shimizu K, Lacza Z, Rajapakse N, Horiguchi T, Snipes J, Busija DW. 2002. MitoK(ATP) opener, diazoxide, reduces neuronal damage after middle cerebral artery occlusion in the rat. *Am. J. Physiol. Heart Circ. Physiol.* 283:H1005–11
72a. Inoue I, Nagase H, Kishi K, Higuti T. 1991. ATP-sensitive K^+ channel in the mitochondrial inner membrane. *Nature* 352:244–47
73. Holmuhamedov EL, Jovanovic S, Dzeja PP, Jovanovic A, Terzic A. 1998. Mitochondrial ATP-sensitive K^+ channels modulate cardiac mitochondrial function. *Am. J. Physiol.* 275:H1567–76
74. Liu Y, Fiskum G, Schubert D. 2002. Generation of reactive oxygen species by the mitochondrial electron transport chain. *J. Neurochem.* 80:780–87
75. St-Pierre J, Buckingham JA, Roebuck SJ, Brand MD. 2002. Topology of superoxide production from different sites in the mitochondrial electron transport chain. *J. Biol. Chem.* 277:44784–90
76. Oldenburg O, Cohen MV, Downey JM. 2003. Mitochondrial K(ATP) channels in preconditioning. *J. Mol. Cell Cardiol.* 35:569–75
77. Vanden Hoek T, Becker L, Shao Z, Li C, Schumacker P. 1998. Reactive oxygen species released from mitochondria during brief hypoxia induce preconditioning in cardiomyocytes. *J. Biol. Chem.* 273:18092–98
78. Pain T, Yang XM, Critz SD, Yue Y, Nakano A, et al. 2000. Opening of mitochondrial K-ATP channels triggers the preconditioned state by generating free radicals. *Circ. Res.* 87:460–66
79. Bolli R. 2001. Cardioprotective function of inducible nitric oxide synthase and role of nitric oxide in myocardial ischemia and preconditioning: an overview of a decade of research. *J. Mol. Cell Cardiol.* 33:1897–918
80. Semenza GL. 1999. Regulation of mammalian O_2 homeostasis by hypoxia-inducible factor 1. *Annu. Rev. Cell Dev. Biol.* 15:551–78
81. Soitamo AJ, Rabergh CM, Gassmann M, Sistonen L, Nikinmaa M. 2001. Characterization of a hypoxia-inducible factor (HIF-1α) from rainbow trout. Accumulation of protein occurs at normal venous oxygen tension. *J. Biol. Chem.* 276:19699–705

65. First direct evidence for ion channel arrest in anoxic turtle neurons.

82. Helton R, Cui J, Scheel JR, Ellison JA, Ames C, et al. 2005. Brain-specific knock-out of hypoxia-inducible factor-1α reduces rather than increases hypoxic-ischemic damage. *J. Neurosci.* 25:4099–107
83. Newby AC, Worku CY, Meghi P, Nadazawa M, Skladanowski AC. 1990. Adenosine: a retaliatory metabolite or not? *News Physiol. Sci.* 5:67–70
84. Collis MG, Hourani SM. 1993. Adenosine receptor subtypes. *Trends Pharmacol. Sci.* 14:360–66
85. Nilsson G, Lutz P. 1992. Adenosine release in anoxic turtle brain as a mechanism for anoxic survival. *J. Exp. Biol.* 162:345–51
86. Perez-Pinzon M, Lutz P, Sick T, Rosenthal M. 1993. Adenosine, a "retaliatory" metabolite, promotes anoxia tolerance in turtle brain. *J. Cereb. Blood Flow Metab.* 13:728–32

85, 86. Recognize that adenosine plays a vital role in anoxia adaptation in turtle brain.

87. Sick TJ, Rosenthal M, LaManna JC, Lutz PL. 1982. Brain potassium ion homeostasis, anoxia, and metabolic inhibition in turtles and rats. *Am. J. Physiol.* 243:R281–88
88. Hylland P, Nilsson GE, Lutz PL. 1994. Time course of anoxia-induced increase in cerebral blood flow rate in turtles: evidence for a role of adenosine. *J. Cereb. Blood Flow Metab.* 14:877–81
89. Buck L, Bickler P. 1995. Role of adenosine in NMDA receptor modulation in the cerebral cortex of an anoxia-tolerant turtle (*Chyrsemys picta belli*). *J. Exp. Biol.* 198:1621–28
90. Ghai HS, Buck LT. 1999. Acute reduction in whole cell conductance in anoxic turtle brain. *Am. J. Physiol.* 277:R887–93
91. Krumschnabel G, Biasi C, Wieser W. 2000. Action of adenosine on energetics, protein synthesis and K^+ homeostasis in teleost hepatocytes. *J. Exp. Biol.* 203:2657–65
92. Nilsson GE. 1991. The adenosine receptor blocker aminophylline increases anoxic ethanol excretion in crucian carp. *Am. J. Physiol.* 261:R1057–60
93. Lutz P, Kabler S. 1997. Release of adenosine and ATP in the brain of the freshwater turtle (*Trachemys scripta*) during long-term anoxia. *Brain Res.* 769:281–86
94. Nilsson GE, Hylland P, Lofman CO. 1994. Anoxia and adenosine induce increased cerebral blood-flow in crucian carp. *Am. J. Physiol.* 267:R590–95
95. Soderstrom-Lauritzsen V, Nilsson GE, Lutz PL. 2001. Effect of anoxia and adenosine on cerebral blood flow in the leopard frog (*Rana pipiens*). *Neurosci. Lett.* 311:85–88
96. Soderstrom V, Renshaw GM, Nilsson GE. 1999. Brain blood flow and blood pressure during hypoxia in the epaulette shark *Hemiscyllium ocellatum*, a hypoxia-tolerant elasmobranch. *J. Exp. Biol.* 202:829–35
97. Jackson DC. 2004. Surviving extreme lactic acidosis: the role of calcium lactate formation in the anoxic turtle. *Respir. Physiol. Neurobiol.* 144:173–78
98. Jackson DC. 2004. Acid-base balance during hypoxic hypometabolism: selected vertebrate strategies. *Respir. Physiol. Neurobiol.* 141:273–83

97, 98. Describe the unique role of the shell in the buffering of acidosis during long-term dormancy in turtles.

99. Jackson DC. 2002. Hibernating without oxygen: physiological adaptations of the painted turtle. *J. Physiol.* 543:731–37
100. Jackson DC. 2000. How a turtle's shell helps it survive prolonged anoxic acidosis. *News Physiol. Sci.* 15:181–85
101. Jackson DC, Ramsey AL, Paulson JM, Crocker CE, Ultsch GR. 2000. Lactic acid buffering by bone and shell in anoxic softshell and painted turtles. *Physiol. Biochem. Zool.* 73:290–97
102. Jackson DC. 2000. Living without oxygen: lessons from the freshwater turtle. *Comp. Biochem. Physiol. A* 125:299–315
103. Ramaglia V, Buck LT. 2004. Time-dependent expression of heat shock proteins 70 and 90 in tissues of the anoxic western painted turtle. *J. Exp. Biol.* 207:3775–84

104. Lushchak VI, Lushchak LP, Mota AA, Hermes-Lima M. 2001. Oxidative stress and antioxidant defenses in goldfish *Carassius auratus* during anoxia and reoxygenation. *Am. J. Physiol. Regul. Integr. Comp. Physiol.* 280:R100–7
105. Vig E, Nemcsok J. 1989. The effect of hypoxia and paraquat on the superoxide dismutase activity in different organs of carp, *Cyprinus carpio* L. *J. Fish Biol.* 35:23–25
106. Storey KB. 1996. Metabolic adaptations supporting anoxia tolerance in reptiles: recent advances. *Comp. Biochem. Physiol. B* 113:23–35
107. Dinkelacker SA, Costanzo JP, Iverson JB, Lee REJ. 2005. Survival and physiological responses of hatchling blanding's turtles (*Emydoidea blandingii*) to submergence in normoxic and hypoxic water under simulated winter conditions. *Physiol. Biochem. Zool.* 78:356–63
108. Dinkelacker SA, Costanzo JP, Lee REJ. 2005. Anoxia tolerance and freeze tolerance in hatchling turtles. *J. Comp. Physiol. B* 175:209–17
109. Brockes JP, Kumar A. 2005. Appendage regeneration in adult vertebrates and implications for regenerative medicine. *Science* 310:1919–23
110. Rosenmund C, Westbrook GL. 1993. Calcium-induced actin depolymerization reduced NMDA channel activity. *Neuron* 10:805–14
111. Weissman T, Noctor SC, Clinton BK, Honig LS, Kriegstein AR. 2003. Neurogenic radial glial cells in reptile, rodent and human: from mitosis to migration. *Cereb. Cortex* 13:550–59
112. Birse SC, Leonard RB, Coggeshall RE. 1980. Neuronal increase in various areas of the nervous system of the guppy, Lebistes. *J. Comp. Neurol.* 194:291–301
113. Raymond PA, Easter SSJ. 1983. Postembryonic growth of the optic tectum in goldfish. I. Location of germinal cells and numbers of neurons produced. *J. Neurosci* 3:1077–91
114. Zupanc GK. 1999. Neurogenesis, cell death and regeneration in the adult gymnotiform brain. *J. Exp. Biol.* 202:1435–46
115. Polenov AL, Chetverukhin VK. 1993. Ultrastructural radioautographic analysis of neurogenesis in the hypothalamus of the adult frog, *Rana temporaria*, with special reference to physiological regeneration of the preoptic nucleus. II. Types of neuronal cells produced. *Cell Tissue Res.* 271:351–62
116. Font E, Desfili E, Perez-Canellas MM, Garcia-Verdugo JM. 2001. Neurogenesis and neuronal regeneration in the adult reptilian brain. *Brain Behav. Evol.* 58:276–95
117. Garcia-Verdugo JM, Ferron S, Flames N, Collado L, Desfilis E, Font E. 2002. The proliferative ventricular zone in adult vertebrates: a comparative study using reptiles, birds, and mammals. *Brain Res. Bull.* 57:765–75
118. Lopez-Garcia C, Molowny A, Garcia-Verdugo JM, Ferrer I. 1988. Delayed postnatal neurogenesis in the cerebral cortex of lizards. *Brain Res.* 471:167–74
119. Polenov AL, Chetverukhin VK, Jakovleva IV. 1972. The role of the ependyma of the recessus praeopticus in formation and the physiological regeneration of the nucleus praeopticus in lower vertebrates. *Z. Mikrosk. Anat. Forsch.* 85:513–32
120. Perez-Canellas MM, Font E, Garcia-Verdugo JM. 1977. Postnatal neurogenesis in the telencephalon of turtles: evidence for nonradial migration of new neurons from distant proliferative ventricular zones to the olfactory bulbs. *Brain Res. Dev. Brain Res.* 101:125–37
121. Margotta V, Fonti R, Palladini G, Filoni S, Lauro GM. 1991. Transient expression of glial-fibrillary acidic protein (GFAP) in the ependyma of the regenerating spinal cord in adult newts. *J. Hirnforsch.* 32:485–90

122. Hisatsune C, Umemori H, Inoue T, Michikawa T, Kohda K, et al. 1997. Phosphorylation-dependent regulation of *N*-methyl-D-aspartate receptors by calmodulin. *J. Biol. Chem.* 272:20805–10
123. Shin DS, Wilkie MP, Pamenter ME, Buck LT. 2005. Calcium and protein phosphatase 1/2A attenuate *N*-methyl-D-aspartate receptor activity in the anoxic turtle cortex. *Comp. Biochem. Physiol. A* 142:50–57

Integration of Rapid Signaling Events with Steroid Hormone Receptor Action in Breast and Prostate Cancer

Carol A. Lange,[1] Daniel Gioeli,[2] Stephen R. Hammes,[3] and Paul C. Marker[4]

[1]Departments of Medicine (Division of Hematology, Oncology, and Transplant) and Pharmacology; email: Lange047@umn.edu

[2]Department of Microbiology, University of Virginia Health System, Charlottesville, Virginia 22908; email: dgg3f@virginia.edu

[3]Department of Internal Medicine (Division of Endocrinology and Metabolism), University of Texas Southwestern Medical Center, Dallas, Texas 75390; email: Stephen.Hammes@UTSouthwestern.edu

[4]Department of Genetics, Cell Biology and Development, University of Minnesota Cancer Center, Minneapolis, Minnesota 55455; email: Marke032@umn.edu

Annu. Rev. Physiol. 2007. 69:171–99

First published online as a Review in Advance on October 12, 2006

The *Annual Review of Physiology* is online at http://physiol.annualreviews.org

This article's doi: 10.1146/annurev.physiol.69.031905.160319

0066-4278/07/0315-0171$20.00

Key Words

progesterone receptor, androgen receptor, epidermal growth factor, mitogen-activated protein kinase, cyclin D1

Abstract

Steroid hormone receptors (SRs) are ligand-activated transcription factors and sensors for growth factor–initiated signaling pathways in hormonally regulated tissues, such as the breast or prostate. Recent discoveries suggest that several protein kinases are rapidly activated in response to steroid hormone binding to cytoplasmic SRs. Induction of rapid signaling upon SR ligand binding ensures that receptors and coregulators are appropriately phosphorylated as part of optimal transcription complexes. Alternatively, SR-activated kinase cascades provide additional avenues for SR-regulated gene expression independent of SR nuclear action. We provide an overview of SR and signaling cross talk in breast and prostate cancers, using the human progesterone receptor (PR) and androgen receptor (AR) as models. Kinases are emerging as key mediators of SR action. Cross talk between SR and membrane-initiated signaling events suggests a mechanism for coordinate regulation of gene subsets by mitogenic stimuli in hormonally responsive normal tissues; such cross talk is suspected to contribute to cancer biology.

INTRODUCTION

Steroid hormone receptors (SRs): a subclass of the nuclear receptor superfamily whose members include the progesterone, androgen, estrogen, glucocorticoid, and mineralocorticoid receptors

AR: androgen receptor

PR: progesterone receptor

Despite extensive research on the etiology of hormonally regulated cancers, the basic molecular mechanisms underlying cellular dysregulation remain undefined. In this review, we hypothesize that an integral feature of steroid hormone receptors (SRs),[1] in addition to their nuclear function as transcription factors, is their cytoplasmic interaction with growth factor–initiated signaling pathways in multiple cell types. Recent data indicate that the rapid signaling effects of SRs occur independently of their nuclear actions but play a key role in the regulation of SR nuclear transcriptional events and represent an independent avenue for coordinating gene regulation in response to activation of kinase pathways. This review focuses on genomic and nongenomic signaling cascades initiated by SRs, with the purpose of identifying key SRs and kinases that are involved in the two arms of one integrated pathway in hormonal regulation of cancer cells. We discuss the regulation of integrated SR signaling systems as a means of treatment for patients with breast and prostate cancer. Owing to the similarities in both classical and rapid signaling mechanisms, we consider the integrated actions of the androgen receptor (AR) and progesterone receptor (PR) as model SR family members.

Classical Actions of Steroid Hormone Receptors

All SRs are classically defined as ligand-activated transcription factors that bind directly or indirectly to DNA to control gene regulation. Testicular-derived testosterone and peripherally derived dihydrotestosterone (DHT) are the principal ligands for AR, and the only active form of AR is encoded by a single gene located on the X chromosome at q11-12. Similarly, the action of PR is mediated through binding with the ovarian steroid hormone progesterone. Three isoforms of PR are produced from a single gene on chromosome 11 at q22-23 by use of alternate promoters and internal translational start sites. PR isoforms include the full-length PR-B (116 kDa) and N-terminally truncated PR-A (94 kDa) and PR-C (60 kDa) isoforms. PR-C is not a functional transcription factor but acts as a dominant inhibitor of uterine PR-B in the fundal myometrium during labor (1). In the absence of steroid hormone, SRs are complexed with several chaperone molecules, including heat shock protein (hsp) 90, hsp70, hsp40, Hop, and p23; these interactions are requisite for proper protein folding and assembly of stable SR-hsp90 heterocomplexes that are competent to bind ligand (2). Hsps also function to connect SRs to protein trafficking systems.

After binding to steroid hormone, the receptor undergoes restructuring; upon receptor dimerization and hsp dissociation, the activated receptor binds directly to specific hormone response elements (HREs) in the promoter regions of target genes. For example, PR interacts with progesterone response elements (PREs) and PRE-like sequences in the *c-myc* (3), *fatty acid synthetase* (4), and *MMTV* (*mouse mammary tumor virus*) promoters (5), whereas AR classically targets the androgen response element (ARE)-containing *prostate-specific antigen* (PSA), *maspin*, *p21*, and *fibroblast growth factor 1* promoters (6). However, treatment with steroid hormones also leads to an upregulation of regulatory molecules that lack a classical PRE or ARE in their proximal promoter regions, such as progestin- or androgen-regulated *epidermal growth factor receptor* (7, 8), *c-fos* (9, 10), and *cyclin D1* (11, 12). Without canonical HREs, SR regulation of these genes can occur through indirect DNA-binding mechanisms, as in the case of tethering of PR to specificity protein 1, to promote p21 transcription in the presence of progestin (13). PR may also regulate genes by tethering to activating protein 1 (14) or STATs (signal transducers and activators of transcription) (10, 15). AR regulates genes

[1] Please see the Appendix at the end for a list of abbreviations used in this review.

via interaction with β-catenin and SMAD proteins (reviewed below).

When bound to DNA, either directly or otherwise, all SRs interact with components of the basal transcription machinery, assisted by nuclear receptor coregulatory molecules. Coregulators control the susceptibility of chromatin to transcription (chromatin remodeling) and the recruitment of transcriptional machinery (e.g., RNA polymerase-II). Histone acetyl transferases function as coactivators, whereas histone deacetylases function as corepressors. These enzymes work in concert with other transcriptional regulator proteins, including the ATP-dependent chromatin remodeling complexes (SWI/SNF), arginine methyltransferases (CARM1 and PRMT1), and histone kinases (reviewed in Reference 16).

Nonclassical Actions of Progesterone and Androgen Receptors in Oocytes

SRs also mediate changes in gene expression and cell biology by direct activation of cytoplasmic signaling pathways. Smith & Ecker (17) first described a transcription-independent steroid effect of biological relevance in their 1969 paper demonstrating that steroids promote frog oocyte maturation (i.e., meiotic resumption) independently of transcription. Nuclei were manually removed from *Rana pipiens* or *Xenopus laevis* oocytes, and yet activation of mitogen-activated protein kinase (MAPK) and CDK1 still occurred in response to steroid (18). These studies provided unequivocal evidence that the steroid-triggered signals regulating meiosis could be initiated without genomic machinery. Since these initial discoveries, the *X. laevis* model system has been used extensively to study intracellular signals regulating meiosis, and progesterone was assumed to be the physiological regulator of maturation. Researchers have known for some time that the addition of steroid induces a drop in cAMP and increased translation of Mos [amphibian MAPK/ERK kinase kinase (MEKK)] mRNA. Newly translated Mos then activates MEK, leading to MAPK activation. This cascade results in the activation of the cyclin B–CDK1 complex, which catalyzes entry into M phase of meiosis I (reviewed in Reference 18). Surprisingly, although much was learned about the downstream signals triggered by progesterone in this 1969 paper, very little progress was made in the ensuing 30 years regarding the immediate signaling molecules activated by progesterone. For a long time neither the identity of the *Xenopus* PR nor the signaling mechanism regulating the drop in cAMP was known.

The first breakthrough came when several laboratories demonstrated that G protein signaling might be altered by the addition of progesterone. Specifically, researchers proposed that this occurred via a release of inhibition, whereby constitutive G protein signaling held oocytes in meiotic arrest by stimulating adenylyl cyclase and increasing intracellular cAMP. The addition of steroid attenuated this inhibitory signal, resulting in a drop in cAMP, MAPK activation, and meiotic progression (19). This model system was one of the first to demonstrate a biologically important role for G proteins in nongenomic steroid signaling, a concept that is now well accepted. The second breakthrough came when the classical *Xenopus* PR gene was cloned, but manipulation of its expression only partially affected progesterone-mediated maturation (reviewed in Reference 19). At approximately the same time, in vivo studies revealed that the *Xenopus* ovary made essentially no progesterone in response to gonadotropin but produced extremely high levels of androstenedione and testosterone, both of which are equal or better promoters of maturation in vitro (20). Furthermore, researchers showed that oocytes express high levels of CYP17, which converts progesterone to androstenedione; thus, when adding progesterone to oocytes, investigators were studying both progesterone- and androstenedione-mediated maturation (19). These observations paved the way for more current pharmacological and RNA-interference studies, which have shown that

Coregulators: factors recruited by steroid hormone receptors that potentiate (coactivator) or attenuate (corepressor) signaling by those receptors

Mitogen-activated protein kinases (MAPKs): serine/threonine protein kinases activated by upstream MAPK kinases (MEKs) in response to mitogenic stimuli. In mammalian cells, MAPKs are mediators of cell proliferation, survival, migration, and differentiation

MEKK: MAPK/ERK kinase kinase; also known as MAPK kinase kinase

MEK: MAPK/ERK kinase; also known as MAPK kinase

androgens, via the classical AR, are the likely mediators of maturation in frog oocytes (19, 21).

Evidence now suggests that in mammalian systems steroids promote mouse oocyte maturation, using signaling pathways that are similar to those characterized in frogs. As in frogs, both androgen and progesterone can promote oocyte maturation in mice, and pharmacological studies suggest that they are signaling through their classical receptors (22). However, unlike in frogs, progesterone appears to be the physiologically relevant mediator of meiosis in mice (22).

ANDROGEN-INDEPENDENT ANDROGEN RECEPTOR SIGNALING PATHWAYS IN PROSTATE CANCER

In adult males, AR is essential for maintaining many male-specific organs, including the prostate gland. Blocking the activity of AR results in loss of secretory activity and epithelial apoptosis in the prostate (23). This normal function of AR has been exploited for more than 65 years as the basis for androgen ablation therapy in prostate cancer (24). Although the specific clinical approaches used to achieve androgen ablation have changed over time, androgen ablation remains the main therapy for advanced prostate cancer. However, despite an initial response to the treatment, the therapy eventually fails in virtually all patients, resulting in androgen-independent prostate cancer. A majority of androgen-independent prostate cancers continue to express AR and to transactivate AR-responsive genes. Cross talk with other signaling pathways may allow AR to continue to function in the absence of androgens or in the presence of very low levels of androgens.

Late-stage androgen-independent prostate cancer almost always retains expression of AR, despite androgen ablation (25). A majority of prostate tumors from patients failing androgen ablation therapy overexpress AR, thereby sensitizing AR to low levels of androgen. In experimental systems, AR is often overexpressed or stabilized (26). AR overexpression is often (in 30% of cases) associated with gene amplification (27). Frequently (in 10–40% of cases), AR is mutated in advanced prostate cancers, which often results in ARs that can be activated by nonandrogenic ligands (reviewed in Reference 28). Furthermore, overexpression of transcriptional coactivators often accompanies prostate cancer progression, facilitating AR activity (29). Finally, AR and its cofactors can be activated in response to signal transduction from growth factors (reviewed in Reference 30). These observations strongly suggest the continued involvement of ARs in the growth and survival of androgen-independent prostate cancers.

Most research on cross talk between ARs and androgen-independent signaling pathways has occurred in the context of prostate cancer and therefore may reflect dysfunctional cellular processes. Current data support at least three distinct ways in which ARs engage in cross talk with other signaling pathways. First, kinases from other signal transduction pathways directly phosphorylate AR and modulate its transcriptional activity. Second, kinases from other signal transduction pathways phosphorylate dedicated steroid receptor coactivator/corepressor proteins and regulate their activity. Third, AR participates directly in cross talk via protein-protein interactions with signal transduction intermediates from other signaling pathways. Thus, AR functions not only as a transcription factor but also as a node that integrates multiple extracellular signals.

Posttranslational modifications to AR can either activate it in the absence of hormone or sensitize it to low hormone levels (reviewed in Reference 30). Increases in autocrine and paracrine growth factor loops are among the most commonly reported changes correlated with the progression of prostate cancer from a localized and androgen-dependent disease to a disseminated and androgen-independent disease. Investigators have made similar observations in experimental models of prostate cancer in which androgen-dependent prostate cancer cell lines require exogenous growth factors to efficiently form tumors in athymic mice, whereas androgen-independent prostate cancer cell lines do not (31–33). Moreover, forced overexpression of HER2/neu in androgen-dependent prostate cancer cells drives androgen-independent growth (34). Finally, inhibition of epidermal growth factor receptor (EGFR)/HER2 signaling can inhibit prostate cancer cell growth in vitro and in vivo (35, 36) as well as AR transcriptional activity, protein stability, DNA binding, and phosphorylation on serine 81 (37).

The Ras Pathway and Prostate Cancer Progression

Virtually all the growth factor receptors upregulated in prostate cancer activate Ras for at least a portion of their signal transduction activity. Yet, Ras mutations are infrequent in prostate cancer (38). This is consistent with the hypothesis that wild-type Ras is chronically activated by autocrine and paracrine growth factor stimulation in prostate cancer. Thus, there is no selective advantage for growth of cells with mutationally activated Ras. However, Ras signaling represents a convergence point for numerous, diverse extracellular signals, and thus Ras and its effectors may play a role in regulating prostate cancer progression and AR function.

Activation of Ras is sufficient to induce androgen-independent growth of prostate cancer cells; expression of an activated v-Ha-Ras in androgen-dependent LNCaP cells enabled these cells to grow in the absence of androgen (39). More recently, researchers demonstrated that Ras activation can play a causal role in moving prostate cancer cells toward decreased hormone dependence and increased malignant phenotype. Bakin et al. (40) created an LNCaP cell line that expressed activated V12 H-Ras effector loop mutants; these cells were dependent on androgen for growth in vivo and responsive to androgen in vitro. Some activated Ras mutants dramatically reduced the androgen requirement of LNCaP cells with respect to both growth and PSA expression; the same mutants that caused these biological changes (T35S and E37G) were those that caused an intrinsic activation of the MAP kinase pathway under basal, serum-free conditions. This observation correlates activation of the MAP kinase pathway with changes in androgen dependence in cell culture. Expression of Ras also increased the ability of LNCaP cells to form tumors and to resist regression after castration. Collectively, these findings show that activation of Ras signaling is sufficient for progression of LNCaP cells toward androgen independence. Moreover, tumor progression correlates with activation of MAP kinase signaling.

The necessity of Ras signaling in prostate cancer progression has been shown in at least one model: Expression of a dominant-negative N17 Ha-Ras restored androgen dependence to an androgen-independent cell line. C4-2 cells demonstrate decreased androgen dependence of growth both in vitro and in vivo as well as increased tumorigenicity in vivo compared with the parental LNCaP cells. Additionally, C4-2 cells, unlike the parental LNCaP cells, demonstrate the ability to grow in soft agarose (anchorage independence) (41). Expression of dominant-negative Ras under the control of a tetracycline-inducible promoter in C4-2 prostate cancer cells restored androgen dependence to androgen-independent C4-2 cells (42). When C-42 derivatives were implanted in nude mice or when dominant-negative N17 Ras was induced with Doxycycline, the derivatives continued to grow after castration. However, the tumors regressed, in most cases completely, when the mice were castrated and treated with Doxycycline to induce N17 Ras.

Evidence from clinical samples supports these findings and suggests that the Ras/Raf/MEK/ERK pathway plays a critical role in prostate cancer progression (30). Numerous studies have reported an elevation of MAPK activity in human primary and metastatic prostate tumors and that activated MAP kinase levels increase with increasing Gleason score and tumor stage (43, 44). This suggests a correlation between activation of the MAP kinase pathway and prostate cancer progression.

In summary, growth factor receptor signals can activate AR or sensitize it to reduced levels of ligand. The ability of signaling cascades to influence AR function may play a significant role in the development and progression of prostate cancer where the increase in signal transduction activity has been associated with the acquisition of androgen-independent disease.

Role of Androgen Receptor Phosphorylation

AR phosphorylation (**Figure 1*a***) has been identified at serines 16, 81, 94, 256, 308, 424, and 650 (45–47). Other studies have suggested candidate phosphorylation sites on AR by in vitro phosphorylation reactions and/or by the identification and mutagenesis of kinase consensus sites. These sites include serines 213, 515, and 791 (48–50). Of the sites directly

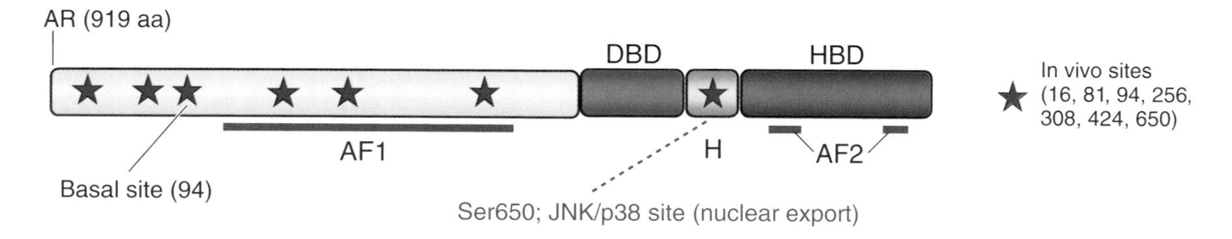

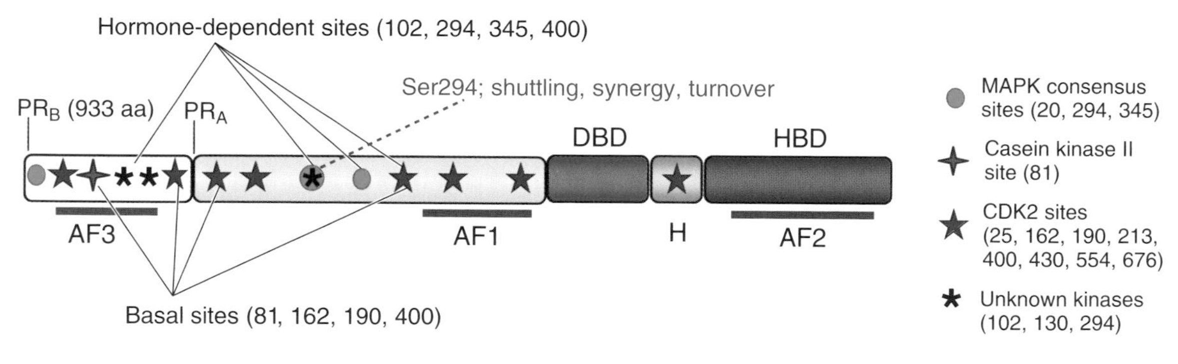

Figure 1

Phosphorylation sites in the human androgen receptor (AR) and progesterone receptor (PR). (*a*) AR phosphorylation. AR phosphorylation sites are shown. AF1 and AF2 (*underlined*) denote activation functional domains; DBD, DNA-binding domain; H, hinge; and HBD, hormone-binding domain. (*b*) PR phosphorylation. Thirteen serine residues and one threonine residue in human PR are shown as in *a*, to represent basal (constitutive) and hormone-induced phosphorylation sites (95), and may contribute to PR regulation by MAPK (55, 93, 94), casein kinase II (92), and CDK2 (90, 95). Individual PR phosphorylation sites may be regulated by multiple protein kinases (55) and/or in a sequential manner (135), illustrating the complexity of PR regulation by phosphorylation.

identified, Ser94 is constitutively phosphorylated. Serines 16, 81, 256, 308, 424, and 650 are regulated in response to androgen, and one of these, Ser650, becomes phosphorylated in response to a number of nonsteroid agonists, including epidermal growth factor (EGF), PMA, forskolin, and anisomycin (45, 47, 51). Recently, the use of phosphospecific antibodies has strongly suggested that AR is phosphorylated on Ser213 (52, 53). Taneja et al. (52) demonstrate that this site is regulated by PI3K-Akt signaling and suggest that Ser213 is phosphorylated in specific developmental contexts. Determining the function of these phosphorylation sites is critical for understanding AR biology, how androgen signaling is integrated with other signal transduction pathways, and for further delineating the role that AR plays in prostate cancer progression.

Studies with Androgen Receptor Ser650

Several investigations have provided a plausible account of the function of AR Ser650 phosphorylation, with important implications for both AR function and prostate cancer biology. Gioeli et al. (51) provide three lines of evidence supporting the conclusion that MKK4/JNK and MKK6/p38 directly regulate Ser650 phosphorylation. First, exogenous expression of stress kinases can increase AR Ser650 phosphorylation. Second, pharmacological inhibition of stress kinases inhibits endogenous AR Ser650 phosphorylation. Third, JNK1 and p38α can directly phosphorylate AR on Ser650 in vitro. These in vitro kinase data, in conjunction with the observation that cotreatment with both p38 and JNK inhibitors (SB203580 and SP600125, respectively) is required to inhibit endogenous LNCaP AR Ser650 phosphorylation, suggest that both JNK and p38 phosphorylate AR in vivo (51). Ser650 phosphorylation negatively regulates AR; siRNA knockdown of either MKK4 or MKK6 increased PSA mRNA levels in both the absence and presence of dihydrotestosterone (DHT). The increase in PSA mRNA levels generated by MKK siRNA was significantly reduced with either the AR antagonist bicalutamide or AR siRNA, suggesting AR dependence. This result was specific for MKK4 and MKK6, as siRNA to MEK1 and MEK2 had no effect on PSA mRNA levels (51). Furthermore, a heterokaryon shuttling assay that assesses nuclear-cytoplasmic shuttling of AR phosphorylation site mutants determined that (*a*) Ser650 phosphorylation was required for optimal nuclear export, (*b*) inhibition of stress kinase signaling reduced nuclear export of AR, and (*c*) stress kinase regulation of AR nuclear-cytoplasmic shuttling is dependent on Ser650 phosphorylation (51). These results are consistent with studies showing that signal transduction pathways often regulate the localization of other steroid receptors (54–56).

EGF: epidermal growth factor

Collectively, the data suggest that MKK4 and MKK6 signaling can antagonize AR transcription by increasing AR phosphorylation on Ser650 and facilitating AR nuclear export. Although this is an important first step toward elucidating the function of one of the multiple AR phosphorylation sites, much remains unknown about AR phosphorylation. Understanding how AR phosphorylation is regulated and modulates AR function is critical for obtaining a full picture of how growth factor and steroid signaling cross talk can contribute to prostate cancer progression.

Interactions Between Androgen Receptor and Steroid Hormone Receptor Coregulatory Proteins

Transcriptional coregulators are frequently overexpressed in advanced prostate cancer, facilitating AR activity. Indeed, several coactivators of AR [e.g., steroid receptor coactivator 1 (SRC-1), cyclic AMP responsive element binding protein (CREB)-binding protein (CBP), p300, amplified in breast cancer 1 (AIB1)] contribute to prostate cancer progression to androgen independence and have been identified as targets of signaling pathways (reviewed in Reference 29). For example, protein kinase A (PKA) and MAPK signaling directly regulate SRC-1 phosphorylation and activity, and EGF potentiates SRC-1-induced SR (chick PR) transcriptional activity (57). Similarly, in androgen-independent prostate cancer cells, EGF increases AR transcriptional activity through MAPK-dependent increases in transcription intermediary factor 2 (TIF2)/glutamate receptor interacting protein 1 (GRIP1) phosphorylation (58). Numerous studies have implicated a direct role for coregulatory proteins in prostate cancer. SRC-1 knockout mice have defects in prostate growth (59), and SRC-1 and TIF2/GRIP1 are overexpressed in recurrent prostate cancers (60). Overexpression of TIF2/GRIP1, AR-associated protein 55 (ARA55), or AR-associated protein 70 increases AR transcriptional activity in response to low-affinity ligands (e.g., DHEA, androstenedione, estradiol) or to low concentrations of DHT (61, 62). Phosphorylation of these coactivators may provide an alternative to overexpression as a mechanism for regulating AR. Consistent with this, p300 mediates interleukin 6 (IL6) activation of AR, and overexpression of p300 can overcome the ability of MEK-inhibition to block the IL6-simulated transactivation (63, 64). In summary, overexpression or activation of every component of the AR signaling pathway (growth factors and growth factor receptors, Ras and Ras effectors, coactivators that include kinase substrates, and AR itself) conspires to decrease the dependence of prostate cancer cells on androgen. This constellation of mutually reinforcing mechanisms toward androgen independence underscores the challenge to develop effective therapies for advanced disease.

Protein-Protein Interactions: Cross Talk Between the Androgen Receptor and WNT/β-Catenin/TCF Pathways

The WNT pathway regulates many key cellular processes, including cell proliferation, apoptosis, migration, and cell polarity, in vertebrates and other organisms (65). Collectively, the ~19 WNT ligands activate several intracellular signal transduction pathways, including a canonical pathway and several noncanonical pathways (66). In the canonical pathway, WNT ligands bind to cell-surface Frizzled receptors and, acting through a series of intermediates, increase the stability of nonmembrane-associated β-catenin. This promotes the accumulation of β-catenin in the nucleus, where β-catenin forms transcriptionally active complexes with members of the TCF/LEF family of transcription factors (65).

In addition to its role in canonical WNT signaling, β-catenin participates in protein-protein interactions with AR (67–69). This interaction modulates signaling by both the AR and WNT pathways. Overexpression of a wild-type β-catenin enhanced the androgen-stimulated activity of AR on multiple androgen-responsive promoters in LNCaP and TSU-Pr1 prostate cancer cell lines (68). This enhancement was more pronounced with a mutant form of β-catenin that is resistant to degradation by the ubiquitin-proteasome pathway. In addition, stable mutant β-catenin greatly enhanced the ability of weak AR ligands, including androstenedione and estradiol, to activate AR-mediated transcription, and partially blocked the activity of the AR antagonist bicalutamide (68).

Verras et al. (70) subsequently confirmed that the activation of β-catenin by the canonical WNT ligand WNT3A can stimulate AR transcriptional activity for both minimal androgen-responsive promoters and endogenous androgen-responsive genes. Interestingly, this study showed that, without the addition of androgen ligands, WNT3A could stimulate nuclear accumulation of β-catenin and an 11-fold induction of transcription from the androgen-responsive 7-kb PSA promoter in LNCaP cells. This observation suggests that WNT3A can bypass the need for androgens in AR activation or that

WNT3A can sensitize AR to the extremely low levels of steroid hormones present in charcoal-stripped serum. This result contrasts with previous studies showing that β-catenin stimulates the transcriptional activity of AR only in the presence of AR agonists (67–69). This discrepancy may be due to effects of WNT3A that are not mediated through β-catenin or by technical differences among the studies. Although activation of WNTs or β-catenin enhances AR transcriptional activity, AR activation inhibits transcription from promoters regulated by the TCF/LEF family of transcription factors, suggesting that these related factors compete with AR for a limited pool of β-catenin to transactivate distinct sets of target genes.

Protein-Protein Interactions: Cross Talk Between the Androgen Receptor and TGF-β/SMAD Pathways

Transforming growth factor-β (TGF-β) is a potent modulator of epithelial cells in the prostate and many other organs. TGF-β inhibits the growth of normal epithelial cells (71), but it can also accelerate cancer progression during late stages of epithelial tumorigenesis (72). Ligands from the TGF-β superfamily bind to heteromeric cell-surface receptors to initiate signal transduction. The receptor complex signals through the SMAD family of intracellular proteins that includes receptor-regulated SMADs (R-Smads) (SMADs 1, 2, 3, 5, and 8), common mediator SMADs (Co-SMADs) (SMAD4), and inhibitory SMADs (I-SMADs) (SMADs 6 and 7) (73). Upon ligand stimulation, R-SMADs are phosphorylated by the receptor complex, associate with Co-SMADs, and translocate to the nucleus, where they associate with DNA and other transcription factors to alter gene expression. I-SMADs antagonize signaling by R- and Co-SMADs.

AR participates in protein-protein interactions with SMAD3 and/or SMAD4 (74). These interactions have been associated with changes in the transcriptional response to both the AR and TGF-β pathways. However, the signaling outcome for interactions between AR and SMADs has differed among current studies. The AR-SMAD3 interaction repressed AR-mediated transcription in some experiments (75, 76) but stimulated AR-mediated transcription in other experiments (77, 78). Additional experiments have shown that SMAD4 can block the ability of SMAD3 to stimulate AR-mediated transcription (78). In cases in which AR and SMAD3 interact, activation of AR also repressed TGF-β-mediated transcription (76, 79). Taken together, these studies support a model in which associations between AR and SMADs modulate the transcriptional responses to both pathways; the outcome of this association for signaling depends on the cellular context.

PROGESTERONE RECEPTOR AND SIGNALING CROSS TALK IN BREAST CANCER

A direct role for PR in breast cancer development or progression remains poorly defined relative to studies with AR in prostate cancer or ERα in breast cancer. However, both ERα and PR are required for normal breast development. Estrogen induces ductal elongation, whereas progestins act directly on the mammary gland to induce ductal sidebranching and alveologenesis (reviewed in Reference 80). These events occur in association with the actions of growth factors. EGF promotes the proliferation of terminal end buds during normal breast development and augments ductal outgrowth and sidebranching induced by estrogen plus progesterone (81). In fact, PR isoform expression cannot be induced in response to estrogen unless EGF is present (82), suggesting the existence of important cross talk between EGFRs and both SRs. Liganded PRs are potent breast mitogens. However, mammary epithelial cells that express PR also express ER, and estrogen is usually required to induce PR expression. For this reason, it has been difficult to separate the effects of progesterone alone from estrogen, itself a potent breast mitogen. Indeed, PR isoforms are grossly understudied relative to ERα with regard to studies in the breast and breast cancer.

In the mature mammary gland, PR and ER are found in a minority population of luminal, nondividing epithelial cells that lie adjacent to proliferating cells. These SR-positive cells represent only approximately 7–10% of the epithelial cell population in the normal adult mammary gland. Estrogen receptor (ER)/PR-positive cells appear to be held in a nonproliferative state by inhibitory molecules such as TGF-β or high levels of p27, the CDK inhibitor (reviewed in Reference 83). Communication between the epithelial and stromal compartments mediates the proliferation of neighboring (adjacent) cells by expression and secretion of proproliferative molecules such as Wnts or IGF-II. Recent data suggest that SR-positive cells in the breast act as "feeder" cells for nearby stem cell–like progenitor cells (84). In contrast to the normal breast, where proliferating cells are clearly devoid of SRs, the majority of newly diagnosed breast cancers (~80%) express ER and PR. This suggests that SR-positive cells undergo an early switch to autocrine signaling mechanisms and/or that SR-positive lineages somehow retain the ability to divide. PR-containing cells also divide in the pregnant mammary gland, where PR-B colocalizes with cyclin D1 in BrdU-stained (dividing) cells (85). Thus, pathways involved in normal mammary gland growth and development are likely reactivated during breast cancer progression.

A more direct role for PRs in breast cancer was perhaps best illustrated by clinical findings showing that progestins increase breast cancer risk when taken with estrogen as part of combined hormone replacement therapy; tumors were larger and of higher grade relative to estrogen alone or placebo (86). Experimental data in model organisms support these findings and suggest a proproliferative role for progestins (87). Progestins are not carcinogens. However, progesterone may induce recently initiated precancerous breast cell populations to inappropriately re-enter the cell cycle or stimulate dormant stem cells to undergo self-renewal (87a, 87b). As breast tumors progress, they become resistant to endocrine-based treatments (anti-estrogens and/or aromatase inhibitors). However, the majority (65%) of resistant breast cancers retain high levels of SRs (ERα and PRs). In these resistant, SR-positive cancers, the rapid action of SRs at the membrane may have gradually become dominant and perhaps inappropriately integrated with the regulation of their classical transcriptional activities. PRs activated by extremely low or subthreshold concentrations of hormone or PRs phosphorylated in the absence of hormone are then capable of activating membrane-associated signaling pathways, including c-Src kinase, EGFR, and the p42/p44 MAPK pathway (discussed below). Elevation of MAPK activity and downstream signaling occurs frequently in breast cancer and provides a strong survival and proliferative stimulus to breast cancer cells (87c, 87d). MAPK signaling downstream of EGFR or Her2 (erbB2) is also associated with resistance to endocrine therapies (88).

Role of Direct Progesterone Receptor Phosphorylation in Breast Cancer Models

As with AR and other SR family members, phosphorylation-dephosphorylation events provide an added level of complexity to PR action (**Figure 1*b***). PR isoforms are heavily phosphorylated by multiple protein kinases; phosphorylation occurs primarily on serine residues within the N termini and, to a lesser degree, on serine residues throughout the receptor. PR contains a total of 14 known phosphorylation sites (reviewed in Reference 89). Serines at positions 81, 162, 190, and 400 were originally defined as basal sites (90) because they appeared to be constitutively phosphorylated in the absence of hormone (**Figure 1**). Serines 102, 294, and 345 are hormone-induced sites that are maximally phosphorylated 1–2 h following progestin treatment (91). Interestingly, hormone-induced phosphorylation of PR includes DNA-dependent and -independent kinase activiites (91a,b). Specific kinases responsible for phosphorylation of selected sites have been identified, whereas others remain unknown. For example, serines at positions 81 and 294 are phosphorylated by casein kinase II (92) and MAPK (93, 94), respectively. Progestins can also stimulate Ser294 phosphorylation independently of MAPKs by activation of an unknown kinase(s) (55). Eight of the total 14 sites (i.e., Thr430 and serines 25, 162, 190, 213, 400, 554, and 676) are known to be phosphorylated by cyclin A/cyclin-dependent protein kinase 2 (CDK2) complexes in vitro (90, 95). Researchers have confirmed five of these sites (i.e., serines 162, 190, 213, 400, 676) as authentic in vivo phosphorylation sites (90, 92, 95).

Although the role of PR phosphorylation is not fully understood, such phosphorylation may influence aspects of transcriptional regulation such as interaction with coregulators, as has been found for ERα (96) and recently for PR (97). Regulation of ligand-dependent (94) and -independent (98, 99) PR nuclear localization, transcriptional activities (99a), and receptor turnover (93) all involve phosphorylation. For example, MAPK-dependent PR Ser294 phosphorylation is required for rapid nuclear translocation of unliganded PR and nuclear export of liganded PR, suggesting that MAPK signaling regulates PR action by altering nucleocytoplasmic shuttling (55). The functional significance of PR nuclear sequestration in response to MAPK activation is unknown. Nuclear localization may serve to concentrate inactive or active receptors, as PRs are degraded in the cytoplasm or upon nuclear export (55). Following a brief (5–15 min) pretreatment with EGF, phosphorylated nuclear PR-B receptors become ultrasensitive to subphysiological progestin levels. But when EGF and progesterone are added to cells simultaneously, the magnitude of PR sensitivity is much diminished relative to EGF pretreatment, suggesting the importance of the temporal order of phosphorylation events in the regulation of hormone sensitivity (100, 101). Additionally, unliganded phosphorylated PR appears to regulate genes via nonclassical mechanisms. For example, PR-B regulation of IRS-2 expression occurs independently of ligand but requires phosphorylation of PR Ser294 in breast cancer cells (101).

CDK2: cyclin-dependent protein kinase 2

Indeed, SRs act as "sensors" for signal transduction pathways. As investigators have reported for ERα (102, 103) and AR (above), phosphorylated PRs are hypersensitive to subphysiological levels of progestins relative to their underphosphorylated counterparts (101). EGF and progestins synergistically upregulate mRNA or protein levels for a number of growth regulatory genes (10), including cyclin D1 and cyclin E (7); the regulation of cyclins by progestins is MAPK dependent. Cyclins, in turn, regulate the progression of cells through the cell cycle by interaction with cyclin-dependent protein kinases (cdks). Progestins activate CDK2 (12), and PRs are predominantly phosphorylated by CDK2 at proline-directed (S/TP) sites (90, 95), perhaps allowing for the coordinate regulation of PR transcriptional activity during cell-cycle progression. In support of this idea, Narayanan and coworkers (97, 104) found that PR activity is highest in S phase and lower in G0/G1 phases of the cell cycle but that this activity is impaired during G2/M phases, concomitant with lowered PR phosphorylation. Overexpression of either cyclin A or CDK2 enhanced both PR and AR transcriptional activity; whereas cyclin A interacts with the N terminus of PR, CDK2 seems to alter PR function indirectly by increasing the phosphorylation and recruitment of SRC-1 to liganded PR.

Multiple Functions of Progesterone Receptor Ser294 Phosphorylation in Breast Cancer Models

PR Ser294 is a ligand-inducible phosphorylation site; it becomes rapidly phosphorylated upon exposure to hormone (91). Ser294 is also a proline-directed or MAPK consensus site (PXXSP). In contrast to progestin-induced Ser294 phosphorylation, which occurs within 30–60 min independently of MAPK activation, growth factor–induced Ser294 phosphorylation occurs within 3–5 min and is MAPK dependent (55). Studies suggest that PR Ser294 is a significant site for PR regulation by multiple kinases (55, 93, 94, 101). Ser294 phosphorylation appears to mediate increased PR nucleocytoplasmic shuttling (55) and is required to sustain transcriptional synergy in the presence of progestins and growth factors (94). Following ligand binding, PR undergoes rapid downregulation (105, 105a). This process is also greatly augmented by the phosphorylation of Ser294, which targets liganded PR for ubiquitination and degradation by the 26S-proteasome pathway (93). Mutant PR, in which Ser294 has been replaced by alanine (S294A), binds ligand and, like wild-type PR, undergoes a characteristic upshift in gel mobility owing to phosphorylation at other sites, enters the nucleus, and binds to PREs (55, 93, 94, 101). However, failure of S294A PR to undergo ubiquitination leaves the receptor stable in the presence of progestins relative to wild-type PR (93). Interestingly, when stably expressed in breast cancer cells, liganded, mutant S294A PR is a weak transcriptional activator and fails to respond to agents that activate MAPK (94). Creation of a Ser294 phosphomimic receptor by replacement of Ser294 with aspartic acid (S294D) reversed these effects, resulting in hyperactive PR with increased turnover relative to wild-type PR (C.A. Lange & A. Daniel, unpublished results). Thus, reversible phosphorylation of PR Ser294 couples increased transcriptional activity to rapid downregulation of PR protein by the ubiquitin-proteasome pathway; the link between these events is under investigation but may involve regulation of transcriptional events by components of the ubiquitin pathway and/or participation of nucleocytoplasmic shuttling factors or chaperones.

Phosphorylation of Ser294 may also mediate some aspects of ligand-independent PR action. Growth factors such as EGF are strong activators of p42/p44 MAPKs and induce rapid phosphorylation of PR Ser294 and nuclear accumulation (55). Mutation of the consensus MAPK site (S294A) abolished EGF-mediated translocation; however, the ability of progestin (R5020) to induce nuclear localization of S294A PR was unaffected. EGF-induced nuclear accumulation of PR suggests a mechanism for ligand-independent transcriptional activation. Labriola et al. (98) reported that the EGF family member heregulin can stimulate nuclear localization, DNA binding, and transcriptional activity of PR in T47D breast cancer cells in the absence of hormone. This was accompanied by activation of MAPK and PR Ser294 phosphorylation. Qiu et al. (55) reported that similar changes mapped to PR Ser294 phosphorylation but that growth factors alone did not stimulate PR transcriptional activity or alter PR downregulation in T47D cell variants (94). Although these events required ligand binding, they were greatly augmented by MAPK activation (55, 94). The reasons for these differences are unknown but may involve differential expression of EGFR family members expressed on the cell surface between T47D cell line clones, leading to differences in the activation of downstream intracellular kinases such as CDK2 (discussed below). In any case, these exciting data (55, 98) suggest a continuum between PR hypersensitivity to extremely low ligand concentrations and complete ligand independence, a phenomenon that is well documented for AR (discussed above). The regulation of PR by alternate signaling pathways, including elevated MAPK activity often exhibited by breast tumors, may contribute to disregulated gene expression and changes in cell growth and/or survival.

CDK2 Regulation of Progesterone Receptor: Role of Ser400

In contrast to Ser294, PR Ser400 is both basally phosphorylated and regulated by ligand in vivo, and Ser400 phosphorylation is mediated by CDK2 in vitro (90). Like mitogens, progestins regulate CDK2 activity (12, 106). Additionally, like progestins, mitogens also induce rapid and robust phosphorylation of PR Ser400 (99), suggesting a mechanism of cell-cycle-dependent regulation of PR. In the presence of activated CDK2, phospho-Ser400 PRs were nuclear, suggesting that the phosphorylation of Ser400 sequesters unliganded PRs (99). CDK2 overexpression increased PR transcriptional activity in the absence or presence of progestin. Mutation of Ser400 to alanine (S400A) selectively blocked ligand-independent PR transcriptional activity, with little effect on transcription induced in response to ligand binding (99). This result suggests that CDK2 positively regulates unliganded PR (i.e., basal transcriptional activity) by acting at Ser400 while still exerting positive effects in the presence of ligand, perhaps in cooperation with other CDK2 sites on PR (107) or its coactivators (97). Interestingly, Ser400 is adjacent to a nine-amino-acid destruction (D)-box motif that may alter PR turnover, suggesting an important means by which PRs are regulated in response to phosphorylation (94). Consistent with this idea, activated CDK2 induced rapid PR downregulation in the presence or absence of progestins, whereas CDK2 inhibitors blocked ligand-induced PR downregulation (99). Similar to PR, phosphorylation of AR regulates receptor location and activity (discussed above), as well as protein turnover (107a); mutation of putative Akt sites in AR (Ser210 and Ser790) prevented Mdm2 E3 ligase–mediated AR ubitquitinylation and proteasome-dependent degradation. The mechanisms linking SR stability/turnover to transcriptional activity require further examination. However, researchers have proposed a model in which selected SRs are regulated by protein kinases that primarily modulate receptor location or shuttling, leading to changes in transcriptional activity and/or protein turnover (99, 101, 108, 109).

SH3: Src homology three domain (interaction with proline-rich regions)

NOVEL EXTRANUCLEAR ACTIONS OF PROGESTERONE AND ANDROGEN RECEPTORS

The extranuclear actions of classical SR transcription factors have become an intense area of investigation, particularly in breast and prostate cancer cell models. Migliaccio et al. (110) were the first to report p60–Src kinase and MAPK activation by estradiol in MCF-7 cells and interaction between PR, ERα, and p60–Src kinase in T47D cells (111). Steroid hormone treatment of breast or prostate cancer cells causes a rapid and transient activation of MAPK signaling that is SR dependent but independent of SR transcriptional activity (111, 112). Whereas the genomic effects of steroid hormone treatment are delayed by several minutes to hours (i.e., following transcription and translation), the extranuclear, or nongenomic, effects occur rapidly, in only a few minutes. For example, following progestin treatment, maximal activation of p60–Src kinase occurs within 2–5 min, and downstream activation of p42/p44 MAPKs, within 5–10 min (111, 112). Human PR contains a proline-rich (PXXP) motif that mediates direct binding to the Src homology three (SH3) domains of signaling molecules in the p60–Src kinase family in a ligand-dependent manner (112). Purified liganded PR-A and PR-B activated the Src-related protein kinase, HcK, in vitro. Mutation of the PXXP sequence in PR abolished the ability of progestins to bind c-Src and activate both c-Src (or HcK) and p42/p44 MAPKs, indicating that MAPKs are likely activated by a c-Src-dependent mechanism involving Ras activation via phosphorylation of the c-Src substrate adaptor proteins p190 and Shc, followed by Grb-2 and Sos binding. Mutation of the PR DNA-binding

SH2: Src homology two domain (interaction with phosphotyrosine residues)

LXXLL motif (NR box): an amphipathic alpha helical motif that appears in single or multiple copies in many coactivators and scaffold molecules such as MNAR; contributes to ligand-dependent interactions of coregulators with AF2 domains of nuclear receptors

mPR: membrane G protein–coupled receptor that binds progesterone

domain abolished PR transcriptional activity without having an effect on c-Src or MAPK activation. A polyproline-rich sequence in AR interacts with c-Src by a similar mechanism in prostate cancer cells (113). Treatment with either steroid hormone (androgen or estrogen) or EGF triggered the association of AR with c-Src and either ERβ in prostate cancer cells or ERα in breast cancer cells (113, 114), indicating that, like most cell-surface receptors, multiple SRs can input to MAPK activation.

Ballare et al. (115) recently reported that MAPK activation by progestins is blocked by antiprogestins and anti-estrogens in COS-7 cells transfected with both PR and ERα. They propose that c-Src/MAPK activation by PR is mediated indirectly by the interaction of the Src homology 2 (SH2) domain of c-Src with phosphotyrosine 537 of ERα (115). In their model, the activation of c-Src and the MAPK pathway by progestins depends upon the presence of unliganded ERα, which interacts constitutively with PR-B via two domains that flank PRs' proline-rich sequence. In contrast to the studies conducted by Boonyaratanakornkit et al. (112), Ballare and coworkers (115) found that deletion of either of these two ER-interacting domains in PR-B blocked c-Src/MAPK activation by progestins in the presence of ERα. Mutation of PR-B's PXXP domain had no effect. However, Boonyaratanakornkit et al. (112) found that PR expression increased c-Src activity in COS-7 cells in the absence of progestins and independently of added ER; coexpression of both PR and ERα reduced basal levels of c-Src activity, revealing a clear stimulation of c-Src by progestin. In addition, progestins activated c-Src in PR-null MCF-12A cells transduced with wild-type PR but not PXXP-mutant PR adenoviruses. Both groups found that ERα interacts with the SH2 domain of c-Src, but neither group tested the effects of estrogen on the ability of progesterone to activate c-Src or MAPKs (112, 115).

Although discrepancies between the two models must be resolved, overexpression of SRs in COS-7 cells may lead to concentration-dependent effects, resulting in the formation of different signaling complexes depending on the presence of other signaling and adaptor molecules. In support of this idea, Wong et al. (116) identified an additional ER-interacting adaptor protein, termed MNAR (modulator of nongenomic activity of estrogen receptor), that contains both LXXLL (nuclear receptor–binding) and PXXP (SH3 domain–binding) motifs. MNAR is essential for ER-Src interaction, but it is not required for progestin/PR-dependent activation of c-Src (D.P. Edwards, personal communication). In *Xenopus* oocytes, MNAR interacts with AR and appears to mediate inhibition of meiosis via Gβγ signaling; Xenopus MNAR also enhances AR transcription via c-Src kinase activation in CV1 cells (117). Constitutive signaling via the AR/c-Src/MNAR complex occurs in androgen-independent prostate cancer cells, whereas transient signaling from this complex is regulated by AR ligand binding in androgen-dependent cells (118). A newly described protein, termed DOC-2/DAB2 (differentially expressed in ovarian cancer/disabled 2) was recently shown to antagonize AR-mediated prostate cancer cell growth by disruption of the AR/c-Src complex (119). Taken together, these data indicate that multiple interactions contribute to direct protein kinase activation by SRs and suggest that at least some nongenomic signaling functions of amphibian PR/AR have been conserved in mammals. Interestingly, Zhu et al. (120) have described a separate gene product encoding the putative mammalian homolog of membrane progesterone receptor (mPR), a progesterone-binding G protein–coupled receptor first identified in spotted seatrout oocytes. Fish and mammalian mPRs appear to be structurally unrelated to PR-A or PR-B. Further studies are needed to determine if mPR plays a role in progestin-induced rapid signaling or if mPR interacts with classical PRs. However, studies with mPR underscore the important concept that steroid hormone-binding proteins other than classical SRs may regulate some nongenomic steroid-mediated signaling events.

Integration of Rapid Signaling and Nuclear Steroid Hormone Receptor Actions

Although the role of cytoplasmic SRs in mammalian physiology remains unclear, SR-mediated rapid activation of signaling molecules may serve to potentiate several nuclear functions of activated SRs (**Figure 2**). Amplification of SR nuclear functions may occur through rapid, direct phosphorylation of SRs and/or their coregulators in response to activation of SR-induced cytoplasmic pathways that coincide with ligand binding. Such a positive feedback loop would explain the dramatic influence of activated signaling pathways on both PR and AR nuclear function. For example, several progestin-dependent functions of PR are MAPK dependent, including upregulation of cyclins D1 and E, CDK2 activation, and S-phase entry (7, 94, 99, 121).

Following ligand binding, most SRs stimulate a transient (3–10 min) activation of MAPKs. However, mitogenic signaling requires sustained (hours to days) MAPK activation in fibroblast cell models (122). Recently, Faivre (C.A. Lange & E.J. Faivre, unpublished results) found that, in addition to rapid and transient activation of MAPK by progestin/PR-B (5–15 min), progestin-bound PR-B induced subsequent oscillations in MAPK activity culminating in a sustained (hours to days) phase of MAPK activation that was c-Src kinase and EGFR dependent. Further studies revealed the creation of an autocrine signaling loop in which PR-B triggered transcriptional upregulation of Wnt-1, leading to activation of Frizzled-receptor dependent MMP9 and shedding of EGF ligands from the cell surface. Consistent with Wnt-1 dependent transactivation of EGFR by progestins, progestin-induced cyclin D1 protein upregulation, S-phase entry, and soft-agar growth of T47D breast cancer cells were blocked by either shRNA targeted to Wnt-1 or by inhibitors of MAPK, c-Src, and EGFR. Furthermore, progestins failed to stimulate S-phase entry in MCF-7 cells that (*a*) stably express a PXXP-mutant PR-B unable to bind to the SH3 domain of c-Src and (*b*) activate MAPK (121). These data indicate that the integrated actions of cytoplasmic and nuclear PRs, acting via both the PR-B and c-Src interaction and PR-B-driven transcriptional events, coopt EGFR signaling to induce breast cancer cell proliferation. Other important effectors downstream of EGFR signaling include STATs. Progestins induce the tyrosine phosphorylation and nuclear translocation of Stat5 (10) and Stat3 (15). Proietti et al. (15) demonstrated that Stat3 phosphorylation and activation by the nongenomic actions of PR are critical events for breast cancer cell growth; T47D cell growth and tumor growth of progestin-induced mammary adenocarcinomas in BALB/c mice depend on PR activation of Jak1 and Jak2, c-Src, and Stat3.

The extranuclear actions of SRs may contribute to deregulated breast cancer cell growth (121) and/or increased breast cancer risk (86), perhaps by linking steroid hormone action to the regulation of MAPK-regulated genes (i.e., transcription factor targets of MAPK; see MAPKs Regulate Cyclin D1 Expression). The extranuclear actions of liganded ERα are thought to induce a state of adaptive hypersensitivity during endocrine therapy in which growth factor signaling pathways are coopted by upregulated ERα (123). In this model of ER-dependent MAPK activation, liganded cell membrane–associated ERα molecules interact with the adapter protein Shc and induce its phosphorylation, leading to the recruitment of Grb-2 and Sos, followed by the activation of Ras and the Raf-1/MEK/MAPK module. ERα activation of MAPK may explain why many tumors respond well to aromatase inhibitors but fail to respond to selective estrogen receptor modulators (SERMS) designed to inhibit ER transcriptional activity. SERMs can act as partial transcriptional agonists of phosphorylated receptors and may not block ER-dependent

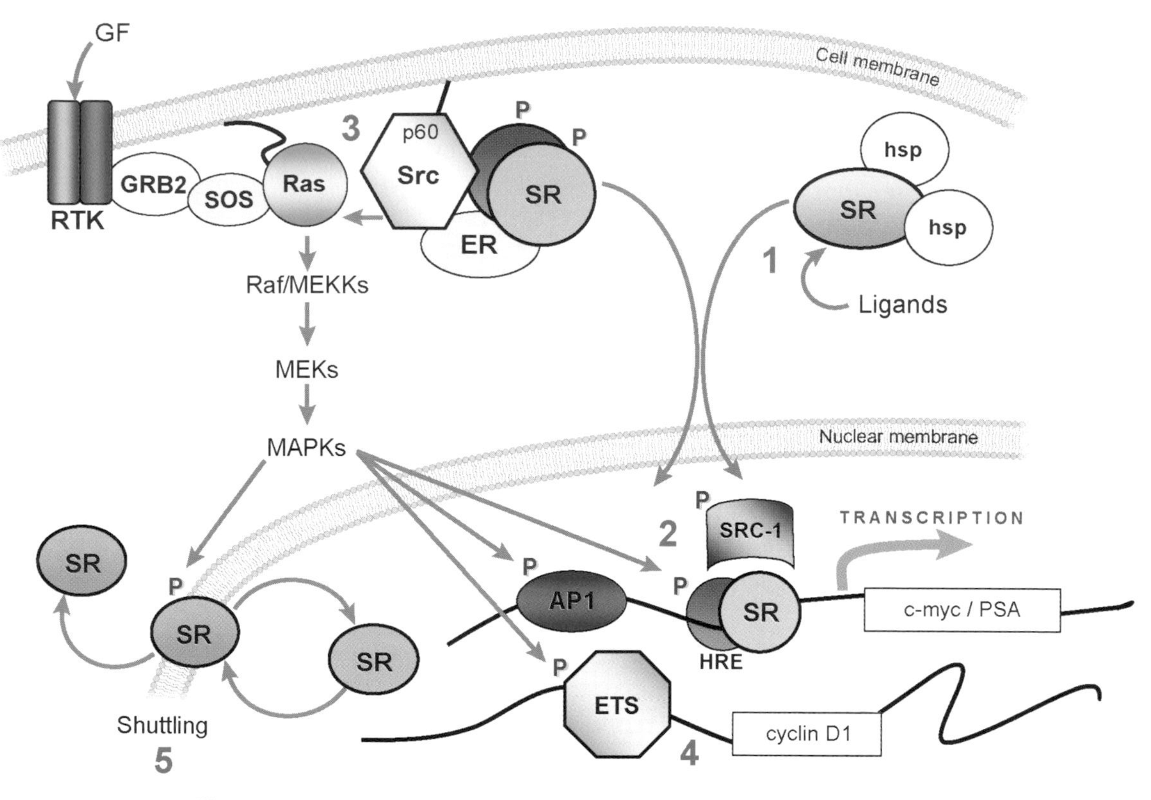

Figure 2

Functional significance of SR phosphorylation. Phosphorylation (P) of specific sites in ARs or PRs (SR) couples multiple SR functions, including transcriptional synergy in the presence of steroid hormones and growth factors predicted to activate MAPK and/or CDK2 as well as nuclear import or export (shuttling) in response to MAPK activation. In the case of PRs, rapid ligand-dependent downregulation by the ubiquitin-proteasome pathway (degradation) occurs upon nuclear export. (*1*) Ligand binding mediates dissociation of heat shock proteins (hsp) and nuclear accumulation of SR dimers. (*2*) Nuclear SRs mediate gene regulation via the classical pathway; phosphorylated SRs may recruit regulatory molecules that are phosphoproteins, and function in one or more interconnected processes (transcription, localization, and turnover), perhaps linked by a common cellular machinery. (*3*) SRs and growth factors activate MAPKs independently via a c-Src kinase–dependent pathway. This may result in positive regulation of SR action via feedback regulation (i.e., direct phosphorylation of liganded SRs or coactivators) occurring in both the absence and presence of steroid hormone ligands and on hormone response element (HRE)-containing or other SR-regulated gene promoters. (*4*) Activation of MAPKs by SRs provides for regulation of gene targets whose promoters do not contain HREs and are otherwise independent of SR-transcriptional activities but utilize SR-activated MAPKs, such as regulation of the cyclin D1 promoter by ETS factors. (*5*) MAPK regulation of SRs mediates nuclear accumulation/shuttling (for PRs) and nuclear export (for ARs) that is coupled to regulation of SR transcriptional events. Other abbreviations used: AP1, activating protein 1; GF, growth factor; RTK; receptor tyrosine kinase.

MAPK activation (123). PR-B or ARs in SR-positive breast cancers may participate in MAPK-activating complexes, perhaps bypassing anti-estrogen therapies. Few groups have studied membrane-associated or cytoplasmic signaling complexes containing both ERα and PR-B or ARs (113, 114). However, ARs are frequently (70% of the time) expressed in metastatic breast cancer (124), and the expression of functional ARs defines a subset of ER/PR-negative breast cancers (125). These studies suggest that it will be important to target SRs that may substitute for ERα in the activation of c-Src-dependent mitogenic signaling cascades.

Integrated Steroid Hormone Receptor Actions in Gene Expression

An important end point of MAPK signaling is upregulation of cyclin D1 (see MAPKs Regulate Cyclin D1 Expression). Cyclin D1 mRNA and protein levels increase in response to estrogen, progesterone, or androgen treatment (12, 126, 127), and cyclin D1 is frequently elevated in breast and prostate cancers (128, 129). Interestingly, the D1a isoform of cyclin D1 acts as an androgen-induced transcriptional repressor of AR via direct binding to the AR N terminus (130). However, the cyclin D1b variant acts to promote androgen-induced prostate cancer proliferation and is frequently overexpressed relative to cyclin D1a in prostate cancer cell lines and tumors (131). Recent evidence suggests that SRs are often recruited to distal enhancer regions far upstream or downstream of hormone-regulated gene proximal promoters; distal HRE-containing elements in association with proteins that bind nearby termed pioneer factors function to recruit and tether the distant SR complex to the proximal promoter via the creation of a chromatin loop (132, 133). Thus, SR recruitment to distant enhancer sites provides a mechanism for the direct regulation of genes such as cyclin D1 via the classical pathway (e.g., via SR binding at putative distant HRE sites). As SR-driven tumors progress, membrane SRs may begin to function dominantly, leading to a switch in promoter regulation to MAPK-dependent induction via proximal promoter sites. This may explain how tumors escape the action of SR antagonists primarily blocking transcriptional events but may fail to inhibit the signaling functions of these receptors. In support of this idea, multiple SRs regulate cyclin D1 expression, perhaps via distant sites. However, transcriptional regulation of the cyclin D1 proximal promoter region by steroids (i.e., progestins or estrogens) is MAPK dependent (121, 134). Thus, the activation of cytoplasmic signaling pathways by liganded SRs not only provides enhanced SR action at

MAPKs REGULATE CYCLIN D1 EXPRESSION

Growth factor stimulation of MAPKs results in the phosphorylation and activation of numerous transcription factors, including c-Ets-2, a member of the large family of ETS factors. ETS proteins promote G1-phase progression by the regulation of genes involved in the control of cell proliferation, including the cell-cycle regulator cyclin D1 (136). Cyclin D1 activates and defines the substrate specificity of cdks 4 and 6, whose activity is required for progression from G1 to S phase (137, 138). Cdks are inactive in the absence of cyclins; the availability of cyclin D1 is a major determinant for cell-cycle progression through G1 in response to diverse mitogens, including steroid hormones and EGF (139). MAPKs can also function posttranslationally to regulate the assembly and activation of cyclin D1/cdk complexes (140). Cyclin D1 is elevated in ~45% of breast cancers (128) and ~25–37% of prostate cancers (129). AR-positive (141) and androgen-independent (142) prostate cancers are most likely to overexpress cyclin D1. Interestingly, cyclin D1–null mice exhibit deficiencies in mammary gland development, including specific defects in alveolar growth (143, 144), a phenotype similar to adult female mice lacking PR-B (145). The oncogenic effects of cyclin D1 overexpression are mediated by interaction with the transcription factor CCAAT enhancer binding protein (C/EBPβ) (146). C/EBPβ knockout mice also exhibit similarly compromised mammary alveolar development (147), indicating the importance of coordinate expression of these gene products (PR-B, cyclin D1, and C/EBPβ).

specific SR-regulated genes via HRE sequences, but couples this to the regulation of additional genes whose promoters clearly use SRs and can also utilize SR-activated MAPK pathways independently of SR transcriptional activity (**Figure 2**).

CONCLUDING REMARKS

Rather than acting in an obligatory or switch-like manner, phosphorylation events are generally considered to exert subtle effects on nuclear SRs; kinase inputs act primarily as a "rheostat" for a continuum of SR transcriptional activities. However, this conclusion is based largely on observations made with liganded receptors in the absence of controlled inhibition or activation of alternate signaling pathways. In fact, studies with human PR or AR reviewed herein suggest that the effects of phosphorylation are quite profound in the context of multiple signaling inputs. We conclude that the phosphorylation status of a particular SR serves as a direct sensor of cellular kinase activities to coordinate responses to growth factors and steroid hormones. In the absence of alternate stimuli, independent activation of MAPKs by extranuclear liganded SRs may result in positive regulation of receptor action via feedback regulation by direct phosphorylation of SRs or their coregulatory partners. This may occur in both the presence and absence of steroid hormone ligands and on diverse gene promoters and via distant sites in chromatin. In addition, activation of cytoplasmic kinase cascades including MAPK modules by liganded receptors provides for the regulation of gene targets whose promoters can function entirely independently of SR transcriptional activities but rely on the activity of MAPK-targeted transcription factors such as the Ets family members, Elk-1, c-myc, fos, and jun (components of AP-1). This important linkage provides for well-integrated control of a large number of genes or gene subsets coordinately regulated in response to convergence of growth factor and SR signaling. Finally, the newly discovered ability of SRs to activate kinase pathways classically defined as key regulators of cell growth underscores the concept that the activation of signal transduction pathways is an integral feature of SR action. This aspect of SR function is likely to play an important role in cancer progression toward the development of resistance to endocrine therapies (123). Targeting the relevant protein kinases (c-Src, MAPKs, and CDKs) as an integral feature of SR action should provide significant improvements over the use of traditional SR blocking strategies for advanced or progressive breast or prostate cancers.

SUMMARY POINTS

1. Steroid hormone receptors (SRs) act as sensors for other signaling pathways, including growth factor–initiated kinase cascades such as the Ras/Raf/MEK/MAPK module; kinase signaling is an integral feature of SR regulation and function.
2. SRs are capable of rapidly activating signaling pathways independently of their function as ligand-activated transcription factors.
3. Rapid activation of kinases by liganded SRs serves as a direct input to regulate SR-directed transcription positively or negatively via direct phosphorylation of SRs and/or their coregulatory partners and other functional components of transcription complexes.
4. Kinase pathways activated by liganded SRs provide alternate routes of gene regulation that are independent of SR nuclear action, but such pathways act via direct phosphorylation of other transcription factor substrates.

5. In the presence of activated kinase pathways, SRs often function as hypersensitive receptors responsive to extremely low ligand concentrations or function completely independently of ligand.
6. The ability of SRs to activate multiple mitogenic kinase pathways (c-Src, MAPKs, and CDKs) likely contributes to the cancer biology (increased growth and survival, progression to steroid hormone independence) of hormonally regulated tissues.
7. SRs and their associated kinase signaling pathways should be targeted as part of combined chemotherapy approaches aimed at blocking these two arms of SR action.

APPENDIX OF ABBREVIATIONS USED

AR, androgen receptor; CBP, cyclic AMP responsive element binding protein (CREB)-binding protein; CDK2, cyclin-dependent protein kinase 2; C/EBPβ, CCAAT enhancer binding protein; EGF, epidermal growth factor; ER, estrogen receptor; HREs, hormone response elements; Hsp, heat shock protein; IL6, interleukin 6; MAPK, mitogen-activated protein kinase; MEK, MAPK/ERK kinase; MEKK, MAPK/ERK kinase kinase; MMTV, mouse mammary tumor virus; MNAR, modulator of nongenomic activity of estrogen receptor; mPR, membrane progesterone receptor; PR, progesterone receptor; PRE, progesterone response element; PSA, prostate-specific antigen; SERM, selective estrogen receptor modulator; SH2, Src homology two domain (interaction with phospho-tyrosine residues); SH3, Src homology three domain (interaction with proline-rich regions); SR, steroid hormone receptor; SRC, steroid receptor coactivator; TIFs, transcription intermediary factors.

ACKNOWLEDGMENTS

Studies contributed by the authors' laboratories were supported by NIH grant DK53825/CA123763 (to C.A.L.), Department of Defense (DOD/CDMRP) grant W81XWH-04-1-0112 (to D.G.), NIH grants DK069662 and AG024278 and DOD/CDMRP grant PC050617 (to P.M.), and NIH grant DK59913 (to S.R.H.).

LITERATURE CITED

1. Condon JC, Hardy DB, Kovaric K, Mendelson CR. 2006. Up-regulation of the progesterone receptor (PR)-C isoform in laboring myometrium by activation of nuclear factor-κB may contribute to the onset of labor through inhibition of PR function. *Mol. Endocrinol.* 20:764–75
2. Pratt WB, Toft DO. 2003. Regulation of signaling protein function and trafficking by the hsp90/hsp70-based chaperone machinery. *Exp. Biol. Med.* 228:111–33
3. Moore MR, Zhou JL, Blankenship KA, Strobl JS, Edwards DP, Gentry RN. 1997. A sequence in the 5′ flanking region confers progestin responsiveness on the human c-myc gene. *J. Steroid Biochem. Mol. Biol.* 62:243–52
4. Chalbos D, Chambon M, Ailhaud G, Rochefort H. 1987. Fatty acid synthetase and its mRNA are induced by progestins in breast cancer cells. *J. Biol. Chem.* 262:9923–26

5. Krusekopf S, Chauchereau A, Milgrom E, Henderson D, Cato AC. 1991. Co-operation of progestational steroids with epidermal growth factor in activation of gene expression in mammary tumor cells. *J. Steroid Biochem. Mol. Biol.* 40:239–45
6. Dehm SM, Tindall DJ. 2006. Molecular regulation of androgen action in prostate cancer. *J. Cell Biochem.* 99:333–34
7. Lange CA, Richer JK, Shen T, Horwitz KB. 1998. Convergence of progesterone and epidermal growth factor signaling in breast cancer. Potentiation of mitogen-activated protein kinase pathways. *J. Biol. Chem.* 273:31308–16
8. Brass AL, Barnard J, Patai BL, Salvi D, Rukstalis DB. 1995. Androgen up-regulates epidermal growth factor receptor expression and binding affinity in PC3 cell lines expressing the human androgen receptor. *Cancer Res.* 55:3197–203
9. Church DR, Lee E, Thompson TA, Basu HS, Ripple MO, et al. 2005. Induction of AP-1 activity by androgen activation of the androgen receptor in LNCaP human prostate carcinoma cells. *Prostate* 63:155–68
10. Richer JK, Lange CA, Manning NG, Owen G, Powell R, Horwitz KB. 1998. Convergence of progesterone with growth factor and cytokine signaling in breast cancer. Progesterone receptors regulate signal transducers and activators of transcription expression and activity. *J. Biol. Chem.* 273:31317–26
11. Gregory CW, Johnson RTJ, Presnell SC, Mohler JL, French FS. 2001. Androgen receptor regulation of G1 cyclin and cyclin-dependent kinase function in the CWR22 human prostate cancer xenograft. *J. Androl.* 22:537–48
12. Groshong SD, Owen GI, Grimison B, Schauer IE, Todd MC, et al. 1997. Biphasic regulation of breast cancer cell growth by progesterone: role of the cyclin-dependent kinase inhibitors, p21 and p27 (Kip1). *Mol. Endocrinol.* 11:1593–607
13. Owen GI, Richer JK, Tung L, Takimoto G, Horwitz KB. 1998. Progesterone regulates transcription of the p21 (WAF1) cyclin-dependent kinase inhibitor gene through Sp1 and CBP/p300. *J. Biol. Chem.* 273:10696–701
14. Tseng L, Tang M, Wang Z, Mazella J. 2003. Progesterone receptor (hPR) upregulates the fibronectin promoter activity in human decidual fibroblasts. *DNA Cell Biol.* 22:633–40
15. Proietti C, Salatino M, Rosemblit C, Carnevale R, Pecci A, et al. 2005. Progestins induce transcriptional activation of signal transducer and activator of transcription 3 (Stat3) via a Jak- and Src-dependent mechanism in breast cancer cells. *Mol. Cell Biol.* 25:4826–40
16. McKenna NJ, O'Malley BW. 2002. Combinatorial control of gene expression by nuclear receptors and coregulators. *Cell* 108:465–74
17. Smith LD, Ecker RE. 1969. Role of the oocyte nucleus in physiological maturation in *Rana pipiens*. *Dev. Biol.* 19:281–309
18. Masui Y. 2001. From oocyte maturation to the in vitro cell cycle: the history of discoveries of Maturation-Promoting Factor (MPF) and Cytostatic Factor (CSF). *Differentiation* 69:1–17
19. Hammes SR. 2004. Steroids and oocyte maturation—a new look at an old story. *Mol. Endocrinol.* 18:769–75
20. **Lutz LB, Cole LM, Gupta MK, Kwist KW, Auchus RJ, Hammes SR. 2001. Evidence that androgens are the primary steroids produced by *Xenopus laevis* ovaries and may signal through the classical androgen receptor to promote oocyte maturation. *Proc. Natl. Acad. Sci. USA* 98:13728–33**
21. White SN, Jamnongjit M, Gill A, Lutz LB, Hammes SR. 2005. Specific modulation of nongenomic androgen signaling in the ovary. *Steroids* 70:352–60

20. Androgens, acting via the classical AR, are the physiological mediators of *Xenopus* oocyte maturation.

22. Jamnongjit M, Hammes SR. 2005. Oocyte maturation: the coming of age of a germ cell. *Semin. Reprod. Med.* 23:234–41
23. Staack A, Kassis AP, Olshen A, Wang Y, Wu D, et al. 2003. Quantitation of apoptotic activity following castration in human prostatic tissue in vivo. *Prostate* 54:212–19
24. Denis LJ, Griffiths K. 2000. Endocrine treatment in prostate cancer. *Semin. Surg. Oncol.* 18:52–74
25. Hobisch A, Culig Z, Radmayr C, Bartsch G, Klocker H, Hittmair A. 1995. Distant metastases from prostatic carcinoma express androgen receptor protein. *Cancer Res.* 55:3068–72
26. Chen CD, Welsbie DS, Tran C, Baek SH, Chen R, et al. 2004. Molecular determinants of resistance to antiandrogen therapy. *Nat. Med.* 10:33–39
27. Linja MJ, Savinainen KJ, Saramaki OR, Tammela TL, Vessella RL, Visakorpi T. 2001. Amplification and overexpression of androgen receptor gene in hormone-refractory prostate cancer. *Cancer Res.* 61:3550–55
28. Edwards J, Bartlett JM. 2005. The androgen receptor and signal-transduction pathways in hormone-refractory prostate cancer. Part 1: Modifications to the androgen receptor. *BJU Int.* 95:1320–26
29. Edwards J, Bartlett JM. 2005. The androgen receptor and signal-transduction pathways in hormone-refractory prostate cancer. Part 2: Androgen-receptor cofactors and bypass pathways. *BJU Int.* 95:1327–35
30. Gioeli D. 2005. Signal transduction in prostate cancer progression. *Clin. Sci.* 108:293–308
31. Thalmann GN, Anezinis PE, Chang SM, Zhau HE, Kim EE, et al. 1994. Androgen-independent cancer progression and bone metastasis in the LNCaP model of human prostate cancer. *Cancer Res.* 54:2577–81
32. Gleave ME, Hsieh JT, von Eschenbach AC, Chung LW. 1992. Prostate and bone fibroblasts induce human prostate cancer growth in vivo: implications for bidirectional tumor-stromal cell interaction in prostate carcinoma growth and metastasis. *J. Urol.* 147:1151–59
33. Pietrzkowski Z, Mulholland G, Gomella L, Jameson BA, Wernicke D, Baserga R. 1993. Inhibition of growth of prostatic cancer cell lines by peptide analogues of insulin-like growth factor 1. *Cancer Res.* 53:1102–6
34. Craft N, Shostak Y, Carey M, Sawyers CL. 1999. A mechanism for hormone-independent prostate cancer through modulation of androgen receptor signaling by the HER-2/neu tyrosine kinase. *Nat. Med.* 5:280–85
35. Mellinghoff IK, Tran C, Sawyers CL. 2002. Growth inhibitory effects of the dual ErbB1/ErbB2 tyrosine kinase inhibitor PKI-166 on human prostate cancer xenografts. *Cancer Res.* 62:5254–59
36. Agus DB, Akita RW, Fox WD, Lewis GD, Higgins B, et al. 2002. Targeting ligand-activated ErbB2 signaling inhibits breast and prostate tumor growth. *Cancer Cell* 2:127–37
37. Mellinghoff IK, Vivanco I, Kwon A, Tran C, Wongvipat J, Sawyers CL. 2004. HER2/neu kinase-dependent modulation of androgen receptor function through effects on DNA binding and stability. *Cancer Cell* 6:517–27
38. Carter BS, Epstein JI, Isaacs WB. 1990. ras gene mutations in human prostate cancer. *Cancer Res.* 50:6830–32
39. Voeller HJ, Wilding G, Gelmann EP. 1991. v-rasH expression confers hormone-independent in vitro growth to LNCaP prostate carcinoma cells. *Mol. Endocrinol.* 5:209–16

40. Bakin RE, Gioeli D, Sikes RA, Bissonette EA, Weber MJ. 2003. Constitutive activation of the Ras/mitogen-activated protein kinase signaling pathway promotes androgen hypersensitivity in LNCaP prostate cancer cells. *Cancer Res.* 63:1981–89
41. Thalmann GN, Sikes RA, Wu TT, Degeorges A, Chang SM, et al. 2000. LNCaP progression model of human prostate cancer: androgen-independence and osseous metastasis. *Prostate* 44:91–103
42. Bakin RE, Gioeli D, Bissonette EA, Weber MJ. 2003. Attenuation of Ras signaling restores androgen sensitivity to hormone-refractory C4-2 prostate cancer cells. *Cancer Res.* 63:1975–80
43. Price DT, Rocca GD, Guo C, Ballo MS, Schwinn DA, Luttrell LM. 1999. Activation of extracellular signal-regulated kinase in human prostate cancer. *J. Urol.* 162:1537–42
44. Royuela M, Arenas MI, Bethencourt FR, Sanchez-Chapado M, Fraile B, Paniagua R. 2002. Regulation of proliferation/apoptosis equilibrium by mitogen-activated protein kinases in normal, hyperplastic, and carcinomatous human prostate. *Hum. Pathol.* 33:299–306
45. Gioeli D, Ficarro SB, Kwiek JJ, Aaronson D, Hancock M, et al. 2002. Androgen receptor phosphorylation. Regulation and identification of the phosphorylation sites. *J. Biol. Chem.* 277:29304–14
46. Zhu Z, Becklin RR, Desiderio DM, Dalton JT. 2001. Identification of a novel phosphorylation site in human androgen receptor by mass spectrometry. *Biochem. Biophys. Res. Commun.* 284:836–44
47. Zhou ZX, Kemppainen JA, Wilson EM. 1995. Identification of three proline-directed phosphorylation sites in the human androgen receptor. *Mol. Endocrinol.* 9:605–15
48. Yeh S, Lin HK, Kang HY, Thin TH, Lin MF, Chang C. 1999. From HER2/Neu signal cascade to androgen receptor and its coactivators: a novel pathway by induction of androgen target genes through MAP kinase in prostate cancer cells. *Proc. Natl. Acad. Sci. USA* 96:5458–63
49. Lin HK, Yeh S, Kang HY, Chang C. 2001. Akt suppresses androgen-induced apoptosis by phosphorylating and inhibiting androgen receptor. *Proc. Natl. Acad. Sci. USA* 98:7200–5
50. Wen Y, Hu MC, Makino K, Spohn B, Bartholomeusz G, et al. 2000. HER-2/neu promotes androgen-independent survival and growth of prostate cancer cells through the Akt pathway. *Cancer Res.* 60:6841–45
51. Gioeli D, Black BE, Gordon V, Spencer A, Kesler CT, et al. 2006. Stress kinase signaling regulates androgen receptor phosphorylation, transcription, and localization. *Mol. Endocrinol.* 20:503–15

51. JNK/p38 MAPKs mediate AR nuclear export and transrepression via phosphorylation of AR Ser650.

52. Taneja SS, Ha S, Swenson NK, Huang HY, Lee P, et al. 2005. Cell-specific regulation of androgen receptor phosphorylation in vivo. *J. Biol. Chem.* 280:40916–24
53. Lin HK, Hu YC, Yang L, Altuwaijri S, Chen YT, et al. 2003. Suppression versus induction of androgen receptor functions by the phosphatidylinositol 3-kinase/Akt pathway in prostate cancer LNCaP cells with different passage numbers. *J. Biol. Chem.* 278:50902–7
54. Itoh M, Adachi M, Yasui H, Takekawa M, Tanaka H, Imai K. 2002. Nuclear export of glucocorticoid receptor is enhanced by c-Jun N-terminal kinase-mediated phosphorylation. *Mol. Endocrinol.* 16:2382–92
55. Qiu M, Olsen A, Faivre E, Horwitz KB, Lange CA. 2003. Mitogen-activated protein kinase regulates nuclear association of human progesterone receptors. *Mol. Endocrinol.* 17:628–42

55. PR Ser294 phosphorylation mediates PR nuclear-cytoplasmic shuttling.

56. Lee H, Bai W. 2002. Regulation of estrogen receptor nuclear export by ligand-induced and p38-mediated receptor phosphorylation. *Mol. Cell Biol.* 22:5835–45

57. Rowan BG, Weigel NL, O'Malley BW. 2000. Phosphorylation of steroid receptor coactivator-1. Identification of the phosphorylation sites and phosphorylation through the mitogen-activated protein kinase pathway. *J. Biol. Chem.* 275:4475–83
58. Gregory CW, Fei X, Ponguta LA, He B, Bill HM, et al. 2004. Epidermal growth factor increases coactivation of the androgen receptor in recurrent prostate cancer. *J. Biol. Chem.* 279:7119–30
59. Xu J, Qiu Y, DeMayo FJ, Tsai SY, Tsai MJ, O'Malley BW. 1998. Partial hormone resistance in mice with disruption of the steroid receptor coactivator-1 (SRC-1) gene. *Science* 279:1922–25
60. Gregory CW, He B, Johnson RT, Ford OH, Mohler JL, et al. 2001. A mechanism for androgen receptor-mediated prostate cancer recurrence after androgen deprivation therapy. *Cancer Res.* 61:4315–19
61. Heinlein CA, Chang C. 2002. Androgen receptor (AR) coregulators: an overview. *Endocr. Rev.* 23:175–200
62. Yeh S, Kang HY, Miyamoto H, Nishimura K, Chang HC, et al. 1999. Differential induction of androgen receptor transactivation by different androgen receptor coactivators in human prostate cancer DU145 cells. *Endocrine* 11:195–202
63. Debes JD, Schmidt LJ, Huang H, Tindall DJ. 2002. P300 mediates androgen-independent transactivation of the androgen receptor by interleukin 6. *Cancer Res.* 62:5632–36
64. Huang H, Tindall DJ. 2002. The role of the androgen receptor in prostate cancer. *Crit. Rev. Eukaryot. Gene Expr.* 12:193–207
65. Nusse R. 2003. Wnts and Hedgehogs: lipid-modified proteins and similarities in signaling mechanisms at the cell surface. *Development* 130:5297–305
66. Pandur P, Maurus D, Kuhl M. 2002. Increasingly complex: New players enter the Wnt signaling network. *Bioessays* 24:881–84
67. Yang F, Li X, Sharma M, Sasaki CY, Longo DL, et al. 2002. Linking beta-catenin to androgen-signaling pathway. *J. Biol. Chem.* 277:11336–44
68. Truica CI, Byers S, Gelmann EP. 2000. Beta-catenin affects androgen receptor transcriptional activity and ligand specificity. *Cancer Res.* 60:4709–13
69. Mulholland DJ, Cheng H, Reid K, Rennie PS, Nelson CC. 2002. The androgen receptor can promote beta-catenin nuclear translocation independently of adenomatous polyposis coli. *J. Biol. Chem.* 277:17933–43
70. Verras M, Brown J, Li X, Nusse R, Sun Z. 2004. Wnt3a growth factor induces androgen receptor-mediated transcription and enhances cell growth in human prostate cancer cells. *Cancer Res.* 64:8860–66
71. Massague J. 1990. The transforming growth factor-beta family. *Annu. Rev. Cell Biol.* 6:597–641
72. Cui W, Fowlis DJ, Bryson S, Duffie E, Ireland H, et al. 1996. TGFβ1 inhibits the formation of benign skin tumors, but enhances progression to invasive spindle carcinomas in transgenic mice. *Cell* 86:531–42
73. Massague J. 1998. TGF-β signal transduction. *Annu. Rev. Biochem.* 67:753–91
74. Zhu B, Kyprianou N. 2005. Transforming growth factor β and prostate cancer. *Cancer Treat. Res.* 126:157–73
75. Hayes SA, Zarnegar M, Sharma M, Yang F, Peehl DM, et al. 2001. SMAD3 represses androgen receptor-mediated transcription. *Cancer Res.* 61:2112–18
76. van der Poel HG. 2005. Androgen receptor and TGFβ1/Smad signaling are mutually inhibitory in prostate cancer. *Eur. Urol.* 48:1051–58

77. Kang HY, Lin HK, Hu YC, Yeh S, Huang KE, Chang C. 2001. From transforming growth factor-β signaling to androgen action: identification of Smad3 as an androgen receptor coregulator in prostate cancer cells. *Proc. Natl. Acad. Sci. USA* 98:3018–23
78. Kang HY, Huang KE, Chang SY, Ma WL, Lin WJ, Chang C. 2002. Differential modulation of androgen receptor-mediated transactivation by Smad3 and tumor suppressor Smad4. *J. Biol. Chem.* 277:43749–56
79. Chipuk JE, Cornelius SC, Pultz NJ, Jorgensen JS, Bonham MJ, et al. 2002. The androgen receptor represses transforming growth factor-β signaling through interaction with Smad3. *J. Biol. Chem.* 277:1240–48
80. Hovey RC, Trott JF, Vonderhaar BK. 2002. Establishing a framework for the functional mammary gland: from endocrinology to morphology. *J. Mammary Gland Biol. Neoplasia* 7:17–38
81. Haslam SZ, Counterman LJ, Nummy KA. 1993. Effects of epidermal growth factor, estrogen, and progestin on DNA synthesis in mammary cells in vivo are determined by the developmental state of the gland. *J. Cell Physiol.* 155:72–78
82. **Ankrapp DP, Bennett JM, Haslam SZ. 1998. Role of epidermal growth factor in the acquisition of ovarian steroid hormone responsiveness in the normal mouse mammary gland. *J. Cell Physiol.* 174:251–60**

82. EGF is required for the acquisition of ovarian hormone responsiveness in the mammary gland; evidence is presented for the existence of important cross talk in vivo.

83. Robinson GW, Hennighausen L, Johnson PF. 2000. Side-branching in the mammary gland: the progesterone-Wnt connection. *Genes Dev.* 14:889–94
84. Li Y, Rosen JM. 2005. Stem/progenitor cells in mouse mammary gland development and breast cancer. *J. Mammary Gland Biol. Neoplasia* 10:17–24
85. Aupperlee MD, Smith KT, Kariagina A, Haslam SZ. 2005. Progesterone receptor isoforms A and B: temporal and spatial differences in expression during murine mammary gland development. *Endocrinology* 146:3577–88
86. Chlebowski RT, Hendrix SL, Langer RD, Stefanick ML, Gass M, et al. 2003. Influence of estrogen plus progestin on breast cancer and mammography in healthy postmenopausal women: the Women's Health Initiative Randomized Trial. *JAMA* 289:3243–53
87. Haslam SZ, Osuch JR, Raafat AM, Hofseth LJ. 2002. Postmenopausal hormone replacement therapy: effects on normal mammary gland in humans and in a mouse postmenopausal model. *J. Mammary Gland Biol. Neoplasia* 7:93–105
87a. Booth BW, Smith GH. 2006. Estrogen receptor-α and progesterone receptor are expressed in label-retaining mammary epithelial cells that divide asymmetrically and retain their template DNA strands. *Breast Cancer Res.* 8:R49
87b. Clarke RB. 2006. Ovarian steroids and the human breast: regulation of stem cells and cell proliferation. *Maturitas* 54: 327–34
87c. Dunn KL, Espino PS, Drobic B, He S, Davie JR. 2005. The Ras-MAPK signal transduction pathway, cancer and chromatin remodeling. *Biochem. Cell Biol.* 83:1–14
87d. Shen Q, Brown PH. 2003. Novel agents for the prevention of breast cancer: targeting transcription factors and signal transduction pathways. *J. Mammary Gland Biol. Neoplasia* 8:45–73
88. Schiff R, Massarweh SA, Shou J, Bharwani L, Arpino G, et al. 2005. Advanced concepts in estrogen receptor biology and breast cancer endocrine resistance: implicated role of growth factor signaling and estrogen receptor coregulators. *Cancer Chemother. Pharmacol.* 56(Suppl. 1):10–20
89. Lange CA. 2004. Making sense of cross-talk between steroid hormone receptors and intracellular signaling pathways: Who will have the last word? *Mol. Endocrinol.* 18:269–78

90. Zhang Y, Beck CA, Poletti A, Clement JP, Prendergast P, et al. 1997. Phosphorylation of human progesterone receptor by cyclin-dependent kinase 2 on three sites that are authentic basal phosphorylation sites in vivo. *Mol. Endocrinol.* 11:823–32
91. Zhang Y, Beck CA, Poletti A, Edwards DP, Weigel NL. 1995. Identification of a group of Ser-Pro motif hormone-inducible phosphorylation sites in the human progesterone receptor. *Mol. Endocrinol.* 9:1029–40
91a. Takimoto GS, Tasset DM, Eppert AC, and Horwitz KB. 1992. Hormone-induced progesterone receptor phosphorylation consists of sequential DNA-independent and DNA-dependent stages: analysis with zinc-finger mutants and the progesterone antagonist ZK89299. *Proc. Natl. Acad. Sci. USA* 89:3050–54
91b. Sartorius CA, Takimoto GS, Richer JK, Tung L, Horwitz KB. 2000. Association of the Ku autoantigen/DNA-dependent protein kinase holoenzyme and poly(ADP-ribose) polymerase with the DNA binding domain of progesterone receptors. *J. Mol. Endocrinol.* 24:165–82
92. Zhang Y, Beck CA, Poletti A, Edwards DP, Weigel NL. 1994. Identification of phosphorylation sites unique to the B form of human progesterone receptor. In vitro phosphorylation by casein kinase II. *J. Biol. Chem.* 269:31034–40
93. Lange CA, Shen T, Horwitz KB. 2000. Phosphorylation of human progesterone receptors at serine-294 by mitogen-activated protein kinase signals their degradation by the 26S proteasome. *Proc. Natl. Acad. Sci. USA* 97:1032–37
94. Shen T, Horwitz KB, Lange CA. 2001. Transcriptional hyperactivity of human progesterone receptors is coupled to their ligand-dependent down-regulation by mitogen-activated protein kinase-dependent phosphorylation of serine 294. *Mol. Cell Biol.* 21:6122–31
95. Knotts TA, Orkiszewski RS, Cook RG, Edwards DP, Weigel NL. 2001. Identification of a phosphorylation site in the hinge region of the human progesterone receptor and additional amino-terminal phosphorylation sites. *J. Biol. Chem.* 276:8475–83
96. Font de Mora J, Brown M. 2000. AIB1 is a conduit for kinase-mediated growth factor signaling to the estrogen receptor. *Mol. Cell Biol.* 20:5041–47
97. Narayanan R, Adigun AA, Edwards DP, Weigel NL. 2005. Cyclin-dependent kinase activity is required for progesterone receptor function: novel role for cyclin A/Cdk2 as a progesterone receptor coactivator. *Mol. Cell Biol.* 25:264–77
98. Labriola L, Salatino M, Proietti CJ, Pecci A, Coso OA, et al. 2003. Heregulin induces transcriptional activation of the progesterone receptor by a mechanism that requires functional ErbB-2 and mitogen-activated protein kinase activation in breast cancer cells. *Mol. Cell Biol.* 23:1095–111
99. Pierson-Mullany LK, Lange CA. 2004. Phosphorylation of progesterone receptor serine 400 mediates ligand-independent transcriptional activity in response to activation of cyclin-dependent protein kinase 2. *Mol. Cell Biol.* 24:10542–57
99a. Takimoto GS, Hovland AR, Tasset DM, Melville MY, Tung L, Horwitz KB. 1996. Role of phosphorylation on DNA binding and transcriptional functions of human progesterone receptors. *J. Biol. Chem.* 271:13308–16
100. Simons SSJ. 2006. How much is enough? Modulation of dose-response curve for steroid receptor-regulated gene expression by changing concentrations of transcription factor. *Curr. Top Med. Chem.* 6:271–85
101. Qiu M, Lange CA. 2003. MAP kinases couple multiple functions of human progesterone receptors: degradation, transcriptional synergy, and nuclear association. *J. Steroid Biochem. Mol. Biol.* 85:147–57

93. PRs are degraded by the ubiquitin-proteasome pathway; PR Ser294 phosphorylation augments ligand-dependent PR turnover.

94. PR Ser294 phosphorylation mediates transcriptional synergy in the presence of progestins and agents that activate MAPK; PR transcription and stability are inversely related and appear to be coupled events.

102. Migliaccio A, Di Domenico M, Green S, de Falco A, Kajtaniak EL, et al. 1989. Phosphorylation on tyrosine of in vitro synthesized human estrogen receptor activates its hormone binding. *Mol. Endocrinol.* 3:1061–69

103. Ali S, Metzger D, Bornert JM, Chambon P. 1993. Modulation of transcriptional activation by ligand-dependent phosphorylation of the human oestrogen receptor A/B region. *EMBO J.* 12:1150–60

104. PR transcriptional activity is highest in S phase of the cell cycle.

104. Narayanan R, Edwards DP, Weigel NL. 2005. Human progesterone receptor displays cell cycle-dependent changes in transcriptional activity. *Mol. Cell Biol.* 25:2885–98

105. Nardulli AM, Katzenellenbogen BS. 1988. Progesterone receptor regulation in T47D human breast cancer cells: analysis by density labeling of progesterone receptor synthesis and degradation and their modulation by progestin. *Endocrinology* 122:1532–40

105a. Sheridan PL, Krett NL, Gordon JA, Horwitz KB. 1988. Human progesterone receptor transformation and nuclear down-regulation are independent of phosphorylation. *Mol. Endo.* 2:1329–42

106. Musgrove EA, Swarbrick A, Lee CS, Cornish AL, Sutherland RL. 1998. Mechanisms of cyclin-dependent kinase inactivation by progestins. *Mol. Cell Biol.* 18:1812–25

107. Edwards DP, Weigel NL, Nordeen SK, Beck CA. 1993. Modulators of cellular protein phosphorylation alter the trans-activation function of human progesterone receptor and the biological activity of progesterone antagonists. *Breast Cancer Res. Treat.* 27:41–56

107a. Lin HK, Wang L, Hu YC, Altuwaijri S, Chang C. 2002. Phosphorylation-dependent ubiquitinylation and degradation of androgen receptor by Akt require Mdm2 E3 ligase. *EMBO J.* 21:4037–48

108. Muratani M, Tansey WP. 2003. How the ubiquitin-proteasome system controls transcription. *Nat. Rev. Mol. Cell Biol.* 4:192–201

109. Lonard DM, Nawaz Z, Smith CL, O'Malley BW. 2000. The 26S proteasome is required for estrogen receptor-α and coactivator turnover and for efficient estrogen receptor-α transactivation. *Mol. Cell* 5:939–48

110. Migliaccio A, Di Domenico M, Castoria G, de Falco A, Bontempo P, et al. 1996. Tyrosine kinase/p21ras/MAP-kinase pathway activation by estradiol-receptor complex in MCF-7 cells. *EMBO J.* 15:1292–300

111. PR, ER, and c-Src exist in a signaling complex in breast cancer cells.

111. Migliaccio A, Piccolo D, Castoria G, Di Domenico M, Bilancio A, et al. 1998. Activation of the Src/p21ras/Erk pathway by progesterone receptor via cross-talk with estrogen receptor. *EMBO J.* 17:2008–18

112. PR-B interacts with the SH3 domain of c-Src kinase via a polyproline-rich region of the PR N terminus and directly activates c-Src kinases.

112. Boonyaratanakornkit V, Scott MP, Ribon V, Sherman L, Anderson SM, et al. 2001. Progesterone receptor contains a proline-rich motif that directly interacts with SH3 domains and activates c-Src family tyrosine kinases. *Mol. Cell* 8:269–80

113. Migliaccio A, Castoria G, Di Domenico M, de Falco A, Bilancio A, et al. 2000. Steroid-induced androgen receptor-estradiol receptor β-Src complex triggers prostate cancer cell proliferation. *EMBO J.* 19:5406–17

114. Migliaccio A, Castoria G, Di Domenico M, Ballare C, Beato M, Auricchio F. 2005. The progesterone receptor/estradiol receptor association and the progestin-triggered S-phase entry. *Ernst Schering Res. Found. Workshop* 52:39–54

115. Ballare C, Uhrig M, Bechtold T, Sancho E, Di Domenico M, et al. 2003. Two domains of the progesterone receptor interact with the estrogen receptor and are required for progesterone activation of the c-Src/Erk pathway in mammalian cells. *Mol. Cell Biol.* 23:1994–2008

116. Wong C, McNally C, Nickbarg E, Komm B, Cheskis B. 2002. Estrogen receptor-interacting protein that modulates its nongenomic activity-crosstalk with Src/Erk phosphorylation cascade. *Proc. Natl. Acad. Sci. USA* 99:14783–88
117. Haas D, White SN, Lutz LB, Rasar M, Hammes SR. 2005. The modulator of nongenomic actions of the estrogen receptor (MNAR) regulates transcription-independent androgen receptor-mediated signaling: evidence that MNAR participates in G protein-regulated meiosis in *Xenopus laevis* oocytes. *Mol. Endocrinol.* 19:2035–46
118. Unni E, Sun S, Nan B, McPhaul MJ, Cheskis B, et al. 2004. Changes in androgen receptor nongenotropic signaling correlate with transition of LNCaP cells to androgen independence. *Cancer Res.* 64:7156–68
119. Zhoul J, Hernandez G, Tu SW, Huang CL, Tseng CP, Hsieh JT. 2005. The role of DOC-2/DAB2 in modulating androgen receptor-mediated cell growth via the nongenomic c-Src-mediated pathway in normal prostatic epithelium and cancer. *Cancer Res.* 65:9906–13
120. Zhu Y, Bond J, Thomas P. 2003. Identification, classification, and partial characterization of genes in humans and other vertebrates homologous to a fish membrane progestin receptor. *Proc. Natl. Acad. Sci. USA* 100:2237–42
121. Skildum A, Faivre E, Lange CA. 2005. Progesterone receptors induce cell cycle progression via activation of mitogen-activated protein kinases. *Mol. Endocrinol.* 19:327–39

121. PR-induced S-phase entry is MAPK dependent and mediated by the PR-B interaction with the SH3 domain of c-Src kinase in breast cancer cells.

122. Murphy LO, Blenis J. 2006. MAPK signal specificity: the right place at the right time. *Trends Biochem. Sci.* 31:268–75
123. Santen R, Jeng MH, Wang JP, Song R, Masamura S, et al. 2001. Adaptive hypersensitivity to estradiol: potential mechanism for secondary hormonal responses in breast cancer patients. *J. Steroid Biochem. Mol. Biol.* 79:115–25
124. Schippinger W, Regitnig P, Dandachi N, Wernecke KD, Bauernhofer T, et al. 2006. Evaluation of the prognostic significance of androgen receptor expression in metastatic breast cancer. *Virchows Arch.* 449:24–30
125. Doane AS, Danso M, Lal P, Donaton M, Zhang L, et al. 2006. An estrogen receptor-negative breast cancer subset characterized by a hormonally regulated transcriptional program and response to androgen. *Oncogene* 25:3994–4008
126. Altucci L, Addeo R, Cicatiello L, Dauvois S, Parker MG, et al. 1996. 17β-Estradiol induces cyclin D1 gene transcription, p36D1-p34cdk4 complex activation and p105Rb phosphorylation during mitogenic stimulation of G_1-arrested human breast cancer cells. *Oncogene* 12:2315–24
127. Knudsen KE, Arden KC, Cavenee WK. 1998. Multiple G1 regulatory elements control the androgen-dependent proliferation of prostatic carcinoma cells. *J. Biol. Chem.* 273:20213–22
128. Gillett C, Fantl V, Smith R, Fisher C, Bartek J, et al. 1994. Amplification and overexpression of cyclin D1 in breast cancer detected by immunohistochemical staining. *Cancer Res.* 54:1812–17
129. Kaltz-Wittmer C, Klenk U, Glaessgen A, Aust DE, Diebold J, et al. 2000. FISH analysis of gene aberrations (MYC, CCND1, ERBB2, RB, and AR) in advanced prostatic carcinomas before and after androgen deprivation therapy. *Lab. Invest.* 80:1455–64
130. Knudsen KE, Cavenee WK, Arden KC. 1999. D-type cyclins complex with the androgen receptor and inhibit its transcriptional transactivation ability. *Cancer Res.* 59:2297–301

131. Burd CJ, Petre CE, Morey LM, Wang Y, Revelo MP, et al. 2006. Cyclin D1b variant influences prostate cancer growth through aberrant androgen receptor regulation. *Proc. Natl. Acad. Sci. USA* 103:2190–95
132. Carroll JS, Liu XS, Brodsky AS, Li W, Meyer CA, et al. 2005. Chromosome-wide mapping of estrogen receptor binding reveals long-range regulation requiring the forkhead protein FoxA1. *Cell* 122:33–43
133. Carroll JS, Brown M. 2006. Estrogen receptor target gene: an evolving concept. *Mol. Endocrinol.* 20:1707–14
134. Marino M, Acconcia F, Bresciani F, Weisz A, Trentalance A. 2002. Distinct nongenomic signal transduction pathways controlled by 17β-estradiol regulate DNA synthesis and cyclin D_1 gene transcription in HepG2 cells. *Mol. Biol. Cell* 13:3720–29
135. Clemm DL, Sherman L, Boonyaratanakornkit V, Schrader WT, Weigel NL, Edwards DP. 2000. Differential hormone-dependent phosphorylation of progesterone receptor A and B forms revealed by a phosphoserine site-specific monoclonal antibody. *Mol. Endocrinol.* 14:52–65
136. Albanese C, Johnson J, Watanabe G, Eklund N, Vu D, et al. 1995. Transforming p21ras mutants and c-Ets-2 activate the cyclin D1 promoter through distinguishable regions. *J. Biol. Chem.* 270:23589–97
137. Reed SI. 1997. Control of the G1/S transition. *Cancer Surv.* 29:7–23
138. Sherr CJ. 1996. Cancer cell cycles. *Science* 274:1672–77
139. Lukas J, Bartkova J, Bartek J. 1996. Convergence of mitogenic signaling cascades from diverse classes of receptors at the cyclin D-cyclin-dependent kinase-pRb-controlled G1 checkpoint. *Mol. Cell Biol.* 16:6917–25
140. Cheng M, Sexl V, Sherr CJ, Roussel MF. 1998. Assembly of cyclin D-dependent kinase and titration of p27Kip1 regulated by mitogen-activated protein kinase kinase (MEK1). *Proc. Natl. Acad. Sci. USA* 95:1091–96
141. Kolar Z, Murray PG, Scott K, Harrison A, Vojtesek B, Dusek J. 2000. Relation of Bcl-2 expression to androgen receptor, p21WAF1/CIP1, and cyclin D1 status in prostate cancer. *Mol. Pathol.* 53:15–18
142. Agus DB, Cordon-Cardo C, Fox W, Drobnjak M, Koff A, et al. 1999. Prostate cancer cell cycle regulators: response to androgen withdrawal and development of androgen independence. *J. Natl. Cancer Inst.* 91:1869–76
143. Fantl V, Stamp G, Andrews A, Rosewell I, Dickson C. 1995. Mice lacking cyclin D1 are small and show defects in eye and mammary gland development. *Genes Dev.* 9:2364–72
144. Sicinski P, Donaher JL, Parker SB, Li T, Fazeli A, et al. 1995. Cyclin D1 provides a link between development and oncogenesis in the retina and breast. *Cell* 82:621–30
145. Lydon JP, DeMayo FJ, Conneely OM, O'Malley BW. 1996. Reproductive phenotpes of the progesterone receptor null mutant mouse. *J. Steroid. Biochem. Mol. Biol.* 56:67–77
146. Lamb J, Ramaswamy S, Ford HL, Contreras B, Martinez RV, et al. 2003. A mechanism of cyclin D1 action encoded in the patterns of gene expression in human cancer. *Cell* 114:323–34
147. Seagroves TN, Krnacik S, Raught B, Gay J, Burgess-Beusse B, et al. 1998. C/EBPβ, but not C/EBPα, is essential for ductal morphogenesis, lobuloalveolar proliferation, and functional differentiation in the mouse mammary gland. *Genes Dev.* 12:1917–28

RELATED RESOURCES

1. Shupnik MA. 2004. Crosstalk between steroid receptors and the c-Src-receptor tyrosine kinase pathways: implications for cell proliferation. *Oncogene* 23(48):7979–89
2. Edwards DP. 2005. Regulation of signal transduction pathways by estrogen and progesterone. *Annu. Rev. Physiol.* 67:335–76
3. Ishizawar R, Parsons SJ. 2004. c-Src and cooperating partners in human cancer. *Cancer Cell* 6(3):209–14
4. Verras M, Sun Z. 2006. Roles and regulation of Wnt signaling and beta-catenin in prostate cancer. *Cancer Lett.* 237(1):22–32

Nuclear Receptor Structure: Implications for Function

David L. Bain, Aaron F. Heneghan, Keith D. Connaghan-Jones, and Michael T. Miura

Department of Pharmaceutical Sciences, University of Colorado Health Sciences Center, Denver, Colorado 80262; email: David.Bain@UCHSC.edu, Aaron.Heneghan@UCHSC.edu, Keith.Connaghan-Jones@UCHSC.edu, Michael.Miura@UCHSC.edu

Annu. Rev. Physiol. 2007. 69:201–20

First published online as a Review in Advance on November 30, 2006

The *Annual Review of Physiology* is online at http://physiol.annualreviews.org

This article's doi:
10.1146/annurev.physiol.69.031905.160308

0066-4278/07/0315-0201$20.00

Key Words

DNA-binding domain, ligand-binding domain, activation function, progesterone receptor, crystallography, NMR

Abstract

Small lipophilic molecules such as steroidal hormones, retinoids, and free fatty acids control many of the reproductive, developmental, and metabolic processes in eukaryotes. The mediators of these effects are nuclear receptor proteins, ligand-activated transcription factors capable of regulating the expression of complex gene networks. This review addresses the structure and structural properties of nuclear receptors, focusing on the well-studied ligand-binding and DNA-binding domains as well as our still-emerging understanding of the largely unstructured N-terminal regions. To emphasize the allosteric interdependence among these subunits, a more detailed inspection of the structural properties of the human progesterone receptor is presented. Finally, this work is placed in the context of developing a quantitative and mechanistic understanding of nuclear receptor function.

INTRODUCTION

PR: progesterone receptor

ER: estrogen receptor

GR: glucocorticoid receptor

AR: androgen receptor

TR: thyroid receptor

RXR: retinoic X receptor

Nuclear receptors are intracellular transcription factors that regulate the activity of complex gene networks (1, 2). The proteins define a superfamily responsible for major aspects of eukaryotic development, differentiation, reproduction, and metabolic homeostasis. This superfamily is typically subdivided into three families, or classes. The steroid receptor family (class I) includes the progesterone receptor (PR), estrogen receptor (ER), glucocorticoid receptor (GR), androgen receptor (AR), and mineralocorticoid receptor. The thyroid/retinoid family (class II) includes the thyroid receptor (TR), vitamin D receptor (VDR), retinoic acid receptor (RAR), and peroxisome proliferator–activated receptor (PPAR). The third class of receptors has long been termed the orphan receptor family: It defines a set of proteins identified by comparative sequence analysis as belonging to the nuclear receptor superfamily but for which the cognate ligand is unknown. The orphan receptors have received particular attention of late because some of the newly identified ligands (e.g., bile acids) implicate their receptors in a variety of metabolic processes, including lipid homeostasis (3).

Although all nuclear receptors regulate gene expression, among the three classes there are subtle differences in the biochemical mechanisms by which the receptors carry out this function. With regard to the steroid receptor family, traditional models posit that, upon binding their hormonal ligand, the receptors release heat shock proteins, translocate into the nucleus, and bind as homodimers to imperfect palindromic (i.e., inverted repeat) response elements at upstream promoter sites. ER recognizes a consensus sequence of AGGTCA, and the remaining steroid receptors recognize a consensus AGAACA sequence. DNA binding is coupled to the recruitment of coactivator proteins and subsequent transcriptional activation. The steroid receptors bind as head-to-head dimers: One protomer of the dimer binds to the highly conserved hexanucleotide half-site, and the second protomer binds to a less conserved hexanucleotide sequence. The two half-sites are separated by an invariant three nucleotides, although the type of nucleotide is not highly conserved.

The class II receptor proteins typically function as heterodimers; TR, VDR, RAR, and PPAR associate with the retinoic X receptor (RXR) and bind as a dimeric complex to direct repeat response elements. The protomers within the heterodimer recognize a hexanucleotide consensus sequence of AGGTCA. Unlike the steroid receptors, these proteins bind in a head-to-tail orientation and can accommodate small changes in the number of nucleotides between the half-sites necessary for combinatorial specificity (4). From a functional perspective, the heterodimers tend to stay bound to their response elements regardless of whether agonist ligands are present: In the absence of ligand, gene activation is prevented by corepressor interactions with the DNA-bound heterodimer. Upon binding ligand, corepressor proteins are released and coactivators are recruited, leading to transcriptional activation.

A complete understanding of orphan receptor function is still developing, but current evidence indicates that orphan receptors either can heterodimerize with RXR or can bind as monomers at response elements to carry out their function. Typically these proteins recognize similar hexanucleotide sequences, as do the class II receptors, but because some of the orphan receptors also bind as monomers, the DNA sequences flanking the recognition sites are also critical.

The above models have served as a central framework for understanding nuclear receptor action. However, their limitations are evident in work demonstrating that the ability of nuclear receptors to regulate gene expression depends upon allosterically mediated structural transitions both within the receptor and among coactivating proteins and the promoter DNA. Some of these transitions involve localized changes in helical orientation, whereas others involve global disorder-order transitions. The physical mechanisms underlying allosterically driven structural rearrangements are largely unknown. Adding to the complexity, nuclear receptor–coactivator interactions at the promoter are not static in character but instead involve highly unstable, time-dependent processes involving upward of 50 or so proteins (5, 6). Keeping this in mind, we present here an overview of nuclear receptor structure-function relationships, followed by an analysis of the structure and structural properties of nuclear receptor subunits. Finally, we discuss the structural and physical properties of the human PR, both of its subunits and of the holoprotein, as a case study for pointing out the strengths and weaknesses in our understanding of receptor function.

OVERVIEW OF NUCLEAR RECEPTOR STRUCTURE-FUNCTION RELATIONSHIPS

Early biochemical studies of partially purified receptors demonstrated that nuclear receptor proteins contained two structural subunits: a moderately conserved C-terminal ligand-binding domain (LBD) and a highly conserved and centrally located DNA-binding domain (DBD) (see **Figure 1**) (7, 8). The LBD serves a number of critical functions. First, as indicated by its name, the LBD contains an interior binding pocket specific for its cognate hormone or ligand. Second, the domain contains a ligand-regulated transcriptional activation function (AF-2) necessary for recruiting various coactivating proteins. The coactivators are then capable of interacting with chromatin-remodeling proteins and the general transcriptional activation machinery (9). Finally, the LBD is the primary mediator of solution self-assembly reactions (e.g., dimerization or tetramerization) necessary for high-affinity DNA response element binding (10).

The second identified subunit, the DBD, docks the receptor to the hexanucleotide

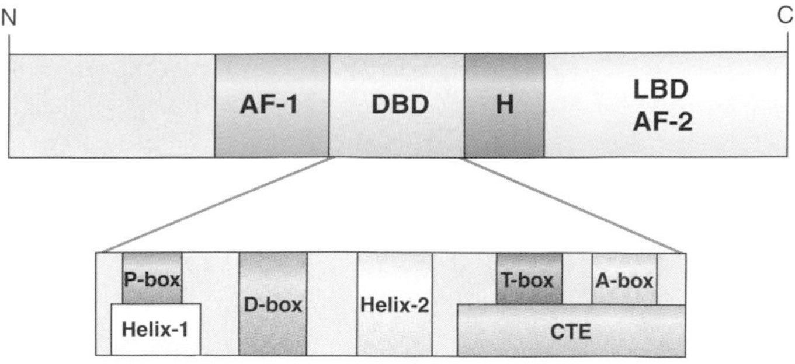

Figure 1

Schematic layout of nuclear receptor structure. AF, activation function; DBD, DNA-binding domain; H, hinge; LBD, ligand-binding domain. Shown below the schematic of the full-length receptor is an expanded view of the DBD, indicating the relative locations of the P-box, D-box, T-box, A-box, helix 1, helix 2, and the C-terminal extension (CTE).

response elements located within nuclear receptor–regulated promoters. As discussed in more detail below, the DBD also serves as an allosteric transmitter of information to other regions of the receptor molecule. The DBD is connected to the LBD via a short amino acid sequence termed the hinge. The complete functional properties of the hinge sequence are still unclear, although it can be phosphorylated, and phosphorylation is coupled to increased transcriptional activation (11, 12).

Most nuclear receptors contain amino acid sequences N-terminal to the DBD. These residues contain a transcriptional activation function termed AF-1. In contrast to the moderately conserved AF-2 sequence embedded within the LBD, the AF-1 sequence shows weak conservation (<15%) across the nuclear receptor superfamily and even within subgroups such as the steroid receptor family. This lack of sequence homology (both within AF-1 and in the greater N-terminal region) may be critical in explaining, for example, how closely related steroid receptors can bind to similar response elements in vitro yet differentially regulate gene promoters containing those same binding sequences in vivo. Finally, although the AF-1 sequence can function as a ligand-independent transcriptional activator, it also can functionally synergize with AF-2 (13).

LBD: ligand-binding domain

DBD: DNA-binding domain

AF: activation function

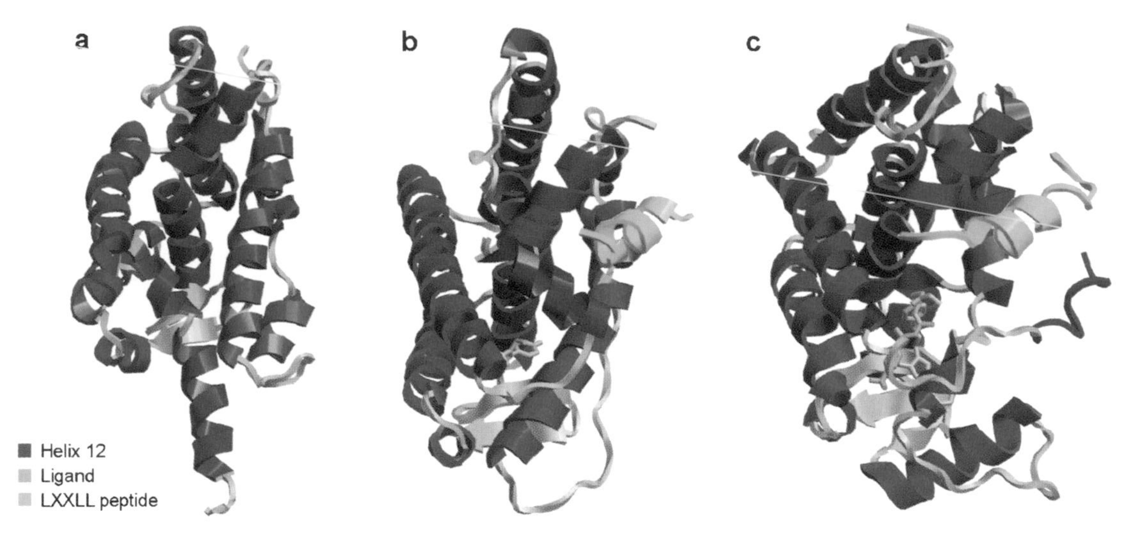

Figure 2

Representative ribbon diagrams of ligand-binding domain (LBD) structures for the three classes of nuclear receptors. (*a*) Apo-RXR (retinoic X receptor) structure (20). (*b*) Estrogen receptor alpha (ERα) LBD bound to agonist ligand and coactivator LXXLL peptide (28). (*c*). Peroxisome proliferator–activated receptor (PPAR) bound to antagonist ligand and corepressor LXXLL peptide (31).

The functional properties of residues located between the far N terminus and AF-1 are the least understood of any of the nuclear receptor sequences. For at least some receptors, these residues play a role in attenuating or modulating transcriptional activation properties (14, 15). Additionally, serine and threonine residues located within this sequence are phosphorylated in response to ligand binding and the cell cycle and thus function as signal transduction sensors (16, 17). It is not clear whether phosphorylation or other posttranslational modifications (18) are linked to changes in nuclear receptor structure, although it is well documented in model peptide systems that phosphorylation of amino acid side chains can alternatively stabilize or destabilize secondary structure (19).

NUCLEAR RECEPTOR STRUCTURE

The Ligand-Binding Domain

The first high-resolution structure of a nuclear receptor LBD was revealed in 1995, when the apo-RXR structure was determined using a crystallographic approach (see **Figure 2*a***) (20). Since then, approximately half of the nuclear receptor LBD structures have been determined either by crystallography or by NMR spectroscopy (for a summary, see Reference 21). Without exception, each structure reveals a globular domain made up of approximately 12 α-helices. At the tertiary level, the helices typically form three antiparallel helical sheets that combine to make what is often described as an α-helical sandwich. The ligand-binding pocket for each receptor is located in the interior of the structure and is formed by a subset of the surrounding helices.

The strength and specificity of LBD-ligand complexes are based largely on hydrophobic interactions, extensive hydrogen-bonding networks, and the steric size and shape of the binding pocket. In the case of steroid receptors, for example, hydrophobic regions within the pocket closely contour the shape of the ligand, and polar groups serve to specifically bind and orient it (22–26). Taken together, these contributions allow the receptors to discriminate among closely related steroidal structures. With regard to the entire nuclear receptor superfamily, the overall size and shape of the binding pockets apparently correlate with receptor function (21). Thus, steroid receptors that maintain high affinity toward only a small number of ligands have smaller volumes within their binding pockets but extensive polar side chains that can precisely hydrogen bond with the natural ligand. By contrast, orphan receptors that interact with diverse metabolic ligands tend to have larger-volume binding pockets that can accommodate a number of different structures.

The LBD includes a ligand-dependent activation function (AF-2) capable of recruiting proteins such as the steroid receptor coactivator (SRC) family (9). The structural interface for this function localizes to a hydrophobic groove formed by several helices of the LBD, including helix 12 (also called the AF-2 helix). Coactivators that contain helical LXXLL motifs bind via hydrophobic interactions in the groove. Interaction specificity is conferred by charge-clamp electrostatic interactions between the LBD and coactivator residues that cap each end of the two-turn LXXLL helix.

Agonist ligands regulate LBD-coactivator interactions by modulating the conformational mobility of both the LBD and helix 12. In the absence of agonist, the LBD is inactive either because helix 12 is positioned away from the LBD core structure, and thus unable to complete formation of the hydrophobic groove (**Figure 2*a***) (20), or the domain exists as a broad ensemble of conformations, only a few of which are in the active form (27). Upon binding an activating ligand, helix 12 is stabilized against the surface of the LBD, allowing formation of the hydrophobic binding groove and coactivator recruitment (**Figure 2*b***) (28). Stabilization can occur through direct contacts between the ligand and helix 12 (22), by interactions with ligand-stabilized helices near the binding pocket (28), or by long-range interactions between the ligand-binding and coactivator-binding pockets (29). Another means of regulation occurs at the quaternary level: RXR proteins maintain a tetrameric state that physically prevents coactivator binding. Agonist binding is coupled to dissociation of tetramers to dimers, resulting in the concomitant formation and exposure of the hydrophobic groove (30).

How do antagonist ligands and corepressor proteins prevent coactivator recruitment? Antagonists can inhibit coactivator binding by sterically blocking the ability of helix 12 to approach the core LBD structure (23, 31) or by inducing helix 12 to bind in the hydrophobic groove and thus unproductively mimic the coactivator (28). For nuclear receptors such as PPAR, corepressors can also prevent coactivator recruitment (32–34). Like coactivators, these molecules contain LXXLL-binding motifs and thus are able to recognize the hydrophobic residues that make up AF-2. However, unlike coactivators, the corepressor-binding motif forms a lengthy three-turn helix that sterically blocks helix 12 from acquiring an active conformation and thus prevents formation of the hydrophobic binding cleft (**Figure 2*c***) (31). The LBD-corepressor complex can be further stabilized by the presence of the antagonist ligand, which not only prevents helix 12 from forming the active conformation but, importantly, creates a larger binding surface for the corepressor LXXLL motif.

The DNA-Binding Domain

The glucocorticoid receptor (GR) DBD was one of the first structures determined for any receptor subunit and still serves as a representative model for the nuclear receptor superfamily (**Figure 3*a***) (35–37). The GR DBD folds into a globular domain made up of two nonequivalent zinc-finger structures; each zinc atom is coordinated by four cysteine residues. The atoms are necessary to retain stable domain structure and function because removal of the zinc ion leads to protein unfolding and loss of DNA-binding activity (38). Also prominent in the structure are two α-helices; the N-terminal helix (helix 1) directly interacts with the major groove of each

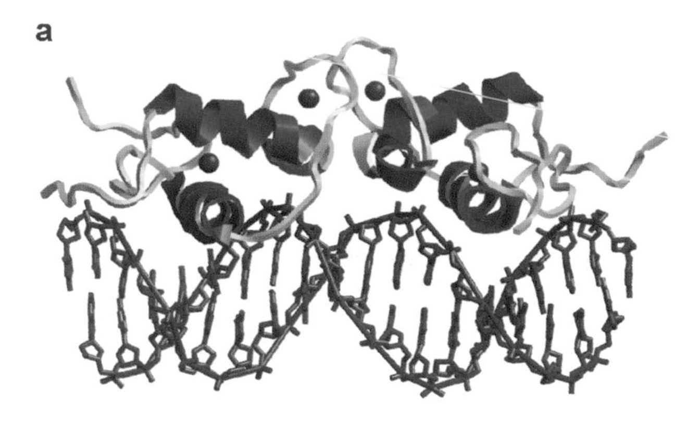

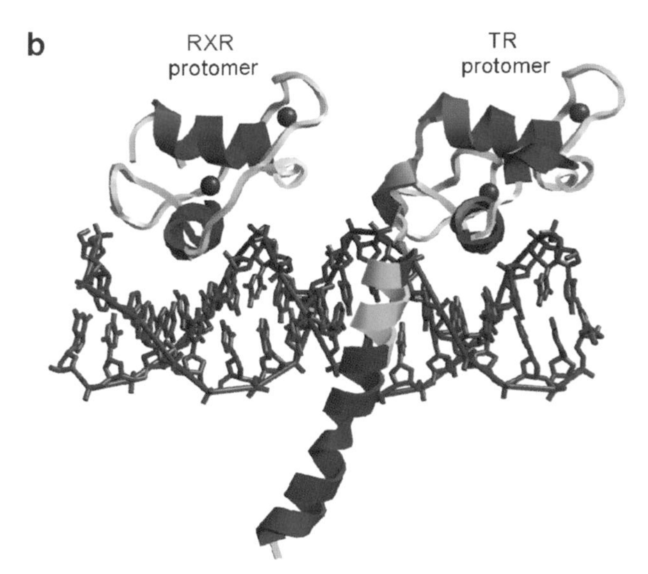

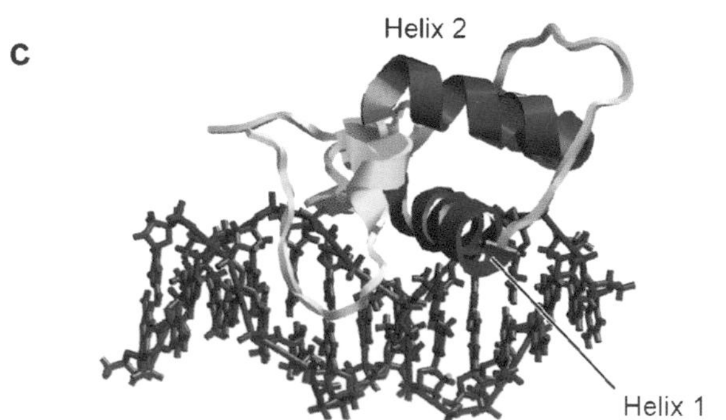

DNA half-site, making base-specific contacts, whereas the C-terminal helix (helix 2) overlays helix 1 in a perpendicular fashion and contributes to stabilization of the overall protein structure.

The GR DBD is monomeric in solution but can undergo DNA-induced dimerization upon binding a palindromic (inverted repeat) response element. The residues that make up the dimer interface are located within the C-terminal zinc finger and define the D-box; the residues critical to sequence-specific DNA binding are located within helix 1 and are defined as the P-box. A comparison of the free GR structure with the DNA-bound structure indicates that interactions between the P-box and the DNA half-site are coupled to conformational changes in the D-box necessary for cooperative recruitment of the second monomer (39). Thus, the specific binding sequence acts as an allosteric effector of function. Similar results were seen for the Class II family discussed below.

The RXR-TR heterodimer bound to a direct repeat sequence is representative of class II receptor structure (**Figure 3*b***) (40). The overall tertiary fold of each protomer of the heterodimer, including the two zinc fingers and α-helices, is similar to that of the GR structure. Moreover, analogous to what was seen in the GR DBD structure, DNA binding is coupled to structural changes necessary for heterodimer formation and high-affinity

Figure 3

Ribbon diagrams of DNA-binding domain (DBD) structures complexed to their response elements for the three classes of nuclear receptors. (*a*) The glucocorticoid (GR) DBD homodimer bound to an inverted repeat response element (37). (*b*) The retinoic X receptor (RXR)–thyroid receptor (TR) heterodimer bound to a direct repeat response element (40). The RXR protomer is shown to the left, and the TR protomer is shown to the right. (*c*) The estrogen-related receptor 2 (ERR2) orphan receptor monomer bound to a half-site response element (43). The axis of helix 1 is oriented toward the viewer; helix 2 is above and perpendicular to helix 1.

binding. However, in the case of RXR, DNA binding is linked to both helix formation and helix loss (41): Helical ordering occurs in the second zinc-finger region and is required for heterodimerization. Helical unfolding occurs in the C-terminal extension (CTE) of the DBD and appears necessary for heterodimer formation. [By contrast, modeling analysis indicates that this helix must be present to allow RXR homodimer formation (41).] In the case of TR, two helices within the CTE play critical roles in DNA binding and heterodimer assembly: (*a*) A helix in the CTE T-box sequence is necessary to complete the heterodimeric interface with the second zinc-finger helix of RXR, and (*b*) a helix within an A-box sequence makes extensive contacts with the DNA. Whether these two helices are induced upon DNA binding is unclear in the absence of a TR structure unliganded to DNA.

A complete picture of the structural basis of DNA binding by orphan receptors is still coming into focus (42). However, for at least a subset of these receptors, DNA binding occurs through a monomer-binding reaction rather than heterodimerization with RXR. Like RXR and TR, orphan receptors such as the estrogen-related receptor 2 protein (ERR2) generate high-affinity binding through the use of the CTE of the core DBD (43). However, in the absence of DNA, the CTE of ERR2 is largely unstructured. Upon binding a hexanucleotide half-site, the CTE becomes ordered by interacting with the DNA minor groove at sequences flanking the canonical response element (see **Figure 3*c***). The interacting sequences localize to an A-box within the CTE, and the contacts contribute significantly to the enhanced stability of binding. Additionally, the CTE folds back on itself to form a pseudo–dimer interface with the core DBD structure. This latter interaction not only adds greater stability to the protein-DNA assembly but suggests that receptor dimerization arose from a progenitor monomer-binding species.

Sequences N-Terminal to the DNA-Binding Domain

Our knowledge of the structural properties of sequences N-terminal to the DBD continues to lag behind that of the LBD and DBD. However, various studies on steroid receptor N-terminal regions have revealed that these sequences tend to be unstructured or weakly folded in isolation and yet can acquire significant secondary and tertiary structure by exposure to interacting macromolecules (either proteins or DNA) or by perturbation of solution conditions. It is not clear whether these phenomena can be extended to the N-terminal sequences of type II and type III receptors because the latter have not been as intensively investigated. However, analysis of the RXR N-terminal sequence suggests that this receptor shares, at least qualitatively, some of the characteristics seen in the steroid receptors (44). A summary and analysis of the results found for several steroid receptor proteins are below.

CTE: C-terminal extension

ERR: estrogen-related receptor

Evidence for minimal folded structure in steroid receptor N-terminal regions dates back to early proteolysis studies (7, 8). Consistent with this, work on the isolated GR AF-1 (also known as tau-1) revealed that it was largely random coil in solution (45). However, spectroscopic analysis of this AF in the presence of the naturally occurring osmolyte TMAO demonstrated that the sequence could acquire secondary and tertiary structure, thus making it competent to interact with transcription factors, including TBP and SRC-1 (46). Additionally, studies using a GR construct containing the N-terminal sequence linked to its DBD demonstrated that binding to DNA response elements could also increase structure within AF-1, thus implicating the DBD as an allosteric mediator of communication between the response elements and activation function (47).

In parallel with this work, limited proteolysis studies on the human progesterone receptor (PR) demonstrated that the N-terminal regions contained structure but that structural stability was highly dependent on the presence of the DBD (48, 49) (discussed in more detail below). Furthermore, these studies were the first to map the locations of N-terminal structural instability and found that these locations corresponded to previously identified phosphorylation motifs and the AF-1 functional domain boundary. Additionally, analysis of PR when it was bound to a palindromic response element revealed DNA-dependent changes in structure that localized to AF-1 and to the hinge.

Studies on the AR AF-1 revealed properties like those seen in the GR studies described above. The isolated AF contained only a low degree of secondary structure in the absence of stabilizing agents, but the addition of TMAO induced a more structured and α-helical conformation (50). Additionally, fluorescence studies indicated that DNA response element binding was coupled to changes in N-terminal structure (51) and that interactions with the general transcription factor TFIIF triggered changes in AF-1 structure (52).

Finally, CD and NMR analyses of the N-terminal regions of ER-α and ER-β demonstrated that both sequences were unstructured in vitro (53). However, only the ER-α construct could interact with the TATA-binding protein, and interaction was coupled to induction of structure. Evidence of an induced-folding reaction within ER-α was consistent with the results of limited proteolysis studies of ER-β, which demonstrated that promoter binding was linked to structural changes in the holoprotein (54). Furthermore, the types of structural changes were specific to the type of promoter sequence and coupled to differential interactions of coactivators.

The effect that DNA has on the N-terminal structure of nuclear receptors is in agreement with a broader theme that eukaryotic promoter-binding sites do not simply act as docking stations to allow the localized assembly of transcription factors but that the sites also function as ligands capable of influencing receptor structure via allosteric mechanisms (55). Unfortunately, the physical mechanism(s) by which these structural changes occur in nuclear receptors remains unclear. Some insight into the origins of allosteric coupling may be found in studies of the glucocorticoid receptor DBD: Analysis of DBD mutants shown to be constitutive transcriptional activators found that the mutations mapped near regions known to undergo DNA-dependent conformational changes (39, 56). Moreover, the structures of the mutant DBDs in the absence of DNA were largely identical to those found for the wild-type DBD when bound to DNA (57). In other words, the mutations decoupled a normally DNA-induced structural switch. How exactly this structural change might lead to activation remains unclear—the structural changes themselves may form a surface capable of directly interacting with coactivators or may trigger conformational changes in AF-1, thus leading to subsequent coactivator recruitment. A variation on this theme may come from recent work demonstrating that the JDP-2 protein, analogous to a DNA response element, can bind to the PR DBD and induce structural changes within the N-terminal AF-1 (58).

Steroid receptor N-terminal sequences have a number of common properties, such as having a partially or totally unfolded structure and the ability to modulate that structure allosterically. However, it is still not clear whether these phenomena occur by a similar physical folding mechanism and whether they apply to other nuclear receptor N-terminal sequences. The use of different techniques and approaches for analyzing each steroid receptor is a contributing factor, as is our still-developing understanding of the physical mechanisms underlying protein folding and disorder-order transitions within protein subunits. Thus, side-by-side comparative and quantitative analysis will be necessary to ascertain whether steroid receptors use similar approaches to mediate folding and function. Although a common folding mechanism would be an appealing result, it may be an unlikely one, given the lack of N-terminal sequence

identity among nuclear receptors. In fact, the structural properties and folding mechanisms of N-terminal steroid receptor sequences may differ precisely to generate receptor-specific functional responses via differential recruitment of coactivators, for example (59).

Another unresolved question is why the N-terminal sequences exist as unfolded or partially folded conformational ensembles. In the past few years, an increasing number of proteins have been found to be natively unfolded or intrinsically disordered (60). It is not yet clear whether nuclear receptor N-terminal sequences are truly natively unfolded, although they share some of the defining characteristics (61). If these sequences are indeed unfolded in vivo, the functional basis likely is to increase the specificity of receptor-coactivator interactions via only folding (and thus interacting) with appropriate target proteins. Additionally, a linked folding-binding reaction will result in a decrease in the intrinsic energetics of the interaction owing to an entropic penalty to folding; this weakened affinity may allow for a more functionally nimble gene regulatory switch.

In conclusion, the above structural studies have given us enormous insight into the properties of nuclear receptor subunits, yet it is still not clear how the parts of the receptors interact to generate holoprotein function. To emphasize the strengths and limitations of our understanding of these relationships, we present below a more detailed analysis of the PR subunit and holoprotein properties.

A CASE STUDY: THE HUMAN PROGESTERONE RECEPTOR

As noted in the Introduction, PR belongs to the steroid receptor family of nuclear receptors. PR mediates the action of the hormone progesterone, a key ligand in reproduction and pregnancy. Understanding PR function is complicated in part because the receptor exists in vivo as two functionally distinct isoforms, PR-A and PR-B (62). As shown in **Figure 4*a***, the two isoforms are identical, except the A-

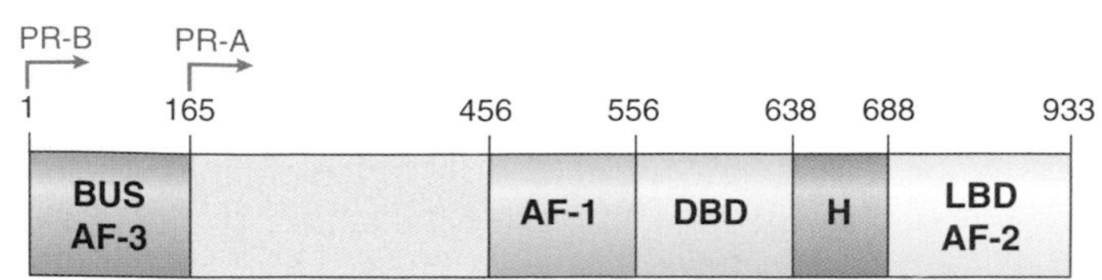

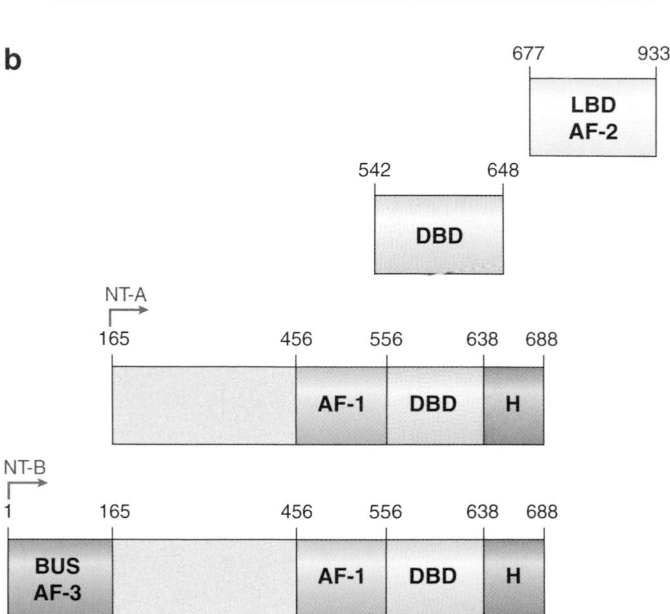

Figure 4

Schematic of full-length progesterone receptor (PR) isoform structure and of experimentally analyzed deletion constructs. (*a*) Schematic layout of the two full-length isoforms. (*b*) Amino acid composition of the deletion constructs discussed in the text. Abbreviations used: AF, activation function; BUS, B-unique sequence; DBD, DNA-binding domain; H, hinge; LBD, ligand-binding domain.

receptor is lacking 164 amino acids at its N terminus. As seen in all other nuclear receptors, the isoforms contain a C-terminal LBD and a centrally located DBD. The two domains are linked by a 50-amino-acid hinge sequence of unclear function in PR. Transcriptional activation functions are located N-terminal to the DBD (AF-1) and within the LBD (AF-2). In apparent contrast to all other nuclear receptors, there is a third activation function (AF-3) located within the 164-residue B-unique sequence (BUS). (63). This AF is not a traditional transcriptional activator: It can function only in the context of its homologous DBD,

PR-B: progesterone receptor B-isoform

suggesting that direct contact between the two surfaces is necessary to carry out function.

Despite their high degree of sequence identity, the two isoforms display significantly different functional properties on natural and synthetic promoters: (*a*) PR-B is typically a much stronger transcriptional activator than PR-A (63). (*b*) The antiprogestin RU486 acts as a partial antagonist toward the B-receptor, although it is a pure antagonist toward the A-receptor (64). (*c*) PR-A gene knockout mice develop uterine dysplasia and abnormal ovaries, whereas PR-B gene knockouts affect the mammary glands, causing premature ductal growth arrest and incomplete lobular-alveolar differentiation (65, 66). (*d*) Microarray studies have demonstrated that the two isoforms regulate different subsets of genes (67). The mechanistic origins of these differences have yet to be elucidated.

The Ligand-Binding Domain

Williams & Sigler (26) in 1998 elucidated the first high-resolution structure of the PR LBD complexed to its ligand, progesterone (**Figure 4*b***). As observed in all other receptor LBDs, the PR structure is composed of a globular, antiparallel α-helical sandwich. Although the structure retains the canonical tertiary fold (**Figure 2**), it is made up of only ten helices: Helix 2 is entirely missing in the PR structure, and helices 10 and 11 are condensed into a single contiguous unit. Helix 12, most critical to regulating coactivator recruitment, is longer than in other LBDs and spans across the ligand-binding pocket. Progesterone binds in what Williams and Sigler term the lower half of the LBD, forming highly specific hydrogen bonds and van der Waals contacts with residues from a short β-turn and helices 3, 5, 7, 11, and 12. Additional hydrophobic interactions between the ligand and the walls of the binding pocket contribute to the stability and specificity of the binding reaction.

Because the PR LBD has not yet been crystallized in the presence of an LXXLL coactivator peptide, the exact nature of the receptor-coactivator interaction is unclear. However, structural comparison to other LBD structures reveals that helix 12 participates in forming a hydrophobic groove likely necessary for binding a coactivating protein (68). Stabilization of the helix in the presence of progesterone comes from hydrophobic interactions with helix 4 and by hydrogen bonding with the agonist hormone. By comparison, manual positioning of the progesterone antagonist RU486 results in steric clashes with helix 12 and tryptophan 755, which contacts helix 12 in the progesterone-bound structure (26). These observations suggest that helix 12 would have to be displaced in the presence of antagonist and thus would be unable to form the interface necessary for coactivator recruitment.

Several additional PR LBD structures, complexed with various steroidal and nonsteroidal ligands, have been solved since the first structure was determined (69–71). Regardless of ligand type, the LBD maintains the same tertiary fold as that discussed above. However, close inspection of the interactions at the binding pocket reveals that receptor-ligand contacts are adjusted depending on ligand type. As a consequence, the volume of the PR LBD binding pocket is modulated from 490 $Å^3$ to 730 $Å^3$, depending on the ligand. The volume changes are accommodated by readjustments of main-chain and side-chain orientations at the binding interface. These studies illuminate the ability of the PR LBD to adjust its local conformation to accommodate a range of ligand types and structures. These studies also may slightly blur the distinction between seemingly highly specific steroid receptor–ligand interactions and the less specific nature of orphan receptor–ligand interactions (21).

The DNA-Binding Domain

Recently, the structure of the PR DBD complexed to a palindromic response element was determined using X-ray crystallography (see **Figure 4*b***; 72). As seen for other steroid receptor structures, PR binds the DNA as a dimer, using a head-to-head orientation. The overall tertiary fold of the core DBD structure is also quite similar to that of other nuclear receptor DBDs (see **Figure 3*a***). If we use the nomenclature defined above, helix 1 of each DBD monomer lies within the major groove of the hexanucleotide half-site, and helix 2 lies above and perpendicular to helix 1. C-terminal to helix 2 is a short helix termed helix 2′. Analysis of the protein-DNA interactions reveals that amino acid side-chain contacts with bases in the major groove are almost identical to those seen for the closely related glucocorticoid and androgen receptors. However, the minor groove within the trinucleotide spacer between each half-site is highly compressed relative to that of the GR- and AR-DNA complexes. The functional implications of this latter observation are still under investigation.

Despite the many similarities to other receptor DBD structures, several surprising observations may have implications for steroid receptor promoter recognition and functional specificity. Most notably, the CTE of the DBD interacted directly with the minor groove of the DNA at bases outside each hexanucleotide response element. Furthermore, mutational analysis revealed that the CTE contributed to overall PR binding affinity and to the receptor's ability to bind isolated half-sites. The contribution of the CTE is not due to nonspecific interactions, as a binding site-selection assay revealed that DBD binding is coupled to sequence preferences that flank the canonical hexanucleotide sequence.

What might these results mean? As discussed above, the CTEs of class II and class III receptors are important for high-affinity DNA binding. In particular, the CTE plays a significant role in the binding of orphan receptor monomers to half-sites. The results described for the PR DBD may suggest a similar role: PR-regulated promoters tend not to exhibit clearly recognizable palindromic response elements. Rather, the promoters are often composed of poorly conserved palindromic sites or clustered half-sites, suggesting that the PR CTE helps to stabilize binding at these promoters. Moreover, because steroid receptors can recognize the same response elements yet differentially regulate promoters containing those response elements, the nonconserved CTE may play a role in the selective binding of steroid receptors at target promoters.

Residues N-Terminal to the DNA-Binding Domain

Using biochemical and biophysical approaches, Horwitz and coworkers (48, 49) demonstrated that PR isoforms lacking only the LBD (denoted as NTA and NTB; see **Figure 4*b***) maintain considerable structure in the N-terminal region. However, analysis indicated that this "structure" exists in solution as an ensemble of extended conformations rather than as any unique globular-type fold. Mapping studies revealed that the regions of lowest structural stability corresponded to previously identified phosphorylation motifs and the functional boundaries of the AF-1 subunit; regions of high stability were interspersed among these sites. However, these results do not imply that the folded sequences are intrinsically stable, as PR fragments that no longer contained the DBD were immediately degraded by proteases. Taken together, these observations indicated that the DBD can stabilize and influence N-terminal structure. These results may also explain why nuclear receptor activation functions are typically unfolded when studied in isolation.

Upon binding to a palindromic response element, both NTA and NTB underwent changes in conformation. These changes were localized to the AF-1 region and the hinge and thus demonstrated that DNA binding is coupled to allosteric structural transitions mediated through the DBD. DNA-dependent changes in N-terminal structure appear to be a common theme for nuclear receptors, although the functional implications are not entirely clear. Evidence suggests that changes

Free energy change (ΔG): is related to the equilibrium dissociation constant (K_d) through the standard expression $\Delta G = -RT \ln 1/K_d$

are necessary for recruitment of coactivating proteins to the promoter (73). However, the hypothesis that DNA-bound receptors truly recruit coactivators—that is, increase the binding energetics of the AF-coactivator interaction—has yet to be rigorously tested (cf. 74).

Biochemical analysis of NTA and NTB revealed that the macroscopic structural properties of residues common to both isoforms were largely identical. In contrast, analytical ultracentrifugation analysis demonstrated that the two constructs showed differences in their ensemble distribution of hydrodynamic conformations and that these differences occurred at the level of either secondary or tertiary structure. This observation led to the hypothesis that isoform-specific functional differences were not due to macroscopic structural differences between the two isoforms but rather that the BUS could stabilize a more functionally active set of conformers within the PR-B ensemble relative to that existing in the PR-A ensemble.

The Progesterone Receptor Holoprotein

As of this writing, neither of the PR isoforms (nor any other full-length nuclear receptor) has yielded to crystallographic or NMR-based structural analysis. The likely culprits are lack of tightly folded structure in the N-terminal region and difficulties in generating functionally homogeneous and highly concentrated receptor preparations. In spite of these issues, it has been possible to carry out rigorous thermodynamic and hydrodynamic studies of the PR isoforms, and these studies may offer some insight into the gross structural properties of full-length nuclear receptors.

Analytical ultracentrifugation analysis of highly purified and functionally homogeneous PR-B demonstrates that the receptor undergoes self-association in the micromolar range (75). This affinity is considerably weaker than the nanomolar dimerization constants semiquantitatively estimated for full-length GR (76, 77) and ER (78 and references therein) and is consistent with the small dimerization interface seen in the PR LBD crystal structure (26). However, the isolated PR LBD and PR constructs lacking the LBD are purely monomeric in solution (26, 48, 49). Therefore, the ability of the holoprotein to dimerize indicates that self-association is a global property of PR—it cannot be attributed solely to the LBD, even if that domain provides the sole dimerization interface. Thus, from a structural perspective, these results indicate that sequences and structures outside of the interface allosterically contribute to the dimerization energetics.

A functional consequence of a micromolar dimerization constant is that, at the nanomolar concentrations of PR-B necessary to initiate DNA binding, the receptor is almost entirely (99.9%) monomeric in solution (79). This observation raises the question as to whether preformed PR dimers are the active DNA-binding species (80, 81). In other words, may it be the case that the monomer is instead the active species? This question takes on greater weight when one examines the energetics of PR binding to its response elements: Thermodynamic analysis of PR-B:DNA interactions reveals that the PR-B dimer has an intrinsic dissociation constant of 81 pM (equivalent to a -12.8 kcal mol^{-1} free energy change), indicative of extremely high binding affinity and in stark contrast to the apparent nanomolar binding affinities seen by visual inspection of binding curves (79, 82). By comparison, analysis of the binding data, through the use of a model in which only solution monomers bind to DNA half-sites, resolves an intrinsic dissociation constant of 39 nM (-9.4 kcal mol^{-1} binding free energy). Thus, successive monomer binding to a palindromic binding site is thermodynamically favored over preformed dimer binding by $+6.0$ kcal mol^{-1}. Furthermore, if PR-B assembles at its response elements only as a dimer, this betrays the notion that receptors have only moderate DNA-binding affinity and makes it somewhat unclear why the dimer would need accessory proteins to create a highly stable protein-DNA complex (82). By contrast, if monomer:half-site interactions are dominant, they would be perfectly positioned to take advantage of enhanced binding via accessory proteins. Finally, the thermodynamic studies suggest that caution should be observed in the experimental design of DNA-binding techniques such as gel shift assays. Because these assays typically use short, labeled oligonucleotides at low nanomolar concentrations, the picomolar affinity of PR dimers (or other nuclear receptors) may cause the resulting binding data to reflect the stoichiometric addition of bound protein rather than any meaningful equilibrium binding affinity.

Regardless of how PR assembles at response elements, the interactions are coupled to conformational transitions outside the DBD, most notably in the AF-1 and hinge sequences (48, 49). Because these changes can include dramatic disorder-to-order folding transitions, investigators have presumed that folding must be energetically costly. However, it was not until recently that the amount of this cost was actually experimentally estimated. Thermodynamic analysis revealed an enormous +6 kcal mol^{-1} unfavorable contribution to PR dimer assembly at a palindromic response element (79). This penalty correlates with structural changes in both the DNA and the protein, and thus assignment of the energetics to specific sequences and residues is not yet possible. However, the size of the penalty indicates that there are large hurdles associated with rearranging receptor-DNA structure (see side bar, Macromolecular Interactions and Effective Concentrations). Taken together with the picomolar dimer binding affinity, the energetics of PR-DNA interactions reflect a balance of very strong favorable and unfavorable forces, perhaps analogous to that seen in protein folding. Finally, this view is not limited to interactions at individual binding sites: A strong cooperative interaction (99–400-fold increase in overall stability) between adjacently bound PR dimers is coupled to energetically unfavorable deformation of the promoter DNA (79). A challenge for structural biology will be to connect more directly the observed thermodynamics with receptor-DNA structural features.

MACROMOLECULAR INTERACTIONS AND EFFECTIVE CONCENTRATIONS

The size of a +6 kcal mol^{-1} energetic penalty to PR-B binding can be put in perspective by considering the relationship of the penalty size to the receptor's effective concentration at the DNA-binding site (83, 84). Effective concentrations are typically used to emphasize the localized increase in reactant concentration upon complex formation. For example, the architecture of an enzymatic active site serves to orient reactive side chains in close proximity to a bound substrate. The consequence of this interaction is that the effective concentration of the substrate, relative to its concentration in bulk solution, can be increased greater than 10,000-fold, leading to increased rates of catalysis. By contrast, the large penalty to PR-B binding results in the more than 10,000-fold reduction in effective concentration of a receptor protomer at the DNA-binding site relative to the receptor concentration in bulk solution. That PR-B nonetheless forms a stable complex with the response element is a testament to the strong favorable forces underlying the picomolar intrinsic binding affinity.

The self-assembly energetics of PR-B are driven, at least in part, by electrostatic interactions (75). Thus, at high salt concentrations, the PR-B dimer dissociates to a hydrodynamically homogeneous monomer. This result allows one to assign some gross structural properties to the receptor isoform. For example, under these conditions the PR-B monomer has a Stokes radius of 64 Å and sediments as a highly asymmetric structure, having a frictional ratio of 1.67. If modeled as a hydrated prolate ellipsoid, PR-B has a major-to-minor-axis ratio of approximately 13:1. Limited proteolysis of the monomeric receptor indicates that this asymmetry is not due to an actual ellipsoid or rod-like conformation but instead arises from a nonglobular [but roughly spherical (48)] structure, with sequences N-terminal to the DBD existing as an unfolded

Stokes radius (r_S): represents the hydrodynamic radius of a macromolecule when modeled as a rigid sphere or partially folded ensemble of conformations. Comparison to previous work on the B-isoform lacking its C-terminal LBD (49) reveals that the holoprotein maintains different rates and patterns of proteolysis within the AF-1 and hinge sequences (84a), thus indicating that the LBD intramolecularly modulates the conformation(s) of these regions. This difference may have implications for understanding the functional synergy seen in the holoreceptor when compared with various deletion constructs (13).

Clearly, our resolution of the structural properties of full-length PR-B falls well short of the atomic-level detail available for the isolated DBD and LBD. Nonetheless, it is still possible to compare directly the results of each of the studies to obtain a more comprehensive perspective of PR structure. **Figure 5** shows a schematic of the experimentally determined hydrodynamic volumes of the full-length PR-B monomer [Stokes radius of 64 Å (75)] relative to the calculated volumes of the monomeric LBD and DBD (Stokes radii of 25 Å and 19 Å, respectively) when modeled as rigid spheres. The size of the full-length receptor dwarfs that of the individual domains. In particular, there is a 38-fold increase in volume when comparing the full-length receptor (regardless of its detailed atomic structure) with that of the isolated DBD. This disparity arises in part simply because the DBD makes up less than 10% of the molecular mass of the holoprotein. However, the primary reason for the volume difference is the low packing density of the conformationally extended N-terminal peptide chain. Similar results are seen when comparing the holoprotein with its

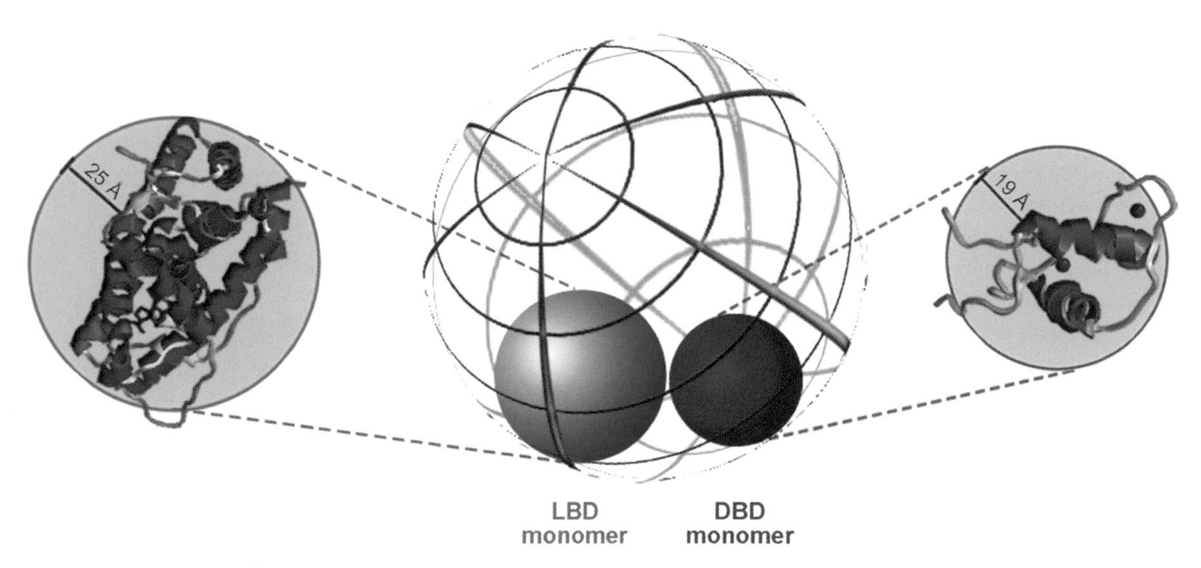

Figure 5

Schematic representation of the hydrodynamic properties of the full-length progesterone receptor B-isoform (PR-B) monomer, a monomeric core DNA-binding domain (DBD), and a monomeric ligand-binding domain (LBD) when modeled as rigid spheres. Shown as a black and white wire structure is the predicted volume of the PR-B monomer as determined from the Stokes radius (r_s = 64 Å; $V = 4/3\pi r^3$). Inset in blue and red are the predicted volumes of the PR LBD monomer and DBD monomer, respectively (r_s = 25 Å for the LBD and 19 Å for the DBD). For perspective, ribbon diagrams of the LBD and DBD structures are overlaid. The Stokes radii of the monomeric species either were experimentally determined, using combined sedimentation velocity and sedimentation equilibrium analysis (PR-B; 75), or calculated from the atomic coordinates of the PR LBD and GR DBD crystal structures (26, 37), using bead modeling (87, 88). The schematic is drawn to scale. (Because the coordinates of the PR DBD structure had not been released at the time of this writing, the coordinates of the essentially identical GR DBD structure were used for the calculation.)

isolated LBD: There is a 17-fold increase in volume. These observations, although lacking in atomic-level precision, are entirely accurate at the hydrodynamic level. We hope that the comparisons emphasize the challenges that lie ahead in obtaining a clearer picture of receptor structure-function relationships.

UNANSWERED QUESTIONS AND FUTURE DIRECTIONS

As described in this review, although nuclear receptors have a modular functional and structural layout, the various parts of the protein (domains, activation functions, recognition motifs, binding interfaces, etc.) do not act independently. Instead, receptor function arises through a nonlinear sum of interactions both within the protein and between various transcription factors and promoter response elements. Moreover, many of these interactions are coupled to complex structural rearrangements or large-scale folding reactions. A major challenge in understanding nuclear receptor function will be in defining the quantitative mechanisms by which the interactions among these parts translate into the "system" behavior of the intact holoprotein (and by extension, the intact transcriptional activation complex). For example, even as the physical and chemical forces necessary to induce folding of AF-1 sequences are beginning to be understood, the molecular mechanisms by which these events occur naturally, through allosteric communication via the DBD and DNA, are largely unknown. Similarly, although the stereochemical basis for understanding LBD-coactivator interactions is well characterized, this viewpoint offers minimal insight into the mechanisms by which these interactions are coupled to large-scale folding of the coactivating protein (85). Our understanding of these issues will be greatly advanced if a high-resolution structure of a full-length receptor or receptor-coactivator complex is determined. However, it is doubtful that only structural analysis will reveal the mechanisms underlying function. Rather, any quantitative and predictive understanding of receptor function will require structural and biochemical analyses to be integrated with rigorous thermodynamic and kinetic dissections of the macromolecular interactions. Much remains to be learned about the molecular origins of nuclear receptor structure-function relationships, and a fundamental understanding will require a multidisciplinary approach.

SUMMARY POINTS

1. Nuclear receptors are made up of a modular structure that includes a C-terminal ligand-binding domain, a centrally located DNA-binding domain, and multiple transcriptional activation functions.
2. Although modular in structure, the receptor subunits do not act independently in the context of the holoprotein; allosteric interactions between the subunits are critical to the ability of nuclear receptors to function as efficient gene regulatory switches.
3. Allosteric interactions are coupled to both subtle and dramatic changes in structure. The physical mechanisms underlying allosteric communication are yet to be determined.

ACKNOWLEDGMENTS

We thank Dr. Mair Churchill for generously sharing results of the PR DBD structural analysis prior to publication. This work was supported in part by NIH grants R01-DK061933 (to D.L.B.) and F32-DK070519 (to A.F.H.).

LITERATURE CITED

1. Mangelsdorf DJ, Thummel C, Beato M, Herrlich P, Schutz G, et al. 1995. The nuclear receptor superfamily: the second decade. *Cell* 83:835–39
2. Tsai MJ, O'Malley BW. 1994. Molecular mechanisms of action of steroid/thyroid receptor superfamily members. *Annu. Rev. Biochem.* 63:451–86
3. Chawla A, Repa JJ, Evans RM, Mangelsdorf DJ. 2001. Nuclear receptors and lipid physiology: opening the X-files. *Science* 294:1866–70
4. Umesono K, Murakami KK, Thompson CC, Evans RM. 1991. Direct repeats as selective response elements for the thyroid hormone, retinoic acid, and vitamin D3 receptors. *Cell* 65:1255–66
5. Metivier R, Penot G, Hubner MR, Reid G, Brand H, et al. 2003. Estrogen receptor-α directs ordered, cyclical, and combinatorial recruitment of cofactors on a natural target promoter. *Cell* 115:751–63
6. Nagaich AK, Walker DA, Wolford R, Hager GL. 2004. Rapid periodic binding and displacement of the glucocorticoid receptor during chromatin remodeling. *Mol. Cell* 14:163–74
7. Birnbaumer M, Schrader WT, O'Malley BW. 1983. Assessment of structural similarities in chick oviduct progesterone receptor subunits by partial proteolysis of photoaffinity-labeled proteins. *J. Biol. Chem.* 258:7331–37
8. Wrange O, Okret S, Radojcic M, Carlstedt-Duke J, Gustafsson JA. 1984. Characterization of the purified activated glucocorticoid receptor from rat liver cytosol. *J. Biol. Chem.* 259:4534–41
9. Xu J, Li Q. 2003. Review of the in vivo functions of the p160 steroid receptor coactivator family. *Mol. Endocrinol.* 17:1681–92
10. Kumar V, Chambon P. 1988. The estrogen receptor binds tightly to its responsive element as a ligand-induced homodimer. *Cell* 55:145–56
11. Knotts TA, Orkiszewski RS, Cook RG, Edwards DP, Weigel NL. 2001. Identification of a phosphorylation site in the hinge region of the human progesterone receptor and additional amino-terminal phosphorylation sites. *J. Biol. Chem.* 276:8475–83
12. Lee YK, Choi YH, Chua S, Park YJ, Moore DD. 2006. Phosphorylation of the hinge domain of the nuclear hormone receptor LRH-1 stimulates transactivation. *J. Biol. Chem.* 281:7850–55
13. Takimoto GS, Tung L, Abdel-Hafiz H, Abel MG, Sartorius CA, et al. 2003. Functional properties of the N-terminal region of progesterone receptors and their mechanistic relationship to structure. *J. Steroid Biochem. Mol. Biol.* 85:209–19
14. Giangrande PH, Pollio G, McDonnell DP. 1997. Mapping and characterization of the functional domains responsible for the differential activity of the A and B isoforms of the human progesterone receptor. *J. Biol. Chem.* 272:32889–900
15. Hovland AR, Powell RL, Takimoto GS, Tung L, Horwitz KB. 1998. An N-terminal inhibitory function, IF, suppresses transcription by the A-isoform but not the B-isoform of human progesterone receptors. *J. Biol. Chem.* 273:5455–60
16. Narayanan R, Edwards DP, Weigel NL. 2005. Human progesterone receptor displays cell cycle-dependent changes in transcriptional activity. *Mol. Cell Biol.* 25:2885–98
17. Takimoto GS, Hovland AR, Tasset DM, Melville MY, Tung L, Horwitz KB. 1996. Role of phosphorylation in DNA binding and transcriptional functions of human progesterone receptors. *J. Biol. Chem.* 271:13308–16
18. Abdel-Hafiz H, Takimoto GS, Tung L, Horwitz KB. 2002. The inhibitory function in human progesterone receptor N termini binds SUMO-1 protein to regulate autoinhibition and transrepression. *J. Biol. Chem.* 277:33950–56

19. Mestas SP, Lumb KJ. 1999. Electrostatic contribution of phosphorylation to the stability of the CREB-CBP activator-coactivator complex. *Nat. Struct. Biol.* 6:613–14
20. Bourguet W, Ruff M, Chambon P, Gronemeyer H, Moras D. 1995. Crystal structure of the ligand-binding domain of the human nuclear receptor RXR-α. *Nature* 375:377–82
21. Li Y, Lambert MH, Xu HE. 2003. Activation of nuclear receptors: a perspective from structural genomics. *Structure* 11:741–46
22. Bledsoe RK, Montana VG, Stanley TB, Delves CJ, Apolito CJ, et al. 2002. Crystal structure of the glucocorticoid receptor ligand binding domain reveals a novel mode of receptor dimerization and coactivator recognition. *Cell* 110:93–105
23. Brzozowski AM, Pike AC, Dauter Z, Hubbard RE, Bonn T, et al. 1997. Molecular basis of agonism and antagonism in the oestrogen receptor. *Nature* 389:753–58
24. He B, Gampe RTJ, Kole AJ, Hnat AT, Stanley TB, et al. 2004. Structural basis for androgen receptor interdomain and coactivator interactions suggests a transition in nuclear receptor activation function dominance. *Mol. Cell* 16:425–38
25. Li Y, Suino K, Daugherty J, Xu HE. 2005. Structural and biochemical mechanisms for the specificity of hormone binding and coactivator assembly by mineralocorticoid receptor. *Mol. Cell* 19:367–80
26. Williams SP, Sigler PB. 1998. Atomic structure of progesterone complexed with its receptor. *Nature* 393:392–96
27. Johnson BA, Wilson EM, Li Y, Moller DE, Smith RG, Zhou G. 2000. Ligand-induced stabilization of PPARγ monitored by NMR spectroscopy: implications for nuclear receptor activation. *J. Mol. Biol.* 298:187–94
28. Shiau AK, Barstad D, Loria PM, Cheng L, Kushner PJ, et al. 1998. The structural basis of estrogen receptor/coactivator recognition and the antagonism of this interaction by tamoxifen. *Cell* 95:927–37
29. Nettles KW, Sun J, Radek JT, Sheng S, Rodriguez AL, et al. 2004. Allosteric control of ligand selectivity between estrogen receptors α and β: implications for other nuclear receptors. *Mol. Cell* 13:317–27
30. Gampe RTJ, Montana VG, Lambert MH, Wisely GB, Milburn MV, Xu HE. 2000. Structural basis for autorepression of retinoid X receptor by tetramer formation and the AF-2 helix. *Genes Dev.* 14:2229–41
31. Xu HE, Stanley TB, Montana VG, Lambert MH, Shearer BG, et al. 2002. Structural basis for antagonist-mediated recruitment of nuclear co-repressors by PPARα. *Nature* 415:813–17
32. Chen JD, Evans RM. 1995. A transcriptional co-repressor that interacts with nuclear hormone receptors. *Nature* 377:454–57
33. Horlein AJ, Naar AM, Heinzel T, Torchia J, Gloss B, et al. 1995. Ligand-independent repression by the thyroid hormone receptor mediated by a nuclear receptor co-repressor. *Nature* 377:397–404
34. Xu L, Glass CK, Rosenfeld MG. 1999. Coactivator and corepressor complexes in nuclear receptor function. *Curr. Opin. Genet. Dev.* 9:140–47
35. Freedman LP. 1992. Anatomy of the steroid receptor zinc finger region. *Endocr. Rev.* 13:129–45
36. Hard T, Kellenbach E, Boelens R, Maler BA, Dahlman K, et al. 1990. Solution structure of the glucocorticoid receptor DNA-binding domain. *Science* 249:157–60
37. Luisi BF, Xu WX, Otwinowski Z, Freedman LP, Yamamoto KR, Sigler PB. 1991. Crystallographic analysis of the interaction of the glucocorticoid receptor with DNA. *Nature* 352:497–505

38. Freedman LP, Luisi BF, Korszun ZR, Basavappa R, Sigler PB, Yamamoto KR. 1988. The function and structure of the metal coordination sites within the glucocorticoid receptor DNA binding domain. *Nature* 334:543–46
39. Baumann H, Paulsen K, Kovacs H, Berglund H, Wright AP, et al. 1993. Refined solution structure of the glucocorticoid receptor DNA-binding domain. *Biochemistry* 32:13463–71
40. Rastinejad F, Perlmann T, Evans RM, Sigler PB. 1995. Structural determinants of nuclear receptor assembly on DNA direct repeats. *Nature* 375:203–11
41. Holmbeck SM, Foster MP, Casimiro DR, Sem DS, Dyson HJ, Wright PE. 1998. High-resolution solution structure of the retinoid X receptor DNA-binding domain. *J. Mol. Biol.* 281:271–84
42. Ingraham HA, Redinbo MR. 2005. Orphan nuclear receptors adopted by crystallography. *Curr. Opin. Struct. Biol.* 15:708–15
43. Gearhart MD, Holmbeck SM, Evans RM, Dyson HJ, Wright PE. 2003. Monomeric complex of human orphan estrogen related receptor-2 with DNA: a pseudo-dimer interface mediates extended half-site recognition. *J. Mol. Biol.* 327:819–32
44. Folkers GE, van Heerde EC, van der Saag PT. 1995. Activation function 1 of retinoic acid receptor $\beta 2$ is an acidic activator resembling VP16. *J. Biol. Chem.* 270:23552–59
45. Dahlman-Wright K, Baumann H, McEwan IJ, Almlof T, Wright AP, et al. 1995. Structural characterization of a minimal functional transactivation domain from the human glucocorticoid receptor. *Proc. Natl. Acad. Sci. USA* 92:1699–703
46. Kumar R, Lee JC, Bolen DW, Thompson EB. 2001. The conformation of the glucocorticoid receptor af1/tau1 domain induced by osmolyte binds co-regulatory proteins. *J. Biol. Chem.* 276:18146–52
47. Kumar R, Baskakov IV, Srinivasan G, Bolen DW, Lee JC, Thompson EB. 1999. Interdomain signaling in a two-domain fragment of the human glucocorticoid receptor. *J. Biol. Chem.* 274:24737–41
48. Bain DL, Franden MA, McManaman JL, Takimoto GS, Horwitz KB. 2000. The N-terminal region of the human progesterone A-receptor. Structural analysis and the influence of the DNA binding domain. *J. Biol. Chem.* 275:7313–20
49. Bain DL, Franden MA, McManaman JL, Takimoto GS, Horwitz KB. 2001. The N-terminal region of human progesterone B-receptors: biophysical and biochemical comparison to A-receptors. *J. Biol. Chem.* 276:23825–31
50. Reid J, Kelly SM, Watt K, Price NC, McEwan IJ. 2002. Conformational analysis of the androgen receptor amino-terminal domain involved in transactivation. Influence of structure-stabilizing solutes and protein-protein interactions. *J. Biol. Chem.* 277:20079–86
51. Brodie J, McEwan IJ. 2005. Intra-domain communication between the N-terminal and DNA-binding domains of the androgen receptor: modulation of androgen response element DNA binding. *J. Mol. Endocrinol.* 34:603–15
52. Kumar R, Betney R, Li J, Thompson EB, McEwan IJ. 2004. Induced α-helix structure in AF1 of the androgen receptor upon binding transcription factor TFIIF. *Biochemistry* 43:3008–13
53. Warnmark A, Wikstrom A, Wright AP, Gustafsson JA, Hard T. 2001. The N-terminal regions of estrogen receptor α and β are unstructured in vitro and show different TBP binding properties. *J. Biol. Chem.* 276:45939–44
54. Wood JR, Likhite VS, Loven MA, Nardulli AM. 2001. Allosteric modulation of estrogen receptor conformation by different estrogen response elements. *Mol. Endocrinol.* 15:1114–26
55. Lefstin JA, Yamamoto KR. 1998. Allosteric effects of DNA on transcriptional regulators. *Nature* 392:885–88

56. Lefstin JA, Thomas JR, Yamamoto KR. 1994. Influence of a steroid receptor DNA-binding domain on transcriptional regulatory functions. *Genes Dev.* 8:2842–56
57. van Tilborg MA, Lefstin JA, Kruiskamp M, Teuben J, Boelens R, et al. 2000. Mutations in the glucocorticoid receptor DNA-binding domain mimic an allosteric effect of DNA. *J. Mol. Biol.* 301:947–58
58. Wardell SE, Kwok SC, Sherman L, Hodges RS, Edwards DP. 2005. Regulation of the amino-terminal transcription activation domain of progesterone receptor by a cofactor-induced protein folding mechanism. *Mol. Cell Biol.* 25:8792–808
59. Li X, Wong J, Tsai SY, Tsai MJ, O'Malley BW. 2003. Progesterone and glucocorticoid receptors recruit distinct coactivator complexes and promote distinct patterns of local chromatin modification. *Mol. Cell Biol.* 23:3763–73
60. Dyson HJ, Wright PE. 2002. Coupling of folding and binding for unstructured proteins. *Curr. Opin. Struct. Biol.* 12:54–60
61. Uversky VN. 2002. Natively unfolded proteins: a point where biology waits for physics. *Protein Sci.* 11:739–56
62. Kastner P, Krust A, Turcotte B, Stropp U, Tora L, et al. 1990. Two distinct estrogen-regulated promoters generate transcripts encoding the two functionally different human progesterone receptor forms A and B. *EMBO J.* 9:1603–14
63. Sartorius CA, Melville MY, Hovland AR, Tung L, Takimoto GS, Horwitz KB. 1994. A third transactivation function (AF3) of human progesterone receptors located in the unique N-terminal segment of the B-isoform. *Mol. Endocrinol.* 8:1347–60
64. Meyer ME, Pornon A, Ji JW, Bocquel MT, Chambon P, Gronemeyer H. 1990. Agonistic and antagonistic activities of RU486 on the functions of the human progesterone receptor. *EMBO J.* 9:3923–32
65. Mulac-Jericevic B, Lydon JP, DeMayo FJ, Conneely OM. 2003. Defective mammary gland morphogenesis in mice lacking the progesterone receptor B isoform. *Proc. Natl. Acad. Sci. USA* 100:9744–49
66. Mulac-Jericevic B, Mullinax RA, DeMayo FJ, Lydon JP, Conneely OM. 2000. Subgroup of reproductive functions of progesterone mediated by progesterone receptor-B isoform. *Science* 289:1751–54
67. Richer JK, Jacobsen BM, Manning NG, Abel MG, Wolf DM, Horwitz KB. 2002. Differential gene regulation by the two progesterone receptor isoforms in human breast cancer cells. *J. Biol. Chem.* 277:5209–18
68. Tanenbaum DM, Wang Y, Williams SP, Sigler PB. 1998. Crystallographic comparison of the estrogen and progesterone receptor's ligand binding domains. *Proc. Natl. Acad. Sci. USA* 95:5998–6003
69. Madauss KP, Deng SJ, Austin RJ, Lambert MH, McLay I, et al. 2004. Progesterone receptor ligand binding pocket flexibility: crystal structures of the norethindrone and mometasone furoate complexes. *J. Med. Chem.* 47:3381–87
70. Matias PM, Donner P, Coelho R, Thomaz M, Peixoto C, et al. 2000. Structural evidence for ligand specificity in the binding domain of the human androgen receptor. Implications for pathogenic gene mutations. *J. Biol. Chem.* 275:26164–71
71. Zhang Z, Olland AM, Zhu Y, Cohen J, Berrodin T, et al. 2005. Molecular and pharmacological properties of a potent and selective novel nonsteroidal progesterone receptor agonist tanaproget. *J. Biol. Chem.* 280:28468–75
72. Roemer SC, Donham DC, Sherman L, Pon VH, Edwards DP, Churchill ME. 2006. Structure of the progesterone receptor-DNA complex: novel interactions required for binding to half-site response elements. *Mol. Endocrinol.* 20:3042–52

73. Kumar R, Thompson EB. 2005. Gene regulation by the glucocorticoid receptor: structure:function relationship. *J. Steroid Biochem. Mol. Biol.* 94:383–94
74. Librizzi MD, Brenowitz M, Willis IM. 1998. The TATA element and its context affect the cooperative interaction of TATA-binding protein with the TFIIB-related factor, TFIIIB70. *J. Biol. Chem.* 273:4563–68
75. Heneghan AF, Berton N, Miura MT, Bain DL. 2005. Self-association energetics of an intact, full-length nuclear receptor: The B-isoform of human progesterone receptor dimerizes in the micromolar range. *Biochemistry* 44:9528–37
76. Perlmann T, Eriksson P, Wrange O. 1990. Quantitative analysis of the glucocorticoid receptor-DNA interaction at the mouse mammary tumor virus glucocorticoid response element. *J. Biol. Chem.* 265:17222–29
77. Segard-Maurel I, Rajkowski K, Jibard N, Schweizer-Groyer G, Baulieu EE, Cadepond F. 1996. Glucocorticosteroid receptor dimerization investigated by analysis of receptor binding to glucocorticosteroid responsive elements using a monomer-dimer equilibrium model. *Biochemistry* 35:1634–42
78. Tamrazi A, Carlson KE, Daniels JR, Hurth KM, Katzenellenbogen JA. 2002. Estrogen receptor dimerization: Ligand binding regulates dimer affinity and dimer dissociation rate. *Mol. Endocrinol.* 16:2706–19
79. Heneghan AF, Connaghan-Jones KD, Miura MT, Bain DL. 2006. Cooperative DNA binding by the B-isoform of human progesterone receptor: Thermodynamic analysis reveals strongly favorable and unfavorable contributions to assembly. *Biochemistry* 45:3285–96
80. DeMarzo AM, Beck CA, Onate SA, Edwards DP. 1991. Dimerization of mammalian progesterone receptors occurs in the absence of DNA and is related to the release of the 90-kDa heat shock protein. *Proc. Natl. Acad. Sci. USA* 88:72–76
81. Rodriguez R, Weigel NL, O'Malley BW, Schrader WT. 1990. Dimerization of the chicken progesterone receptor in vitro can occur in the absence of hormone and DNA. *Mol. Endocrinol.* 4:1782–90
82. Onate SA, Prendergast P, Wagner JP, Nissen M, Reeves R, et al. 1994. The DNA-bending protein HMG-1 enhances progesterone receptor binding to its target DNA sequences. *Mol. Cell Biol.* 14:3376–91
83. Jencks WP. 1981. On the attribution and additivity of binding energies. *Proc. Natl. Acad. Sci. USA* 78:4046–50
84. Creighton TE. 1983. *Proteins, Structure and Molecular Principles*. New York: W. H. Freeman and Co. 512 pp.

84a. Connaghan-Jones KD, Heneghan AF, Miura MT, Bain DL. 2006. Hydrodynamic analysis of the human progesterone receptor A-isoform reveals that self-association occurs in the micromolar range. *Biochemistry*. 45:12090–99

85. Demarest SJ, Martinez-Yamout M, Chung J, Chen H, Xu W, et al. 2002. Mutual synergistic folding in recruitment of CBP/p300 by p160 nuclear receptor coactivators. *Nature* 415:549–53
86. Deleted in proof
87. Garcia de la Torre J, Navarro S, Lopez Martinez MC, Diaz FG, Lopez Cascales JJ. 1994. HYDRO: a computer program for the prediction of hydrodynamic properties of macromolecules. *Biophys. J.* 67:530–31
88. Garcia DL Torre J, Huertas ML, Carrasco B. 2000. Calculation of hydrodynamic properties of globular proteins from their atomic-level structure. *Biophys. J.* 78:719–30

Regulation of Intestinal Cholesterol Absorption

David Q.-H. Wang

Department of Medicine, Liver Center and Gastroenterology Division, Beth Israel Deaconess Medical Center, Harvard Medical School and Harvard Digestive Diseases Center, Boston, Massachusetts 02115; email: dqwang@caregroup.harvard.edu

Annu. Rev. Physiol. 2007. 69:221–48

First published online as a Review in Advance on September 20, 2006

The *Annual Review of Physiology* is online at http://physiol.annualreviews.org

This article's doi:
10.1146/annurev.physiol.69.031905.160725

0066-4278/07/0315-0221$20.00

Key Words

bile salt, transporter, chylomicron, nutrition, sitosterol

Abstract

The identification of defective structures in the ATP-binding cassette (ABC) transporters ABCG5 and ABCG8 in patients with sitosterolemia suggests that these two proteins are an apical sterol export pump promoting active efflux of cholesterol and plant sterols from enterocytes back into the intestinal lumen for excretion. The newly identified Niemann-Pick C1–like 1 (NPC1L1) protein is also expressed at the apical membrane of enterocytes and plays a crucial role in the ezetimibe-sensitive cholesterol absorption pathway. These findings indicate that cholesterol absorption is a multistep process that is regulated by multiple genes at the enterocyte level and that the efficiency of cholesterol absorption may be determined by the net effect between influx and efflux of intraluminal cholesterol molecules crossing the brush border membrane of the enterocyte. Combination therapy using cholesterol absorption (NPC1L1) inhibitor (ezetimibe) and 3-hydroxy-3-methylglutaryl-CoA (HMG-CoA) reductase inhibitors (statins) provides a powerful novel strategy for the prevention and treatment of hypercholesterolemia.

INTRODUCTION

Bile acids: the end products of cholesterol degradation in the liver that are the major constituents of bile and that are essential for the absorption of cholesterol from the small intestine

Plant sterol (phytosterol): sterols present in plants

Stenol: sterols with one or more nuclear double bonds

ACAT: acyl-CoA:cholesterol acyltransferase

ATP-binding cassette (ABC) transporters: a superfamily of highly conserved proteins involved in the membrane transport of a variety of substrates to various compartments

Sterol: any unsaponifiable steroid alcohol with an aliphatic side chain of 8–10 carbon atoms and a hydroxyl group at C3

Because increased plasma cholesterol levels are one of the most important risk factors for coronary heart disease, the National Cholesterol Education Program Adult Treatment Panel III guidelines and the results of the Heart Protection Study have provided a stronger rationale to treat high-risk patients to a low-density lipoprotein (LDL) cholesterol goal of <100 mg dl^{-1} (1). In particular, individuals at substantial risk for atherosclerosis or patients with cardiovascular diseases should meet defined targets for LDL cholesterol levels. The setting of these targets has greatly increased the number of individuals who need cholesterol-lowering therapy.

The cholesterol carried in LDL is derived principally from de novo synthesis and absorption from the diet. In humans, there is a significant and positive correlation between the level of plasma LDL cholesterol and the efficiency of intestinal cholesterol absorption (2). Thus, the restriction of dietary calories, cholesterol, and saturated fat has been recommended as the primary initial therapeutic intervention for the treatment of patients with dyslipidemia (3, 4). Despite significant restrictions in dietary intake, the reduction of dietary cholesterol is often not associated with a significant decrease in circulating LDL cholesterol levels. Therefore, pharmacological modulation of cholesterol absorption is potentially an effective way of lowering plasma LDL cholesterol levels in the general population.

Because intestinal cholesterol absorption is a complex process that involves multiple interrelated sequential degradative and synthetic pathways, it provides multiple therapeutic targets in the management of patients with hypercholesterolemia. For example, the bile acid sequestrants (e.g., cholestyramine, resins, and colestipol) reduce cholesterol absorption primarily via interruption of the enterohepatic circulation of bile acids and may result in a secondary increase in hepatic LDL receptor activity (5, 6). Dietary plant sterols at a dose of 2 g per day have been recommended as an adjunctive lifestyle treatment for hypercholesterolemia (7). Recent clinical studies have found that plant sterol treatment induces an approximately 8–14% decrease in plasma LDL cholesterol levels in subjects with mild or moderate hypercholesterolemia (8). Specific lipase inhibitors such as orlistat may also suppress cholesterol absorption by blocking the degradative process within the gastrointestinal lumen (9, 10), which results in a decreased solubility of cholesterol during the critical stage of intestinal diffusion. The intestinal acyl-CoA:cholesterol acyltransferase (ACAT) inhibitors (11) and cholesterol ester transfer protein inhibitors (12) are currently being tested in clinical trials, and the potential to alter ATP-binding cassette (ABC) transporter activity in the intestine is also being investigated. Moreover, the discovery and development of ezetimibe (13, 14)—a novel, selective, and potent inhibitor that effectively blocks intestinal absorption of dietary and biliary cholesterol—open a new door to the treatment of hypercholesterolemia (15–18). Ezetimibe, which can be administered either alone or in combination with statins, is a safe and efficacious treatment for hypercholesterolemia and potentially enables more patients to reach recommended LDL cholesterol goals.

Pharmacological inhibitors of cholesterol absorption should ideally be targeted for use in individuals who demonstrate quantitatively enhanced cholesterol transport from the intestinal tract to the circulation. In particular, two recently identified intestinal sterol transporters provide further insights in the regulation of intestinal cholesterol absorption. Mutations in the genes encoding either ABCG5 or ABCG8 result in sitosterolemia (19, 20), which is characterized by increased intestinal absorption and diminished biliary secretion of plant sterols, inducing a significant increase in plasma concentrations of plant sterols (21). These findings suggest that

ABCG5 and ABCG8 work as sterol efflux pumps in the small intestine and liver. The discovery of ezetimibe greatly helped to reveal the role of the Niemann-Pick C1–like 1 (NPC1L1) protein in the intestinal uptake of cholesterol and plant sterols (22, 23). As shown in **Figure 1**, intestinal cholesterol absorption is a multistep process that is regulated by multiple genes in the enterocyte. This review highlights the recent progress in elucidating the genetic mechanisms of intestinal cholesterol absorption, the molecular biology of intestinal sterol transporters, and the pharmacological approaches by which plant sterols and ezetimibe inhibit the absorption process.

Niemann-Pick C1–like 1 (NPC1L1) protein: has 50% amino acid homology to NPC1, functions in intracellular cholesterol trafficking, and is defective in the cholesterol storage disorder, i.e., the Niemann-Pick type C disease

Sitosterolemia: a rare autosomal-recessive inherited disease induced by mutations of either the *ABCG5* or *ABCG8* gene and characterized by increased plasma plant sterol levels, xanthomas, and premature onset of atherosclerosis

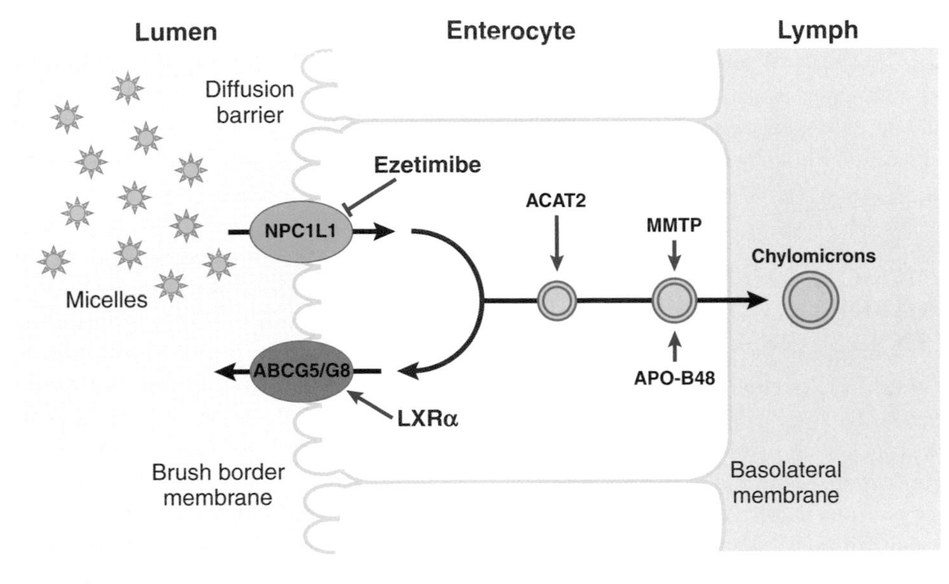

Figure 1

Within the intestinal lumen, the micellar solubilization of sterols allows them to move through the diffusion barrier overlying the surface of the absorptive cells. In the absence of bile acids, individual sterol molecules must diffuse across the diffusion barrier overlying the brush border of the enterocyte. Hence, uptake of the sterols is largely diffusion limited. In the presence of bile acids, large amounts of the sterol molecules are delivered to the aqueous-membrane interface so that the uptake rate is greatly increased. The principal mechanism whereby hydrophilic bile acids inhibit cholesterol absorption appears to be via diminution of intraluminal micellar cholesterol solubilization. Plant sterols and plant stanols have a higher affinity to mixed micelles than does cholesterol, and they thereby displace cholesterol from these micelles and reduce cholesterol absorption. The Niemann-Pick C1–like 1 (NPC1L1) protein, a newly identified sterol influx transporter, is located at the apical membrane of the enterocyte, which may actively facilitate the uptake of cholesterol by promoting the passage of sterols across the brush border membrane of the enterocyte. Most likely, ezetimibe reduces cholesterol absorption by inhibiting the activity of intestinal NPC1L1. In contrast, ABCG5 and ABCG8 promote active efflux of cholesterol and plant sterols from the enterocyte into the intestinal lumen for excretion. Liver X receptor α (LXRα) may be essential for the upregulation of the *ABCG5* and *ABCG8* genes in response to high dietary cholesterol. The combined regulatory effects of NPC1L1, ABCG5, and ABCG8 may play a critical role in modulating the amount of cholesterol that reaches the lymph from the intestinal lumen. In addition, several proteins involved in other steps in the absorption process, e.g., acyl-CoA:cholesterol acyltransferase isoform 2 (ACAT2), apolipoprotein B48 (APO-B48), and microsomal triglyceride transfer protein (MTTP), involve esterification of cholesterol and its incorporation into nascent chylomicrons that are subsequently secreted into the lymph. These intracellular events may also exert major influences on cholesterol absorption. Therefore, intestinal cholesterol absorption is a multistep process that is regulated by multiple genes. Modified from Reference 93, with permission.

DISTINGUISHING INTESTINAL CHOLESTEROL ABSORPTION AND INTESTINAL CHOLESTEROL UPTAKE

Conceptually, intestinal absorption of cholesterol has to be distinguished from its uptake by the enterocyte. Intestinal absorption of cholesterol is most accurately defined as the transfer of intraluminal cholesterol into intestinal or thoracic duct lymph. Intestinal uptake of cholesterol refers to its entry from the lumen into intestinal absorptive cells. As can be inferred from these definitions, intestinal cholesterol absorption is a multistep process that is regulated by multiple genes, and any factors that change the transportation of cholesterol from the intestinal lumen to the lymph can influence the efficiency of intestinal cholesterol absorption.

Micelles: small globular polymolecular aggregates of bile acids that form by apposition of their hydrophobic surfaces in an aqueous solution and that incorporate cholesterol and phospholipids, enhancing their solubility

PHYSICAL CHEMISTRY AND PHYSIOLOGY OF INTESTINAL CHOLESTEROL ABSORPTION

Cholesterol that enters the small intestinal lumen to be absorbed by the enterocytes derives mainly from three sources: diet, bile, and intestinal epithelial sloughing. The average intake of cholesterol in the Western diet is approximately 300–500 mg per day. Bile is estimated to contribute nearly 800–1200 mg of cholesterol per day to the intraluminal pool. A third source of intraluminal cholesterol comes from the turnover of intestinal mucosal epithelium, which provides roughly 300 mg of cholesterol per day. Although the entire length of the small intestine can absorb cholesterol from the lumen, the major sites of absorption are in the upper part of the small intestine, i.e., the duodenum and proximal jejunum. Thus, because intestinal sloughing occurs throughout the intestinal tract and cholesterol absorption seems to be confined to the very proximal small intestine, the intestinal sloughing pool may not contribute greatly to cholesterol absorption.

Cholesterol absorption begins in the stomach, where dietary ingredients are mixed with lingual and gastric enzymes, resulting in partial fat digestion by preduodenal lipases and emulsification by peristalsis. The stomach also regulates the delivery of gastric chyme to the duodenum, where it is mixed with bile and pancreatic juice. The major lipases and proteins secreted by the pancreas into the intestinal lumen in response to a meal include carboxyl ester lipase (CEL), pancreatic triglyceride lipase, and the Group 1B phospholipase A_2, as well as pancreatic lipase–related protein-1 and -2. Only unesterified cholesterol can be incorporated into bile acid micelles and transported to the brush border of enterocyte, so an extremely important step is de-esterification of intestinal cholesteryl esters. Additionally, because the pool of unesterified cholesterol (mainly biliary source) in the intestinal lumen is relatively much larger than the esterified dietary pool of cholesterol, inhibition or loss of some of the pancreatic lipolytic enzyme activities is not likely to be effective in reducing cholesterol absorption. These observations may partly explain why targeted disruption of the *CEL* gene has only a slight inhibitory effect on intestinal cholesterol absorption in mice (24, 25). Interestingly, the lack of triglyceride hydrolytic activity in the intestinal lumen significantly reduces dietary cholesterol absorption but does not influence triglyceride digestion or fat absorption in pancreatic triglyceride lipase knockout mice (26). The regulatory effects of the Group 1B phospholipase A_2 as well as of pancreatic lipase–related protein-1 and -2 on intestinal cholesterol absorption have not yet been defined.

In general, most of the cholesterol in food exists in the unesterified form—<15% of cholesterol is in the form of cholesteryl esters—and dietary cholesterol ingestion is often associated with fat consumption. Dietary cholesterol appearing in the intestinal lumen is usually mixed with triglycerides and phospholipids in the form of lipid emulsion. Digestion of the phospholipids and triglycerides in the surface and core, respectively, of the lipid emulsion particles is necessary to liberate the dietary cholesterol to phospholipid vesicles and bile acid micelles for its transport to the brush border of enterocyte for absorption.

Because some lipolytic products (e.g., cholesterol) are poorly soluble in a pure aqueous environment, they must depend on the solubilizing properties of bile acids. Luminal bile acids are derived from hepatic secretion and reabsorbed from the intestinal lumen back to the liver in a process termed the enterohepatic circulation of bile acids. The detergency of bile acids is obligatory for intestinal cholesterol uptake through micellar solubilization of the intraluminal sterol. Bile acids are a biological amphipathic detergent, and their monomers can aggregate spontaneously to form simple micelles when their concentrations exceed the critical micellar concentration. Simple micelles (3 nm in diameter) are small, thermodynamically stable aggregates that can solubilize a minimal amount of cholesterol. In contrast, phospholipids, monoacylglycerides, and free fatty acids are readily soluble.

Together with ionized and nonionized fatty acids, monoacylglycerides, and lysophospholipids, bile acids form mixed micelles, which significantly enhances the solubility of cholesterol. Mixed micelles (4–8 nm in diameter) are large, thermodynamically stable aggregates, and their sizes vary principally depending on the relative proportion of bile acids and phospholipids. They function as a concentrated reservoir and transport vehicle for cholesterol across the unstirred water layer toward the brush border of the small intestine to facilitate uptake of monomeric cholesterol by the enterocyte. Furthermore, excess lipids not dissolved in the micellar phase can be maintained as a stable emulsion by bile acids, phospholipids, monoacylglycerides, and fatty acids in the intestinal lumen.

During lipolysis, a liquid crystalline phase composed of multilamellar products of lipid digestion forms at the surface of the emulsion droplets. Vesicles are unilamellar spherical structures that contain phospholipids and cholesterol and few, if any, bile acids. Thus, vesicles (40–100 nm in diameter) are substantially larger than either simple or mixed micelles but much smaller than liquid crystals (500 nm in diameter) that are composed of multilamellar spherical structures. This liquid crystalline phase provides an accessible source of cholesterol and other lipids for continuous formation and modification of mixed micelles in the presence of bile acids. Within the intestinal lumen, the presence of hydrophilic bile acids may reduce solubility of cholesterol by inducing phase separation of the sterol from mixed micelles to a coexisting liquid crystalline vesicle phase (27). Most likely, hydrophilic bile acids facilitate incorporation of cholesterol molecules into a stable liquid crystalline/vesicle phase from which they are poorly absorbed by enterocytes. In contrast, hydrophobic bile acids markedly increase micellar cholesterol solubility (27, 28) and thereby augment cholesterol absorption. This suggests that the hydrophobic bile acids are more effective promoters of cholesterol absorption than are the hydrophilic bile acids.

Before cholesterol molecules in the small intestinal lumen can interact with the newly identified intestinal sterol transporter NPC1L1 for uptake and subsequent transport across the brush border of the enterocyte, they must pass through a diffusion barrier that is located at the intestinal lumen–membrane interface, which may alter the kinetics of cholesterol absorption. Moreover, diffusion through the unstirred water layer is a relatively slow process for cholesterol that is virtually insoluble in aqueous systems. Therefore, the unstirred water layer, a series of water lamellae at the interface between the bulk water phase of the lumen and the apical membrane of the enterocyte, is considered an important barrier through which a cholesterol molecule in the bulk phase must pass to be absorbed. Additionally, the intestinal mucous coat has an important role as a diffusion-limiting barrier, as cholesterol molecules may be extensively bound to surface mucins prior to being transferred into the enterocyte. Physiological levels of the epithelial mucin encoded by the *MUC1* gene are necessary for normal intestinal uptake and absorption of cholesterol in mice (29). Because cholesterol absorption

Intestinal diffusion barrier: an unstirred water layer and a surface mucous coat on the apical membrane of the enterocyte

Chylomicrons: the major transport vehicles of cholesterol and triglycerides from the intestine to the liver

MTTP: microsomal triglyceride transfer protein

HMG-CoA: 3-hydroxy-3-methylglutaryl-CoA

efficiency is reduced by 50% in MUC1-deficient mice, there may be alternative pathways for cholesterol absorption. Furthermore, uptake and absorption of cholesterol but not fatty acids are decreased in MUC1 knockout mice because the movement of big rigid molecules, such as cholesterol, across the cell membrane is different from that of smaller, less rigid, and space-occupying molecules, such as fatty acids. As the lipid-protein interaction and structural assembly of proteins may influence the kinetics of net cholesterol movement across the cell membrane of enterocyte, it is crucial to investigate how structural protein integrity or assembly at the cell membrane level is maintained during the intestinal absorption of cholesterol.

Another potential step for sorting/regulation is when the absorbed cholesterol molecules reach the endoplasmic reticulum, where an enzyme usually known as ACAT2 esterifies cholesterol (30). ACAT2 is highly specific for cholesterol and does not appreciably esterify plant sterols. Cholesteryl esters are then pumped by a key step in chylomicron biogenesis; microsomal triglyceride transfer protein (MTTP) transfers neutral lipids into nascent chylomicrons, allowing them to mature and exit the endoplasmic reticulum for eventual secretion as chylomicron particles into the lymph (31). During cholesterol absorption, there is little increase in the cholesterol content of the small intestinal wall, demonstrating that the absorbed cholesterol can be rapidly processed and exported from the enterocyte into the intestinal or thoracic duct lymph. Essentially all cholesterol molecules that move from the intestinal lumen into enterocytes are unesterified; however, cholesterol exported into intestinal lymph following a cholesterol-rich meal is approximately 70–80% esterified, suggesting that esterification is an important step for bulk entry into the nascent chylomicrons. Therefore, the enterocyte's cholesterol-esterifying activity may be an important regulator of intestinal cholesterol absorption because re-esterification of the absorbed cholesterol within the enterocyte enhances the diffusion gradient for intraluminal cholesterol entry into the cell. The inhibition of ACAT by pharmacological intervention significantly reduces transmucosal transport of cholesterol in rats (32, 33), and deletion of the *ACAT2* gene decreases intestinal cholesterol absorption in mice (34). Moreover, the inhibition of intestinal 3-hydroxy-3-methylglutaryl-CoA (HMG-CoA) reductase by pharmacological treatment with statins also diminishes intestinal cholesterol absorption in humans (37) and other animals (35, 36).

In addition, intestinal cholesterol absorption is significantly inhibited in apolipoprotein-B48 knockout mice (38) and in "apolipoprotein-B100-only" mice that synthesize exclusively apolipoprotein-B100 (39) because of a failure in the assembly and/or delivery of chylomicrons into the intestinal lymph. Intestinal MTTP transfers neutral lipids into newly formed chylomicrons in the endoplasmic reticulum (40), and *MTTP* mutations cause abetalipoproteinemia in humans, which is characterized by severe steatorrhea, neurological symptoms, fatty liver, and very low plasma cholesterol levels (41). Targeted disruption of the *CEL* gene induces a significant decrease in the number of chylomicron particles produced by the enterocyte after a lipid meal; most of the intestinal lipoproteins produced by CEL knockout mice are VLDL-sized particles (24). The exact mechanism by which CEL participates in chylomicron assembly is currently unknown, but indirect evidence suggests that CEL has an important effect on intracellular lipid trafficking. Although intestinal apolipoproteins AI/CIII/AIV may play a role in the regulation of cholesterol absorption (42), the regulatory effects of these proteins remain to be defined. Nevertheless, all these observations on chylomicron assembly suggest that the later steps in the cholesterol absorption process are also critically important.

Finally, cholesterol and bile acids that escape intestinal reabsorption are excreted as fecal neutral and acidic sterols. This represents the major route for sterol elimination from the body.

FACTORS INFLUENCING INTESTINAL CHOLESTEROL ABSORPTION EFFICIENCY

Any factor that changes the transportation of cholesterol from the intestinal lumen to the lymph can influence the efficiency of cholesterol absorption because intestinal cholesterol absorption is a multistep process. **Table 1** summarizes dietary, pharmacological, biliary, cellular, and luminal factors that may influence intestinal cholesterol absorption. When dietary conditions are controlled, biliary factors may exert a major influence on the efficiency of cholesterol absorption, any changes in which may partly explain interindividual and interstrain differences in cholesterol absorption efficiency in humans and other animals. For example, hepatic output and pool size of biliary bile acids are markedly reduced in mice with homozygous disruption of the cholesterol 7α-hydroxylase (*CYP7A1*) gene that encodes the key enzyme of the neutral pathway of bile acid synthesis (43). As a result, the mice absorb only trace amounts of cholesterol because of bile acid deficiency in bile. Similarly, upon deletion of the sterol 27-hydroxylase (*CYP27*) gene, which encodes the main enzyme of the alternative pathway of bile acid synthesis, the knockout mice display significantly reduced bile acid synthesis and pool size. Consequently, intestinal cholesterol absorption decreases from 54% to 4%, whereas fecal neutral sterol excretion increases 2.5-fold (44). However, in both knockout strains, cholesterol absorption is restored by feeding a diet containing cholic acid (43, 44).

These findings confirm that hepatic output and pool size of biliary bile acids play a critical role in intestinal cholesterol absorption by the regulation of intraluminal bile acid micellar concentrations. Hydrophilic and hydrophobic bile acid feeding studies in mice (27, 28) show that changes in the hydrophilic-hydrophobic balance of biliary bile acid pool influence cholesterol absorption. In an alloxan-induced mouse model of diabetes, percentages of cholesterol absorption are significantly increased. This is because the biosynthesis of hydrophilic tauro-β-muricholic acid in the liver is reduced and the biosynthesis of cholic acid is augmented so that the hydrophilic-hydrophobic index of bile acid pool is increased remarkably (45). These alterations in turn induce biliary cholesterol hypersecretion and cholesterol gallstone formation (45). Moreover, targeted deletion of the *ABCB4* gene encoding the canalicular phosphatidylcholine flippase of the hepatocyte abolishes biliary secretion of phospholipids, which significantly suppresses intestinal cholesterol absorption (46, 47). Studies of homozygous and heterozygous *ABCB4*-deficient mice suggest that physiological levels of biliary phospholipid outputs are necessary for normal intestinal cholesterol absorption (47). Disruption of the ileal bile acid transporter (IBAT) eliminates enterohepatic cycling of bile acids in mice (48). However, there is a mild modulation in cholesterol absorption efficiency under chow feeding conditions, which can be explained by the facts that cholesterol absorption occurs predominantly in the proximal intestine and that IBAT-mediated uptake of bile acids occurs primarily in the distal intestine. In addition, the bile acid pool size in the proximal intestine is not different between control and IBAT inhibitor–treated animals, primarily because of the compensatory increase in bile acid synthesis by the liver in the latter group of animals.

IBAT: ileal bile acid transporter

Several human and animal studies have found that changes in small intestinal transit rate influence the efficiency of cholesterol absorption. Mice with deletion of the cholecystokinin-1 receptor (*CCK-1R*) gene show a significant increase in intestinal cholesterol absorption, which correlates with a significantly slower small intestinal transit rate (49). This in turn induces biliary cholesterol hypersecretion and cholesterol

Table 1 Possible factors influencing intestinal cholesterol absorption[a,b]

Factors[c]	Effects on percent cholesterol absorption and type of study	Mouse Chr[d]	cM	Human ortholog	References
Dietary factors					
↑ Cholesterol	(–) Animal feeding studies				93
↑ Monounsaturated	↓ African green monkey feeding studies				140
↑ ω-3 polyunsaturated	↓ African green monkey feeding studies				140
↑ Fish oils	↓ Rat lymphatic transport studies				141
↑ Sphingomyelin	↓ Animal feeding studies				142
↑ Fiber	↓ Human and animal feeding studies				143
↑ Plant sterols (phytosterols)	↓ Human and animal feeding studies				123, 129
Pharmacological factors					
↑ Hydrophilic bile acids	↓ Human and animal feeding studies				27
↑ Hydrophobic bile acids	↑ Human and animal feeding studies				27, 28
↑ Ezetimibe	↓ Human and animal feeding studies				98, 131
↑ ACAT inhibitors	↓ Human and animal feeding studies				11, 32, 33
↑ Statins	↓ Human and animal feeding studies				35–37
↑ Bile acid sequestrants	↓ Human and animal feeding studies				5, 6
↑ Intestinal lipase inhibitors	↓ Human and animal feeding studies				9, 10
↑ Estrogen	↑ Animal feeding studies				55
Biliary factors					
↓ Biliary bile salt output	↓ Cholesterol 7α-hydroxylase (–/–) mice				43, 144
↓ Size of bile salt pool	↓ Cholesterol 7α-hydroxylase (–/–) mice				43, 144
↓ Biliary phospholipid output	↓ *Abcb4* (–/–) mice				46, 47
↑ Biliary cholesterol output	↑ Animal studies				52, 77
↑ Cholesterol content of bile	↑ Animal studies				45, 52, 77
↑ HI of bile salt pool	↑ Animal studies				27, 45
Cellular factors					
↓ ACAT2	↓ ACAT2 inhibitors and *Acat2* (–/–) mice	15	61.7	12q13.13	11, 32–34
↓ HMG-CoA reductase	↓ HMG-CoA reductase inhibitors in human and mouse studies	13	49.0	5q13.3-q14	35–37
↓ ABCA1	↓↑ *Abca1* (–/–) mice[e]	4	23.1	9q31.1	116–118
↓ ABCG5 and ABCG8	↑ *Abcg5/g8* (–/–) mice and *Abcg5/g8* transgenic mice; and sitosterolemia	17	54.5	2p21	19–21, 78–80, 89
↓ NPC1L1	↓ *Npc1l1* (–/–) mice and ezetimibe feeding studies	11	ND	7p13	22, 23, 98, 131
Aminopeptidase N	To be identified	7	ND	15q25-q26	
↓ SR-BI	(–) ↓ *Sr-b1* (–/–) mice and *Sr-b1* transgenic mice[e]	5	68.0	12q24.31	113–115
↓ IBAT	(–) *Ibat* (–/–) mice	8	20	13q33	48
↓ Caveolin 1	(–) Caveolin 1 (–/–) mice	6	A2[f]	7q31.1	145
Caveolin 2	To be identified	6	A2	7q31.1	
MTTP	To be identified	3	66.2	4q24	
SCP2	To be identified	4	52.0	1p32	
OSBP	To be identified	19	7.0	11q12-q13	

(*Continued*)

Table 1 (*Continued*)

Factors[c]	Effects on percent cholesterol absorption and type of study	Mouse Chr[d]	cM	Human ortholog	References
↓ APO-B48	↓ *ApoB48* (−/−) mice and "Apo-B100-only" mice	12	2.0	2p24-p23	38, 39
APO-AI	To be identified	9	27.0	11q23-q24	
↓ APO-AIV	(–) *Apo-AIV* (−/−) mice and *Apo-AIV* transgenic mice	9	27.0	11q23	146, 147
APO-CIII	To be identified	9	27.0	11q23.1-q23.2	
↑ Estrogen receptor α	↑ Animal studies	10	12.0	6q25.1	55
Estrogen receptor β	To be identified	12	33.0	14q23.2	
NR1H4 (FXR)	To be identified	10	50.0	12q23.1	
NR1H3 (LXRα)	To be identified	2	40.4	11p11.2	
NR1H2 (LXRβ)	To be identified	7	ND	19q13.3	
↑ NR2B1 (RXRα)	↓ RXR agonist and mouse study	2	17.0	9q34.3	116
↑ NR1C1 (PPARα)	↓ PPARα agonist and PPARα (−/−) mice	15	48.8	22q13.31	148
↑ NR1C2 (PPARδ)	↓ PPARδ agonist and mouse study	17	13.5	6p21.2-p21.1	149
NR1C3 (PPARγ)	To be identified	6	52.7	3p25	
Luminal factors					
↑ Small intestinal transit time	↑ Cck-1 receptor (−/−) mice				49
↑ Gastric emptying time	↑ Inbred strains of mice				150
↓ MUC1 mucin	↓ *Muc1* (−/−) mice	3	44.8	1q21	29
MUC2 mucin	To be identified	7	69.0	11p15.5	
MUC3 mucin	To be identified	5	75.0	7q22	
MUC4 mucin	To be identified	16	ND	3q29	
MUC5ac mucin	To be identified	7	69.0	11p15.5	
MUC5b mucin	To be identified	7	69.0	11p15.5	
MUC6 mucin	To be identified	7	69.0	11p15.5	
↓ Carboxyl ester lipase	(–) ↓ Carboxyl ester lipase (−/−) mice	2	16.0	9q34.3	24, 25
↓ Pancreatic triglyceride lipase	↓ Pancreatic triglyceride lipase (−/−) mice	19	29.0	10q26.1	26
Sphingomyelinase	To be identified	4	ND	8q12-q13	

[a] Table is modified from Reference 52, with permission.

[b] Abbreviations: ABC, ATP-binding cassette (transporter); ACAT2, acyl-CoA:cholesterol acyltransferase, isoform 2; APO, apolipoprotein; CCK, cholecystokinin; Chr, Chromosome; cM, centimorgan; FXR, farnesoid X receptor; HI, hydrophobicity index; HMG, 3-hydroxy-3-methylglutaryl; IBAT, ileal bile acid transporter; LXR, liver X receptor; MTTP, microsomal triglyceride transfer protein; MUC, mucin gene; ND, not determined; NPC1L1, Niemann-Pick C1 like 1 (protein); NR, nuclear receptor; OSBP, oxysterol-binding protein; p, short arm of the Chr; q, long arm of the Chr; RXR, retinoid X receptor; PPAR, peroxisomal proliferator activated receptor; SCP2, sterol carrier protein 2; SR-BI, scavenger receptor class B type 1.

[c] ↑ represents increase, ↓ decrease, and (–) no effect.

[d] Map position is based on conserved homology between mouse and human genomes and assigned indirectly from localization in other species. Information on homologous regions was retrieved from the mouse/human homology databases maintained at the Jackson Laboratory (**http://www.informatics.jax.org/searches/marker_form.shtml**) and the National Center for Biotechnology Information (**http://www.ncbi.nlm.nih.gov/HomoloGene**).

[e] Contradictory results were reported by different groups (see text for details).

[f] As inferred from conserved map locations in mouse and human genomes, the mouse gene may be localized on proximal Chr 2 band A2.

gallstone formation (49). In contrast, guinea pigs resistant to systemic effects of dietary cholesterol display shorter small intestinal transit times than do hypercholesterolemic guinea pigs (50). Furthermore, acceleration of small intestine transit induced by pharmacological intervention is consistently associated with decreased cholesterol absorption in humans (51). However, it is surprising that there are essentially similar small intestinal transit times, lengths, and weights among low-, middle-, and high-cholesterol-absorbing inbred mouse strains (52). These findings suggest that under normal physiological conditions, luminal factors apparently do not account for the major differences in the efficiency of intestinal cholesterol absorption in diverse healthy inbred strains of mice.

In addition, there are well-known gender differences in the efficiency of cholesterol absorption in humans and other animals (53–56). Estrogen significantly increases hepatic output of biliary lipids and bile acid–dependent bile flow rate (57–59). These biliary factors markedly promote cholesterol absorption in animals and humans, especially in those exposed to high levels of estrogen. In addition, estrogen likely regulates expression of the sterol transporter genes in the intestine via the estrogen receptor pathway (55). Furthermore, the efficiency of intestinal cholesterol absorption increases markedly with age (53–55), suggesting that cholesterol absorption is modified by aging. Aging significantly increases secretion rate of biliary lipids and cholesterol content of bile as well as the size and hydrophobicity index of the bile acid pool (53, 60–62). These biliary factors together may have a major effect on increased efficiency of cholesterol absorption with age (53–55). It is imperative to explore whether aging per se enhances intestinal cholesterol absorption by the mechanism whereby *Longevity* (aging) genes may influence expression of the intestinal sterol transporter genes.

GENETIC ANALYSIS OF INTESTINAL CHOLESTEROL ABSORPTION

Epidemiological investigations and animal studies show that there are significant interindividual differences and interstrain variations in intestinal cholesterol absorption efficiency in primates (63, 64), including humans (2, 65–68), as well as in inbred strains of mice (52, 69–72), rats (73), and rabbits (74, 75). These observations strongly suggest that intestinal cholesterol absorption is regulated by multiple genes, as diet, the key environmental factor, is controlled in these studies. The question arises, however, as to which cellular step(s) in the intestinal absorption of cholesterol might be inherently different. As examined by a fecal dual-isotope ratio method in probands with very high and low plasma cholestanol levels (a plant sterol that correlates with cholesterol absorption) and their siblings (76), siblings of the higher-absorbing probands display significantly higher cholesterol absorption efficiency ($49 \pm 2\%$) than do siblings of the lower-absorbing probands ($37 \pm 3\%$). Furthermore, there are significant differences in cholesterol absorption efficiency measured by plasma, fecal, and lymphatic methods among 12 inbred strains of mice: <25% in AKR/J, C3H/J, and A/J strains; 25–30% in SJL/J, DBA/2J, BALB/cJ, SWR/J, and SM/J strains; and 31–40% in C57L/J, C57BL/6J, FVB/J, and 129/SvJ strains (52). In particular, when dietary factors are controlled by feeding a normal rodent chow diet containing trace (<0.02%) amounts of cholesterol, the efficiency of cholesterol absorption in the C57L/J strain with intact enterohepatic circulation of bile acids is significantly higher than in the AKR/J strain, as measured by four independent methods, i.e., the plasma and the fecal dual-isotope ratio methods, the lymphatic transport of cholesterol, and the mass balance method (52, 77). When these studies were repeated in mice with chronic biliary fistulae but in the setting of duodenal infusion of taurocholate and egg yolk lecithin, the marked differences in the efficiency of intestinal cholesterol absorption still persist between these two mouse strains (52, 77). Furthermore, cholesterol absorption in (AKR/J × C57L/J)F_1 progeny mimics that in the higher-absorbing parental strain C57L/J, suggesting that high cholesterol absorption is a dominant trait in mice (52). These systematic studies suggest that the genetic factors at the enterocyte level are crucial in determining the variations of intestinal cholesterol absorption efficiency (52).

With quantitative trait locus (QTL) mapping techniques, genetic loci that determine cholesterol absorption efficiency can be identified by genome-wide linkage studies in experimental crosses of inbred mouse strains. In backcrossing progeny of (AKR/J × DBA/2J)F_1 to the lower-absorbing DBA/2J parental strain (72), a QTL that influences cholesterol absorption efficiency is detected on chromosome 2 at 64 cM with a significant LOD score of 3.5 (designated cholesterol absorption gene locus 1, *Chab1*), and a second suggestive QTL (*Chab2*) is on chromosome 10 at 24 cM with a LOD score of 1.9. Three additional loci are identified in 21 different DBA/2J × AKR/J recombinant inbred strains with LOD scores between 1.6 and 2.0: *Chab3* on chromosome 6 at 51 cM, *Chab4* on chromosome 15 at 58 cM, and *Chab5* on chromosome 19 at 16 cM. In contrast, when (129P3/J × SJL/J)F_1 progeny are backcrossed to the higher-absorbing 129P3/J parental strain, QTL analysis reveals a locus named *Chab6* on chromosome 1 at 57 cM with a LOD score of 2.1 and a second locus named *Chab7* on chromosome 5 at 57 cM with a LOD score of 3.3. However, the individual genes underlying each of these cholesterol absorption QTLs remain to be identified. **Figure 2** shows physiological relevant genes as well as quantitative trait loci (QTL) and their candidate genes on chromosomes of the mouse genome, which may be involved in the regulation of intestinal cholesterol absorption.

IDENTIFICATION OF INTESTINAL STEROL TRANSPORTERS AND THEIR MOLECULAR BIOLOGY

Although accumulated evidence has clearly established the importance of cholesterol transfer to bile acid micelles before the transport of cholesterol to the brush border membrane of enterocytes for absorption, the mechanism by which the cholesterol molecules in micelles are taken up across the brush border membrane independently of bile acid uptake is still under extensive investigation. The long-standing hypothesis suggests that cholesterol absorption is an energy-independent, simple passive diffusion process in which micellar cholesterol is in equilibrium with monomolecular cholesterol in solution and the monomeric cholesterol is absorbed to the brush border membrane down a concentration gradient. The intestinal sterol uptake and transport process has been assumed to be controlled mainly by two enzymes: ACAT2, which enhances intracellular cholesterol esterification, and MTTP, which is responsible for intestinal chylomicron assembly.

However, much new evidence supports the notion that a transporter-facilitated mechanism is involved in cholesterol uptake by the enterocyte. First, interindividual differences and interstrain variations in the efficiency of intestinal cholesterol absorption occur in humans and animals, suggesting that intestinal cholesterol absorption is regulated by multiple genes. Second, patients with sitosterolemia mainly display excess plant sterol absorption, indicating that they have lost the ability to discriminate between plant sterols and cholesterol (21, 78–80). Third, structurally related plant sterols such as sitosterol and campesterol, which differ from cholesterol only in the degree of saturation of the sterol nucleus or in the nature of the side chain at C24, are less efficiently absorbed than cholesterol (81, 82). Fourth, intestinal cholesterol absorption can be specifically inhibited by cholesterol absorption inhibitors such as

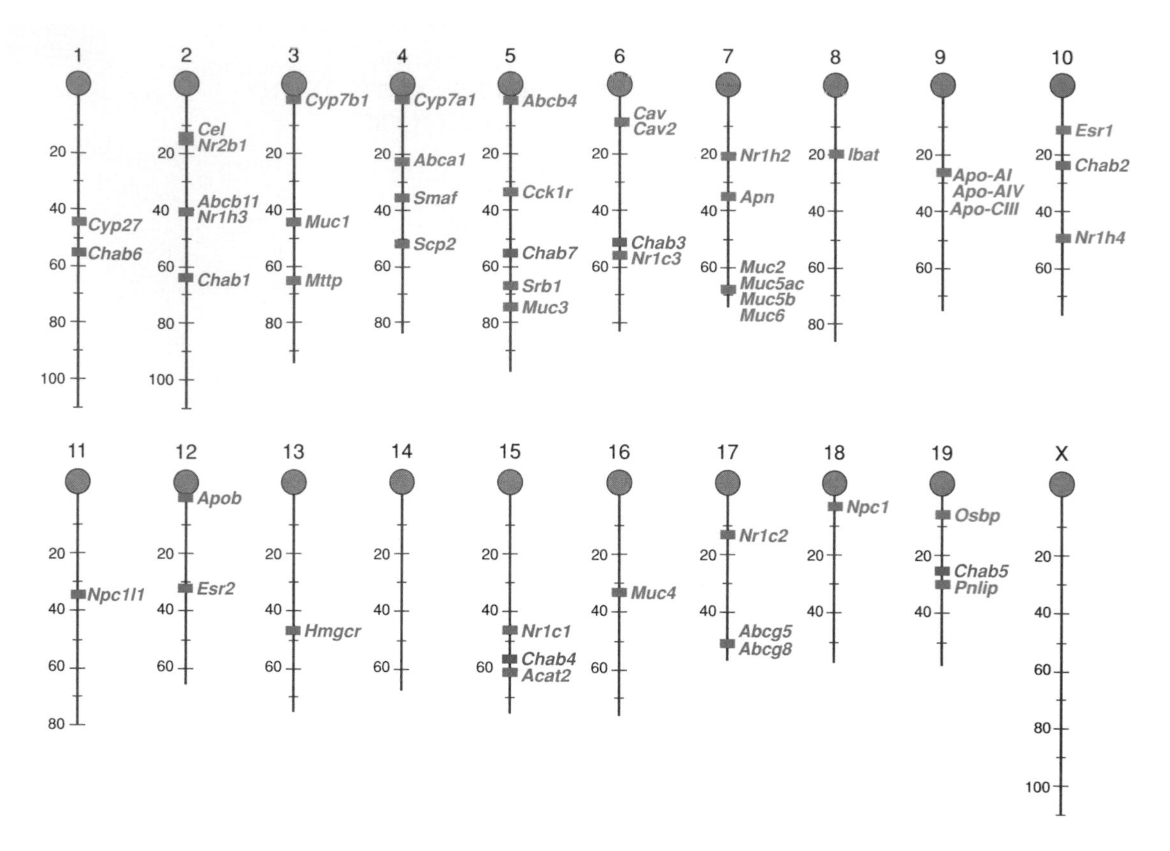

Figure 2

Composite map of the genes for intestinal sterol transporters and lipid metabolism, and of quantitative trait loci (QTLs) for cholesterol absorption (*Chab*) genes, as well as of candidate genes for the regulation of cholesterol absorption on chromosomes representing the entire mouse genome. A vertical line represents each chromosome, with the centromere at the top; genetic distances from the centromere (*horizontal lines*) are indicated to the left of the chromosomes in centimorgans (cM). Chromosomes are drawn to scale, based on the estimated cM position of the most distally mapped locus taken from Mouse Genome Database. The locations of the genes for intestinal sterol transporters and lipid metabolism as well as of candidate genes for the regulation of cholesterol absorption are represented by horizontal blue lines. QTLs (*Chab* genes) are indicated by horizontal red lines with the gene symbols to the right. Abbreviations of genes, followed by the protein encoded: *Acat*, acyl-CoA:cholesterol transferase; *Apn*, aminopeptidase N; *Apo*, apolipoprotein; *Cav*, Caveolin; *Cck1r*, cholecystokinin 1 receptor; *Cel*, carboxyl ester lipase; *Cyp7a1*, cholesterol 7α-hydroxylase; *Cyp7b1*, oxysterol 7α-hydroxylase; *Cyp27*, sterol 27α-hydroxylase; *Esr1*, estrogen receptor α; *Esr2*, estrogen receptor β; *Hmgcr*, HMG-CoA reductase; *Ibat*, ileal bile acid transporter; *Muc*, mucin; *Npc1*, Niemann-Pick C1 (protein); *Nr*, nuclear receptor; *Osbp*, oxysterol-binding protein; *Pnlip*, pancreatic triglyceride lipase; *Scp*, sterol carrier protein; *Smaf*, sphingomyelinase; *Srb*, scavenger receptor class B.

2-azetidinones (83). Finally, transport of cholesterol from mixed micelles or small unilamellar vesicles to brush border membrane vesicles follows second-order kinetics and is sensitive to proteases (84, 85).

Major progress in the search for the intestinal sterol transporters came from the discovery that mutations in the genes encoding human ABCG5 and ABCG8 transporters cause sitosterolemia. Patel et al. (86) first mapped sitosterolemia to a special gene locus on the human chromosome 2p21, between *D2S2294* and *D2S2298*. Using a microarray analysis of cDNAs from intestines and livers of mice treated with a liver X receptor (LXR) agonist, as well as a positional cloning approach, two groups of scientists (19, 20) independently identified the two adjacent genes *ABCG5* and *ABCG8* encoding transporters expressed in the liver and intestine. Unlike other ABC transporter genes that encode proteins with 12 transmembrane domains, *ABCG5* and *ABCG8* each encode a protein with 6 transmembrane domains, and the heterodimerization of the resulting two proteins to form a 12-transmembrane protein complex is required for transport activity (87).

Patients with sitosterolemia absorb 20–30% of dietary sitosterol intake, in contrast to the typical <5% sitosterol absorption rate in normal individuals (21, 78–80). Interestingly, sitosterolemic patients also absorb a greater fraction of dietary cholesterol and excrete less cholesterol into the bile as compared with normal subjects, resulting in hypercholesterolemia. Indeed, the results from the studies of transgenic and knockout mice as well as in vitro systems (88–90) show that ABCG5 and ABCG8 are localized in the apical brush border membrane of enterocytes and the canalicular membrane of hepatocytes. These transporters may represent an efficient efflux pump system for both cholesterol and plant sterols in the small intestine and liver and may transport sterols out of the cell either back into the intestinal lumen or into the bile, thereby regulating the intestinal absorption and biliary secretion of cholesterol and plant sterols (88–91). Also, several polymorphisms in the *ABCG5* and *ABCG8* genes that may moderately influence plasma sterol levels have been identified (92). These findings explain, in part, why cholesterol absorption is a selective process, in which plant sterols and other noncholesterol sterols are absorbed poorly or not at all. The structure-function relationship between (*a*) cholesterol and plant sterols such as sitosterol, campesterol, and stigmasterol and (*b*) shellfish sterols should be further studied. More recently, researchers found that in inbred strains of mice, there is a negative correlation between the efficiency of cholesterol absorption and the expression levels of ABCG5 and ABCG8 in the jejunum and ileum but not in the duodenum (93). This suggests that under normal physiological conditions, the jejunal and ileal ABCG5 and ABCG8 play a major regulatory role in modulating the amount of cholesterol that is absorbed from the intestine (93).

LXR: liver X receptor

As found in LXRα und LXRβ knockout mice, LXRs are essential for diet-induced upregulation of the *ABCG5* and *ABCG8* genes (94). Also, studies in ABCG5 and ABCG8 double-knockout mice show that stimulation of cholesterol excretion by the nonsteroidal synthetic LXR agonist T0901317 requires intact ABCG5 and ABCG8 transporters (94). These studies suggest that mRNA expression for the *ABCG5* and *ABCG8* genes is activated by dietary cholesterol via LXRs. The observation that incubation with mixed micelles enriched with sitostanol, the Δ^5-saturated homolog of sitosterol, led to an increase in ABCA1 expression, which is also LXR mediated, in Caco-2 cells raises the question of whether plant sterols themselves or plant sterol derivatives have the capability to activate LXRs. LXRs are activated by endogenous oxysterols such as 22(*R*)-hydroxycholesterol, 24(*S*),25-epoxycholesterol, and 24*S*- and 27-hydroxycholesterol. Whereas plant sterols themselves do not exert any relevant LXR activation, several plant sterol derivatives show similar LXR activation capacities as does 24(*S*),25-epoxycholesterol (95). Despite these observations, it is imperative to investigate whether plant sterols, under conditions of high dietary plant sterol feeding, upregulate expression levels of the *ABCG5* and *ABCG8* genes via the LXR pathway. An upregulation of these transporters may promote transport of plant sterols from enterocytes back into the intestinal lumen and function as a gatekeeper to avoid a rapid increase in plasma levels of plant sterols.

SR-BI: scavenger receptor class B type I

The discovery of ezetimibe as the specific and potent inhibitor of intestinal cholesterol absorption has focused attention on a putative sterol influx transporter that may be a target for ezetimibe. The inhibition of cholesterol absorption by ezetimibe is not mediated via changes in either the size or composition of the intestinal bile acid pool or in the mRNA expression levels of *ABCG5*, *ABCG8*, *ABCA1*, and scavenger receptor class B type I (*SR-BI*). Rather, the mechanism of inhibition may involve disruption of the uptake of luminal sterol across the brush border membrane (96). Furthermore, ezetimibe neither inhibits pancreatic lipolytic enzyme activities in the intestinal lumen nor affects bile acid micelle solubilization of cholesterol (97). [^{3}H]-labeled ezetimibe is localized to the brush border membrane of the enterocyte (98) and appears to inhibit directly the activity of the putative brush border membrane transporter(s) that actively mediates the uptake of cholesterol. Using a genomic-bioinformatics approach, Altmann and coworkers (22, 23) identified transcripts containing expression patterns and structural characteristics anticipated in cholesterol transporters (e.g., sterol-sensing and transmembrane domains, extracellular signal peptides), and these researchers established NPC1L1 as a strong candidate for the ezetimibe-sensitive cholesterol transporter. Moreover, ezetimibe-treated mice and mice with targeted inactivation of the *NPC1L1* gene display similar dietary cholesterol absorption characteristics (22), suggesting that NPC1L1 is the ezetimibe-inhibitable cholesterol transporter. The NPC1L1 protein has 50% amino acid homology to NPC1, which is defective in the cholesterol storage disease Niemann-Pick type C and functions in intracellular cholesterol trafficking (99). However, in contrast to *NPC1*, which is expressed in many tissues, *NPC1L1* is expressed predominantly in the intestine, with peak expression in the proximal jejunum (22, 23), which parallels the efficiency of cholesterol absorption along the gastrocolic axis. Subfractionation of the brush border membrane suggests that NPC1L1 is associated with the apical membrane fraction of enterocytes. A rat homolog of the human *NPC1L1* gene has been found to be the only gene that encodes a protein that contains an extracellular signal peptide, transmembrane sequences, N-linked glycosylation sites, and a sterol-sensing domain (100). More recently, investigators discovered that the binding affinities of ezetimibe and several key analogs to recombinant NPC1L1 are virtually identical to those observed for native enterocyte membranes and that ezetimibe no longer binds to membranes from NPC1L1 knockout mice, suggesting that NPC1L1 is the direct molecular target of ezetimibe (101, 102).

However, preliminary attempts to reconstitute cholesterol transport activity in nonenterocyte cells by overexpression of NPC1L1 have been unsuccessful; thus, additional proteins may be required to reconstitute a fully functional cholesterol transporter. In particular, these may include caveolin 1, which can form a heterocomplex with annexin 2 (and cyclophilins) in zebrafish and mouse intestines (103). A stable 55-kDa complex of annexin 2 and caveolin 1 seems involved in intracellular sterol trafficking. The resistance of the intestinal caveolin 1–annexin 2 heterocomplex to boiling in sodium dodecyl sulfate, reducing conditions, and ether extraction implies that a covalent interaction other than disulfide cross-linking is involved in formation of the complex (103). Moreover, incubation with ezetimibe leads to complete disruption of the caveolin 1–annexin 2 complex in the early state of zebrafish embryos. Pharmacological treatment of mice with ezetimibe disrupts the complex only in hypercholesterolemic mice, as induced by a high-cholesterol and high-fat diet or by LDL-receptor gene knockout (103), indicating that the caveolin 1 heterocomplexes may represent additional ezetimibe targets that regulate intestinal cholesterol transport. The immunoprecipitation experiments suggest that ezetimibe disrupts the caveolin 1–annexin 2 complex most likely by direct interaction with the caveolin 1 protein. Nevertheless, the exact molecular mechanism by which NPC1L1 regulates cholesterol absorption remains to be defined.

Over the past decade, several groups have been searching for cholesterol transporters that are located at the apical brush border membrane of the enterocyte (104–107). Kramer and colleagues (108) investigated potential target structures of ezetimibe with the help of photoreactive derivatives of 2-azetidinones and a photoreactive cholesterol derivative photocholesterol, i.e., [3α-^{3}H]-6-azi-5α-cholestan-3β-ol. They identified an 80-kDa and a 145-kDa integral membrane protein as putative components of the intestinal cholesterol transporters (108). They found that the photoreactive analog of ezetimibe binds to another 145-kDa integral membrane protein. The 80-kDa cholesterol-binding protein does not interact with cholesterol absorption inhibitors and vice versa, and neither cholesterol nor plant sterols interfere with the 145-kDa molecular target for cholesterol absorption inhibitors (108). Interestingly, there is no competition in the binding of the 2-azetidinone at the 145-kDa protein with cholesterol. However, binding of photocholesterol to the 80-kDa cholesterol-binding protein may be inhibited in a concentration-dependent manner by cholesterol, sitosterol, and campesterol, whereas cholesterol absorption inhibitors, such as ezetimibe, have no influence on the binding of photocholesterol to the putative 80-kDa cholesterol-binding protein. Both proteins are different from the above-described candidate proteins for the intestinal cholesterol transporters (ABCA1, ABCG5, ABCG8, NPC1L1, and SR-BI). More recently, the 145-kDa ezetimibe-binding protein was purified by three different methods, and the protein sequencing reveals its identity with the membrane-bound ectoenzyme aminopeptidase N (APN) (109, 110). Because APN has a role in endocytotic processes, binding of ezetimibe to APN may block endocytosis of cholesterol-rich membrane microdomains, thereby inhibiting intestinal cholesterol absorption (109, 110). However, researchers have yet to pinpoint the exact mechanism whereby APN influences intestinal cholesterol absorption.

APN: aminopeptidase N

Immunoblotting data show that the SR-BI protein is expressed in brush border membrane preparations and in Caco-2 cells, and preincubation with an anti-SR-BI antibody partially inhibits cholesterol and cholesteryl ester uptake by brush border membrane vesicles and Caco-2 cells, in contrast to results in control incubations without antibody (111). These in vitro experiments suggest that SR-BI is a cholesterol transporter in the intestine and involved in the absorption of dietary cholesterol. The distribution of SR-BI along the gastrocolic axis and on the apical membrane of the enterocyte is also consistent with its participation in cholesterol absorption (112). However, targeted disruption of the *SR-BI* gene appears to have little effect on intestinal cholesterol absorption in mice (113–115). More importantly, the cholesterol absorption inhibitor ezetimibe, which labels SR-BI in the enterocyte, also inhibits cholesterol absorption in *SR-BI* knockout mice (114), suggesting that *SR-BI* cannot be the ezetimibe-sensitive target gene responsible for intestinal cholesterol absorption.

Some nuclear hormone receptor RXR and LXR agonists upregulate ABCA1 expression levels, with a concomitant decrease in cholesterol absorption (116), suggesting that the intestinal ABCA1 transporter serves to efflux cholesterol from the enterocyte back into the intestinal lumen for excretion. However, there are controversial results from the ABCA1 knockout mouse studies: One study shows only a marginal increase in cholesterol absorption (117), whereas another study shows decreased cholesterol absorption (118). Subsequent characterization of mice expressing no ABCA1 (i.e., *ABCA1*-deficient mice) (119) and of the Wisconsin hypoalpha mutant (WHAM) chicken with spontaneously occurring ABCA1 dysfunction (120, 121) revealed no impairment in percent cholesterol absorption, fecal neutral steroid excretion, or biliary cholesterol secretion, even after challenge with a synthetic LXR agonist. Furthermore, in situ hybridization techniques revealed that ABCA1 is predominantly in cells present in the lamina propria in mice (94) and occasionally in the enterocyte in the primate (122), which is inconsistent with the view that it plays a major role in cholesterol absorption. ABCA1 may be involved in the transfer of cholesterol from enterocytes into lymph and/or to blood macrophages and promote efficient cholesterol efflux from enterocytes to plasma HDL, but additional studies are required to clarify these observations.

Besides ABCG5, ABCG8, and NPC1L1, other sterol transporters in the intestine may be involved in the regulation of intestinal cholesterol absorption, although these transporters have only been proposed and not identified. It will be important to investigate the molecular and genetic mechanisms underlying the dominant rate-limiting step/factor in intestinal cholesterol absorption.

INHIBITORS OF INTESTINAL CHOLESTEROL ABSORPTION

The use of cholesterol absorption inhibitors for treating hypercholesterolemia has a long history, and several classes of compounds have been established and developed. Here we focus discussion on plant sterols and plant stanols (phytosterols) and ezetimibe, as well as their inhibitory actions on intestinal cholesterol absorption, because they markedly lower plasma total and LDL cholesterol levels in humans.

Plant Sterols and Stanols (Phytosterols)

Plant sterols are naturally occurring sterols and structurally related to cholesterol. Their chemical structure is very similar to that of cholesterol, with a Δ^5 double bond and a 3β-hydroxyl group but with structural modifications of the side chain. Plant sterols have the same basic importance in plants as cholesterol in animals; i.e., they play a crucial role in cell membrane function. Campesterol and sitosterol, the 24-methyl and the 24-ethyl analog of cholesterol, respectively, are the most abundant plant sterols, and they are found at low concentrations in human plasma. They are part of the diet and are exclusively taken up from the intestine. Over the past decade, plant sterols as ingredients in functional foods have been found to reduce plasma cholesterol levels (7, 8, 123). The effective doses are 1.5–3 g per day, which leads to a 8–16% reduction in plasma LDL cholesterol concentrations. The dietary intake of cholesterol and plant sterols is almost equal, but plant sterols are absorbed poorly or not at all. For example, the absorption efficiencies of sitosterol and campesterol are 5–8% and 9–18%, respectively (124). Most of the plant sterols that do enter the enterocyte most likely are pumped back rapidly into the intestinal lumen for excretion. In addition to poor net absorption, researchers have observed rapid biliary excretion of plant sterols. These mechanisms keep plasma plant sterol levels to <1 mg dl^{-1} in humans. Furthermore, plant sterols are poorly soluble in aqueous systems and require formulation for bioactivity. To increase their lipid solubility, it is imperative to esterify plant sterols and to dissolve them at high concentrations in the triglyceride phase of margarines (125). Cholesterol absorption from dietary and biliary sources is significantly reduced in the presence of plant sterols, and consequently the unabsorbed cholesterol excreted in the feces is increased markedly. The commonly accepted basic mechanisms of inhibitory action of these compounds are that plant sterols can become efficiently incorporated into micelles in the intestinal lumen, displace the cholesterol, and lead to its precipitation with other, nonsolubilized plant sterols (126–129). Moreover, competition between cholesterol and plant sterols for incorporation into micelles and for transfer into the brush border membrane, as well as competition within the enterocyte for ACAT, may partly explain the inhibitory effect of large amounts of plant sterols on cholesterol absorption. This process reduces both hepatic cholesterol and triglyceride contents, which is compensated for by two different mechanisms: an increase in cholesterol synthesis and an increase in LDL receptor levels. In contrast, Δ^{22}-sterols (stigmasterol) markedly reduce cholesterol synthesis via competitive inhibition of sterol Δ^{24}-reductase, which is an interesting secondary mechanism for future research (130).

Ezetimibe

Ezetimibe (SCH 58235), 1-(4-fluorophenyl)-(3*R*)-[3-(4-fluorophenyl)-(3*S*)-hydroxypro pyl]-(4*S*)-(4-hydroxyphenyl)-2-azetidinone, and an analog, SCH 48461, (3*R*)-(3-phenyl propyl)-1,(4*S*)-bis(4-methoxyphenyl)-2-azet idinone, are highly selective intestinal cholesterol absorption inhibitors that effectively and potently prevent the absorption of cholesterol by inhibiting the uptake of dietary and biliary cholesterol across the brush border membrane of the enterocyte. The high potency of these compounds is reflected by a 50% inhibition dose of 0.0005 mg kg^{-1} and 0.05 mg kg^{-1} in a series of different animal models (98, 131). Following oral administration, ezetimibe undergoes rapid glucuronidation in the enterocyte during its first pass (131–133). Both ezetimibe and its glucuronide are circulated enterohepatically and repeatedly delivered back to the site of action in the intestine, resulting in multiple peaks of the drug and accounting for an elimination half-life of approximately 22 hours (134). This may explain why ezetimibe displays a longer duration of action and the effort of treatment persists for several days after its cessation; thus, once-daily dosing should be sufficient for an adequate therapeutic effect. Furthermore, after oral administration of the glucuronide (SCH-60663), >95% of the compound remains in the intestine (131). That the glucuronide is more potent than ezetimibe in inhibiting cholesterol absorption confirms that ezetimibe acts directly in the intestine, as does glucuronide (131). Because ezetimibe and its analogs are relatively small molecular structures, they do not change the physical-chemical nature of the intraluminal environment, nor do they affect the enterohepatic flux of bile acids. Additionally, ezetimibe does not affect absorption of triglycerides, fatty acids, bile acids, or fat-soluble vitamins, including vitamins A, D, and E and α- and β-carotenes.

In addition, during ezetimibe treatment, there is a marked compensatory increase in cholesterol synthesis in the liver, but not in the peripheral organs, and an accelerated loss of cholesterol in the feces, with little or no change in the rate of conversion of cholesterol to bile acids. Thus, the combination of ezetimibe with HMG-CoA reductase inhibitors such as atorvastatin or simvastatin is a powerful new therapeutic approach with which to control plasma LDL cholesterol levels in the general population as well as to provide a complementary treatment strategy for high-risk patients, e.g., patients with primary hypercholesterolemia (15–18, 135, 136), homozygous familial hypercholesterolemia (137), or sitosterolemia (138).

Among the most frequently reported adverse events of combination therapy of ezetimibe and statins are elevated plasma aminotransferase and γ-glutamyltransferase activities (131). The Federal Drug Agency in Germany has recently released a cautious note on ezetimibe with respect to the number of myopathies and increases of liver enzymes reported (at least 15 patients, most of whom were on combination therapy with statins) (139). Because systemic concentrations of ezetimibe are maintained continually during treatment, the side effects of ezetimibe need further investigation in patients undergoing long-term therapy before this regimen is widely recommended. In addition, on the basis of its mechanism of action, ezetimibe may provide additional reduction in plasma total and LDL cholesterol concentrations when used in combination with fibrate, niacin, or bile acid–binding resins, but clinical trial data are not yet available. Because intestinal cholesterol absorption shows a wide variation in the general population, it may be reasonable to distinguish prospective responders from nonresponders to therapy with ezetimibe so as to optimize cholesterol-lowering therapy in the future.

CONCLUSIONS

A strong genetic factor in the regulation of intestinal cholesterol absorption has been established, indicating that intestinal cholesterol absorption is a complex and a multistep process that is regulated by multiple genes at the enterocyte level. Also, this process provides multiple therapeutic targets for preventing the absorption of cholesterol from dietary and biliary sources by suppressing uptake and transport of cholesterol through the enterocytes. A better understanding of the molecular genetics of intestinal cholesterol absorption will lead to novel approaches for the prevention and treatment of hypercholesterolemia.

SUMMARY POINTS

1. There are interindividual differences and interstrain variations in intestinal cholesterol absorption efficiency in humans and other animals. Genetic factors at the enterocyte level are crucial in determining the variations of intestinal cholesterol absorption efficiency, and high cholesterol absorption is a dominant trait.
2. The ABCG5 and ABCG8 transporters represent apical sterol export pumps that promote active efflux of cholesterol and plant sterols from enterocytes back into the intestinal lumen for excretion. This explains why cholesterol absorption is a selective process; plant sterols and other noncholesterol sterols are absorbed poorly or not at all.
3. The complete insensitivity of *Npc1l1*-deficient mice to ezetimibe indicates that NPC1L1 plays a critical role in the ezetimibe-sensitive cholesterol absorption pathway.
4. Cholesterol absorption is a multistep process that is regulated by multiple genes at the enterocyte level (**Figure 1**). Any factors that change the transportation of cholesterol from the intestinal lumen to the lymph can influence the efficiency of intestinal cholesterol absorption (**Table 1** and **Figure 2**).
5. The absorption efficiency of cholesterol is most likely determined by the net effect between influx and efflux of intraluminal cholesterol molecules across the brush border of the enterocyte.
6. Combination therapy using a novel, specific, and potent cholesterol absorption (NPC1L1) inhibitor (ezetimibe) and HMG-CoA reductase inhibitors (statins) offers an efficacious new approach to the prevention and treatment of hypercholesterolemia.
7. The detergency of bile acids through micellar solubilization of intraluminal sterols is obligatory for intestinal cholesterol absorption.
8. Plant sterols reduce cholesterol absorption by displacing cholesterol from micelles, primarily because they have higher affinity to micelles than does cholesterol.

ACKNOWLEDGMENTS

The author thanks Dr. Patrick Tso (University of Cincinnati School of Medicine) for reading the manuscript as well as for his critical comments and suggestions. This work is supported in part by a research grant DK54012 from the National Institutes of Health (U.S. Public Health Service).

LITERATURE CITED

1. Grundy SM, Cleeman JI, Merz CN, Brewer HBJ, Clark LT, et al. 2004. Implications of recent clinical trials for the National Cholesterol Education Program Adult Treatment Panel III guidelines. *Arterioscler. Thromb. Vasc. Biol.* 24:e149–61
2. Kesäniemi YA, Miettinen TA. 1997. Cholesterol absorption efficiency regulates plasma cholesterol level in the Finnish population. *Eur. J. Clin. Invest.* 17:391–95
3. Knopp RH, Walden CE, Retzlaff BM, McCann BS, Dowdy AA, et al. 1997. Long-term cholesterol-lowering effects of 4 fat-restricted diets in hypercholesterolemic and combined hyperlipidemic men. The Dietary Alternatives Study. *J. Amer. Med. Assoc.* 278:1509–15
4. De Lorgeril M, Salen P, Martin JL, Monjaud I, Delaye J, Mamelle N. 1999. Mediterranean diet, traditional risk factors, and the rate of cardiovascular complications after myocardial infarction: final report of the Lyon Diet Heart Study. *Circulation* 99:779–85
5. Davidson MH, Dillon MA, Gordon B, Jones P, Samuels J, et al. 1999. Colesevelam hydrochloride (cholestagel): a new, potent bile acid sequestrant associated with a low incidence of gastrointestinal side effects. *Arch. Intern. Med.* 159:1893–1900
6. Insull WJ Toth P, Mullican W, Hunninghake D, Burke S, et al. 2001. Effectiveness of colesevelam hydrochloride in decreasing LDL cholesterol in patients with primary hypercholesterolemia: a 24-week randomized controlled trial. *Mayo Clin. Proc.* 76:971–82
7. Maki KC, Davidson MH, Umporowicz DM, Schaefer EJ, Dicklin MR, et al. 2001. Lipid responses to plant-sterol-enriched reduced-fat spreads incorporated into a National Cholesterol Education Program Step I diet. *Am. J. Clin. Nutr.* 74:33–43
8. Miettinen TA, Puska P, Gylling H, Vanhanen H, Vartiainen E. 1995. Reduction of serum cholesterol with sitostanol-ester margarine in a mildly hypercholesterolemic population. *N. Engl. J. Med.* 333:1308–12
9. Mittendorfer B, Ostlund REJ, Patterson BW, Klein S. 2001. Orlistat inhibits dietary cholesterol absorption. *Obes. Res.* 9:599–604
10. Muls E, Kolanowski J, Scheen A, Van Gaal L. 2001. ObelHyx Study Group. The effects of orlistat on weight and on serum lipids in obese patients with hypercholesterolemia: a randomized, double-blind, placebo-controlled, multicentre study. *Int. J. Obes. Relat. Metab. Disord.* 25:1713–21
11. Insull WJ Koren M, Davignon J, Sprecher D, Schrott H, et al. 2001. Efficacy and short-term safety of a new ACAT inhibitor, avasimibe, on lipids, lipoproteins, and apolipoproteins, in patients with combined hyperlipidemia. *Atherosclerosis* 157:137–44
12. Davidson MH, Maki K, Umporowicz D, Wheeler A, Rittershaus C, Ryan U. 2003. The safety and immunogenicity of a CETP vaccine in healthy adults. *Atherosclerosis* 169:113–20
13. **Rosenblum SB, Huynh T, Afonso A, Davis HRJ, Yumibe N, et al. 1998. Discovery of 1-(4-fluorophenyl)-(3R)-[3-(4-fluorophenyl)-(3S)-hydroxypropyl]-(4S)-(4-hydroxyphenyl)-2-azetidinone (SCH 58235): a designed, potent, orally active inhibitor of cholesterol absorption. *J. Med. Chem.* 41:973–80**

13. This fundamentally important work reports the effectively and potently inhibitory effects of ezetimibe on intestinal cholesterol absorption.

14. Burnett DA, Caplen MA, Davis HRJ, Burrier RE, Clader JW. 1994. 2-Azetidinones as inhibitors of cholesterol absorption. *J. Med. Chem.* 37:1733–36
15. Knopp RH, Dujovne CA, Le Beaut A, Lipka LJ, Suresh R, Veltri EP. 2003. Ezetimbe Study Group. Evaluation of the efficacy, safety, and tolerability of ezetimibe in primary hypercholesterolaemia: a pooled analysis from two controlled phase III clinical studies. *Int. J. Clin. Pract.* 57:363–68
16. Dujovne CA, Bays H, Davidson MH, Knopp R, Hunninghake DB, et al. 2001. Reduction of LDL cholesterol in patients with primary hypercholesterolemia by SCH 48461: results of a multicenter dose-ranging study. *J. Clin. Pharmacol.* 41:70–78
17. Bays HE, Moore PB, Drehobl MA, Rosenblatt S, Toth PD, et al. 2001. Ezetimibe Study Group. Effectiveness and tolerability of ezetimibe in patients with primary hypercholesterolemia: pooled analysis of two phase II studies. *Clin. Ther.* 23:1209–30
18. Dujovne CA, Ettinger MP, McNeer JF, Lipka LJ, LeBeaut AP, et al. 2002. Esetimibe Study Group. Efficacy and safety of a potent new selective cholesterol absorption inhibitor, ezetimibe, in patients with primary hypercholesterolemia. *Am. J. Cardiol.* 90:1092–97
19. **Berge KE, Tian H, Graf GA, Yu L, Grishin NV, et al. 2000. Accumulation of dietary cholesterol in sitosterolemia caused by mutations in adjacent ABC transporters. *Science* 290:1771–75**
20. **Lee MH, Lu K, Hazard S, Yu H, Shulenin S, et al. 2001. Identification of a gene, *ABCG5*, important in the regulation of dietary cholesterol absorption. *Nat. Genet.* 27:79–83**
21. Bhattacharyya AK, Connor WE. 1974. β-sitosterolemia and xanthomatosis. A newly described lipid storage disease in two sisters. *J. Clin. Invest.* 53:1033–43
22. **Altmann SW, Davis HRJ, Zhu LJ, Yao X, Hoos LM, et al. 2004. Niemann-Pick C1 like 1 protein is critical for intestinal cholesterol absorption. *Science* 303:1201–4**
23. Davis HR, Zhu LJ, Hoos LM, Tetzloff G, Maguire M, et al. 2004. Niemann-Pick C1 Like 1 (NPC1L1) is the intestinal phytosterol and cholesterol transporter and a key modulator of whole-body cholesterol homeostasis. *J. Biol. Chem.* 279:33586–92
24. Kirby RJ, Zheng S, Tso P, Howles PN, Hui DY. 2002. Bile salt-stimulated carboxyl ester lipase influences lipoprotein assembly and secretion in intestine. A process mediated via ceramide hydrolysis. *J. Biol. Chem.* 277:4104–9
25. Weng W, Li L, Van Bennekum AM, Potter SH, Harrison EH, et al. 1999. Intestinal absorption of dietary cholesteryl ester is decreased but retinyl ester absorption is normal in carboxyl ester lipase knockout mice. *Biochemistry* 38:4143–49
26. Huggins KW, Camarota LM, Howles PN, Hui DY. 2003. Pancreatic triglyceride lipase deficiency minimally affects dietary fat absorption but dramatically decreases dietary cholesterol absorption in mice. *J. Biol. Chem.* 278:42899–905
27. Wang DQH, Tazuma S, Cohen DE, Carey MC. 2003. Feeding natural hydrophilic bile acids inhibits intestinal cholesterol absorption: studies in the gallstone-susceptible mouse. *Am. J. Physiol.* 285:G494–502
28. Wang DQH, Lammert F, Cohen DE, Paigen B, Carey MC. 1999. Cholic acid aids absorption, biliary secretion, and phase transitions of cholesterol in murine cholelithogenesis. *Am. J. Physiol.* 276:G751–60
29. Wang HH, Afdhal NH, Gendler SJ, Wang DQH. 2004. Lack of the intestinal Muc1 mucin impairs cholesterol uptake and absorption but not fatty acid uptake in *Muc1*$^{-/-}$ mice. *Am. J. Physiol.* 287:G547–54
30. Lee RG, Willingham MC, Davis MA, Skinner KA, Rudel LL. 2000. Differential expression of ACAT1 and ACAT2 among cells within liver, intestine, kidney, and adrenal of nonhuman primates. *J. Lipid Res.* 41:1991–2001

19. Reports that mutations in either *ABCG5* or *ABCG8* result in sitosterolemia, shows these genes' relative mRNA expression levels in human tissues, and explores expression levels of these two genes in response to high dietary cholesterol in the livers and small intestines of mice.

20. Presents evidence that sitosterolemia results from mutations in *ABCG5* in a study of nine patients.

22. This fundamentally important paper reports that the newly identified Niemann-Pick C1–like 1 (NPC1L1) protein is also expressed at the apical membrane of enterocytes and plays a crucial role in the ezetimibe-sensitive cholesterol absorption pathway.

31. van Greevenbroek MM, Robertus-Teunissen MG, Erkelens DW, de Bruin TW. 1998. Participation of the microsomal triglyceride transfer protein in lipoprotein assembly in Caco-2 cells: interaction with saturated and unsaturated dietary fatty acids. *J. Lipid Res.* 39:173–85
32. Bennett Clark S, Tercyak AM. 1984. Reduced cholesterol transmucosal transport in rats with inhibited mucosal acyl CoA:cholesterol acyltransferase and normal pancreatic function. *J. Lipid Res.* 25:148–59
33. Heider JG, Pickens CE, Kelly LA. 1983. Role of acyl CoA:cholesterol acyltransferase in cholesterol absorption and its inhibition by 57–118 in the rabbit. *J. Lipid Res.* 24:1127–34
34. **Buhman KK, Accad M, Novak S, Choi RS, Wong JS, et al. 2000. Resistance to diet-induced hypercholesterolemia and gallstone formation in ACAT2-deficient mice. *Nat. Med.* 6:1341–47**
35. Nielsen LB, Stender S, Kjeldsen K. 1993. Effect of lovastatin on cholesterol absorption in cholesterol-fed rabbits. *Pharmacol. Toxicol.* 72:148–51
36. Hajri T, Ferezou J, Laruelle C, Lutton C. 1995. Crilvastatin, a new 3-hydroxy-3-methylglutaryl-coenzyme A reductase inhibitor, inhibits cholesterol absorption in genetically hypercholesterolemic rats. *Eur. J. Pharmacol.* 286:131–36
37. Vanhanen H, Kesaniemi YA, Miettinen TA. 1992. Pravastatin lowers serum cholesterol, cholesterol-precursor sterols, fecal steroids, and cholesterol absorption in man. *Metabolism* 41:588–95
38. **Young SG, Cham CM, Pitas RE, Burri BJ, Connolly A, et al. 1995. A genetic model for absent chylomicron formation: mice producing apolipoprotein B in the liver, but not in the intestine. *J. Clin. Invest.* 96:2932–46**
39. Wang HH, Wang DQH. 2005. Reduced susceptibility to cholesterol gallstone formation in mice that do not produce apolipoprotein B48 in the intestine. *Hepatology* 42:894–904
40. Gordon DA, Jamil H, Gregg RE, Olofsson SO, Boren J. 1996. Inhibition of the microsomal triglyceride transfer protein blocks the first step of apolipoprotein B lipoprotein assembly but not the addition of bulk core lipids in the second step. *J. Biol. Chem.* 271:33047–53
41. Wetterau JR, Aggerbeck LP, Bouma ME, Eisenberg C, Munck A, et al. 1992. Absence of microsomal triglyceride transfer protein in individuals with abetalipoproteinemia. *Science* 258:999–1001
42. Ordovas JM, Schaefer EJ. 2000. Genetic determinants of plasma lipid response to dietary intervention: the role of the APOA1/C3/A4 gene cluster and the APOE gene. *Br. J. Nutr.* 83:S127–36
43. Schwarz M, Russell DW, Dietschy JM, Turley SD. 2001. Alternate pathways of bile acid synthesis in the cholesterol 7α-hydroxylase knockout mouse are not upregulated by either cholesterol or cholestyramine feeding. *J. Lipid Res.* 42:1594–603
44. Repa JJ, Lund EG, Horton JD, Leitersdorf E, Russell DW, et al. 2000. Disruption of the sterol 27-hydroxylase gene in mice results in hepatomegaly and hypertriglyceridemia. Reversal by cholic acid feeding. *J. Biol. Chem.* 275:39685–92
45. Akiyoshi T, Uchida K, Takase H, Nomura Y, Takeuchi N. 1986. Cholesterol gallstones in alloxan-diabetic mice. *J. Lipid Res.* 27:915–24
46. Voshol PJ, Havinga R, Wolters H, Ottenhoff R, Princen HM, et al. 1998. Reduced plasma cholesterol and increased fecal sterol loss in multidrug resistance gene 2 P-glycoprotein-deficient mice. *Gastroenterology* 114:1024–34
47. Wang DQH, Lammert F, Cohen DE, Paigen B, Carey MC. 1998. Hyposecretion of biliary phospholipids significantly decreases the intestinal absorption of cholesterol in Mdr2 (−/−) and (+/−) mice. *Gastroenterology* 114:A913 (Abstr.)

34. This important study provides evidence showing the impact of ACAT2 deficiency on intestinal cholesterol absorption and hepatic cholesterol metabolism in mice fed low-cholesterol or lithogenic diets.

38. Presents evidence for the importance of the intestinal apolipoprotein B in the assembly and secretion of chylomicrons as well as in intestinal cholesterol absorption.

48. Dawson PA, Haywood J, Craddock AL, Wilson M, Tietjen M, et al. 2003. Targeted deletion of the ileal bile acid transporter eliminates enterohepatic cycling of bile acids in mice. *J. Biol. Chem.* 278:33920–27
49. Wang DQH, Schmitz F, Kopin AS, Carey MC. 2004. Targeted disruption of the murine cholecystokinin-1 receptor promotes intestinal cholesterol absorption and susceptibility to cholesterol cholelithiasis. *J. Clin. Invest.* 114:521–28
50. Traber MG, Ostwald R. 1978. Cholesterol absorption and steroid excretion in cholesterol-fed guinea pigs. *J. Lipid Res.* 19:448–56
51. Ponz de Leon M, Iori R, Barbolini G, Pompei G, Zaniol P, Carulli N. 1982. Influence of small-bowel transit time on dietary cholesterol absorption in human beings. *N. Engl. J. Med.* 307:102–3
52. **Wang DQH, Paigen B, Carey MC. 2001. Genetic factors at the enterocyte level account for variations in intestinal cholesterol absorption efficiency among inbred strains of mice. *J. Lipid Res.* 42:1820–30**

52. Documents that interstrain differences in intestinal cholesterol absorption efficiency, which are determined by genetic factors at the enterocyte level, exist among inbred strains of mice and that high cholesterol absorption is a dominant trait in mice.

53. Wang DQH. 2002. Aging per se is an independent risk factor for cholesterol gallstone formation in gallstone susceptible mice. *J. Lipid Res.* 43:1950–59
54. Hollander D, Morgan D. 1979. Increase in cholesterol intestinal absorption with aging in the rat. *Exp. Gerontol.* 14:201–4
55. Duan P, Wang HH, Ohashi A, Wang DQH. 2006. Role of intestinal sterol transporters Abcg5, Abcg8, and Npc1l1 in cholesterol absorption in mice: gender and age effects. *Am. J. Physiol.* 290:G269–76
56. Turley SD, Schwarz M, Spady DK, Dietschy JM. 1998. Gender-related differences in bile acid and sterol metabolism in outbred CD-1 mice fed low- and high-cholesterol diets. *Hepatology* 28:1088–94
57. Wang HH, Afdhal NH, Wang DQH. 2004. Estrogen receptor α, but not β, plays a major role in 17β-estradiol-induced murine cholesterol gallstones. *Gastroenterology* 127:239–49
58. Bennion LJ, Ginsberg RL, Gernick MB, Bennett PH. 1976. Effects of oral contraceptives on the gallbladder bile of normal women. *N. Engl. J. Med.* 294:189–92
59. Henriksson P, Einarsson K, Eriksson A, Kelter U, Angelin B. 1989. Estrogen-induced gallstone formation in males. Relation to changes in serum and biliary lipids during hormonal treatment of prostatic carcinoma. *J. Clin. Invest.* 84:811–16
60. Uchida K, Nomura Y, Kadowaki M, Takase H, Takano K, Takeuchi N. 1978. Age-related changes in cholesterol and bile acid metabolism in rats. *J. Lipid Res.* 19:544–52
61. Einarsson K, Nilsell K, Leijd B, Angelin B. 1985. Influence of age on secretion of cholesterol and synthesis of bile acids by the liver. *N. Engl. J. Med.* 313:277–82
62. Valdivieso V, Palma R, Wünkhaus R, Antezana C, Severín C, Contreras A. 1978. Effect of aging on biliary lipid composition and bile acid metabolism in normal Chilean women. *Gastroenterology* 74:871–74
63. Bhattacharyya AK, Eggen DA. 1980. Cholesterol absorption and turnover in rhesus monkey as measured by two methods. *J. Lipid Res.* 21:518–24
64. Lofland HBJ, Clarkson TB, St. Clair RW, Lehner NDM. 1972. Studies on the regulation of plasma cholesterol levels in squirrel monkeys of two genotypes. *J. Lipid Res.* 13:39–47
65. Sehayek E, Nath C, Heinemann T, McGee M, Seidman CE, et al. 1998. U-shape relationship between change in dietary cholesterol absorption and plasma lipoprotein responsiveness and evidence for extreme interindividual variation in dietary cholesterol absorption in humans. *J. Lipid Res.* 39:2415–22
66. McNamara DJ, Kolb R, Parker TS, Batwin H, Samuel P, et al. 1987. Heterogeneity of cholesterol homeostasis in man. Response to changes in dietary fat quality and cholesterol quantity. *J. Clin. Invest.* 79:1729–39

67. Bosner MS, Lange LG, Stenson WF, Ostlund REJ. 1999. Percent cholesterol absorption in normal women and men quantified with dual stable isotopic tracers and negative ion mass spectrometry. *J. Lipid Res.* 40:302–8
68. Miettinen TA, Tilvis RS, Kesaniemi YA. 1990. Serum plant sterols and cholesterol precursors reflect cholesterol absorption and synthesis in volunteers of a randomly selected male population. *Am. J. Epidemiol.* 131:20–31
69. Kirk EA, Moe GL, Caldwell MT, Lernmark JA, Wilson DL, LeBoeuf RC. 1995. Hyper- and hyporesponsiveness to dietary fat and cholesterol among inbred mice: searching for level and variability genes. *J. Lipid Res.* 36:1522–32
70. Carter CP, Howles PN, Hui DY. 1997. Genetic variation in cholesterol absorption efficiency among inbred strains of mice. *J. Nutr.* 127:1344–48
71. Jolley CD, Dietschy JM, Turley SD. 1999. Genetic differences in cholesterol absorption in 129/Sv and C57BL/6 mice: effect on cholesterol responsiveness. *Am. J. Physiol.* 276:G1117–24
72. **Schwarz M, Davis DL, Vick BR, Russell DW. 2001. Genetic analysis of intestinal cholesterol absorption in inbred mice. *J. Lipid Res.* 42:1801–11**

72. Describes a genetic mapping strategy to identify regions in the chromosome containing candidate genes that may influence intestinal cholesterol absorption in inbred mice.

73. Van Zutphen LFM, Den Bieman MGCW. 1981. Cholesterol response in inbred strains of rats, *Rattus norvegicus*. *J. Nutr.* 111:1833–38
74. Beynen AC, Meijer GW, Lemmens AG, Glatz JFC, Versluis A, et al. 1989. Sterol balance and cholesterol absorption in inbred strains of rabbits hypo- or hyperresponsive to dietary cholesterol. *Atherosclerosis* 77:151–57
75. van Zutphen LFM, Fox RR. 1977. Strain differences in response to dietary cholesterol by JAX rabbits: correlation with esterase patterns. *Atherosclerosis* 28:435–46
76. Gylling H, Miettinen TA. 2002. Inheritance of cholesterol metabolism of probands with high or low cholesterol absorption. *J. Lipid Res.* 43:1472–76
77. **Wang DQH, Carey MC. 2003. Measurement of intestinal cholesterol absorption by plasma and fecal dual-isotope ratio, mass balance, and lymph fistula methods in the mouse: an analysis of direct versus indirect methodologies. *J. Lipid Res.* 44:1042–59**

77. Provides quantifiable methodologies for exploring genetic mechanisms of cholesterol absorption and for investigating the assembly and secretion of chylomicrons as well as intestinal lipoprotein metabolism.

78. Miettinen TA. 1980. Phytosterolaemia, xanthomatosis and premature atherosclerotic arterial disease: a case with high plant sterol absorption, impaired sterol elimination and low cholesterol synthesis. *Eur. J. Clin. Invest.* 10:27–35
79. Salen G, Shore V, Tint GS, Forte T, Shefer S, et al. 1989. Increased sitosterol absorption, decreased removal, and expanded body pools compensate for reduced cholesterol synthesis in sitosterolemia with xanthomatosis. *J. Lipid Res.* 30:1319–30
80. Salen G, Tint GS, Shefer S, Shore V, Nguyen L. 1992. Increased sitosterol absorption is offset by rapid elimination to prevent accumulation in heterozygotes with sitosterolemia. *Arterioscler. Thromb.* 12:563–68
81. Piironen V, Lindsay DG, Miettinen TA, Toivo J, Lampi AM. 2000. Plant sterols: biosynthesis, biological function and their importance to human nutrition. *J. Sci. Food Agric.* 80:939–66
82. Moreau RA, Whitaker BD, Hicks KB. 2002. Phytosterols, phytostanols, and their conjugates in foods: structural diversity, quantitative analysis, and health-promoting uses. *Prog. Lipid Res.* 41:457–500
83. Clader JW. 2004. The discovery of ezetimibe: a view from outside the receptor. *J. Med. Chem.* 47:1–9
84. Thurnhofer H, Hauser H. 1990. Uptake of cholesterol by small intestinal brush border membrane is protein-mediated. *Biochemistry* 29:2142–48

85. Compassi S, Werder M, Boffelli D, Weber FE, Hauser H, Schulthess G. 1995. Cholesteryl ester absorption by small intestinal brush border membrane is protein-mediated. *Biochemistry* 34:16473–82
86. Patel SB, Salen G, Hidaka H, Kwiterovich PO, Stalenhoef AF, et al. 1998. Mapping a gene involved in regulating dietary cholesterol absorption. The sitosterolemia locus is found at chromosome 2p21. *J. Clin. Invest.* 102:1041–44
87. Graf GA, Li WP, Gerard RD, Gelissen I, White A, et al. 2002. Coexpression of ATP-binding cassette proteins ABCG5 and ABCG8 permits their transport to the apical surface. *J. Clin. Invest.* 110:659–69
88. Yu L, Hammer RE, Li-Hawkins J, von Bergmann K, Lütjohann D, et al. 2002. Disruption of Abcg5 and Abcg8 in mice reveals their crucial role in biliary cholesterol secretion. *Proc. Natl. Acad. Sci. USA* 99:16237–42
89. Yu L, Li-Hawkins J, Hammer RE, Berge KE, Horton JD, et al. 2002. Overexpression of ABCG5 and ABCG8 promotes biliary cholesterol secretion and reduces fractional absorption of dietary cholesterol. *J. Clin. Invest.* 110:671–80
90. Yu L, York J, von Bergmann K, Lütjohann D, Cohen JC, Hobbs HH. 2003. Stimulation of cholesterol excretion by the liver X receptor agonist requires ATP-binding cassette transporters G5 and G8. *J. Biol. Chem.* 278:15565–70
91. Klett EL, Lu K, Kosters A, Vink E, Lee MH, et al. 2004. A mouse model of sitosterolemia: absence of Abcg8/sterolin-2 results in failure to secrete biliary cholesterol. *BMC Med.* 2:5–26
92. Berge KE, von Bergmann K, Lütjohann D, Guerra R, Grundy SM, et al. 2002. Heritability of plasma noncholesterol sterols and relationship to DNA sequence polymorphism in ABCG5 and ABCG8. *J. Lipid Res.* 43:486–94
93. Duan P, Wang HH, Wang DQH. 2004. Cholesterol absorption is mainly regulated by the jejunal and ileal ATP-binding cassette sterol efflux transporters Abcg5 and Abcg8 in mice. *J. Lipid Res.* 45:1312–23
94. Repa JJ, Berge KE, Pomajzl C, Richardson JA, Hobbs HH, Mangelsdorf DJ. 2002. Regulation of ATP-binding cassette sterol transporters ABCG5 and ABCG8 by the liver X receptors α and β. *J. Biol. Chem.* 277:18793–800
95. Kaneko E, Matsuda M, Yamada Y, Tachibana Y, Shimomura I, Makishima M. 2003. Induction of intestinal ATP-binding cassette transporters by a phytosterol-derived liver X receptor agonist. *J. Biol. Chem.* 278:36091–98
96. Repa JJ, Dietschy JM, Turley SD. 2002. Inhibition of cholesterol absorption by SCH 58053 in the mouse is not mediated via changes in the expression of mRNA for ABCA1, ABCG5, or ABCG8 in the enterocyte. *J. Lipid Res.* 43:1864–74
97. van Heek M, Farley C, Compton DS, Hoos L, Davis HR. 2001. Ezetimibe selectively inhibits intestinal cholesterol absorption in rodents in the presence and absence of exocrine pancreatic function. *Br. J. Pharmacol.* 134:409–17
98. van Heek M, Farley C, Compton DS, Hoos L, Alton KB, et al. 2000. Comparison of the activity and disposition of the novel cholesterol absorption inhibitor, SCH58235, and its glucuronide, SCH60663. *Br. J. Pharmacol.* 129:1748–54
99. Carstea ED, Morris JA, Coleman KG, Loftus SK, Zhang D, et al. 1997. Niemann-Pick C1 disease gene: homology to mediators of cholesterol homeostasis. *Science* 277:228–31
100. Davies JP, Levy B, Ioannou YA. 2000. Evidence for a Niemann-Pick C (NPC) gene family: identification and characterization of NPC1L1. *Genomics* 65:137–45
101. Garcia-Calvo M, Lisnock J, Bull HG, Hawes BE, Burnett DA, et al. 2005. The target of ezetimibe is Niemann-Pick C1-like 1 (NPC1L1). *Proc. Natl. Acad. Sci. USA* 102:8132–37

102. Iyer SPN, Yao X, Crona JH, Hoos LM, Tetzloff G, et al. 2005. Characterization of the putative native and recombinant rat sterol transporter Niemann-Pick C1 Like 1 (NPC1L1) protein. *Biochim. Biophys. Acta* 1722:282–92
103. Smart EJ, de Rose RA, Farber SA. 2004. Annexin 2-caveolin 1 complex is a target of ezetimibe and regulates intestinal cholesterol transport. *Proc. Natl. Acad. Sci. USA* 101:3450–55
104. Sparrow CP, Patel S, Baffic J, Chao YS, Hernandez M, et al. 1999. A fluorescent cholesterol analog traces cholesterol absorption in hamsters and is esterified in vivo and in vitro. *J. Lipid Res.* 40:1747–57
105. Kramer W, Glombik H, Petry S, Heuer H, Schafer H, et al. 2000. Identification of binding proteins for cholesterol absorption inhibitors as components of the intestinal cholesterol transporter. *FEBS Lett.* 487:293–97
106. Hernandez M, Montenegro J, Steiner M, Kim D, Sparrow C, et al. 2000. Intestinal absorption of cholesterol is mediated by a saturable, inhibitable transporter. *Biochim. Biophys. Acta* 1486:232–42
107. Detmers PA, Patel S, Hernandez M, Montenegro J, Lisnock JM, et al. 2000. A target for cholesterol absorption inhibitors in the enterocyte brush border membrane. *Biochim. Biophys. Acta* 1486:243–52
108. Kramer W, Girbig F, Corsiero D, Burger K, Fahrenholz F, et al. 2003. Intestinal cholesterol absorption: identification of different binding proteins for cholesterol and cholesterol absorption inhibitors in the enterocyte brush border membrane. *Biochim. Biophys. Acta* 1633:13–26
109. Kramer W, Girbig F, Corsiero D, Pfenninger A, Frick W, et al. 2005. Aminopeptidase N (CD13) is a molecular target of the cholesterol absorption inhibitor ezetimibe in the enterocyte brush border membrane. *J. Biol. Chem.* 280:1306–20
110. Orsó E, Werner T, Wolf Z, Bandulik S, Kramer W, Schmitz G. 2006. Ezetimib influences the expression of raft-associated antigens in human monocytes. *Cytometry* 69A:206–8
111. Hauser H, Dyer JH, Nandy A, Vega MA, Werder M, et al. 1998. Identification of a receptor mediating absorption of dietary cholesterol in the intestine. *Biochemistry* 37:17843–50
112. Cai SF, Kirby RJ, Howles PN, Hui DY. 2001. Differentiation-dependent expression and localization of the class B type I scavenger receptor in intestine. *J. Lipid Res.* 42:902–9
113. Mardones P, Quinones V, Amigo L, Moreno M, Miquel JF, et al. 2001. Hepatic cholesterol and bile acid metabolism and intestinal cholesterol absorption in scavenger receptor class B type I-deficient mice. *J. Lipid Res.* 42:170–80
114. Altmann SW, Davis HRJ Yao X, Laverty M, Compton DS, et al. 2002. The identification of intestinal scavenger receptor class B, type I (SR-BI) by expression cloning and its role in cholesterol absorption. *Biochim. Biophys. Acta* 1580:77–93
115. Wang DQH, Carey MC. 2002. Susceptibility to murine cholesterol gallstone formation is not affected by partial disruption of the HDL receptor SR-BI. *Biochim. Biophys. Acta* 1583:141–50
116. Repa JJ, Turley SD, Lobaccaro JA, Medina J, Li L, et al. 2000. Regulation of absorption and ABC1-mediated efflux of cholesterol by RXR heterodimers. *Science* 289:1524–29
117. McNeish J, Aiello RJ, Guyot D, Turi T, Gabel C, et al. 2000. High density lipoprotein deficiency and foam cell accumulation in mice with targeted disruption of ATP-binding cassette transporter-1. *Proc. Natl. Acad. Sci. USA* 97:4245–50

118. Drobnik W, Lindenthal B, Lieser B, Ritter M, Weber TC, et al. 2001. ATP-binding cassette transporter A1 (ABCA1) affects total body sterol metabolism. *Gastroenterology* 120:1203–11
119. Plosch T, Kok T, Bloks VW, Smit MJ, Havinga R, et al. 2002. Increased hepatobiliary and fecal cholesterol excretion upon activation of the liver X receptor is independent of ABCA1. *J. Biol. Chem.* 277:33870–77
120. Mulligan JD, Flowers MT, Tebon A, Bitgood JJ, Wellington C, et al. 2003. ABCA1 is essential for efficient basolateral cholesterol efflux during the absorption of dietary cholesterol in chickens. *J. Biol. Chem.* 278:13356–66
121. Attie AD, Hamon Y, Brooks-Wilson AR, Gray-Keller MP, MacDonald ML, et al. 2002. Identification and functional analysis of a naturally occurring E89K mutation in the ABCA1 gene of the WHAM chicken. *J. Lipid Res.* 43:1610–17
122. Lawn RM, Wade DP, Couse TL, Wilcox JN. 2001. Localization of human ATP-binding cassette transporter 1 (ABC1) in normal and atherosclerotic tissues. *Arterioscler. Thromb. Vasc. Biol.* 21:378–85
123. Ostlund REJ, Racette SB, Okeke A, Stenson WF. 2002. Phytosterols that are naturally present in commercial corn oil significantly reduce cholesterol absorption in humans. *Am. J. Clin. Nutr.* 75:1000–4
124. Heinemann T, Axtmann G, von Bergmann K. 1993. Comparison of intestinal absorption of cholesterol with different plant sterols in man. *Eur. J. Clin. Invest.* 23:827–31
125. Hallikainen MA, Sarkkinen ES, Gylling H, Erkkila AT, Uusitupa MI. 2000. Comparison of the effects of plant sterol ester and plant stanol ester-enriched margarines in lowering serum cholesterol concentrations in hypercholesterolaemic subjects on a low-fat diet. *Eur. J. Clin. Nutr.* 54:715–25
126. Nissinen M, Gylling H, Vuoristo M, Miettinen TA. 2002. Micellar distribution of cholesterol and phytosterols after duodenal plant stanol ester infusion. *Am. J. Physiol.* 282:G1009–15
127. Ikeda I, Tanabe Y, Sugano M. 1989. Effects of sitosterol and sitostanol on micellar solubility of cholesterol. *J. Nutr. Sci. Vitaminol.* 35:361–69
128. Ikeda I, Tanaka K, Sugano M, Vahouny GV, Gallo LL. 1988. Discrimination between cholesterol and sitosterol for absorption in rats. *J. Lipid Res.* 29:1583–91
129. Ikeda I, Tanaka K, Sugano M, Vahouny GV, Gallo LL. 1988. Inhibition of cholesterol absorption in rats by plant sterols. *J. Lipid Res.* 29:1573–82
130. Plat J, Mensink RP. 2002. Effects of plant stanol esters on LDL receptor protein expression and on LDL receptor and HMG-CoA reductase mRNA expression in mononuclear blood cells of healthy men and women. *FASEB J.* 16:258–60
131. Sudhop T, von Bergmann K. 2002. Cholesterol absorption inhibitors for the treatment of hypercholesterolaemia. *Drugs* 62:2333–47
132. Clader JW, Burnett DA, Caplen MA, Domalski MS, Dugar S, et al. 1996. 2-Azetidinone cholesterol absorption inhibitors: structure-activity relationships on the heterocyclic nucleus. *J. Med. Chem.* 39:3684–93
133. van Heek M, France CF, Compton DS, McLeod RL, Yumibe NP, et al. 1997. In vivo metabolism-based discovery of a potent cholesterol absorption inhibitor, SCH58235, in the rat and rhesus monkey through the identification of the active metabolites of SCH48461. *J. Pharmacol. Exp. Ther.* 283:157–63
134. Ezzet F, Krishna G, Wexler DB, Statkevich P, Kosoglou T, Batra VK. 2001. A population pharmacokinetic model that describes multiple peaks due to enterohepatic recirculation of ezetimibe. *Clin. Ther.* 23:871–85

135. Melani L, Mills R, Hassman D, Lipetz R, Lipka L, et al. 2003. Ezetimibe Study Group. Efficacy and safety of ezetimibe coadministered with pravastatin in patients with primary hypercholesterolemia: a prospective, randomized, double-blind trial. *Eur. Heart J.* 24:717–28
136. Feldman T, Koren M, Insull WJ, McKenney J, Schrott H, et al. 2004. Treatment of high-risk patients with ezetimibe plus simvastatin coadministration versus simvastatin alone to attain National Cholesterol Education Program Adult Treatment Panel III low-density lipoprotein cholesterol goals. *Am. J. Cardiol.* 93:1481–86
137. Gagne C, Gaudet D, Bruckert E. 2002. Ezetimibe Study Group. Efficacy and safety of ezetimibe coadministered with atorvastatin or simvastatin in patients with homozygous familial hypercholesterolemia. *Circulation* 105:2469–75
138. Salen G, von Bergmann K, Lutjohann D, Kwiterovich P, Kane J, et al. 2004. Multicenter Sitosterolemia Study Group. Ezetimibe effectively reduces plasma plant sterols in patients with sitosterolemia. *Circulation* 109:966–71
139. Arzneimittelkommission der deutschen Äerzteschaft. 2004. Myopathien bzw. Leberreaktionen unter Ezetimib (Ezetrol). *Deutsches. Aerzteblatt.* 101:A959 (Abstr.)
140. Johnson FL, St Clair RW, Rudel LL. 1985. Effects of the degree of saturation of dietary fat on the hepatic production of lipoproteins in the African green monkey. *J. Lipid Res.* 26:403–17
141. Chen IS, Hotta SS, Ikeda I, Cassidy MM, Sheppard AJ, Vahouny GV. 1987. Digestion, absorption and effects on cholesterol absorption of menhaden oil, fish oil concentrate and corn oil by rats. *J. Nutr.* 117:1676–80
142. Eckhardt ER, Wang DQH, Donovan JM, Carey MC. 2002. Dietary sphingomyelin suppresses intestinal cholesterol absorption by decreasing thermodynamic activity of cholesterol monomers. *Gastroenterology* 122:948–56
143. Vahouny GV, Roy T, Gallo LL, Story JA, Kritchevsky D, et al. 1978. Dietary fiber and lymphatic absorption of cholesterol in the rat. *Am. J. Clin. Nutr.* 31:S208–10
144. Schwarz M, Russell DW, Dietschy JM, Turley SD. 1998. Marked reduction in bile acid synthesis in cholesterol 7α-hydroxylase-deficient mice does not lead to diminished tissue cholesterol turnover or to hypercholesterolemia. *J. Lipid Res.* 39:1833–43
145. Valasek MA, Weng J, Shaul PW, Anderson RG, Repa JJ. 2005. Caveolin-1 is not required for murine intestinal cholesterol transport. *J. Biol. Chem.* 280:28103–9
146. Weinstock PH, Bisgaier CL, Hayek T, Aalto-Setala K, Sehayek E, et al. 1997. Decreased HDL cholesterol levels but normal lipid absorption, growth, and feeding behavior in apolipoprotein A-IV knockout mice. *J. Lipid Res.* 38:1782–94
147. Aalto-Setala K, Bisgaier CL, Ho A, Kieft KA, Traber MG, et al. 1994. Intestinal expression of human apolipoprotein A-IV in transgenic mice fails to influence dietary lipid absorption or feeding behavior. *J. Clin. Invest.* 93:1776–86
148. Knight BL, Patel DD, Humphreys SM, Wiggins D, Gibbons GF. 2003. Inhibition of cholesterol absorption associated with a PPARα-dependent increase in ABC binding cassette transporter A1 in mice. *J. Lipid Res.* 44:2049–58
149. van der Veen JN, Kruit JK, Havinga R, Baller JF, Chimini G, et al. 2005. Reduced cholesterol absorption upon PPARδ activation coincides with decreased intestinal expression of NPC1L1. *J. Lipid Res.* 46:526–34
150. Kirby RJ, Howles PN, Hui DY. 2004. Rate of gastric emptying influences dietary cholesterol absorption efficiency in selected inbred strains of mice. *J. Lipid Res.* 45:89–98

RELATED RESOURCES

Lammert F, Wang DQ-H. 2005. New insights into the genetic regulation of intestinal cholesterol absorption. *Gastroenterology* 129:718–34

Turley SD, Dietschy JM. 2003. Sterol absorption by the small intestine. *Curr. Opin. Lipid.* 14:233–40

Tso P. 1994. Intestinal lipid absorption. In *Physiology of the Gastrointestinal Tract*, ed. LR Johnson, DH Alpers, J Christensen, ED Jacobson, JH Walsh, pp. 1867–1907. New York: Raven

Why Does Pancreatic Overstimulation Cause Pancreatitis?

Ashok K. Saluja,[1] Markus M. Lerch,[2] Phoebe A. Phillips,[1] and Vikas Dudeja[1]

[1]Department of Surgery, University of Minnesota, Minneapolis, Minnesota 55455; email: asaluja@umn.edu

[2]Department of Gastroenterology, Endocrinology and Nutrition, Ernst-Moritz-Arndt-Universität Greifswald, Greifswald, Germany

Annu. Rev. Physiol. 2007. 69:249–69

First published online as a Review in Advance on October 20, 2006

The *Annual Review of Physiology* is online at http://physiol.annualreviews.org

This article's doi:
10.1146/annurev.physiol.69.031905.161253

0066-4278/07/0315-0249$20.00

Key Words

lysosome, heat shock protein, zymogen, exocrine pancreas, cholecystokinin, cytokine

Abstract

Many animal models are available to investigate the pathogenesis of pancreatitis, an inflammatory disorder of the pancreas. However, the secretagogue hyperstimulation model of pancreatitis is the most commonly used. Animals infused with high doses of cholecystokinin (CCK) exhibit hyperamylasemia, pancreatic edema, and acinar cell injury, which closely mimic pancreatitis in humans. Intra-acinar zymogen activation is an essential early event in the pathogenesis of secretagogue-induced pancreatitis. Early in the course of pancreatitis, lysosomal hydrolases colocalize with digestive zymogens and activate them. These activated zymogens then cause acinar cell injury and necrosis, a characteristic of pancreatitis. Besides being the site of initiation of injury in pancreatitis, acinar cells also synthesize and release cytokines and chemokines very early in the course of pancreatitis, which then attract and activate inflammatory cells and initiate the disease's systemic phase.

INTRODUCTION

CDE diet: choline-deficient, ethionine-supplemented diet

CCK: cholecystokinin

Acute pancreatitis is an inflammatory disorder of the pancreas that varies in severity from mild to severe and whose pathogenesis is not well understood. The pancreas is an enzyme factory that secretes large amounts of digestive enzymes, many of which are proenzymes known as zymogens. For activation these zymogens require cleavage of their activation peptide by a protease. In the small intestine, enterokinase cleaves the pancreatic zymogen trypsinogen to form trypsin. Trypsin then activates all other pancreatic zymogens (e.g., proelastase, procarboxypeptidase). In healthy organisms, pancreatic proteases remain inactive during their synthesis, secretion from acinar cells, and transport through the pancreatic duct. Their activation occurs only upon their arrival in the small intestine lumen. In contrast, in pancreatitis premature intrapancreatic zymogen activation may result in autodigestion of the pancreas (1).

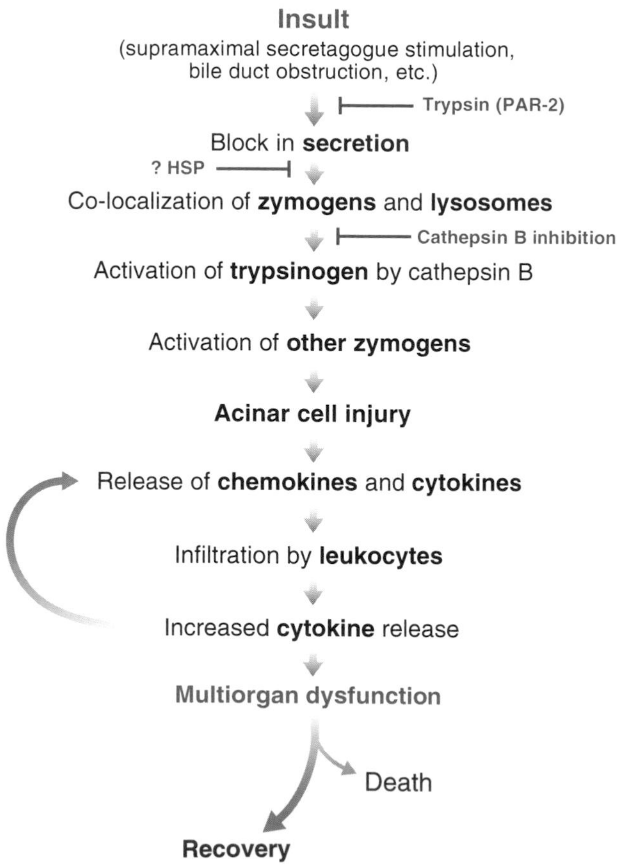

A number of studies have presented various animal models of pancreatitis in different species, including the bile duct obstruction model of pancreatitis and the choline-deficient, ethionine-supplemented diet (CDE diet) model of pancreatitis in mice. However, most studies evaluating the pathogenesis of pancreatitis use the supramaximal stimulation model of pancreatitis, in which a supramaximal concentration of the secretagogue cholecystokinin (CCK) or its analog caerulein is given to rodents, which induces acute pancreatitis closely resembling that in humans. In addition, in vitro studies using isolated acini have shed light on early events in pancreatitis. When exposed to the gastrointestinal hormone CCK, the pancreas and isolated acini respond similarly.

Under the current paradigm for CCK-induced pancreatitis and other forms of pancreatitis, after the initial insult there are early events including (*a*) a block in secretion, (*b*) the colocalization of zymogens and lysosomal enzymes, (*c*) the activation of trypsinogen and other zymogens, and (*d*) acinar cell injury (**Figure 1**). As a result of injury the acinar cells release chemokines and cytokines, which initiate the late events in pancreatitis, including the recruitment of inflammatory cells into the tissue.

This review focuses on the mechanism whereby overstimulation of the pancreas with secretagogues causes pancreatitis. We discuss the relevant intracellular events (such as activation of zymogens, block in secretion, disruption of acinar cytoskeleton, increased cytosolic calcium, cytokine production, and activation of kinases) as well as extracellular events, including the dissociation of cell-cell contacts, the infiltration of inflammatory cells, and systemic events.

Insult
(supramaximal secretagogue stimulation, bile duct obstruction, etc.)
Trypsin (PAR-2)
Block in **secretion**
? HSP
Co-localization of **zymogens** and **lysosomes**
Cathepsin B inhibition
Activation of **trypsinogen** by cathepsin B
Activation of **other zymogens**
Acinar cell injury
Release of **chemokines** and **cytokines**
Infiltration by **leukocytes**
Increased **cytokine** release
Multiorgan dysfunction
Death
Recovery

Figure 1

Schematic representation of the sequence of events involved in the pathogenesis of pancreatitis and ensuing multiorgan dysfunction.

CCK and its Receptors

In 1928, Ivy & Oldberg (2) first discovered CCK, a major gastrointestinal hormone with a central role in pancreatic secretion. Physiological doses of CCK mediate pancreatic secretion and growth and do not cause pancreatitis. In contrast, supramaximal doses of CCK (a concentration exceeding the dose at which maximal amylase secretion is observed) inhibit pancreatic secretion, resulting in zymogen activation and acinar cell injury, which trigger a series of events leading to acute pancreatitis.

Rodent pancreatic acinar cells express receptors for CCK (3–6). In mice and rats, the CCK receptors on acinar cells are directly involved in pancreatic secretion. However, the same may not hold true for human pancreatic acinar cells, in which the CCK receptors have low levels of expression (7). Also, CCK does not cause any changes in human acinar cell function [as assessed by calcium release, amylase secretion, and extracellular-regulated kinase (ERK) phosphorylation] (7). Supporting that CCK_A receptors on human pancreatic acini do not cause enzyme secretion, Adler et al. (8) reported that a cholinergic mechanism underlies CCK-mediated pancreatic secretion in humans. In contrast, in experimental rodent models of acute pancreatitis, CCK may function via CCK_A receptors (9–12). Support for this is that administration of large doses of CCK analogs in rats and mice cause pancreatitis (12–14) and that, in the CDE model of pancreatitis in mice, coadministration of CCK makes the pancreatitis more severe (11). In addition, OLETF rats that possess a selective defect in the CCK_A receptor develop less severe pancreatitis induced by taurocholate intraductal infusion, by a closed duodenal loop, or by intraperitoneal L-arginine administration (15).

INTRA-ACINAR EVENTS IN PANCREATITIS

The site and mechanism of initiation of pancreatitis have been enduring mysteries. Initially, researchers believed that the pancreatic juice leaking from the pancreatic duct was responsible for the onset of pancreatitis and that the disease began in the periductal region (16). Later, the observation of pancreatic fat necrosis at the time of autopsy in the patients dying of pancreatitis led to the hypothesis that the initial event was the release of active pancreatic lipase from the acinar cells, leading to peripancreatic fat necrosis (17). Subsequent controlled studies performed in animal models that simulate the human disease suggested that the acinar cell was the initial site of morphological damage (18). Today there is consensus that the events that culminate into pancreatitis are initiated in acinar cells. Thus, besides animal models of pancreatitis, isolated pancreatic acini are considered a valid model with which to investigate the pathogenesis of pancreatitis. The pathophysiological events that occur within acinar cells are outlined below.

Supramaximal dose: a dose of the secretagogue exceeding the dose at which maximal amylase secretion is elicited

OLETF rat: Otsuka Long Evans Tokushima Fatty rat

TAP: trypsinogen activation peptide

Intra-Acinar Cell Zymogen Activation

Acinar cells produce a profusion of digestive enzymes, which, to prevent auto-digestion, are synthesized and stored as inactive zymogen precursors. A low level of zymogen activation occurs within the acinar cells even under physiological conditions, but intracellular protective mechanisms such as the presence of trypsin inhibitors, nonoptimal pH conditions, and proteases degrading the activated enzymes normally prevent cell damage.

In vivo and in vitro studies have demonstrated the importance of zymogen activation in the pathogenesis of pancreatitis. The activation of trypsinogen and other pancreatic zymogens occurs in the pancreatic homogenate from animals with caerulein-induced pancreatitis (19–21). This zymogen activation appears to be an early event. Trypsin activity is detected as early as 10 min after supramaximal stimulation by caerulein in rats and increases over time (20, 22). Similarly, the levels of trypsinogen activation peptide (TAP),

CA$_1$: carboxypeptidase A$_1$

which is cleaved from trypsinogen and active enzymes such as carboxypeptidase A$_1$ (CA$_1$), increase as early as 15 min after initiation of supramaximal stimulation (20, 22). All other markers of pancreatitis, e.g., hyperamylasemia, pancreatic edema, and acinar cell vacuolization, can only be detected after 30 min of supramaximal stimulation (20). This sequence of events strongly supports the paradigm that zymogen activation is the cause and not the result of pancreatitis. Moreover, pretreatment with protease inhibitors reduces the severity of hyperstimulation pancreatitis in animals (23).

In vitro studies on pancreatic acini have confirmed the findings from animal studies. In analogy to the pancreatic necrosis observed in the in vivo models, various biochemical parameters such as cytosolic lactic dehydrogenase release, propidium iodide (PI) incorporation, and trypan blue retention have been used as markers of cellular injury in in vitro models. Using those surrogate markers, Grady et al. (24) and Saluja et al. (25) observed that supramaximal stimulation by CCK (concentration $>10^{-10}$ M) results in zymogen activation and acinar cell injury. No activation occurred at CCK concentrations below 10^{-10} M (24, 25).

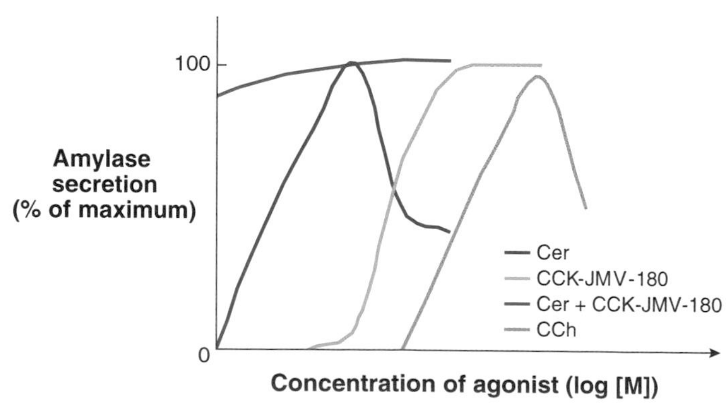

Figure 2

A schematic representation showing amylase secretion from pancreatic acinar cells stimulated with various secretagogues. Supramaximal doses of either caerulein (Cer) and carbachol (CCh) inhibit secretion, whereas CCK-JMV-180 does not cause high-dose inhibition. Administration of CCK-JMV-180 at 1 μM with caerulein prevents the high-dose inhibition induced by caerulein alone. Modified from Saluja et al. (27).

Schmid et al. (26) also reported a similar pattern of zymogen activation when acini were stimulated with supramaximal concentrations (>0.1 μM) of carbachol (a cholinergic receptor agonist). Pretreatment of acinar cells with cell-permeable protease inhibitors (like benzamidine and pefabloc) prior to supramaximal stimulation by caerulein prevented both zymogen activation and acinar cell injury (24, 25). These findings emphasize the role of zymogen activation in acinar cell injury caused by supramaximal secretagogue stimulation.

How does a physiological secretory stimulus of a secretagogue turn, at a supramaximal dose, into a pathological signal with zymogen activation and cell injury? A comparison of the dose response curves of the various secretagogues provides some clues (**Figure 2**). The dose response curve of caerulein for pancreatic secretion is biphasic, with maximal stimulation at 0.2 μg kg^{-1} hr^{-1} in rats (27). Increasing the dose to 5 μg kg^{-1} hr^{-1} paradoxically inhibits the secretion. Only infusion of this supramaximal dose of caerulein induces pancreatitis in the animals, as indicated by hyperamylasemia and edema of the pancreas (28). Similar dose response curves are obtained for caerulein and carbachol in vitro, using acinar cells. The maximal stimulatory concentration of caerulein for amylase secretion is 10^{-10} M in vitro; higher concentrations lead to a progressive decrease in amylase secretion and zymogen activation (**Figure 2**) (27, 29). In contrast to caerulein, CCK-JMV-180, another CCK analog, has a monophasic dose response curve for amylase secretion (**Figure 2**) (27). When pancreatic acini are stimulated with CCK-JMV-180, the maximal rate of amylase secretion is observed at concentrations of ~1 μM, whereas higher concentrations neither increase nor decrease this rate of secretion. Similarly, the administration of CCK-JMV-180 to animals at 0.2 mg kg^{-1} hr^{-1} causes a maximal rate of amylase secretion, and infusion of a higher dose (unlike for caerulein) does not inhibit secretion (27). Remarkably, a supramaximal dose of CCK-JMV-180 does not cause pancreatitis (27). These observations indicate that high-dose inhibition of secretion induced by supramaximal concentrations of CCK, caerulein, or carbachol has a role in inducing zymogen activation, cellular injury, and pancreatitis.

Studies using bombesin, which has a monophasic dose response curve like that of CCK-JMV-180 and unlike that of caerulein, further clarify the operation of supramaximal inhibition of secretion by secretagogues. Infusion of a supramaximal dose (500 $\mu g\ kg^{-1}\ hr^{-1}$) of bombesin in rats does not lead to pancreatitis (24, 30). Similarly, a supramaximal concentration of bombesin does not lead to pancreatic acinar cell injury in vitro (24). Surprisingly, a supramaximal dose of bombesin stimulates zymogen activation, i.e., the processing of procarboxypeptidase A_1 (PCA_1) to CA_1. However, the CA_1 produced is secreted into the medium, as bombesin does not inhibit secretion at supramaximal doses. In contrast, hyperstimulation by caerulein and carbachol leads to inhibition of secretion at supramaximal doses. The activated zymogens thus remain inside the acinar cells, causing cellular injury, as lactic dehydrogenase release and trypan blue retention confirm (24, 31). Both the activation of zymogens and their retention inside the acinar cell therefore are necessary for cellular injury in pancreatitis.

Pancreatic acinar cells express the G protein–coupled transmembrane receptor known as proteinase-activated receptor-2 (PAR-2), which is activated by trypsin. PAR-2 is activated during acute pancreatitis (32). Although the physiological role of PAR-2 in the pancreas is unknown, the activation of PAR-2 by trypsin (which is prematurely activated and released into the interstitial space) protects against pancreatitis. Supporting evidence for this is that the administration of the PAR-2-activating peptide SLIGRL decreases the severity of CCK-induced pancreatitis (33, 34). Furthermore, injury from CCK-induced pancreatitis is more extensive in PAR-2 knockout mice than in wild-type mice (35).

PCA_1: procarboxypeptidase A_1

PAR: proteinase-activated receptor

Supramaximal Inhibition of Secretion: Role of High-Affinity and Low-Affinity CCK Receptors

Studies employing CCK-JMV-180 have been fundamental to our understanding of the molecular mechanism of the high-dose inhibition induced by CCK and its analog caerulein. The CCK receptors exist in two affinity states: high and low. CCK-JMV-180 can act as a partial agonist on high-affinity CCK receptors and as an antagonist on low-affinity CCK receptors (36). As discussed above, CCK-JMV-180 has a monophasic dose response curve for amylase secretion and does not inhibit secretion at supramaximal doses. When given along with caerulein, it also prevents the inhibition of secretion induced by supramaximal doses of caerulein (**Figure 2**) (28). This indicates that the stimulation of enzyme secretion by both CCK-JMV-180 and physiological doses of CCK or caerulein occurs via stimulation of high-affinity CCK receptors. In contrast, the inhibition of amylase secretion induced by high concentrations of CCK and caerulein appears to be mediated by the low-affinity receptors. This action can be prevented by CCK-JMV-180, which acts as an antagonist on low-affinity receptors. The role of low-affinity receptors in the inhibition of secretion is further demonstrated by the finding that caerulein, when administered in supramaximal doses along with bombesin, also inhibits bombesin-induced secretion (37). By virtue of its ability to prevent high-dose inhibition induced by caerulein, CCK-JMV-180 can prevent pancreatitis induced by high-dose caerulein (28).

Taken together, these findings establish that caerulein-induced high-dose inhibition of secretion is mediated by the stimulation of low-affinity receptors. Furthermore, this high-dose inhibition of secretion is important for initiating acinar cell injury and pancreatitis.

GRAMP: granule membrane protein

Role of the Actin Cytoskeleton in Supramaximal Inhibition of Secretion and Pancreatitis

Pancreatic acinar cells, like many other eukaryotic cells, have a dense web of actin filaments immediately beneath their plasma membrane (38, 39). Ultrastructurally, loss of the terminal actin web and its associated intermediate filaments occurs in pancreatic acinar cells in response to supramaximal concentrations of caerulein, leading to the inhibition of secretion (40, 41). CCK-JMV-180, on the other hand, does not cause any changes in the ultrastructure of the acinar subapical actin web nor, as discussed above, does it lead to inhibition of secretion at supramaximal concentrations (42). Additionally, CCK-JMV-180, when administered along with caerulein, prevents the ablation of apical cytoskeleton and inhibition of secretion induced by supramaximal concentration of caerulein. Therefore, the subapical actin web may be important for the secretory function of acinar cells, and its ablation by caerulein at supramaximal concentrations may be responsible for the inhibition of secretion. These changes in the actin web are mediated through the stimulation of low-affinity receptors. It is unclear how the stimulation of low-affinity receptors leads to ablation of the actin cytoskeleton. However, this most likely involves the activation of phospholipase C by the stimulation of low-affinity receptors (42).

Role of Lysosomes in Pancreatitis

As mentioned above, zymogen activation is the central initiating event in pancreatitis, and a normal pancreas has many protective mechanisms that ensure against such an occurrence. How, then, are these zymogens prematurely activated within the pancreas during early stages of acute pancreatitis? Many studies have attempted to address this question, engendering several hypotheses. One such theory is the colocalization hypothesis, which suggests that, during early stages of pancreatitis, pancreatic zymogens become colocalized with and thus activated by lysosomal enzymes such as cathepsin B to produce active trypsin. This in turn activates other digestive zymogens, resulting in cell injury.

Using subcellular fractionation, Saluja et al. (43) observed the redistribution of cathepsin B activity from the lysosome-rich fraction to the zymogen granule–rich fraction after supramaximal stimulation with caerulein. Immunolocalization studies have confirmed the colocalization of lysosomal and digestive enzymes in the same organelle (44). This redistribution of cathepsin B from the lysosome-enriched to the zymogen-enriched fraction was noted within 15 min of the start of caerulein administration, and trypsinogen activation was observed concurrently (20). On the other hand, hyperamylasemia, pancreatic edema, and acinar cell injury occurred at a later time only, indicating that colocalization is not the result but the cause of cell injury. Similar colocalization of zymogen granules and lysosomal enzymes has been observed in other models of pancreatitis (28, 45, 46), suggesting that this phenomenon is not limited to secretagogue-induced pancreatitis but is a generalized phenomenon applicable to all models as well as human disease. Immunolocalization studies also indicate not only that the zymogens and lysosomal enzymes colocalize but also that trypsinogen activation occurs in cytoplasmic vacuoles containing the lysosomal hydrolase cathepsin B. Antibodies against the lysosomal membrane protein GRAMP 92 as the marker confirm these observations (**Figure 3**) (47).

There have been some concerns about the colocalization hypothesis with regard to the ability of cathepsin B to activate trypsinogen. Of primary concern was that the optimal pH of the lysosomal enzymes is approximately 5, whereas the sites of colocalization appear to have a neutral pH. However, substantial activation of trypsinogen by cathepsin B can occur at neutral pH as well (48–50).

Further evidence for the role of cathepsin B in trypsinogen activation comes from

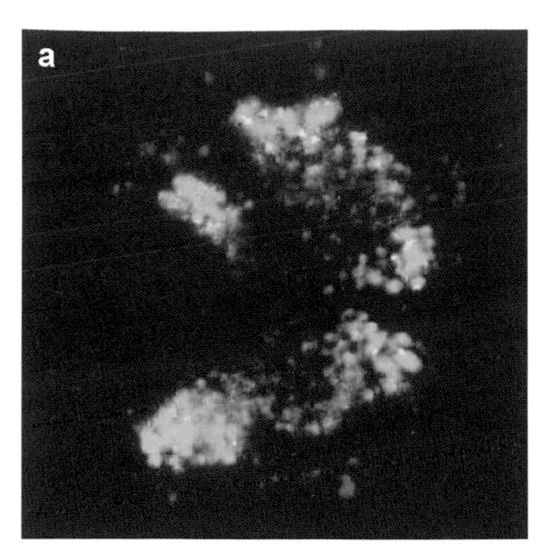

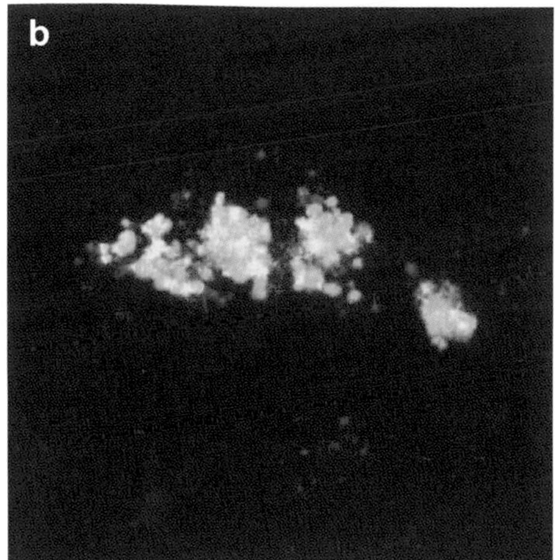

Figure 3

Light microscopy illustrating the colocalization of trypsinogen and lysosomes within 30 min of a supramaximal dose of caerulein. Double-labeling fluorescence microscopy of paraffin-embedded pancreatic sections was carried out with purified trypsinogen antibody/flourescein-conjugated antirabbit IgG (*green*) and with anti-GRAMP92 (lysosomal membrane protein)/Texas red-conjugated antisheep IgG (*red*). (*a*) Before caerulein stimulation. (*b*) After caerulein stimulation. The areas of colocalization appear as yellow dots on merging the images.

evaluations of the effect of cathepsin B inhibition on pancreatitis. Pretreatment of rat pancreatic acini with E-64d, a cell-permeable form of the cathepsin B inhibitor E-64, led to complete inhibition of cathepsin B and prevented caerulein-induced activation of trypsinogen (51). The concentrations of E-64d that led to partial inhibition of cathepsin B in vitro could not prevent caerulein-induced trypsinogen activation. The cell-impermeable E-64 is unable to completely block cathepsin B activity and thus is unable to prevent the activation of trypsinogen by caerulein hyperstimulation (51). This explains the findings of some in vivo studies, in which E-64 did not protect the animals from caerulein hyperstimulation pancreatitis (52). Similarly, treatment of the acini with another cathepsin B inhibitor CA-074me completely prevented caerulein-induced in vitro activation of trypsinogen activation (53). Inhibition in animals of cathepsin B by CA-074me resulted in reduced severity of caerulein-induced pancreatitis (**Figure 4**) (53).

Van Acker et al. (54) tested the importance of cathepsin B in the activation of trypsinogen by using cathepsin B knockout mice in which the cathepsin B gene had been deleted by targeted disruption. After induction of experimental secretagogue-induced pancreatitis, the trypsin activity in the pancreas of cathepsin B knockout mice was more than 80% lower than in the wild-type animals. Also, pancreatic damage, as indicated by various parameters including the extent of acinar tissue necrosis, was substantially lower in the knockout animals.

These experiments prove the involvement of cathepsin B in intrapancreatic trypsinogen activation and the onset of acute pancreatitis. In contrast, trypsinogen autoactivation—as opposed to cathepsin B–induced trypsinogen activation—is not a trigger mechanism for acute pancreatitis (55).

Phosphatidylinositol-3 kinase (PI-3 kinase) (class 3) is involved in the perturbation of normal sorting of the lysosomal hydrolases, thus leading to the colocalization

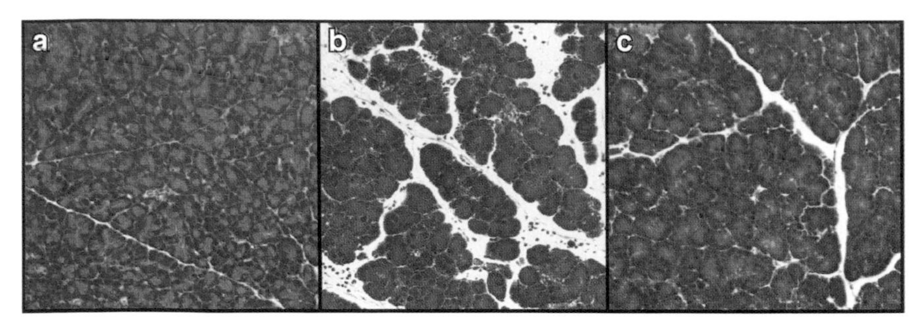

Figure 4

Effect of inhibition of cathepsin B by CA-074me before caerulein administration to mice on the morphological changes of pancreatitis [modified from Van Acker et al. (51)]. (*a–c*) Representative photomicrographs taken of samples from control mice (*a*), mice given supramaximal doses of caerulein for 6 h (*b*), and mice given CA-074me before supramaximal doses of caerulein (*c*). The caerulein-alone sample (*b*) shows marked inflammation, edema, and acinar cell necrosis, which are markedly reduced in a sample taken from an animal given CA-074me before caerulein (*c*).

PI-3 kinase: phosphatidylinositol 3-kinase

IP_3: inositol 1,4,5-triphosphate

and activation of trypsinogen (56). Inhibition of PI-3 kinase activity by either wortmannin or LY294002 prevented the colocalization of zymogen and lysosomal hydrolases and thus decreased activation of trypsinogen and the severity of pancreatitis. Similar effects were seen in another model of pancreatitis involving injection of taurocholate in the bile duct (56).

Investigators have suggested other additional mechanisms of trypsinogen activation in pancreatitis. Gukovskaya et al. (57) indicate that neutrophils recruited into the pancreas during pancreatitis may also be involved in the activation of trypsinogen. Animals whose neutrophils were depleted by antineutrophil serum show reduced trypsin activation as compared with control animals. Mediators generated by NADPH oxidase from the neutrophils appear to cause this trypsinogen activation (57). Similarly, when the ability of neutrophils to enter the pancreas is abolished by inhibiting their neutrophil elastase, an enzyme they require to cleave cell contacts between acinar cells, zymogen activation is also prevented (58).

Role of Calcium in Pancreatitis

Intracellular calcium is involved in the signal-secretion coupling in pancreatic acinar cells. In the resting condition, the concentration of calcium within the acinar cell cytosol is 10^{-7} M, which is much lower than the concentrations in the extracellular fluid (10^{-3} M) and in the intracellular stores (10^{-4} M) (59). This maintenance of low cytoplasmic calcium concentration enables small calcium increases in different regions of the cells to serve as signals to control intracellular events. Moreover, high cytosolic calcium concentrations are toxic to many types of cells (60).

Different secretagogues such as acetylcholine and CCK bind to and activate G protein–linked transmembrane receptors on the basolateral membrane of the acinar cells. G protein activation then leads to the activation of phospholipase C-β, which cleaves membrane-bound phosphatidylinositol-4,5-bisphosphate (PIP_2) into diacylglycerol and free inositol 1,4,5-triphosphate (IP_3). IP_3 then binds to receptors located on the endoplasmic reticulum membrane to induce calcium release. The characteristics of pancreatic calcium signals are specific to a particular early secretagogue as well as to its concentration. CCK, caerulein, CCK-JMV-180, and acetylcholine at physiological concentrations elicit repetitive calcium spikes or oscillations (61). For acetylcholine these oscillations are restricted to the secretory pole of the cell (62), whereas cholecystokinin leads to short-lasting local spikes followed by longer calcium transients that spread to the entire cell. Each oscillation is associated with a burst of exocytotic activity and the release of zymogen into the duct lumen (63). Supramaximal stimulation of acinar cells, in contrast, induces a completely different pattern of calcium signals. Instead of the oscillatory activity observed with physiological doses of CCK, there is a much larger rise followed by a sustained elevation at a lower level (61).

The oscillatory changes in the calcium levels induced by CCK-JMV-180 or by low concentrations of CCK or caerulein appear to be induced by the stimulation of high-affinity CCK receptors. The inability of CCK-JMV-180 to cause calcium changes, as do supramaximal concentrations of CCK, suggests the involvement of low-affinity CCK receptors in these calcium changes (61). Additionally, by virtue of its antagonistic action on low-affinity CCK receptors, CCK-JMV-180, when coadministered with caerulein, prevents the typical intracellular calcium changes induced by supramaximal concentration of caerulein (61).

Substantial direct and indirect evidence suggests the involvement of calcium in the pathogenesis of pancreatitis. That patients with endocrine conditions resulting in hypercalcemia are predisposed to pancreatitis suggests an association between hypercalcemia and pancreatitis (64). Hypercalcemia appears to be the cause of pancreatitis that develops in patients maintained on extracorporeal blood circulation during major cardiac surgery (65). In addition, in animal experiments, hypercalcemia either decreases the threshold level for the onset of pancreatitis or induces morphological alterations reminiscent of pancreatitis. Disturbances in the calcium homeostasis of pancreatic acinar cells occur early in the secretagogue-induced model of pancreatitis (66). The attenuation of the calcium elevation in acinar cells by cytosolic calcium chelator BAPTA-AM prevents zymogen activation, suggesting that calcium is essential for zymogen activation (25, 67, 68). The sustained elevation following the initial spike of intracellular calcium induced by the supramaximal concentration of caerulein is attenuated in the absence of, and thus appears to depend on, extracellular calcium. In the absence of extracellular calcium, the activation of trypsinogen induced by a supramaximal dose of caerulein is also attenuated, suggesting that the initial and transient rise in calcium caused by the release of calcium from the internal stores is not sufficient to permit trypsinogen activation (25, 67). In contrast, interference with high calcium plateaus by using the natural Ca^{2+} antagonist Mg^{2+} (69) or a calcium chelator in vivo abolishes trypsinogen activation as well as pancreatitis (70).

BAPTA-AM: 1,2-bis(*o*-aminophenoxy)ethane-*N*,*N*,*N*′,*N*′-tetraacetic acid tetra(acetoxymethyl) ester

NF-κB: nuclear factor-κB

Although calcium undoubtedly is required to activate pancreatic digestive proteases, some investigators believe that the elevation of intra-acinar calcium, regardless of its source or its stimulatory agent (e.g., thapsigargin, ionomycin), is sufficient to induce protease activation (68). However, studies involving the elevation of intra-acinar calcium by a variety of reagents with different sites of action have shown that the mere elevation of intracellular calcium does not lead to trypsinogen activation. This suggests that calcium, although essential, is not sufficient to activate trypsinogen. Presumably, then, other effects of supramaximal secretagogue are needed to activate trypsinogen and induce pancreatitis.

Husain et al. (71) recently postulated a role of ryanodine-sensitive pools of calcium in zymogen activation. These researchers observed that the trypsinogen activation takes place in the supranuclear compartment of the acinar cell, which overlaps with the intracellular location of ryanodine receptors. Moreover, inhibition of ryanodine receptors reduces basolateral calcium signals and zymogen activation.

Role of NF-κB in Pancreatitis

Investigators have recently obtained a greater understanding of inflammation in the pathogenesis of pancreatitis. Besides the inflammatory cells recruited into the pancreas during an episode of pancreatitis, acinar cells themselves may act as a source of cytokines and modify the course of pancreatitis. Many transcription factors activated early in pancreatitis induce the synthesis and release of various cytokines and chemokines such as tumor necrosis factor (TNF)-α, IL-6, IL-1β, mob-1, MIP-2, and chemokine KC (72–75). These cytokines may then activate and attract inflammatory cells into the pancreas and continue the process of injury (76).

One such transcription factor is nuclear factor-κB (NF-κB). NF-κB, a key regulator of cytokine induction, is a known modulator of inflammation in various organ systems. NF-κB represents a family of proteins sharing the Rel homology domain; these proteins bind to DNA as homo- or heterodimers and activate a multitude of cellular stress-related and early response genes for the cytokines, growth factors, adhesion molecules, and acute-phase proteins (77). Physiologically, NF-κB is kept silent in the cytoplasm via interaction with inhibitory proteins of the IκB family that mask the nuclear localization signal. Various agents ranging from cell-damaging physical factors and viruses to mitogens and cytokines activate NF-κB. Upon activation, IκB proteins become phosphorylated and proteolytically degraded, allowing NF-κB dimers to translocate rapidly into the nucleus, where they bind to DNA and modulate the transcription of various cytokines.

In the rat caerulein–induced pancreatitis model, there is rapid, strong, biphasic activation of NF-κB (78). No NF-κB binding activity was detected in response to a low, physiological dose of caerulein or infusion of the CCK analog CCK-JMV-180. Furthermore, caerulein-induced NF-κB activation was abolished by coinfusion of CCK-JMV-180, which, as discussed above, prevents pancreatitis (78). Protein kinase C isoforms delta and epsilon play a role in this NF-κB activation induced by caerulein (79). Other models of pancreatitis show similar NF-κB activation patterns, suggesting that it is a universal phenomenon (80, 81). NF-κB may be an important regulator of expression of inflammatory mediators (78, 80) in experimental pancreatitis. Caerulein hyperstimulation greatly increases the expression of interleukin (IL)-6 and IL-8 mRNA up to 100-fold (78). Blockage of NF-κB activation (by *N*-acetyl cysteine, for example) prevents this induction of cytokine transcription by caerulein hyperstimulation, thus indicating an involvement of NF-κB in cytokine activation by caerulein (78). These cytokines and other inflammatory mediators thus produced play a role in the activation and recruitment of neutrophils, which promote further injury (82).

The inhibition of NF-κB activation in the caerulein hyperstimulation model of pancreatitis as well as other models of pancreatitis significantly reduces the severity of pancreatitis. When NF-κB activation was prevented by use of the antioxidant *N*-acetyl cysteine or other means, all parameters of rat caerulein pancreatitis were diminished (78, 80). Other authors have used a variety of compounds to inhibit NF-κB activation in other models of pancreatitis and have reported similar findings (83, 84). Except for one study (85), which did not show any beneficial effect of inhibition of NF-κB on experimental pancreatitis, most investigators agree that blocking NF-κB activation reduces the severity of the disease.

The link between pancreatic NF-κB activation and trypsinogen activation in experimental pancreatitis has also been investigated. Although NF-κB and trypsinogen activation in rat pancreatic acini occur in a similar time frame, they are independent events (86, 87). NF-κB activation can occur independently of trypsinogen activation induced by liposaccharide in vivo and by phorbol ester in pancreatic acini in vitro. Similarly, another compound, MG-132, prevents caerulein-induced NF-κB activation but not trypsinogen activation (83). Moreover, secretagogue stimulation of pancreatic acini after 6 h of culture did not lead to intra-acinar trypsinogen activation but caused NF-κB activation. Apparently, trypsinogen activation initiates the cellular injury in pancreatitis, and NF-κB activation, via the production of cytokines and recruitment of various inflammatory cells, causes the continuation and propagation of the injury.

EXTRACELLULAR EVENTS

Inflammatory Cell Infiltration

Events that occur subsequent to acinar cell injury are believed to determine the severity of pancreatitis. Pancreatic acinar cells also synthesize and release cytokines and chemokines, resulting in the recruitment of inflammatory cells such as neutrophils and macrophages. The recruitment and activation of various inflammatory cells leads to further acinar cell injury and causes an elevation of various proinflammatory mediators such as TNF-α; IL-1, IL-2, IL-6, and other chemokines; and anti-inflammatory factors such as IL-10 and IL-1 receptor antagonist (88, 89).

The past several years of research have yielded several key inflammatory mediators that play a role in experimental CCK-induced pancreatitis as well as in human disease. One such mediator is platelet-activating factor (PAF), a potent proinflammatory phospholipid mediator that has an active role in wound healing, physiological inflammation, and apoptosis (90). PAF is synthesized by a variety of inflammatory cells, including macrophages, leukocytes, and eosinophils (91, 92). Murine pancreatic acini also can synthesize PAF in response to caerulein (93). Many studies have shown a link between PAF and the pathogenesis of pancreatitis. We have previously shown that terminating the action of PAF with recombinant PAF-acetylhydrolase (which accelerates the breakdown of PAF) prior to the administration of supramaximal caerulein amelioriates pancreatitis, as evident by a reduction in hyperamylasemia, acinar cell vacuolization, and pancreatic inflammation (94). Several other studies using various models of experimental pancreatitis support these data by employing PAF antagonists (95–98).

CCK STIMULATION AND PANCREATITIS IN HUMANS: ROLE OF SENSITIZATION BY ETHANOL

Secretagogue overstimulation causes pancreatitis in experimental animal models. However, overstimulation does not appear to be the cause of pancreatitis in humans, as in humans CCK levels are not known to rise above physiological concentrations. Another agent such as ethanol may sensitize the acinar cells to injury by CCK, thus allowing physiological CCK levels to induce pancreatitis. This is plausible, as ethanol consumption predisposes one to pancreatitis. Some investigators have evaluated the ethanol-induced sensitization of the pancreas to the effects of CCK. Even physiological doses of CCK can activate zymogens in ethanol-sensitized acinar cells (133). In addition, in animal models chronic ethanol feeding sensitizes the pancreas to injury by physiological concentrations of caerulein (134). In humans some additional factor probably sensitizes the pancreas to injury by physiological doses of CCK.

With regard to PAF and the human disease, clinical trials have used a PAF receptor antagonist (Lexipafant) that binds to and blocks the PAF receptor. Different studies show conflicting results (see review in Reference 99). The initial pilot study found a significantly reduced incidence of organ failure and lower cytokine levels in patients administered Lexipafant compared with placebo results (100). Similarly, McKay et al. (101) examined 50 patients and reported a reduction in organ failure. However, a larger trial involving 290 patients who were randomized to placebo or Lexipafant™ demonstrated no improvement in organ injury compared with placebo (102).

Substance P, the neuropeptide released from afferent nerve endings that is important in inflammatory processes (103), plays an important role in regulating the severity of pancreatitis. Our group showed that substance P is upregulated in caerulein-induced pancreatitis and that the expression of its receptor [neurokinin 1 receptor (NK-1R)] is upregulated on acinar cells (104). Moreover, when

PAF: platelet-activating factor

NK-1R: neurokinin-1 receptor

PPT-A: preprotachykinin-A

PPT-C: preprotachykinin-C

SIRS: systemic inflammatory response syndrome

ARDS: adult respiratory distress syndrome

HSF-1: heat shock factor-1

HSP: heat shock protein

we administered caerulein to NK-1R knockout mice, there was a significant decrease in the parameters that characterize the severity of pancreatitis (104). In support of these data, preprotachykinin-A (PPT-A) gene deletion in mice protects against caerulein-induced pancreatitis and lung injury; substance P and neurokinin are products of the PPT-A gene (105). Furthermore, treatment with a NK-1R antagonist (CP-96345) protects mice against caerulein-induced acute pancreatitis and lung injury (106). Lau & Bhatia (107) recently demonstrated that in caerulein-induced pancreatitis, CP-96345 treatment suppressed the elevation of substance P, PPT-A mRNA expression, and the peprotachykinin-C (PPT-C) gene in the pancreas and lungs.

The significance of neutrophils in the pathogenesis of pancreatitis is illustrated by work investigating the depletion of neutrophils with antineutrophil serum in a model of caerulein-induced pancreatitis. In the absence of neutrophils, the severity of pancreatitis and associated lung injury is blunted (57, 108).

Tissue infiltration by inflammatory cells has traditionally been regarded as occuring late and not early in the course of pancreatitis (18, 109). However, Mayerle et al. (58) reported inflammatory cell infiltration as early as 1 h after the start of caerulein-induced pancreatitis [as assessed by CD45-positive leukocytes, pancreatic tissue myeloperoxidase activity, and pancreatic tissue polymorphonuclear (PMN) elastase expression]. Other researchers have shown that neutrophils are attracted to the site of pancreatic injury by IL-8, TNF-α, and MCP-1, which are released from macrophages, endothelial cells, and epithelial cells (110). In addition, Gukovskaya et al. (73) have demonstrated that as early as 30 min after the start of a supramaximal caerulein infusion, pancreatic acini release significantly increased amounts of TNF-α and therefore can attract neutrophils to the site of inflammation. There is also evidence that incubation of isolated neutrophils with purified pancreatic enzymes leads to the degranulation of the neutrophils and transmigration in a matrigel chamber (111). Based on this evidence, Mayerle et al. (58) postulated that, given that intracellular protease activation is an early event in acute pancreatitis, activated digestive proteases directly induce the degranulation of neutrophils, resulting in a rise in extracellular PMN elastase.

Systemic Effects

In most patients the acinar cell damage and local inflammation associated with pancreatitis resolve. However, for some unfortunate patients the disease progresses to systemic illness. Systemic inflammatory response syndrome (SIRS) is a result of uncontrolled local inflammation. These patients are at a higher risk of multiorgan failure. Pancreatitis-associated lung injury [adult respiratory distress syndrome (ARDS)] is frequently connected with early deaths in patients with severe acute pancreatitis (112). Activated neutrophils may be responsible for the development of this injury and may correlate with the occurrence and severity of experimental ARDS (113, 114). As mentioned above, substance P and its receptor (neurokinin) also play an important role in experimental ARDS induced by hyperstimulation with caerulein.

Heat Shock Proteins and Pancreatitis

Heat shock proteins (HSPs) are protective against the toxic mediators of inflammation, providing a cellular defense against the damaging effects of reactive oxygen species and cytokines (115, 116). At the time of stress the transcriptional activation of heat shock factor-1 (HSF-1) mediates the increased synthesis of HSPs (117). The members of the major cytoprotective HSPs are expressed constitutively or can be induced in the pancreas. Stresses such as heat shock, water immersion, hyperosmolarity, chemicals, or supramaximal doses of CCK induce HSP synthesis in the pancreas.

Studies examining the effects of supramaximal doses of caerulein on the expression of

HSPs in the pancreas suggest that they are upregulated during CCK-induced pancreatitis and protect against injury. Supramaximal stimulation by caerulein induces the synthesis of HSP27 and HSP70 in isolated pancreatic acini as well as in the pancreas of animals with caerulein-induced pancreatitis (118–120). Studies performed with other models of pancreatitis, including the dibutyltin dichloride (DBTC) model (121) and the arginine-induced model, provide further support for the existence of increased HSPs in pancreatitis (122). Caerulein-induced pancreatitis is more severe in HSF-1 knockout mice, which cannot induce synthesis of these HSPs, suggesting that these HSPs may constitute a self-defense mechanism (123). Further support for this role comes from studies examining the effect of HSP overexpression on the severity of pancreatitis. Prior induction of HSPs by water immersion (HSP60 and HSP70) or hyperthermia (HSP70) before caerulein administration leads to decreased severity of pancreatitis (121, 124–128). Overexpression of HSPs by other means, including ischemic preconditioning (129), the HSP70 coinducer BRX-220 (130), or transgenic overexpression (131), protects against pancreatitis.

We provided direct evidence that HSP70 plays an essential role in thermal stress–induced protection against pancreatitis via administration of antisense against HSP70 (118). Antisense but not sense HSP70 reduced the thermal stress–induced HSP70 expression, restored the ability of supramaximal caerulein stimulation to cause intrapancreatic trypsinogen activation, and abolished the protective effect of prior thermal stress against pancreatitis (118).

The exact mechanisms by which HSP70 may act to protect against caerulein-induced pancreatitis are not known. Hwang et al. (132) recently showed that basal trypsin activity and the zymogen:lysosomal ratio of cathepsin B activity before caerulein injection was higher in HSP70 knockout mice than in wild-type mice. After caerulein stimulation, trypsin activity increased twofold in HSP70 knockout mice. Thus, HSP70 expression is probably preventing lysosomal enzyme/digestive zymogen colocalization, and as a result intra-acinar cell trypsinogen activation is reduced and pancreatitis is ameliorated. This may occur in two ways. (*a*) HSP70 blocks the intracellular trafficking changes that lead to zymogen activation and pancreatitis. (*b*) HSP70 may act to prevent the pathological rise in calcium required for trypsinogen activation. In addition, HSPs may also affect other inflammatory mediators such as NF-κB (87, 124).

FUTURE DIRECTIONS

Although much progress has been made with regard to elucidating the pathogenesis of pancreatitis, no specific therapy is yet available. A major problem is that the patients present at a stage at which the initial pancreatic events have already taken place and injury is beginning to progress to a systemic phase. Attempts to target the inflammatory mediators of this systemic phase (like PAF) in the clinical setting have been largely unsuccessful. This suggests that there is an interplay of multiple factors and that a better understanding of these extracellular events is needed for the development of specific therapies to mitigate the severity of pancreatitis.

SUMMARY POINTS

1. Caerulein-induced pancreatitis is the most widely used animal model with which to study pancreatitis.
2. The pancreatic acinar cell is the initiation site of pancreatitis.
3. Intra-acinar zymogen activation is a key early event in the onset of acute pancreatitis.

4. Early in the development of pancreatitis, colocalization of lysosomal enzymes and zymogens leads to premature intra-acinar cell activation of these zymogens.
5. Supramaximal doses of caerulein cause inhibition of secretion. This results in retention of the activated zymogens inside the acinar cell, leading to cellular injury.
6. Calcium is required for zymogen activation, but it is not sufficient to cause pancreatitis.
7. Injured pancreatic acinar cells release cytokines, which result in inflammatory cell infiltration and further pancreatic injury. These inflammatory cells release more cytokines, which are responsible for multiorgan failure and mortality in some unfortunate patients.
8. Heat shock proteins protect against injury in pancreatitis.

LITERATURE CITED

1. Chiari H. 1896. Uber die Selbstverdauung des menschlichen Pankreas. *Z. Helik.* 17:69–96
2. Ivy AC, Oldberg E. 1928. A hormone mechanism for gallbladder contraction and evacuation. *Am. J. Physiol.* 65:599–613
3. Monstein HJ, Nylander AG, Salehi A, Chen D, Lundquist I, Hakanson R. 1996. Cholecystokinin-A and cholecystokinin-B/gastrin receptor mRNA expression in the gastrointestinal tract and pancreas of the rat and man. A polymerase chain reaction study. *Scand. J. Gastroenterol.* 31:383–90
4. Morisset J, Wong H, Walsh JH, Laine J, Bourassa J. 2000. Pancreatic CCK(B) receptors: their potential roles in somatostatin release and delta-cell proliferation. *Am. J. Physiol. Gastrointest. Liver Physiol.* 279:G148–56
5. Wank SA, Pisegna JR, de Weerth A. 1994. Cholecystokinin receptor family. Molecular cloning, structure, and functional expression in rat, guinea pig, and human. *Ann. N.Y. Acad. Sci.* 713:49–66
6. Zhou W, Povoski SP, Bell RHJ. 1995. Characterization of cholecystokinin receptors and messenger RNA expression in rat pancreas: evidence for expression of cholecystokinin-A receptors but not cholecystokinin-B (gastrin) receptors. *J. Surg. Res.* 58:281–89
7. Ji B, Bi Y, Simeone D, Mortensen RM, Logsdon CD. 2001. Human pancreatic acinar cells lack functional responses to cholecystokinin and gastrin. *Gastroenterology* 121:1380–90
8. Adler G, Beglinger C, Braun U, Reinshagen M, Koop I, et al. 1991. Interaction of the cholinergic system and cholecystokinin in the regulation of endogenous and exogenous stimulation of pancreatic secretion in humans. *Gastroenterology* 100:537–43
9. Beglinger C. 1999. Potential role of cholecystokinin in the development of acute pancreatitis. *Digestion* 60(Suppl. 1):61–63
10. Niederau C, Grendell JH. 1999. Role of cholecystokinin in the development and progression of acute pancreatitis and the potential of therapeutic application of cholecystokinin receptor antagonists. *Digestion* 60(Suppl. 1):69–74
11. Niederau C, Liddle RA, Ferrell LD, Grendell JH. 1986. Beneficial effects of cholecystokinin-receptor blockade and inhibition of proteolytic enzyme activity in experimental acute hemorrhagic pancreatitis in mice. Evidence for cholecystokinin as a major factor in the development of acute pancreatitis. *J. Clin. Invest.* 78:1056–63
12. Otsuki M. 2000. Pathophysiological role of cholecystokinin in humans. *J. Gastroenterol. Hepatol.* 15(Suppl.):D71–83

13. Adler G, Hupp T, Kern HF. 1979. Course and spontaneous regression of acute pancreatitis in the rat. *Virchows Arch. A* 382:31–47
14. Niederau C, Ferrell LD, Grendell JH. 1985. Caerulein-induced acute necrotizing pancreatitis in mice: protective effects of proglumide, benzotript, and secretin. *Gastroenterology* 88:1192–204
15. Tachibana I, Shirohara H, Czako L, Akiyama T, Nakano S, et al. 1997. Role of endogenous cholecystokinin and cholecystokinin-A receptors in the development of acute pancreatitis in rats. *Pancreas* 14:113–21
16. Foulis AK. 1980. Histological evidence of initiating factors in acute necrotising pancreatitis in man. *J. Clin. Pathol.* 33:1125–31
17. Kloppel G, Dreyer T, Willemer S, Kern HF, Adler G. 1986. Human acute pancreatitis: its pathogenesis in the light of immunocytochemical and ultrastructural findings in acinar cells. *Virchows Arch. A* 409:791–803
18. Lerch MM, Saluja AK, Dawra R, Ramarao P, Saluja M, Steer ML. 1992. Acute necrotizing pancreatitis in the opossum: earliest morphological changes involve acinar cells. *Gastroenterology* 103:205–13
19. Bialek R, Willemer S, Arnold R, Adler G. 1991. Evidence of intracellular activation of serine proteases in acute cerulein-induced pancreatitis in rats. *Scand. J. Gastroenterol.* 26:190–96
20. Grady T, Saluja A, Kaiser A, Steer M. 1996. Edema and intrapancreatic trypsinogen activation precede glutathione depletion during caerulein pancreatitis. *Am. J. Physiol.* 271:G20–26
21. Luthen R, Niederau C, Grendell JH. 1995. Intrapancreatic zymogen activation and levels of ATP and glutathione during caerulein pancreatitis in rats. *Am. J. Physiol.* 268:G592–604
22. Mithofer K, Fernandez-del Castillo C, Rattner D, Warshaw AL. 1998. Subcellular kinetics of early trypsinogen activation in acute rodent pancreatitis. *Am. J. Physiol.* 274:G71–79
23. Suzuki M, Isaji S, Stanten R, Frey CF, Ruebner B. 1992. Effect of protease inhibitor FUT-175 on acute hemorrhagic pancreatitis in mice. *Int. J. Pancreatol.* 11:59–65
24. Grady T, Mah'Moud M, Otani T, Rhee S, Lerch MM, Gorelick FS. 1998. Zymogen proteolysis within the pancreatic acinar cell is associated with cellular injury. *Am. J. Physiol.* 275:G1010–17
25. Saluja AK, Bhagat L, Lee HS, Bhatia M, Frossard JL, Steer ML. 1999. Secretagogue-induced digestive enzyme activation and cell injury in rat pancreatic acini. *Am. J. Physiol.* 276:G835–42
26. Schmid SW, Modlin IM, Tang LH, Stoch A, Rhee S, et al. 1998. Telenzepine-sensitive muscarinic receptors on rat pancreatic acinar cells. *Am. J. Physiol.* 274:G734–41
27. Saluja AK, Saluja M, Printz H, Zavertnik A, Sengupta A, Steer ML. 1989. Experimental pancreatitis is mediated by low-affinity cholecystokinin receptors that inhibit digestive enzyme secretion. *Proc. Natl. Acad. Sci. USA* 86:8968–71
28. Saluja A, Saluja M, Villa A, Leli U, Rutledge P, et al. 1989. Pancreatic duct obstruction in rabbits causes digestive zymogen and lysosomal enzyme colocalization. *J. Clin. Invest.* 84:1260–66
29. Saluja AK, Dawra RK, Lerch MM, Steer ML. 1992. CCK-JMV-180, an analog of cholecystokinin, releases intracellular calcium from an inositol trisphosphate-independent pool in rat pancreatic acini. *J. Biol. Chem.* 267:11202–7
30. Powers RE, Grady T, Orchard JL, Gilrane TB. 1993. Different effects of hyperstimulation by similar classes of secretagogues on the exocrine pancreas. *Pancreas* 8:58–63

31. Chaudhuri A, Kolodecik TR, Gorelick FS. 2005. Effects of increased intracellular cAMP on carbachol-stimulated zymogen activation, secretion, and injury in the pancreatic acinar cell. *Am. J. Physiol. Gastrointest. Liver Physiol.* 288:G235–43
32. Olejar T, Matej R, Zadinova M, Pouckova P. 2001. Expression of proteinase-activated receptor 2 during taurocholate-induced acute pancreatic lesion development in Wistar rats. *Int. J. Gastrointest. Cancer* 30:113–21
33. Namkung W, Han W, Luo X, Muallem S, Cho KH, et al. 2004. Protease-activated receptor 2 exerts local protection and mediates some systemic complications in acute pancreatitis. *Gastroenterology* 126:1844–59
34. Singh VP, Bhagat L, Navina S, Sharif R, Dawra R, Saluja AK. 2006. PAR-2 protects against pancreatitis by stimulating exocrine secretion. *Gut* In press
35. Sharma A, Tao X, Gopal A, Ligon B, Andrade-Gordon P, et al. 2005. Protection against acute pancreatitis by activation of protease-activated receptor-2. *Am. J. Physiol. Gastrointest. Liver Physiol.* 288:G388–95
36. Matozaki T, Martinez J, Williams JA. 1989. A new CCK analogue differentiates two functionally distinct CCK receptors in rat and mouse pancreatic acini. *Am. J. Physiol.* 257:G594–600
37. Kiehne K, Herzig KH, Otte JM, Folsch UR. 2002. Low-affinity CCK-1 receptors inhibit bombesin-stimulated secretion in rat pancreatic acini—implication of the actin cytoskeleton. *Regul. Pept.* 105:131–37
38. Stock C, Launay JF, Grenier JF, Bauduin H. 1978. Pancreatic acinar cell changes induced by caerulein, vinblastine, deuterium oxide, and cytochalasin B in vitro. *Lab. Invest.* 38:157–64
39. Williams JA. 1977. Effects of cytochalasin B on pancreatic acinar cell structure and secretion. *Cell Tissue Res.* 179:453–66
40. Jungermann J, Lerch MM, Weidenbach H, Lutz MP, Kruger B, Adler G. 1995. Disassembly of rat pancreatic acinar cell cytoskeleton during supramaximal secretagogue stimulation. *Am. J. Physiol.* 268:G328–38
41. O'Konski MS, Pandol SJ. 1990. Effects of caerulein on the apical cytoskeleton of the pancreatic acinar cell. *J. Clin. Invest.* 86:1649–57
42. O'Konski MS, Pandol SJ. 1993. Cholecystokinin JMV-180 and caerulein effects on the pancreatic acinar cell cytoskeleton. *Pancreas* 8:638–46
43. Saluja A, Hashimoto S, Saluja M, Powers RE, Meldolesi J, Steer ML. 1987. Subcellular redistribution of lysosomal enzymes during caerulein-induced pancreatitis. *Am. J. Physiol.* 253:G508–16
44. Watanabe O, Baccino FM, Steer ML, Meldolesi J. 1984. Supramaximal caerulein stimulation and ultrastructure of rat pancreatic acinar cell: early morphological changes during development of experimental pancreatitis. *Am. J. Physiol.* 246:G457–67
45. Hirano T. 1999. Cytokine suppressive agent improves survival rate in rats with acute pancreatitis of closed duodenal loop. *J. Surg. Res.* 81:224–29
46. Hirano T, Manabe T, Imanishi K, Tobe T. 1993. Protective effect of a cephalosporin, Shiomarin, plus a new potent protease inhibitor, E3123, on rat taurocholate-induced pancreatitis. *J. Gastroenterol. Hepatol.* 8:52–59
47. Otani T, Chepilko SM, Grendell JH, Gorelick FS. 1998. Codistribution of TAP and the granule membrane protein GRAMP-92 in rat caerulein-induced pancreatitis. *Am. J. Physiol.* 275: G999–1009
48. Figarella C, Miszczuk-Jamska B, Barrett AJ. 1988. Possible lysosomal activation of pancreatic zymogens. Activation of both human trypsinogens by cathepsin B and spontaneous acid. Activation of human trypsinogen 1. *Biol. Chem. Hoppe Seyler* 369(Suppl.):293–98

49. Greenbaum LM, Hirshkowitz A, Shoichet I. 1959. The activation of trypsinogen by cathepsin B. *J. Biol. Chem.* 234:2885–90
50. Lerch MM, Saluja AK, Dawra R, Saluja M, Steer ML. 1993. The effect of chloroquine administration on two experimental models of acute pancreatitis. *Gastroenterology* 104:1768–79
51. Saluja AK, Donovan EA, Yamanaka K, Yamaguchi Y, Hofbauer B, Steer ML. 1997. Cerulein-induced in vitro activation of trypsinogen in rat pancreatic acini is mediated by cathepsin B. *Gastroenterology* 113:304–10
52. Korsten MA, Dlugosz JW. 1993. Cathepsin B inhibition in two models of acute pancreatitis. *Int. J. Pancreatol.* 14:149–55
53. Van Acker GJ, Saluja AK, Bhagat L, Singh VP, Song AM, Steer ML. 2002. Cathepsin B inhibition prevents trypsinogen activation and reduces pancreatitis severity. *Am. J. Physiol. Gastrointest. Liver Physiol.* 283:G794–800
54. Halangk W, Lerch MM, Brandt-Nedelev B, Roth W, Ruthenbuerger M, et al. 2000. Role of cathepsin B in intracellular trypsinogen activation and the onset of acute pancreatitis. *J. Clin. Invest.* 106:773–81
55. Halangk W, Kruger B, Ruthenburger M, Sturzebecher J, Albrecht E, et al. 2002. Trypsin activity is not involved in premature, intrapancreatic trypsinogen activation. *Am. J. Physiol. Gastrointest. Liver Physiol.* 282:G367–74
56. Singh VP, Saluja AK, Bhagat L, van Acker GJ, Song AM, et al. 2001. Phosphatidylinositol 3-kinase-dependent activation of trypsinogen modulates the severity of acute pancreatitis. *J. Clin. Invest.* 108:1387–95
57. Gukovskaya AS, Vaquero E, Zaninovic V, Gorelick FS, Lusis AJ, et al. 2002. Neutrophils and NADPH oxidase mediate intrapancreatic trypsin activation in murine experimental acute pancreatitis. *Gastroenterology* 122:974–84
58. Mayerle J, Schnekenburger J, Kruger B, Kellermann J, Ruthenburger M, et al. 2005. Extracellular cleavage of E-cadherin by leukocyte elastase during acute experimental pancreatitis in rats. *Gastroenterology* 129:1251–67
59. Petersen OH, Gerasimenko OV, Gerasimenko JV, Mogami H, Tepikin AV. 1998. The calcium store in the nuclear envelope. *Cell Calcium* 23:87–90
60. Nicotera P, Bellomo G, Orrenius S. 1992. Calcium-mediated mechanisms in chemically induced cell death. *Annu. Rev. Pharmacol. Toxicol.* 32:449–70
61. Matozaki T, Goke B, Tsunoda Y, Rodriguez M, Martinez J, Williams JA. 1990. Two functionally distinct cholecystokinin receptors show different modes of action on Ca^{2+} mobilization and phospholipid hydrolysis in isolated rat pancreatic acini. Studies using a new cholecystokinin analog, JMV-180. *J. Biol. Chem.* 265:6247–54
62. Thorn P, Lawrie AM, Smith PM, Gallacher DV, Petersen OH. 1993. Ca^{2+} oscillations in pancreatic acinar cells: spatiotemporal relationships and functional implications. *Cell Calcium* 14:746–57
63. Maruyama Y, Petersen OH. 1994. Delay in granular fusion evoked by repetitive cytosolic Ca^{2+} spikes in mouse pancreatic acinar cells. *Cell Calcium* 16:419–30
64. Mithofer K, Fernandez-del Castillo C, Frick TW, Lewandrowski KB, Rattner DW, Warshaw AL. 1995. Acute hypercalcemia causes acute pancreatitis and ectopic trypsinogen activation in the rat. *Gastroenterology* 109:239–46
65. Fernandez-del Castillo C, Harringer W, Warshaw AL, Vlahakes GJ, Koski G, et al. 1991. Risk factors for pancreatic cellular injury after cardiopulmonary bypass. *N. Engl. J. Med.* 325:382–87

66. Ward JB, Sutton R, Jenkins SA, Petersen OH. 1996. Progressive disruption of acinar cell calcium signaling is an early feature of cerulein-induced pancreatitis in mice. *Gastroenterology* 111:481–91
67. Kruger B, Albrecht E, Lerch MM. 2000. The role of intracellular calcium signaling in premature protease activation and the onset of pancreatitis. *Am. J. Pathol.* 157:43–50
68. Raraty M, Ward J, Erdemli G, Vaillant C, Neoptolemos JP, et al. 2000. Calcium-dependent enzyme activation and vacuole formation in the apical granular region of pancreatic acinar cells. *Proc. Natl. Acad. Sci. USA* 97:13126–31
69. Mooren FC, Turi S, Gunzel D, Schlue WR, Domschke W, et al. 2001. Calcium-magnesium interactions in pancreatic acinar cells. *FASEB J.* 15:659–72
70. Mooren F, Hlouschek V, Finkes T, Turi S, Weber IA, et al. 2003. Early changes in pancreatic acinar cell calcium signaling after pancreatic duct obstruction. *J. Biol. Chem.* 278:9361–69
71. Husain SZ, Prasad P, Grant WM, Kolodecik TR, Nathanson MH, Gorelick FS. 2005. The ryanodine receptor mediates early zymogen activation in pancreatitis. *Proc. Natl. Acad. Sci. USA* 102:14386–91
72. Blinman TA, Gukovsky I, Mouria M, Zaninovic V, Livingston E, et al. 2000. Activation of pancreatic acinar cells on isolation from tissue: cytokine upregulation via p38 MAP kinase. *Am. J. Physiol. Cell Physiol.* 279:C1993–2003
73. Gukovskaya AS, Gukovsky I, Zaninovic V, Song M, Sandoval D, et al. 1997. Pancreatic acinar cells produce, release, and respond to tumor necrosis factor-α. Role in regulating cell death and pancreatitis. *J. Clin. Invest.* 100:1853–62
74. Han B, Logsdon CD. 1999. Cholecystokinin induction of mob-1 chemokine expression in pancreatic acinar cells requires NF-κB activation. *Am. J. Physiol.* 277:C74–82
75. Yu JH, Lim JW, Namkung W, Kim H, Kim KH. 2002. Suppression of cerulein-induced cytokine expression by antioxidants in pancreatic acinar cells. *Lab. Invest.* 82:1359–68
76. Grady T, Liang P, Ernst SA, Logsdon CD. 1997. Chemokine gene expression in rat pancreatic acinar cells is an early event associated with acute pancreatitis. *Gastroenterology* 113:1966–75
77. Wulczyn FG, Krappmann D, Scheidereit C. 1996. The NF-κB/Rel and IκB gene families: mediators of immune response and inflammation. *J. Mol. Med.* 74:749–69
78. Gukovsky I, Gukovskaya AS, Blinman TA, Zaninovic V, Pandol SJ. 1998. Early NF-κB activation is associated with hormone-induced pancreatitis. *Am. J. Physiol.* 275:G1402–14
79. Satoh A, Gukovskaya AS, Nieto JM, Cheng JH, Gukovsky I, et al. 2004. PKC-delta and -epsilon regulate NF-κB activation induced by cholecystokinin and TNF-alpha in pancreatic acinar cells. *Am. J. Physiol. Gastrointest. Liver Physiol.* 287:G582–91
80. Dunn JA, Li C, Ha T, Kao RL, Browder W. 1997. Therapeutic modification of nuclear factor kappa B binding activity and tumor necrosis factor-alpha gene expression during acute biliary pancreatitis. *Am. Surg.* 63:1036–43; discussion 43–44
81. Vaquero E, Gukovsky I, Zaninovic V, Gukovskaya AS, Pandol SJ. 2001. Localized pancreatic NF-κB activation and inflammatory response in taurocholate-induced pancreatitis. *Am. J. Physiol. Gastrointest. Liver Physiol.* 280:G1197–208
82. Zaninovic V, Gukovskaya AS, Gukovsky I, Mouria M, Pandol SJ. 2000. Cerulein upregulates ICAM-1 in pancreatic acinar cells, which mediates neutrophil adhesion to these cells. *Am. J. Physiol. Gastrointest. Liver Physiol.* 279:G666–76
83. Letoha T, Somlai C, Takacs T, Szabolcs A, Rakonczay ZJ, et al. 2005. The proteasome inhibitor MG132 protects against acute pancreatitis. *Free Radic. Biol. Med.* 39:1142–51

84. Virlos I, Mazzon E, Serraino I, Di Paola R, Genovese T, et al. 2003. Pyrrolidine dithiocarbamate reduces the severity of cerulein-induced murine acute pancreatitis. *Shock* 20:544–50
85. Steinle AU, Weidenbach H, Wagner M, Adler G, Schmid RM. 1999. NF-κB/Rel activation in cerulein pancreatitis. *Gastroenterology* 116:420–30
86. Han B, Ji B, Logsdon CD. 2001. CCK independently activates intracellular trypsinogen and NF-κB in rat pancreatic acinar cells. *Am. J. Physiol. Cell Physiol.* 280:C465–72
87. Hietaranta AJ, Singh VP, Bhagat L, van Acker GJ, Song AM, et al. 2001. Water immersion stress prevents caerulein-induced pancreatic acinar cell NF-κB activation by attenuating caerulein-induced intracellular Ca^{2+} changes. *J. Biol. Chem.* 276:18742–47
88. Davies MG, Hagen PO. 1997. Systemic inflammatory response syndrome. *Br. J. Surg.* 84:920–35
89. Makhija R, Kingsnorth AN. 2002. Cytokine storm in acute pancreatitis. *J. Hepatobiliary Pancreat. Surg.* 9:401–10
90. Liu LR, Xia SH. 2006. Role of platelet-activating factor in the pathogenesis of acute pancreatitis. *World J. Gastroenterol.* 12:539–45
91. Ishii S, Shimizu T. 2000. Platelet-activating factor (PAF) receptor and genetically engineered PAF receptor mutant mice. *Prog. Lipid Res.* 39:41–82
92. Stafforini DM, McIntyre TM, Zimmerman GA, Prescott SM. 2003. Platelet-activating factor, a pleiotrophic mediator of physiological and pathological processes. *Crit. Rev. Clin. Lab. Sci.* 40:643–72
93. Zhou W, Levine BA, Olson MS. 1993. Platelet-activating factor: a mediator of pancreatic inflammation during cerulein hyperstimulation. *Am. J. Pathol.* 142:1504–12
94. Hofbauer B, Saluja AK, Bhatia M, Frossard JL, Lee HS, et al. 1998. Effect of recombinant platelet-activating factor acetylhydrolase on two models of experimental acute pancreatitis. *Gastroenterology* 115:1238–47
95. Ais G, Lopez-Farre A, Gomez-Garre DN, Novo C, Romeo JM, et al. 1992. Role of platelet-activating factor in hemodynamic derangements in an acute rodent pancreatic model. *Gastroenterology* 102:181–87
96. Dabrowski A, Gabryelewicz A, Chyczewski L. 1995. The effect of platelet activating factor antagonist (BN 52021) on acute experimental pancreatitis with reference to multiorgan oxidative stress. *Int. J. Pancreatol.* 17:173–80
97. Fujimura K, Kubota Y, Ogura M, Yamaguchi T, Binnaka T, et al. 1992. Role of endogenous platelet-activating factor in caerulein-induced acute pancreatitis in rats: protective effects of a PAF-antagonist. *J. Gastroenterol. Hepatol.* 7:199–202
98. Lane JS, Todd KE, Gloor B, Chandler CF, Kau AW, et al. 2001. Platelet activating factor antagonism reduces the systemic inflammatory response in a murine model of acute pancreatitis. *J. Surg. Res.* 99:365–70
99. Abu-Zidan FM, Windsor JA. 2002. Lexipafant and acute pancreatitis: a critical appraisal of the clinical trials. *Eur. J. Surg.* 168:215–19
100. Kingsnorth AN, Galloway SW, Formela LJ. 1995. Randomized, double-blind phase II trial of Lexipafant, a platelet-activating factor antagonist, in human acute pancreatitis. *Br. J. Surg.* 82:1414–20
101. McKay CJ, Curran F, Sharples C, Baxter JN, Imrie CW. 1997. Prospective placebo-controlled randomized trial of lexipafant in predicted severe acute pancreatitis. *Br. J. Surg.* 84:1239–43

102. Johnson CD, Kingsnorth AN, Imrie CW, McMahon MJ, Neoptolemos JP, et al. 2001. Double blind, randomised, placebo controlled study of a platelet activating factor antagonist, lexipafant, in the treatment and prevention of organ failure in predicted severe acute pancreatitis. *Gut* 48:62–69
103. Bowden JJ, Garland AM, Baluk P, Lefevre P, Grady EF, et al. 1994. Direct observation of substance P-induced internalization of neurokinin 1 (NK1) receptors at sites of inflammation. *Proc. Natl. Acad. Sci. USA* 91:8964–68
104. Bhatia M, Saluja AK, Hofbauer B, Frossard JL, Lee HS, et al. 1998. Role of substance P and the neurokinin 1 receptor in acute pancreatitis and pancreatitis-associated lung injury. *Proc. Natl. Acad. Sci. USA* 95:4760–65
105. Bhatia M, Slavin J, Cao Y, Basbaum AI, Neoptolemos JP. 2003. Preprotachykinin-A gene deletion protects mice against acute pancreatitis and associated lung injury. *Am. J. Physiol. Gastrointest. Liver Physiol.* 284:G830–36
106. Lau HY, Wong FL, Bhatia M. 2005. A key role of neurokinin 1 receptors in acute pancreatitis and associated lung injury. *Biochem. Biophys. Res. Commun.* 327:509–15
107. Lau HY, Bhatia M. 2006. The effect of CP96,345 on the expression of tachykinins and neurokinin receptors in acute pancreatitis. *J. Pathol.* 208:364–71
108. Frossard JL, Saluja A, Bhagat L, Lee HS, Bhatia M, et al. 1999. The role of intercellular adhesion molecule 1 and neutrophils in acute pancreatitis and pancreatitis-associated lung injury. *Gastroenterology* 116:694–701
109. Lerch MM, Saluja AK, Runzi M, Dawra R, Steer ML. 1995. Luminal endocytosis and intracellular targeting by acinar cells during early biliary pancreatitis in the opossum. *J. Clin. Invest.* 95:2222–31
110. Bhatia M, Brady M, Shokuhi S, Christmas S, Neoptolemos JP, Slavin J. 2000. Inflammatory mediators in acute pancreatitis. *J. Pathol.* 190:117–25
111. Keck T, Balcom JH, Fernandez-del Castillo C, Antoniu BA, Warshaw AL. 2002. Matrix metalloproteinase-9 promotes neutrophil migration and alveolar capillary leakage in pancreatitis-associated lung injury in the rat. *Gastroenterology* 122:188–201
112. Robertson CS, Basran GS, Hardy JG. 1988. Lung vascular permeability in patients with acute pancreatitis. *Pancreas* 3:162–65
113. Bhatia M, Saluja AK, Hofbauer B, Lee HS, Frossard JL, Steer ML. 1998. The effects of neutrophil depletion on a completely noninvasive model of acute pancreatitis-associated lung injury. *Int. J. Pancreatol.* 24:77–83
114. O'Donovan DA, Kelly CJ, Bouchier-Hayes DM, Grace P, Redmond HP, et al. 1995. Alpha-1-antichymotrypsin is an effective inhibitor of pancreatitis-induced lung injury. *Eur. J. Gastroenterol. Hepatol.* 7:847–52
115. Jacquier-Sarlin MR, Fuller K, Dinh-Xuan AT, Richard MJ, Polla BS. 1994. Protective effects of hsp70 in inflammation. *Experientia* 50:1031–38
116. Polla BS, Perin M, Pizurki L. 1993. Regulation and functions of stress proteins in allergy and inflammation. *Clin. Exp. Allergy* 23:548–56
117. Morimoto RI. 1993. Cells in stress: transcriptional activation of heat shock genes. *Science* 259:1409–10
118. Bhagat L, Singh VP, Song AM, van Acker GJ, Agrawal S, et al. 2002. Thermal stress-induced HSP70 mediates protection against intrapancreatic trypsinogen activation and acute pancreatitis in rats. *Gastroenterology* 122:156–65
119. Ethridge RT, Ehlers RA, Hellmich MR, Rajaraman S, Evers BM. 2000. Acute pancreatitis results in induction of heat shock proteins 70 and 27 and heat shock factor-1. *Pancreas* 21:248–56

120. Weber CK, Gress T, Muller-Pillasch F, Lerch MM, Weidenbach H, Adler G. 1995. Supramaximal secretagogue stimulation enhances heat shock protein expression in the rat pancreas. *Pancreas* 10:360–67
121. Weber H, Wagner AC, Jonas L, Merkord J, Hofken T, et al. 2000. Heat shock response is associated with protection against acute interstitial pancreatitis in rats. *Dig. Dis. Sci.* 45:2252–64
122. Tashiro M, Schafer C, Yao H, Ernst SA, Williams JA. 2001. Arginine induced acute pancreatitis alters the actin cytoskeleton and increases heat shock protein expression in rat pancreatic acinar cells. *Gut* 49:241–50
123. Bhagat L, van Acker GJ, Manzoor R, Singh VP, Song AM, et al. 2002. Targeted deletion of heat shock factor 1 (HSF-1) exacerbates the severity of sectretagogue-induced pancreatitis. *Pancreas* 25:421
124. Frossard JL, Pastor CM, Hadengue A. 2001. Effect of hyperthermia on NF-κB binding activity in cerulein-induced acute pancreatitis. *Am. J. Physiol. Gastrointest. Liver Physiol.* 280:G1157–62
125. Lee HS, Bhagat L, Frossard JL, Hietaranta A, Singh VP, et al. 2000. Water immersion stress induces heat shock protein 60 expression and protects against pancreatitis in rats. *Gastroenterology* 119:220–29
126. Metzler W, Hofken T, Weber H, Printz H, Goke B, Wagner AC. 1999. Hyperthermia, inducing pancreatic heat-shock proteins, fails to prevent cerulein-induced stress kinase activation. *Pancreas* 19:150–57
127. Otaka M, Itoh H, Kuwabara T, Zeniya A, Fujimori S, et al. 1993. Induction of a 60-kDa heat shock protein in rat pancreas by water-immersion stress. *Int. J. Biochem.* 25:1769–73
128. Rakonczay ZJ, Takacs T, Mandi Y, Ivanyi B, Varga S, et al. 2001. Water immersion pretreatment decreases proinflammatory cytokine production in cholecystokinin-octapeptide-induced acute pancreatitis in rats: possible role of HSP72. *Int. J. Hyperthermia* 17:520–35
129. Warzecha Z, Dembinski A, Ceranowicz P, Konturek SJ, Dembinski M, et al. 2005. Ischemic preconditioning inhibits development of edematous cerulein-induced pancreatitis: involvement of cyclooxygenases and heat shock protein 70. *World J. Gastroenterol.* 11:5958–65
130. Rakonczay ZJ, Ivanyi B, Varga I, Boros I, Jednakovits A, et al. 2002. Nontoxic heat shock protein coinducer BRX-220 protects against acute pancreatitis in rats. *Free Radic. Biol. Med.* 32:1283–92
131. Kubisch C, Dimagno MJ, Tietz AB, Welsh MJ, Ernst SA, et al. 2004. Overexpression of heat shock protein Hsp27 protects against cerulein-induced pancreatitis. *Gastroenterology* 127:275–86
132. Hwang JH, Ryu JK, Yoon YB, Lee KH, Park YS, et al. 2005. Spontaneous activation of pancreas trypsinogen in heat shock protein 70.1 knock-out mice. *Pancreas* 31:332–36
133. Katz M, Carangelo R, Miller LJ, Gorelick F. 1996. Effect of ethanol on cholecystokinin-stimulated zymogen conversion in pancreatic acinar cells. *Am. J. Physiol.* 270:G171–75
134. Pandol SJ, Periskic S, Gukovsky I, Zaninovic V, Jung Y, et al. 1999. Ethanol diet increases the sensitivity of rats to pancreatitis induced by cholecystokinin octapeptide. *Gastroenterology* 117:706–16

Timing and Computation in Inner Retinal Circuitry

Stephen A. Baccus

Department of Neurobiology, Stanford University School of Medicine, Stanford, California 94305; email: baccus@stanford.edu

Annu. Rev. Physiol. 2007. 69:271–90

First published online as a Review in Advance on October 20, 2006

The *Annual Review of Physiology* is online at http://physiol.annualreviews.org

This article's doi: 10.1146/annurev.physiol.69.120205.124451

0066-4278/07/0315-0271$20.00

Key Words

neural circuit, neural coding, amacrine cells, interneurons, motion processing

Abstract

In the vertebrate inner retina, the second stage of the visual system, different components of the visual scene are transformed, discarded, or selected before visual information is transmitted through the optic nerve. This review discusses the connections between higher-level functions of visual processing, mathematical descriptions of the neural code, inner retinal circuitry, and visual computations. In the inner plexiform layer, bipolar cells deliver spatially and temporally filtered input to approximately ten anatomical strata. These layers receive a unique combination of excitation and inhibition, causing cells in different layers to respond with different kinetics to visual input. These distinct temporal channels interact through amacrine cells, a diverse class of inhibitory interneurons, which transmit signals within and between layers. In particular, wide-field amacrine cells transmit transient inhibition over long distances within a layer. These mechanisms and properties are combined into computations to detect the presence of differential motion and suppress the visual effects of eye movements.

Inner plexiform layer (IPL): the second synaptic layer of the vertebrate retina, containing approximately ten strata, most with distinct response kinetics

INTRODUCTION

Explaining how a neural circuit functions requires an understanding of how diverse lower-level mechanisms combine to transform information to perform a higher-level task. In the vertebrate retina, more than a century of anatomical measurements (1), more than half a century of physiological measurements of light responses (2), decades of modeling of the neural code (3), and recent descriptions of new phenomena have begun to lay a foundation for this understanding.

The focus of this review is the vertebrate cone pathway, and anatomical and computational properties of the rod pathway are not addressed here. The retina encodes the visual scene into a sequence of action potentials, using two synaptic layers, one in the outer retina and one in the inner retina. Bipolar cells serve as both the output of the first layer and input to the second layer (**Figure 1**). In the outer plexiform layer, photoreceptor signals are transmitted to bipolar cells and are modified by up to several types of horizontal cells. In the inner plexiform layer (IPL), bipolar cell signals are transmitted to ganglion cells and are modified by up to 30 types of amacrine cells (4). Consequently, the neural circuitry and synaptic organization of the inner retina are substantially more complex than in the outer retina.

The work of Kuffler (5) presents the current classical view of retinal signaling: Retinal ganglion cells respond with an excitatory center and an opposing, inhibitory surround. These pioneering experiments were codified by early mathematical models of the retina, which described the output of ganglion cells as a weighted sum over space and time of light intensity in the receptive field center and surround (3). In this view, for an On-type ganglion cell, each photon in the center region leads to an increase in firing, each photon in the surround leads to a delayed decrease in firing, and the result of this sum is thresholded to yield a positive rate. Accordingly, the canonical picture of the retina is that this simple linear weighted sum accounts for the output of the entire retina.

However, much of this linear center-surround organization derives from outer retinal processing and can be seen in the responses of bipolar cells (6). A strong deviation from this linear processing is the property of light adaptation, whereby the retina adjusts its sensitivity and kinetics, depending on the ambient light level (7). Once again, properties of photoreceptors and the outer retina can account for much (but not all) of this adaptive processing.

What then is the function of the inner retina's complex circuitry? Recent results have

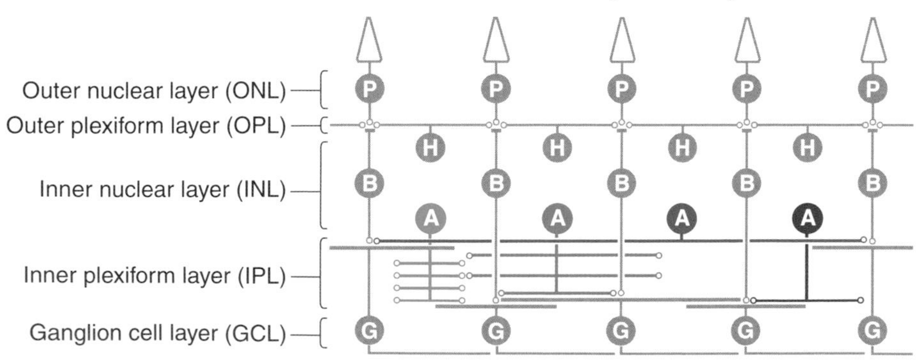

Figure 1

The vertebrate retina. A schematic diagram of the main cell types and general circuit arrangement in the vertebrate retina. Cell types: P, photoreceptor; H, horizontal cell; B, bipolar cell; A, amacrine cell; G, ganglion cell. This figure highlights the diverse population of amacrine cells, including monostratified cells, diffuse cells that span multiple strata in the IPL, narrow-field cells that cover a small range of visual space, and wide-field cells that extend their processes over long distances in the retinal plane.

shown that the output of photoreceptors is processed and transformed in ways that cannot be accounted for by a simple thresholded, weighted sum over space and time. These findings have been gathered from different species, including salamander, fish, rabbit, guinea pig, cat, and primate. Although species differences occur, there is often great commonality in most basic phenomena, and findings are pooled freely here. The following phenomena of visual processing arise at least in part from circuitry in the inner retina.

NEURAL CODING IN THE INNER RETINA

Concerted Signaling

Retinal ganglion cells do not sample independently from the visual scene but coordinate their outputs to encode the stimulus. The most evident aspect of this property is that nearby cells of similar functional type fire synchronously far more often than expected by chance (8–11). Combinations of several ganglion cells fire in concerted patterns that signal the presence of common visual input (11). These firing patterns arise owing to common chemical and electrical synaptic input (12). Some of this shared input arises in the inner retina from electrical synapses between amacrine and ganglion cells or between ganglion cells.

Contrast Adaptation

The early visual system adapts in multiple ways to use its dynamic range more efficiently. Light adaptation in phototransduction (7, 13) and retinal circuitry (14, 15) allows the visual system to operate across a large range of mean luminance. Contrast adaptation is a process that adjusts sensitivity in the inner retina; changes in the range of light intensity fluctuations cause changes in retinal sensitivity and response kinetics (16–21). This process allows a neuron to avoid saturation and to use its dynamic range more efficiently by decreasing its sensitivity in response to a large input (22). This adaptive process is inherited by higher-level circuits in the lateral geniculate nucleus, particularly in the primate magnocellular pathway (23), and can be observed in the cortex (24) and in visual perception (25). Adaptation to contrast occurs over multiple timescales, ranging from <0.1 s, termed fast contrast adaptation (19) or contrast gain control (26), to tens of seconds, termed slow contrast adaptation (16). During slow contrast adaptation, changes in sensitivity occur in part at the bipolar-to-ganglion-cell synapse, prior to summation within the ganglion cell (27), and in part within the ganglion cell (18, 19, 28).

Dynamic Predictive Coding

Recently, Hosoya et al. (29) found that the retina also adapts to spatio-temporal correlations, termed pattern adaptation. The hardwired properties of center-surround receptive fields and biphasic temporal responses reduce spatial and temporal correlations, respectively. However, the statistics of the visual scene are not static, and correlations between nearby points in space and time can change. For example, in a forest, tree trunks viewed from a distance cause image points to be correlated vertically but not horizontally. Viewed up close, bark and leaves present a more uncorrelated environment. An efficient encoder of the visual scene should adjust to these different predictable statistics of the stimulus. In fact, retinal ganglion cells adapt over a few seconds to encode the current visual environment more efficiently (29). When exposed to either horizontal or vertical lines, correlations in time, or spatio-temporal correlations, ganglion cell receptive fields change shape so as to decrease sensitivity to the dominant correlation structure of the visual scene. This reduction in sensitivity to current spatio-temporal correlations implements a dynamic version of a strategy called predictive coding (30). This process decreases sensitivity to predictable features of the stimulus and increases

DS: direction selective

Object motion sensitive (OMS) ganglion cells: a class of ganglion cells that respond to local, object motion but not global, background motion

sensitivity to novel features, thus increasing the efficiency of the representation of the visual scene. This type of retinal adaptation to correlations requires inhibitory transmission in the inner retina. According to one hypothesis, pattern adaptation may result from anti-Hebbian plasticity at inhibitory synapses, but such a mechanism has yet to be identified.

Direction Selectivity

Some ganglion cells have a strong selectivity for the direction of motion (31, 32). Although most often studied in the rabbit, they have been identified in diverse mammalian and nonmammalian species (for references, see Reference 33). For these direction-selective (DS) ganglion cells, motion in the most preferred direction strongly activates the cell, whereas motion in the opposite direction evokes almost no activity. Although most research focuses on these two extremes of the cell's response, DS ganglion cells have a broad tuning curve, gradually varying their response to encode the direction of motion (31, 32). One important component of the circuit underlying directional sensitivity is a type of amacrine cell known as the starburst amacrine cell, which is required for direction selectivity (34, 35). Individual dendrites of starburst amacrine cells are themselves directionally selective, and selective input from individual dendrites of these amacrine cells may convey directionally selective inhibition onto DS ganglion cells (35–37). Multiple presynaptic and postsynaptic mechanisms combine to produce direction selectivity, and this phenomenon and circuit are among the most thoroughly studied aspects of inner retinal visual processing. More detailed knowledge of the circuit for direction selectivity exists than can be addressed here, and the topic is reviewed in Reference 38.

Motion Anticipation

Changes in light intensity are encoded with a delay that arises from multiple mechanisms, including phototransduction and neurotransmission. Moving objects change position during this delay, creating the potential problem that the neural representation of a moving object will always lag behind the object's location. The visual system compensates for these delays by displacing the representation of a continuously moving object in the direction of motion, effectively representing the object in advance of its immediate location. Psychophysical experiments show that subjects report the apparent position of a moving object as shifted relative to a stationary object (39). This process begins in the retina as the population of retinal ganglion cells activated by a moving object shifts in the direction of motion (40).

The necessary ingredients to produce this phenomenon are simpler than one might expect: a broad receptive field and fast adjustable gain such that input to one region decreases the gain of the entire receptive field. These properties are well known in the retina. Ganglion cells sum over multiple bipolar cells, and fast contrast adaptation (contrast gain control) quickly reduces the gain of the entire ganglion cell receptive field if one region is strongly activated (41). Although retinal mechanisms do not produce the entire perceptual shift of anticipated motion, this same set of underlying mechanisms likely is present in many brain regions. Broad receptive fields and negative feedback occur in many neural circuits, with the consequence that motion across the receptive field is expected to shift the population representation in the direction of motion. This process may also underlie diverse related perceptual phenomena involving changing stimuli of different modalities (42).

Object Motion Sensitivity

Certain types of retinal ganglion cells, termed object motion sensitive (OMS) cells, respond very sensitively to differential motion between the receptive field center and the periphery, as produced by an object moving over the background, but are strongly suppressed by background motion, as produced by fixational eye movements (43). Suppression of background motion involves timed, transient inhibition that blanks out coincident transient excitation. As a result, when the entire visual field moves together, motion signals are silenced in these cells, whereas if one region experiences motion different from the background, that motion signal is transmitted.

Global Shift Suppression

During fast gaze-shifting eye movements, known as saccades, visual sensitivity to changing stimuli decreases. Retinal motion alone is sufficient to produce this perceptual suppression (44), although some component may arise from the brain's motor command centers (45, 46). When visual stimuli that mimic saccades are presented to the isolated retina, certain retinal ganglion cells are strongly inhibited with brief, well-timed inhibition (47). Both object motion sensitivity and global shift suppression involve timed inhibition from globally correlated shifting stimuli, and the two phenomena may share common circuitry.

MODELING OF RETINAL FUNCTION

To make the connection between higher-level functions such as contrast adaptation and lower-level mechanisms such as synaptic transmission, great benefit has come from mathematical modeling of retinal responses toward the goal of creating a precise description of the neural code. A feature of the retina that makes this possible is the precision to which it responds, the full extent of which has only recently been appreciated. Experiments using naturalistic stimuli with a broad temporal bandwidth produce bursts of ganglion cell spikes reproducible to within one millisecond or less, although this precision depends on the stimulus (52) (**Figure 2*a–d***). A complete description of how the retina encodes natural visual scenes is beyond the scope of current models. However, simple models have successfully described retinal responses to some simple stimuli.

A stimulus is expected to cause a cell (or an animal) to respond with some latency and for some duration. Typically, the response may have a threshold and also be limited by some maximal level. These intuitive notions can be made precise by a simple model of sensory responses, termed a linear-nonlinear (LN) model, in which the stimulus is passed through two stages (**Figure 2*e***). The stimulus intensity $s(t)$ is first passed through a linear temporal filter $F(t)$, which captures the time course of the response. The output is then transformed by a nonlinear function N into the model's output $r'(t)$.

$$r'(t) = N\left(\int s(\tau)F(t-\tau)d\tau\right). \qquad 1.$$

This model is often calculated by correlating the response with a Gaussian white-noise stimulus, which contains a broad temporal bandwidth and thus many temporal patterns, in a method known as white-noise analysis (49, 50). This model closely approximates the response of a retinal ganglion cells under a small but important range of stimuli, such as a uniform, randomly flickering stimulus. The model also can be used to fit all other classes of retinal neurons, using different filters and nonlinearities (17, 20, 50).

The operation of passing the stimulus through the linear temporal filter can be considered from two points of view. In one view, each photon yields an output equal to the time course of the filter. To compute the overall output, at each point in time the results of all photons are summed together according to a convolution of the filter with the stimulus (Equation 1). Thus, the term linear is appropriate; the response to two photons is equal to the sum of the responses to each single photon. In this view, the filter is the impulse response function of the system.

In an alternative but mathematically equivalent view, instead of computing the output by summing the responses to single photons, one can start with the goal of computing the

Global shift suppression: the inhibitory effect produced in some ganglion cells by fast, global visual shifts that mimic saccades

Linear-nonlinear (LN) model: a simple model of a cell's response; consists of a linear filter followed by a static nonlinearity

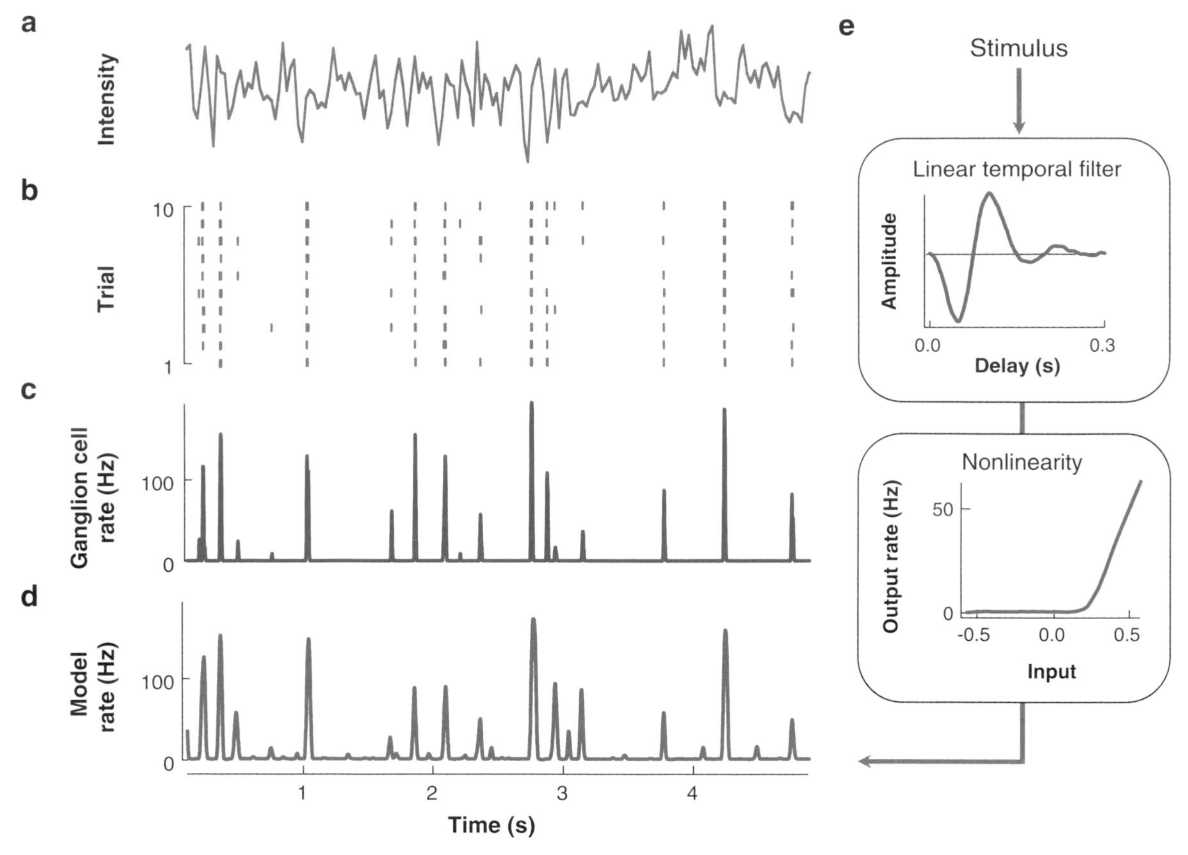

Figure 2

Precise, sparse firing in the retina captured by a simple model. (*a*) Random flicker sequence of light intensity drawn from a Gaussian distribution. (*b*) Response of a single salamander ganglion cell to repeated presentations of the sequence, recorded with a multielectrode array. (*c*) Peristimulus time histogram showing the cell's firing rate. (*d*) Firing rate predicted by a linear-nonlinear (LN) model. (*e*) LN model consisting of the stimulus intensity weighted over time by a linear temporal filter and then passed through a static nonlinearity.

output at a single point in time. To do so, one simply computes a weighted sum over the recent stimulus intensity; the weighting is equal to the reverse time course of the filter. This time-reversed filter is equal to the temporal receptive field, for which the separate elements of the receptive field are separate time points of the recent stimulus, weighted according to the sensitivity of the cell to the recent stimulus intensity. In the case of a spatially modulated stimulus, the LN model can be generalized to include a spatio-temporal linear filter, representing the spatio-temporal receptive field, for which a separate temporal filter exists for each spatial location (51).

Static nonlinearity: a generally nonlinear function that does not depend on time and can capture properties such as thresholding or saturation

Of course, biological systems are not truly linear—increasing the light intensity by a factor of one million does not yield membrane potential responses in the kilovolt range. Thus, the nonlinearity corrects for the fact that neurons have a threshold, are sensitive over a limited dynamic range, and saturate. (51).

The combined operations of the temporal filter and threshold nonlinearity produce several important consequences. The biphasic filter and threshold together emphasize rare excursions of the stimulus far away from its mean value. Thus, ganglion cell firing in response to naturalistic stimuli is sparse, composed of brief periods of firing separated by longer periods of silence (52) (**Figure 2**).

Figure 3 illustrates how the filter and nonlinearity of different cells act to produce the firing pattern of a ganglion cell. Two cells are compared, using filters of different kinetics. For each cell, a flash, the equivalent of an impulse, leads to a signal equal to the time course of the filter. A step in light intensity, when passed through the biphasic filter, leads to a transient signal that emphasizes the onset or offset of the step. Although both filters are biphasic, one filter transmits the sustained component of the stimulus more than the

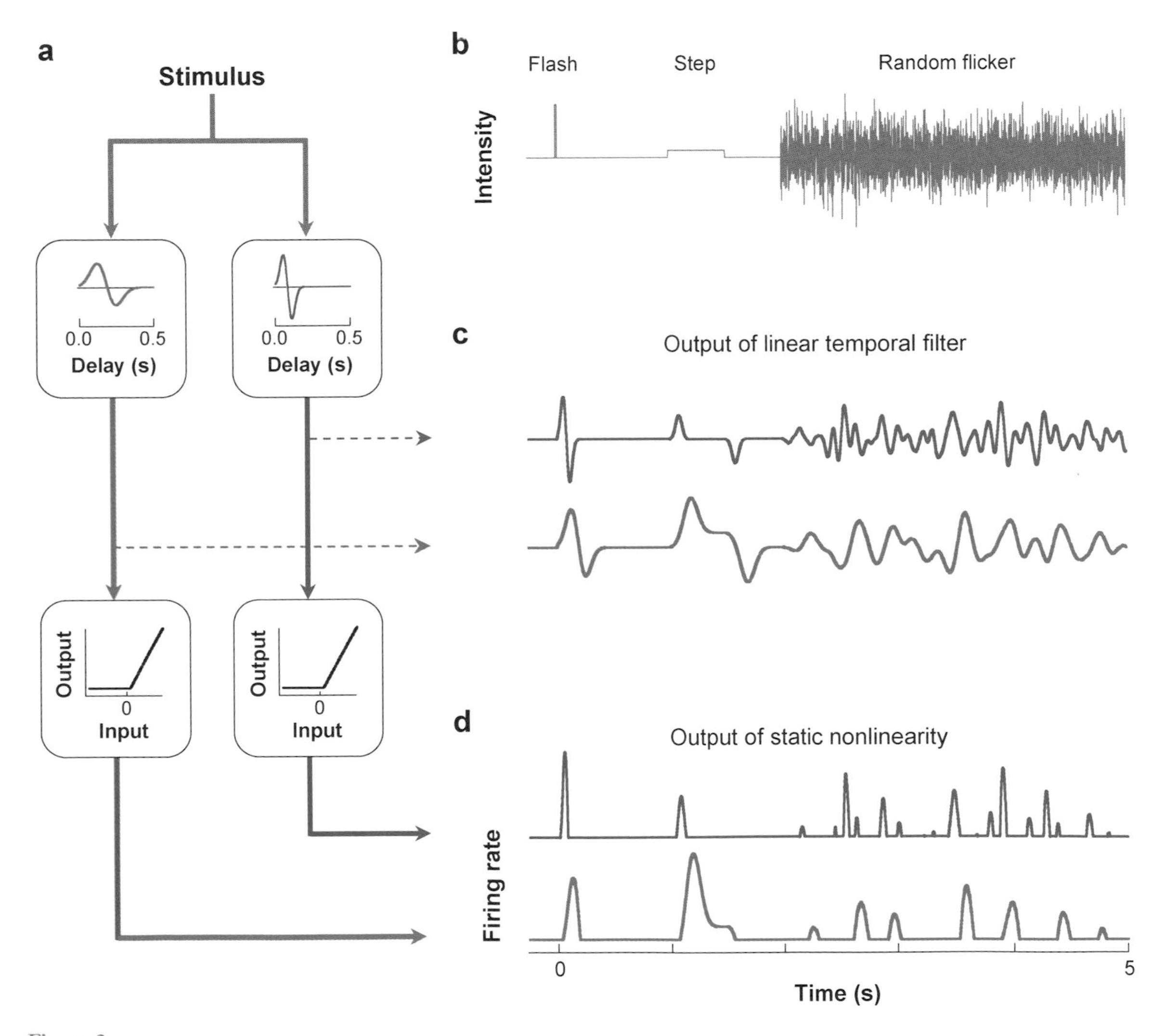

Figure 3

Effect of kinetics on response timing in a linear-nonlinear (LN) model. (*a*) Two model cells with different kinetics. (*b*) Different stimuli—a flash, a step, or random flicker—are presented. (*c*) Flash responses yield a filtered stimulus equivalent to the temporal filter. The response to a step approximates the integral of the filter. (*d*) After thresholding, a more naturalistic stimulus, random flicker, yields responses that occur at different times in the two cells.

GCL: ganglion cell layer

other does. When a more naturalistic stimulus is applied, the different time courses of the filter produce different filtered versions of the stimulus. After the filtered stimulus passes through the threshold nonlinearity, only the extreme positive values produce an output. The brief periods of firing are distinct for the two cells because the filters produce large positive values of the stimulus at different times. It is instructive to compare the results of different types of stimuli. Brief flashes give only a partial description of the response because the threshold hides information about the cell's full kinetics. Similarly, a step increase in intensity may not reveal the sparseness of the response to an ongoing, rapidly varying stimulus.

Although the LN model is quite successful at predicting the firing rate of ganglion cells in response to a uniform field stimulus, this simple model fails to capture some aspects of firing. These include the timing of individual spikes, fast negative feedback, and the variability of ganglion cell spiking. These properties have been incorporated in an accurate spiking model that captures the responses and variability of the ganglion cell response to a uniform field stimulus (53).

The spatio-temporal filter of an LN model refines the classical notion of retinal signaling. Each cell integrates over a small window of the visual field, creating a linear sum of the light intensity in that region. Kuffler (5) showed in the cat retina that ganglion cells weight this sum with an opposite sign in the receptive field center and surrounding regions. However, researchers have long known that ganglion cells are influenced by visual information a long distance away from the classic receptive field center and surround.

McIlwain first discussed that stimuli distant from the center may produce excitation, termed the periphery effect (54) and later the shift effect (55). He further identified that this effect was not a simple linear summation from the classical surround; changes in intensity of either sign in the periphery could excite a cell (56). Changing intensity far away from the classical receptive field can have varying effects. In the absence of central input, peripheral stimulation can be excitatory (55). However, Werblin & Copenhagen (57) showed that peripheral motion can also be inhibitory, suppressing the response to central stimulation. These effects are more complex than simpler linear summation, as can be accounted for by the circuitry of the outer retina. Dowling & Boycott (58) suggested that amacrine cells potentially transmit signals carried by the long-distance periphery effect.

INNER RETINAL CELLULAR PROPERTIES

Laminar organization is common in the nervous system. In the spinal cord, thalamus, hippocampus, and cortex, cells make connections within defined layers (59). In the retina, an overall laminar architecture is formed by three layers of cell bodies and two intervening synaptic layers (**Figure 1**). This laminar structure is most intricate in the second layer of synaptic processing, the IPL, where neurites of bipolar, amacrine, and ganglion cells form a dense synaptic network. Studies of the morphology of these neurons have shown that many cells restrict their processes to narrow layers. Cells range from monostratified, with dendrites or axons branching in a layer only 1 or 2 μm thick, to diffuse, spreading their processes across the entire 30-μm thickness of the IPL (60). Inventories of the stratification of different cell types have indicated that there are at least ten strata within the IPL (61).

It has long been known that these different strata in the IPL represent distinct functional pathways. Synaptic terminals of On and Off bipolar cells branch in different layers of the IPL; On bipolar cells terminate closer to the ganglion cell layer (GCL) (62, 63). Ganglion cells with DS responses arborize in specific layers of the IPL (64, 65). More recently, several systematic studies have examined different cell types and compared their fine stratification in the IPL with their light responses (61, 69, 70, 79). These studies found that different layers carry signals that have distinct kinetics.

Roska & Werblin (61) performed a detailed study in the rabbit to measure how excitatory and inhibitory input is combined within ganglion cells. For each ganglion cell, they compared synaptic currents and spiking with the layer of dendritic arborization. The authors concluded that in each layer, a ganglion cell's dendrites integrate a unique set of excitatory and inhibitory synaptic currents, leading to distinct patterns of spiking. As a result, ganglion cells with dendrites in different layers responded to a visual stimulus with different kinetics. Overall, measurements with a simple flashing square reveal approximately three to four distinct Off-type kinetics and three to four distinct On-type kinetics in the rabbit retina (**Figure 4*a***).

The visual scene maps onto the retina by a simple projection onto the two-dimensional photoreceptor layer. However, the finding that distinct layers have distinct kinetics indicates a different type of transformation: A different stimulus attribute, namely temporal selectivity, is mapped onto the retina's axial dimension. There may be some organization to this mapping because in rabbit, cells with faster kinetics have dendrites that are furthest from the ganglion cell layer (61). However, this requires further study and may depend on the species. An analogy can be made to other neural circuits for which similar stimulus features are mapped to nearby brain locations. Other examples include frequency-selective domains in the auditory cortex (66) and the visual cortex, wherein cells within a cortical column respond to similar features (67). Notably, although columns in the visual cortex have long been known, a full understanding of their functional significance has been elusive without a detailed knowledge of the neural circuitry (68).

In the retina, the anatomy and basic light responses of the circuitry spanning different sublamina of the IPL have been studied. Studies of bipolar cells with the most narrowly stratified axon terminals have indicated that

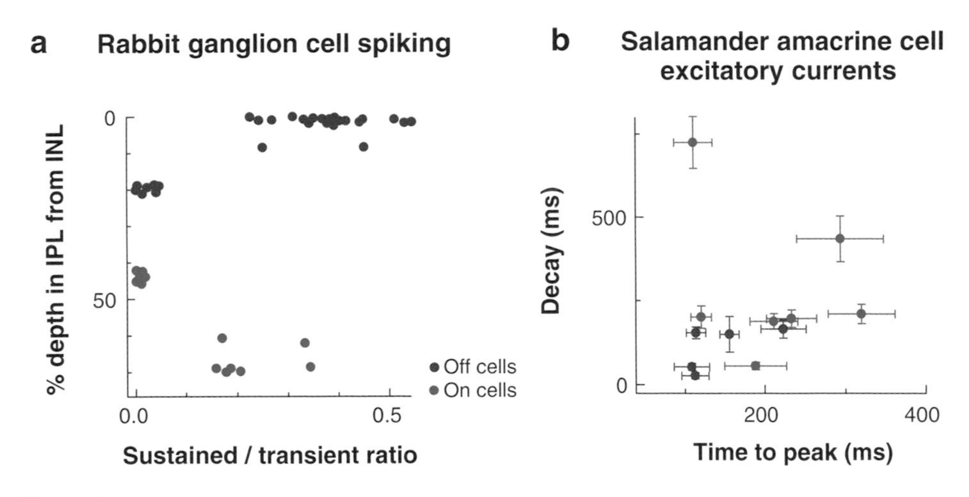

Figure 4

Different kinetics in different layers of the retina. (*a*) Kinetics of different types of rabbit monostratified ganglion cells responding to a 1-s flash, calculated as the ratio of the peak firing rate (transient component) to the response after 1 s (sustained component). The result is plotted against the depth of their dendrites in the inner plexiform layer (IPL). INL, inner nuclear layer. Adapted from Reference 61. (*b*) Kinetics of different groups of salamander monostratified amacrine cells responding to a flash, plotted as the decay time constant of excitatory currents versus the time to peak. Each point is from a group of cells arborizing in a different stratum in the IPL. Adapted from tabular data in Reference 79.

there are approximately ten distinct layers in the IPL (69, 70). Cells of a single layer share very similar light response kinetics, whereas those in different layers have different kinetics. Different excitatory currents arise in part from different types of glutamate receptors on bipolar cell dendrites and are delivered to specific layers in the IPL. There, the excitatory currents mix with distinct inhibitory currents that likely arise from amacrine transmission. Thus, there is a match between a bipolar cell's input and its output, such that visual information of a specific kinetic type relative to the visual stimulus is delivered to a specific layer of processing.

These studies do not involve a formal cluster analysis of the cells that are claimed to comprise different classes. Thus, it is not settled whether examples of cells are sampled sparsely from a large continuum or whether they are elements of distinct classes.

AMACRINE CIRCUITRY

The simple notion of bipolar cells delivering different temporal dimensions of the stimulus to different ganglion cell dendrites is complicated by the fact that amacrine cells mix these signals, making connections within layers and between layers. Of the basic retinal cell types, the least is known about amacrine cells, an extremely diverse class of inhibitory cells (71). These cells vary in their light responses, morphology, molecular profiles, and synaptic connections (71, 72). However, much more is known about these mechanistic details than about their specific functional contributions to retinal processing. The few cell types whose functions are known appear to have very specific roles, such as mediating directionally selective responses to motion (35, 36), carrying signals from rod photoreceptors (73, 74), and adjusting retinal responses to the average light level by releasing dopamine as a neuromodulator (75, 76). More general amacrine functions include effects on spatial and temporal processing. Some amacrine cells transmit inhibitory signals over long distances using Na action potentials, thus contributing a suppressive input that can extend beyond the classical receptive field surround (77). In addition, some amacrine cells affect the kinetics of ganglion cell responses, truncating the ganglion cell response to a prolonged stimulus (78). But for the remaining 20 or more distinguished morphological types (4), their general classification as lateral inhibitory neurons summarizes much of our present level of understanding.

Through extensive study of amacrine cell physiology and anatomy, investigators have made various observations of properties relevant to their function. Within the IPL, amacrine cells arborize differently: Some are narrowly stratified and restricted to a single layer, and some branch diffusely across many layers (60, 79). Aside from their layer of stratification in the retina, amacrine cells vary in their extent in the retinal plane. They range from narrow field, <100 microns, to wide field, >400 microns, sometimes spanning many millimeters across the retina. Experiments using simple flashes have shown that amacrine cells also have different temporal responses and range from transient to sustained (6). In the face of this apparent complexity, some rules linking the anatomy of amacrine cells and their light responses have emerged.

Monostratified Amacrine Cells in Different Layers Have Distinct Kinetics

As for bipolar cells and ganglion cells, the responses of amacrine cells correspond to the layers in which they arborize. Pang et al. (79) performed a systematic survey of amacrine cell light responses and morphology. An examination of the detailed list of data indicates that monostratified amacrine cells comprise approximately eight distinct kinetic classes (**Figure 4*b***). Their light responses arise as a combination of synaptic input from both bipolar cells and other amacrine cells.

Small, Narrow-Field Amacrine Cells Have a Greater Diversity than Do Wide-Field Cells

Small amacrine cells include both monostratified and diffuse cells, receiving input from either one or many retinal layers. Many narrow-field cells give sustained responses to a step increase in light (80). These findings suggest a number of implications that remain to be tested. For diffuse cells spanning multiple layers, a sustained response may arise from the summation of multiple timescales of synaptic input from the different layers. One would also expect that cells spanning multiple layers influence more types of ganglion cells. It is also tempting to think that the greater diversity of smaller amacrine cells results in more types of spatio-temporal processing of visual input across a fine spatial scale.

Wide-Field Amacrine Cells Are Transient and Narrowly Stratified

Although small amacrine cells can be either monostratified or diffuse, sustained or transient, large wide-field amacrine cells appear always to be narrowly stratified (71, 79). In addition, they appear always to have transient responses (79). The property of transience in wide-field amacrine cells arises from multiple mechanisms, in part from inhibitory input from other amacrine cells (81), in part from intrinsic conductances (82), and in part from the desensitization of glutamate receptors (83). The transient output of these amacrine cells is widespread in the retina; nearly all ganglion cell types receive wide-field transient inhibition from amacrine cells (84). One type of wide-field amacrine cell, the polyaxonal amacrine cell, receives central input and produces wide-field output. Although the name amacrine was chosen because the cells were thought to lack axons (1), polyaxonal amacrine cells have multiple axons that produce tetrodotoxin-sensitive action potentials (85, 86). These cells distribute inhibition across a wide region of space but a narrow slice of time to distinct temporal parallel channels. This organization can be contrasted with narrow-field sustained diffuse cells, which deliver inhibition to affect visual signals arising across a broad region of time but a narrow region of space.

Wide-field amacrine cell: an inhibitory amacrine cell that has a large dendritic or axonal arbor. These are nearly always transient and narrowly stratified in the IPL

Polyaxonal amacrine cell: a type of wide-field spiking amacrine cell with multiple axons that transmits transient inhibition long distances across the retina

Linear spatio-temporal filter: a mathematical operation that performs a weighted sum of input over space and time

NEURAL COMPUTATIONS

How are these circuit and physiological properties combined to produce a computation? Recent studies have shown how spatial filtering, timing, amacrine transmission, retinal layers, and nonlinearities combine together to perform computations.

Object Motion Sensitivity

Animals are very sensitive to retinal motion produced by moving objects but almost completely insensitive to retinal motion produced by eye movements. In particular, fixational drift eye movements continually scan the image across the retina. During natural posture, these eye movements can be surprisingly large, up to 0.5° in amplitude (the width of the full moon) and up to ~1° per second (88). Somewhere in the visual system, retinal motion from fixational eye movements is discarded, whereas retinal motion from moving objects is preserved. A recent study identifies OMS ganglion cells as having the requisite properties to signal the presence of a moving object: selectivity for differential motion while failing to respond to global motion (43).

The heart of this computation is a model consisting of several stages. The central excitatory input to the cell consists of multiple nonlinear subunits (**Figure 5*a***). When an object moves across this receptive field (**Figure 5*b***), in the first stage, each subunit in each location applies a simple linear spatio-temporal filter. A subunit weights the stimulus over a nearby region of space and the preceding interval of time, applying a biphasic temporal filter that emphasizes changes in intensity. The result encodes both positive and negative changes in intensity at each location

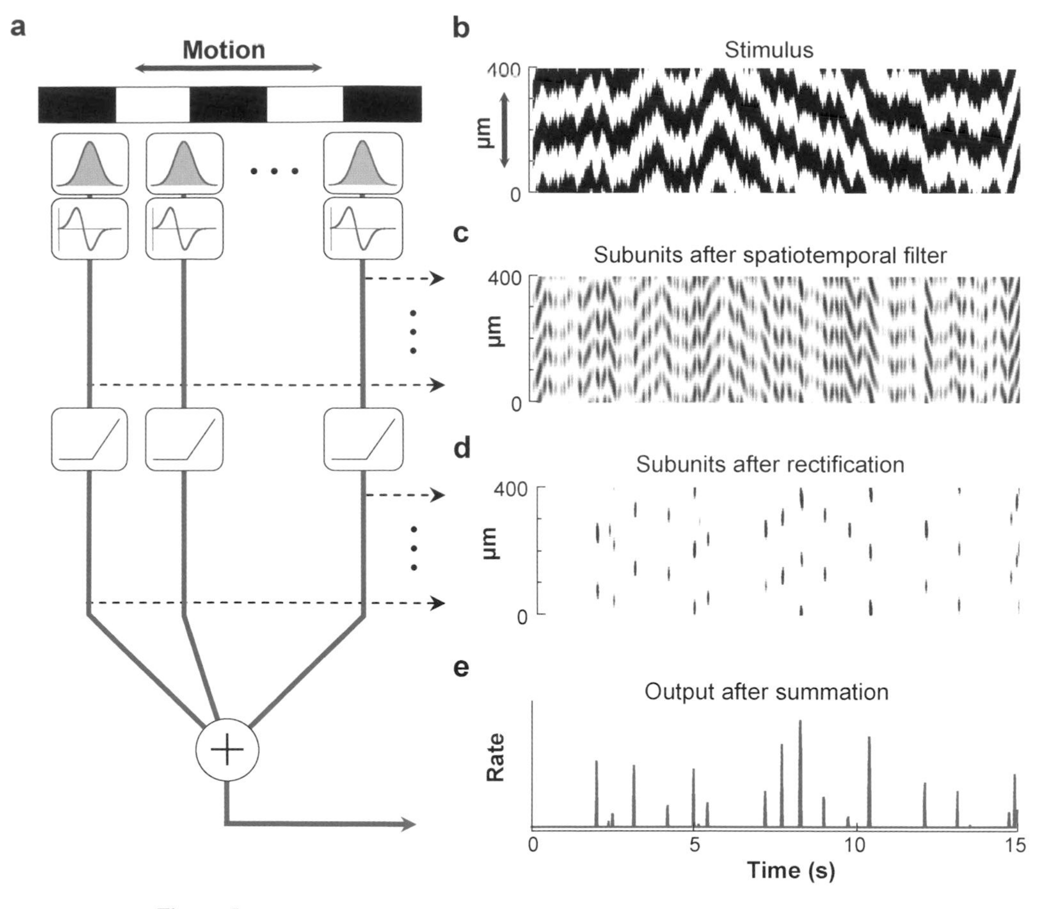

Figure 5

(*a*) Object motion sensitive (OMS) ganglion cell response converts motion into a sparse firing pattern. Model of the OMS cell central input (43), consisting of the stimulus passed through a set of subunits, each of which computes a linear sum of stimulus intensity over space and time. The output of each subunit is thresholded, and then all subunits are summed. (*b*) Stimulus consisting of a jittering one-dimensional striped grating. The vertical axis represents space, and the horizontal axis represents time. (*c*) After the stimulus passes through the subunits, one at each spatial location on the vertical axis, the stimulus is temporally and spatially smoothed. Blue and red denote a positive and negative subunit response, respectively. (*d*) Thresholding of the signal at each spatial location preserves only the strongest positive signals of those subunits that detected a change in intensity. (*e*) After summation, the stimulus trajectory is converted into a firing rate output.

(**Figure 5*c***). At the next stage, the output of each linear subunit is rectified by a threshold nonlinearity. As a result, only large positive changes in intensity are transmitted (**Figure 5*d***). Finally, the positive changes are summed, yielding a single firing rate (**Figure 5*e***). The biphasic temporal filter and the high threshold conspire to create an output that is sparse in time. Thus, the motion trajectory is converted into a temporal firing pattern.

The central excitatory part of the OMS cell model shares properties with the cat Y cell, which also sums over multiple nonlinear subunits (89, 90). Although this response is constructed from the same basic pieces as the LN model, linear filtering, summation, and static nonlinearities, the order of these operations leads to an interesting property. Although the motion trajectory is encoded, the specific spatial pattern is not. The key to this property is that individual subunits are rectified before they are summed. If summation across the entire receptive field occurred prior to the threshold, subunits under a light patch would cancel out subunits under a dark patch. However, the rectifying step eliminates the contribution of those subunits with negative responses, preventing this cancellation.

Much discussion has followed the idea that a critical aspect of higher-level processing is to maintain sensitivity to an important property, such as object identity, while becoming invariant to certain unimportant properties of a stimulus (91). In the OMS computation, this type of stimulus generalization (32) is achieved by pooling over spatial locations, causing the response to be invariant to the spatial pattern (43). The step of rectification preserves the sensitivity to each component of the pattern during this pooling step.

Differential motion detection is then accomplished by comparing the motion trajectory at different locations. To make this comparison, inhibitory input is delivered from the surrounding background region (**Figure 6*a***). This inhibitory input has the same nonlinear summation and temporal selectivity as the central excitatory input. Because the motion trajectory is converted into a distinct sequence of synaptic input, excitation and inhibition are translated into sparse sequences according to the same rules. Thus, if central excitation and background inhibition are driven by the same trajectory, as occurs during global motion owing to eye movements, excitation and inhibition will be coincident and therefore will cancel (**Figure 6*b***). In the case of differential motion, different trajectories in the object and background regions will yield different temporal sequences of excitation and inhibition. Because those sequences are both sparse, inhibition will have minimal influence on excitatory input (**Figure 6*c***).

What circuitry might underlie this comparison between temporal sequences? If excitation and inhibition have the same temporal selectivity, then this comparison can be achieved simply by connecting excitatory and inhibitory cells in the same stratum in the IPL. Wide-field amacrine cells, which are nearly always transient and narrowly stratified, are suited to deliver precisely timed inhibition to bipolar cell axon terminals or ganglion cell dendrites with similar temporal selectivity. Polyaxonal amacrine cells are especially suited to this task, as their axons can span a large fraction of the retina. In support of this idea, some polyaxonal amacrine cells in the salamander have the same kinetics as OMS ganglion cells. In response to the same motion trajectory, the two cell classes have nearly synchronous spiking responses (43). Polyaxonal, or axon-bearing, amacrine cells have been described in many species, including rabbit, salamander, and primate (85, 92, 93). It remains to be seen whether polyaxonal amacrine cells and OMS ganglion cells occupy the same stratum in the IPL and whether these amacrine cells make direct connections to OMS ganglion cells or their bipolar cell inputs within the same retinal layer.

Global Shift Suppression

Saccadic eye movements of humans and other animals cause brief, high-speed retinal motion (94). One might expect that such large retinal shifts cause great excitation across the entire retina, but humans do not perceptually report this effect. In fact, during saccades, the visual threshold to detect changing stimuli is elevated. In primates, this suppression appears to be selective for the magnocellular pathway (95). Global retinal image motion by itself can produce this suppressive effect (44), although some additional suppression may be generated by the motor command for saccades (45, 46).

Recently, Roska & Werblin (47) analyzed synaptic inputs from the rabbit peripheral retina that could underlie the suppressive component of saccadic global shifts. Using natural movies with shifts that approximated

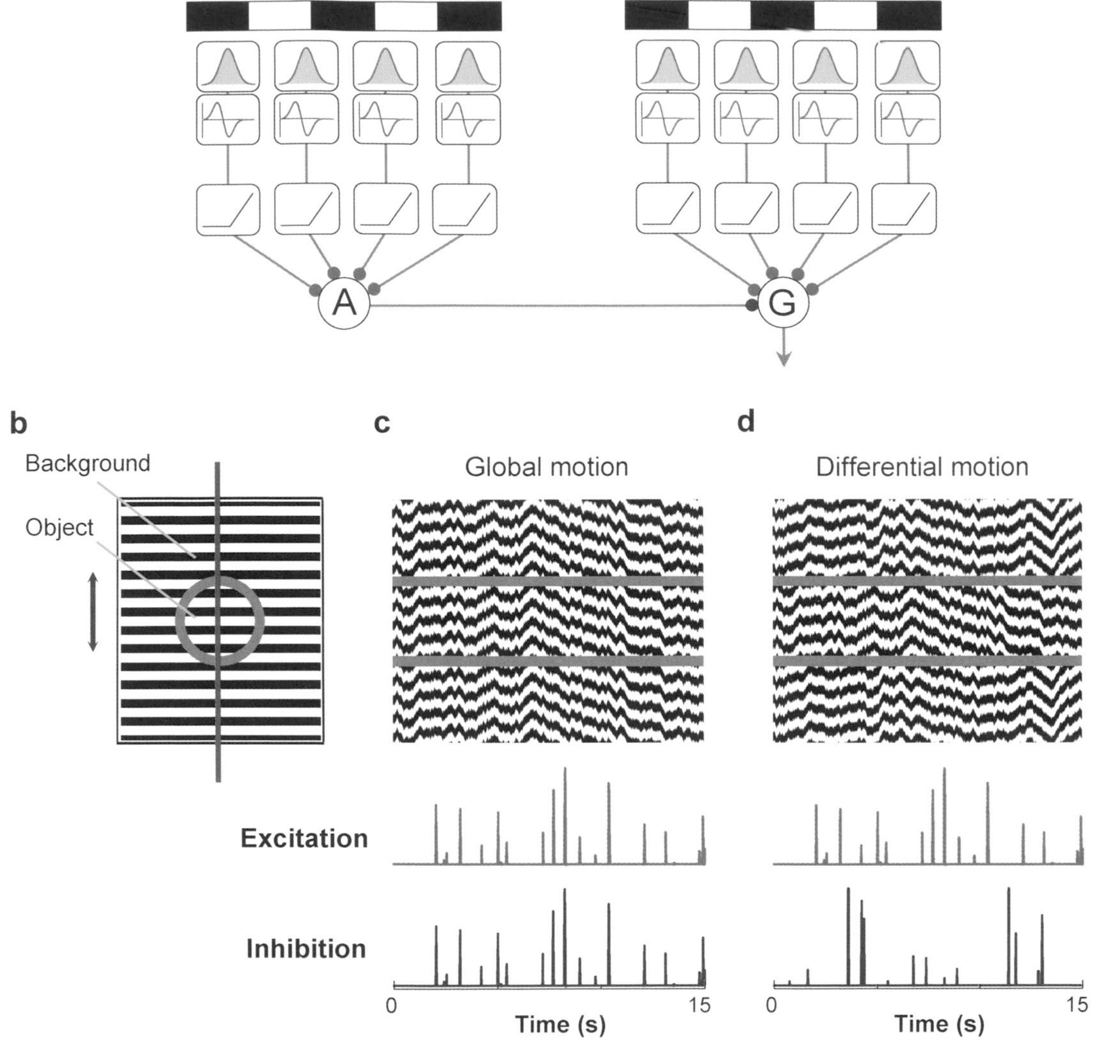

Figure 6

Object motion sensitivity (OMS) computation. (*a*) Model of OMS computation consisting of nonlinear excitatory input from the central object region and similar nonlinear inhibitory input from the surrounding background region. Blue circles indicate excitatory input, and the red circle indicates inhibitory input, not necessarily monosynaptic. A, amacrine cell; G, ganglion cell. (*b*) Differential motion stimulus consisting of an object region and background region; each region can be moved independently. The vertical blue line indicates the location of the one-dimensional cross section of stimuli shown in *c* and *d*. (*c*) (*Top*) Global motion stimulus, for which object and background regions move together coherently. The vertical axis represents space, and the horizontal axis represents time. (*Bottom*) Excitation and inhibition patterns produced by the motion trajectories in each region. Excitation arises from the central object region, and inhibition arises from the background. (*d*) (*Top*) Differential motion stimulus, for which object and background regions move independently. (*Bottom*) Patterns of excitation and inhibition for the two regions.

saccades, they found that large, fast, global motion silences a class of ganglion cells. Global shifts produce both excitatory and inhibitory input to a ganglion cell. For the ganglion cell types silenced by the shifts, precisely timed inhibition from amacrine cells nullified coincident excitation. Roska & Werblin further found that the ganglion cells that were suppressed by global shifts had their dendrites in a restricted layer of the IPL.

Thus, as for the OMS circuit, the precise, coincident timing of inhibition is important for suppressing global motion. Moreover, that dendrites of suppressed ganglion cells are within a narrow region of the IPL supports the idea that connections within retinal layers transmit inhibition that carries a specific temporal selectivity. Again, wide-field, transient, narrowly stratified amacrine cells are the prime candidates to produce this suppression. The context is somewhat different from the OMS circuit—large, fast, saccadic eye movements instead of small, drifting eye movements. But the concept is very similar, and global shift suppression may share components with the OMS circuit.

CONCLUSION

The wealth of anatomical, physiological, mathematical, and biochemical information about retinal mechanisms and function, combined with psychophysical and behavioral measurements of visual function, has enabled the first steps toward understanding how the inner retina translates the visual scene. The natural human intuition for visual processing and the high precision of retinal responses greatly facilitate experimental and theoretical study. Common features are shared between the inner retina and other parts of the brain. Other brain regions, including the cortex and hippocampus, also show a laminar organization, diverse inhibitory neurons, and spatial proximity of neurons that carry similar information. As such, the inner retina will likely serve as a guide to understanding how the nervous system performs complex, adaptive computations.

SUMMARY POINTS

1. The inner retina generates a number of aspects of visual processing, including concerted signaling, adaptation to contrast and spatio-temporal correlations, anticipation of moving stimuli, selectivity for directional motion and object motion, and the suppression of fast global motion.
2. Ganglion cell responses to naturalistic stimuli of broad temporal bandwidth are precise and sparse in time. Responses to these stimuli can sometimes be captured by a relatively simple model consisting of a linear spatio-temporal filter and time-independent (static) nonlinear function. These models can be used to determine how synapses and circuits give rise to higher-level computations.
3. There are approximately ten synaptic strata in the inner plexiform layer (IPL). Different strata, or layers, respond with distinct kinetics, forming separate but interacting temporal channels. Synaptic input is delivered by bipolar cells to individual layers, transformed by amacrine cells that transmit signals within and between layers, and conveyed to ganglion cell dendrites.
4. Amacrine cells are an extremely diverse group of inhibitory interneurons, consisting of approximately 30 types. Wide-field amacrine cells are narrowly stratified and transient, thus delivering brief inhibition across a wide range of space within a temporal channel.

5. Object motion sensitive (OMS) ganglion cells are sensitive to local motion produced by a moving object but are silenced by global motion as produced by fixational eye movements. The properties of sensitivity to motion but invariance to spatial pattern can be explained by central, transient excitatory input that is pooled over nonlinear subunits. Selectivity for local motion over global motion can be explained by wide-field transient inhibitory input that is pooled from similar subunits and transmitted across the retina by precisely timed wide-field amacrine cells, cancelling the excitatory response during global motion.
6. Human visual perception to some types of changing visual stimuli is suppressed during gaze-shifting saccades. This suppression has a visual component and may arise owing to the suppression of certain types of ganglion cells. Some ganglion cells receive strong, precisely timed inhibition during fast, shifting global stimuli that approximate saccades. These ganglion cells have their dendrites in a small subset of layers in the IPL and may receive inhibition from specific wide-field amacrine cells.

FUTURE ISSUES

1. There likely are unknown functions of retinal visual processing. Some possibilities include selectivity for specific features of natural visual scenes or adaptation to the more complex, undescribed statistics of natural scenes. Although natural scenes are complex, a full understanding of retinal function will benefit from an effort to study reduced components of natural scenes, as has already occurred for the visual contribution of eye movements.
2. Simple models have had success in predicting the responses of single neurons to simple stimuli, but they do not approach the benchmark of predicting the response of a population of retinal ganglion cells to natural visual scenes. Although phenomena such as contrast adaptation, pattern adaptation, and concerted signaling have been described, there are no explicit models that predict the responses for single neurons or populations across a wide range of stimuli that produce these phenomena.
3. The functional connectivity of the inner retina is largely unknown. It will be important to learn which of nearly 30 amacrine cell types influence the approximately 10 bipolar cell types and 10 ganglion cell types. Consequently, the specific contributions of each of the amacrine cell types, and how those cell types change and modify ganglion cell light responses, are unknown.

ACKNOWLEDGMENTS

I thank Markus Meister and Bence Ölveczky for many invaluable discussions. Support was provided by the National Institutes of Health (National Eye Institute) and the Pew Charitable Trusts.

LITERATURE CITED

1. Cajal SR. 1972. *The Structure of the Retina*. Springfield, IL: Thomas

2. Adrian ED, Matthews R. 1928. The action of light on the eye. Part III. The interaction of retinal neurones. *J. Physiol.* 65:273–98
3. Rodieck RW. 1965. Quantitative analysis of cat retinal ganglion cell response to visual stimuli. *Vis. Res.* 5:583–601
4. Masland RH. 2001. The fundamental plan of the retina. *Nat. Neurosci.* 4:877–86
5. Kuffler SW. 1953. Discharge patterns and functional organization of mammalian retina. *J. Neurophysiol.* 16:37–68
6. Kaneko A. 1970. Physiological and morphological identification of horizontal, bipolar and amacrine cells in goldfish retina. *J. Physiol.* 207:623–33
7. Burns ME, Baylor DA. 2001. Activation, deactivation, and adaptation in vertebrate photoreceptor cells. *Annu. Rev. Neurosci.* 24:779–805
8. Mastronarde DN. 1983. Interactions between ganglion cells in cat retina. *J. Neurophysiol.* 49:350–65
9. Mastronarde DN. 1983. Correlated firing of cat retinal ganglion cells. I. Spontaneously active inputs to X- and Y-cells. *J. Neurophysiol.* 49:303–24
10. Meister M, Lagnado L, Baylor DA. 1995. Concerted signaling by retinal ganglion cells. *Science* 270:1207–10
11. Schnitzer MJ, Meister M. 2003. Multineuronal firing patterns in the signal from eye to brain. *Neuron* 37:499–511
12. Brivanlou IH, Warland DK, Meister M. 1998. Mechanisms of concerted firing among retinal ganglion cells. *Neuron* 20:527–39
13. Pugh EN, Nikonov S, Lamb TD. 1999. Molecular mechanisms of vertebrate photoreceptor light adaptation. *Curr. Opin. Neurobiol.* 9:410–18
14. Barlow HB, Levick WR. 1969. Changes in the maintained discharge with adaptation level in the cat retina. *J. Physiol.* 202:699–718
15. Enroth-Cugell C, Lennie P. 1975. The control of retinal ganglion cell discharge by receptive field surrounds. *J. Physiol.* 247:551–78
16. Smirnakis SM, Berry MJ, Warland DK, Bialek W, Meister M. 1997. Adaptation of retinal processing to image contrast and spatial scale. *Nature* 386:69–73
17. Baccus SA, Meister M. 2002. Fast and slow contrast adaptation in retinal circuitry. *Neuron* 36:909–19
18. Kim KJ, Rieke F. 2001. Temporal contrast adaptation in the input and output signals of salamander retinal ganglion cells. *J. Neurosci.* 21:287–99
19. Kim KJ, Rieke F. 2003. Slow Na^+ inactivation and variance adaptation in salamander retinal ganglion cells. *J. Neurosci.* 23:1506–16
20. Rieke F. 2001. Temporal contrast adaptation in salamander bipolar cells. *J. Neurosci.* 21:9445–54
21. Chander D, Chichilnisky EJ. 2001. Adaptation to temporal contrast in primate and salamander retina. *J. Neurosci.* 21:9904–16
22. Laughlin SB. 1989. The role of sensory adaptation in the retina. *J. Exp. Biol.* 146:39–62
23. Solomon SG, Peirce JW, Dhruv NT, Lennie P. 2004. Profound contrast adaptation early in the visual pathway. *Neuron* 42:155–62
24. Ohzawa I, Sclar G, Freeman RD. 1985. Contrast gain control in the cat's visual system. *J. Neurophysiol.* 54:651–67
25. Blakemore C, Campbell FW. 1969. On the existence of neurones in the human visual system selectively sensitive to the orientation and size of retinal images. *J. Physiol.* 203:237–60
26. Victor JD. 1987. The dynamics of the cat retinal X cell center. *J. Physiol.* 386:219–46

27. Manookin MB, Demb JB. 2006. Presynaptic mechanism for slow contrast adaptation in mammalian retinal ganglion cells. *Neuron* 50:453–64
28. Zaghloul KA, Boahen K, Demb JB. 2005. Contrast adaptation in subthreshold and spiking responses of mammalian Y-type retinal ganglion cells. *J. Neurosci.* 25:860–68
29. Hosoya T, Baccus SA, Meister M. 2005. Dynamic predictive coding by the retina. *Nature* 436:71–77
30. Srinivasan MV, Laughlin SB, Dubs A. 1982. Predictive coding: a fresh view of inhibition in the retina. *Proc. R. Soc. London Ser. B* 216:427–59
31. Barlow HB, Hill RM, Levick WR. 1964. Retinal ganglion cells responding selectively to direction and speed of image motion in the rabbit. *J. Physiol.* 173:377–407
32. Barlow HB, Levick WR. 1965. The mechanism of directionally selective units in rabbit's retina. *J. Physiol.* 178:477–504
33. Wyatt HJ, Daw NW. 1975. Directionally sensitive ganglion cells in the rabbit retina: specificity for stimulus direction, size, and speed. *J. Neurophysiol.* 38:613–26
34. Yoshida K, Watanabe D, Ishikane H, Tachibana M, Pastan I, Nakanishi S. 2001. A key role of starburst amacrine cells in originating retinal directional selectivity and optokinetic eye movement. *Neuron* 30:771–80
35. Fried SI, Munch TA, Werblin FS. 2002. Mechanisms and circuitry underlying directional selectivity in the retina. *Nature* 420:411–14
36. Euler T, Detwiler PB, Denk W. 2002. Directionally selective calcium signals in dendrites of starburst amacrine cells. *Nature* 418:845–52
37. Fried SI, Munch TA, Werblin FS. 2005. Directional selectivity is formed at multiple levels by laterally offset inhibition in the rabbit retina. *Neuron* 46:117–27
38. Vaney DI, Taylor WR. 2002. Direction selectivity in the retina. *Curr. Opin. Neurobiol.* 12:405–10
39. Nijhawan R. 2002. Neural delays, visual motion and the flash-lag effect. *Trends Cogn. Sci.* 6:387–93
40. Berry MJ, Brivanlou IH, Jordan TA, Meister M. 1999. Anticipation of moving stimuli by the retina. *Nature* 398:334–38
41. Benardete EA, Kaplan E. 1999. The dynamics of primate M retinal ganglion cells. *Vis. Neurosci.* 16:355–68
42. Sheth BR, Nijhawan R, Shimojo S. 2000. Changing objects lead briefly flashed ones. *Nat. Neurosci.* 3:489–95
43. Olveczky BP, Baccus SA, Meister M. 2003. Segregation of object and background motion in the retina. *Nature* 423:401–8
44. Mackay DM. 1970. Elevation of visual threshold by displacement of retinal image. *Nature* 225:90–92
45. Riggs LA, Merton PA, Morton HB. 1974. Suppression of visual phosphenes during saccadic eye movements. *Vis. Res.* 14:997–1011
46. Diamond MR, Ross J, Morrone MC. 2000. Extraretinal control of saccadic suppression. *J. Neurosci.* 20:3449–55
47. Roska B, Werblin F. 2003. Rapid global shifts in natural scenes block spiking in specific ganglion cell types. *Nat. Neurosci.* 6:600–8
48. Deleted in proof
49. Hunter IW, Korenberg MJ. 1986. The identification of nonlinear biological systems: Wiener and Hammerstein cascade models. *Biolog. Cybernet.* 55:135–44
50. Sakai HM, Naka K, Korenberg MJ. 1988. White-noise analysis in visual neuroscience. *Vis. Neurosci.* 1:287–96
51. Chichilnisky EJ. 2001. A simple white noise analysis of neuronal light responses. *Network* 12:199–213

43. Describes object motion sensitive (OMS) ganglion cells, gives a computational model of the response, and shows the likely role of polyaxonal amacrine cells.

47. Shows that some rabbit ganglion cells are suppressed by fast, global shifts that mimic saccadic eye movements and measures the stratification of these cells.

50. Describes the Gaussian white-noise approach of modeling a cell's response by using cross-correlation of the stimulus with the response.

51. Describes the spatio-temporal linear-nonlinear (LN) model and its calculation, using reverse correlation.

52. **Berry MJ, Warland DK, Meister M. 1997. The structure and precision of retinal spike trains. *Proc. Natl. Acad. Sci. USA* 94:5411–16**
53. Keat J, Reinagel P, Reid RC, Meister M. 2001. Predicting every spike: a model for the responses of visual neurons. *Neuron* 30:803–17
54. McIlwain JT. 1964. Receptive fields of optic tract axons and lateral geniculate cells: peripheral extent and barbiturate sensitivity. *J. Neurophysiol.* 27:1154–73
55. Fischer B, Kruger J, Droll W. 1975. Quantitative aspects of the shift-effect in cat retinal ganglion cells. *Brain Res.* 83:391–403
56. McIlwain JT. 1966. Some evidence concerning the physiological basis of the periphery effect in the cat's retina. *Exp. Brain Res.* 1:265–71
57. Werblin FS, Copenhagen DR. 1974. Control of retinal sensitivity. 3. Lateral interactions at the inner plexiform layer. *J. Gen. Physiol.* 63:88–110
58. Dowling JE, Boycott BB. 1966. Organization of the primate retina: electron microscopy. *Proc. R. Soc. London Ser. B* 166:80–111
59. Sanes JR, Yamagata M. 1999. Formation of lamina-specific synaptic connections. *Curr. Opin. Neurobiol.* 9:79–87
60. MacNeil MA, Masland RH. 1998. Extreme diversity among amacrine cells: implications for function. *Neuron* 20:971–82
61. **Roska B, Werblin F. 2001. Vertical interactions across ten parallel, stacked representations in the mammalian retina. *Nature* 410:583–87**
62. Famiglietti EVJ, Kolb H. 1976. Structural basis for ON-and OFF-center responses in retinal ganglion cells. *Science* 194:193–95
63. Nelson R, Famiglietti EVJ, Kolb H. 1978. Intracellular staining reveals different levels of stratification for on- and off-center ganglion cells in cat retina. *J. Neurophysiol.* 41:472–83
64. Amthor FR, Oyster CW, Takahashi ES. 1984. Morphology of on-off direction-selective ganglion cells in the rabbit retina. *Brain Res.* 298:187–90
65. Amthor FR, Takahashi ES, Oyster CW. 1989. Morphologies of rabbit retinal ganglion cells with complex receptive fields. *J. Comp. Neurol.* 280:97–121
66. Schreiner CE, Read HL, Sutter ML. 2000. Modular organization of frequency integration in primary auditory cortex. *Annu. Rev. Neurosci.* 23:501–29
67. Blakemore C, Price DJ. 1987. The organization and postnatal development of area 18 of the cat's visual cortex. *J. Physiol.* 384:263–92
68. Horton JC, Adams DL. 2005. The cortical column: a structure without a function. *Philos. Trans. R. Soc. London B* 360:837–62
69. **Wu SM, Gao F, Maple BR. 2000. Functional architecture of synapses in the inner retina: segregation of visual signals by stratification of bipolar cell axon terminals. *J. Neurosci.* 20:4462–70**
70. Pang JJ, Gao F, Wu SM. 2004. Stratum-by-stratum projection of light response attributes by retinal bipolar cells of *Ambystoma*. *J. Physiol.* 558:249–62
71. MacNeil MA, Heussy JK, Dacheux RF, Raviola E, Masland RH. 1999. The shapes and numbers of amacrine cells: matching of photofilled with Golgi-stained cells in the rabbit retina and comparison with other mammalian species. *J. Comp. Neurol.* 413:305–26
72. Yang CY, Lukasiewicz P, Maguire G, Werblin FS, Yazulla S. 1991. Amacrine cells in the tiger salamander retina: morphology, physiology, and neurotransmitter identification. *J. Comp. Neurol.* 312:19–32
73. Vaney DI, Young HM, Gynther IC. 1991. The rod circuit in the rabbit retina. *Vis. Neurosci.* 7:141–54
74. Veruki ML, Hartveit E. 2002. Electrical synapses mediate signal transmission in the rod pathway of the mammalian retina. *J. Neurosci.* 22:10558–66

52. Shows that retinal ganglion cells fire with high precision when presented with broad temporal bandwidth stimuli.

61. Systematic survey comparing the excitatory and inhibitory currents, spiking, and anatomical stratification of rabbit ganglion cells.

69. Systematic survey comparing the anatomy and light responses of salamander bipolar cells.

75. Puopolo M, Hochstetler SE, Gustincich S, Wightman RM, Raviola E. 2001. Extrasynaptic release of dopamine in a retinal neuron: activity dependence and transmitter modulation. *Neuron* 30:211–25
76. Gustincich S, Feigenspan A, Wu DK, Koopman LJ, Raviola E. 1997. Control of dopamine release in the retina: a transgenic approach to neural networks. *Neuron* 18:723–36
77. Cook PB, Lukasiewicz PD, McReynolds JS. 1998. Action potentials are required for the lateral transmission of glycinergic transient inhibition in the amphibian retina. *J. Neurosci.* 18:2301–8
78. Nirenberg S, Meister M. 1997. The light response of retinal ganglion cells is truncated by a displaced amacrine circuit. *Neuron* 18:637–50
79. Pang JJ, Gao F, Wu SM. 2002. Segregation and integration of visual channels: layer-by-layer computation of ON-OFF signals by amacrine cell dendrites. *J. Neurosci.* 22:4693–701

79. Systematic survey comparing the anatomy and light responses of salamander amacrine cells.

80. Werblin F, Maguire G, Lukasiewicz P, Eliasof S, Wu SM. 1988. Neural interactions mediating the detection of motion in the retina of the tiger salamander. *Vis. Neurosci.* 1:317–29
81. Roska B, Nemeth E, Werblin FS. 1998. Response to change is facilitated by a three-neuron disinhibitory pathway in the tiger salamander retina. *J. Neurosci.* 18:3451–59
82. Barnes S, Werblin F. 1987. Direct excitatory and lateral inhibitory synaptic inputs to amacrine cells in the tiger salamander retina. *Brain Res.* 406:233–37
83. Tran MN, Higgs MH, Lukasiewicz PD. 1999. AMPA receptor kinetics limit retinal amacrine cell excitatory synaptic responses. *Vis. Neurosci.* 16:835–42
84. Wunk DF, Werblin FS. 1979. Synaptic inputs to the ganglion cells in the tiger salamander retina. *J. Gen. Physiol.* 73:265–86
85. Cook PB, Werblin FS. 1994. Spike initiation and propagation in wide field transient amacrine cells of the salamander retina. *J. Neurosci.* 14:3852–61
86. Maguire G. 1999. Spatial heterogeneity and function of voltage- and ligand-gated ion channels in retinal amacrine neurons. *Proc. R. Soc. London Ser. B* 266:987–92
87. Deleted in proof
88. Skavenski AA, Hansen RM, Steinman RM, Winterson BJ. 1979. Quality of retinal image stabilization during small natural and artificial body rotations in man. *Vis. Res.* 19:675–83
89. Hochstein S, Shapley RM. 1976. Linear and nonlinear spatial subunits in Y cat retinal ganglion cells. *J. Physiol.* 262:265–84
90. Victor JD, Shapley RM. 1979. The nonlinear pathway of Y ganglion cells in the cat retina. *J. Gen. Physiol.* 74:671–89

90. Describes the model of a Y-type cell, consisting of nonlinear subunits first rectified, then summed.

91. Wallis G, Rolls ET. 1997. Invariant face and object recognition in the visual system. *Prog. Neurobiol.* 51:167–94
92. Famiglietti EV. 1992. Polyaxonal amacrine cells of rabbit retina: size and distribution of PA1 cells. *J. Comp. Neurol.* 316:406–21
93. Stafford DK, Dacey DM. 1997. Physiology of the A1 amacrine: a spiking, axon-bearing interneuron of the macaque monkey retina. *Vis. Neurosci.* 14:507–22
94. Ross J, Morrone MC, Goldberg ME, Burr DC. 2001. Changes in visual perception at the time of saccades. *Trends Neurosci.* 24:113–21
95. Ross J, Burr D, Morrone C. 1996. Suppression of the magnocellular pathway during saccades. *Behav. Brain Res.* 80:1–8

Understanding Circuit Dynamics Using the Stomatogastric Nervous System of Lobsters and Crabs

Eve Marder[1] and Dirk Bucher[2]

[1]Volen Center and Biology Department, Brandeis University, Waltham, Massachusetts 02454; email: marder@brandeis.edu

[2]The Whitney Laboratory for Marine Bioscience, University of Florida, St. Augustine, Florida 32080; email: bucher@whitney.ufl.edu

Annu. Rev. Physiol. 2007. 69:291–316

First published online as a Review in Advance on September 29, 2006

The *Annual Review of Physiology* is online at http://physiol.annualreviews.org

This article's doi:
10.1146/annurev.physiol.69.031905.161516

0066-4278/07/0315-0291$20.00

Key Words

central pattern generator, neuronal oscillators, neuromodulation, pyloric rhythm, gastric mill rhythm

Abstract

Studies of the stomatogastric nervous systems of lobsters and crabs have led to numerous insights into the cellular and circuit mechanisms that generate rhythmic motor patterns. The small number of easily identifiable neurons allowed the establishment of connectivity diagrams among the neurons of the stomatogastric ganglion. We now know that (*a*) neuromodulatory substances reconfigure circuit dynamics by altering synaptic strength and voltage-dependent conductances and (*b*) individual neurons can switch among different functional circuits. Computational and experimental studies of single-neuron and network homeostatic regulation have provided insight into compensatory mechanisms that can underlie stable network performance. Many of the observations first made using the stomatogastric nervous system can be generalized to other invertebrate and vertebrate circuits.

STNS: stomatogastric nervous system

STG: stomatogastric ganglion

INTRODUCTION

Recent years have seen a rebirth of interest in understanding how neural circuits generate behavior. Therefore, it is a particularly good time to review and critically examine what we know about the stomatogastric nervous system (STNS), one of the premier systems for analyzing how circuit dynamics arise from the properties of its neurons and their connections. The process of understanding how the STNS generates movements of the crustacean foregut has involved multiple cycles of revisiting many of the same issues over the years, as each decade has revealed "new generation" insights into how even this small nervous system generates rhythmic motor patterns.

The STNS was developed as an experimental preparation almost 40 years ago, in the early days of circuit analysis (1), to understand the generation of rhythmic motor patterns. Over time this system has revealed numerous general principles relevant to central pattern generators (CPGs) and other large and small circuits in both invertebrates and vertebrates. As we look forward to understanding the larger and more complex circuits in the vertebrate brain, lessons learned in small circuits can help pose more precisely the issues crucial for circuit analysis in all systems.

The 40 years of work on the STNS have generated a considerable literature now nearing almost 1000 original journal articles, many reviews (e.g., References 2–4), and two books (5, 6). Navigating through this literature can be a daunting task, made more difficult because studies of the STNS have employed a number of different crustacean species, including spiny lobsters (*Panulirus argus*, *Panulirus interruptus*), clawed lobsters (*Homarus americanus* and *Homarus gammarus*), a variety of crabs (*Cancer borealis* most commonly), crayfish, and shrimp. Although all big-picture conclusions that have arisen from STNS studies hold for all species, some details do vary across species (7).

Features of the Stomatogastric Nervous System that Facilitate Circuit Analysis

The STNS has important attributes that have been crucial in making it a useful preparation for circuit analysis:

1. When the STNS is removed from the animal and placed in a saline-filled dish, it continues to produce fictive motor patterns that resemble closely those recorded in vivo (8–12).
2. The neurons of the stomatogastric ganglion (STG) are unambiguously identifiable from preparation to preparation.
3. Intracellular recordings from the somata of the STG neurons reveal large-amplitude synaptic potentials and other underlying subthreshold changes in membrane potential.
4. Unlike most CPGs that consist of interneurons that drive motor neurons (13), most of the synaptic connections important for the generation of rhythmic motor patterns in the STG occur among the motor neurons. Thus, recordings from the motor neurons provide, at the same time, recordings of the output as well as of the operation of the circuit.
5. It is routinely possible to obtain simultaneous recordings of most, if not all, relevant circuit neurons, using a combination of intracellular and extracellular recordings. Routine STNS experiments include 4 simultaneous intracellular recordings and 8–12 extracellular nerve recordings.
6. The large neuronal somata allow hand dissection of individual neurons for biochemical and molecular characterization at the single-neuron level (14–16).
7. The in vitro preparations are routinely stably active for 18–24 h and can be maintained for days and weeks if required (17).

Today, as we look at attempts to understand vertebrate spinal cord, brainstem, and brain circuits, work is hampered because of the lack of one or more of the above attributes. For this reason, attempts to identify neurons unambiguously in vertebrates are critical (18, 19).

ORGANIZATION OF THE CRUSTACEAN STOMATOGASTRIC NERVOUS SYSTEM

The STNS controls the movements of four regions of the crustacean foregut, or stomach. **Figure 1*a*** shows a side view of a lobster and indicates the position of the stomach, the heart, and the main portions of the nervous system. The STG is found in the dorsal artery, where it is a direct target for hormones released from the pericardial organs and other sources. The crustacean stomach (**Figure 1*b***) is a complex mechanical device that grinds and filters food, using the movements made by more than 40 pairs of striated muscles (20). The stomach muscles, which move the gastric mill and pylorus, are innervated by motor neurons in the STG. Although the general features of the stomach are conserved across decapods, the shape of the stomach is quite different in the oblong lobster and the flat crab. These animals also show anatomical differences in the nervous system and in the STG.

The STNS consists of a group of four linked ganglia, the paired commissural ganglia (CoGs), the unpaired esophageal ganglion (OG), and the STG (**Figure 1*b*,*c***). Each of the CoGs contains approximately 400 neurons, and the OG contains approximately 18 neurons. Together the CoGs and OG contain many descending modulatory neurons that control STG activity (21, 22). The STG consists of ~30 neurons [the exact number varies from species to species and, for some cell types, from animal to animal within a species (23, 24)] and contains the motor neurons that move the muscles of the gastric mill and pyloric regions of the stomach (**Figure 1*b***) (20).

Figure 1*c* is a diagram of the STNS dissected free from the stomach as it is routinely prepared for in vitro electrophysiological recordings. Intracellular glass microelectrodes are used to record from the somata of STG neurons, and extracellular nerve recordings are used to identify STG neurons and to characterize motor patterns. In **Figure 1*d***, simultaneous extracellular recordings of all the motor neurons show the pyloric and gastric mill rhythms in the lobster *H. americanus*. As we discuss below, the pyloric rhythm is faster than the gastric mill rhythm, and although they can usually be separately characterized, there are also strong interactions between them.

In addition to the gastric mill and pyloric rhythms, the STNS also generates the cardiac sac and esophageal rhythms. The generation of these latter two rhythms depends on neurons not found in the STG, and the circuits responsible for them are not known.

CoG: commissural ganglion

OG: esophageal ganglion

GM neuron: gastric mill neuron

PY neuron: pyloric neuron

PD neuron: pyloric dilator neuron

THE STRUCTURE OF THE SOMATOGASTRIC GANGLION AND ITS NEURONS

The number of STG neurons varies from 25–26 in the crab *C. borealis* (23), 28–30 in *P. interruptus* (20), and 29–32 in *H. americanus* (24). Differences in the number of two of the neuron types in the STG, the gastric mill (GM) and pyloric (PY) neurons, account for much of this variability, whereas all other neurons are found invariantly as either single copies or pairs of neurons (24).

STG neurons have a large soma (typically 50–100 μm) and complex branching patterns. **Figure 2** shows dye fills of a single pyloric dilator (PD) neuron in the three species indicated. These fills illustrate both the structures of the individual neurons and the differences in ganglion shape and size of the adult animals routinely used for physiological analyses. Many STG neurons have major neurites as large as 15–20 μm in diameter, with fine diameter processes that ramify extensively through much of the neuropil. Synaptic profiles are found on these finer processes (23–26).

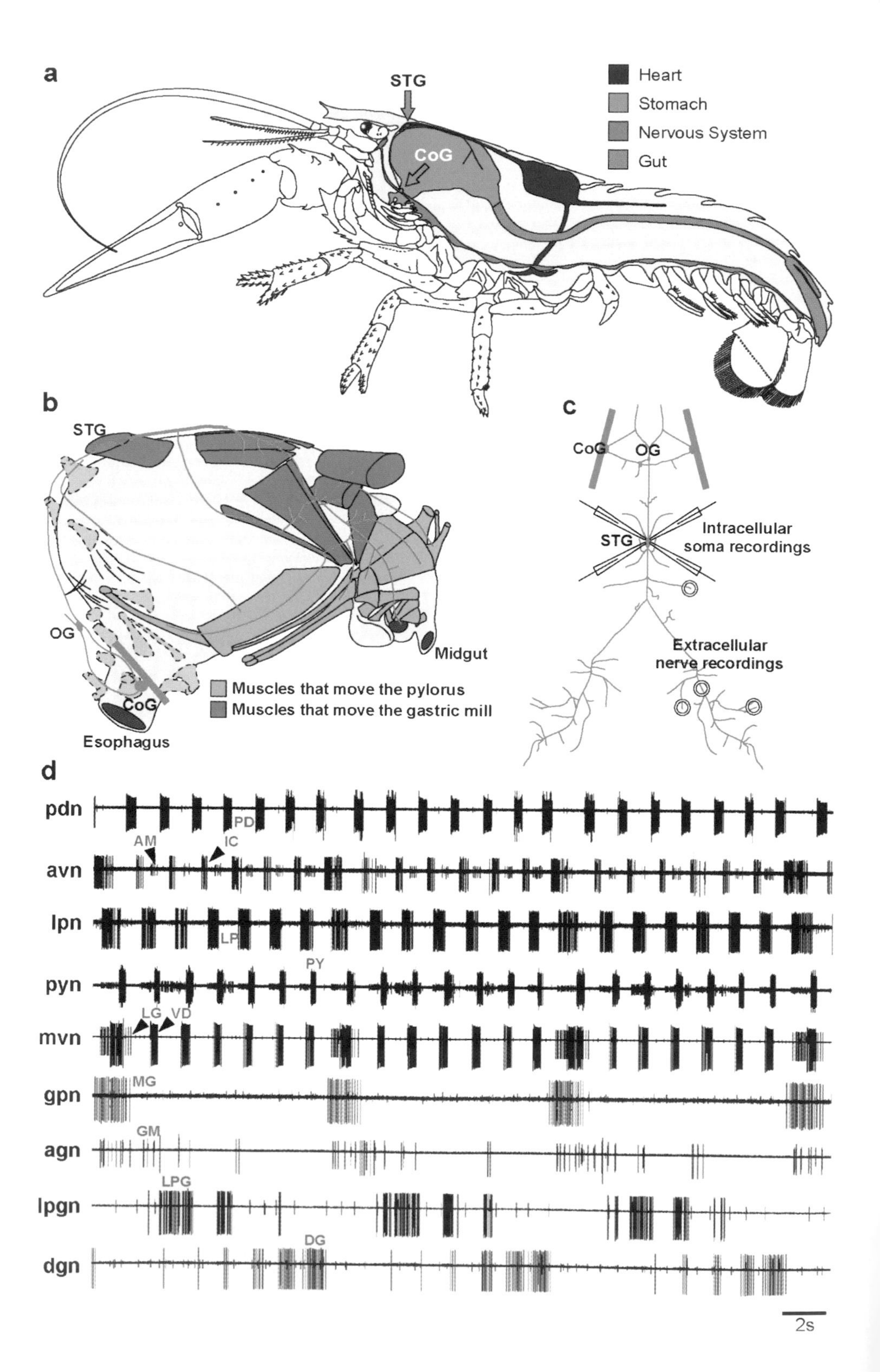
a
STG
CoG
Heart
Stomach
Nervous System
Gut
b
STG
OG
CoG
Esophagus
Midgut
Muscles that move the pylorus
Muscles that move the gastric mill
c
CoG
OG
STG
Intracellular soma recordings
Extracellular nerve recordings
d
pdn
PD
AM
IC
avn
lpn
LP
PY
pyn
LG
VD
mvn
MG
gpn
GM
agn
LPG
lpgn
DG
dgn
2s

THE STOMATOGASTRIC GANGLION IS MULTIPLY MODULATED

Many studies have identified the neurotransmitters used by the STG neurons themselves (14, 27, 28) and the neuromodulators that are released into the neuropil of the STG as a consequence of the actions of sensory neurons and descending modulatory projection neurons (22, 29–31). Additionally, researchers have characterized the substances found in the pericardial organs and other neurosecretory structures (32, 33). Thirty years ago, these studies employed biochemical and histochemical methods (34, 35). Subsequently, enormous progress was made with immunocytochemical methods (36–38). Most recently, the introduction of mass spectroscopy for peptide identification has accelerated the pace of neuropeptide identification in the crustacean nervous system (39, 40).

Figure 3 summarizes much of what is known about the neuromodulatory control pathways to the STG. An important take-home message for workers in other circuits is that the STG is multiply modulated. Consequently, no single substance or several substances are the major source of neuromodulatory inputs to the STG. Rather, the challenge is to understand which of these substances are colocalized in specific neurons (31, 41), to understand the extent to which these substances act at the same time or different times to regulate the networks of the STG, and to uncover the mechanisms by which each of these substances regulates STG motor patterns.

Many of the same substances are released into the hemolymph to act as circulating hormones and are released directly into the STG neuropil from descending modulatory projections (**Figure 3**). Presumably, many of the circulating hormones are released in the context of specific behavioral states such as feeding (42) or molting (43). However, a detailed understanding of how neuromodulatory hemolymph concentrations fluctuate according to the animal's behavioral state (44) is still lacking for most of the substances listed in **Figure 3** (left panel).

THE PYLORIC RHYTHM

Electromyographic recordings made in vivo show that the pyloric rhythm is almost always continuously expressed in the intact animal, although its frequency and intensity vary with the animal's physiological status (8, 9). The pyloric rhythm recorded in vitro closely

Figure 1

The stomatogastric nervous system (STNS). (*a*) Side view of a lobster showing the position of the stomach and the STNS. CoG, commissural ganglion; STG, stomatogastric ganglion. (*b*) Side view of the lobster stomach showing the muscles that move the pylorus and gastric mill, the ganglia of the STNS, and the location of the major motor nerves innervating the stomach muscles. OG, esophageal ganglion. (*c*) Schematic of the STNS as it is usually studied in vitro. The nerves and ganglia are dissected off free of the stomach. Extracellular recordings are made with pin electrodes placed in Vaseline wells around the motor nerves. Intracellular recordings are made with glass microelectrodes from ganglia somata. (*d*) Simultaneous extracellular recordings from nine different nerves (pdn, avn, lpn, pyn, mvn gpn, agn, lpgn, and dgn, where the "n" refers to "nerve"). Recordings show the activity of each of the STG motor neurons during ongoing gastric mill and pyloric rhythms in the lobster *H. americanus*. The pyloric rhythm is the faster rhythm and is seen as the alternating activity of the pyloric dilator (PD), lateral pyloric (LP), pyloric (PY), ventricular dilator (VD), and inferior cardiac (IC) neurons. The gastric mill rhythm is slower and is seen as the bursts of activity in the medial gastric (MG), dorsal gastric (DG), gastric mill (GM), lateral posterior gastric (LPG), and lateral gastric (LG) neurons. The DG and LPG neuron bursts are interrupted in time with the pyloric rhythm. agn, anterior gastric nerve; AM, anterior median neuron, avn, anterior ventricular nerve; IC, inferior cardiac neuron; mvn, median ventricular neuron; gpn, gastropyloric nerve. *a* is modified from Reference 189, and *b–d* are modified, with permission, from Reference 140.

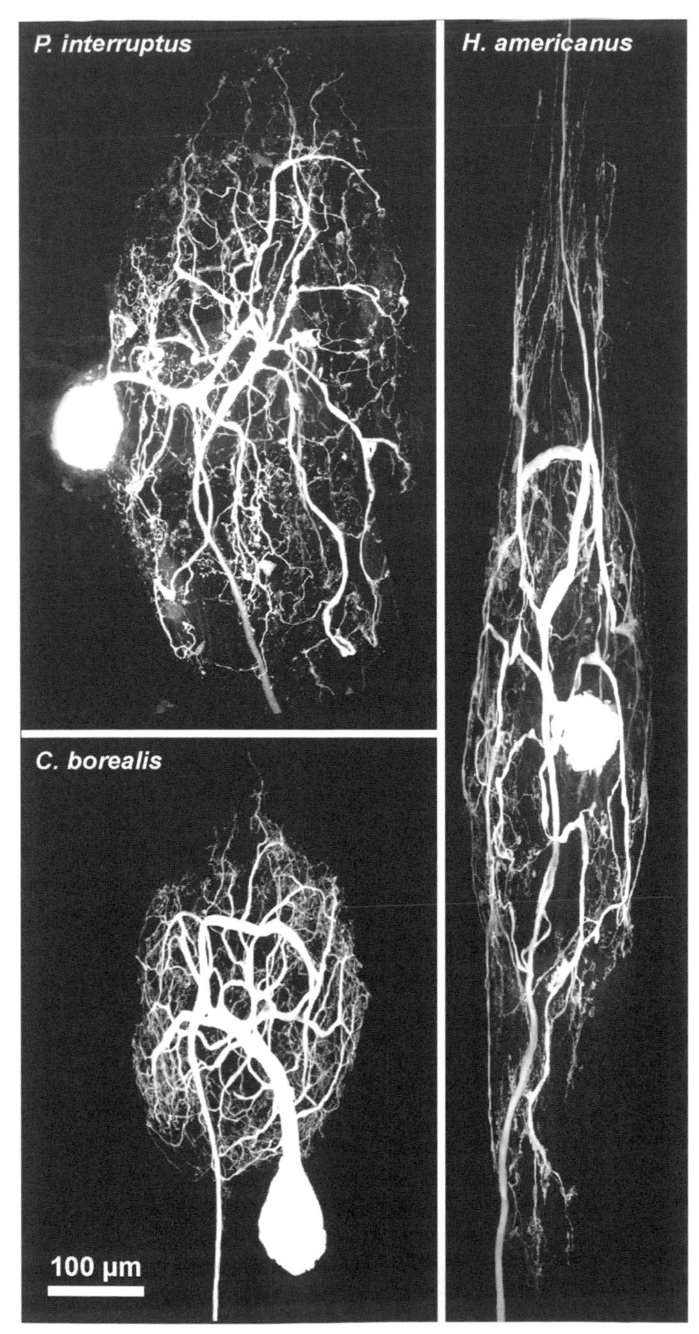

Figure 2

Structure of the pyloric dilator neurons in three crustacean species. In each case, the neurons were filled with Alexa 568 hydrazide and imaged with a confocal microscope. Scale bar is the same for all three images. *H. americanus* fill is used, with permission, from Reference 140, *P. interruptus* fill is courtesy of J. Thuma & S.L. Hooper (unpublished work), and *C. borealis* fill is from D. Bucher (unpublished work).

resembles those recorded in vivo (30). The pyloric rhythm is a triphasic motor pattern with a period of ~1–2 s (**Figure 4**). The canonical pyloric rhythm consists of bursts of action potentials in the PD neurons, followed by bursts of action potentials in the lateral pyloric (LP) neuron, then by bursts in the PY neurons. The inferior cardiac (IC) neuron fires often with the LP neuron burst, and the ventricular dilator (VD) neuron commonly fires with the PY neurons. The anterior burster (AB) neuron is an interneuron that projects anteriorly through the stomatogastric nerve to the CoGs and is electrically coupled to the PD neurons.

Early research on the pyloric rhythm (1, 45–49) focused on several fundamental questions. (*a*) What is the underlying mechanism for rhythm production? (*b*) What determines the specific phase relationships or timing of the elements within the pattern? (*c*) What determines the frequency of the rhythm? Researchers are asking these same questions today of vertebrate spinal cord and respiratory circuits (50–53).

The first steps in answering those questions were to determine the intrinsic membrane properties of each of the pyloric network neurons and to determine their connectivity. There are two major classes of inputs to the pyloric network neurons: inputs from other neurons within the pyloric network itself and inputs from the anterior CoGs and OG. Therefore, the intrinsic properties of the pyloric neurons have been studied under two conditions: (*a*) with impulse activity from the anterior inputs blocked or removed and STG presynaptic inputs also blocked and (*b*) with anterior inputs left intact but STG level presynaptic inputs removed. To remove synaptic inputs from other STG neurons, researchers (*a*) block the glutamatergic inhibitory synaptic inputs with 10^{-5} M picrotoxin (27, 54, 55) and (*b*) remove other inputs, including electrically coupled neurons, by photoinactivation after injection with a fluorescent dye such as Lucifer yellow (56). In the presence of the descending modulatory inputs, all STG neurons show some evidence

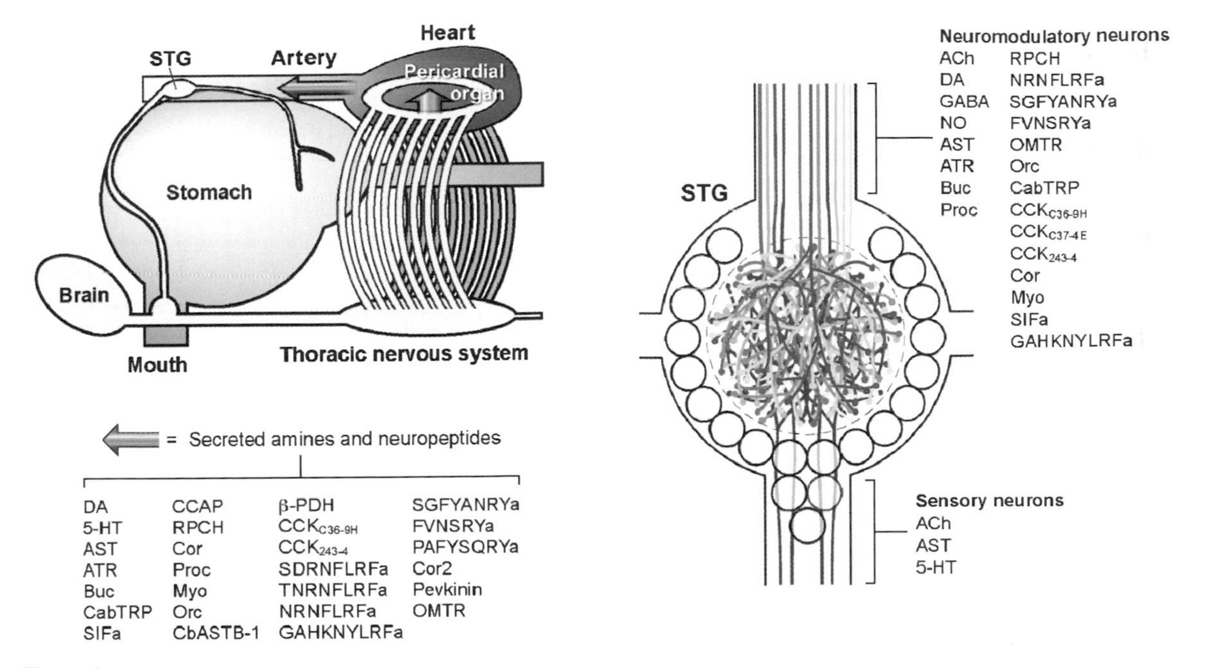

Figure 3

Neuromodulatory control of the stomatogastric ganglion (STG). (*Left*) The STG is shown in the dorsal artery, directly anterior to the heart. The pericardial organs are neurosecretory structures that release many amines and neuropeptides directly into the circulatory system at the level of the heart. (*Right*) The STG is directly modulated by terminals of descending neuromodulatory neurons and ascending sensory neurons. These direct neural projections also release many small molecules and neuropeptides into the neuropil of the STG. Modified from Reference 190.

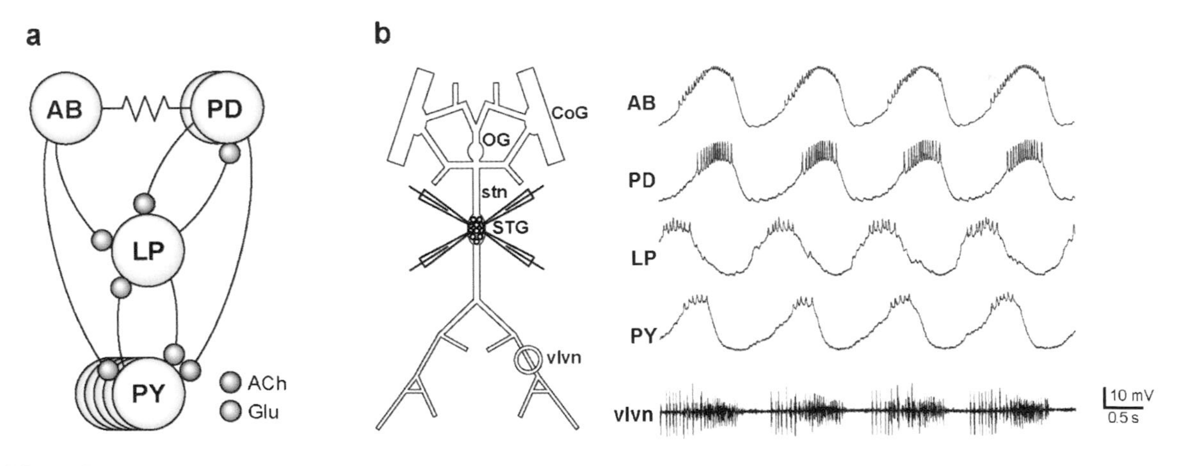

Figure 4

The pyloric rhythm. (*a*) Simplified connectivity diagram of the pyloric circuit [without the ventricular dilator (VD) and inferior cardiac (IC) neurons]. A resistor symbol indicates an electrical coupling between an anterior burster (AB) neuron and a pyloric dilator (PD) neuron. Circles indicate chemical inhibitory connections (ACh, cholinergic; Glu, glutamatergic). (*b*) Schematic of the stomatogastric nervous system (STNS) and simultaneous intracellular (*top four traces*) and extracellular (*bottom trace*) recordings from *H. americanus* that show the typical triphasic pattern. CoG, commissural ganglion; OG, esophageal ganglion; stn, stomatogastric nerve; vlvn, ventral branch, lateral ventricular nerve.

LP neuron: lateral pyloric neuron

IC neuron: inferior cardiac neuron

VD neuron: ventricular dilator neuron

AB neuron: anterior burster neuron

of plateau or bursting behavior (49, 57, 58). However, after the descending inputs are removed, only the AB neuron retains its ability to burst (48, 59), whereas the other neurons fire tonically or fall silent.

Investigators determined the connectivity among the pyloric network neurons, using a combination of simultaneous intracellular recordings and cell kills (1, 47, 55). **Figure 4*a*** shows (*a*) a simplified connectivity diagram of the pyloric network neurons that supports the generation of the pyloric rhythm and (*b*) the neurotransmitters that mediate these synaptic connections. **Figure 4*b*** shows simultaneous intracellular recordings from the somata of these neurons in *H. americanus*.

A first-approximation description of the pyloric rhythm is as follows. The AB neuron is an intrinsic oscillator that, by virtue of its electrical coupling with the PD neurons, causes them to fire bursts of action potentials. Together the PD and AB neurons inhibit the LP and PY neurons, forcing them to fire in alternation with the PD neurons. The LP neuron rebounds from inhibition before the PY neurons, because of various factors, and in turn inhibits the PY neurons. When the PY neurons finally rebound from inhibition, they terminate the LP neuron burst. In this scenario, the rhythm depends strongly on the intrinsic pacemaker properties of the AB neuron, and the phase of the pattern depends on a variety of factors that govern the time at which the LP and PY neurons rebound from inhibition (45, 60–62).

Neuromodulation of the Stomatogastric Ganglion Circuits

Many different substances, including amines, neuropeptides, and gases, modulate the pyloric circuit directly (**Figure 2**) (36, 38, 63–74). Exogenous application of these substances results in changes in the frequency, phase relationships, and number of spikes per burst in different network neurons so that the same network can be reconfigured into various different output patterns (22, 75, 76). Intensive study over the past 20 years has provided some important insights into the mechanisms by which neuromodulatory control is effected in the STG:

(a) Some neuromodulators act on several different voltage-gated channels in the same neuron (77, 78) (**Figure 5*a***).
(b) A number of different neuromodulators converge onto the same voltage-dependent conductance (74) (**Figure 5*b***).
(c) Every neuron in the pyloric circuit is subject to neuromodulation by multiple substances (65, 74, 79).
(d) Every synapse in the pyloric circuit is subject to neuromodulation (78, 80).
(e) The same modulator can influence different synapses in opposing directions (80).

Figure 5*a* summarizes the multiple actions of dopamine on the pyloric neurons in *P. interruptus* (60, 78, 81–83). All pyloric neurons have dopamine receptors, and in each cell type dopamine modulates a different subset of ion channels. Dopamine action on the same channel type can have a different sign in different neurons. Dopamine also modulates the majority of synapses in the pyloric circuits (not shown in **Figure 5*a***).

All STG neurons have receptors for multiple transmitters and neuromodulators. **Figure 5*b*** depicts the known complement of receptors in the *C. borealis* LP neuron. These include receptors to classical transmitters, amines, and a range of neuropeptides. The neuropeptide proctolin was the first to be described to activate a voltage-gated cation current in STG neurons (84), now referred to as the proctolin current. However, later work showed that many of the excitatory neuropeptide receptors found in a given pyloric neuron converge onto the same current (73, 74). Thus, differential circuit modulation by different peptides is the result of different complements of receptors in different neurons. Together, these findings demonstrate that all components of the circuit are subject to neuromodulatory control. This raises a number of

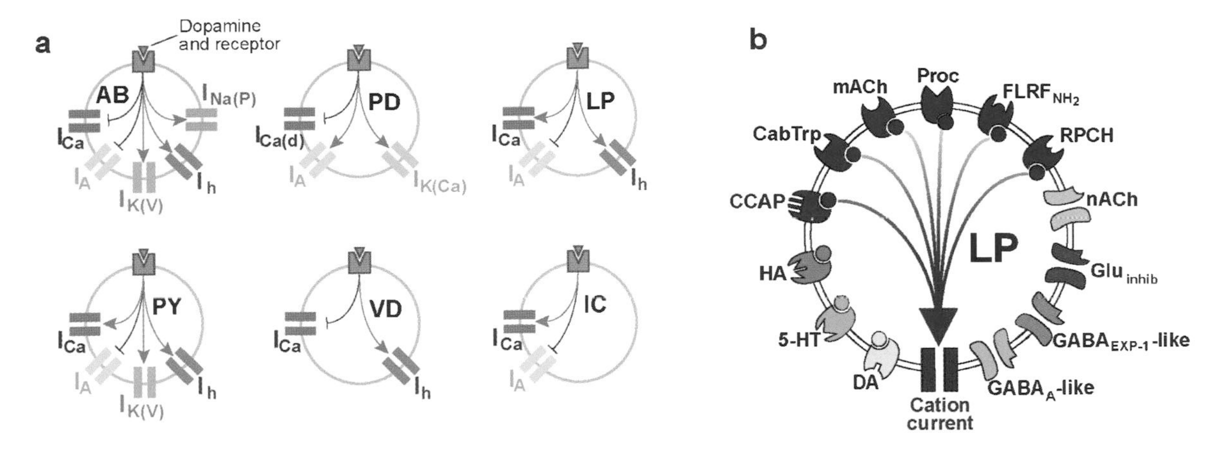

Figure 5

Multiple neuromodulatory mechanisms. (*a*) Dopamine receptors are on all pyloric neurons in *P. interruptus*. In each cell type, dopamine modulates a different subset of ion channels. Dopamine action on the same channel type can have a different sign in different neurons. These schematics summarize data contained in numerous publications from the Harris-Warrick laboratory (60, 78, 81–83). AB, anterior burster neuron; IC, inferior cardiac neuron; LP, lateral pyloric neuron; PD, pyloric dilator neuron; PY, pyloric neuron; VD, ventricular dilator neuron. (*b*) The LP neuron in *C. borealis* has receptors for more than 10 neurotransmitters and modulators, many of which converge on the same cation current. Summarizes data from References 73 and 74 and unpublished data.

questions relevant to maintaining stable circuit function, as it is difficult to understand how it is possible to alter every parameter controlling network function while retaining many of the essential features of the circuit performance.

The Descending Modulatory Projection Neurons

The existence of descending pathways that could influence STG motor patterns was established early (85–87) in spiny lobsters. The anterior pyloric modulator (APM) neuron was an early example of a modulatory neuron that changed the excitability and plateau properties of its target neurons as well as altered the phase relationships of the STG motor patterns (88, 89). Subsequently, the most extensive studies of modulatory projection neurons and their interactions with STG target circuits have been done in *C. borealis*, which has approximately 25 pairs of descending projection neurons to the STG (21).

Some of the descending projection neurons receive synaptic connections from their target neurons in the STG, creating a local circuit among these terminals and STG neurons (90, 91). Consequently, tonic projection neuron activity can be locally configured into rhythmic transmitter release by presynaptic actions.

Most, if not all, of the descending modulatory neurons contain multiple cotransmitters (31, 41, 92), which can evoke a variety of postsynaptic actions and act on different target neurons (93, 94). Sensory inputs can activate these modulatory projection neurons (95–99) to evoke specific sets of motor patterns from the STG circuits.

Maintenance of Constant Phase While Frequency Varies Depends on Synaptic Depression and I_A

One of the remarkable features of the pyloric rhythm is that the phase at which the follower neurons fire is relatively constant while the frequency varies (61, 100–105). This is at first glance puzzling, as all membrane and synaptic currents that play a role in determining rebound properties have fixed time constants. Nonetheless, recent work has provided some insight into how this may occur. The synapses among STG neurons have both spike-mediated and graded components (106, 107), and the graded component of many of the synapses depresses (108, 109) such that the synaptic current decreases at higher frequencies and increases at lower burst frequencies. The transient outward current I_A and the hyperpolarization-activated inward current I_h play important roles in determining when a follower neuron recovers from inhibition (3, 60, 83, 110). Because I_A requires hyperpolarization to remove inactivation, a short interburst interval decreases I_A. Thus, the effects of frequency on synaptic depression and on I_A interact to promote phase constancy (104).

Frequency Control in a Pacemaker-Driven Network

How is frequency controlled in pacemaker-driven networks? To a first approximation, the AB neuron is the pacemaker for the pyloric network. However, it is electrically coupled to the two PD neurons, and the PD neurons are inhibited by the LP neuron. To what extent do these interactions influence the frequency of the pyloric rhythm?

The AB neuron and the PD neurons burst synchronously, but they release different neurotransmitters (27), respond to different neuromodulators (59), and have different intrinsic membrane properties (49). Under some neuromodulatory conditions, the frequency of the entire pacemaker kernel is significantly slower than that of the isolated AB neuron (65). Motivated by this result, Kepler et al. (111) constructed simple, two-neuron, electrically coupled circuits in which one neuron was an oscillator and the second neuron, nonoscillatory. This study demonstrated that, depending on the nature of the oscillator, a nonoscillatory neuron could either increase or decrease the frequency of the oscillator to which it was electrically coupled. A recent modeling study (112) of the electrically coupled PD and AB neurons suggests that the coupling between two dissimilar neurons may extend the range of frequencies over which the coupled network is stable.

The only feedback to the electrically coupled pacemaker kernel from the rest of the pyloric circuit comes from a synapse from the LP neuron to the PD neurons. The role of this synapse in controlling the frequency of the pyloric rhythm has been somewhat elusive. Hyperpolarizing the LP neuron, or otherwise removing it, sometimes had little effect and other times increased the frequency of the pyloric rhythm (113). This is explained by the phase-response curve of the PD neurons (114–117), which is flat at the phase of the pyloric cycle during which the LP neuron usually fires. Because of this, neuromodulators that strongly potentiate the strength of this synapse can nonetheless have relatively little effect on the frequency of the pyloric rhythm (117).

Many modulators influence the frequency of the pyloric rhythm (59, 65–67, 79, 118), and many of these act directly on the AB neuron (59, 65, 79, 118). Although there are few voltage clamp data available on the isolated AB neuron, Harris-Warrick & Flamm (119) showed that whereas the slow wave that sustains bursting persists in the presence of TTX in dopamine, all slow-wave activity is lost in TTX in octopamine and serotonin. This suggests that a different balance of voltage-dependent currents underlies bursting in different modulators. This intuition is strengthened by modeling studies (120–123) that show that various combinations of conductance densities can sustain similar-looking bursting activity.

THE GASTRIC MILL RHYTHM

The gastric mill rhythm controls the movements of the two lateral teeth and single medial tooth in the inside of the stomach (**Figure 6*a***). Unlike the pyloric rhythm, which in vivo is continuously expressed, the

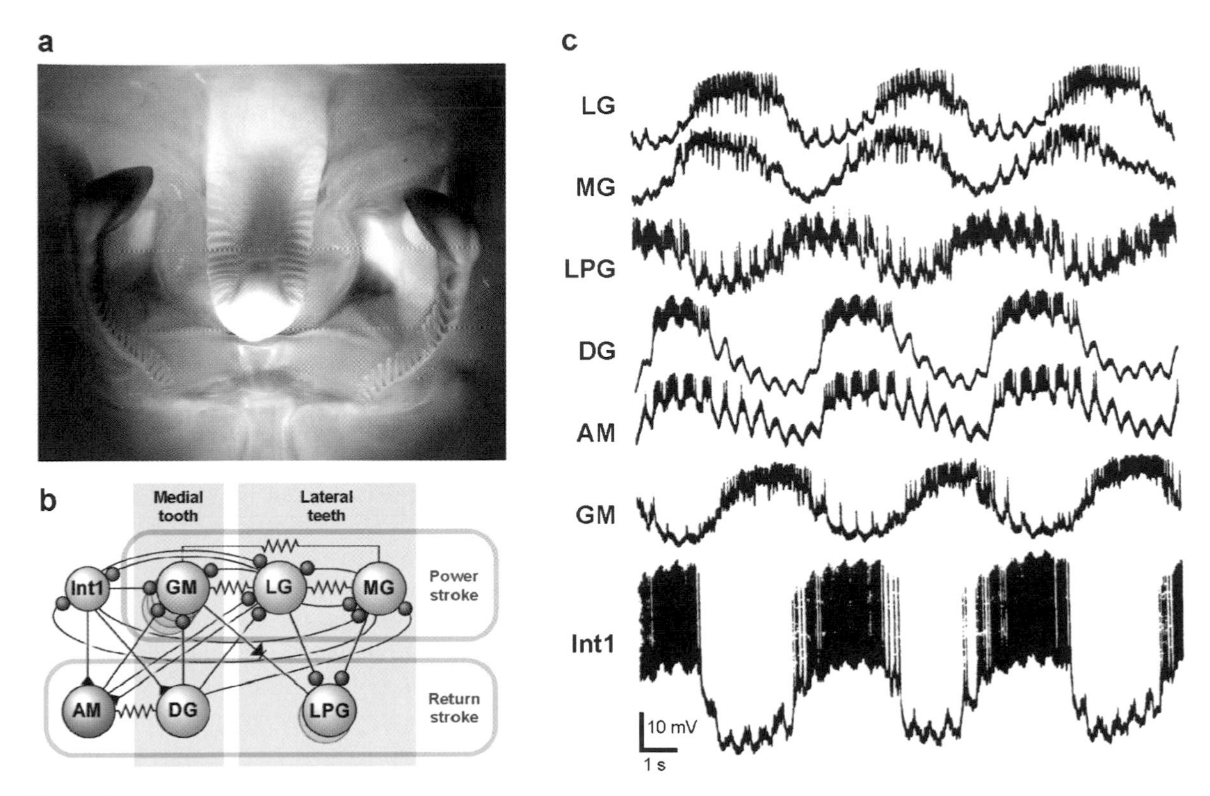

Figure 6

The gastric mill rhythm. (*a*) Photograph of the gastric mill teeth inside the stomach of *P. interruptus*. (*b*) Diagram of the gastric mill circuit. Circles indicate inhibitory connections, triangles indicate excitatory connections, resistor symbols indicate electrical coupling, and the diode symbol indicates a rectifying electrical synapse. Neurons are grouped according to which teeth they control and in which phase they are active. The anterior median (AM) neuron is shown in a different color because it innervates a cardiac sac muscle. Modified with permission from Reference 191. DG, dorsal gastric neuron; GM, gastric mill neuron; Int1, interneuron 1; LG, lateral gastric neuron; LPG, lateral posterior gastric neuron; MG, medial gastric neuron. (*c*) Intracellular recordings of gastric mill neurons in *P. interruptus*. Alternating activity is seen between the LG, MG, and GM neurons in one phase and the LPG, DG, AM, and Int1 neurons in the other phase. The membrane potentials show substantial modulation in pyloric time. Modified with permission from Reference 12.

gastric rhythm in vivo is intermittently active (8) in response to feeding (124, 125) and has a highly variable period, most often approximately 8–20 s. The gastric mill rhythm is less stereotyped than the pyloric rhythm: It displays a number of different forms, depending on how it is activated, that are characterized by different phase relations among the participating muscles and the neurons that innervate them (8, 11, 99, 126, 127). Unlike the pyloric rhythm, the gastric mill rhythm does not have a single pacemaker neuron within the STG, but as a first approximation, the gastric mill rhythm is an emergent property of the reciprocal inhibition among the participating network neurons (128, 129) and interactions with ascending and descending projection neurons (90, 99, 126, 127, 130).

Mulloney & Selverston (128, 129, 131) generated the first connectivity diagram for the gastric rhythm (**Figure 6*b***), using the lobster *P. interruptus*. Heinzel (10, 11) and Heinzel & Selverston (12) further characterized extensively the *P. interrruptus* gastric mill rhythms, using a combination of endoscopy and physiology. **Figure 6*c*** shows

DG neuron: dorsal gastric neuron

LG neuron: lateral gastric neuron

MG neuron: medial gastric neuron

LPG neuron: lateral posterior gastric neuron

simultaneous intracellular recordings from all neurons that participate in the gastric mill rhythm in *P. interruptus*. Note the alternation between dorsal gastric (DG) neurons and GM neurons, which control the medial tooth movements, and the alternation between the lateral gastric (LG) neuron/medial gastric (MG) neurons and the lateral posterior gastric (LPG) neurons, which control the movements of the lateral teeth.

Subsequently, a great deal of work has been done on the gastric mill rhythms activated by specific modulatory projection neurons in the crab *C. borealis* (41, 90, 98, 132, 133). **Figure 7*a*** shows a provisional connectivity diagram for the *C. borealis* STG. Although lobsters and crabs share many features of the gastric mill activity, there also appear to be some important differences. This raises the interesting possibility that different patterns of connectivity have evolved to produce similar motor patterns in different species. Moreover, at least in *C. borealis*, identified modulatory projection neurons are part of the circuitry that generates the gastric mill rhythm (22, 90, 126).

Work from the Nusbaum laboratory has characterized the effects of many different modulatory projection neurons in *C. borealis* (22, 41, 90, 93–98, 134, 135). Many projection neurons influence both the gastric and pyloric rhythms. For example, the modulatory commissural neuron 1 (MCN1) elicits a gastric mill rhythm whose period is an integer multiple of the pyloric rhythm period (132, 136). Moreover, under normal physiological conditions, interactions from the pyloric rhythm regulate MCN1's activity, thus establishing two different mechanisms by which the pyloric and gastric mill activity are coordinated (135).

NEURONS CAN SWITCH BETWEEN DIFFERENT CIRCUITS

For many years researchers thought that the pyloric and gastric mill circuits were separate circuits that showed relatively weak interactions among them, despite the presence of extensive synaptic interactions between pyloric and gastric neurons. However, we now know that, in the absence of gastric mill activity in crabs and some lobsters, neurons that usually are part of the gastric mill circuit fire in time with the pyloric rhythm (137, 138) (**Figure 7*b***). When these neurons fire in pyloric time, they can entrain and reset the pyloric rhythm (139). When the gastric mill rhythm is active and the neurons fire in time with the gastric mill rhythm, they can entrain and reset the gastric mill rhythm (139). Thus, gastric neurons genuinely switch between being members of the pyloric or gastric CPGs.

Interactions also exist when both rhythms are active. **Figure 7*c*** shows recordings from the same experiment as **Figure 7*b***, in this case during ongoing gastric activity. When both rhythms are active, pyloric neuron firing patterns can show substantial modulation over the gastric cycle, as seen in the prolonged interburst interval of the PD neuron at the onset of the LG burst (red bars in **Figure 7*c***). Gastric neurons can show substantial membrane potential modulation in pyloric time (blue arrow in **Figure 7*c***; see also **Figure 6**). Therefore, STG neurons express both patterns simultaneously to different degrees (139, 140). Such a description may be particularly useful because both gastric and pyloric muscles express both rhythms in the contractions they produce (137, 141).

In addition, neuromodulators can recruit neurons into new circuit configurations. In the spiny lobster, the neuromodulator red pigment concentrating hormone strongly potentiates the synapses between the IV neurons of the cardiac sac rhythm and neurons of the gastric mill rhythm. This results in a new network in which cardiac sac and gastric neurons are coordinately active (142). Similarly, activity of the PS neurons, the IV homologs in the lobster *H. gammarus*, produces a novel rhythm in which members of the gastric mill and pyloric networks participate (143, 144).

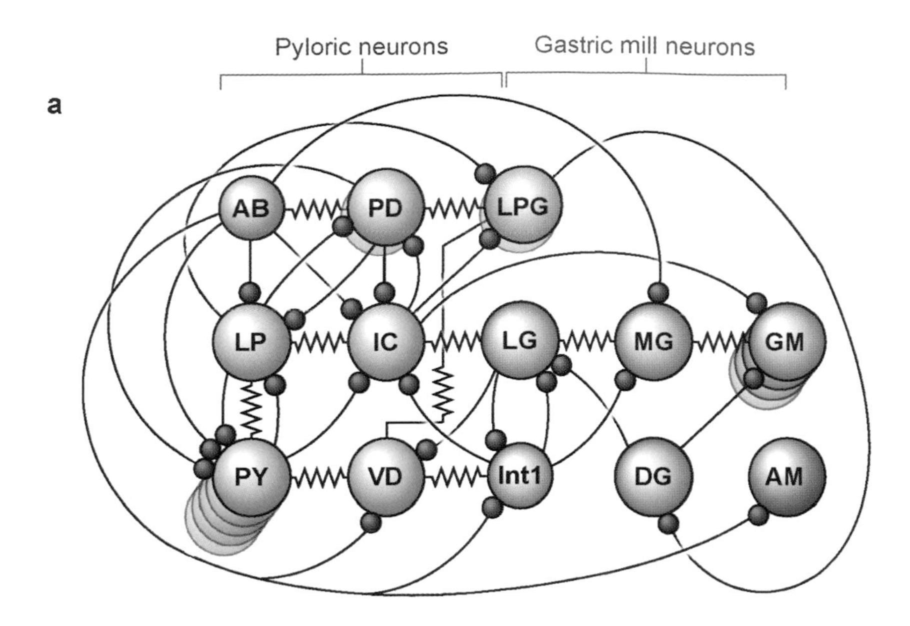

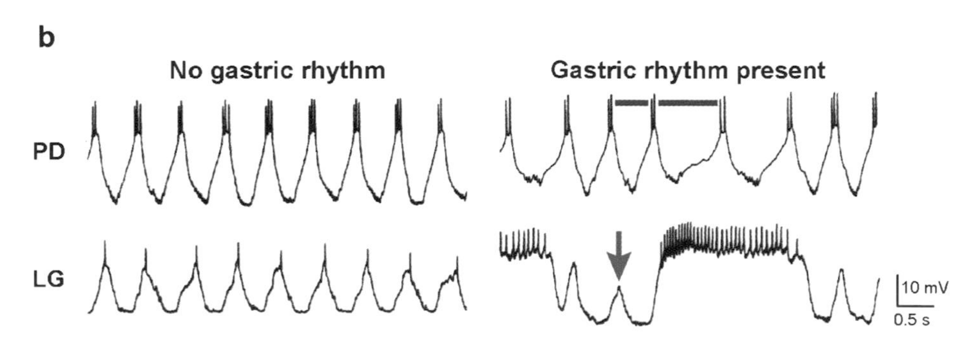

Figure 7

Interactions between the gastric mill and pyloric rhythms. (*a*) Provisional connectivity diagram of the STG neurons in *C. borealis* that shows the substantial synaptic connections between gastric mill circuit and pyloric circuit neurons. Neurons that innervate muscles of the pylorus [and the anterior burster (AB) interneuron] are shown in red, and neurons that innervate muscles of the gastric mill [and interneuron 1 (Int1)] are shown in blue. The lateral posterior ganglion (LPG) neurons innervate pyloric and gastric mill muscles, and the anterior median (AM) neuron innervates a cardiac sac muscle. Circles indicate inhibitory connections, and resistor symbols indicate electrical coupling. Modified from Reference 22. DG, dorsal gastric neuron; IC, inferior cardiac neuron; LG, lateral gastric neuron; LP, lateral pyloric neuron; MG, medial gastric neuron; PD, pyloric dilator neuron; VD, ventricular dilator neuron. (*b*) Intracellular recordings of the PD and LG neurons in *C. borealis*. In the absence of gastric mill activity, the LG neuron fires in time with the pyloric rhythm. In the presence of gastric mill activity, LG is active in gastric time but still shows membrane potential modulation in pyloric time (*blue arrow*). The pyloric rhythm slows down during the LG burst (*red bars*). Modified with permission from Reference 138.

SENSORY INPUT TO THE STOMATOGASTRIC GANGLION

Sensory input plays an important role in shaping the output of CPGs, and the STNS is no different from other motor systems in this regard. Various sensory neurons whose activity alters the pyloric and gastric mill neurons have been identified. Sensory feedback in motor systems is usually studied with respect to the control of timing and magnitude that it exerts through connections with either motor neurons or CPG neurons (145). However, an important role of sensory feedback in the STNS seems to be the activation of and interaction with different descending neuromodulatory pathways (86, 96).

The posterior stomach receptors were the first mechanoreceptors to be described (146). These neurons influence the gastric and pyloric rhythms, presumably acting in the CoGs and other sites. The gastropyloric receptor (GPR) neurons are stretch receptors that innervate stomach muscles (29, 98, 147–153). They synapse directly on STG neurons (150) as well as project anteriorly to the CoGs, where they activate specific sets of descending projection neurons (98). Another stretch receptor, the anterior gastric neuron, has a bipolar soma just posterior to the STG and monitors stretch in the large gastric mill muscles but projects anteriorly without synapsing in the STG (99, 127, 154, 155). The ventral cardiac neurons are a recently described set of sensory neurons that act directly on modulatory projection neurons in the CoGs (95, 97). Interestingly, ventral cardiac neurons and GPRs elicit distinct gastric mill rhythms, although they both activate the same descending projection neurons, MCN1 and CPN2 (97, 98).

Substances present in the hemolymph, including serotonin and the neuropeptide allatostatin (147, 148), modulate the responses of the GPR neurons to muscle stretch. These substances not only alter the stretch responses but also influence the precision of their spiking.

DEVELOPMENT, MATURATION, AND GROWTH OF THE STOMATOGASTRIC GANGLION

In lobsters, the STG is present with its full constellation of neurons before the midpoint of embryonic development (156). At this time it is spontaneously active (156, 157) and generates a rhythm that drives the muscles of the embryonic stomach, including some that eventually become pyloric region muscles and others that eventually become gastric mill region muscles (156, 158). The existence of this seemingly conjoint rhythm, which combines members of the future gastric mill and pyloric networks during embryonic time, has several possible explanations. At one extreme, this may be another example of the neuromodulatory reconfiguration of the STG networks, and the embryonic system may be essentially similar to that of the adult, but in a different neuromodulatory state. Alternatively, the differences between the embryonic rhythm and those generated in the adult may result from differences in the synaptic and intrinsic properties of the neurons at these early stages.

Much of the neuromodulatory complement is formed quite early in development (159–163), although some neuromodulators do not become detectable until larval stages. The receptors for most, if not all, of the modulators that act in the adult appear to be present in the embryo (163–165; K. Rehm, unpublished observations).

Based on modeling work, the Meyrand laboratory has suggested that strong electrical coupling among the neurons in the embryonic network accounts for the fact that the future gastric mill neurons tend to fire in time with the future pyloric neurons (166). A descending projection in the embryo may be responsible for maintaining a high level of coupling (158, 166), which is later inhibited as the animals go through metamorphosis.

Lobsters undergo a final metamorphosis after their three larval stages. At this point they are still quite small, but these juveniles resemble the adult in body and stomach structure. Despite considerable changes in body and ganglion size, the STNS isolated from juveniles produces pyloric rhythms virtually indistinguishable from those seen in adults (105). Thus, there must be mechanisms in place to assure that network stability is maintained as individual neurons are adding membranes and synthesizing and inserting channels and as distances between synapses are changing.

HOMEOSTASIS AND RECOVERY OF FUNCTION

As described above, when the STG is isolated from its descending modulatory inputs, the gastric mill rhythm stops, and the pyloric rhythm usually also stops. Nonetheless, if the preparations are maintained in sterile conditions in vitro after one to several days, the pyloric rhythm resumes (**Figure 8**), now in the absence of neuromodulatory inputs (17, 102, 167–170), after a period in which the preparations generate bouts of intermittent activity (102). This recovery is blocked by inhibitors of mRNA synthesis (169), and is associated with increased excitability in the PD neurons (170). This recovery is consistent with the interpretation that neurons in the STG have a target activity level and mechanisms by which they can sense that activity and alter their channel densities accordingly (167, 171–174). STG neurons also may respond directly to the loss of neuromodulatory signals, which may trigger the changes that allow recovery of activity (168).

The hypothesis that neurons and networks have a target activity level and homeostatic mechanisms tending to maintain stable neuronal and network function predicts that overexpression of a current may result in compensatory changes in one or more other currents (171, 172). Injection of *shal* mRNA encoding I_A (15, 175) into PD neurons resulted in enhanced I_A current measured in voltage clamp but no obvious change in the firing pattern of the PD neurons (176). A compensatory upregulation of I_h explained the lack of change in excitability (176). However, the same upregulation of I_h occurred when an inactive form of the *shal* gene was expressed, arguing that a direct molecular mechanism, and not activity, mediated the compensatory change (176, 177). This coupling of the two currents was not reciprocal, as injection of the mRNA encoding I_h failed to alter I_A, and did result in altered patterns of activity (178).

Correlated expression of I_A and I_h was also seen in recent single-neuron, real-time PCR experiments from identified crab STG neurons (16). In this study, the values of these currents were tightly correlated with each other in a given PD neuron, but the values across preparations were highly variable. However, the values of these currents were very similar in the two electrically coupled PD neurons within a preparation. This may indicate that a small molecular metabolite is important for controlling the expression of these currents or that the two electrically coupled neurons

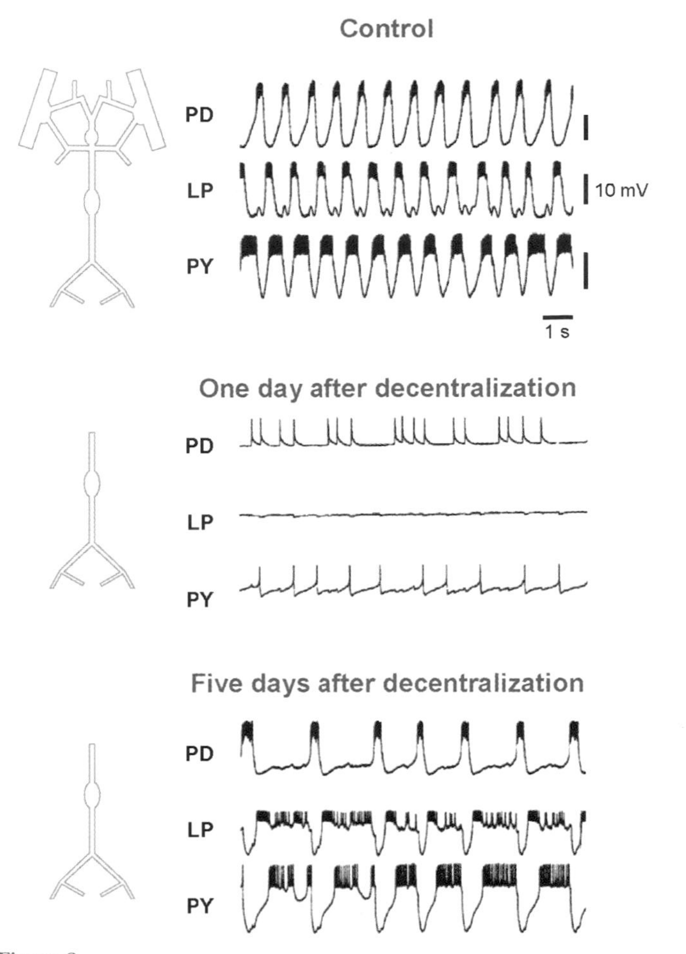

Figure 8

Recovery of the pyloric rhythm after decentralization in the lobster *Jasus lalandii*. The triphasic rhythmic activity seen in control lobsters (*top panel*) ceases after descending modulatory input is removed (*middle panel*). Rhythmic activity resumes after several days in the absence of modulatory input (*lower panel*). LP, lateral pyloric neuron; PD, pyloric dilator neuron; PY, pyloric neuron. Modified with permission from Reference 170.

within a given animal have very similar histories of activity (16).

The molecular identification of the genes for channels and receptors (15, 175, 179–188) opens the possibility of determining where ion channels and receptors are found over the complex structures of STG neurons. For example, different genes appear to contribute to A-type K^+ currents in different regions of neurons (179), and antibodies raised against different Ca^{2+} channels apparently also show differential distribution of labeling (187). Nonetheless, these studies provide only a starting point for understanding the extent to which STG neurons specifically localize both ion channels and receptors and to which this localization is modified by experience.

SUMMARY POINTS

1. Neurons and neural circuits are modulated by many substances that may act singly or in concert to reconfigure neuronal networks. The targets for neuromodulation include all neurons within a circuit and all the neurons' synapses.
2. A pacemaker kernel of three neurons drives the pyloric rhythm, which depends on a series of mechanisms to maintain approximately constant firing phases over a range of frequencies.
3. The gastric mill rhythm emerges from the connectivity among the gastric mill neurons and the activity of descending modulatory inputs. Different patterns of connectivity may subserve the production of similar gastric mill motor patterns in different species, and within the same species, different mechanisms may generate similar gastric mill motor patterns.
4. Individual neurons may fire in time with more than one rhythm and may switch among different pattern-generating circuits. Consequently, activity alone is not a sufficient criterion for neuronal identification, thus complicating the identification of circuit neurons in larger vertebrate systems.
5. Sensory modification of the STG motor patterns results both from direct projections to the STG and from the activation of specific sets of descending modulatory projections. Sensory neurons themselves are subject to neuromodulation that alters their response to stretch.
6. The STG is fully formed early in development but generates motor patterns different from those in the adult.
7. If the STG is deprived of its normal constellation of neuromodulatory inputs for 24–72 h, activity resumes independently of those neuromodulatory inputs. Thus, there are mechanisms that maintain stable network output under different physiological conditions.
8. There may be multiple combinations of conductance densities consistent with the activity patterns of individual identified neurons. The expression of some channel genes may be coupled, resulting in mechanisms for compensation for changes in some currents.

FUTURE ISSUES

1. What kinds of mechanisms stabilize constant performance of the pyloric and gastric mill networks over many years, despite ongoing channel turnover and considerable growth of the neurons and the biomechanical plant that they drive?
2. If every synapse and neuron in a network is subject to neuromodulation, what prevents overmodulation and keeps networks in the appropriate operating range?
3. What combinations of membrane conductances give rise to the specific properties of the different classes of identified neurons? What specifies neuronal identity at the molecular level?

LITERATURE CITED

1. Maynard DM. 1972. Simpler networks. *Ann. N.Y. Acad. Sci.* 193:59–72
2. Harris-Warrick RM, Marder E. 1991. Modulation of neural networks for behavior. *Annu. Rev. Neurosci.* 14:39–57
3. Hartline DK, Russell DF, Raper JA, Graubard K. 1988. Special cellular and synaptic mechanisms in motor pattern generation. *Comp. Biochem. Physiol.* 91C:115–31
4. Selverston AI, Russell DF, Miller JP, King DG. 1976. The stomatogastric nervous system: structure and function of a small neural network. *Prog. Neurobiol.* 7:215–90
5. Harris-Warrick RM, Marder E, Selverston AI, Moulins M. 1992. *Dynamic Biological Networks. The Stomatogastric Nervous System.* Cambridge, MA: MIT Press. 328 pp.
6. Selverston AI, Moulins M, eds. 1987. *The Crustacean Stomatogastric System.* Berlin: Springer-Verlag. 338 pp.
7. Meyrand P, Faumont S, Simmers J, Christie AE, Nusbaum MP. 2000. Species-specific modulation of pattern-generating circuits. *Eur. J. Neurosci.* 12:2585–96
8. Clemens S, Combes D, Meyrand P, Simmers J. 1998. Long-term expression of two interacting motor pattern-generating networks in the stomatogastric system of freely behaving lobster. *J. Neurophysiol.* 79:1396–408
9. Rezer E, Moulins M. 1983. Expression of the crustacean pyloric pattern generator in the intact animal. *J. Comp. Physiol. A* 153:17–28
10. Heinzel HG. 1988. Gastric mill activity in the lobster. II. Proctolin and octopamine initiate and modulate chewing. *J. Neurophysiol.* 59:551–65
11. Heinzel HG. 1988. Gastric mill activity in the lobster. I. Spontaneous modes of chewing. *J. Neurophysiol.* 59:528–50
12. Heinzel HG, Selverston AI. 1988. Gastric mill activity in the lobster. III. Effects of proctolin on the isolated central pattern generator. *J. Neurophysiol.* 59:566–85
13. Marder E, Calabrese RL. 1996. Principles of rhythmic motor pattern generation. *Physiol. Rev.* 76:687–717
14. Marder E. 1976. Cholinergic motor neurones in the stomatogastric system of the lobster. *J. Physiol.* 257:63–86
15. Baro DJ, Levini RM, Kim MT, Willms AR, Lanning CC, et al. 1997. Quantitative single-cell-reverse transcription-PCR demonstrates that A-current magnitude varies as a linear function of *shal* gene expression in identified stomatogastric neurons. *J. Neurosci.* 17:6597–10
16. Schulz DJ, Goaillard JM, Marder E. 2006. Variable channel expression in identified single and electrically coupled neurons in different animals. *Nat. Neurosci.* 9:356–62

17. Mizrahi A, Dickinson PS, Kloppenburg P, Fénelon V, Baro DJ, et al. 2001. Long-term maintenance of channel distribution in a central pattern generator neuron by neuromodulatory inputs revealed by decentralization in organ culture. *J. Neurosci.* 21:7331–39
18. Kiehn O, Butt SJ. 2003. Physiological, anatomical and genetic identification of CPG neurons in the developing mammalian spinal cord. *Prog. Neurobiol.* 70:347–61
19. Sugino K, Hempel CM, Miller MN, Hattox AM, Shapiro P, et al. 2006. Molecular taxonomy of major neuronal classes in the adult mouse forebrain. *Nat. Neurosci.* 9:99–107
20. Maynard DM, Dando MR. 1974. The structure of the stomatogastric neuromuscular system in *Callinectes sapidus*, *Homarus americanus* and *Panulirus argus* (decapoda crustacea). *Philos. Trans. R. Soc. London Ser. B* 268:161–220
21. Coleman MJ, Nusbaum MP, Cournil I, Claiborne BJ. 1992. Distribution of modulatory inputs to the stomatogastric ganglion of the crab, *Cancer borealis*. *J. Comp. Neurol.* 325:581–94
22. Nusbaum MP, Beenhakker MP. 2002. A small-systems approach to motor pattern generation. *Nature* 417:343–50
23. Kilman VL, Marder E. 1996. Ultrastructure of the stomatogastric ganglion neuropil of the crab, *Cancer borealis*. *J. Comp. Neurol.* 374:362–75
24. Bucher D, Johnson CD, Marder E. 2006. Neuronal morphology and neuropil structure in the stomatogastric ganglion of the lobster, *Homarus americanus*. *J. Comp. Neurol.* In press
25. King DG. 1976. Organization of crustacean neuropil. I. Patterns of synaptic connections in lobster stomatogastric ganglion. *J. Neurocytol.* 5:207–37
26. King DG. 1976. Organization of crustacean neuropil. II. Distribution of synaptic contacts on identified motor neurons in lobster stomatogastric ganglion. *J. Neurocytol.* 5:239–66
27. Marder E, Eisen JS. 1984. Transmitter identification of pyloric neurons: electrically coupled neurons use different neurotransmitters. *J. Neurophysiol.* 51:1345–61
28. Lingle C. 1980. The sensitivity of decapod foregut muscles to acetylcholine and glutamate. *J. Comp. Physiol.* 138:187–99
29. Katz PS, Eigg MH, Harris-Warrick RM. 1989. Serotonergic/cholinergic muscle receptor cells in the crab stomatogastric nervous system. I. Identification and characterization of the gastropyloric receptor cells. *J. Neurophysiol.* 62:558–70
30. Marder E, Bucher D. 2001. Central pattern generators and the control of rhythmic movements. *Curr. Biol.* 11:R986–96
31. Nusbaum MP, Blitz DM, Swensen AM, Wood D, Marder E. 2001. The roles of cotransmission in neural network modulation. *Trends Neurosci.* 24:146–54
32. Christie AE, Cain SD, Edwards JM, Clason TA, Cherny E, et al. 2004. The anterior cardiac plexus: an intrinsic neurosecretory site within the stomatogastric nervous system of the crab *Cancer productus*. *J. Exp. Biol.* 207:1163–82
33. Christie AE, Skiebe P, Marder E. 1995. Matrix of neuromodulators in neurosecretory structures of the crab, *Cancer borealis*. *J. Exp. Biol.* 198:2431–39
34. Barker DL, Kushner PD, Hooper NK. 1979. Synthesis of dopamine and octopamine in the crustacean stomatogastric nervous system. *Brain Res.* 161:99–113
35. Kushner PD, Maynard EA. 1977. Localization of monoamine fluorescence in the stomatogastric nervous system of lobsters. *Brain Res.* 129:13–28
36. Beltz B, Eisen JS, Flamm R, Harris-Warrick RM, Hooper S, Marder E. 1984. Serotonergic innervation and modulation of the stomatogastric ganglion of three decapod crustaceans (*Panulirus interruptus*, *Homarus americanus* and *Cancer irroratus*). *J. Exp. Biol.* 109:35–54

37. Nusbaum MP, Marder E. 1989. A modulatory proctolin-containing neuron (MPN). I. Identification and characterization. *J. Neurosci.* 9:1591–99
38. Skiebe P, Schneider H. 1994. Allatostatin peptides in the crab stomatogastric nervous system: inhibition of the pyloric motor pattern and distribution of allatostatin-like immunoreactivity. *J. Exp. Biol.* 194:195–208
39. Li L, Kelley WP, Billimoria CP, Christie AE, Pulver SR, et al. 2003. Mass spectrometric investigation of the neuropeptide complement and release in the pericardial organs of the crab, *Cancer borealis*. *J. Neurochem.* 87:642–56
40. Stemmler EA, Provencher HL, Guiney ME, Gardner NP, Dickinson PS. 2005. Matrix-assisted laser desorption/ionization fourier transform mass spectrometry for the identification of orcokinin neuropeptides in crustaceans using metastable decay and sustained off-resonance irradiation. *Anal. Chem.* 77:3594–606
41. Blitz DM, Christie AE, Coleman MJ, Norris BJ, Marder E, Nusbaum MP. 1999. Different proctolin neurons elicit distinct motor patterns from a multifunctional neuronal network. *J. Neurosci.* 19:5449–63
42. Turrigiano GG, Selverston AI. 1990. A cholecystokinin-like hormone activates a feeding-related neural circuit in lobster. *Nature* 344:866–68
43. Dircksen H. 1998. Conserved crustacean cardioactive peptide: neural networks and function in arthropod evolution. In *Arthropod Endocrinology: Perspectives and Recent Advances*, ed. GM Coast, SG Webster, pp. 302–33. Cambridge, UK: Cambridge Univ. Press
44. Keller R. 1992. Crustacean neuropeptides: structures, functions and comparative aspects. *Experientia* 48:439–48
45. Hartline DK. 1979. Pattern generation in the lobster (*Panulirus*) stomatogastric ganglion. II. Pyloric network simulation. *Biol. Cybern.* 33:223–36
46. Hartline DK, Gassie DV Jr. 1979. Pattern generation in the lobster (*Panulirus*) stomatogastric ganglion. I. Pyloric neuron kinetics and synaptic interactions. *Biol. Cybern.* 33:209–22
47. Miller JP, Selverston AI. 1982. Mechanisms underlying pattern generation in lobster stomatogastric ganglion as determined by selective inactivation of identified neurons. IV. Network properties of pyloric system. *J. Neurophysiol.* 48:1416–32
48. Miller JP, Selverston AI. 1982. Mechanisms underlying pattern generation in lobster stomatogastric ganglion as determined by selective inactivation of identified neurons. II. Oscillatory properties of pyloric neurons. *J. Neurophysiol.* 48:1378–91
49. Selverston AI, Miller JP. 1980. Mechanisms underlying pattern generation in the lobster stomatogastric ganglion as determined by selective inactivation of identified neurons. I. Pyloric neurons. *J. Neurophysiol.* 44:1102–21
50. Grillner S. 2003. The motor infrastructure: from ion channels to neuronal networks. *Nat. Rev. Neurosci.* 4:573–86
51. Rekling JC, Feldman JL. 1998. PreBötzinger complex and pacemaker neurons: hypothesized site and kernel for respiratory rhythm generation. *Annu. Rev. Physiol.* 60:385–405
52. Ramirez JM, Tryba AK, Pena F. 2004. Pacemaker neurons and neuronal networks: an integrative view. *Curr. Opin. Neurobiol.* 14:665–74
53. Kiehn O. 2006. Locomotor circuits in the mammalian spinal cord. *Annu. Rev. Neurosci.* 29:279–306
54. Marder E, Paupardin-Tritsch D. 1978. The pharmacological properties of some crustacean neuronal acetylcholine, gamma-aminobutyric acid and L-glutamate responses. *J. Physiol.* 280:213–36

55. Eisen JS, Marder E. 1982. Mechanisms underlying pattern generation in lobster stomatogastric ganglion as determined by selective inactivation of identified neurons. III. Synaptic connections of electrically coupled pyloric neurons. *J. Neurophysiol.* 48:1392–415
56. Miller JP, Selverston A. 1979. Rapid killing of single neurons by irradiation of intracellularly injected dye. *Science* 206:702–4
57. Elson RC, Huerta R, Abarbanel HD, Rabinovich MI, Selverston AI. 1999. Dynamic control of irregular bursting in an identified neuron of an oscillatory circuit. *J. Neurophysiol.* 82:115–22
58. Szucs A, Pinto RD, Rabinovich MI, Abarbanel HD, Selverston AI. 2003. Synaptic modulation of the interspike interval signatures of bursting pyloric neurons. *J. Neurophysiol.* 89:1363–77
59. Marder E, Eisen JS. 1984. Electrically coupled pacemaker neurons respond differently to the same physiological inputs and neurotransmitters. *J. Neurophysiol.* 51:1362–74
60. Harris-Warrick RM, Coniglio LM, Levini RM, Gueron S, Guckenheimer J. 1995. Dopamine modulation of two subthreshold currents produces phase shifts in activity of an identified motoneuron. *J. Neurophysiol.* 74:1404–20
61. Eisen JS, Marder E. 1984. A mechanism for production of phase shifts in a pattern generator. *J. Neurophysiol.* 51:1375–93
62. Rabbah P, Nadim F. 2005. Synaptic dynamics do not determine proper phase of activity in a central pattern generator. *J. Neurosci.* 25:11269–78
63. Marder E, Thirumalai V. 2002. Cellular, synaptic and network effects of neuromodulation. *Neural Netw.* 15:479–93
64. Flamm RE, Harris-Warrick RM. 1986. Aminergic modulation in lobster stomatogastric ganglion. I. Effects on motor pattern and activity of neurons within the pyloric circuit. *J. Neurophysiol.* 55:847–65
65. Hooper SL, Marder E. 1987. Modulation of the lobster pyloric rhythm by the peptide proctolin. *J. Neurosci.* 7:2097–112
66. Nusbaum MP, Marder E. 1988. A neuronal role for a crustacean red pigment concentrating hormone-like peptide: neuromodulation of the pyloric rhythm in the crab, *Cancer borealis*. *J. Exp. Biol.* 135:165–81
67. Weimann JM, Skiebe P, Heinzel H-G, Soto C, Kopell N, et al. 1997. Modulation of oscillator interactions in the crab stomatogastric ganglion by crustacean cardioactive peptide. *J. Neurosci.* 17:1748–60
68. Turrigiano GG, Selverston AI. 1989. Cholecystokinin-like peptide is a modulator of a crustacean central pattern generator. *J. Neurosci.* 9:2486–501
69. Weimann JM, Marder E, Evans B, Calabrese RL. 1993. The effects of SDRNFLRFamide and TNRNFLRFamide on the motor patterns of the stomatogastric ganglion of the crab *Cancer borealis*. *J. Exp. Biol.* 181:1–26
70. Scholz NL, de Vente J, Truman JW, Graubard K. 2001. Neural network partitioning by NO and cGMP. *J. Neurosci.* 21:1610–18
71. Christie AE, Stemmler EA, Peguero B, Messinger DI, Provencher HL, et al. 2006. Identification, physiological actions, and distribution of VYRKPPFNGSIFamide (Val1-SIFamide) in the stomatogastric nervous system of the American lobster *Homarus americanus*. *J. Comp. Neurol.* 496:406–21
72. Claiborne B, Selverston A. 1984. Histamine as a neurotransmitter in the stomatogastric nervous system of the spiny lobster. *J. Neurosci.* 4:708–21
73. Swensen AM, Golowasch J, Christie AE, Coleman MJ, Nusbaum MP, Marder E. 2000. GABA and responses to GABA in the stomatogastric ganglion of the crab *Cancer borealis*. *J. Exp. Biol.* 203:2075–92

74. Swensen AM, Marder E. 2000. Multiple peptides converge to activate the same voltage-dependent current in a central pattern-generating circuit. *J. Neurosci.* 20:6752–59
75. Marder E, Hooper SL. 1985. Neurotransmitter modulation of the stomatogastric ganglion of decapod crustaceans. In *Model Neural Networks and Behavior*, ed. AI Selverston, pp. 319–37. New York: Plenum Press
76. Marder E, Weimann JM. 1992. Modulatory control of multiple task processing in the stomatogastric nervous system. In *Neurobiology of Motor Progamme Selection*, ed. J Kien, C McCrohan, B Winlow, pp. 3–19. New York: Pergamon Press
77. Kiehn O, Harris-Warrick RM. 1992. 5-HT modulation of hyperpolarization-activated inward current and calcium-dependent outward current in a crustacean motor neuron. *J. Neurophysiol.* 68:496–508
78. Harris-Warrick RM, Johnson BR, Peck JH, Kloppenburg P, Ayali A, Skarbinski J. 1998. Distributed effects of dopamine modulation in the crustacean pyloric network. *Ann. N.Y. Acad. Sci.* 860:155–67
79. Flamm RE, Harris-Warrick RM. 1986. Aminergic modulation in lobster stomatogastric ganglion. II. Target neurons of dopamine, octopamine, and serotonin within the pyloric circuit. *J. Neurophysiol.* 55:866–81
80. Johnson BR, Peck JH, Harris-Warrick RM. 1995. Distributed amine modulation of graded chemical transmission in the pyloric network of the lobster stomatogastric ganglion. *J. Neurophysiol.* 174:437–52
81. Gruhn M, Guckenheimer J, Land B, Harris-Warrick RM. 2005. Dopamine modulation of two delayed rectifier potassium currents in a small neural network. *J. Neurophysiol.* 94:2888–900
82. Kloppenburg P, Levini RM, Harris-Warrick RM. 1999. Dopamine modulates two potassium currents and inhibits the intrinsic firing properties of an identified motor neuron in a central pattern generator network. *J. Neurophysiol.* 81:29–38
83. Harris-Warrick RM, Coniglio LM, Barazangi N, Guckenheimer J, Gueron S. 1995. Dopamine modulation of transient potassium current evokes phase shifts in a central pattern generator network. *J. Neurosci.* 15:342–58
84. Golowasch J, Marder E. 1992. Proctolin activates an inward current whose voltage dependence is modified by extracellular Ca^{2+}. *J. Neurosci.* 12:810–17
85. Dando MR, Selverston AI. 1972. Command fibres from the supraesophageal ganglion to the stomatogastric ganglion in *Panulirus argus*. *J. Comp. Physiol.* 78:138–75
86. Sigvardt KA, Mulloney B. 1982. Sensory alteration of motor patterns in the stomatogastric nervous system of the spiny lobster *Panulirus interruptus*. *J. Exp. Biol.* 97:137–52
87. Sigvardt KA, Mulloney B. 1982. Properties of synapses made by IVN command-interneurones in the stomatogastric ganglion of the spiny lobster *Panulirus interruptus*. *J. Exp. Biol.* 97:153–68
88. Dickinson PS, Nagy F. 1983. Control of a central pattern generator by an identified modulatory interneurone in crustacea. II. Induction and modification of plateau properties in pyloric neurones. *J. Exp. Biol.* 105:59–82
89. Nagy F, Dickinson PS. 1983. Control of a central pattern generator by an identified modulatory interneurone in crustacea. I. Modulation of the pyloric motor output. *J. Exp. Biol.* 105:33–58
90. Coleman MJ, Meyrand P, Nusbaum MP. 1995. A switch between two modes of synaptic transmission mediated by presynaptic inhibition. *Nature* 378:502–5
91. Coleman MJ, Nusbaum MP. 1994. Functional consequences of compartmentalization of synaptic input. *J. Neurosci.* 14:6544–52

92. Christie AE, Stein W, Quinlan JE, Beenhakker MP, Marder E, Nusbaum MP. 2004. Actions of a histaminergic/peptidergic projection neuron on rhythmic motor patterns in the stomatogastric nervous system of the crab *Cancer borealis*. *J. Comp. Neurol.* 469:153–69
93. Blitz DM, Nusbaum MP. 1999. Distinct functions for cotransmitters mediating motor pattern selection. *J. Neurosci.* 19:6774–83
94. Wood DE, Stein W, Nusbaum MP. 2000. Projection neurons with shared cotransmitters elicit different motor patterns from the same neuronal circuit. *J. Neurosci.* 20:8943–53
95. Beenhakker MP, Blitz DM, Nusbaum MP. 2004. Long-lasting activation of rhythmic neuronal activity by a novel mechanosensory system in the crustacean stomatogastric nervous system. *J. Neurophysiol.* 91:78–91
96. Beenhakker MP, DeLong ND, Saideman SR, Nadim F, Nusbaum MP. 2005. Proprioceptor regulation of motor circuit activity by presynaptic inhibition of a modulatory projection neuron. *J. Neurosci.* 25:8794–806
97. Beenhakker MP, Nusbaum MP. 2004. Mechanosensory activation of a motor circuit by coactivation of two projection neurons. *J. Neurosci.* 24:6741–50
98. Blitz DM, Beenhakker MP, Nusbaum MP. 2004. Different sensory systems share projection neurons but elicit distinct motor patterns. *J. Neurosci.* 24:11381–90
99. Combes D, Meyrand P, Simmers J. 1999. Motor pattern specification by dual descending pathways to a lobster rhythm-generating network. *J. Neurosci.* 19:3610–19
100. Hooper SL. 1997. Phase maintenance in the pyloric pattern of the lobster (*Panulirus interruptus*) stomatogastric ganglion. *J. Comput. Neurosci.* 4:191–205
101. Hooper SL. 1997. The pyloric pattern of the lobster (*Panulirus interruptus*) stomatogastric ganglion comprises two phase maintaining subsets. *J. Comput. Neurosci.* 4:207–19
102. Luther JA, Robie AA, Yarotsky J, Reina C, Marder E, Golowasch J. 2003. Episodic bouts of activity accompany recovery of rhythmic output by a neuromodulator- and activity-deprived adult neural network. *J. Neurophysiol.* 90:2720–30
103. Manor Y, Bose A, Booth V, Nadim F. 2003. Contribution of synaptic depression to phase maintenance in a model rhythmic network. *J. Neurophysiol.* 90:3513–28
104. Greenberg I, Manor Y. 2005. Synaptic depression in conjunction with A-current channels promote phase constancy in a rhythmic network. *J. Neurophysiol.* 93:656–77
105. Bucher D, Prinz AA, Marder E. 2005. Animal-to-animal variability in motor pattern production in adults and during growth. *J. Neurosci.* 25:1611–19
106. Graubard K. 1978. Synaptic transmission without action potentials: input-output properties of a nonspiking presynaptic neuron. *J. Neurophysiol.* 41:1014–25
107. Graubard K, Raper JA, Hartline DK. 1980. Graded synaptic transmission between spiking neurons. *Proc. Natl. Acad. Sci. USA* 77:3733–35
108. Manor Y, Nadim F, Abbott LF, Marder E. 1997. Temporal dynamics of graded synaptic transmission in the lobster stomatogastric ganglion. *J. Neurosci.* 17:5610–21
109. Mamiya A, Manor Y, Nadim F. 2003. Short-term dynamics of a mixed chemical and electrical synapse in a rhythmic network. *J. Neurosci.* 23:9557–64
110. Tierney AJ, Harris-Warrick RM. 1992. Physiological role of the transient potassium current in the pyloric circuit of the lobster stomatogastric ganglion. *J. Neurophysiol.* 67:599–609
111. Kepler TB, Marder E, Abbott LF. 1990. The effect of electrical coupling on the frequency of model neuronal oscillators. *Science* 248:83–85
112. Soto-Trevino C, Rabbah P, Marder E, Nadim F. 2005. Computational model of electrically coupled, intrinsically distinct pacemaker neurons. *J. Neurophysiol.* 94:590–604
113. Nadim F, Manor Y, Kopell N, Marder E. 1999. Synaptic depression creates a switch that controls the frequency of an oscillatory circuit. *Proc. Natl. Acad. Sci. USA* 96:8206–11

114. Ayali A, Harris-Warrick RM. 1999. Monoamine control of the pacemaker kernel and cycle frequency in the lobster pyloric network. *J. Neurosci.* 19:6712–22
115. Ayers JL, Selverston AI. 1979. Monosynaptic entrainment of an endogenous pacemaker network: a cellular mechanism for von Holt's magnet effect. *J. Comp. Physiol.* 129:5–17
116. Prinz AA, Thirumalai V, Marder E. 2003. The functional consequences of changes in the strength and duration of synaptic inputs to oscillatory neurons. *J. Neurosci.* 23:943–54
117. Thirumalai V, Prinz AA, Johnson CD, Marder E. 2006. Red pigment concentrating hormone strongly enhances the strength of the feedback to the pyloric rhythm oscillator but has little effect on pyloric rhythm period. *J. Neurophysiol.* 95:1762–70
118. Bal T, Nagy F, Moulins M. 1994. Muscarinic modulation of a pattern-generating network: control of neuronal properties. *J. Neurosci.* 14:3019–35
119. Harris-Warrick RM, Flamm RE. 1987. Multiple mechanisms of bursting in a conditional bursting neuron. *J. Neurosci.* 7:2113–28
120. Epstein IR, Marder E. 1990. Multiple modes of a conditional neural oscillator. *Biol. Cybern.* 63:25–34
121. Goldman MS, Golowasch J, Marder E, Abbott LF. 2001. Global structure, robustness, and modulation of neuronal models. *J. Neurosci.* 21:5229–38
122. Guckenheimer J, Gueron S, Harris-Warrick RM. 1993. Mapping the dynamics of a bursting neuron. *Philos. Trans. R. Soc. London Ser. B* 341:345–59
123. Taylor AL, Hickey TJ, Prinz AA, Marder E. 2006. Structure and visualization of high-dimensional conductance spaces. *J. Neurophysiol.* 96:891–905
124. Clemens S, Massabuau JC, Legeay A, Meyrand P, Simmers J. 1998. In vivo modulation of interacting central pattern generators in lobster stomatogastric ganglion: influence of feeding and partial pressure of oxygen. *J. Neurosci.* 18:2788–99
125. Clemens S, Meyrand P, Simmers J. 1998. Feeding-induced changes in temporal patterning of muscle activity in the lobster stomatogastric system. *Neurosci. Lett.* 254:65–68
126. Norris BJ, Coleman MJ, Nusbaum MP. 1994. Recruitment of a projection neuron determines gastric mill motor pattern selection in the stomatogastric nervous system of the crab, *Cancer borealis*. *J. Neurophysiol.* 72:1451–63
127. Combes D, Meyrand P, Simmers J. 1999. Dynamic restructuring of a rhythmic motor program by a single mechanorecptor neuron in lobster. *J. Neurosci.* 19:3620–28
128. Mulloney B, Selverston AI. 1974. Organization of the stomatogastric ganglion in the spiny lobster. I. Neurons driving the lateral teeth. *J. Comp. Physiol.* 91:1–32
129. Mulloney B, Selverston AI. 1974. Organization of the stomatogastric ganglion in the spiny lobster. III. Coordination of the two subsets of the gastric system. *J. Comp. Physiol.* 91:53–78
130. Dickinson PS, Nagy F, Moulins M. 1988. Control of central pattern generators by an identified neurone in crustacea: activation of the gastric mill motor pattern by a neurone known to modulate the pyloric network. *J. Exp. Biol.* 136:53–87
131. Selverston AI, Mulloney B. 1974. Organization of the stomatogastric ganglion of the spiny lobster. II. Neurons driving the medial tooth. *J. Comp. Physiol.* 91:33–51
132. Bartos M, Manor Y, Nadim F, Marder E, Nusbaum MP. 1999. Coordination of fast and slow rhythmic neuronal circuits. *J. Neurosci.* 19:6650–60
133. Bartos M, Nusbaum MP. 1997. Intercircuit control of motor pattern modulation by presynaptic inhibition. *J. Neurosci.* 17:2247–56
134. Blitz DM, Nusbaum MP. 1997. Motor pattern selection via inhibition of parallel pathways. *J. Neurosci.* 17:4965–75
135. Wood DE, Manor Y, Nadim F, Nusbaum MP. 2004. Intercircuit control via rhythmic regulation of projection neuron activity. *J. Neurosci.* 24:7455–63

136. Nadim F, Manor Y, Nusbaum MP, Marder E. 1998. Frequency regulation of a slow rhythm by a fast periodic input. *J. Neurosci.* 18:5053–67
137. Heinzel HG, Weimann JM, Marder E. 1993. The behavioral repertoire of the gastric mill in the crab, *Cancer pagurus*: an in situ endoscopic and electrophysiological examination. *J. Neurosci.* 13:1793–803
138. Weimann JM, Meyrand P, Marder E. 1991. Neurons that form multiple pattern generators: identification and multiple activity patterns of gastric/pyloric neurons in the crab stomatogastric system. *J. Neurophysiol.* 65:111–22
139. Weimann JM, Marder E. 1994. Switching neurons are integral members of multiple oscillatory networks. *Curr. Biol.* 4:896–902
140. Bucher D, Taylor AL, Marder E. 2006. Central pattern generating neurons simultaneously express fast and slow rhythmic activities in the stomatogastric ganglion. *J. Neurophysiol.* 95:3617–32
141. Thuma JB, Morris LG, Weaver AL, Hooper SL. 2003. Lobster (*Panulirus interruptus*) pyloric muscles express the motor patterns of three neural networks, only one of which innervates the muscles. *J. Neurosci.* 23:8911–20
142. Dickinson PS, Mecsas C, Marder E. 1990. Neuropeptide fusion of two motor pattern generator circuits. *Nature* 344:155–58
143. Meyrand P, Simmers J, Moulins M. 1991. Construction of a pattern-generating circuit with neurons of different networks. *Nature* 351:60–63
144. Meyrand P, Simmers J, Moulins M. 1994. Dynamic construction of a neural network from multiple pattern generators in the lobster stomatogastric nervous system. *J. Neurosci.* 14:630–44
145. Büschges A. 2005. Sensory control and organization of neural networks mediating coordination of multisegmental organs for locomotion. *J. Neurophysiol.* 93:1127–35
146. Dando MR, Laverack MS. 1969. The anatomy and physiology of the posterior stomach nerve (p.s.n.) in some decapod crustacea. *Proc. R. Soc. London Ser. B* 171:465–82
147. Billimoria CP, DiCaprio RA, Birmingham JT, Abbott LF, Marder E. 2006. Neuromodulation of spike-timing precision in sensory neurons. *J. Neurosci.* 26:5910–19
148. Birmingham JT, Billimoria CP, DeKlotz TR, Stewart RA, Marder E. 2003. Differential and history-dependent modulation of a stretch receptor in the stomatogastric system of the crab, *Cancer borealis*. *J. Neurophysiol.* 90:3608–16
149. Birmingham JT, Szuts Z, Abbott LF, Marder E. 1999. Encoding of muscle movement on two time scales by a sensory neuron that switches between spiking and burst modes. *J. Neurophysiol.* 82:2786–97
150. Katz PS, Harris-Warrick RM. 1989. Serotonergic/cholinergic muscle receptor cells in the crab stomatogastric nervous system. II. Rapid nicotinic and prolonged modulatory effects on neurons in the stomatogastric ganglion. *J. Neurophysiol.* 62:571–81
151. Katz PS, Harris-Warrick RM. 1990. Neuromodulation of the crab pyloric central pattern generator by serotonergic/cholinergic proprioceptive afferents. *J. Neurosci.* 10:1495–512
152. Katz PS, Harris-Warrick RM. 1990. Actions of identified neuromodulatory neurons in a simple motor system. *Trends Neurosci.* 13:367–73
153. Katz PS, Harris-Warrick RM. 1991. Recruitment of crab gastric mill neurons into the pyloric motor pattern by mechanosensory afferent stimulation. *J. Neurophysiol.* 65:1442–51
154. Combes D, Simmers AJ, Moulins M. 1995. Structural and functional characterization of a muscle tendon proprioceptor in lobster. *J. Comp. Neurol.* 363:221–34
155. Combes D, Simmers AJ, Moulins M. 1997. Conditional dendritic oscillators in a lobster mechanoreceptor neurone. *J. Physiol.* 499:161–77

156. Casasnovas B, Meyrand P. 1995. Functional differentiation of adult neural circuits from a single embryonic network. *J. Neurosci.* 15:5703–18
157. Richards KS, Miller WL, Marder E. 1999. Maturation of the rhythmic activity produced by the stomatogastric ganglion of the lobster, *Homarus americanus*. *J. Neurophysiol.* 82:2006–9
158. Le Feuvre Y, Fénelon VS, Meyrand P. 1999. Unmasking of multiple adult neural networks from a single embryonic circuit by removal of neuromodulatory inputs. *Nature* 402:660–64
159. Fénelon V, Casasnovas B, Faumont S, Meyrand P. 1998. Ontogenetic alteration in peptidergic expression within a stable neuronal population in lobster stomatogastric nervous system. *J. Comp. Neurol.* 399:289–305
160. Fénelon VS, Kilman V, Meyrand P, Marder E. 1999. Sequential developmental acquisition of neuromodulatory inputs to a central pattern-generating network. *J. Comp. Neurol.* 408:335–51
161. Kilman VL, Fénelon V, Richards KS, Thirumalai V, Meyrand P, Marder E. 1999. Sequential developmental acquisition of cotransmitters in identified sensory neurons of the stomatogastric nervous system of the lobsters, *Homarus americanus* and *Homarus gammarus*. *J. Comp. Neurol.* 408:318–34
162. Pulver SR, Marder E. 2002. Neuromodulatory complement of the pericardial organs in the embryonic lobster, *Homarus americanus*. *J. Comp. Neurol.* 451:79–90
163. Pulver SR, Thirumalai V, Richards KS, Marder E. 2003. Dopamine and histamine in the developing stomatogastric system of the lobster *Homarus americanus*. *J. Comp. Neurol.* 462:400–14
164. Richards KS, Marder E. 2000. The actions of crustacean cardioactive peptide on adult and developing stomatogastric ganglion motor patterns. *J. Neurobiol.* 44:31–44
165. Richards KS, Simon DJ, Pulver SR, Beltz BS, Marder E. 2003. Serotonin in the developing stomatogastric system of the lobster, *Homarus americanus*. *J. Neurobiol.* 54:380–92
166. Bem T, Le Feuvre Y, Simmers J, Meyrand P. 2002. Electrical coupling can prevent expression of adult-like properties in an embryonic neural circuit. *J. Neurophysiol.* 87:538–47
167. Golowasch J, Casey M, Abbott LF, Marder E. 1999. Network stability from activity-dependent regulation of neuronal conductances. *Neural Comput.* 11:1079–96
168. Thoby-Brisson M, Simmers J. 1998. Neuromodulatory inputs maintain expression of a lobster motor pattern-generating network in a modulation-dependent state: evidence from long-term decentralization in vitro. *J. Neurosci.* 18:212–25
169. Thoby-Brisson M, Simmers J. 2000. Transition to endogenous bursting after long-term decentralization requires de novo transcription in a critical time window. *J. Neurophysiol.* 84:596–99
170. Thoby-Brisson M, Simmers J. 2002. Long-term neuromodulatory regulation of a motor pattern-generating network: maintenance of synaptic efficacy and oscillatory properties. *J. Neurophysiol.* 88:2942–53
171. LeMasson G, Marder E, Abbott LF. 1993. Activity-dependent regulation of conductances in model neurons. *Science* 259:1915–17
172. Liu Z, Golowasch J, Marder E, Abbott LF. 1998. A model neuron with activity-dependent conductances regulated by multiple calcium sensors. *J. Neurosci.* 18:2309–20
173. Turrigiano G, Abbott LF, Marder E. 1994. Activity-dependent changes in the intrinsic properties of cultured neurons. *Science* 264:974–77
174. Turrigiano GG, LeMasson G, Marder E. 1995. Selective regulation of current densities underlies spontaneous changes in the activity of cultured neurons. *J. Neurosci.* 15:3640–52

175. Baro DJ, Coniglio LM, Cole CL, Rodriguez HE, Lubell JK, et al. 1996. Lobster *shal*: comparison with *Drosophila shal* and native potassium currents in identified neurons. *J. Neurosci.* 16:1689–701
176. MacLean JN, Zhang Y, Johnson BR, Harris-Warrick RM. 2003. Activity-independent homeostasis in rhythmically active neurons. *Neuron* 37:109–20
177. MacLean JN, Zhang Y, Goeritz ML, Casey R, Oliva R, et al. 2005. Activity-independent coregulation of I_A and I_h in rhythmically active neurons. *J. Neurophysiol.* 94:3601–17
178. Zhang Y, Oliva R, Gisselmann G, Hatt H, Guckenheimer J, Harris-Warrick RM. 2003. Overexpression of a hyperpolarization-activated cation current (I_h) channel gene modifies the firing activity of identified motor neurons in a small neural network. *J. Neurosci.* 23:9059–67
179. Baro DJ, Ayali A, French L, Scholz NL, Labenia J, et al. 2000. Molecular underpinnings of motor pattern generation: differential targeting of *shal* and *shaker* in the pyloric motor system. *J. Neurosci.* 20:6619–30
180. Baro DJ, Cole CL, Harris-Warrick RM. 1996. RT-PCR analysis of *shaker*, *shab*, *shaw*, and *shal* gene expression in single neurons and glial cells. *Recept. Channels* 4:149–59
181. Baro DJ, Cole CL, Harris-Warrick RM. 1996. The lobster *shaw* gene: cloning, sequence analysis and comparison to fly *shaw*. *Gene* 170:267–70
182. Baro DJ, Cole CL, Zarrin AR, Hughes S, Harris-Warrick RM. 1994. *Shab* gene expression in identified neurons of the pyloric network in the lobster stomatogastric ganglion. *Recept. Channels* 2:193–205. Erratum. 1994. *Recept. Channels* 2(4):350
183. Baro DJ, Harris-Warrick RM. 1998. Differential expression and targeting of K^+ channel genes in the lobster pyloric central pattern generator. *Ann. N.Y. Acad. Sci.* 860:281–95
184. Baro DJ, Quinones L, Lanning CC, Harris-Warrick RM, Ruiz M. 2001. Alternate splicing of the *shal* gene and the origin of I_A diversity among neurons in a dynamic motor network. *Neuroscience* 106:419–32
185. Clark MC, Baro DJ. 2006. Molecular cloning and characterization of crustacean type-one dopamine receptors: D1αPan and D1βPan. *Comp. Biochem. Physiol. B* 143:294–301
186. Clark MC, Dever TE, Dever JJ, Xu P, Rehder V, et al. 2004. Arthropod 5-HT2 receptors: a neurohormonal receptor in decapod crustaceans that displays agonist independent activity resulting from an evolutionary alteration to the DRY motif. *J. Neurosci.* 24:3421–35
187. French LB, Lanning CC, Harris-Warrick RM. 2002. The localization of two voltage-gated calcium channels in the pyloric network of the lobster stomatogastric ganglion. *Neuroscience* 112:217–32
188. French LB, Lanning CC, Matly M, Harris-Warrick RM. 2004. Cellular localization of *Shab* and *Shaw* potassium channels in the lobster stomatogastric ganglion. *Neuroscience* 123:919–30
189. Herrick FH. 1909. Natural history of the American lobster. *Bull. U.S. Bur. Fish.* 29:plateXXXIIII
190. Marder E, Bucher D, Schulz DJ, Taylor AL. 2005. Invertebrate central pattern generation moves along. *Curr. Biol.* 15:R685–99
191. Krenz WD, Nguyen D, Perez-Acevedo NL, Selverston AI. 2000. Group I, II, and III mGluR compounds affect rhythm generation in the gastric circuit of the crustacean stomatogastric ganglion. *J. Neurophysiol.* 83:1188–201

Molecular Mechanisms of Renal Ammonia Transport

I. David Weiner[1,2] and L. Lee Hamm[3]

[1]Nephrology Section, North Florida/South Georgia Veterans Health System, and [2]Division of Nephrology, Hypertension and Transplantation, University of Florida, Gainesville, Florida 32608; email: weineid@ufl.edu

[3]Department of Medicine, Tulane University School of Medicine, New Orleans, Louisiana 70112; email: lhamm@tulane.edu

Annu. Rev. Physiol. 2007. 69:317–40

First published online as a Review in Advance on September 5, 2006

The *Annual Review of Physiology* is online at http://physiol.annualreviews.org

This article's doi:
10.1146/annurev.physiol.69.040705.142215

0066-4278/07/0315-0317$20.00

Key Words

acid-base homeostasis, aquaporin, Rh glycoproteins, Na^+-K^+-ATPase, Na^+-K^+-$2Cl^-$ cotransport, Na^+/H^+ exchange, H^+-K^+-ATPase

Abstract

Acid-base homeostasis to a great extent relies on renal ammonia metabolism. In the past several years, seminal studies have generated important new insights into the mechanisms of renal ammonia transport. In particular, the theory that ammonia transport occurs almost exclusively through nonionic NH_3 diffusion and NH_4^+ trapping has given way to a model postulating that a variety of proteins specifically transport NH_3 and NH_4^+ and that this transport is critical for normal ammonia metabolism. Many of these proteins transport primarily H^+ or K^+ but also transport NH_4^+. Nonerythroid Rh glycoproteins transport ammonia and may represent critical facilitators of ammonia transport in the kidney. This review discusses the underlying aspects of renal ammonia transport as well as specific proteins with important roles in renal ammonia transport.

INTRODUCTION

Acid-base homeostasis depends on renal ammonia[1] metabolism. Large amounts of bicarbonate are filtered by the glomerulus, but under normal circumstances renal tubules reabsorb essentially all filtered bicarbonate. The generation of new bicarbonate is necessary to replace the alkali consumed in the buffering of endogenous and exogenous acids. Under normal conditions 60–70% of the new bicarbonate formed by the kidney is due to renal ammonia metabolism, and in response to chronic metabolic acidosis or acid loads, an even larger proportion of new bicarbonate results from increased renal ammonia metabolism (1–3).

Recent studies have yielded important new insights into the mechanisms of renal ammonia transport. In particular, the theory that ammonia transport occurs almost exclusively through nonionic NH_3 diffusion has been replaced by the observation that a variety of proteins specifically transport NH_3 and NH_4^+ and that this transport is critical for normal ammonia metabolism. In this review, we discuss the underlying aspects of renal ammonia transport and then discuss in more detail the specific proteins that mediate important roles in renal ammonia transport.

AMMONIA CHEMISTRY

Ammonia exists in biological solutions in two molecular forms, NH_3 and NH_4^+. The relative amounts of each are governed by the buffer reaction $NH_3 + H^+ \leftrightarrow NH_4^+$. This reaction occurs essentially instantaneously and has a pKa′ under biologically relevant conditions of ~9.1–9.3. Accordingly, in most biological fluids of pH 7.4 or less, most ammonia is present as NH_4^+. At pH 7.4 only ~1% of total ammonia is present as NH_3. Moreover, because most biological fluids exist at a pH substantially below the pKa′ of this buffer reaction, small changes in pH result in exponential changes in relative NH_3 concentration with almost no change in NH_4^+ concentration. Because NH_3 is a relatively small, uncharged molecule, it can diffuse across most lipid bilayers. However, its membrane permeability is not infinite, and thus transepithelial NH_3 gradients can occur in the presence of high rates of either NH_4^+ or H^+ transport. NH_4^+ is a cation with very limited permeability across lipid bilayers in the absence of specific transport proteins. However, in aqueous solutions, NH_4^+ and K^+ have nearly identical biophysical characteristics (**Table 1**), which enables NH_4^+ transport at the K^+ transport site of many proteins.

RENAL AMMONIA METABOLISM

Renal ammonia metabolism and transport involve integrated responses of multiple portions of the kidney (**Figure 1**). Essentially none of urinary ammonia is filtered by the glomerulus, making urinary ammonia excretion unique among the major compounds present in the urine. Instead, ammonia is produced by the kidney and is then selectively transported into either the urine or into the renal vein. Ammonia excreted into the urine as NH_4^+ results in equimolar new bicarbonate formation, whereas ammonia that returns to the systemic circulation through the renal vein is metabolized by the liver to urea and glutamine (4). Under normal conditions, approximately two-thirds of hepatic ammonia metabolism result in urea formation; hepatic ureagenesis utilizes equimolar bicarbonate, thereby consuming bicarbonate produced in the proximal tubule in ammoniagenesis and resulting in no net new bicarbonate formation (5). The selective transport of ammonia into either the urine or the renal vein includes integrated transport mechanisms in the proximal tubule, thick ascending limb of the loop of Henle, and collecting duct. We do not review all data regarding the development of this

[1]Ammonia exists in two molecular forms, NH_3 and NH_4^+. In this review, we use the terms "ammonia" or "total ammonia" to refer to the sum of these two molecular species. When referring to either of these molecular species, we use specifically either "NH_3" or "NH_4^+."

Table 1 Biophysical characteristics of common cations[a]

Cation	Atomic weight	Ionic radius (126, 127) (Å)	Hydrodynamic radius (126, 127) (Å)	Mobility in H_2O (126, 127) (10^{-4} cm^2 s^{-1} V^{-1})	T_i^{H2O} (126, 128)
Li^+	6	0.060	1.73	4.01	0.33
Na^+	23	0.095	1.67	5.19	0.39
NH_4^+	18	0.133	1.14	7.60	0.49
K^+	39	0.143	1.14	7.62	0.49

[a]References, where applicable, are in parentheses in column headers.

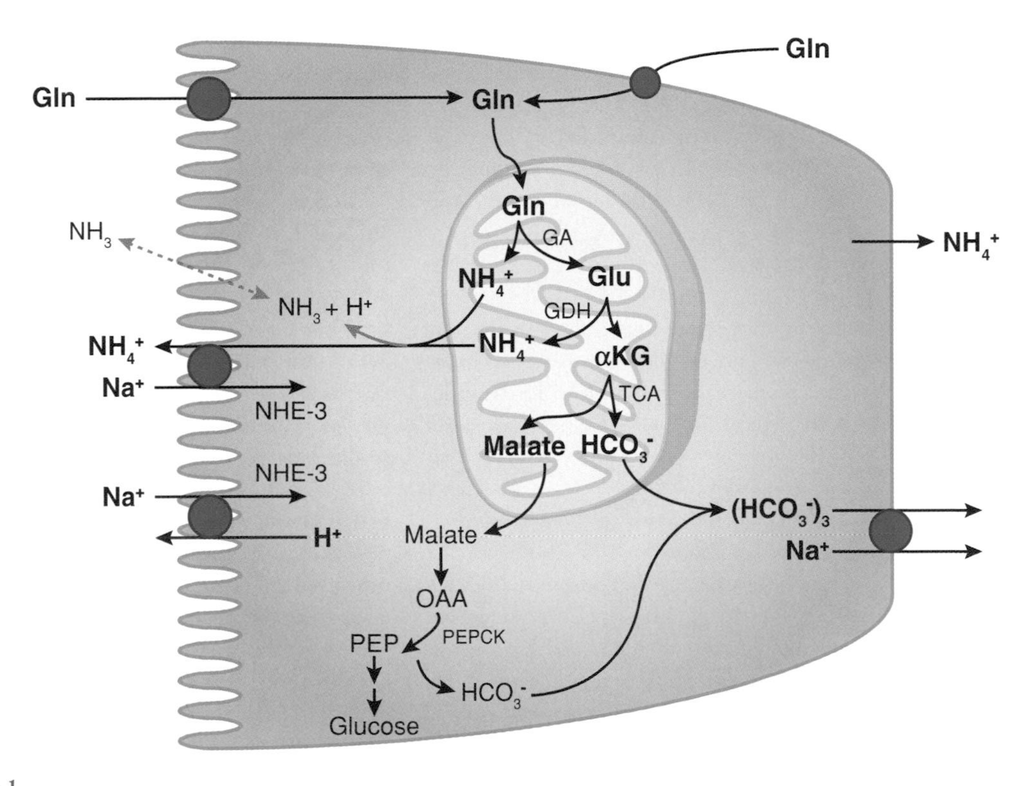

Figure 1

Proximal tubule ammoniagenesis. Ammonia is produced from glutamine (Gln) as a result of proximal tubule ammoniagenesis. Glutamine is transported across both the apical and basolateral plasma membranes and then transported into mitochondria. The enzyme glutaminase (GA) is the first step in ammoniagenesis, and glutamate dehydrogenase (GDH) results in the production of the second NH_4^+ molecule. Metabolism of α-ketoglutarate (αKG) leads to the production of the first of two HCO_3^- ions. Further metabolism in the cytoplasm results in the production of a second HCO_3^-. Thus, complete metabolism of each glutamine produces two NH_4^+ and two HCO_3^- ions. Blue circles denote transport proteins. Dotted gray line indicates the minor component of transport, and the solid gray line indicates the major component of transport. Other abbreviations used: NHE-3, Na^+/H^+ exchanger type 3; OAA, oxaloacetic acid; PEP, phosphoenolpyruvate; PEPCK, phosphoenolpyruvate carboxykinase; TCA, tricarboxylic acid cycle.

model, as several excellent reviews can be consulted for more detail (1, 6–8).

NHE-3: Na^+/H^+ exchanger type 3

The proximal tubule segments metabolize the amino acid glutamine to produce equal numbers of NH_4^+ and HCO_3^- molecules (**Figure 1**) (see References 9 and 10 for a review of the biochemistry of ammonia production). The proximal tubule secretes ammonia into the tubule lumen, but there is also some transport across the basolateral membrane and ultimately into the renal veins (~24–45% of ammonia produced, depending on acid-base homeostasis) (11). Ammonia is secreted into the luminal fluid as NH_4^+ through a mechanism involving transport via the apical Na^+/H^+ exchanger NHE-3 (12, 13) and to a lesser degree through an apical Ba^{2+}-sensitive K^+ channel (13, 14). In addition, a significant component of NH_3 diffusion across the apical membrane occurs, although the exact proportion is unresolved. **Figure 2** summarizes proximal tubule ammonia metabolism.

In the thick ascending limb of the loop of Henle, multiple proteins contribute to luminal ammonia reabsorption. The furosemide-sensitive apical Na^+-K^+-$2Cl^-$ cotransporter NKCC2 mediates the majority of NH_4^+ reabsorption; an apical K^+/NH_4^+ antiporter and an amiloride-sensitive NH_4^+ conductance also contribute (15–17). Some of the ammonia absorbed by the medullary thick ascending limb of the loop of Henle undergoes recycling into the thin descending limb of the loop of Henle; this involves passive NH_3 diffusion, with a small component of passive NH_4^+ transport (18, 19). Secondarily active ammonia absorption by the thick ascending limb of the loop of Henle and passive ammonia secretion into the thin descending limb of the loop of Henle result in the increasing axial ammonia concentrations in the medullary interstitium that parallel the hypertonicity gradient. Moreover, because of ammonia absorption by the thick ascending limb, ammonia delivery to the distal tubule is only ~20% of final urinary ammonia content (1, 2).

The secretion of ammonia by the collecting duct accounts for the majority (~80%) of urinary ammonia content. Several studies using in vitro–microperfused collecting duct segments have shown that collecting duct ammonia secretion involves parallel NH_3 and H^+ secretion, with little to no NH_4^+ permeability in collecting duct segments (1, 7). However, the findings of parallel NH_3 and H^+ secretion are also consistent with parallel NH_4^+/H^+ exchange and H^+ transport. In the absence of luminal carbonic anhydrase activity in most segments of the collecting duct, a luminal disequilibrium pH amplifies the NH_3 gradient and increases the rate of transepithelial ammonia secretion (1, 7). Thus, the fundamental mechanisms of ammonia transport in the collecting duct differ from those in the proximal tubule and the thick ascending limb of the loop of Henle.

Recent studies have examined the mechanism of the NH_3 transport observed in the collecting duct. NH_3 is a small, neutral molecule that is relatively permeable across lipid bilayers (20). As a result, collecting duct NH_3 transport may reflect either diffusive or transporter-mediated NH_3 transport. One way to differentiate between these two possibilities is to examine the correlation between extracellular ammonia concentration and ammonia transport; with diffusive transport the transport rate parallels the concentration gradient, whereas transporter-mediated uptake exhibits saturable kinetics. Researchers addressed this question using a cultured renal collecting duct epithelial cell line, mouse inner medullary collecting duct (mIMCD)-3 cells, grown on permeable support membranes to enable separate study of apical and basolateral transport mechanisms. Methylammonia (MA) was used as an ammonia analog, enabling the use of the radiolabeled molecule analog [^{14}C]-MA for quantification of transport.

Studies examining basolateral ammonia/MA uptake showed that the renal collecting duct cell line mIMCD-3 exhibited basolateral MA transport (21). Transport was due to a combination of both a saturable, transporter-mediated component and

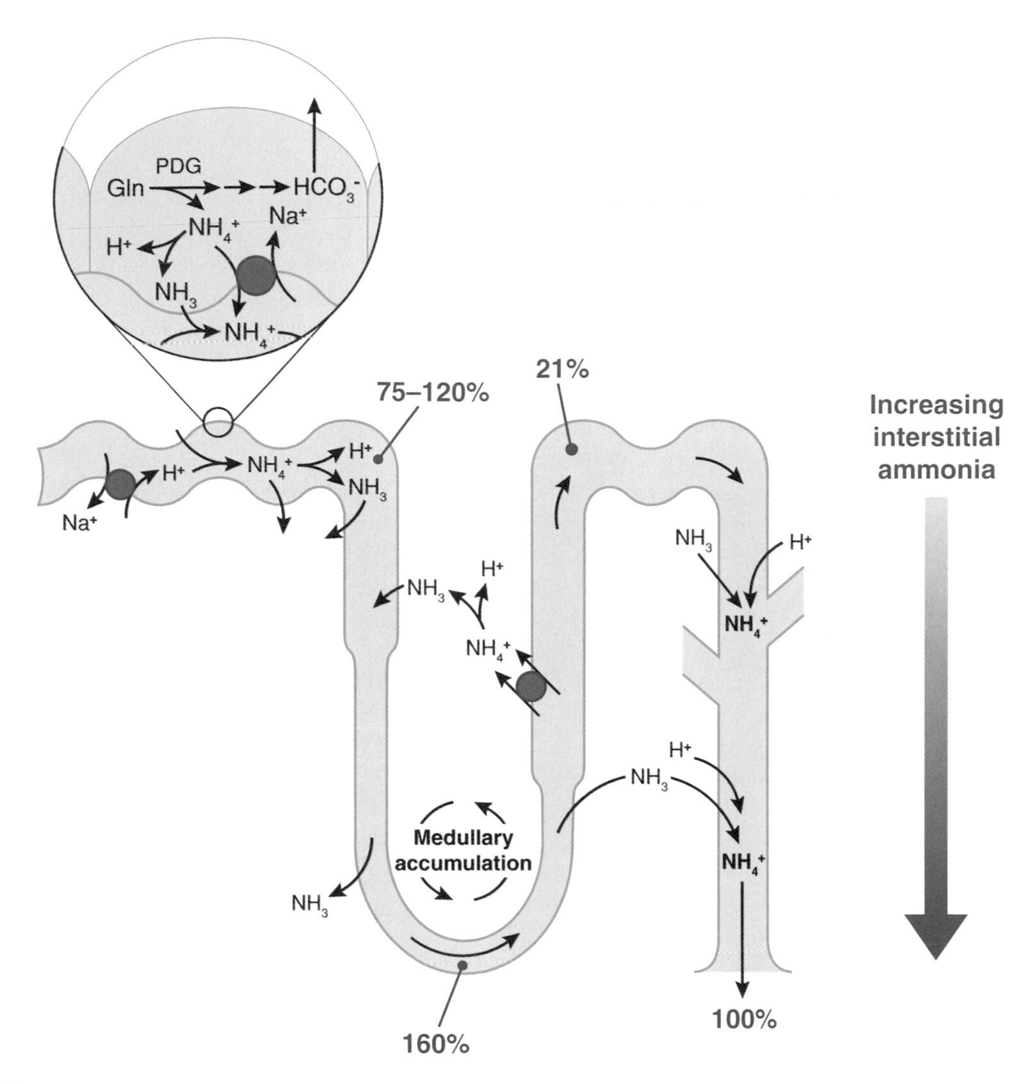

Figure 2

Summary of renal ammonia metabolism. Ammonia is produced in the proximal tubule as a result of metabolism of glutamine (Gln) to NH_4^+ and HCO_3^-; ammonia is preferentially secreted into the luminal fluid as either NH_4^+ or NH_3. Bicarbonate produced from ammoniagenesis is preferentially secreted across the basolateral membrane. Ammonia is secreted by the proximal tubule into the luminal fluid and then undergoes recycling by the loop of Henle, resulting in medullary accumulation. Ammonia delivery to the bend of the loop of Henle is ~160% of that which is excreted in the final urine. Ammonia is reabsorbed in the thick ascending limb of the loop of Henle through multiple mechanisms and then is secreted in the collecting duct. Collecting duct ammonia secretion involves parallel NH_3 and H^+ secretion. Numbers in red represent delivery as a percentage of final urinary ammonia excretion. Blue circles denote transport proteins. PDG, phosphate-dependent glutaminase.

a nonsaturable, diffusive component; the transporter-mediated component predominated at concentrations below 7 mM. With a K_i of ~2 mM, ammonia competitively inhibited MA transport activity. Functional characterization of the saturable transport activity demonstrated that it did not involve extracellular Na^+ or K^+; was not inhibited by chemical inhibitors of Na^+-K^+-ATPase, Na^+-K^+-$2Cl^-$ cotransporters NKCC-1 or NKCC-2,

KCC proteins: K^+-Cl^- cotransporting proteins

Rhbg: Rh B glycoprotein

Rhcg: Rh C glycoprotein

K^+ channels, or KCC proteins; and was not inhibited by changes in membrane voltage. Varying intracellular and extracellular pH demonstrated that transport activity paralleled the transmembrane H^+ gradient. Thus, the renal collecting duct mIMCD-3 cell exhibits basolateral, electroneutral, facilitated NH_3 transport activity, an ammonia transport activity not previously described in mammalian cells. The transport activity observed is also consistent with NH_4^+/H^+ exchange activity. Because these cells expressed basolateral Rh B glycoprotein (Rhbg), this transport activity likely reflects Rhbg-mediated transport. However, Rhbg-specific inhibitors were unavailable, and manipulation of Rhbg protein expression was not performed, preventing definitive proof of the role of Rhbg in basolateral NH_3 transport.

Similar studies examined apical ammonia transport mechanisms. Apical transport exhibited both a saturable, transporter-mediated component and a nonsaturable, diffusive component (22). The transporter-mediated component exhibited (*a*) Na^+ and K^+ independence, (*b*) modulation by extracellular and intracellular pH, and (*c*) lack of inhibition by changes in membrane voltage or extracellular K^+. The apparent affinity of this transport activity for ammonia, measured as the K_i of ammonia, ~4 mM, was slightly less than that observed for the basolateral ammonia transport activity, consistent with the higher luminal than peritubular ammonia concentrations to which collecting duct cells are exposed. Inhibitors of apical Na^+-K^+-ATPase, H^+-K^+-ATPase, and Na^+/H^+ exchange did not alter transport. Thus, the collecting duct mIMCD-3 cell exhibits an apical electroneutral transport activity functionally consistent with either NH_4^+/H^+ exchange or facilitated NH_3 transport. These cells express apical Rh C glycoprotein (Rhcg), suggesting that Rhcg mediates this transport activity. Although these studies (21, 22) establish facilitated ammonia transport, the relative extents of lipid solubility diffusion of NH_3 and protein-mediated facilitated diffusion of NH_3 in the apical and basolateral membranes of collecting duct cells in vivo remain to be determined.

Metabolic acidosis, which increases renal net acid excretion and new bicarbonate formation predominantly through increases in ammonia production and excretion, is associated with the stimulation of the key components of ammonia metabolism and transport. These include augmented proximal tubule ammoniagenesis, which involves increases in the following: activity and expression of key ammoniagenic enzymes (9, 23); proximal tubule luminal NH_4^+ secretion (24, 25) due, at least in part, to greater NHE-3 expression (26); ammonia reabsorption in the thick ascending limb of the loop of Henle (27); and collecting duct ammonia secretion (2, 3).

SPECIFIC PROTEINS INVOLVED IN RENAL AMMONIA TRANSPORT

NHE-3

NH_4^+ produced in the proximal tubule is preferentially transported across the apical membrane into the luminal fluid, as opposed to transport across the basolateral plasma membrane into the peritubular space. Apical ammonia secretion is inhibited by the combination of low luminal Na^+ and luminal amiloride, suggesting that this process may involve Na^+/NH_4^+ exchange via apical NHE-3 (28). Consistent with this interpretation is that proximal tubule brush border vesicles exhibit Na^+/NH_4^+ exchange activity (29). Although parallel Na^+/H^+ exchange and NH_3 secretion may account for some of this transport, luminal acidification in the absence of apical Na^+/H^+ exchange activity does not result in equivalent rates of net ammonia secretion (28). These observations suggest that apical NHE-3 has an important role in the preferential secretion of cytosolic ammonia into the luminal fluid in the proximal tubule.

NHE-3 is also present in the apical membrane of the thick ascending limb of the loop of Henle. However, because this transporter secretes NH_4^+ and the thick ascending limb of the loop of Henle reabsorbs NH_4^+, NHE-3 is unlikely to have an important role in loop of Henle ammonia reabsorption.

The regulation of proximal tubule apical NHE-3 is important in the regulation of proximal tubule ammonia transport. Chronic metabolic acidosis, changes in extracellular K^+, and angiotensin II (AII) regulate proximal tubule ammonia secretion, apparently through changes in apical Na^+/H^+ exchange activity (25, 30). In chronic metabolic acidosis and hypokalemia, parallel changes in NHE-3 expression were observed (26, 31). In the S3 segment, chronic metabolic acidosis increases AII-stimulated apical ammonia secretion, and the Na^+/H^+ exchange inhibitor amiloride in the presence of low luminal Na^+ concentration blocks secretion (32). This suggests that chronic metabolic acidosis increases AII-stimulated, Na^+/H^+ exchange–mediated apical ammonia secretion in the S3 segment.

K^+ Channels

At a molecular level, K^+ and NH_4^+ have nearly identical biophysical characteristics, including bare and hydrated radii (**Table 1**). This molecular mimicry enables NH_4^+ to substitute for K^+ at the K^+-binding site of most K^+ transporters, including K^+ channels (**Table 2**). K^+ channel–mediated transport of NH_4^+ has been shown for various K^+ channel families, including strong and weak inward rectifying, voltage-gated, Ca^{2+}-activated, delayed rectifier, and L-type transient K^+ channels (summarized in Reference 33). In general, the relative conductance of K^+ channels for NH_4^+ is 10–20% of that observed for K^+ (33). This decreased affinity probably reflects a lower affinity for NH_4^+ than for K^+ at binding sites within the channel's pore (33).

AII: angiotensin II

In vitro microperfusion studies show that barium, a nonspecific K^+ channel inhibitor, blocks a component of proximal tubule ammonia transport (13). Several K^+ channels, including KCNA10, TWIK-1, and KCNQ1, are present in the apical membrane of the proximal tubule (34–38); which of these mediate ammonia transport is currently not known. Because the negative intracellular electrical potential favors NH_4^+ uptake, apical K^+ channels are more likely to contribute to NH_4^+ absorption that can occur in the proximal straight tubule than to NH_4^+ secretion that predominates in the proximal convoluted tubule. Basolateral barium-sensitive transport is also present in the proximal tubule (39).

In the medullary thick ascending limb of the loop of Henle, apical K^+ channels can contribute to NH_4^+ transport, particularly when apical Na^+-K^+-$2Cl^-$ cotransport is inhibited (17). The primary apical K^+ channels in the medullary thick ascending limb of the loop of Henle are ATP sensitive. These proteins appear primarily to enable recycling of K^+ reabsorbed by the apical Na^+-K^+-$2Cl^-$ cotransporter and do not seem to have important roles in ammonia transport in the absence of inhibition of apical Na^+-K^+-$2Cl^-$ cotransport (40).

The electrochemical gradient for NH_4^+ transport by K^+ channels normally favors

Table 2 Comparison of selectivity properties of several K^+ channels[a]

Channel	γ, pS				P_X/P_K, bi-ionic		
	K^+	Rb^+	NH_4^+	Tl^+	Rb^+	NH_4^+	Tl^+
KcsA (129)	55.6 ± 0.6	23.2 ± 0.8	24.0 ± 0.3	16.4 ± 0.1	0.78 ± 0.07	0.20 ± 0.01	3.2 ± 0.1
Shaker (130)	18	9	14	—	0.66	0.09	—
BK (131)	230	20	41	111	0.7	0.1	1.3
$K_{ir}2.1$ (33)	29	3	17	24	0.49	0.1	1.8

[a]References are in parentheses under the first column.

CCD: cortical collecting duct

OMCD: outer medullary collecting duct

IMCD: inner medullary collecting duct

NH_4^+ uptake from the extracellular into the intracellular compartment, primarily owing to the negative intracellular electrical potential. Thus, basolateral K^+ channels in the collecting duct may contribute to peritubular NH_4^+ uptake. At present, no studies have examined this possibility. However, the observation that prolonged ammonia incubation acidifies CCD intercalated cells (41) suggests that significant NH_4^+ transport, possibly due to K^+ channels, may occur in the cortical collecting duct (CCD).

Na^+-K^+-$2Cl^-$ Cotransport

Researchers have identified two isoforms of the Na^+-K^+-$2Cl^-$ transporter, termed NKCC-1 and NKCC-2 (42). These isoforms exist in a wide variety of tissues, where they mediate multiple physiological functions, including ion transport, cell volume regulation, blood pressure regulation, and saliva and endolymph formation (42–44). NKCC-1, also known as BSC-2 and as the secretory isoform, is expressed at the basolateral membrane of many cells involved in secretory functions. In the kidney, NKCC-1 is present almost exclusively in the basolateral membrane of A-type intercalated cells in the outer and inner medullary collecting ducts (OMCD and IMCD, respectively) (45). However, when bumetanide, a highly specific inhibitor, is added to the peritubular solution, it does not alter ammonia secretion by in vitro–microperfused rat OMCD (46). Thus, it is unlikely that NKCC-1 contributes to OMCD ammonia secretion. Studies using cultured IMCD cells suggest that basolateral Na^+-K^+-$2Cl^-$ cotransport in the IMCD, presumably involving NKCC-1, contributes to basolateral NH_4^+ uptake (47, 48). However, other studies using in vitro–microperfused rat terminal IMCD segments have not confirmed this observation (49).

NKCC-2 is a kidney-specific Na^+-K^+-$2Cl^-$ cotransporter isoform specifically expressed in the apical plasma membrane of the thick ascending limb of the loop of Henle (50, 51), where it is the major mechanism for ammonia reabsorption (8, 40). The affinity of NKCC-2 for ammonia is ~2 mM, enabling effective competition of NH_4^+ with K^+ at the K^+-binding site for transport (52). Hyperkalemia suppresses urinary ammonia excretion. This likely occurs, at least in part, because higher luminal K^+ inhibits NH_4^+ reabsorption by NKCC-2, thereby inhibiting development of the medullary ammonia concentration gradient (40).

Changes in NKCC-2 protein expression are a major component of the increase in MTAL ammonia absorption that occurs with chronic metabolic acidosis (27, 53). Increased mRNA expression precedes increased protein expression (53); the elevation in systemic glucocorticoids that occurs with chronic metabolic acidosis may mediate the increases in mRNA and protein expression (54). Changes in cell-surface expression of NKCC-2 can regulate NKCC-2 ion transport in response to other stimuli such as cAMP (55); whether this mechanism is important for the regulation of ammonia transport is unknown.

Na^+-K^+-ATPase

Na^+-K^+-ATPase is a family of heterodimeric proteins present in essentially all nucleated cells. The alpha subunit comprises the catalytic and ion transport domain. The function of the beta subunit is not completely clear but may include directing membrane localization (56). Na^+-K^+-ATPase is present in the basolateral plasma membrane of essentially all kidney cells, but its expression appears greatest in the medullary thick ascending limb of the loop of Henle, with lesser expression in the cortical thick ascending limb, distal convoluted tubule (DCT), CCD, medullary collecting duct (MCD), and the proximal tubule (57).

NH_4^+ competes with K^+ at the K^+-binding site of Na^+-K^+-ATPase, thereby enabling net Na^+-NH_4^+ exchange (58, 59). The affinity of the K^+-binding site of

Na^+-K^+-ATPase for NH_4^+ is similar to that for K^+ (58, 59), but at least in the renal cortex, interstitial ammonia concentrations are lower than K^+ concentrations. Moreover, in vitro–microperfused tubule studies show that, even in the presence of elevated peritubular ammonia concentrations, ouabain, a Na^+-K^+-ATPase inhibitor, does not alter transepithelial ammonia secretion in the CCD (60). Accordingly, basolateral Na^+-K^+-ATPase is not likely to transport NH_4^+ to a significant extent in the CCD (58).

In contrast, interstitial ammonia concentrations are substantially higher in the inner medulla, enabling Na^+-K^+-ATPase-mediated basolateral ammonia uptake to contribute significantly to transepithelial ammonia secretion by the IMCD (58, 59). Direct studies show that NH_4^+ competes with K^+ as a substrate for hydrolytic activity and that basolateral Na^+-K^+-ATPase mediates NH_4^+ uptake, increases apical proton secretion, and is important for transepithelial ammonia secretion (49, 59, 61). Using measured values for Na^+-K^+-ATPase K_m and V_{max} and assuming that vasa recta K^+ and NH_4^+ concentrations are similar to interstitial concentrations, Wall and colleagues (62) have shown that rates of NH_4^+ uptake via Na^+-K^+-ATPase are remarkably similar to measured rates of NH_4^+ secretion. Furthermore, physiological changes in interstitial K^+ concentrations associated with hypokalemia predict a two- to threefold increase in NH_4^+ uptake (62). Direct studies examining the effects of hypokalemia on Na^+-K^+-ATPase-mediated ammonia transport have confirmed this model (63). Moreover, the increase in ammonia uptake is almost totally attributable to changes in interstitial K^+ concentration and does not involve changes in Na^+-K^+-ATPase hydrolytic activity (63). Thus, the increased rate of ammonia secretion that occurs during hypokalemia may be due, at least in part, to the decreased interstitial K^+ concentration, which results in more effective NH_4^+ transport by basolateral Na^+-K^+-ATPase.

K^+/NH_4^+ (H^+) Exchanger

An apical K^+/NH_4^+ (H^+) transporter that is present in the thick ascending limb of the loop of Henle mediates most of the ammonia reabsorption in this nephron segment that is not mediated by NKCC-2 (8). This transport activity is electroneutral and inhibited by barium and verapamil (16, 64). Neither the gene product nor the protein that correlates with this transport activity has been identified.

AQP1, 3, 8, 9: aquaporins 1, 3, 8, 9

H^+-K^+-ATPase

NH_4^+ also is a substrate for nongastric (or colonic) H^+-K^+-ATPase in the collecting duct and colon. This has been demonstrated in heterologous expression systems, colonic apical membranes, and IMCDs perfused in vitro (65–68). Although the exact substrates for colonic H^+-K^+-ATPase are still not completely resolved or reconciled with in vivo transport, we know that NH_4^+ can substitute for K^+. However, recent data suggest that NH_4^+ cannot substitute for the H^+ site (68). Because colonic H^+-K^+-ATPase in vivo is induced during K^+ deficiency, it may facilitate increased ammonia excretion (67). Ammonia also stimulates H^+ secretion and bicarbonate reabsorption in the CCD, possibly via several mechanisms (69), including stimulation of, or coupling with, H^+-K^+-ATPase (41).

Aquaporins

Aquaporins are an extended family of proteins that mediate facilitated transmembrane water transport (70, 71). Because H_2O and NH_3 have similar molecular sizes and charge distribution, an increasing number of studies are examining the role of aquaporins in transmembrane NH_3 transport, in particular ammonia transport by the aquaporins AQP1, AQP3, AQP8, and AQP9.

The first study investigating the possible role of aquaporins as ammonia transporters examined AQP1 (72). There, the addition of extracellular ammonia to control,

Rhag: Rh A glycoprotein

H_2O-injected oocytes that did not express AQP1 resulted in intracellular acidification and cellular depolarization. These results indicate NH_4^+ entry coupled with a relatively low NH_3 permeability (72). In contrast, significantly less intracellular acidification occurred in AQP1-expressing oocytes. Thus, AQP1 expression either inhibited endogenous NH_4^+ permeability or increased NH_3 permeability. Because AQP1 expression did not alter ammonia-dependent membrane depolarization, AQP1 expression is unlikely to alter NH_4^+ permeability (72). Moreover, increasing extracellular pH from 7.5 to 8.0 in the continuous presence of extracellular ammonia, which increases extracellular NH_3 but does not substantially change extracellular NH_4^+, caused more rapid intracellular alkalinization in AQP1-expressing oocytes than in control oocytes. Thus, AQP1 appeared to enable facilitated NH_3 transport (72). Although it is tempting to speculate that AQP1 may contribute to NH_3 permeability in the thin descending limb of the loop of Henle, it is unknown whether AQP1 knockout mice show altered ammonia transport there. However, not all studies have confirmed that AQP1 can transport NH_3 (73, 74).

In the basolateral membrane of collecting duct principal cells, AQP3 plays an important role in H_2O transport (75). When expressed in *Xenopus* oocytes, AQP3 increases membrane NH_3 permeability, suggesting that AQP3 can function as an NH_3 transporter (73). Whether AQP3 contributes to renal principal cell basolateral NH_3 transport has not been reported.

AQP8, similar to AQP3, increases NH_3 permeability in *Xenopus* oocytes; it also increases permeability to the ammonia analogs formamide and MA (73). In the kidney, AQP8 is weakly expressed in the cytoplasm of the proximal tubule, CCD, and OMCD, with no evidence of plasma membrane expression (76). Acid-base homeostasis, urine ammonia concentration, and urine pH are similar in AQP8 knockout and wild-type mice under both basal- and acid-loaded conditions, suggesting that AQP8 deletion does not alter basal- or acid-stimulated net acid excretion (77, 78). However, AQP8 knockout mice do exhibit minor changes in hepatic ammonia accumulation, renal excretion of infused ammonia, and intrarenal ammonia concentrations (78).

AQP9 expression, similar to AQP8, increases apparent NH_3 permeability and permeability to the analogs formamide and MA in *Xenopus* oocytes (73). However, AQP9 does not appear to be expressed in the kidney.

The molecular specificity of ammonia transport by aquaporins may relate to specific amino acids in the aromatic/arginine constriction region of these proteins. This region is located below the channel mouth and may be narrower than the NPA constriction (79). Site-directed mutagenesis shows that specific amino acids, including Phe-56, His-180, and Arg-195 of AQP1, are important for ammonia permeability but not for water permeability (74). The central NPA constriction, although important for aquaporin water permeability, does not appear critical for aquaporin ammonia permeability (74).

Rh Glycoproteins

The identification that the Rh glycoprotein family proteins can transport ammonia has been the most recent addition to our understanding of the molecular mechanisms of renal ammonia transport. These proteins are mammalian orthologs of Mep/AMT proteins, the ammonia transporter families present in yeast, plants, bacteria, and many other organisms (80, 81). Three mammalian Rh glycoproteins—Rh A glycoprotein (RhAG/Rhag), Rh B glycoprotein (RhBG/Rhbg), and Rh C glycoprotein (RhCG/Rhcg)—have been identified to date (82, 83). By convention, Rh A glycoprotein in human tissue is termed RhAG, and in nonhumans it is termed Rhag; similar terminologies apply for RhBG/Rhbg and RhCG/Rhcg.

RhAG/Rhag. The Rh complex in erythrocytes consists of RhAG/Rhag in association with the nonglycosylated Rh proteins RhD and RhCE in humans or with Rh30 in nonhuman mammals. Marini et al. (84) predicted in 1997 that Rh proteins would have structural similarity to Mep and AMT proteins and might function as ammonia transporters. Subsequent studies established that RhAG and RhCG expression restored growth defects of yeast deficient in endogenous ammonia transporters and that RhAG mediated efflux of the ammonia analog MA (85), suggesting that RhAG functioned as a mammalian ammonia transporter.

Various studies have examined RhAG/Rhag-mediated ammonia transport characteristics. Heterologous expression studies show that RhAG transports the ammonia analog MA, with an EC_{50} of ~1.6 mM, and that transport is electroneutral and coupled to H^+ gradients (86). Transport is not coupled to either Na^+ or K^+, nor is it affected by a wide variety of inhibitors of other transporter families. Moreover, RhAG-mediated transport is bidirectional (85, 87). Studies comparing ammonia transport in erythrocytes from Rh_{null} individuals, who do not express erythroid RhAG, with transport in erythrocytes with normal RhAG expression show that NH_3 transport parallels RhAG expression (88). These characteristics functionally identify that RhAG mediates either facilitated NH_3 transport or NH_4^+/H^+ exchange activity. However, at least one study has suggested that RhAG, when expressed in HeLa cells, transports both NH_3 and NH_4^+ (89).

RhAG/Rhag is an erythrocyte- and erythroid-precursor-specific protein (90, 91). At present, there is no evidence that RhAG/Rhag contributes to renal ammonia metabolism and/or transport.

The nonglycosylated Rh protein RhCE apparently transports neither ammonia nor its analog MA, nor does it alter transport by RhAG (86). Moreover, structural models, using the *Escherichia coli* AmtB structure as a template, suggest that the arrangement of key amino acids is sufficiently different in RhCE and RhD that RhCE and RhD either do not transport ammonia or do so by mechanisms that differ from those used by AmtB, RhAG, RhBG, and RhCG (92).

RhBG/Rhbg. RhBG/Rhbg is the second member of the mammalian Rh glycoprotein family. RhBG/Rhbg is expressed in a wide variety of organs involved in ammonia metabolism, including the kidneys, liver, skin, stomach, and gastrointestinal tract (82, 93). The kidneys express basolateral Rhbg immunoreactivity in the distal convoluted tubule (DCT), connecting segment (CNT), initial collecting tubule (ICT), CCD, OMCD, and IMCD (94, 95). Both intercalated and principal cells in the DCT, CNT, ICT, and CCD express Rhbg, and expression is greater in intercalated cells than in principal cells; this is consistent with intercalated cells having a greater role in acid-base homeostasis than do principal cells. In the IMCD, only intercalated cells express Rhbg. The only exception to this pattern is that the CCD B-type intercalated cell does not express Rhbg immunoreactivity detectable by light microscopy (95). This is consistent with a primary role of the B-type intercalated cell in bicarbonate secretion and Cl^- reabsorption. The basolateral expression of Rhbg is stabilized through specific interactions of the cytoplasmic carboxy terminus with ankyrin-G (96).

RhBG/Rhbg is expressed in multiple extrarenal sites involved in ammonia transport and metabolism. In the liver, Rhbg is expressed in the sinusoidal membrane of perivenous hepatocytes (97), which is the basolateral membrane. Perivenous hepatocytes have an affinity for ammonia of ~110 μM and are responsible for high-affinity hepatic ammonia metabolism (98). In the skin Rhbg mRNA is expressed in hair follicles and sweat glands (82). Although the role of Rhbg in sweat formation is undefined at present, sweat ammonia concentrations average ~3 mM and can exceed 11 mM (99, 100); these concentrations are greater than in all other body fluids except for urine. The gastrointestinal tract is another major site of ammonia transport and metabolism, and villous epithelial cells, the major site of ion transport, from the duodenum through the colon express basolateral Rhbg immunoreactivity (93).

Multiple heterologous expression studies demonstrate that RhBG/Rhbg can transport ammonia and the ammonia analog MA. However, studies differ as to the exact molecular species, NH_3 or NH_4^+, transported. Some studies show that ammonia and MA transport is electroneutral, coupled to proton gradients, independent of Na^+ or K^+, and unaffected by changes in membrane voltage, indicating electroneutral, cation-independent NH_3 transport (101–103). However, other data suggest that Rhbg mediates electrogenic NH_4^+ transport (104). In all these studies, the affinity of Rhbg for ammonia is ~2–4 mM. Why different studies identify seemingly differing molecular transport mechanisms is currently a mystery.

Several studies have examined the physiological role of Rhbg in renal ammonia metabolism. Seshardri et al. (105) induced chronic metabolic acidosis, which increases renal ammonia excretion, in normal Sprague-Dawley rats. These researchers observed no detectable changes in Rhbg protein expression, either by immunoblot analysis or by immunohistochemistry. Two possibilities thus exist. First, Rhbg is not involved in the increased renal ammonia excretion in response to chronic metabolic acidosis. Alternatively, if Rhbg contributes to increased collecting duct ammonia transport, then the increased transport occurs through mechanisms independent of changes in protein expression detectable by immunoblot analysis or by light microscopy (105).

In another study, genetic deletion of pendrin, an apical Cl^-/HCO_3^- exchanger present in B-type intercalated cells and non-A, non-B cells, decreased Rhbg expression (106). This adaptation may reflect increased urinary acidification in response to pendrin deletion, which increases the gradient for NH_3 transport, thus decreasing the necessity for ammonia transport via Rhbg.

Finally, Chambrey et al. (107) have studied an Rhbg knockout mouse. Their extensive physiological studies show normal basal acid-base balance, a normal response to chronic acid loading, and normal basolateral NH_3 and NH_4^+ transport in in vitro–microperfused CCD segments (107). These results suggest either that Rhbg is not involved in renal ammonia metabolism or that certain mechanisms activated in response to Rhbg deletion fully compensate for the lack of Rhbg (107). However, there were no differences in Rhcg, H^+-ATPase, or AE1 expression, likely candidate proteins if compensatory adaptations were present (107).

At present, the specific role of RhBG/Rhbg in renal ammonia metabolism is unclear. It does not appear to be required for renal ammonia metabolism, at least under basal conditions or in response to metabolic acidosis. Its specific role in acid-base homeostasis will undoubtedly be an important area of investigation.

RhCG/Rhcg. RhCG/Rhcg is the third member of the mammalian Rh glycoprotein family. It is expressed in multiple ammonia-transporting/metabolizing organs, including the kidneys, central nervous system, testes, skeletal muscle and liver, and gastrointestinal tract (83, 85, 93, 97, 108). In the kidney, Rhcg expression parallels that of Rhbg; i.e., it occurs in the DCT, CNT, ICT, CCD, OMCD, and IMCD (95, 109). The cellular expression of Rhcg is similar to that of Rhbg, with intercalated cell expression exceeding principal cell expression in the DCT, CNT, ICT, CCD, and OMCD and expression only in intercalated cells in the IMCD (95, 109).

However, the subcellular location of Rhcg differs from the exclusive basolateral location observed for Rhbg. In the mouse kidney, Rhcg immunoreactivity is apical (95, 109). One study in the rat showed that Rhcg immunoreactivity was exclusively apical (109) whereas other rat studies show both apical and basolateral expression (105, 110). Recent studies in the human kidney show both apical and basolateral RhCG expression (111). Quantitative analysis of rat Rhcg localization, using immunogold electron microscopy to identify the specific cellular location of RhCG, shows that basolateral plasma membrane Rhcg expression exceeds apical plasma membrane expression, at least in the OMCD in the inner stripe (110). In addition, rat renal Rhcg is also present in intracellular sites, including cytoplasmic vesicles, suggesting that vesicular trafficking regulates plasma membrane Rhcg expression (**Figure 3*a***) (110). The presence of both apical and basolateral plasma membrane RhCG/Rhcg expression in the human and rat

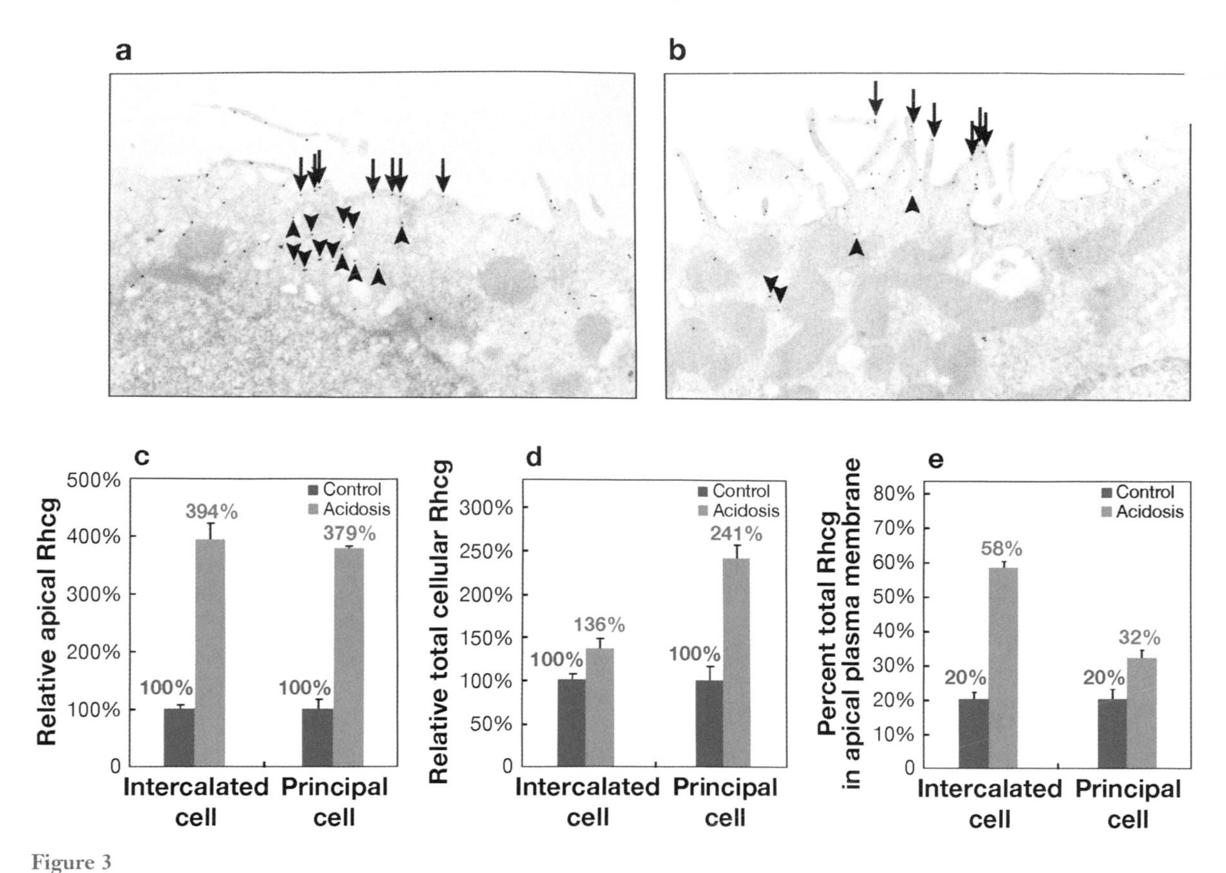

Figure 3

Rhcg expression in the OMCD in response to chronic metabolic acidosis. (*a*) Under basal conditions Rhcg is present both in the apical plasma membrane (*arrows*) and in intracellular compartments (*arrowheads*) in intercalated cells in the OMCD. (*b*) Chronic metabolic acidosis increases the proportion of Rhcg present in the apical plasma membrane (*arrows*) and decreases Rhcg present in the intracellular compartment (*arrowheads*). (*c*) Chronic metabolic acidosis increases the amount of Rhcg in the apical plasma membrane in both the OMCD intercalated cell and the principal cell as compared with control conditions (100%). The relative increase is similar in intercalated cells and principal cells. Absolute expression (not shown) is greater in the intercalated cell than in the principal cell under both control and acidosis conditions (110). (*d*) Total cellular Rhcg in intercalated and principal cells in response to chronic metabolic acidosis, expressed relative to total cellular Rhcg in control conditions (100%). Although Rhcg expression increases in both the intercalated cell and principal cell, the relative increase is greater in the principal cell than in the intercalated cell. (*e*) The proportions of total cellular Rhcg present in the apical plasma membrane in the intercalated cell and principal cell under control conditions and in response to chronic metabolic acidosis. The increase in response to metabolic acidosis is greater in the intercalated cell than in the principal cell. *a* and *b* are reprinted from Reference 110, with permission. *c*, *d*, and *e* use data published previously in Reference 110.

kidneys raises the possibility that Rhcg contributes to both apical and basolateral plasma membrane ammonia transport.

As for Rhbg, multiple studies have addressed and differed as to the molecular ammonia species transported by RhCG/Rhcg. Some studies suggest that RhCG/Rhcg mediates electroneutral NH_3 transport (101–103), whereas others report both NH_3 and NH_4^+ transport (112) or only electrogenic NH_4^+ transport (113). The explanation for these differing observations is currently unknown. However, only electroneutral NH_3 transport likely contributes to apical membrane ammonia secretion in the collecting duct. The negative intracellular electrical potential and high luminal NH_4^+ concentrations would favor luminal NH_4^+ reabsorption via an electrogenic NH_4^+ transport mechanism (22, 114).

Studies examining the regulation of Rhcg in response to chronic metabolic acidosis have yielded important observations. Chronic metabolic acidosis increases Rhcg protein expression in the OMCD and the IMCD but not in the cortex (105). The OMCD and IMCD changes appear to occur through posttranscriptional mechanisms because Rhcg steady-state mRNA expression is not detectably altered. In the IMCD, the increase in expression is specific to the IMCD intercalated cell.

In the OMCD in the inner stripe, at least two mechanisms are involved in the response of Rhcg to chronic metabolic acidosis (110). Quantitative analysis of the subcellular location of Rhcg, via immunoelectron microscopy and morphometric analysis, shows that chronic metabolic acidosis increases Rhcg expression in both the intercalated cell and the principal cell (**Figure 3**). Although the absolute levels of Rhcg expression are greater in the intercalated cell in the principal cell, the relative increase in cellular expression was substantially greater in the principal cell (↑~240%) than in the intercalated cell (↑~35%).

A second regulatory mechanism operates through changes in the subcellular location of Rhcg (110). Under basal conditions, Rhcg is located in both the apical and basolateral plasma membrane and in subapical sites in both the principal cell and intercalated cell. In response to chronic metabolic acidosis, particularly in the intercalated cell, there is a dramatic increase in apical plasma membrane expression and a decrease in cytoplasmic Rhcg expression (110). These results suggest translocation of Rhcg from cytoplasmic to apical plasma membrane sites. A similar response, although one of substantially less magnitude, occurs in the principal cell (110).

Chronic metabolic acidosis also increases basolateral plasma membrane Rhcg expression (110). This increase parallels the change in total cellular expression and does not appear to involve redistribution from cytoplasmic sites to the basolateral plasma membrane.

These observations suggest that at least two mechanisms regulate Rhcg expression: changes in total cellular expression and changes in the subcellular distribution of Rhcg (110). Moreover, the relative roles of these two mechanisms differ in adjacent cell types, the principal cell and the intercalated cell. These changes are also membrane-specific, as the increase in basolateral plasma membrane Rhcg expression in the principal cell parallels changes in total cellular expression and does not involve changes in the relative subcellular distribution of Rhcg. The mechanisms involved in the cell- and membrane-specific differences in Rhcg expression in response to specific physiological stimuli are unknown at present.

Tertiary structure of Rh glycoproteins. Researchers are beginning to unravel the molecular mechanisms by which Rh glycoproteins transport ammonia and its analog MA. The related *E. coli* protein AmtB has been crystallized, and its tertiary structure has been identified (**Figure 4**) (115–117). AmtB is expressed as a trimeric complex. The individual subunits of the trimeric complex span the plasma membrane 11 times and have two structurally related halves with opposite

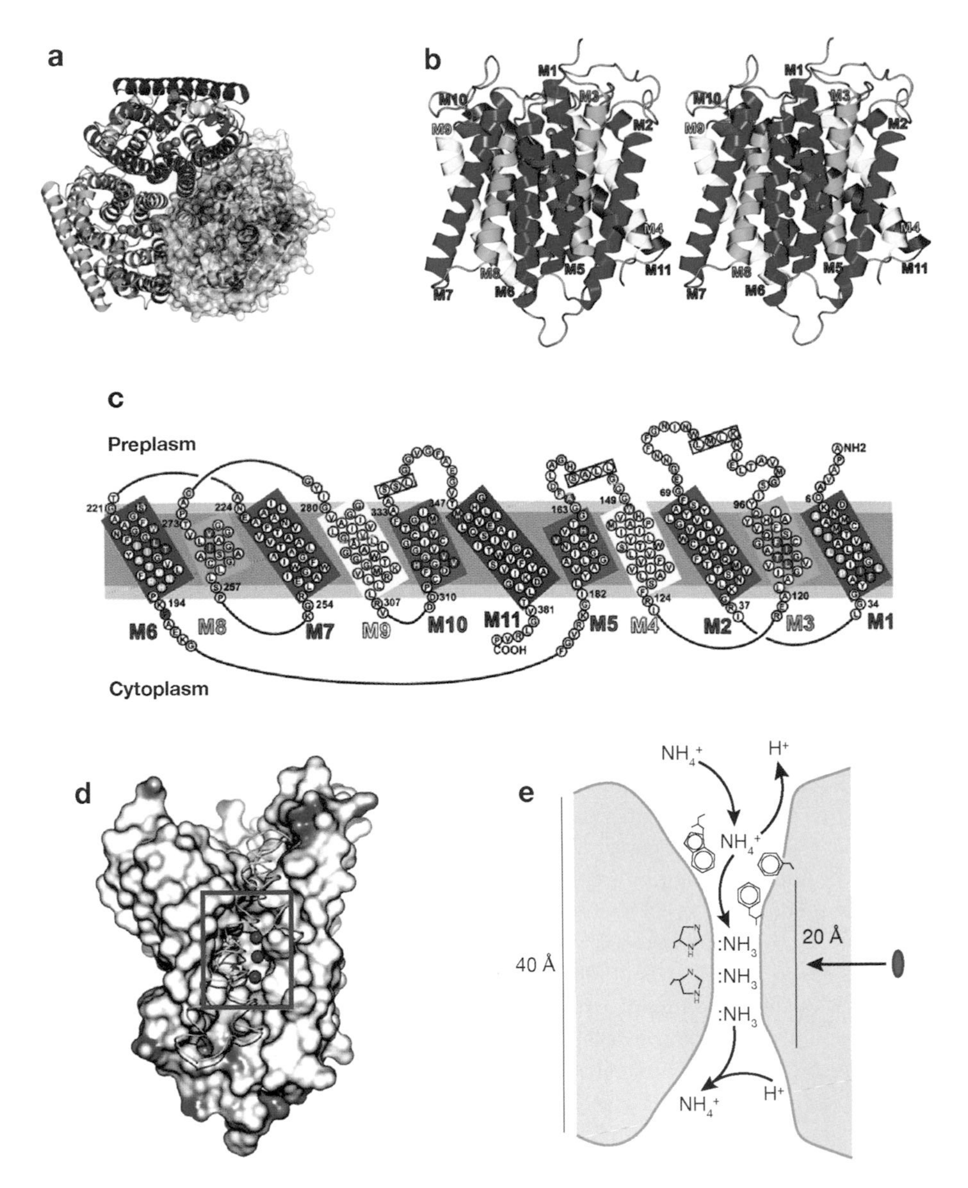

Figure 4

Tertiary structure of *E. coli* AmtB. (*a*) A ribbon representation of AmtB viewed from the extracellular surface demonstrates the trimeric structure of AmtB. In the top monomer, homologous transmembrane segments are shown in the same color. In the right monomer, a solvent-accessible transparent surface is colored according to the electrostatic potential (*red* for negative and *blue* for positive). Potential ammonia-binding sites are shown as the blue sphere for NH_3 and red sphere for NH_4^+. (*b*) A stereoview of the monomeric structure. The extracellular surface is uppermost. NH_3-binding sites are shown as blue spheres. (*c*) The amino acid sequences of the transmembrane extracellular and intracellular loops. Homologous sequences are shown in similar colors. (*d*) The ammonia-conducting portion of AmtB after removal of portions of helices M8, M9, and M10. The lumen surface is colored according to the electrostatic potential (*red* for negative and *blue* for positive). The histidines near the three NH_3 sites (*blue spheres*) are shown in yellow. Narrow hydrophobic regions through the conducting channel lie above and below the NH_3 sites. (*e*) The deduced mechanism of ammonia transport through AmtB. The blue oval indicates the 20-Å hydrophobic channel of the protein. Figure from Reference 116, with permission from AAAS.

polarity. A vestibule, present in both the extracellular and intracellular regions of the protein, recruits NH_4^+. In AmtB, NH_4^+ and MA^+ appear to be stabilized in the vestibule through interactions with Trp-148, Phe-103, and Ser-219. Although mutation of Asp-160 abolishes transport activity (118), this amino acid does not directly interact with ammonia or MA; instead, it orients the carbonyl groups of Asp-160, Phe-161, and Ala-162 (116). Transport then occurs through a 20-Å hydrophobic channel that spans the membrane. Trp-148, Phe-103, Phe-161, and Tyr-140 form the first hydrophobic barrier to transport through this channel; the diameter of this pore is 1.2 Å, suggesting that transport requires movement of the side chains of Phe-107 and Phe-215 (116). In the channel pore, two conserved histidines, His-168 and His-318, appear to interact with and stabilize NH_3 at specific locations, Am2, Am3, and Am4 (116, 117). The unprotonated Nδ1s of His-168 and His-H318 are fixed by hydrogen bonds to each other, thereby providing two Cε1-Hs to the nitrogens of Am2 and Am3 and one Nε acceptor that interact with the N-H bond of Am4. Because the Cε1-Hs can donate, but cannot accept, N–H bonds, specificity for NH_3 transport rather than NH_4^+ transport is generated. As a result, NH_4^+ is concentrated at the Am1 site, progressively desolvated as it progresses through the channel pore to Am2, and deprotonated at Am3. NH_3 and the methyl derivative CH_3NH_2, but not NH_4^+ and $CH_3NH_3^+$, are transported through the pore of AmtB (116). The tertiary structure of AmtB is a novel structural motif without significant similarity to other known proteins (116, 117).

The degree to which mammalian Rh glycoproteins mimic the structure of AmtB is an active area of investigation. Homology models suggest that Rhag, Rhbg, and Rhcg have a channel architecture similar to AmtB (92). Preliminary studies suggest that mutations at Phe-74, Val-137, and Phe-235 of RhCG, equivalent to Ile-28, Leu-114, and Phe-215, respectively, in AmtB, inhibit NH_3 transport (119). This suggests that the molecular mechanism of ammonia transport by Rhcg is similar to those used by AmtB. Finally, the reported K_ms of mammalian glycoproteins for ammonia and MA, >1 mM, are >100-fold greater than those reported for AmtB. This is associated with differences in the π-cation-stabilizing rings; changes in Trp-148 to leucine or valine and Phe/Tyr-103 to isoleucine occur in the Rh glycoproteins (118).

Alternative functions of Rh glycoproteins. There are important controversies regarding the functions of the mammalian Rh glycoproteins. As noted above, genetic deletion of Rhbg does not detectably alter renal ammonia metabolism under basal conditions or in response to chronic metabolic acidosis, nor does it alter basal plasma ammonia levels, suggesting that Rhbg deletion does not impact hepatic ammonia metabolism (107). Thus, either Rhbg does not function as an ammonia transporter or other transport mechanisms can compensate in the absence of Rhbg.

Recent studies suggest that Rh glycoproteins transport CO_2. The green algae *Chlamydomonas reinhardtii* express both Amt proteins homologous to Mep/Amt proteins and Rh proteins that are relatively more homologous to mammalian Rh glycoproteins (120). Inhibiting expression of the Rh glycoprotein homolog Rh1 does not alter extracellular ammonia uptake (121). Interestingly, elevating media CO_2 content increases Rh1 expression and inhibiting Rh1 expression suppresses the normal increase in growth rates that occur in response to elevated CO_2 (121, 122). One possibility, as yet unproven, is that Rh1 is involved in CO_2 uptake; the CO_2 is then used for carbon fixation, increased nutrient utilization, and increased growth rates and is not involved in extracellular ammonia uptake.

Evolutionary analysis of AMT proteins and Rh glycoproteins suggests that Rh glycoprotein genes arose after the development of AMT proteins and that the development of Rh glycoproteins was associated with the loss of AMT proteins (123). In organisms expressing both AMT and Rh proteins, Rh proteins appeared to initially diverge rapidly away from Amt and then, over a longer period, to evolve more slowly. Furthermore, functionally divergent amino acid sites exist in the transmembrane segments surrounding the NH_3-conducting lumen identified in AmtB, suggesting the presence of differing substrate specificity (123).

Two recent studies have examined whether mammalian RhAG/Rhag transports CO_2. These studies used erythrocytes from normal individuals and Rh_{null} individuals, who do not express erythrocyte RhAG, and from wild-type and Rhag knockout mice. In one study, erythrocyte CO_2 permeability was decreased in erythrocytes lacking RhAG/Rhag (124), whereas in the other, RhAG/Rhag absence did not alter CO_2 permeability (125). The reason for these differing results is not known but may reflect use of differing analytical techniques—mass spectrometric measurement of $C^{18}O^{16}O$ permeability versus fluorometric analysis of CO_2-induced intracellular pH changes—and in their use of intact erythrocytes versus erythrocyte ghosts (124, 125).

CONCLUSIONS

Fundamental changes in our understanding of the molecular mechanisms of renal ammonia transport have occurred in the past several years. The specific transport of NH_3 and NH_4^+ by specific proteins appears critical for normal acid-base homeostasis.

SUMMARY POINTS

1. Ammonia metabolism is a primary component of acid-base homeostasis through its role in new bicarbonate generation.
2. Models of ammonia transport solely involving passive lipid-phase diffusion of NH_3 and diffusion trapping of NH_4^+ cannot explain many aspects of ammonia metabolism.
3. Specific proteins that have important roles in transepithelial ammonia transport are expressed in different regions of the kidney, including the proximal tubule, thick ascending limb of the loop of Henle, distal convoluted tubule, connecting segment, initial collecting tubule, and collecting duct.

FUTURE ISSUES

1. The role of NHE-3 and other transporters in proximal tubule ammonia secretion should be investigated.
2. The role(s) of aquaporin family members in renal ammonia transport deserves further research.
3. The transport characteristics of mammalian Rh glycoproteins, i.e., facilitated NH_3 transport versus NH_4^+/H^+ exchange versus electrogenic NH_4^+ transport, should be elucidated.
4. The relative roles of Rhbg and Rhcg in distal ammonia secretion should be explored.

ACKNOWLEDGMENTS

The preparation of this review was supported by funds from the NIH (grants DK-45788 and NS-47624).

LITERATURE CITED

1. DuBose TD Jr, Good DW, Hamm LL, Wall SM. 1991. Ammonium transport in the kidney: new physiological concepts and their clinical implications. *J. Am. Soc. Nephrol.* 1:1193–203
2. Hamm LL, Simon EE. 1987. Roles and mechanisms of urinary buffer excretion. *Am. J. Physiol. Renal Physiol.* 253:F595–605
3. Sajo IM, Goldstein MB, Sonnenberg H, Stinebaugh BJ, Wilson DR, Halperin ML. 1982. Sites of ammonia addition to tubular fluid in rats with chronic metabolic acidosis. *Kidney Int.* 20:353–58
4. Haussinger D, Lamers WH, Moorman AF. 1992. Hepatocyte heterogeneity in the metabolism of amino acids and ammonia. *Enzyme* 46:72–93
5. Haussinger D. 1990. Nitrogen metabolism in liver: structural and functional organization and physiological relevance. *Biochem. J.* 267:281–90
6. Good DW, Knepper MA. 1985. Ammonia transport in the mammalian kidney. *Am. J. Physiol. Renal Physiol.* 248:F459–71
7. Knepper MA. 1991. NH_4^+ transport in the kidney. *Kidney Int.* 40:S95–102
8. Karim Z, Attmane-Elakeb A, Bichara M. 2002. Renal handling of NH_4^+ in relation to the control of acid-base balance by the kidney. *J. Nephrol.* 15(Suppl. 5):S128–34
9. Tannen RL, Sahai A. 1990. Biochemical pathways and modulators of renal ammoniagenesis. *Miner. Electrolyte Metab.* 16:249–58
10. Tannen RL. 1978. Ammonia metabolism. *Am. J. Physiol. Renal Physiol.* 235:F265–77
11. Tizianello A, Deferrari G, Garibotto G, Robaudo C, Acquarone N, Ghiggeri GM. 1982. Renal ammoniagenesis in an early stage of metabolic acidosis in man. *J. Clin. Invest.* 69:240–50
12. Nagami GT. 1989. Ammonia production and secretion by the proximal tubule. *Am. J. Kidney Dis.* 14:258–61
13. Simon EE, Merli C, Herndon J, Cragoe EJ Jr, Hamm LL. 1992. Effects of barium and 5-(*N*-ethyl-*N*-isopropyl)-amiloride on proximal tubule ammonia transport. *Am. J. Physiol. Renal Physiol.* 262:F36–39
14. Hamm LL, Simon EE. 1990. Ammonia transport in the proximal tubule. *Miner. Electrolyte Metab.* 16:283–90
15. Kikeri D, Sun A, Zeidel ML, Hebert SC. 1989. Cell membranes impermeable to NH_3. *Nature* 339:478–80
16. Amlal H, Paillard M, Bichara M. 1994. NH_4^+ transport pathways in cells of medullary thick ascending limb of rat kidney. NH_4^+ conductance and K^+/NH_4^+ (H^+) antiport. *J. Biol. Chem.* 269:21962–71
17. Attmane-Elakeb A, Amlal H, Bichara M. 2001. Ammonium carriers in medullary thick ascending limb. *Am. J. Physiol. Renal Physiol.* 280:F1–9
18. Flessner MF, Mejia R, Knepper MA. 1993. Ammonium and bicarbonate transport in isolated perfused rodent long-loop thin descending limbs. *Am. J. Physiol.* 264:F388–96
19. Mejia R, Flessner MF, Knepper MA. 1993. Model of ammonium and bicarbonate transport along LDL: implications for alkalinization of luminal fluid. *Am. J. Physiol. Renal Physiol.* 264:F397–403

20. Walter A, Gutknecht J. 1986. Permeability of small nonelectrolytes through lipid bilayer membranes. *J. Membr. Biol.* 90:207–17
21. Handlogten ME, Hong SP, Westhoff CM, Weiner ID. 2004. Basolateral ammonium transport by the mouse inner medullary collecting duct cell (mIMCD-3). *Am. J. Physiol. Renal Physiol.* 287:F628–38
22. Handlogten ME, Hong SP, Westhoff CM, Weiner ID. 2005. Apical ammonia transport by the mouse inner medullary collecting duct cell (mIMCD-3). *Am. J. Physiol. Renal Physiol.* 289:F347–58
23. DiGiovanni SR, Madsen KM, Luther AD, Knepper MA. 1994. Dissociation of ammoniagenic enzyme adaptation in rat S1 proximal tubules and ammonium excretion response. *Am. J. Physiol. Renal Physiol.* 267:F407–14
24. Nagami GT. 2002. Enhanced ammonia secretion by proximal tubules from mice receiving NH_4Cl: role of angiotensin II. *Am. J. Physiol. Renal Physiol.* 282:F472–77
25. Nagami GT, Sonu CM, Kurokawa K. 1986. Ammonia production by isolated mouse proximal tubules perfused in vitro: effect of metabolic acidosis. *J. Clin. Invest.* 78:124–29
26. Ambuhl PM, Amemiya M, Danczkay M, Lotscher M, Kaissling B, et al. 1996. Chronic metabolic acidosis increases NHE3 protein abundance in rat kidney. *Am. J. Physiol. Renal Physiol.* 271:F917–25
27. Good DW. 1990. Adaptation of HCO3- and NH_4^+ transport in rat MTAL: effects of chronic metabolic acidosis and Na^+ intake. *Am. J. Physiol. Renal Physiol.* 258:F1345–53
28. Nagami GT. 1988. Luminal secretion of ammonia in the mouse proximal tubule perfused in vitro. *J. Clin. Invest.* 81:159–64
29. Kinsella JL, Aronson PS. 1981. Interaction of NH_4^+ and Li^+ with the renal microvillus membrane Na^+-H^+ exchanger. *Am. J. Physiol. Cell Physiol.* 241:C220–26
30. Nagami GT. 1990. Effect of bath and luminal potassium concentration on ammonia production and secretion by mouse proximal tubules perfused in vitro. *J. Clin. Invest.* 86:32–39
31. Elkjar ML, Kwon TH, Wang W, Nielsen J, Knepper MA, et al. 2002. Altered expression of renal NHE3, TSC, BSC-1, and ENaC subunits in potassium-depleted rats. *Am. J. Physiol. Renal Physiol.* 283:F1376–88
32. Nagami GT. 2004. Ammonia production and secretion by S3 proximal tubule segments from acidotic mice: role of ANG II. *Am. J. Physiol. Renal Physiol.* 287:F707–12
33. Choe H, Sackin H, Palmer LG. 2000. Permeation properties of inward-rectifier potassium channels and their molecular determinants. *J. Gen. Physiol.* 115:391–404
34. Vallon V, Grahammer F, Volkl H, Sandu CD, Richter K, et al. 2005. KCNQ1-dependent transport in renal and gastrointestinal epithelia. *Proc. Natl. Acad. Sci. USA* 102:17864–69
35. Nie X, Arrighi I, Kaissling B, Pfaff I, Mann J, et al. 2005. Expression and insights on function of potassium channel TWIK-1 in mouse kidney. *Pflügers Arch.* 451:479–88
36. Merot J, Bidet M, Le MS, Tauc M, Poujeol P. 1989. Two types of K^+ channels in the apical membrane of rabbit proximal tubule in primary culture. *Biochim. Biophys. Acta* 978:134–44
37. Yao X, Tian S, Chan HY, Biemesderfer D, Desir GV. 2002. Expression of KCNA10, a voltage-gated K channel, in glomerular endothelium and at the apical membrane of the renal proximal tubule. *J. Am. Soc. Nephrol.* 13:2831–39
38. Cluzeaud F, Reyes R, Escoubet B, Fay M, Lazdunski M, et al. 1998. Expression of TWIK-1, a novel weakly inward rectifying potassium channel in rat kidney. *Am. J. Physiol. Cell Physiol.* 275:C1602–9
39. Volkl H, Lang F. 1991. Electrophysiology of ammonia transport in renal straight proximal tubules. *Kidney Int.* 40:1082–89

40. Good DW. 1994. Ammonium transport by the thick ascending limb of Henle's loop. *Annu. Rev. Physiol.* 56:623–47
41. Frank AE, Wingo CS, Weiner ID. 2000. Effects of ammonia on bicarbonate transport in the cortical collecting duct. *Am. J. Physiol. Renal Physiol.* 278:F219–26
42. Haas M, Forbush B. 1998. The Na-K-Cl cotransporters. *J. Bioenerg. Biomembr.* 30:161–72
43. Wall SM, Knepper MA, Hassell KA, Fischer MP, Shodeinde A, et al. 2006. Hypotension in NKCC1 null mice: role of the kidneys. *Am. J. Physiol. Renal Physiol.* 290:F409–16
44. Meyer JW, Flagella M, Sutliff RL, Lorenz JN, Nieman ML, et al. 2002. Decreased blood pressure and vascular smooth muscle tone in mice lacking basolateral Na^+-K^+-$2Cl^-$ cotransporter. *Am. J. Physiol. Heart Circ. Physiol.* 283:H1846–55
45. Ginns SM, Knepper MA, Ecelbarger CA, Terris J, He X, et al. 1996. Immunolocalization of the secretory isoform of Na-K-Cl cotransporter in rat renal intercalated cells. *J. Am. Soc. Nephrol.* 7:2533–42
46. Wall SM, Fischer MP. 2002. Contribution of the Na^+-K^+-$2Cl^-$ cotransporter (NKCC1) to transepithelial transport of H^+, NH_4^+, K^+, and Na^+ in rat outer medullary collecting duct. *J. Am. Soc. Nephrol.* 13:827–35
47. Wall SM, Trinh HN, Woodward KE. 1995. Heterogeneity of NH_4^+ transport in mouse inner medullary collecting duct cells. *Am. J. Physiol. Renal Physiol.* 269:F536–44
48. Glanville M, Kingscote S, Thwaites DT, Simmons NL. 2001. Expression and role of sodium, potassium, chloride cotransport (NKCC1) in mouse inner medullary collecting duct (mIMCD-K2) epithelial cells. *Pflügers Arch.* 443:123–31
49. Wall SM. 1997. Ouabain reduces net acid secretion and increases pH_i by inhibiting NH_4^+ uptake on rat tIMCD Na^+-K^+-ATPase. *Am. J. Physiol. Renal Physiol.* 273:F857–68
50. Ecelbarger CA, Terris J, Hoyer JR, Nielsen S, Wade JB, Knepper MA. 1996. Localization and regulation of the rat renal Na^+-K^+-$2Cl^-$ cotransporter, BSC-1. *Am. J. Physiol. Renal Physiol.* 271:F619–28
51. Plotkin MD, Kaplan MR, Verlander JW, Lee WS, Brown D, et al. 1996. Localization of the thiazide sensitive Na-Cl cotransporter, rTSC1 in the rat kidney. *Kidney Int.* 50:174–83
52. Kinne R, Kinne-Saffran E, Schutz H, Schölermann B. 1986. Ammonium transport in medullary thick ascending limb of rabbit kidney: involvement of the Na^+,K^+,Cl^--cotransporter. *J. Membr. Biol.* 94:279–84
53. Attmane-Elakeb A, Mount DB, Sibella V, Vernimmen C, Hebert SC, Bichara M. 1998. Stimulation by in vivo and in vitro metabolic acidosis of expression of rBSC-1, the Na^+-K^+(NH_4^+)-$2Cl^-$ cotransporter of the rat medullary thick ascending limb. *J. Biol. Chem.* 273:33681–91
54. Attmane-Elakeb A, Sibella V, Vernimmen C, Belenfant X, Hebert SC, Bichara M. 2000. Regulation by glucocorticoids of expression and activity of rBSC1, the Na^+-K^+(NH_4^+)-$2Cl^-$ cotransporter of medullary thick ascending limb. *J. Biol. Chem.* 275:33548–53
55. Ortiz PA. 2006. cAMP increases surface expression of NKCC2 in rat thick ascending limbs: role of VAMP. *Am. J. Physiol. Renal Physiol.* 290:F608–16
56. Martin DW. 2005. Structure-function relationships in the Na^+,K^+-pump. *Semin. Nephrol.* 25:282–91
57. Jorgensen PL. 1980. Sodium and potassium ion pump in kidney tubules. *Physiol. Rev.* 60:864–917
58. Kurtz I, Balaban RS. 1986. Ammonium as a substrate for Na^+-K^+-ATPase in rabbit proximal tubules. *Am. J. Physiol. Renal Physiol.* 250:F497–502
59. Wall SM, Koger LM. 1994. NH_4^+ transport mediated by Na^+-K^+-ATPase in rat inner medullary collecting duct. *Am. J. Physiol. Renal Physiol.* 267:F660–70

60. Knepper MA, Good DW, Burg MB. 1984. Mechanism of ammonia secretion by cortical collecting ducts of rabbits. *Am. J. Physiol. Renal Physiol.* 247:F729–38
61. Wall SM. 1996. NH_4^+ augments net acid-secretion by a ouabain-sensitive mechanism in isolated-perfused inner medullary collecting ducts. *Am. J. Physiol. Renal Physiol.* 270:F432–39
62. Wall SM. 2003. Mechanisms of NH_4^+ and NH_3 transport during hypokalemia. *Acta Physiol. Scand.* 179:325–30
63. Wall SM, Fischer MP, Kim GH, Nguyen BM, Hassell KA. 2002. In rat inner medullary collecting duct, NH_4^+ uptake by the Na,K-ATPase is increased during hypokalemia. *Am. J. Physiol. Renal Physiol.* 282:F91–102
64. Attmane-Elakeb A, Boulanger H, Vernimmen C, Bichara M. 1997. Apical location and inhibition by arginine vasopressin of K^+/H^+ antiport of the medullary thick ascending limb of rat kidney. *J. Biol. Chem.* 272:25668–77
65. Cougnon M, Bouyer P, Jaisser F, Edelman A, Planelles G. 1999. Ammonium transport by the colonic H^+-K^+-ATPase expressed in *Xenopus* oocytes. *Am. J. Physiol. Cell Physiol.* 277:C280–87
66. Codina J, Pressley TA, DuBose TD Jr. 1999. The colonic H^+,K^+-ATPase functions as a Na^+-dependent $K^+(NH_4^+)$-ATPase in apical membranes from rat distal colon. *J. Biol. Chem.* 274:19693–98
67. Nakamura S, Amlal H, Galla JH, Soleimani M. 1999. NH_4^+ secretion in inner medullary collecting duct in potassium deprivation: role of colonic H^+-K^+-ATPase. *Kidney Int.* 56:2160–67
68. Swarts HGP, Koenderink JB, Willems PHGM, De Pont JJ. 2005. The non-gastric H,K-ATPase is oligomycin-sensitive and can function as an H^+, NH_4^+-ATPase. *J. Biol. Chem.* 280:33115–22
69. Frank AE, Wingo CS, Andrews PM, Ageloff S, Knepper MA, Weiner ID. 2002. Mechanisms through which ammonia regulates cortical collecting duct net proton secretion. *Am. J. Physiol. Renal Physiol.* 282:F1120–28
70. King LS, Kozono D, Agre P. 2004. From structure to disease: the evolving tale of aquaporin biology. *Nat. Rev. Mol. Cell Biol.* 5:687–98
71. Knepper MA. 1994. The aquaporin family of molecular water channels. *Proc. Natl. Acad. Sci. USA* 91:6255–58
72. Nakhoul NL, Hering-Smith KS, Abdulnour-Nakhoul SM, Hamm LL. 2001. Transport of NH_3/NH in oocytes expressing aquaporin-1. *Am. J. Physiol. Renal Physiol.* 281:F255–63
73. Holm LM, Jahn TP, Moller AL, Schjoerring JK, Ferri D, et al. 2005. NH_3 and NH_4^+ permeability in aquaporin-expressing *Xenopus* oocytes. *Pflügers Arch.* 450:415–28
74. Beitz E, Wu B, Holm LM, Schultz JE, Zeuthen T. 2006. Point mutations in the aromatic/arginine region in aquaporin 1 allow passage of urea, glycerol, ammonia, and protons. *Proc. Natl. Acad. Sci. USA* 103:269–74
75. Ma T, Song Y, Yang B, Gillespie A, Carlson EJ, et al. 2000. Nephrogenic diabetes insipidus in mice lacking aquaporin-3 water channels. *Proc. Natl. Acad. Sci. USA* 97:4386–91
76. Elkjar ML, Nejsum LN, Gresz V, Kwon TH, Jensen UB, et al. 2001. Immunolocalization of aquaporin-8 in rat kidney, gastrointestinal tract, testis, and airways. *Am. J. Physiol. Renal Physiol.* 281:F1047–57
77. Yang B, Song Y, Zhao D, Verkman AS. 2005. Phenotype analysis of aquaporin-8 null mice. *Am. J. Physiol Cell Physiol.* 288:C1161–70
78. Yang B, Zhao D, Solenov E, Verkman AS. 2006. Evidence from knockout mice against physiologically significant aquaporin 8-facilitated ammonia transport. *Am. J. Physiol. Cell Physiol.* 291:C417–23

79. de Groot BL, Grubmuller H. 2001. Water permeation across biological membranes: mechanism and dynamics of aquaporin-1 and GlpF. *Science* 294:2353–57
80. von Wiren N, Gazzarrini S, Gojon A, Frommer WB. 2000. The molecular physiology of ammonium uptake and retrieval. *Curr. Opin. Plant Biol.* 3:254–61
81. Gazzarrini S, Lejay L, Gojon A, Ninnemann O, Frommer WB, von Wiren N. 1999. Three functional transporters for constitutive, diurnally regulated, and starvation-induced uptake of ammonium into Arabidopsis roots. *Plant Cell* 11:937–48
82. Liu Z, Peng J, Mo R, Hui C, Huang CH. 2001. Rh type B glycoprotein is a new member of the Rh superfamily and a putative ammonia transporter in mammals. *J. Biol. Chem.* 276:1424–33
83. Liu Z, Chen Y, Mo R, Hui C, Cheng JF, et al. 2000. Characterization of human RhCG and mouse Rhcg as novel nonerythroid Rh glycoprotein homologues predominantly expressed in kidney and testis. *J. Biol. Chem.* 275:25641–51
84. Marini AM, Urrestarazu A, Beauwens R, Andre B. 1997. The Rh (rhesus) blood group polypeptides are related to NH_4^+ transporters. *Trends Biochem. Sci.* 22:460–61
85. Marini AM, Matassi G, Raynal V, Andre B, Cartron JP, Cherif-Zahar B. 2000. The human Rhesus-associated RhAG protein and a kidney homologue promote ammonium transport in yeast. *Nat. Genet.* 26:341–44
86. Westhoff CM, Ferreri-Jacobia M, Mak DD, Foskett JK. 2002. Identification of the erythrocyte Rh-blood group glycoprotein as a mammalian ammonium transporter. *J. Biol. Chem.* 277:12499–502
87. Westhoff CM, Siegel DL, Burd CG, Foskett JK. 2004. Mechanism of genetic complementation of ammonium transport in yeast by human erythrocyte Rh-associated glycoprotein (RhAG). *J. Biol. Chem.* 279:17443–48
88. Ripoche P, Bertrand O, Gane P, Birkenmeier C, Colin Y, Cartron JP. 2004. Human Rhesus-associated glycoprotein mediates facilitated transport of NH_3 into red blood cells. *Proc. Natl. Acad. Sci. USA* 101:17222–27
89. Benjelloun F, Bakouh N, Fritsch J, Hulin P, Lipecka J, et al. 2005. Expression of the human erythroid Rh glycoprotein (RhAG) enhances both NH_3 and NH_4^+ transport in HeLa cells. *Pflügers Arch.* 450:155–67
90. Cartron JP. 1999. RH blood group system and molecular basis of Rh-deficiency. *Baillieres Best Pract. Res. Clin. Haematol.* 12:655–89
91. Liu Z, Huang CH. 1999. The mouse Rhl1 and Rhag genes: sequence, organization, expression, and chromosomal mapping. *Biochem. Genet.* 37:119–38
92. Conroy MJ, Bullough PA, Merrick M, Avent ND. 2005. Modelling the human rhesus proteins: implications for structure and function. *Br. J. Haematol.* 131:543–51
93. Handlogten ME, Hong SP, Zhang L, Vander AW, Steinbaum ML, et al. 2005. Expression of the ammonia transporter proteins, Rh B Glycoprotein and Rh C Glycoprotein, in the intestinal tract. *Am. J. Physiol. Gastrointest. Liver Physiol.* 288:G1036–47
94. Quentin F, Eladari D, Cheval L, Lopez C, Goossens D, et al. 2003. RhBG and RhCG, the putative ammonia transporters, are expressed in the same cells in the distal nephron. *J. Am. Soc. Nephrol.* 14:545–54
95. Verlander JW, Miller RT, Frank AE, Royaux IE, Kim YH, Weiner ID. 2003. Localization of the ammonium transporter proteins, Rh B Glycoprotein and Rh C glycoprotein, in the mouse kidney. *Am. J. Physiol. Renal Physiol.* 284:F323–37
96. Lopez C, Metral S, Eladari D, Drevensek S, Gane P, et al. 2004. The ammonium transporter RhBG: requirement of a tyrosine-based signal and ankyrin-G for basolateral targeting and membrane anchorage in polarized kidney epithelial cells. *J. Biol. Chem.* 280:8221–28

97. Weiner ID, Miller RT, Verlander JW. 2003. Localization of the ammonium transporters, Rh B Glycoprotein and Rh C Glycoprotein in the mouse liver. *Gastroenterology* 124:1432–40
98. Kaiser S, Gerok W, Haussinger D. 1988. Ammonia and glutamine metabolism in human liver slices: new aspects on the pathogenesis of hyperammonaemia in chronic liver disease. *Eur. J. Clin. Invest.* 18:535–42
99. Czarnowski D, Gorski J. 1991. Sweat ammonia excretion during submaximal cycling exercise. *J. Appl. Physiol.* 70:371–74
100. Brusilow SW, Gordes EH. 1968. Ammonia secretion in sweat. *Am. J. Physiol.* 214:513–17
101. Mak DD, Dang B, Weiner ID, Foskett JK, Westhoff CM. 2006. Characterization of transport by the kidney Rh glycoproteins, RhBG and RhCG. *Am. J. Physiol. Renal Physiol.* 290:F297–305
102. Ludewig U. 2004. Electroneutral ammonium transport by basolateral Rhesus B glycoprotein. *J. Physiol.* 559:751–59
103. Zidi-Yahiaoui N, Mouro-Chanteloup I, D'Ambrosio AM, Lopez C, Gane P, et al. 2005. Human Rhesus B and Rhesus C glycoproteins: properties of facilitated ammonium transport in recombinant kidney cells. *Biochem. J.* 391:33–40
104. Nakhoul NL, DeJong H, Abdulnour-Nakhoul SM, Boulpaep EL, Hering-Smith K, Hamm LL. 2004. Characteristics of renal Rhbg as an NH_4^+ transporter. *Am. J. Physiol. Renal Physiol.* 288:F170–81
105. Seshadri RM, Klein JD, Kozlowski S, Sands JM, Kim YH, et al. 2006. Renal expression of the ammonia transporters, Rhbg and Rhcg, in response to chronic metabolic acidosis. *Am. J. Physiol. Renal Physiol.* 290:F397–408
106. Kim YH, Verlander JW, Matthews SW, Kurtz I, Shin WK, et al. 2005. Intercalated cell H^+/OH^- transporter expression is reduced in *Slc26a4* null mice. *Am. J. Physiol. Renal Physiol.* 289:F1262–72
107. Chambrey R, Goossens D, Bourgeois S, Picard N, Bloch-Faure M, et al. 2005. Genetic ablation of Rhbg in mouse does not impair renal ammonium excretion. *Am. J. Physiol. Renal Physiol.* 289:F1281–90
108. Weiner ID. 2006. Expression of the non-erythroid Rh glycoproteins in mammalian tissues. *Transfus. Clin. Biol.* 13:159–63
109. Eladari D, Cheval L, Quentin F, Bertrand O, Mouro I, et al. 2002. Expression of RhCG, a new putative NH_3/NH_4^+ transporter, along the rat nephron. *J. Am. Soc. Nephrol.* 13:1999–2008
110. Seshadri RM, Klein JD, Smith T, Sands JM, Handlogten ME, et al. 2006. Changes in the subcellular distribution of the ammonia transporter Rhcg, in response to chronic metabolic acidosis. *Am. J. Physiol. Renal Physiol.* 290:F1443–52
111. Han KH, Croker BP, Clapp WL, Werner D, Sahni M, et al. 2006. Expression of the ammonia transporter, Rh C glycoprotein, in normal and neoplastic human kidney. *J. Am. Soc. Nephrol.* In press. doi:10.1681/ASN.2006020160
112. Bakouh N, Benjelloun F, Hulin P, Brouillard F, Edelman A, et al. 2004. NH_3 is involved in the NH_4^+ transport induced by the functional expression of the human Rh C glycoprotein. *J. Biol. Chem.* 279:15975–83
113. Nakhoul NL, Palmer SA, Abdulnour-Nakhoul S, Hering-Smith K, Hamm LL. 2002. Ammonium transport in oocytes expressing Rhcg. *FASEB J.* 16(4):A53 (Abstr.)
114. Weiner ID. 2004. The Rh gene family and renal ammonium transport. *Curr. Opin. Nephrol. Hypertens.* 13:533–40

115. Conroy MJ, Jamieson SJ, Blakey D, Kaufmann T, Engel A, et al. 2004. Electron and atomic force microscopy of the trimeric ammonium transporter AmtB. *EMBO Rep.* 5:1153 58
116. Khademi S, O'Connell J III, Remis J, Robles-Colmenares Y, Miercke LJ, Stroud RM. 2004. Mechanism of ammonia transport by Amt/MEP/Rh: structure of AmtB at 1.35 Å. *Science* 305:1587–94
117. Zheng L, Kostrewa D, Berneche S, Winkler FK, Li XD. 2004. The mechanism of ammonia transport based on the crystal structure of AmtB of *Escherichia coli*. *Proc. Natl. Acad. Sci. USA* 101:17090–95
118. Javelle A, Severi E, Thornton J, Merrick M. 2004. Ammonium sensing in *Escherichia coli*: role of the ammonium transporter AmtB and AmtB-GlnK complex formation. *J. Biol. Chem.* 279:8530–38
119. Zidi-Yahiaoui N, Ripoche P, Le Van KC, Gane P, D'Ambrosio AM, et al. 2006. Ammonium transport properties of HEK293 cells expressing RhCG mutants: preliminary analysis of structure/function by site-directed mutagenesis. *Transfus. Clin. Biol.* 13:128–31
120. Kim KS, Feild E, King N, Yaoi T, Kustu S, Inwood W. 2005. Spontaneous mutations in the ammonium transport gene *AMT4* of *Chlamydomonas reinhardtii*. *Genetics* 170:631–44
121. Soupene E, Inwood W, Kustu S. 2004. Lack of the Rhesus protein Rh1 impairs growth of the green alga *Chlamydomonas reinhardtii* at high CO_2. *Proc. Natl. Acad. Sci. USA* 101:7787–92
122. Soupene E, King N, Feild E, Liu P, Niyogi KK, et al. 2002. Rhesus expression in a green alga is regulated by CO_2. *Proc. Natl. Acad. Sci. USA* 99:7769–73
123. Huang CH, Peng J. 2005. Evolutionary conservation and diversification of Rh family genes and proteins. *Proc. Natl. Acad. Sci. USA* 102:15512–17
124. Endeward V, Cartron JP, Ripoche P, Gros G. 2006. Red cell membrane CO_2 permeability in normal human blood and in blood deficient in various blood groups, and effect of DIDS. *Transfus. Clin. Biol.* 13:123–27
125. Ripoche P, Goossens D, Devuyst O, Gane P, Colin Y, et al. 2006. Role of RhAG and AQP1 in NH_3 and CO_2 gas transport in red cell ghosts: a stopped-flow analysis. *Transfus. Clin. Biol.* 13:117–22
126. Mudry B, Guy RH, Gado-Charro MB. 2006. Transport numbers in transdermal iontophoresis. *Biophys. J.* 90:2822–30
127. Atkins PW. 1978. Molecules in motion: ion transport and molecular diffusion. In *Physical Chemistry*, ed. PW Atkins, pp. 819–48. Oxford, UK: Oxford Univ. Press
128. Falk KG. 1929. Transference numbers of electrolytes in aqueous solutions. In *International Critical Tables of Numerical Data, Physics, Chemistry and Technology*, ed. EW Washburn, VI:309–11. New York: McGraw-Hill
129. LeMasurier M, Heginbotham L, Miller C. 2001. KcsA: It's a potassium channel. *J. Gen. Physiol.* 118:303–14
130. Heginbotham L, MacKinnon R. 1993. Conduction properties of the cloned Shaker K^+ channel. *Biophys. J.* 65:2089–96
131. Eisenman G, Latorre R, Miller C. 1986. Multi-ion conduction and selectivity in the high-conductance Ca^{++}-activated K^+ channel from skeletal muscle. *Biophys. J.* 50:1025–34

Phosphatonins and the Regulation of Phosphate Homeostasis

Theresa Berndt and Rajiv Kumar

Nephrology and Hypertension Research, Departments of Medicine, Biochemistry, and Molecular Biology, Mayo Clinic College of Medicine, Rochester, Minnesota, 55905; email: rkumar@mayo.edu

Annu. Rev. Physiol. 2007. 69:341–59

First published online as a Review in Advance on September 5, 2006

The *Annual Review of Physiology* is online at http://physiol.annualreviews.org

This article's doi: 10.1146/annurev.physiol.69.040705.141729

0066-4278/07/0315-0341$20.00

Key Words

phosphatonins, fibroblast growth factor-23, secreted frizzled related protein-4, matrix extracellular phosphoglycoprotein, fibroblast growth factor-7, sodium-phosphate cotransporters, parathyroid hormone, 1α,25-dihydroxyvitamin D

Abstract

Inorganic phosphate (P_i) is required for energy metabolism, nucleic acid synthesis, bone mineralization, and cell signaling. The activity of cell-surface sodium-phosphate (Na^+-P_i) cotransporters mediates the uptake of P_i from the extracellular environment. Na^+-P_i cotransporters and organ-specific P_i absorptive processes are regulated by peptide and sterol hormones, such as parathyroid hormone (PTH) and 1α,25-dihydroxyvitamin D (1α,25$(OH)_2D_3$), which interact in a coordinated fashion to regulate P_i homeostasis. Recently, several phosphaturic peptides such as fibroblast growth factor-23 (FGF-23), secreted frizzled related protein-4 (sFRP-4), matrix extracellular phosphoglycoprotein, and fibroblast growth factor-7 have been demonstrated to play a pathogenic role in several hypophosphatemic disorders. By inhibiting Na^+-P_i transporters in renal epithelial cells, these proteins increase renal P_i excretion, resulting in hypophosphatemia. FGF-23 and sFRP-4 inhibit 25-hydroxyvitamin D 1α-hydroxylase activity, reducing 1α,25$(OH)_2D_3$ synthesis and thus intestinal P_i absorption. This review examines the role of these factors in P_i homeostasis in health and disease.

Na^+-P_i **cotransporters:** sodium-phosphate cotransporters

$1\alpha,25(OH)_2D_3$: 1α,25-dihydroxyvitamin D

PTH: parathyroid hormone

THE IMPORTANCE OF PHOSPHOROUS IN BIOLOGICAL PROCESSES

Phosphorus and inorganic phosphate (P_i) play important roles in a variety of biological processes such as cell signaling, nucleic acid synthesis, energy metabolism, membrane function, and bone mineralization (1–5). P_i is required for optimal cellular growth, and serum P_i concentrations, as well as renal P_i reabsorption, are higher in rapidly growing animals than in adults (6, 7). Owing to the involvement of P_i in diverse biological processes, decrements in serum P_i concentrations and a negative P_i balance can result in serious disease. Acute decreases in serum P_i concentrations can result in myopathy, cardiac dysfunction, abnormal neutrophil function, platelet dysfunction, and red-cell membrane fragility (8, 9). Chronic serum P_i deficiency results in impaired bone mineralization, rickets, and osteomalacia because the rate of bone matrix mineralization depends on the availability of phosphorus and calcium (5). Elevated serum P_i concentrations contribute to the pathogenesis of secondary hyperparathyroidism in patients with chronic renal failure (10–14).

MECHANISMS OF CELLULAR P_i UPTAKE

Because a cell's interior is electronegative relative to the exterior, the movement of P_i into the cell does not occur by simple diffusion (15). Various H^+- or Na^+-coupled P_i cotransporters mediate the transport of P_i across cell membranes (15). The structure and function of these have been extensively reviewed, and the reader is directed to other publications in this regard (15–20). The Na^+-coupled P_i cotransporters that are important in P_i uptake in vertebrates belong to two large families, the Na-P_i type II and the Na-P_i type III families. **Figure 1*a*** shows the different families of Na-P_i transporters, and **Figure 1*b*** summarizes the relationships of Na-P_i type II protein sequences to one another (15). The Na-P_i transporters are highly homologous. Na^+-P_i IIa (Npt2) transporters are the most abundant in the kidney and contribute approximately 85% of proximal tubule Na-P_i reabsorption (19, 20). Researchers recently identified, in rat and human kidney, a Na-P_i type IIc transporter that is maximally upregulated in *Npt2*$^{-/-}$ mice and is thought to account for the residual P_i transport in the *Npt2* knockout mice (21, 22).

PHOSPHATE HOMEOSTASIS IN HUMANS AND MAMMALS

In humans, absorption and reabsorption of P_i occur primarily in the intestine and kidney, respectively (see **Figure 2**). In states of neutral P_i balance, the amount of P_i absorbed in the intestine (approximately 1–1.5 g per 24 h) is equivalent to the amount excreted in the urine. 1α,25-Dihydroxyvitamin D ($1\alpha,25(OH)_2D_3$) increases the efficiency of P_i absorption in the intestine, although there is evidence for a $1\alpha,25(OH)_2D_3$-independent increase in P_i transport during P_i deprivation that occurs in the absence of the $1\alpha,25(OH)_2D_3$ receptor (23–28). In the kidney, P_i is reabsorbed along the proximal convoluted and proximal straight tubules (29, 30). Various factors, most importantly parathyroid hormone (PTH), influence the efficiency of renal P_i reabsorption (29, 31–33).

The vitamin D endocrine system and PTH interact to regulate P_i absorption in the intestine and reabsorption in the kidney, as demonstrated in **Figure 3** (30). Animals fed a low-P_i diet have decreased serum P_i concentrations that are associated with a reciprocal increase in circulating plasma calcium concentrations. The increase in plasma calcium concentrations inhibits PTH release, which in turn reduces the renal excretion of P_i. Additionally, a low-P_i diet and reductions in serum P_i are associated with increased $1\alpha,25(OH)_2D_3$ synthesis as a result of stimulation of 25-hydroxyvitamin D 1α-hydroxylase activity (27, 34). Conversely, when animals are fed a high-P_i diet, serum

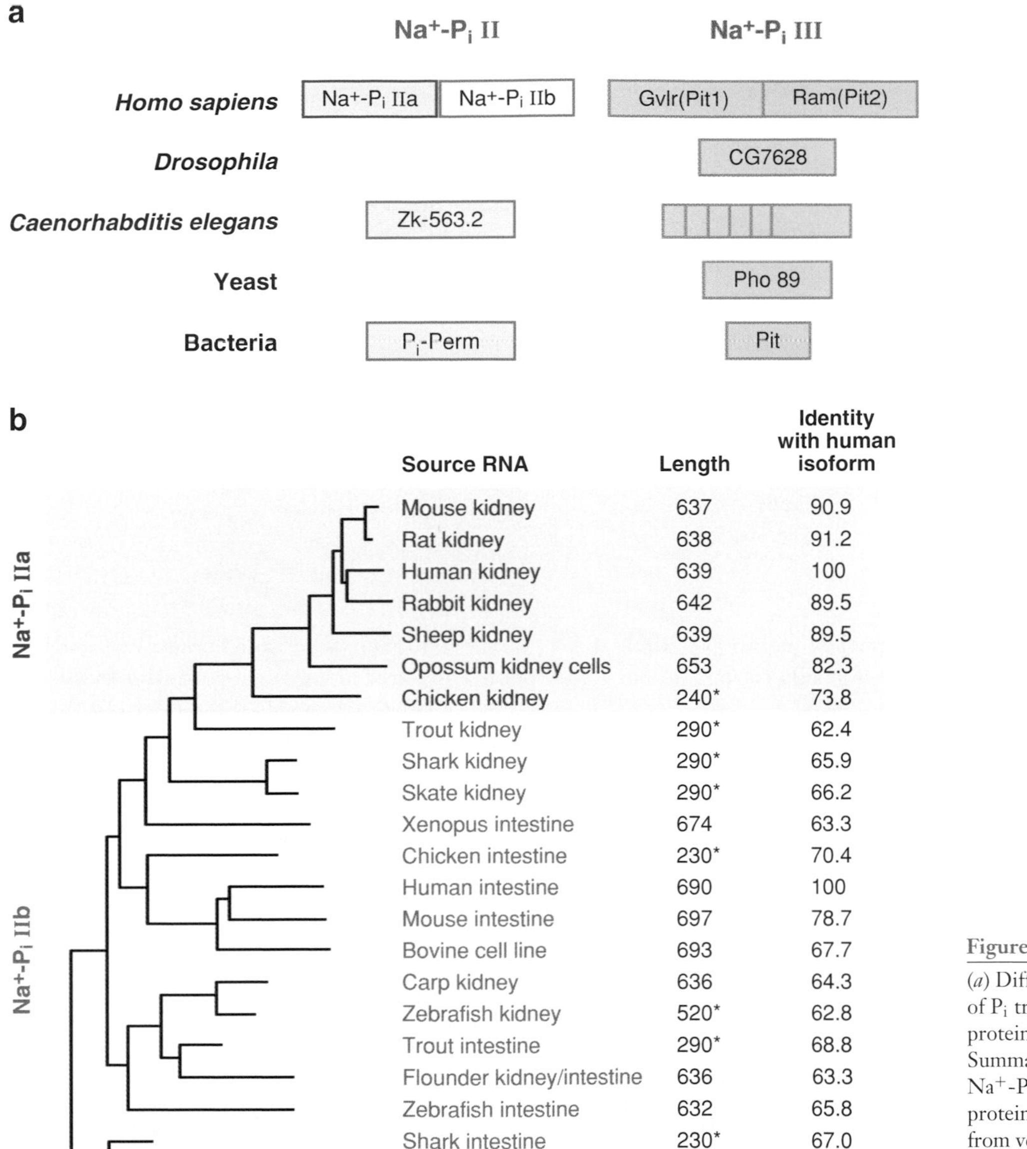

a
Na+-Pi II
Na+-Pi III
Homo sapiens
Na+-Pi IIa
Na+-Pi IIb
Gvlr(Pit1)
Ram(Pit2)
Drosophila
CG7628
Caenorhabditis elegans
Zk-563.2
Yeast
Pho 89
Bacteria
Pi-Perm
Pit
b
Na+-Pi IIa
Na+-Pi IIb
Source RNA
Length
Identity with human isoform
Mouse kidney 637 90.9
Rat kidney 638 91.2
Human kidney 639 100
Rabbit kidney 642 89.5
Sheep kidney 639 89.5
Opossum kidney cells 653 82.3
Chicken kidney 240* 73.8
Trout kidney 290* 62.4
Shark kidney 290* 65.9
Skate kidney 290* 66.2
Xenopus intestine 674 63.3
Chicken intestine 230* 70.4
Human intestine 690 100
Mouse intestine 697 78.7
Bovine cell line 693 67.7
Carp kidney 636 64.3
Zebrafish kidney 520* 62.8
Trout intestine 290* 68.8
Flounder kidney/intestine 636 63.3
Zebrafish intestine 632 65.8
Shark intestine 230* 67.0
Skate intestine 230* 63.0
*Fragment

Figure 1

(*a*) Different families of P_i translocating proteins. (*b*) Summary of Na^+-P_i-related protein sequences from vertebrates. Both *a* and *b* are from Reference 15 with permission.

calcium concentrations decrease, and PTH release is increased. Recent data suggest that PTH release can occur in the absence of changes in serum P_i or calcium concentrations. Following the instillation of high-P_i diets directly into the intestine, changes in PTH secretion in response to dietary P_i occur rapidly (within 10 min) and independently of changes in serum P_i or calcium (35, 36). A signal emanating from the intestine, similar to that which has been documented for Na^+ (37–39), may affect P_i reabsorption by the kidney (40, 41). An elevation in serum P_i after a high-P_i meal reduces 25-hydroxyvitamin D 1α-hydroxylase activity and results in reduced circulating $1\alpha,25(OH)_2D_3$ and diminished

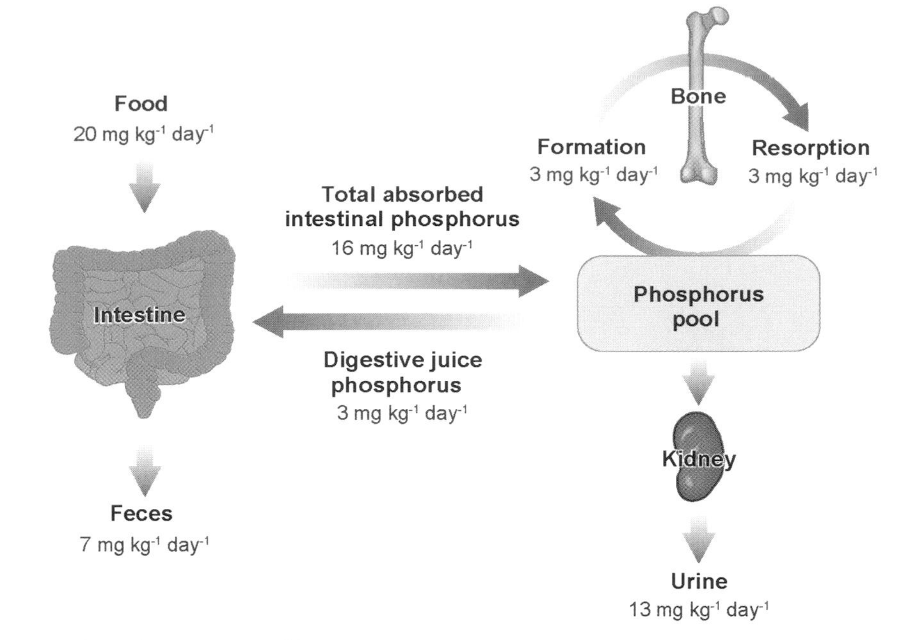

Figure 2

Quantitative aspects of phosphorus homeostasis in humans. Based on Reference 30. The intestine and kidney play major roles in phosphate (P_i) homeostasis by regulating absorption and excretion, respectively. The movement of P_i into bone and soft tissue also determines serum P_i concentrations.

intestinal P_i absorption. **Table 1** lists various factors influencing phosphate absorption in the intestine.

Given the importance of P_i ions in cellular function, various other mechanisms have evolved to maintain P_i balance. Although PTH and the vitamin D endocrine system are involved in the regulation of P_i absorption in the intestine and reabsorption in the kidney, other factors such as arterial CO_2 ($paCO_2$), adrenergic agents, and dopamine can rapidly modulate the phosphaturic effect of PTH to

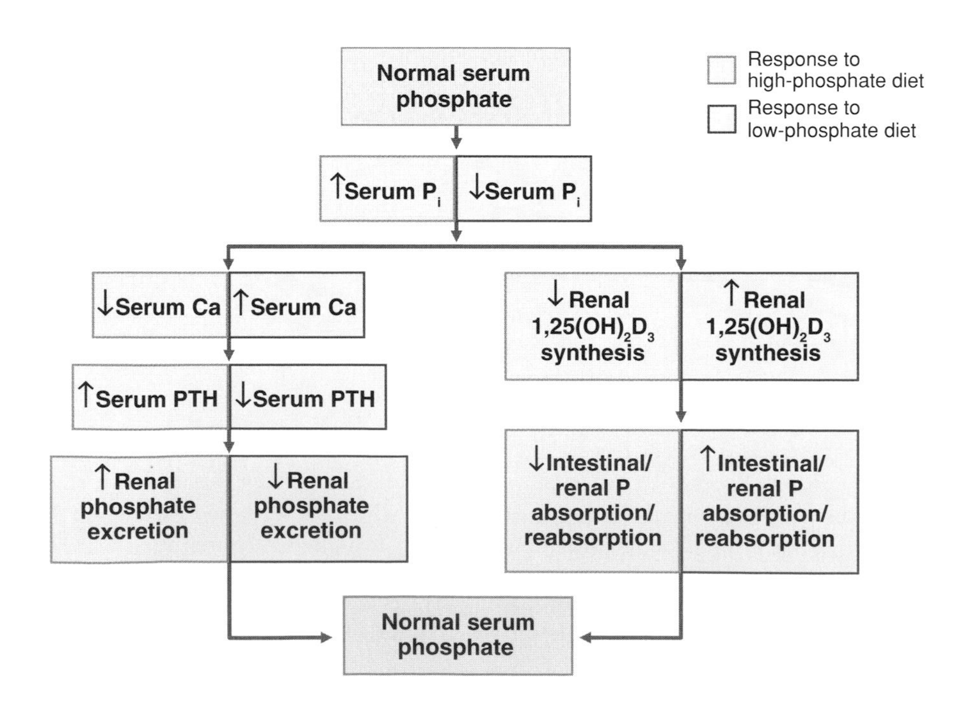

Figure 3

Mechanisms by which phosphate (P_i) homeostasis is maintained. In response to dietary phosphate intake, the vitamin D endocrine system and parathyroid hormone (PTH) interact to regulate phosphate absorption in the intestine and reabsorption in the kidney. From Reference 30 with permission.

alter P_i homeostasis (29). The phosphaturic response to PTH is progressively attenuated by a low-P_i diet. P_i deprivation for two days or fewer results in an attenuated phosphaturic response to PTH that is restored by propanolol infusion, suggesting that stimulation of β adrenoreceptors occurs during short-term P_i deprivation (42). However, P_i deprivation for three days or more results in complete resistance to the phosphaturic response to PTH.

The enhanced renal P_i reabsorption as a result of selective P_i deprivation is rapidly reversed by total fasting through mechanisms that are not understood (43, 44). Rapid changes in renal P_i reabsorption occur independently of changes in serum PTH as well as in the absence of PTH. Cultured proximal tubule cells exposed to low P_i concentrations in the culture medium rapidly increase P_i transport, demonstrating an intrinsic ability of these proximal tubule cells to sense P_i and adapt by increasing P_i transport (45). Serum P_i concentrations are also altered as a result of the movement of P_i from the extracellular fluid space into bone and soft tissues, and the rate at which this occurs is influenced by a number of factors, including blood pH and various peptide hormones. Taken together, these observations suggest the existence of autocrine and paracrine factors that acutely alter the phosphaturic response to PTH and/or directly alter renal P_i reabsorption. **Table 2** summarizes various factors influencing phosphate reabsorption in the kidney.

Table 1 Factors influencing phosphate absorption in the intestine

Factors that increase P_i absorption
Reduced dietary intake of phosphate
Elevated serum 1α,25-dihydroxyvitamin D
Factors that reduce P_i absorption
Reduced serum 1α,25-dihydroxyvitamin D
Elevated concentrations of calcium salts or phosphate binders, e.g., aluminum hydroxide and sevelamer, in intestinal lumen

THE PHOSPHATONINS AND THE REGULATION OF RENAL P_i REABSORPTION

In 1994, our laboratory described the existence of a factor or factors (called the phosphatonins) that induced renal P_i wasting in patients with tumor-induced osteomalacia (TIO) (46, 47). Patients with TIO typically exhibit low serum P_i concentrations, normal or slightly low serum calcium concentrations, normal PTH concentrations, inappropriately low serum $1\alpha,25(OH)_2D_3$ concentrations, renal P_i wasting, and a defect in bone mineralization (47–49). This phenotype is also observed in patients with autosomal-dominant hypophosphatemic rickets (ADHR) and X-linked hypophosphatemic rickets (XLH) (50–53). We demonstrated that conditioned medium from a tumor associated with TIO produced a substance or substances that inhibited Na^+-dependent P_i transport in cultured opossum kidney cells (46). This heat-labile substance(s) was of M_r 10–30,000 and inhibited Na^+-dependent P_i transport by a process independent of cyclic AMP. The activity of the substance was not blocked by a PTH receptor antagonist. This suggested that the factor was

Table 2 Factors influencing phosphate reabsorption in the kidney

Factors that increase P_i reabsorption
Phosphate depletion
Parathyroidectomy
$1\alpha,25(OH)_2D_3$
Volume contraction
Hypocalcemia
Hypocapnia
Factors that decrease P_i reabsorption
Phosphate loading
Parathyroid hormone and cyclic AMP
Volume expansion
Hypercalcemia
Carbonic anhydrase inhibitors
Dopamine
Glucose and alanine
Acid-base disturbances (increased bicarbonate, hypercapnia)
Metabolic inhibitors (e.g., arsenate)
FGF-23
SFRP-4
MEPE
FGF-7

not one already known to influence P_i transport in epithelia.

Several laboratories subsequently showed that factors such as fibroblast growth factor-23 (FGF-23), secreted frizzled related protein-4 (sFRP-4), fibroblast growth factor 7 (FGF-7), and matrix extracellular phosphoglycoprotein (MEPE) are present in these tumors and contribute to the phosphaturia associated with this syndrome (30, 53–61). Identification of these phosphatonin molecules has led to the recognition that these proteins are also involved in other pathophysiological conditions associated with P_i wasting and may contribute to the physiological regulation of renal P_i reabsorption.

Phosphatonins: phosphaturic substances responsible for the pathogenesis of disorders such as tumor-induced osteomalacia, X-linked hypophosphatemic rickets, and autosomal-dominant hypophosphatemic rickets

Tumor-induced osteomalacia (TIO): a syndrome associated with mesenchymal tumors and characterized by hypophosphatemia, renal phosphate wasting, diminished or inappropriately low serum $1\alpha,25(OH)_2D_3$ concentrations, and rickets or osteomalacia

Autosomal-dominant hypophosphatemic rickets (ADHR): an inherited disorder with phenotype similar to that seen in TIO and XLH

X-linked hypophosphatemic rickets (XLH): an inherited disorder with phenotype similar to that seen in TIO

Secreted frizzled related protein-4 (sFRP-4): a Wnt antagonist

MEPE: matrix extracellular phosphoglycoprotein

Fibroblast Growth Factor-23

Biological properties of FGF-23 and its role in the pathogenesis of ADHR, TIO, and XLH. Patients with ADHR typically exhibit low serum P_i concentrations, normal or slightly low serum calcium concentrations, normal PTH concentrations, low or inappropriately normal serum $1\alpha,25(OH)_2D_3$ concentrations, renal P_i wasting, and a defect in bone mineralization. Using positional cloning methods, the ADHR Consortium (53) identified the mutant gene responsible for the disease in these patients. The mutant gene encodes a novel fibroblast growth factor, FGF-23. The *FGF-23* mutation results in the production of a stable and long-lived form of FGF-23. The wild-type FGF-23 protein contains a furin proconvertase site (176 RHTR 179), and proteolysis results in the formation of the amino-terminal fragment of approximately 16 kDa and a smaller fragment of approximately 12 kDa (62). In ADHR patients, the FGF-23 expressed from the mutant gene lacks a normal furin proconvertase site (176 QHTR 179; 176 RHTW/Q 179) (53, 63).

Shimada et al. (64) showed that FGF-23 was overexpressed in a tumor of a patient with TIO. Recombinant FGF-23 administered intraperitoneally to mice induced hypophosphatemia but did not alter serum calcium concentrations. When Chinese hamster ovary cells transfected with an FGF-23 expression plasmid were implanted in nude mice, these animals became hypophosphatemic, and the urinary fractional excretion of P_i was increased within 10 days. Radiological and histological signs of rickets in the long bones were seen after several weeks. Decreased mRNA for the 25-hydroxyvitamin D 1α-hydroxylase cytochrome P450 was observed in the kidneys of these nude mice. In support of these studies, Bowe et al. (65) demonstrated that recombinant FGF-23 inhibited Na^+-dependent P_i transport in opossum kidney cells. Additionally, these authors showed that FGF-23 was the subject of proteolysis by recombinant PHEX (phosphate-regulating gene with homology to endopeptidases on the X chromosome), the endopeptidase mutated in patients with XLH (52). Intravenous infusion of recombinant FGF-23 into mice caused a rapid, dose-dependent increase in the fractional excretion of P_i with little or no change in Na^+ excretion, suggesting that FGF-23 has direct actions on renal P_i transport (**Figure 4**) (60). Thus, FGF-23 is at least one of the phosphatonins responsible for the pathogenesis of hypophosphatemia, renal P_i wasting, and reduced serum $1\alpha,25(OH)_2D_3$ concentrations in patients with TIO. Cure of the disease phenotype and decreases in FGF-23 serum concentrations after tumor removal in patients with TIO support the role of FGF-23 in the pathogenesis of this condition (66–69).

Researchers have generated transgenic animals overexpressing FGF-23 to elucidate further the ability of FGF-23 to decrease serum P_i concentrations, increase renal P_i excretion, and inhibit 25-hydroxyvitamin D 1α-hydroxylase activity (70, 71). As expected, these mice are hypophosphatemic, exhibit excessive renal P_i excretion, and have rickets and either reduced serum $1\alpha,25(OH)_2D_3$ concentrations or 25-hydroxyvitamin D 1α-hydroxylase activity. Conversely, mice in which the *FGF-23* gene has been ablated demonstrate hyperphosphatemia, reduced renal P_i excretion, and elevated serum

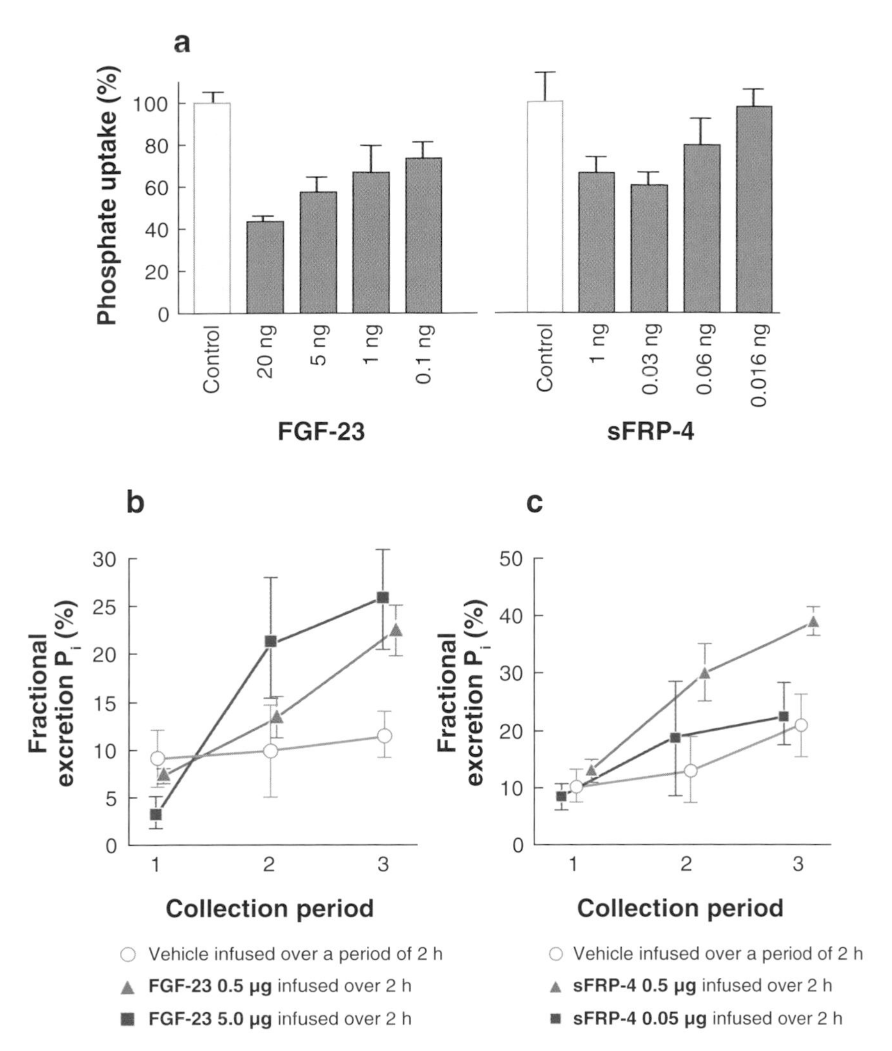

Figure 4

Biological effects of FGF-23 and sFRP-4 in opossum kidney cells in vitro (*a*) and in normal mice in vivo (*b*). (*a*) Opossum kidney cells were maintained in culture, and Na^+-dependent P_i transport was measured following the addition of either FGF-23 (*left panel*) or sFRP-4 (*right panel*). (*b*) The effect of infused FGF-23 on the fractional excretion of P_i in mice. (*c*) The effect of infused sFRP-4 on the fractional excretion of P_i in mice. Figure from Reference 30 with permission.

$1\alpha,25(OH)_2D_3$ concentrations and renal 25-hydroxyvitamin D 1α-hydroxylase mRNA expression (72).

In patients with XLH, mutations of the gene encoding the endopeptidase PHEX are believed to be responsible for the disease phenotype (50–52). Parabiosis and kidney cross-transplantation have clearly shown that there is a circulating hypophosphatemia-inducing factor present in the serum of *Hyp* mice (the mouse homolog of human XLH) (73–75). PHEX may be responsible for the degradation of a phosphatonin (65, 76). FGF-23 is likely the phosphatonin degraded by PHEX. Studies by Bowe et al. (65) with an FGF-23 and recombinant PHEX and the studies of Campos et al. (76) with FGF-23 peptides support a role for PHEX in processing and inactivating FGF-23. Recently, Liu et al. (77) demonstrated that deletion of the *FGF-23* gene in PHEX-deficient mice results in the abolition of the hypophosphatemic phenotype seen in mice deficient in PHEX alone. Furthermore, individuals with XLH and *Hyp* mice have elevated FGF-23 concentrations (67).

Thus, FGF-23—by virtue of its ability to increase renal P_i excretion, inhibit 25-hydroxyvitamin D 1α-hydroxylase activity and $1\alpha,25(OH)_2D_3$ synthesis, and reduce intestinal P_i absorption—likely is involved in

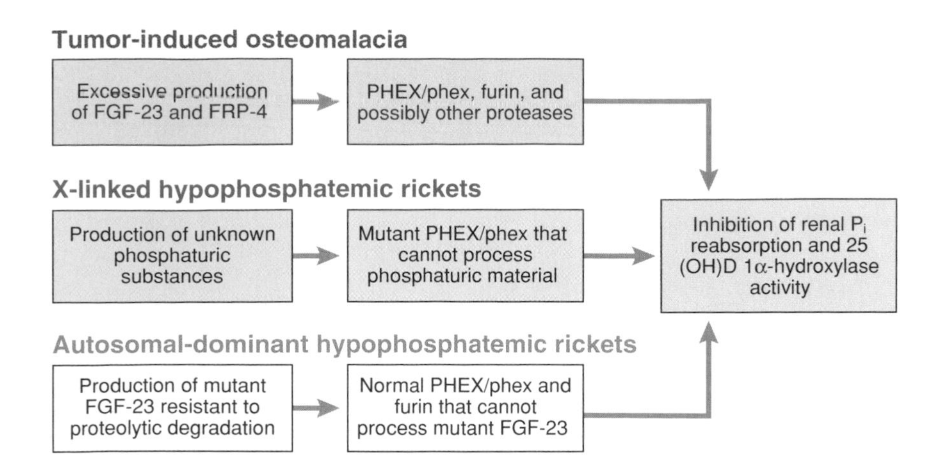

Figure 5
Mechanisms by which hypophosphatemia occurs in tumor-induced osteomalacia (TIO), X-linked hypophosphatemic rickets (XLH), and autosomal-dominant hypophosphatemic rickets (ADHR).

Fibroblast growth factor-23 (FGF-23) and fibroblast growth factor-7 (FGF-7): growth factors known to cause hypophosphatemia and the inhibition of sodium-phosphate cotransport in renal epithelial cells. FGF-23 also inhibits 25-hydroxyvitamin D 1α-hydroxylase activity, thereby reducing $1\alpha,25(OH)_2D_3$ synthesis

Tumoral calcinosis (TC): a disorder that is the mirror image of TIO, XLH, and ADHR and characterized by hyperphosphatemia, reduced renal phosphate excretion, and elevated or inappropriately high serum $1\alpha,25(OH)_2D_3$ concentrations

the pathogenesis of at least three hypophosphatemic conditions, namely, ADHR, TIO, and XLH. The mechanisms by which the disease phenotype occurs in each of these conditions are summarized in **Figure 5**.

The study of patients with a rare disorder, tumor calcinosis (TC), has also yielded interesting information concerning the biological properties of FGF-23. Patients with TC have a biochemical phenotype that is the opposite of that seen in patients with TIO, XLH, and ADHR. TC subjects have hyperphosphatemia, reduced renal P_i excretion, and elevated $1\alpha,25(OH)_2D_3$ concentrations. Two different types of mutations account for this syndrome. The first type of mutation is in the gene *GALNT3*, which encodes a glycosyltransferase responsible for initiating mucin-type O-glycosylation (79). Patients with this syndrome have elevated concentrations of FGF-23, as measured by an assay that detects the carboxyl-terminal portion of the molecule. These patients have relatively low normal concentrations of FGF-23, as measured by an assay that detects only the intact protein. The defect in glycosylation may interfere with the processing of FGF-23, and low concentrations of the intact peptide may be responsible for the biochemical phenotype. The second type of mutation seen in patients with tumoral calcinosis occurs within the *FGF-23* gene itself (80–83). Two recessive mutations in FGF-23, serine 71/glycine (S71G) and serine 129/phenylalanine (S129F), have been identified in patients with TC. These patients have elevated FGF-23 concentrations, as measured by an assay that detects carboxyl-terminal fragments and the carboxyl-terminal portion of the protein. However, FGF-23 concentrations measured by the full-length protein assay are apparently low to normal. Defects in the processing of these proteins in the Golgi apparatus may be responsible for the differences in FGF-23 concentrations measured by the two assays.

Mechanism of action of FGF-23. In vitro binding studies suggest that FGF-23 binds and signals through one of the known FGF receptors. Yamashita et al. (84) demonstrated that FGF-23 binds with high affinity to the FGF receptor 3c, which is mainly expressed in opossum kidney cells. Yu et al. (85) showed that FGF-23 is bound to the c splice isoform of FGFRs 1–4. Yan et al. (86) showed that FGF-23 binds FGFR2 in opossum kidney cells but also presented evidence for a novel FGF-23 receptor on the basolateral surface of the cells. Kurosu et al. (87) recently demonstrated that the binding of FGF-23 to FGFRs 1c, 3c, and 4 is enhanced in the presence of klotho, a membrane protein that shares sequence similarity with the β-glucosidase enzymes. The published data suggest that FGF-23 binds to the c splice isoforms of FGFRs and that the interaction is enhanced by klotho. That other, novel receptors for FGF-23 may be present in cells that respond to FGF-23 cannot be excluded.

FGF-23 activates the mitogen-activated protein kinase (MAPK) pathway, which is the major intracellular signaling pathway of FGF-23 (84). An inhibitor for tyrosine kinases of the FGF-23 receptor, SU 5402, blocks the activity of FGF-23 (84). Additionally, inhibitors of the MAPK pathway, PD98059 and SB203580, also block the activity of FGF-23. Given that klotho increases the affinity of FGF-23 for FGFRs, it is interesting that klotho significantly enhances the ability of FGF-23 to induce phosphorylation of a FGF receptor substrate and extracellular signal–regulated kinase (ERK) in various types of cells (87). FGF-23 causes a redistribution and internalization of Na^+-P_i IIa cotransporters on the surface of renal epithelial cells (71, 86, 88).

Regulation of FGF-23 by P_i. From a physiological perspective, it would be appropriate for FGF-23 concentrations to be regulated by the intake of dietary phosphorus and by serum P_i concentrations. Studies in humans as well as animal models have examined FGF-23 concentrations following changes in dietary P_i intake. In humans, short-term alterations in dietary P_i intake do not influence FGF-23 concentrations. Larsson et al. (78) fed human subjects normal, high-P_i, or low-P_i diets for 72 h. FGF-23 concentrations did not change substantially, suggesting that dietary P_i does not regulate FGF-23 concentrations. In a subsequent study, Ferrari et al. (89) administered a high- or a low-P_i diet to humans, with concomitant changes in dietary calcium designed to minimize changes in PTH concentrations (89). Modest decreases or increases within normal range in FGF-23 concentrations were observed following the administration of a low- or high-P_i diet, respectively. In neither of these studies were short-term changes in urinary P_i excretion studied to determine whether temporal changes in the renal excretion of P_i directly correlated with temporal changes in FGF-23 concentrations. Thus, in humans, dietary variation in P_i intake apparently has no effect, or at most an extremely modest effect, on P_i excretion in the kidney. A recent study has shown that the administration to humans of a high-P_i meal is associated with an increase in the fractional excretion of P_i within an hour of ingestion of that meal (90). In this study, there were modest changes in PTH concentrations and no changes in FGF-23 concentrations until eight hours after eating the meal, suggesting that changes in P_i excretion by the kidney were unrelated to changes in FGF-23 concentrations. These data suggest that early and rapid changes in renal P_i excretion occur following a high-P_i meal and are independent of FGF-23 concentration.

Perwad et al. (91) have shown that in mice dietary P_i intake influences FGF-23 concentrations. Within five days, a high-P_i diet increased, and a low-P_i diet decreased, serum FGF-23 concentrations in these animals. The changes in serum FGF-23 concentrations in these mice were correlated with changes in serum P_i concentrations. Similar studies from our laboratory suggest that, within 24 h, serum FGF-23 and PTH concentrations increase in response to increased dietary P_i intake and decrease in response to low P_i intake in rodents. However, serum FGF-23 concentrations did not correlate with serum P_i in animals fed a high-P_i diet (92).

Regulation of FGF-23 by $1\alpha,25(OH)_2D_3$. $1\alpha,25(OH)_2D_3$ regulates FGF-23 synthesis (93–96). Increasing doses of $1\alpha,25(OH)_2D_3$ proportionately increase FGF-23 serum concentrations. $1\alpha,25(OH)_2D_3$ administration is also associated with increases in serum P_i concentrations. The elevated concentrations of serum P_i may directly inhibit the synthesis of $1\alpha,25(OH)_2D_3$ in the kidney, but increasing FGF-23 concentrations also may inhibit $1\alpha,25(OH)_2D_3$ synthesis. Thus, there may be a negative feedback loop inhibiting $1\alpha,25(OH)_2D_3$ synthesis that involves both P_i and FGF-23. These relationships are summarized in **Figure 6**.

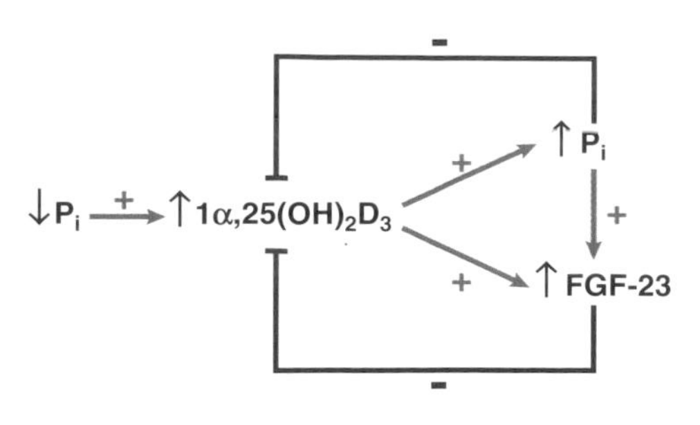

Figure 6

Relationships between concentrations of $1\alpha,25(OH)_2D_3$, serum P_i, and FGF-23.

Secreted Frizzled Related Protein-4

We used serial analysis of gene expression to detect genes that were consistently overexpressed in tumors of patients with TIO (57). sFRP-4 was among the most consistently overexpressed genes found in these tumors. To determine whether sFRP-4 played a role in the pathogenesis of this disorder, we expressed sFRP-4, using recombinant methods, and infused the protein intravenously into rats or mice (30, 56). We also tested the properties of recombinant sFRP-4 by adding it to the medium of opossum kidney cells and examining whether there was an inhibition of Na^+-P_i uptake (56). The intravenous infusion of sFRP-4 into rats increased P_i excretion at both 2 and 8 h; the phosphaturia observed at 8 h was associated with hypophosphatemia. **Figure 4** shows representative results obtained following the intravenous administration of sFRP-4 in mice. There is a dose-dependent increase in the renal fractional excretion of phosphorus in mice given intravenous sFRP-4 at 60 min (clearance 2) and 120 min (clearance 3) after initiation of the sFRP-4 infusion. The effects of sFRP-4 on 25-hydroxyvitamin D 1α-hydroxylase cytochrome P450 messenger RNA concentrations were determined following an 8-h infusion of the protein in rats (56). As noted above, serum P_i concentrations decreased. However, the expected upregulation of 25-hydroxyvitamin D 1α-hydroxylase cytochrome P450 mRNA did not occur, suggesting that sFRP-4 blocked the compensatory upregulation of 25-hydroxyvitamin D 1α-hydroxylase activity and $1\alpha,25(OH)_2D_3$ synthesis.

Mechanism of action of sFRP-4. The secreted frizzled related proteins function as antagonists of the Wnt proteins. Wnt proteins signal in cells by binding to the seven-transmembrane frizzled receptor and to its coreceptor, LRP 5/6. In cells not activated by Wnt, a complex between β-catenin, Axin, APC, and GSK3 causes phosphorylation of β-catenin and its consequent destruction (97–100). Following the binding of Wnt to frizzled receptors and LRP 5/6, phosphorylation of β-catenin is inhibited, and unphosphorylated β-catenin enters the nucleus to activate a variety of genes (97–100). When the secreted frizzled related proteins antagonize Wnt activity, the amount of phosphorylated β-catenin is increased. We demonstrated that the infusion of sFRP-4 into rats was associated with phosphaturia, a concomitant increase in the amount of phospho-β-catenin, and a decrease in the amount of nonphosphorylated β-catenin (56). Thus, sFRP-4 antagonizes Wnt signaling in the kidney. Subsequent studies demonstrated that infusion of sFRP-4 decreased Na^+-P_i cotransporter abundance in the brush border membrane of the proximal tubule and reduced surface expression of the Na^+-P_i IIa cotransporter in the proximal tubules as well as on the surface of opossum kidney cells (**Figures 7** and **8**).

Regulation of sFRP-4 by dietary P_i. sFRP-4 protein concentrations were increased in the homogenates from kidneys of rats that were fed a high-P_i diet for two weeks but not in animals fed a low-P_i diet. This suggests a possible role for sFRP-4 during increases in P_i intake (92).

MEPE

MEPE is also among the most abundantly overexpressed mRNA species found in

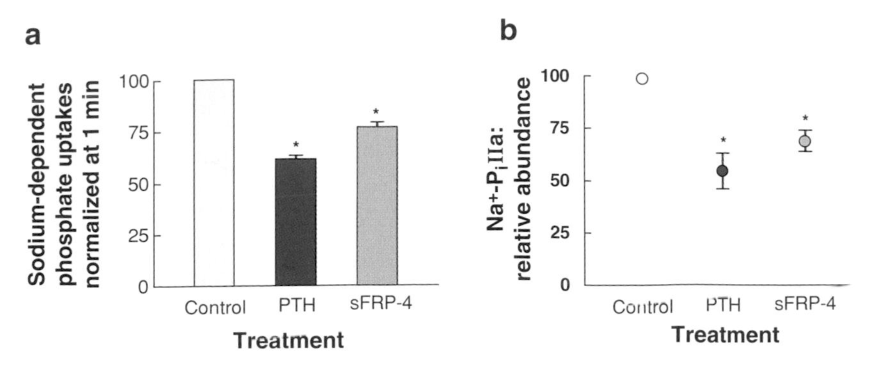

Figure 7

(*a*) Na^+-dependent P_i uptakes into brush border membrane vesicles (BBMV) and Na^+-P_i IIa abundance. Na^+-dependent Pi uptake into renal BBMV prepared from rats infused with vehicle, PTH, or sFRP-4 was measured. Data show uptake rates at 1 min in the PTH- and sFRP-4-treated groups. All data were normalized against the respective control (vehicle-infused) groups. Na^+-dependent P_i uptake following PTH or sFRP-4 treatment is statistically significant, $P < 0.001$, when compared with control. (*b*) Relative abundance of Na^+-P_i IIa protein (Na^+-P_i IIa/actin ratio) in BBMV from the same group of animals used for uptake measurements. Data are normalized and expressed as percentages relative to vehicle-infused rats ± SEM (control group = 100%); $^*P < 0.001$ for PTH compared with control; $^*P < 0.01$ for sFRP-4 versus control as analyzed by ANOVA and Bonferroni's multiple comparison test; $n = 5$–9. From Reference 105 with permission.

tumors associated with renal P_i wasting and osteomalacia (54). Recombinant MEPE expressed in insect cells induces phosphaturia and decreases serum P_i concentrations when administered to mice in vivo (55). Additionally, inhibition of Na^+-dependent P_i uptake was noted in opossum kidney cells incubated in the presence of the recombinant protein. MEPE also inhibits bone mineralization in vitro, and MEPE-null mice have increased bone mineralization (101). Thus, MEPE may be important in the pathogenesis of hypophosphatemia in renal P_i wasting observed in patients with TIO. However, MEPE infusion does not recapitulate the defect in vitamin D metabolism seen in patients with TIO. As noted above, patients with TIO have low serum P_i concentrations and inappropriately reduced or normal concentrations of serum $1\alpha,25(OH)_2D_3$. Infusion of MEPE reduces serum P_i concentrations, and serum $1\alpha,25(OH)_2D_3$ concentrations increase following MEPE infusion, as would be expected in the face of hypophosphatemia (55). Thus, in patients with TIO, it is likely that MEPE contributes to the hypophosphatemia but that other products such as FGF-23 and sFRP-4 inhibit $1\alpha,25(OH)_2D_3$ concentrations by inhibiting the activity of the 25-hydroxyvitamin D 1α-hydroxylase.

MEPE may play a role in the pathogenesis of XLH, in which there is P_i wasting and evidence for a mineralization defect that is independent of low P_i concentrations in the extracellular fluid (50, 51). Recent evidence suggests that MEPE concentration is increased in the bones of mice with the *Hyp* mutation (102). Under normal circumstances, MEPE is proteolyzed to release a peptide containing an ASARM (acidic serine-aspartate-rich motif) sequence. The latter peptide acts as an inhibitor of mineralization. MEPE may be a substrate for PHEX, and PHEX may prevent proteolysis of MEPE and release of the protease-resistant MEPE-ASARM peptide, an inhibitor of mineralization (minhibin) (102). In patients with XLH and in mice with the *Hyp* mutation, PHEX is mutated and therefore cannot bind to either MEPE or the ASARM peptide. This results in the release of MEPE into the circulation, thereby causing hypophosphatemia in renal P_i wasting.

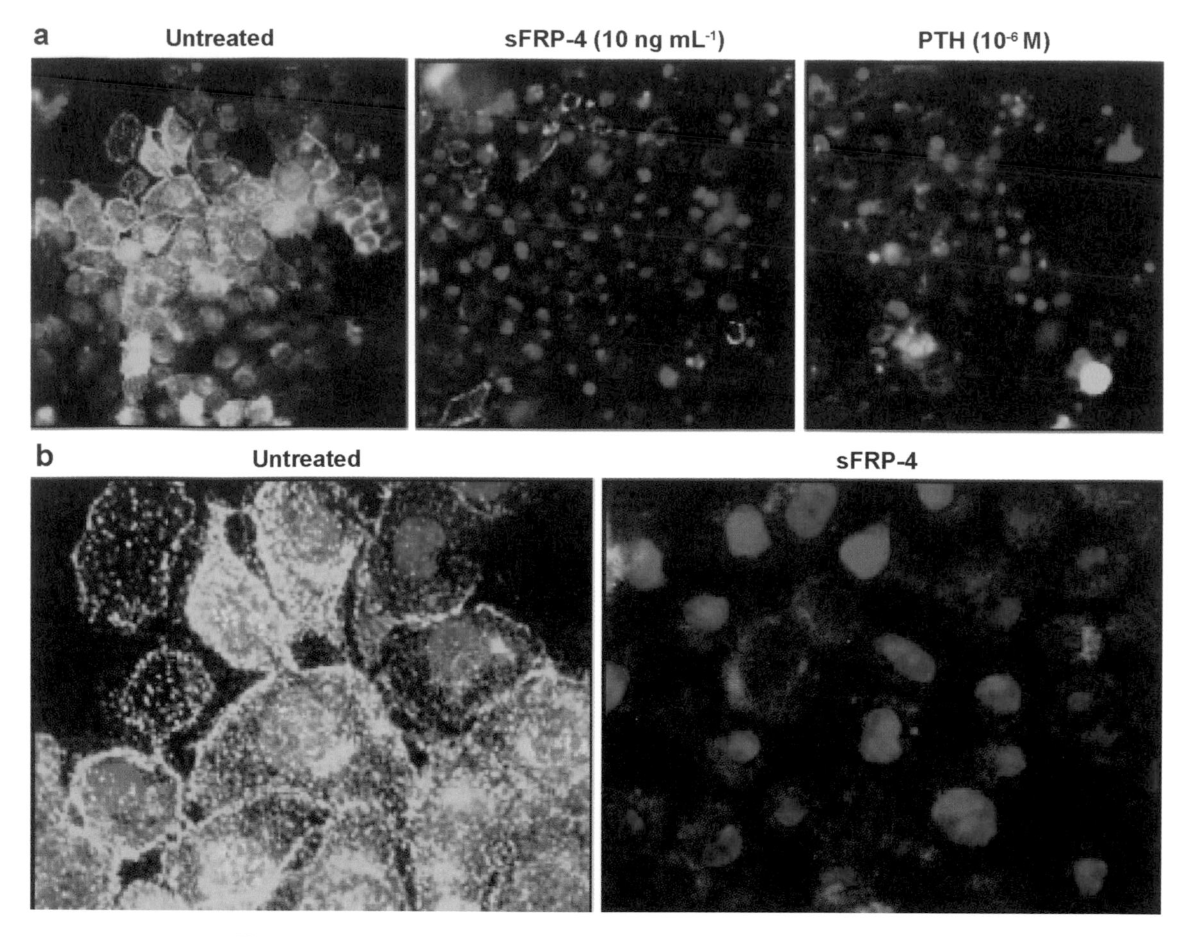

Figure 8

(*a*) Effect of addition of sFRP-4 (10 ng ml^{-1}) or PTH (10^{-6} M) on Na^+-P_i IIa distribution in opossum kidney cells expressing a chimeric Na^+-P_i IIa–V5 transporter (200 times magnification). sFRP-4 or PTH were added to cells in the concentrations indicated for 3 h. Na^+-P_i IIa distribution was examined using an antibody directed against the V5 epitope. (*b*) Effect of sFRP-4 on Na^+-P_i IIa protein distribution in opossum kidney cells expressing a chimeric Na^+-P_i IIa–V5 transporter (400 times magnification). Cells were treated with sFRP-4 (10 ng ml^{-1}) for a period of 3 h. From Reference 105 with permission.

Increased constellations of MEPE-ASARM peptide have been measured in humans with XLH and in *Hyp* mice (103). MEPE concentrations have been measured in normal humans, and concentrations of the protein appear to correlate positively with bone mineral density and serum P_i concentrations (104).

FGF-7

FGF-7, also known as keratinocyte growth factor, is overexpressed in tumors associated with osteomalacia and renal P_i wasting (58). FGF-7 protein inhibited Na^+-dependent P_i transport in opossum kidney cells. Anti-FGF-7 antibodies attenuated the inhibitory effect of tumor supernatants on Na^+-dependent P_i transport. In this study (58), low concentrations of FGF-23 were present in the conditioned medium of tumor cells. FGF-7 is present in normal plasma; however, whether it is elevated in the plasma of subjects with TIO or in response to alterations in dietary P_i intake has not been determined. Nevertheless, the report does point to the complexity of factors involved in the pathogenesis of TIO.

FUTURE ISSUES

Three key issues need to be addressed; they relate to how mammalian organisms respond to changes in P_i intake. A low-P_i dietary state is somewhat artificial because virtually all foods contain substantial amounts of P_i. Indeed, renal P_i reabsorption responses obtained in the fasting state are considerably different than those obtained in situations in which only P_i is (by artificial means) removed from the diet. The adaptation of greatest consequence is that which is required following the ingestion of a high-P_i diet. The compensatory changes that occur should facilitate the excretion of excessive P_i from the body. It would be extremely important to define these mechanisms. In this regard, it is becoming increasingly clear that short-term rapid responses are sensed by, and mediated through, the intestine and that these mechanisms are key to the ability of the mammalian organism to excrete excessive amounts of P_i via the kidney. We believe that the delineation of this "enteric-renal" P_i regulatory pathway requires further definition and study.

A second area of investigation needed is to define the interactions between the phosphatonins. Do these factors act independently, or do they influence one another?

The third area of investigation that will yield considerable insights into the manner in which P_i is transported across epithelia relates to the structure of the Na^+-P_i cotransporter proteins. Structural studies that delineate the three-dimensional topology of these proteins and the manner in which they move P_i across the lipid bilayer will undoubtedly yield significant information about the transport of this important ion.

SUMMARY POINTS

1. P_i homeostasis is preserved during alterations in P_i intake by a variety of phosphaturic peptides.
2. PTH is a key hormone in the regulation of P_i homeostasis.
3. Phosphatonin molecules, initially identified as a result of the study of patients with rare disorders associated with renal P_i excretion, may contribute to the physiological regulation of renal P_i reabsorption.
4. FGF-23, sFRP- 4, MEPE, and FGF-7 all inhibit renal P_i reabsorption.
5. FGF-23 and sFRP-4 synthesis may be regulated by the intake of dietary P_i.
6. Unidentified P_i regulatory factors likely mediate the rapid changes in P_i reabsorption by the kidney in response to alterations in dietary P_i intake.

ACKNOWLEDGMENTS

Work in Dr. Kumar's laboratory is supported by NIH grants DK 65830, DK 73369, and DK 58546.

LITERATURE CITED

1. Cohen P. 1989. The structure and regulation of protein phosphatases. *Annu. Rev. Biochem.* 58:453–508
2. Hubbard SR, Till JH. 2000. Protein tyrosine kinase structure and function. *Annu. Rev. Biochem.* 69:373–98

3. Hunter T, Cooper JA. 1985. Protein-tyrosine kinases. *Annu. Rev. Biochem.* 54:897–930
4. Krebs EG, Beavo JA. 1979. Phosphorylation-dephosphorylation of enzymes. *Annu. Rev. Biochem.* 48:923 59
5. Neuman W. 1980. Bone material and calcification mechanisms. In *Fundamental and Clinical Bone Physiology*, ed. M Urist, pp. 83–107. Philadelphia: J.B. Lippincott Co.
6. Haramati A, Mulroney SE, Webster SK. 1988. Developmental changes in the tubular capacity for phosphate reabsorption in the rat. *Am. J. Physiol.* 255:F287–91
7. Mulroney SE, Lumpkin MD, Haramati A. 1989. Antagonist to GH-releasing factor inhibits growth and renal Pi reabsorption in immature rats. *Am. J. Physiol.* 257:F29–34
8. Knochel JP. 1977. The pathophysiology and clinical characteristics of severe hypophosphatemia. *Arch. Intern. Med.* 137:203–20
9. Knochel JP, Barcenas C, Cotton JR, Fuller TJ, Haller R, Carter NW. 1978. Hypophosphatemia and rhabdomyolysis. *Trans. Assoc. Am. Physiol.* 91:156–68
10. Slatopolsky E. 2003. New developments in hyperphosphatemia management. *J. Am. Soc. Nephrol.* 14:5297–99
11. Slatopolsky E, Bricker NS. 1973. The role of phosphorus restriction in the prevention of secondary hyperparathyroidism in chronic renal disease. *Kidney Int.* 4:141–45
12. Slatopolsky E, Caglar S, Gradowska L, Canterbury J, Reiss E, Bricker NS. 1972. On the prevention of secondary hyperparathyroidism in experimental chronic renal disease using "proportional reduction" of dietary phosphorus intake. *Kidney Int.* 2:147–51
13. Slatopolsky E, Gradowska L, Kashemsant C, Keltner R, Manley C, Bricker NS. 1966. The control of phosphate excretion in uremia. *J. Clin. Invest.* 45:672–77
14. Slatopolsky E, Robson AM, Elkan I, Bricker NS. 1968. Control of phosphate excretion in uremic man. *J. Clin. Invest.* 47:1865–74
15. Werner A, Kinne RK. 2001. Evolution of the Na-P(i) cotransport systems. *Am. J. Physiol. Regul. Integr. Comp. Physiol.* 280:R301–12
16. Murer H. 1992. Homer Smith Award. Cellular mechanisms in proximal tubular Pi reabsorption: some answers and more questions. *J. Am. Soc. Nephrol.* 2:1649–65
17. Murer H, Biber J. 1996. Molecular mechanisms of renal apical Na/phosphate cotransport. *Annu. Rev. Physiol.* 58:607–18
18. Murer H, Hernando N, Forster I, Biber J. 2000. Proximal tubular phosphate reabsorption: molecular mechanisms. *Physiol. Rev.* 80:1373–409
19. Murer H, Hernando N, Forster I, Biber J. 2003. Regulation of Na/Pi transporter in the proximal tubule. *Annu. Rev. Physiol.* 65:531–42
20. Hernando N, Biber J, Forster I, Murer H. 2005. Recent advances in renal phosphate transport. *Ther. Apher. Dial.* 9:323–27
21. Madjdpour C, Bacic D, Kaissling B, Murer H, Biber J. 2004. Segment-specific expression of sodium-phosphate cotransporters NaPi-IIa and -IIc and interacting proteins in mouse renal proximal tubules. *Pflügers Arch.* 448:402–10
22. Ohkido I, Segawa H, Yanagida R, Nakamura M, Miyamoto K. 2003. Cloning, gene structure and dietary regulation of the type-IIc Na/Pi cotransporter in the mouse kidney. *Pflügers Arch.* 446:106–15
23. Tanaka Y, Deluca HF. 1974. Role of 1,25-dihydroxyvitamin D3 in maintaining serum phosphorus and curing rickets. *Proc. Natl. Acad. Sci. USA* 71:1040–44
24. Tanaka Y, Frank H, DeLuca HF. 1972. Role of 1,25-dihydroxycholecalciferol in calcification of bone and maintenance of serum calcium concentration in the rat. *J. Nutr.* 102:1569–77
25. Tanaka Y, Frank H, DeLuca HF. 1973. Biological activity of 1,25-dihydroxyvitamin D3 in the rat. *Endocrinology* 92:417–22

26. Tanaka Y, Frank H, DeLuca HF. 1973. Intestinal calcium transport: stimulation by low phosphorus diets. *Science* 181:564–66
27. Tanaka Y, Deluca HF. 1973. The control of 25-hydroxyvitamin D metabolism by inorganic phosphorus. *Arch. Biochem. Biophys.* 154:566–74
28. Segawa H, Kaneko I, Yamanaka S, Ito M, Kuwahata M, et al. 2004. Intestinal Na-P_i cotransporter adaptation to dietary P_i content in vitamin D receptor null mice. *Am. J. Physiol. Renal Physiol.* 287:F39–47
29. Berndt T, Knox F. 1992. Renal regulation of phosphate excretion. In *The Kidney: Physiology and Pathophysiology*, ed. D Seldin, G Giebisch, pp. 2511–32. New York: Raven Press
30. Berndt TJ, Schiavi S, Kumar R. 2005. "Phosphatonins" and the regulation of phosphorus homeostasis. *Am. J. Physiol. Renal Physiol.* 289:F1170–82
31. Chase LR, Aurbach GD. 1967. Parathyroid function and the renal excretion of 3′5′-adenylic acid. *Proc. Natl. Acad. Sci. USA* 58:518–25
32. Aurbach GD, Heath DA. 1974. Parathyroid hormone and calcitonin regulation of renal function. *Kidney Int.* 6:331–45
33. Aurbach GD, Keutmann HT, Niall HD, Tregear GW, O'Riordan JL, et al. 1972. Structure, synthesis, and mechanism of action of parathyroid hormone. *Recent Prog. Horm. Res.* 28:353–98
34. Condamine L, Menaa C, Vrtovsnik F, Friedlander G, Garabedian M. 1994. Local action of phosphate depletion and insulin-like growth factor 1 on in vitro production of 1,25-dihydroxyvitamin D by cultured mammalian kidney cells. *J. Clin. Invest.* 94:1673–79
35. Martin DR, Ritter CS, Slatopolsky E, Brown AJ. 2005. Acute regulation of parathyroid hormone by dietary phosphate. *Am. J. Physiol. Endocrinol. Metab.* 289:E729–34
36. Brown AJ, Koch MJ, Coyne DW. 2006. Oral feeding acutely down-regulates serum PTH in hemodialysis patients. *Nephron. Clin. Pract.* 103:c106–13
37. Lorenz JN, Nieman M, Sabo J, Sanford LP, Hawkins JA, et al. 2003. Uroguanylin knockout mice have increased blood pressure and impaired natriuretic response to enteral NaCl load. *J. Clin. Invest.* 112:1244–54
38. Fan X, Wang Y, London RM, Eber SL, Krause WJ, et al. 1997. Signaling pathways for guanylin and uroguanylin in the digestive, renal, central nervous, reproductive, and lymphoid systems. *Endocrinology* 138:4636–48
39. Potthast R, Ehler E, Scheving LA, Sindic A, Schlatter E, Kuhn M. 2001. High salt intake increases uroguanylin expression in mouse kidney. *Endocrinology* 142:3087–97
40. Landsman A, Lichtstein D, Ilani A. 2005. Distinctive features of dietary phosphate supply. *J. Appl. Physiol.* 99:1214–19
41. Landsman A, Lichtstein D, Bacaner M, Ilani A. 2001. Dietary phosphate-dependent growth is not mediated by changes in plasma phosphate concentration. *Br. J. Nutr.* 86:217–23
42. Rybczynska A, Hoppe A, Knox FG. 1990. Propranolol restores phosphaturic effect of PTH in short-term phosphate deprivation. *Am. J. Physiol.* 258:R120–23
43. Beck N, Webster SK, Reineck HJ. 1979. Effect of fasting on tubular phosphorus reabsorption. *Am. J. Physiol.* 237:F241–46
44. Kempson SA, Shah SV, Werness PG, Berndt T, Lee PH, et al. 1980. Renal brush border membrane adaptation to phosphorus deprivation: effects of fasting versus low-phosphorus diet. *Kidney Int.* 18:36–47
45. Markovich D, Verri T, Sorribas V, Forgo J, Biber J, Murer H. 1995. Regulation of opossum kidney (OK) cell Na/Pi cotransport by Pi deprivation involves mRNA stability. *Pflügers Arch.* 430:459–63

46. Cai Q, Hodgson SF, Kao PC, Lennon VA, Klee GG, et al. 1994. Brief report: inhibition of renal phosphate transport by a tumor product in a patient with oncogenic osteomalacia. *N. Engl. J. Med.* 330:1645–49
47. Econs MJ, Drezner MK. 1994. Tumor-induced osteomalacia—unveiling a new hormone. *N. Engl. J. Med.* 330:1679–81
48. Kumar R. 1997. Phosphatonin—a new phosphaturetic hormone? (lessons from tumor-induced osteomalacia and X-linked hypophosphataemia). *Nephrol. Dial. Transplant.* 12:11–13
49. Kumar R. 2000. Tumor-induced osteomalacia and the regulation of phosphate homeostasis. *Bone* 27:333–38
50. Drezner MK. 2003. Hypophosphatemic rickets. *Endocr. Dev.* 6:126–55
51. Drezner MK. 2000. PHEX gene and hypophosphatemia. *Kidney Int.* 57:9–18
52. Francis F, Hennig S, Korn B, Reinhardt R, de Jong P, et al. 1995. A gene (PEX) with homologies to endopeptidases is mutated in patients with X-linked hypophosphatemic rickets. *Nat. Genet.* 11:130–36
53. ADHR Consortium. 2000. Autosomal dominant hypophosphataemic rickets is associated with mutations in FGF23. *Nat. Genet.* 26:345–48
54. Rowe PS, de Zoysa PA, Dong R, Wang HR, White KE, et al. 2000. MEPE, a new gene expressed in bone marrow and tumors causing osteomalacia. *Genomics* 67:54–68
55. Rowe PS, Kumagai Y, Gutierrez G, Garrett IR, Blacher R, et al. 2004. MEPE has the properties of an osteoblastic phosphatonin and minhibin. *Bone* 34:303–19
56. Berndt T, Craig TA, Bowe AE, Vassiliadis J, Reczek D, et al. 2003. Secreted frizzled-related protein 4 is a potent tumor-derived phosphaturic agent. *J. Clin. Invest.* 112:785–94
57. De Beur SM, Finnegan RB, Vassiliadis J, Cook B, Barberio D, et al. 2002. Tumors associated with oncogenic osteomalacia express genes important in bone and mineral metabolism. *J. Bone Miner. Res.* 17:1102–10
58. Carpenter TO, Ellis BK, Insogna KL, Philbrick WM, Sterpka J, Shimkets R. 2005. FGF7: an inhibitor of phosphate transport derived from oncogenic osteomalacia-causing tumors. *J. Clin. Endocrinol. Metab.* 90:1012–20
59. Deleted in proof
60. Schiavi SC, Kumar R. 2004. The phosphatonin pathway: new insights in phosphate homeostasis. *Kidney Int.* 65:1–14
61. Schiavi SC, Moe OW. 2002. Phosphatonins: a new class of phosphate-regulating proteins. *Curr. Opin. Nephrol. Hypertens.* 11:423–30
62. Shimada T, Muto T, Urakawa I, Yoneya T, Yamazaki Y, et al. 2002. Mutant FGF-23 responsible for autosomal dominant hypophosphatemic rickets is resistant to proteolytic cleavage and causes hypophosphatemia in vivo. *Endocrinology* 143:3179–82
63. White KE, Carn G, Lorenz-Depiereux B, Benet-Pages A, Strom TM, Econs MJ. 2001. Autosomal-dominant hypophosphatemic rickets (ADHR) mutations stabilize FGF-23. *Kidney Int.* 60:2079–86
64. Shimada T, Mizutani S, Muto T, Yoneya T, Hino R, et al. 2001. Cloning and characterization of FGF23 as a causative factor of tumor-induced osteomalacia. *Proc. Natl. Acad. Sci. USA* 98:6500–5
65. Bowe AE, Finnegan R, Jan de Beur SM, Cho J, Levine MA, et al. 2001. FGF-23 inhibits renal tubular phosphate transport and is a PHEX substrate. *Biochem. Biophys. Res. Commun.* 284:977–81
66. Takeuchi Y, Suzuki H, Ogura S, Imai R, Yamazaki Y, et al. 2004. Venous sampling for fibroblast growth factor-23 confirms preoperative diagnosis of tumor-induced osteomalacia. *J. Clin. Endocrinol. Metab.* 89:3979–82

67. Yamazaki Y, Okazaki R, Shibata M, Hasegawa Y, Satoh K, et al. 2002. Increased circulatory level of biologically active full-length FGF-23 in patients with hypophosphatemic rickets/osteomalacia. *J. Clin. Endocrinol. Metab.* 87:4957–60
68. Larsson T, Zahradnik R, Lavigne J, Ljunggren O, Juppner H, Jonsson KB. 2003. Immunohistochemical detection of FGF-23 protein in tumors that cause oncogenic osteomalacia. *Eur. J. Endocrinol.* 148:269–76
69. Jonsson KB, Zahradnik R, Larsson T, White KE, Sugimoto T, et al. 2003. Fibroblast growth factor 23 in oncogenic osteomalacia and X-linked hypophosphatemia. *N. Engl. J. Med.* 348:1656–63
70. Larsson T, Marsell R, Schipani E, Ohlsson C, Ljunggren O, et al. 2004. Transgenic mice expressing fibroblast growth factor 23 under the control of the alpha1(I) collagen promoter exhibit growth retardation, osteomalacia, and disturbed phosphate homeostasis. *Endocrinology* 145:3087–94
71. Shimada T, Urakawa I, Yamazaki Y, Hasegawa H, Hino R, et al. 2004. FGF-23 transgenic mice demonstrate hypophosphatemic rickets with reduced expression of sodium phosphate cotransporter type IIa. *Biochem. Biophys. Res. Commun.* 314:409–14
72. Shimada T, Kakitani M, Yamazaki Y, Hasegawa H, Takeuchi Y, et al. 2004. Targeted ablation of Fgf23 demonstrates an essential physiological role of FGF23 in phosphate and vitamin D metabolism. *J. Clin. Invest.* 113:561–68
73. Meyer RAJ, Tenenhouse HS, Meyer MH, Klugerman AH. 1989. The renal phosphate transport defect in normal mice parabiosed to X-linked hypophosphatemic mice persists after parathyroidectomy. *J. Bone Miner. Res.* 4:523–32
74. Meyer RAJ, Meyer MH, Gray RW. 1989. Parabiosis suggests a humoral factor is involved in X-linked hypophosphatemia in mice. *J. Bone Miner. Res.* 4:493–500
75. Nesbitt T, Coffman TM, Griffiths R, Drezner MK. 1992. Crosstransplantation of kidneys in normal and Hyp mice. Evidence that the Hyp mouse phenotype is unrelated to an intrinsic renal defect. *J. Clin. Invest.* 89:1453–59
76. Campos M, Couture C, Hirata IY, Juliano MA, Loisel TP, et al. 2003. Human recombinant endopeptidase PHEX has a strict S1′ specificity for acidic residues and cleaves peptides derived from fibroblast growth factor-23 and matrix extracellular phosphoglycoprotein. *Biochem. J.* 373:271–79
77. Liu S, Zhou J, Tang W, Jiang X, Rowe DW, Quarles LD. 2006. Pathogenic role of FGF23 in *Hyp* mice. *Am. J. Physiol. Endocrinol. Metab.* 298:E38–49
78. Larsson T, Nisbeth U, Ljunggren O, Juppner H, Jonsson KB. 2003. Circulating concentration of FGF-23 increases as renal function declines in patients with chronic kidney disease, but does not change in response to variation in phosphate intake in healthy volunteers. *Kidney Int.* 64:2272–79
79. Topaz O, Shurman DL, Bergman R, Indelman M, Ratajczak P, et al. 2004. Mutations in GALNT3, encoding a protein involved in O-linked glycosylation, cause familial tumoral calcinosis. *Nat. Genet.* 36:579–81
80. Araya K, Fukumoto S, Backenroth R, Takeuchi Y, Nakayama K, et al. 2005. A novel mutation in fibroblast growth factor (FGF)23 gene as a cause of tumoral calcinosis. *J. Clin. Endocrinol. Metab.* 90:5523–37
81. Larsson T, Davis SI, Garringer HJ, Mooney SD, Draman MS, et al. 2005. Fibroblast growth factor-23 mutants causing familial tumoral calcinosis are differentially processed. *Endocrinology* 146:3883–91
82. Benet-Pages A, Orlik P, Strom TM, Lorenz-Depiereux B. 2005. An FGF23 missense mutation causes familial tumoral calcinosis with hyperphosphatemia. *Hum. Mol. Genet.* 14:385–90

83. Larsson T, Yu X, Davis SI, Draman MS, Mooney SD, et al. 2005. A novel recessive mutation in fibroblast growth factor-23 causes familial tumoral calcinosis. *J. Clin. Endocrinol. Metab.* 90:2424–27
84. Yamashita T, Konishi M, Miyake A, Inui K, Itoh N. 2002. Fibroblast growth factor (FGF)-23 inhibits renal phosphate reabsorption by activation of the mitogen-activated protein kinase pathway. *J. Biol. Chem.* 277:28265–70
85. Yu X, Ibrahimi OA, Goetz R, Zhang F, Davis SI, et al. 2005. Analysis of the biochemical mechanisms for the endocrine actions of fibroblast growth factor-23. *Endocrinology* 146:4647–56
86. Yan X, Yokote H, Jing X, Yao L, Sawada T, et al. 2005. Fibroblast growth factor 23 reduces expression of type IIa Na^+/P_i cotransporter by signaling through a receptor functionally distinct from the known FGFRs in opossum kidney cells. *Genes Cells* 10:489–502
87. Kurosu H, Ogawa Y, Miyoshi M, Yamamoto M, Nandi A, et al. 2006. Regulation of fibroblast growth factor-23 signaling by klotho. *J. Biol. Chem.* 281:6120–23
88. Baum M, Schiavi S, Dwarakanath V, Quigley R. 2005. Effect of fibroblast growth factor-23 on phosphate transport in proximal tubules. *Kidney Int.* 68:1148–53
89. Ferrari SL, Bonjour JP, Rizzoli R. 2005. Fibroblast growth factor-23 relationship to dietary phosphate and renal phosphate handling in healthy young men. *J. Clin. Endocrinol. Metab.* 90:1519–24
90. Nishida Y, Yamanaka-Okumura H, Taketani Y, Sato T, Nashiki K, et al. 2005. *Postprandial changes of serum fibroblast growth factor 23 (FGF23) levels on high phosphate diet in healthy men*. Presented at Annu. Meet. ASBMR, 27th, Nashville
91. Perwad F, Azam N, Zhang MY, Yamashita T, Tenenhouse HS, Portale AA. 2005. Dietary and serum phosphorus regulate fibroblast growth factor 23 expression and 1,25-dihydroxyvitamin D metabolism in mice. *Endocrinology* 146:5358–64
92. Sommer S, Berndt T, Craig TA, Kumar R. 2006. The phosphatonins and the regulation of phosphate transport and vitamin D metabolism. *J. Steroid Biochem. Mol. Biol.* In press
93. Saito H, Maeda A, Ohtomo S, Hirata M, Kusano K, et al. 2005. Circulating FGF-23 is regulated by 1α,25-dihydroxyvitamin D3 and phosphorus in vivo. *J. Biol. Chem.* 280:2543–49
94. Kolek OI, Hines ER, Jones MD, LeSueur LK, Lipko MA, et al. 2005. 1α,25-Dihydroxyvitamin D3 upregulates FGF23 gene expression in bone: the final link in a renal-gastrointestinal-skeletal axis that controls phosphate transport. *Am. J. Physiol. Gastrointest. Liver Physiol.* 289:G1036–42
95. Collins MT, Lindsay JR, Jain A, Kelly MH, Cutler CM, et al. 2005. Fibroblast growth factor-23 is regulated by 1α,25-dihydroxyvitamin D. *J. Bone Miner. Res.* 20:1944–50
96. Liu S, Tang W, Zhou J, Stubbs JR, Luo Q, et al. 2006. Fibroblast growth factor 23 is a counter-regulatory phosphaturic hormone for vitamin D. *J. Am. Soc. Nephrol.* 17:1305–15
97. Nusse R. 2005. Cell biology: relays at the membrane. *Nature* 438:747–49
98. Nusse R. 2005. Wnt signaling in disease and in development. *Cell Res.* 15:28–32
99. Logan CY, Nusse R. 2004. The Wnt signaling pathway in development and disease. *Annu. Rev. Cell Dev. Biol.* 20:781–810
100. Nusse R. 2003. Wnts and Hedgehogs: lipid-modified proteins and similarities in signaling mechanisms at the cell surface. *Development* 130:5297–305
101. Gowen LC, Petersen DN, Mansolf AL, Qi H, Stock JL, et al. 2003. Targeted disruption of the osteoblast/osteocyte factor 45 gene (OF45) results in increased bone formation and bone mass. *J. Biol. Chem.* 278:1998–2007

102. Rowe PS, Garrett IR, Schwarz PM, Carnes DL, Lafer EM, et al. 2005. Surface plasmon resonance (SPR) confirms that MEPE binds to PHEX via the MEPE-ASARM motif: a model for impaired mineralization in X-linked rickets (HYP). *Bone* 36:33–46
103. Bresler D, Bruder J, Mohnike K, Fraser WD, Rowe PS. 2004. Serum MEPE-ASARM-peptides are elevated in X-linked rickets (HYP): implications for phosphaturia and rickets. *J. Endocrinol.* 183:R1–9
104. Jain A, Fedarko NS, Collins MT, Gelman R, Ankrom MA, et al. 2004. Serum levels of matrix extracellular phosphoglycoprotein (MEPE) in normal humans correlate with serum phosphorus, parathyroid hormone and bone mineral density. *J. Clin. Endocrinol. Metab.* 89:4158–61
105. Berndt TJ, Bielesz B, Craig TA, Tebben PJ, Bacic D, et al. 2006. Secreted frizzled related protein-4 reduces sodium-phosphate co-transporter abundance and activity in proximal tubule cells. *Pflügers Arch.* 451(4):579–87

Specificity and Regulation of Renal Sulfate Transporters

Daniel Markovich[1] and Peter S. Aronson[2]

[1]Department of Physiology and Pharmacology, School of Biomedical Sciences, University of Queensland, Brisbane, QLD 4072 Australia; email: d.markovich@uq.edu.au

[2]Department of Internal Medicine & Department of Cellular and Molecular Physiology, Yale University School of Medicine, New Haven, Connecticut 06520-8029; email: peter.aronson@yale.edu

Annu. Rev. Physiol. 2007. 69:361–75

First published online as a Review in Advance on September 14, 2006

The *Annual Review of Physiology* is online at http://physiol.annualreviews.org

This article's doi: 10.1146/annurev.physiol.69.040705.141319

0066-4278/07/0315-0361$20.00

Key Words

oxalate transport, renal reabsorption, anion exchange, NaS1, sat1, cfex, SLC13A1, SLC26A1, SLC26A6

Abstract

Sulfate is essential for normal cellular function. The kidney plays a major role in sulfate homeostasis. Sulfate is freely filtered and then undergoes net reabsorption in the proximal tubule. The apical membrane Na^+/sulfate cotransporter NaS1 (SLC13A1) has a major role in mediating proximal tubule sulfate reabsorption, as demonstrated by the findings of hyposulfatemia and hypersulfaturia in Nas1-null mice. The anion exchanger SAT1 (SLC26A1), the founding member of the SLC26 sulfate transporter family, mediates sulfate exit across the basolateral membrane to complete the process of transtubular sulfate reabsorption. Another member of this family, CFEX (SLC26A6), is present at the apical membrane of proximal tubular cells. It also can transport sulfate by anion exchange, which probably mediates backflux of sulfate into the lumen. Knockout mouse studies have demonstrated a major role of CFEX as an apical membrane Cl^-/oxalate exchanger that contributes to NaCl reabsorption in the proximal tubule. Several additional SLC26 family members mediate sulfate transport and show some level of renal expression (e.g., SLC26A2, SLC26A7, SLC26A11). Their roles in mediating renal tubular sulfate transport are presently unknown. This paper reviews current data available on the function and regulation of three sulfate transporters (NaS1, SAT1, and CFEX) and their physiological roles in the kidney.

INTRODUCTION

All cells require inorganic sulfate (SO_4^{2-}) for normal function. SO_4^{2-} is among the most important macronutrients in cells and is the fourth most abundant anion in human plasma (≈300 μM). SO_4^{2-} is the major sulfur source in many organisms. Because SO_4^{2-} is a hydrophilic anion that cannot passively cross the lipid bilayer of cell membranes, all cells require a mechanism for SO_4^{2-} influx and efflux to ensure an optimal supply of SO_4^{2-} in the body. SO_4^{2-} transporters are the class of proteins involved in moving SO_4^{2-} across the plasma membranes of cells. The most widely studied organ in terms of SO_4^{2-} transport has been the mammalian kidney. Early studies demonstrated that SO_4^{2-} was freely filtered and extensively reabsorbed by the kidneys (1). Stop-flow experiments proposed that the kidney proximal tubule is the major site of active SO_4^{2-} reabsorption (2). The kidney reabsorbs most filtered SO_4^{2-}; fractional excretion is approximately 10% (1). The active SO_4^{2-} reabsorption process is capacity limited and saturable (3). Under physiological conditions, tubular SO_4^{2-} reabsorption is near the maximal rate (4), whereas if plasma SO_4^{2-} increases, the filtered load of SO_4^{2-} quickly exceeds the maximal tubular reabsorption, and SO_4^{2-} is excreted in the urine. Renal clearance of SO_4^{2-} increases with increasing serum SO_4^{2-} concentrations, reaching approximately the glomerular filtration rate (GFR). At physiological serum SO_4^{2-} concentrations of 700–1000 μM in the rat (5, 6), SO_4^{2-} renal clearance is approximately 30% of the GFR.

In proximal tubule cells, SO_4^{2-} transport systems have been extensively characterized on the luminal brush-border membrane (BBM) and the contraluminal basolateral membrane (BLM) by microperfusion studies in vivo (7–15), with isolated tubules (16), and by membrane vesicle studies (17–22). The principal pathway for secondary active SO_4^{2-} uptake across the BBM is sodium-coupled sulfate transport (Na^+/SO_4^{2-} cotransport) driven by the luminal membrane Na^+ gradient (21–24). The influx of SO_4^{2-} into the proximal tubular cell generates an outward SO_4^{2-} gradient. Early studies demonstrated that exit of SO_4^{2-} across the BLM occurs via a SO_4^{2-}/HCO_3^- exchanger (16, 18, 20, 25), which completes the process of transcellular SO_4^{2-} reabsorption in the proximal tubule. As a result, HCO_3^- is driven into the proximal tubular cell across the BLM, thereby contributing to intracellular pH regulation (i.e., raising the pH). The same scenario occurs across the BBM: Backflux of SO_4^{2-} from the proximal tubular cell to the tubular lumen via a SO_4^{2-}/HCO_3^- exchanger (26–28) for HCO_3^- also drives HCO_3^- into the cell, thereby raising intracellular pH. The exit of HCO_3^- from the proximal tubule cell, by Na^+/HCO_3^- cotransport across the BLM, counteracts the gain in intracellular HCO_3^- (29).

RENAL SULFATE TRANSPORT PROCESSES

Renal Brush-Border Membrane SO_4^{2-} Transport

Early transport studies using radiotracer ^{35}S-sulfate in brush-border membrane vesicles (BBMVs) demonstrated the presence of a Na^+-dependent SO_4^{2-} transport system (8, 21, 22). In rat renal BBMVs, the affinity (K_m) for this transport system was estimated to be 600 μM for SO_4^{2-} and 36 mM for Na^+, with a Hill-coefficient (n) of approximately 1.6 (21). This system was kinetically characterized to transport two Na^+ ions with one SO_4^{2-} ion, a stoichiometry consistent with an electroneutral transport mechanism (21, 22). This electroneutral kinetic model was favored for many years until the molecular isolation of this protein structure in 1993 (30), when a more detailed functional characterization observed this protein to be electrogenic (31). Transport characterization of the cloned renal BBM Na^+/SO_4^{2-} cotransporter

(NaS1) (30) by electrophysiological recordings (31) demonstrated a kinetic model favoring the transport of three Na^+ ions for one SO_4^{2-} ion. Regarding substrate specificity, the BBMV Na^+/SO_4^{2-} cotransport activity was inhibited by thiosulfate, molybdate, and other tetra-oxyanions (including chromate and selenate) but not by phosphate or tungstate (8), suggesting that this transport system was unable to mediate phosphate transport.

In addition to the Na^+/SO_4^{2-} cotransport system present on the BBM, SO_4^{2-} can be transported by an anion exchange mechanism across the luminal membrane of BBMVs isolated from rat (26), bovine (27), and rabbit (28, 32) kidney cortex. This transport system mediates an electroneutral anion exchange and transports HCO_3^- and oxalate in addition to SO_4^{2-} (28). Chloride also is capable of exchanging with SO_4^{2-}, although not as well as is HCO_3^- (26, 32). Upon its cloning, this anion exchanger, CFEX (SLC26A6), had affinity for SO_4^{2-}, HCO_3^-, Cl^-, and oxalate as substrates (33–36) and was localized on the BBM of proximal tubular cells (37, 38). The overall importance and contribution of this transport activity to SO_4^{2-} reabsorption in proximal tubular cells are still unknown.

Renal Basolateral Membrane Sulfate Transport

The proximal tubular BLM possesses an anion exchange SO_4^{2-} transport system that transports SO_4^{2-} out of the cell into the systemic circulation and exchanges SO_4^{2-} with thiosulfate, hydroxyl (OH^-) ions, HCO_3^-, oxalate, and various organic ions (16, 18, 20, 25). In rat and rabbit renal basolateral membrane vesicles (BLMVs), the SO_4^{2-} anion exchanger shows specificity for the exchange of HCO_3^-, OH^-, thiosulfate, and SO_4^{2-} ions but not Cl^-, phosphate, lactate, or PAH (para-amino hippurate) (20, 25, 39). Furthermore, stilbene derivatives DIDS (4,4′-diisothiocyanato-2,2′-disulfonate) and SITS (4-acetamido-4′-isothiocyanato-2,2′-disulfonate), which are inhibitors of anion exchangers (7, 16, 20, 23, 40, 41) and other proteins, may inhibit SO_4^{2-} uptake into BLMVs (17, 23, 39). This BLM transport system has different transport kinetics and a different sensitivity to inhibitors than the BBM SO_4^{2-} anion exchanger (7), suggesting that the two systems are not encoded by the same protein. The apparent K_i (inhibition constant) for DIDS, probenecid, phenol red, and oxalate (among others) on SO_4^{2-} uptake was more than threefold higher for the BBM SO_4^{2-} anion exchanger than for the BLM SO_4^{2-} anion exchanger. This suggests that the BLM transporter is more sensitive to inhibition by these compounds than is the BBM SO_4^{2-} transporter (7). Furthermore, upon cloning of the BLM SO_4^{2-} anion exchanger 1 (sat1) (41), antibodies raised against this protein showed immunocytological staining only on the BLM, suggesting that this protein is structurally (in addition to functionally) different from the BBM SO_4^{2-} anion exchanger (CFEX).

Despite the identification of several different SO_4^{2-} transport systems in renal proximal tubules, it is unclear which of these plays the predominant role in SO_4^{2-} homeostasis. Because the active reabsorption process depends on normal Na^+/K^+ATPase activity in the proximal tubular cells (42), a Na^+ gradient may be more important than anion exchange in SO_4^{2-} uptake, and the Na^+/SO_4^{2-} cotransporter on the BBM may play the predominant role in maintaining such a gradient (43). In the lower urinary tract (i.e., human ureteral epithelial cells), there is evidence for a DIDS-sensitive SO_4^{2-}/Cl^- anion exchanger but not for a Na^+-dependent SO_4^{2-} transporter (44), suggesting that active SO_4^{2-} uptake does not occur in the lower urinary tract but is restricted to the renal proximal tubule. For a model of the SO_4^{2-} transport systems present in the renal proximal tubule, see **Figure 1**.

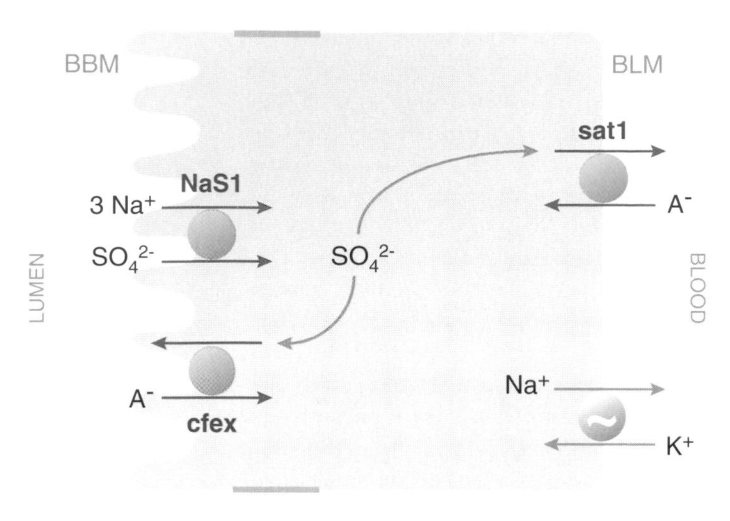

Figure 1

Sulfate transporters in renal proximal tubule. BBM, brush-border membrane; BLM, basolateral membrane.

CLONING AND FUNCTIONAL CHARACTERIZATION OF RENAL SULFATE TRANSPORTERS

The renal proximal tubule is the site of expression of three functionally and structurally distinct SO_4^{2-} transporters: (*a*) the Na^+/SO_4^{2-} (NaS1; SLC13A1) cotransporter, (*b*) the SO_4^{2-} anion transporter 1 (Sat1; SLC26A1), and (*c*) the CFEX (PAT1; SLC26A6) anion exchanger (**Figure 1**).

Na^+/SO_4^{2-} Cotransporters (NaS1; SLC13A1)

Three mammalian SLC13A1 orthologs (45) have been cloned so far: human NaS1 (hNaS1) (46), rat NaS1 (rNaS1) (30), and mouse NaS1 (mNaS1) (47). Northern blotting and reverse transcription polymerase chain reaction (RT-PCR) detected hNaS1 only in the kidney, whereas rNaS1 mRNA was detected by Northern blotting in the kidney and small intestine (30), and mNaS1 was detected by Northern blotting and RT-PCR in the kidney, ileum, duodenum/jejunum, and colon, with weaker signals in the cecum, testis, adrenal gland, and adipose tissue (46). NaS1 protein is localized to the BBM of renal proximal tubular cells (48, 49). hNaS1 encodes a protein of 595 amino acids and 13 putative transmembrane domains (46). The hNaS1 gene consists of 15 exons (spanning more than 83 kb) on human chromosome 7q31–7q32 (46).

NaS1 gene encodes an electrogenic, pH-insensitive, high-affinity, Na^+-dependent SO_4^{2-} transporter (Na^+/SO_4^{2-} cotransporter/symporter) with substrate preferences for the anions SO_4^{2-} (K_m = 93 μM), thiosulfate (K_m = 84 μM), and selenate (K_m = 580 μM) and the cation Na^+ ($K_{mNa}{}^+$ = 16–24 mM) (50). Significant *cis* inhibition for NaS1-induced SO_4^{2-} transport was observed with thiosulfate, selenate, tungstate, and molybdate (for all NaS1 orthologs) and with succinate and citrate (for hNaS1 only) (46).

NaS1 mRNA and protein levels are downregulated in the renal cortex by a high-SO_4^{2-} diet (51), hypothyroidism (52), vitamin D depletion (53), glucocorticoids (54), hypokalemia (49), metabolic acidosis (55), and nonsteroidal anti-inflammatory drugs (NSAIDs) (56). Conversely, NaS1 mRNA and protein levels are upregulated by a low-SO_4^{2-} diet (56), thyroid hormone (58), vitamin D supplementation (53), growth hormone (52), and chronic renal failure (59) and during postnatal growth (60). The mNaS1 gene (*Nas1*) is significantly upregulated by vitamin D and thyroid hormone by binding to vitamin D response element (VDRE) (at −525 bp) and thyroid hormone response element (T_3RE) (at −436 bp), respectively, on the *Nas1* promoter (61). *Nas1* is downregulated by glucocorticoids (47). The hNaS1 gene (*NAS1*) is significantly upregulated by xenobiotic 3-methycholanthrene by binding to a xenobiotic response element (XRE) (at −2052 bp) on the *NAS1* promoter (62). rNaS1-induced SO_4^{2-} transport in *Xenopus* oocytes was completely inhibited by pharmacological activators of protein kinase A (PKA) (8br-cAMP) and protein kinase C (PKC) [di-octinoyl glycerol (DOG)] (63, 64). rNaS1-induced SO_4^{2-} transport in *Xenopus* oocytes was inhibited by the heavy metals mercury, lead, chromium, and cadmium (51), suggesting that heavy metals inhibit renal brush-border Na^+/SO_4^{2-} cotransport via the NaS1 protein through various mechanisms and that this blockade may be responsible for sulfaturia following heavy metal intoxication.

To determine the functional role of NaS1 in SO_4^{2-} homeostasis and the physiological consequences of its absence, we recently generated a mouse lacking a functional NaS1 gene (65). Nas1-null (*Nas1−/−*) mice exhibit increased urinary SO_4^{2-} excretion and are therefore hyposulfatemic; serum SO_4^{2-} concentrations are reduced by >75% in *Nas1−/−* mice as compared with wild-type *Nas1+/+* mice (65). Using BBMV uptake studies, we showed that *Nas1−/−* mice have reduced renal and intestinal Na^+/SO_4^{2-} cotransport. *Nas1−/−* mice have general growth retardation: Body weight is reduced by >20% as compared with *Nas1+/+* and *Nas1+/−* littermates at two weeks of age and remains so throughout adulthood (65). *Nas1−/−* females have lowered fertility, with a 60% reduction in litter size. Spontaneous clonic seizures were observed in *Nas1−/−* mice from eight months of age. Behavioral studies of *Nas1−/−* mice demonstrated impaired memory and olfactory performance (66) as well as decreased object-induced anxiety and decreased locomotor activity in these mice (67). The precise mechanisms of these features are currently being investigated. However, these data do demonstrate that NaS1 is essential for maintaining SO_4^{2-} homeostasis and that its expression is necessary for a wide range of physiological functions.

Sulfate Anion Transporter 1

The first member of the SO_4^{2-} transporter SLC26 gene family (68), sat1 (SLC26A1) (69), was identified using a *Xenopus* oocyte expression cloning system (60). Recently, mouse (70) and human (71) sat1 orthologs (named msat1 and hsat1, respectively) were identified using a homology screening approach with the rat sat1 (rsat1) cDNA (69). msat1 and hsat1 proteins share 94% and 77% amino acid identities, respectively, with the rsat1 protein (64). The coding region of the rsat1 cDNA predicts a protein of 703 amino acids and a calculated molecular mass of 75.4 kDa. Hydrophobicity analysis of rsat1 protein suggests 12 putative transmembrane domains (64). Interestingly, there are two consensus regions that are highly conserved between the sat1 proteins and other members of the SLC26 gene family. These are the phosphopantetheine attachment site (PROSITE PS00012) motif at residues 415–430 of rsat1 (QTQLSSVVSAAVVLLV) and the SO_4^{2-} transporter signature (PROSITE PS01130) motif at residues 98–119 of both rsat1 and msat1 (PIYSLYTSFFANLIYFLMGTSR). Furthermore, there is a SO_4^{2-} transporter and antisigma antagonist (STAS) domain in the C-terminal cytoplasmic domain of the sat1 protein, which appears conserved in all members of the SLC26 family (68).

Northern blot analysis detected a single mRNA (3.8-kb) transcript for rsat1 very strongly in rat liver and kidney, with weaker signals in skeletal muscle and the brain (69). rsat1 mRNA expression levels in the kidney were identical to expression levels in the liver (the tissue from which it was cloned), prompting us to examine the role of sat1 in the kidney. Rat kidney mRNA, when injected into *Xenopus* oocytes, induced a Na^+-independent SO_4^{2-} transport activity that was inhibitable by DIDS, probenecid, and phenol red (72). The degree of inhibition by these compounds was closely correlated with the inhibition pattern of the renal BLM SO_4^{2-} anion exchanger (7) as well as with rsat1 cRNA–induced activity in *Xenopus* oocytes (69). Furthermore, using a hybrid depletion strategy, sat1 antisense oligonucleotides led to a complete abolition of the kidney mRNA–induced, Na^+-independent SO_4^{2-} transport activity (in *Xenopus* oocytes), confirming that rsat1 encodes the Na^+-independent SO_4^{2-} transporter, whose function most closely correlates with the proximal tubular BLM SO_4^{2-}/HCO_3^- anion exchanger (72). Researchers confirmed this by immunocytochemistry, using rsat1 polyclonal antibodies showing sat1 protein localization restricted to the BLM of kidney proximal tubules (73).

msat1 mRNA expression was strongest in the kidney and liver, with lower levels observed in the cecum, calvaria, brain, heart, and skeletal muscle (70). Two distinct transcripts were expressed in kidney and liver owing to alternative utilization of the first intron, corresponding to an internal portion of the 5′-untranslated region. hsat1 mRNA expression was most abundant in the kidney and liver, with lower levels in the pancreas, testis, brain, small intestine, colon, and lung (71). The mouse sat1 gene (*Sat1*; Slc26a1) is approximately ~6 kb in length and consists of four exons, with an optional intron 1. The *Sat1* promoter is ~52% G + C rich and contains a number of well-characterized *cis*-acting elements, including sequences resembling thyroid hormone response elements (T_3REs) and VDREs. *Sat1* promoter drove basal transcription in renal opossum kidney (OK) cells and was stimulated by tri-iodothyronine, or thyroid hormone (T_3), but not by 1,25-$(OH)_2$ vitamin D_3 (vitamin D). Site-directed mutagenesis of an imperfect T_3RE sequence at −454 bp led to a loss of *Sat1* promoter inducibility by T_3. This suggests that position −454 in the *Sat1* promoter contains a functional T_3RE that is responsible for transcriptional activation of *Sat1* by T_3.

The human sat1 gene (*SAT1*; SLC26A1) is localized to human chromosome 4p16.3 and is composed of four exons stretching approximately 6 kb in length, with an alternative splice site formed from an optional exon (exon 2) (71). The *SAT1* promoter is ~60% G + C rich and contains a number of well-characterized *cis*-acting elements but lacks any canonical TATA or CAAT boxes. The *SAT1* 5′ flanking region led to basal promoter activity in renal OK and LLC-PK1 cells. Via use of *SAT1* 5′ flanking region truncations, the first 135 bp was sufficient for basal promoter activity in both cells lines. Unlike in the mouse *Sat1* promoter, neither T_3 nor vitamin D can *trans*-stimulate *SAT1* promoter activity, despite the presence of sequences in its promoter resembling T_3REs and VDREs. Mutation of the activator protein-1 (AP-1) site at position −52 in the *SAT1* promoter led to loss of transcriptional activity, suggesting that AP-1 is required for *SAT1* basal transcription. Unlike NaS1, very little information is available on the regulation of sat1 in vivo.

Functional characterization in *Xenopus* oocytes, via use of radiotracer SO_4^{2-} measurements, revealed rsat1 to encode a Na^+-independent SO_4^{2-} transporter that had a high affinity ($K_m = 136$ μM) for SO_4^{2-} and that was strongly inhibited by a stilbene derivative, DIDS [4,4′-diisothiocyanato-2,2′-disulfonate] ($IC_{50DIDS} = 28$ μM), and oxalate but not by succinate or cholate (69). These properties correlated closely with the functional activities of the SO_4^{2-}/HCO_3^- exchanger in liver canalicular membrane vesicles (74). msat1 led to a ~20-fold induction in Na^+-independent SO_4^{2-} transport and a ~6-fold induction in oxalate and Cl^- transport but was unable to induce any formate, L-leucine, or succinate uptakes, in contrast to water-injected (control) oocytes (70). This suggests that msat1 transport was restricted to the anions SO_4^{2-}, oxalate, and Cl^-. As shown for rsat1 (69), msat1-induced SO_4^{2-} transport occurred only in the presence of extracellular Cl^-, suggesting its requirement for Cl^- (70). Typical Michaelis-Menten saturation kinetics were observed for msat1-induced SO_4^{2-} uptake, with a calculated $K_m = 0.31 \pm 0.05$ mM and $V_{max} = 143.90 \pm 5.32$ pmol h^{-1} for SO_4^{2-} interaction (70). These values are in close agreement with rsat1-induced SO_4^{2-} transport kinetics ($K_m = 0.14$ mM) (69). msat1-induced SO_4^{2-} transport was significantly inhibited by molybdate, selenate, tungstate, DIDS, thiosulfate, phenol red, and probenecid, whereas citrate and glucose had no effect. These data suggest that msat1 encodes a functional SO_4^{2-}/Cl^-/oxalate anion exchanger whose activity is blocked by anion exchange inhibitors DIDS and phenol red, tetra-oxyanions (selenate, molybdate, and tungstate), and thiosulfate. Similarly, using radiotracer measurements in *Xenopus* oocytes, researchers found that hsat1 led to a ~40-fold induction in Na^+-independent SO_4^{2-} uptake, ~5-fold induction in Cl^- uptake, and ~6-fold induction in oxalate uptake, when compared with the water-injected (control) oocytes (71). However, hsat1 was unable to induce formate uptake (71). This agrees with the transport activities for rsat1 and msat1, both of which transport SO_4^{2-}, oxalate, and Cl^- but not formate (69, 70). Typical Michaelis-Menten saturation kinetics were observed for hsat1-induced SO_4^{2-} uptake in *Xenopus* oocytes, with a calculated K_m for SO_4^{2-} of 0.19 ± 0.066 mM and a V_{max} of 52.35 ± 3.30 pmol/oocyte.hour (71). Various compounds were tested as possible inhibitors of hsat1-induced SO_4^{2-} uptake in *Xenopus* oocytes; the stilbene derivative DIDS, phenol red, thiosulfate, molybdate, tungstate, and selenate all significantly inhibited hsat1-induced SO_4^{2-} transport, whereas citrate and glucose had no effect (71).

SLC26A6 (CFEX, PAT1) Anion Exchanger

SLC26A6 (PAT1, CFEX) was the second member of the SLC26 SO_4^{2-} transporter gene family to be found expressed in the kidney proximal tubule (37, 75, 76). Human and mouse SLC26A6 was initially identified by searches of expressed sequence tags and genomic databases for novel homologs of previously known members of the SLC26 gene family (37, 75, 76). SLC26A6 is relatively ubiquitously expressed in both epithelial and nonepithelial tissues, with prominent expression in the heart, placenta, liver, skeletal muscle, kidney, pancreas, stomach, and small intestine (35, 37, 75, 76). Immunolocalization studies demonstrated SLC26A6 expression on the apical membrane in several epithelia, including in the proximal tubule (37, 38), pancreatic duct (75), and small intestine (35).

The SLC26A6 gene consists of 21 exons on human chromosome 3p21 or mouse chromosome 9 (34, 76). Multiple splice isoforms of SLC26A6 have been identified in human and mouse owing to the use of alternative first exons or alternative splice donor and acceptor sites (34, 36, 75, 76). Differences in transport properties of isoforms functionally expressed successfully have not been detected, although there are conflicting results about whether the SLC26A6c and SLC26A6d isoforms are functional at all (36, 77). Tissue-specific differences in expression among splice isoforms have been demonstrated (36).

The first functional expression studies were performed on mouse SLC26A6 and demonstrated Cl^-/formate exchange activity (37). For this reason the transporter was named CFEX. Because the function of human SLC26A6 was initially not known, human SLC26A6 was deposited in the database as putative anion transporter 1 (PAT1) (75). Although PAT1 has been retained as an alternative name for SLC26A6, the same name has been applied to multiple other gene products, including another transporter, a proton-coupled amino acid transporter (78). Thus, the name PAT1 is not unique and probably should be discontinued as a name for SLC26A6. The name CFEX is unique but arguably less than ideal because Cl^-/formate exchange may not be the most important physiological function of SLC26A6, at least in the proximal tubule.

Following initial description of CFEX as a Cl^-/formate exchanger, subsequent studies demonstrated that the transporter has broad specificity and can function in multiple DIDS-sensitive exchange modes involving SO_4^{2-}, oxalate, Cl^-, formate, HCO_3^-, and OH^- as substrates (33–36, 77). CFEX also transports lactate and the anions of both short- and medium-chain fatty acids (79). PAH is a nontransported inhibitor of CFEX activity (33, 79). Cl^-/oxalate exchange mediated by CFEX is electrogenic, consistent with exchange of divalent oxalate for monovalent Cl^- (33). Interestingly, Cl^-/HCO_3^- exchange mediated by CFEX is also electrogenic (34, 80), consistent with the transport of more than a single HCO_3^- in exchange for Cl^-. Indeed, a careful comparison of Cl^- and HCO_3^- flux rates demonstrated that the $HCO_3^-:Cl^-$ stoichiometry is 2:1 (81).

Comparison of the potency of various anions as inhibitors of CFEX activity indicated that the affinity of CFEX for oxalate is appreciably greater than for Cl^-, HCO_3^-, SO_4^{2-}, or formate (33, 34). Similarly, when tested at equal substrate concentrations, transport of oxalate was twice as high as that of SO_4^{2-} or formate, indicating that among these anions CFEX has the greatest activity as an oxalate transporter (33).

Given the ability of CFEX to function as an anion exchanger with broad specificity for many substrates in addition to SO_4^{2-}, a key issue is the physiological role(s) of CFEX in the proximal tubule as well as in other tissues in which it is expressed. Several of the anion exchange activities mediated by CFEX have been previously described at the apical membrane of proximal tubule cells. These include Cl^-/formate exchange (82), electrogenic Cl^-/oxalate exchange (32), Cl^--OH^-/HCO_3^- exchange (83–85), and multiple modes of SO_4^{2-} anion exchange (26, 28).

Studies in CFEX-null mice have provided partial insight into the role of CFEX in mediating these previously observed anion exchange activities (86). Comparison of wild-type and CFEX-null mice shows that CFEX mediates all Cl^-/oxalate exchange activity, a significant portion of the SO_4^{2-}/oxalate exchange activity, and a minor component of the Cl^-/formate exchange activity in renal BBMVs (86). Moreover, studies in CFEX-null mice demonstrated that CFEX mediates a substantial portion of Cl^--OH^-/HCO_3^- exchange across the apical membrane of proximal tubule cells (87).

Investigators performed tubular microperfusion studies in CFEX-null mice to assess the role of CFEX-mediated anion exchange in transtubular NaCl absorption. Proximal tubules were microperfused in situ with a low-HCO_3^- (5 mM), low-pH (6.7) solution, simulating conditions in the later portions of the proximal tubule after substantial reabsorption of HCO_3^- has taken place. Under these conditions, the rate of volume absorption is a measure of isosmotic NaCl absorption (88). Previous studies had indicated that addition of oxalate to the perfusion solutions markedly stimulates proximal tubule NaCl absorption under these conditions (88, 89). This oxalate-stimulated NaCl transport is abolished when SO_4^{2-} is omitted from the perfusion solutions (88) but is not dependent on sodium hydrogen exchanger 3 (NHE3) activity (88, 90). These findings are consistent with a model in which NaCl absorption across the apical membrane of proximal tubule cells takes place by Cl^-/oxalate exchange operating in parallel with SO_4^{2-}/oxalate exchange and Na^+/SO_4^{2-} cotransport. The increment in volume absorption induced by the addition of oxalate to the perfusion solution was completely abolished in CFEX-null mice (87). These results are consistent with the findings in BBMVs that CFEX mediates all apical membrane Cl^-/oxalate exchange activity in the proximal tubule. In contrast, the increment in volume absorption induced by formate was only partially inhibited in CFEX-null mice (87). Previous work had indicated that formate-stimulated NaCl absorption takes place by Cl^-/formate exchange in parallel with Na^+/H^+ exchange and H^+-coupled formate recycling (88, 89, 91). The modest reduction of formate-stimulated NaCl absorption in CFEX-null mice is therefore consistent with the results in BBMVs indicating that CFEX mediates only a minor fraction of apical membrane Cl^-/formate exchange activity. Finally, the baseline rate of volume reabsorption measured in the absence of added formate and oxalate was not detectably reduced in CFEX-null mice (87). These latter findings suggest that CFEX-mediated Cl^--OH^-/HCO_3^- exchange activity operating in parallel with Na^+/H^+ exchange does not contribute to transtubular NaCl absorption in the superficial proximal tubule segments that were microperfused in these experiments. Because these studies were conducted with SO_4^{2-} in the perfusion solution, the absence of reduction of baseline NaCl absorption in CFEX-null mice indicated that CFEX-mediated Cl^-/SO_4^{2-} exchange activity in parallel with Na^+/SO_4^{2-} cotransport also does not contribute to NaCl absorption. Taken together, these findings demonstrate a major role for CFEX in mediating Cl^-/oxalate exchange as the basis for oxalate-stimulated NaCl absorption in the proximal tubule.

The possible role of CFEX in the process of transtubular SO_4^{2-} transport has not yet been similarly assessed. The net direction of CFEX-mediated SO_4^{2-} anion exchange will depend on the gradients across the apical membrane of SO_4^{2-} and any given exchanging anion. Given the presence in the same membrane of NaS1, which is theoretically capable of generating a cell-to-lumen gradient of SO_4^{2-} larger than the corresponding gradient of any exchange partner, CFEX likely mediates backflux of SO_4^{2-} from cell to lumen, as illustrated in **Figure 1**. Possible alterations in body SO_4^{2-} homeostasis in CFEX-null mice also have not been evaluated. At the level of the intact animal, the most striking abnormality so far observed in CFEX-null mice is in oxalate homeostasis. CFEX-null animals have hyperoxaluria, hyperoxalemia, and a high incidence of calcium oxalate urolithiasis owing to enhanced net absorption of oxalate ingested in the diet (86). This abnormality in oxalate homeostasis has been attributed to an important constitutive role of CFEX in mediating oxalate secretion in the small intestine (86, 92).

Recent studies have begun to define the role of CFEX-associated proteins in regulating transporter expression or function. The C terminus of CFEX has a predicted PDZ-binding motif (37, 75). The C terminus of CFEX can bind the PDZ domain–containing proteins NHERF1, NHERF2, and PDZK1 (NHERF3) (36, 38, 93). Interactions with these proteins were abolished when the predicted C-terminal PDZ-binding motif of CFEX was deleted (36, 38). This PDZ-binding motif is absent in the SLC26A6d splice variant (36). The possible functional significance of the interaction of CFEX with NHERF1 and NHERF2 in the kidney in vivo is not yet known. A dramatic loss of expression of CFEX protein and functional activity was observed in the kidneys of PDZK1-null mice, indicating that this scaffolding protein is essential for the normal expression and function of CFEX in the proximal tubule (38).

Researchers have also demonstrated a functional interaction of CFEX with carbonic anhydrase (CA) II (94). Mutation of the CAII-binding site greatly reduced Cl^-/HCO_3^- exchange activity mediated by CFEX, suggesting that the complex of CFEX with CAII represents a CO_2/HCO_3^- transport metabolon (94). Agonists activating PKC inhibit CFEX-mediated Cl^-/HCO_3^- exchange activity. Activation of PKC dissociates CAII from CFEX. Mutation of a putative PKC phosphorylation site adjacent to the CAII binding site in CFEX prevents inhibition by PKC. Taken together, these findings indicate that the regulated binding of CAII is an important modulator of Cl^-/HCO_3^- activity mediated by CFEX (94). The possible interaction of CFEX with CAII or CAIV in proximal tubule cells has not yet been reported.

A mutual functional interaction between CFEX and CFTR exists (80, 95). Coexpression of CFTR caused marked cAMP-dependent stimulation of CFEX-mediated Cl^-/OH^- exchange activity (80). Conversely, such coexpression of CFEX caused stimulation of channel activity of CFTR (95). The activation of CFTR was greatly stimulated by intracellular application of a fusion protein containing the STAS domain of CFEX, implicating this portion of the CFEX sequence in the functional interaction with CFTR (95). Although CFTR is not known to be expressed with CFEX in the proximal tubule, the functional interaction between CFEX and CFTR is of potential importance for the regulation of Cl^- and HCO_3^- transport in tissues in which the two proteins are coexpressed (i.e., pancreatic duct and intestine) (95).

CONCLUSIONS

There has been great progress in the molecular identification of SO_4^{2-} transporters in the kidney and in the elucidation of their physiological roles in the homeostasis of SO_4^{2-} and other ions. The apical membrane Na^+/SO_4^{2-} cotransporter NaS1 plays a major role in mediating proximal tubule SO_4^{2-} reabsorption, as evidenced by the striking hyposulfatemia and hypersulfaturia observed in Nas1-null mice. The anion exchanger SAT1 (SLC26A1) mediates SO_4^{2-} exit across the basolateral membrane to complete the process of transtubular SO_4^{2-} reabsorption. Ongoing knockout mouse studies will elucidate the contribution of SAT1 to SO_4^{2-} homeostasis. CFEX (SLC26A6) present at the apical membrane of proximal tubular cells is also capable of mediating SO_4^{2-} transport by anion exchange. Investigators currently are assessing in CFEX-null mice the contribution of this transporter to net renal SO_4^{2-} reabsorption and SO_4^{2-} homeostasis. Knockout mouse studies have demonstrated that CFEX has a major role as an apical membrane Cl^-/oxalate exchanger that contributes to NaCl reabsorption in the proximal tubule and oxalate secretion in the small intestine. Finally, researchers have shown that several additional SLC26 family members mediate SO_4^{2-} transport and are expressed in the kidney (e.g., SLC26A2, SLC26A7, SLC26A11) (68). Of these, SLC26A7 has been immunolocalized to the apical domain of proximal tubular cells (96) as well as the BLM of thick ascending limb cells (96) and intercalated cells of the outer medullary collecting duct (97). The precise role of the SLC26 family members in mediating renal tubular SO_4^{2-} transport and their overall contribution to SO_4^{2-} homeostasis have yet to be determined.

ACKNOWLEDGMENTS

Work in the authors' laboratories was supported by grants from the ARC and NHMRC to D.M., and NIH grants R01-DK33793 and P01-DK17433 to P.S.A.

LITERATURE CITED

1. Goudsmit AJ, Power MH, Bollman JL. 1939. The excretion of sulfates by the dog. *Am. J. Physiol.* 125:506–20
2. Hierholzer K, Cade R, Gurd R, Kessler R, Pitts R. 1960. Stop flow analysis of renal reabsorption and excretion of sulfate in the dog. *Am. J. Physiol.* 198:833–37
3. Lin JH, Levy G. 1983. Renal clearance of inorganic sulfate in rats: effect of acetaminophen-induced depletion of endogenous sulfate. *J. Pharm. Sci.* 72:213–17
4. Mudge G, Berndt W, Valtin H. 1973. Tubular transport of urea, phosphate, uric acid, sulfate and thiosulfate. In *Handbook of Physiology*, ed. W Berliner, J Orloff and G Geiger, pp. 587–652. Washington, DC: Williams & Williams
5. Krijgsheld KR, Scholtens E, Mulder GJ. 1980. Serum concentration of inorganic sulfate in mammals: species differences and circadian rhythm. *Comp. Biochem. Physiol.* 67A:683–86
6. Mulder GJ. 1981. Sulfate availability in vivo. In *Sulfation of Drugs and Related Compounds*, ed. GJ Mulder, pp. 32–52. Boca Raton, FL: CRC
7. David C, Ullrich KJ. 1992. Substrate specificity of the luminal Na^+-dependent sulfate transport system in the proximal renal tubule as compared to the contraluminal sulfate exchange system. *Pflügers Arch.* 421:455–65

8. Ullrich KJ, Rumrich G, Kloess S. 1980. Bidirectional active transport of thiosulfate in the proximal convolution of the rat kidney. *Pflügers Arch.* 387:127–32
9. Ullrich KJ, Rumrich G, Kloess S. 1984. Contraluminal sulfate transport in the proximal tubule of the rat kidney. I. Kinetics, effects of K^+, Na^+, Ca^{2+}, H^+ and anions. *Pflügers Arch.* 402:264–71
10. Ullrich KJ, Rumrich G, Kloess S. 1985. Contraluminal sulfate transport in the proximal tubule of the rat kidney. II. Specificity: sulfate-ester, sulfonates and amino sulfonates. *Pflügers Arch.* 404:293–99
11. Ullrich KJ, Rumrich G, Kloess S. 1985. Contraluminal sulfate transport in the proximal tubule of the rat kidney. III. Specificity: disulfonates, di- and tri-carboxylates and sulfo-carboxylates. *Pflügers Arch.* 404:300–6
12. Ullrich KJ, Rumrich G, Kloess S. 1985. Contraluminal sulfate transport in the proximal tubule of the rat kidney. IV. Specificity: Salicylate analogs. *Pflügers Arch.* 404:307–10
13. Ullrich KJ, Rumrich G, Kloess S. 1985. Contraluminal sulfate transport in the proximal tubule of the rat kidney. V. Specificity: phenophthaleins, sulfonphthaleins and other sulfodyes, sulfamoyl compounds and dyphenylamine-2-carboxylates. *Pflügers Arch.* 404:311–18
14. Ullrich KJ, Rumrich G. 1988. Contraluminal transport systems in the proximal renal tubule involved in secretion of organic anions. *Am. J. Physiol.* 254:F453–62
15. Ullrich KJ, Murer H. 1982. Sulphate and phosphate transport in the renal proximal tubule. *Philos. Trans. R. Soc. London Ser. B* 299:549–58
16. Brazy PC, Dennis VW. 1981. Sulfate transport in rabbit proximal convoluted tubules: presence of anion exchange. *Am. J. Physiol.* 241:F300–7
17. Grinstein S, Turner RJ, Silverman M, Rothstein A. 1980. Inorganic anion transport in kidney and intestinal brush border and basolateral membranes. *Am. J. Physiol.* 238:F452–60
18. Low I, Friedrich T, Burckhardt G. 1984. Properties of an anion exchanger in rat renal basolateral membrane vesicles. *Am. J. Physiol.* 246:F334–42
19. Lucke H, Stange G, Murer H. 1979. Sulphate/ion-sodium/ion cotransport by brush border membranes vesicles isolated from rat kidney cortex. *Biochem. J.* 182:223–29
20. Pritchard JB, Renfro JL. 1983. Renal sulfate transport at the basolateral membrane is mediated by anion exchange. *Proc. Natl. Acad. Sci. USA* 80:2603–7
21. Schneider EG, Durham JC, Sacktor B. 1984. Sodium-dependent transport of inorganic sulfate by rabbit renal brush-border membrane vesicles. *J. Biol. Chem.* 259(23):14591–99
22. Turner RJ. 1984. Sodium-dependent sulfate transport in renal outer cortical brush border membrane vesicles. *Am. J. Physiol.* 247:F793–98
23. Bastlein C, Burckhardt G. 1986. Sensitivity of rat renal luminal and contraluminal sulfate transport to DIDS. *Am. J. Physiol.* 250:F226–34
24. Lucke H, Stange G, Murer H. 1981. Sulfate-sodium cotransport by brush-border membrane vesicles isolated from rat ileum. *Gastroenterology* 80:22–30
25. Kuo SM, Aronson PS. 1988. Oxalate transport via the sulfate/HCO_3 exchanger in rabbit renal basolateral membrane vesicles. *J. Biol. Chem.* 263:9710–17
26. Pritchard JB. 1987. Sulfate-bicarbonate exchange in brush-border membranes from rat renal cortex. *Am. J. Physiol.* 252:F346–56
27. Talor Z, Gold RM, Yang WC, Arruda JAL. 1987. Anion exchanger is present in both luminal and basolateral renal membranes. *Eur. J. Biochem.* 164:695–702
28. Kuo SM, Aronson PS. 1996. Pathways for oxalate transport in rabbit renal microvillus membrane vesicles. *J. Biol. Chem.* 271:15491–97

29. Grichtchenko II, Choi I, Zhong X, Bray-Ward P, Russell JM, Boron WF. 2001. Cloning, characterization, and chromosomal mapping of a human electroneutral Na^+-driven Cl-HCO_3 exchanger. *J. Biol. Chem.* 276:8358–63
30. Markovich D, Forgo J, Stange G, Biber J, Murer H. 1993. Expression cloning of rat renal Na^+/SO_4^{2-} cotransport. *Proc. Natl. Acad. Sci. USA* 90:8073–77
31. Busch AE, Waldegger S, Herzer T, Biber J, Markovich D, et al. 1994. Electrogenic cotransport of Na^+ and sulfate in Xenopus oocytes expressing the cloned $Na^+SO_4^{2-}$ transport protein NaSi-1. *J. Biol. Chem.* 269:12407–9
32. Karniski LP, Aronson PS. 1987. Anion exchange pathways for C1-transport in rabbit renal microvillus membranes. *Am. J. Physiol.* 253:F513–21
33. Jiang Z, Grichtchenko II, Boron WF, Aronson PS. 2002. Specificity of anion exchange mediated by mouse Slc26a6. *J. Biol. Chem.* 277:33963–67
34. Xie Q, Welch R, Mercado A, Romero MF, Mount DB. 2002. Molecular characterization of the murine Slc26a6 anion exchanger: functional comparison with Slc26a1. *Am. J. Physiol. Renal Physiol.* 283:F826–38
35. Wang Z, Petrovic S, Mann E, Soleimani M. 2002. Identification of an apical Cl^-/HCO_3^- exchanger in the small intestine. *Am. J. Physiol. Gastrointest. Liver Physiol.* 282:G573–79
36. Lohi H, Lamprecht G, Markovich D, Heil A, Kujala M, et al. 2003. Isoforms of SLC26A6 mediate anion transport and have functional PDZ interaction domains. *Am. J. Physiol. Cell Physiol.* 284:C769–79
37. Knauf F, Yang CL, Thomson RB, Mentone SA, Giebisch G, Aronson PS. 2001. Identification of a chloride-formate exchanger expressed on the brush border membrane of renal proximal tubule cells. *Proc. Natl. Acad. Sci. USA* 98:9425–30
38. Thomson RB, Wang T, Thomson BR, Tarrats L, Girardi A, et al. 2005. Role of PDZK1 in membrane expression of renal brush border ion exchangers. *Proc. Natl. Acad. Sci. USA* 102:13331–36
39. Hagenbuch B, Stange G, Murer H. 1985. Transport of sulfate in rat jejunal and proximal tubular basolateral membrane vesicles. *Pflügers Arch.* 405:202–8
40. Aronson PS. 1989. The renal proximal tubule: a model for diversity of anion exchangers and stilbene-sensitive anion transporters. *Annu. Rev. Physiol.* 51:419–41
41. Bissig M, Hagenbuch B, Stieger B, Koller T, Meier PJ. 1994. Functional expression cloning of the canalicular sulfate transport system of rat hepatocytes. *J. Biol. Chem.* 269(4):3017–21
42. Murer H, Burckhardt G. 1983. Membrane transport of anions across epithelia of the mammalian small intestine and kidney proximal tubule. *Rev. Physiol. Biochem. Pharmacol.* 96:1–51
43. Besseghir K, Roch-Ramel F. 1987. Renal excretion of drugs and other xenobiotics. *Renal Physiol.* 10:221–41
44. Elgavish A, Wille JJ, Rahemtulla F, Debro L. 1991. Carrier-mediated sulfate transport in human ureteral epithelial cells cultured in serum-free medium. *Am. J. Physiol.* 261:C916–26
45. Markovich D, Murer H. 2004. The SLC13 gene family of sodium sulfate/carboxylate cotransporters. *Pflügers Arch.* 447:594–602
46. Lee A, Beck L, Markovich D. 2000. The human renal sodium sulfate cotransporter (SLC13A1; hNaSi-1) cDNA and gene: organization, chromosomal localization, and functional characterization. *Genomics* 70:354–63
47. Beck L, Markovich D. 2000. The mouse Na^+-sulfate cotransporter gene *Nas1*. Cloning, tissue distribution, gene structure, chromosomal assignment, and transcriptional regulation by vitamin D. *J. Biol. Chem.* 275:11880–90
48. Lotscher M, Custer M, Quabius E, Kaissling B, Murer H, Biber J. 1996. Immunolocalization of Na/SO_4^- cotransport (NaSi-1) in rat kidney. *Pflügers Arch.* 432:373–78

49. Markovich D, Wang H, Puttaparthi K, Zajicek H, Rogers T, et al. 1999. Chronic K depletion inhibits renal brush border membrane Na/sulfate cotransport. *Kidney Int.* 55:244–51
50. Busch A, Waldegger S, Herzer T, Biber J, Markovich D. 1994. Electrophysiological analysis of Na^+/P_i cotransport mediated by a transporter cloned from rat kidney and expressed in Xenopus oocytes. *Proc. Natl. Acad. Sci. USA* 91:8205–8
51. Markovich D, Knight D. 1998. Renal Na-Si_i cotransporter NaSi-1 is inhibited by heavy metals. *Am. J. Physiol.* 274:F283–89
52. Sagawa K, Han B, DuBois DC, Murer H, Almon RR, et al. 1999. Age- and growth hormone-induced alterations in renal sulfate transport. *J. Pharmacol. Exp. Ther.* 290:1182–87
53. Fernandes I, Hampson G, Cahours X, Morin P, Coureau C, et al. 1997. Abnormal sulfate metabolism in vitamin D-deficient rats. *J. Clin. Invest.* 100:2196–203
54. Lee HJ, Sagawa K, Shi W, Murer H, Morris ME. 2000. Hormonal regulation of sodium/sulfate cotransport in renal epithelial cells. *Proc. Soc. Exp. Biol. Med.* 225:49–57
55. Puttaparthi K, Markovich D, Halaihel N, Wilson P, Zajicek HK, et al. 1999. Metabolic acidosis regulates rat renal Na-Si cotransport activity. *Am. J. Physiol.* 276:C1398–404
56. Sagawa K, Benincosa LJ, Murer H, Morris ME. 1998. Ibuprofen-induced changes in sulfate renal transport. *J. Pharmacol. Exp. Ther.* 287(3):1092–97
57. Deleted in proof
58. Sagawa K, Murer H, Morris ME. 1999. Effect of experimentally induced hypothyroidism on sulfate renal transport in rats. *Am. J. Physiol.* 276:F164–71
59. Fernandes I, Laouari D, Tutt P, Hampson G, Friedlander G, Silve C. 2001. Sulfate homeostasis, NaSi-1 cotransporter, and SAT-1 exchanger expression in chronic renal failure in rats. *Kidney Int.* 59:210–21
60. Markovich D, Werner A, Murer H. 1999. Expression cloning with *Xenopus* oocytes. In *Techniques in Molecular Medicine*, ed. F Hildebrandt, P Igarashi, pp. 310–18. Heidelberg: Springer Verlag
61. Dawson PA, Markovich D. 2002. Regulation of the mouse Nas1 promoter by vitamin D and thyroid hormone. *Pflügers Arch.* 444:353–59
62. Lee A, Markovich D. 2004. Characterization of the human renal Na^+-sulfate cotransporter gene (NAS1) promoter. *Pflügers Arch.* 448:490–99
63. Markovich D. 2000. Molecular regulation and membrane trafficking of mammalian renal phosphate and sulfate transporters. *European J. Cell Biol.* 79:531–38
64. Markovich D. 2001. Physiological roles and regulation of mammalian sulfate transporters. *Physiol. Rev.* 81:1499–534
65. Dawson PA, Beck L, Markovich D. 2003. Hyposulfatemia, growth retardation, reduced fertility and seizures in mice lacking a functional NaS_i-1 gene. *Proc. Natl. Acad. Sci. USA* 100:13704–9
66. Dawson PA, Steane SE, Markovich D. 2005. Impaired memory and olfactory performance in NaSi-1 sulfate transporter deficient mice. *Behav. Brain Res.* 159(1):15–20
67. Dawson PA, Steane SE, Markovich D. 2004. Behavioural abnormalities of the hyposulfataemic *Nas1* knock-out mouse. *Behav. Brain Res.* 154:457–63
68. Mount DB, Romero MF. 2004. The SLC26 gene family of multifunctional anion exchangers. *Pflügers Arch.* 447:710–21
69. Bissig M, Hagenbuch B, Stieger B, Koller T, Meier PJ. 1994. Functional expression cloning of the canalicular sulfate transport system of rat hepatocytes. *J. Biol. Chem.* 269:3017–21
70. Lee A, Beck L, Markovich D. 2003. The mouse sulfate anion transporter gene *Sat1* (Slc26a1): cloning, tissue distribution, gene structure, functional characterization, and transcriptional regulation thyroid hormone. *DNA Cell Biol.* 22:19–31

71. Regeer RR, Lee A, Markovich D. 2003. Characterization of the human sulfate anion transporter (hsat-1) protein and gene (*SAT1*; SLC26A1). *DNA Cell Biol.* 22:107–17
72. Markovich D, Bissig M, Sorribas V, Hagenbuch B, Meier PJ, Murer H. 1994. Expression of rat renal sulfate transport systems in *Xenopus laevis* oocytes. Functional characterization and molecular identification. *J. Biol. Chem.* 269:3022–26
73. Karniski LP, Lotscher M, Fucentese M, Hilfiker H, Biber J, Murer H. 1998. Immunolocalization of sat-1 sulfate/oxalate/bicarbonate anion exchanger in the rat kidney. *Am. J. Physiol.* 275:F79–87
74. Meier P, Valantinas J, Hugentobler G, Rahm I. 1987. Bicarbonate sulfate exchange in canalicular rat liver plasma membrane vesicles. *Am. J. Physiol.* 253:461–68
75. Lohi H, Kujala M, Kerkela E, Saarialho-Kere U, Kestila M, Kere J. 2000. Mapping of five new putative anion transporter genes in human and characterization of SLC26A6, a candidate gene for pancreatic anion exchanger. *Genomics* 70:102–12
76. Waldegger S, Moschen I, Ramirez A, Smith RJ, Ayadi H, et al. 2001. Cloning and characterization of SLC26A6, a novel member of the solute carrier 26 gene family. *Genomics* 72:43–50
77. Chernova MN, Jiang L, Friedman DJ, Darman RB, Lohi H, et al. 2005. Functional comparison of mouse slc26a6 anion exchanger with human SLC26A6 polypeptide variants: differences in anion selectivity, regulation, and electrogenicity. *J. Biol. Chem.* 280:8564–80
78. Boll M, Foltz M, Rubio-Aliaga I, Kottra G, Daniel H. 2002. Functional characterization of two novel mammalian electrogenic proton-dependent amino acid cotransporters. *J. Biol. Chem.* 277:22966–73
79. Nozawa T, Sugiura S, Hashino Y, Tsuji A, Tamai I. 2004. Role of anion exchange transporter PAT1 (SLC26A6) in intestinal absorption of organic anions. *J. Drug Target.* 12:97–104
80. Ko SB, Shcheynikov N, Choi JY, Luo X, Ishibashi K, et al. 2002. A molecular mechanism for aberrant CFTR-dependent HCO_3^- transport in cystic fibrosis. *EMBO J.* 21:5662–72
81. Shcheynikov N, Wang Y, Park M, Ko SB, Dorwart M, et al. 2006. Coupling modes and stoichiometry of Cl^-/HCO_3^- exchange by slc26a3 and slc26a6. *J. Gen. Physiol.* 127:511–24
82. Karniski LP, Aronson PS. 1985. Chloride/formate exchange with formic acid recycling: a mechanism of active chloride transport across epithelial membranes. *Proc. Natl. Acad. Sci. USA* 82:6362–65
83. Warnock DG, Yee VJ. 1981. Chloride uptake by brush border membrane vesicles isolated from rabbit renal cortex. Coupling to proton gradients and K^+ diffusion potentials. *J. Clin. Invest.* 67:103–15
84. Kurtz I, Nagami G, Yanagawa N, Li L, Emmons C, Lee I. 1994. Mechanism of apical and basolateral Na^+-independent Cl^-/base exchange in the rabbit superficial proximal straight tubule. *J. Clin. Invest.* 94:173–83
85. Sheu JN, Quigley R, Baum M. 1995. Heterogeneity of chloride/base exchange in rabbit superficial and juxtamedullary proximal convoluted tubules. *Am. J. Physiol.* 268:F847–53
86. Jiang Z, Asplin JR, Evan AP, Rajendran VM, Velazquez H, et al. 2006. Calcium oxalate urolithiasis in mice lacking anion transporter Slc26a6. *Nat. Genet.* 38:474–78
87. Wang Z, Wang T, Petrovic S, Tuo B, Riederer B, et al. 2005. Renal and intestinal transport defects in Slc26a6-null mice. *Am. J. Physiol. Cell Physiol.* 288:C957–65
88. Wang T, Egbert ALJ, Abbiati T, Aronson PS, Giebisch G. 1996. Mechanisms of stimulation of proximal tubule chloride transport by formate and oxalate. *Am. J. Physiol.* 271:F446–50
89. Wang T, Giebisch G, Aronson PS. 1992. Effects of formate and oxalate on volume absorption in rat proximal tubule. *Am. J. Physiol.* 263:F37–42

90. Wang T, Yang CL, Abbiati T, Shull GE, Giebisch G, Aronson PS. 2001. Essential role of NHE3 in facilitating formate-dependent NaCl absorption in the proximal tubule. *Am. J. Physiol. Renal Physiol.* 281:F288–92
91. Schild L, Giebisch G, Karniski LP, Aronson PS. 1987. Effect of formate on volume reabsorption in the rabbit proximal tubule. *J. Clin. Invest.* 79:32–38
92. Freel RW, Hatch M, Green M, Soleimani M. 2006. Ileal oxalate absorption and urinary oxalate excretion are enhanced in Slc26a6 null mice. *Am. J. Physiol. Gastrointest. Liver Physiol.* 290:G719–28
93. Gisler SM, Pribanic S, Bacic D, Forrer P, Gantenbein A, et al. 2003. PDZK1: I. a major scaffolder in brush borders of proximal tubular cells. *Kidney Int.* 64:1733–45
94. Alvarez BV, Vilas GL, Casey JR. 2005. Metabolon disruption: a mechanism that regulates bicarbonate transport. *EMBO J.* 24:2499–511
95. Ko SB, Zeng W, Dorwart MR, Luo X, Kim KH, et al. 2004. Gating of CFTR by the STAS domain of SLC26 transporters. *Nat. Cell Biol.* 6:343–50
96. Dudas PL, Mentone S, Greineder CF, Biemesderfer D, Aronson PS. 2006. Immunolocalization of anion transporter Slc26a7 in mouse kidney. *Am. J. Physiol. Renal Physiol.* 290:F937–45
97. Petrovic S, Barone S, Xu J, Conforti L, Ma L, et al. 2004. SLC26A7: a basolateral Cl^-/HCO_3^- exchanger specific to intercalated cells of the outer medullary collecting duct. *Am. J. Physiol. Renal Physiol.* 286:F161–69

Overview of Structure and Function of Mammalian Cilia

Peter Satir[1] and Søren Tvorup Christensen[2]

[1]Department of Anatomy and Structural Biology, Albert Einstein College of Medicine of Yeshiva University, Bronx, New York 10461; email: satir@aecom.yu.edu

[2]Institute of Molecular Biology and Physiology, Department of Biochemistry, University of Copenhagen, DK-2100 Copenhagen OE, Denmark; email: stchristensen@aki.ku.dk

Annu. Rev. Physiol. 2007. 69:377–400

First published online as a Review in Advance on September 29, 2006

The *Annual Review of Physiology* is online at http://physiol.annualreviews.org

This article's doi: 10.1146/annurev.physiol.69.040705.141236

0066-4278/07/0315-0377$20.00

Key Words

ciliary motility, primary ciliary dyskinesia, primary cilia, sensory organelles, signaling, tissue homeostasis, development

Abstract

Cilia are membrane-bounded, centriole-derived projections from the cell surface that contain a microtubule cytoskeleton, the ciliary axoneme, surrounded by a ciliary membrane. Axonemes in multiciliated cells of mammalian epithelia are 9 + 2, possess dynein arms, and are motile. In contrast, single nonmotile 9 + 0 primary cilia are found on epithelial cells, such as those of the kidney tubule, but also on nonepithelial cells, such as chondrocytes, fibroblasts, and neurons. The ciliary membranes of all cilia contain specific receptors and ion channel proteins that initiate signaling pathways controlling motility and/or linking mechanical or chemical stimuli, including sonic hedgehog and growth factors, to intracellular transduction cascades regulating differentiation, migration, and cell growth during development and in adulthood. Unique motile 9 + 0 cilia, found during development at the embryonic node, determine left-right asymmetry of the body.

INTRODUCTION

Cilia: microtubule-based cell organelles extending from a basal body, a centriole, at the apical cell surface, containing 9+2 or 9 + 0 axonemes surrounded by a specialized ciliary membrane

Axoneme: the microtubule cytoskeleton of the cilium, consisting of a ring of nine doublet microtubules surrounding a central pair (9+2) or missing the central pair (9 + 0)

Dynein: an AAA-type ATPase that functions as a motor to move cargo along microtubules to their slow-polymerizing (−) end or assembled as inner and outer rows of projections along the axonemal doublets to move the doublet microtubules with respect to one another, thus powering ciliary movement

Primary cilia: single 9 + 0 nonmotile cilia found on many mammalian cells, including kidney epithelia, chondrocytes, fibroblasts, and neurons

Cilia are membrane-bounded, centriole-derived, microtubule-containing projections from the cell surface. The microtubule cytoskeleton of the cilium, the ciliary axoneme, grows from and continues the ninefold symmetry of the centriole, which is nearly identical to, and often becomes, a ciliary basal body. Mammalian ciliary axonemes, like axonemes elsewhere in the animal kingdom, are formed with two major patterns: 9+2, in which the nine doublet microtubules surround a central pair of singlet microtubules, and 9 + 0, in which the central pair is missing (1, 2). Usually, 9 + 0 cilia are also missing the molecular motors, axonemal dyneins, which are responsible for ciliary movement; such cilia are therefore nonmotile. In contrast, 9+2 cilia are motile (**Figure 1**). In addition, whereas epithelial cells may possess several hundred 9+2 motile cilia, 9 + 0 cilia are usually solitary. Nonmotile 9 + 0 cilia form the basis for various specialized sensory structures, including chemosensitive or proprioceptive sensilla of invertebrates, such as the insect ear. The outer segments of the rods and cones of the eye in mammals are expanded 9 + 0 cilia. Olfactory cells have nonmotile multiple long chemoreceptive cilia that are 9+2 at their base but lose this organization distally.

One aspect that received only the attention of specialists until recently is the widespread distribution of 9+0 cilia among the cells of the body. Such cilia are now called primary cilia. Primary cilia are found on epithelial cells such as the kidney tubule, the bile duct, the endocrine pancreas, and the thyroid but also on nonepithelial cells such as chondrocytes, fibroblasts, smooth muscle cells, neurons, and Schwann cells. In addition, unique motile 9 + 0 cilia, bearing dynein arms, are found at the embryonic node during development. For an up-to-date listing of cell types with primary cilia, see the Primary Cilium Resource Site (**http://www.primary-cilium.co.uk/**).

The structure and distribution of both 9+2 and 9+0 cilia have been widely studied since the advent of biological electron microscopy in the middle of the twentieth century. However, except for those examining photoreceptor visual transduction and olfaction, functional studies were largely confined to motile cilia until recently. Many people considered primary cilia to be vestigial, but motile cilia—which occur, for example, on respiratory epithelium, along the female reproductive tract, and on ependymal cells lining the ventricles of the brain and which move mucus or fluid—were thought to be important for health. In particular, the mucociliary escalator of the respiratory tract seemed important for respiratory clearance and the prevention of bacterial colonization (3, 4), although some ciliary function can be superceded by muscle contraction, such as coughing. The mammalian sperm tail contains a motile 9+2 axoneme surrounded by a specialized set of thick fibers. Until the development of in vitro fertilization procedures, sperm motility was essential for fertility, and because male infertility brought many individuals to the clinic, the first true ciliary disease, now known as primary ciliary dyskinesia (PCD), was diagnosed there (5).

It is instructive to realize how much of our present understanding of mammalian cilia depends on work on model organisms, particularly the ciliated protistans *Tetrahymena* and *Chlamydomonas*, and more recently on the model nematode, *Caenorhabditis elegans*. In each case, recent advances in genomics, combined with mutant analysis, have deepened the understanding of mechanisms that are preserved in mammalian cilia. The ciliary proteome has been defined for *Tetrahymena* (6) and *Chlamydomonas* (7). Genes encoding specific ciliary proteins are absent in yeast. There are mouse and human (8) orthologs for virtually all the structurally or functionally characterized ciliary proteins.

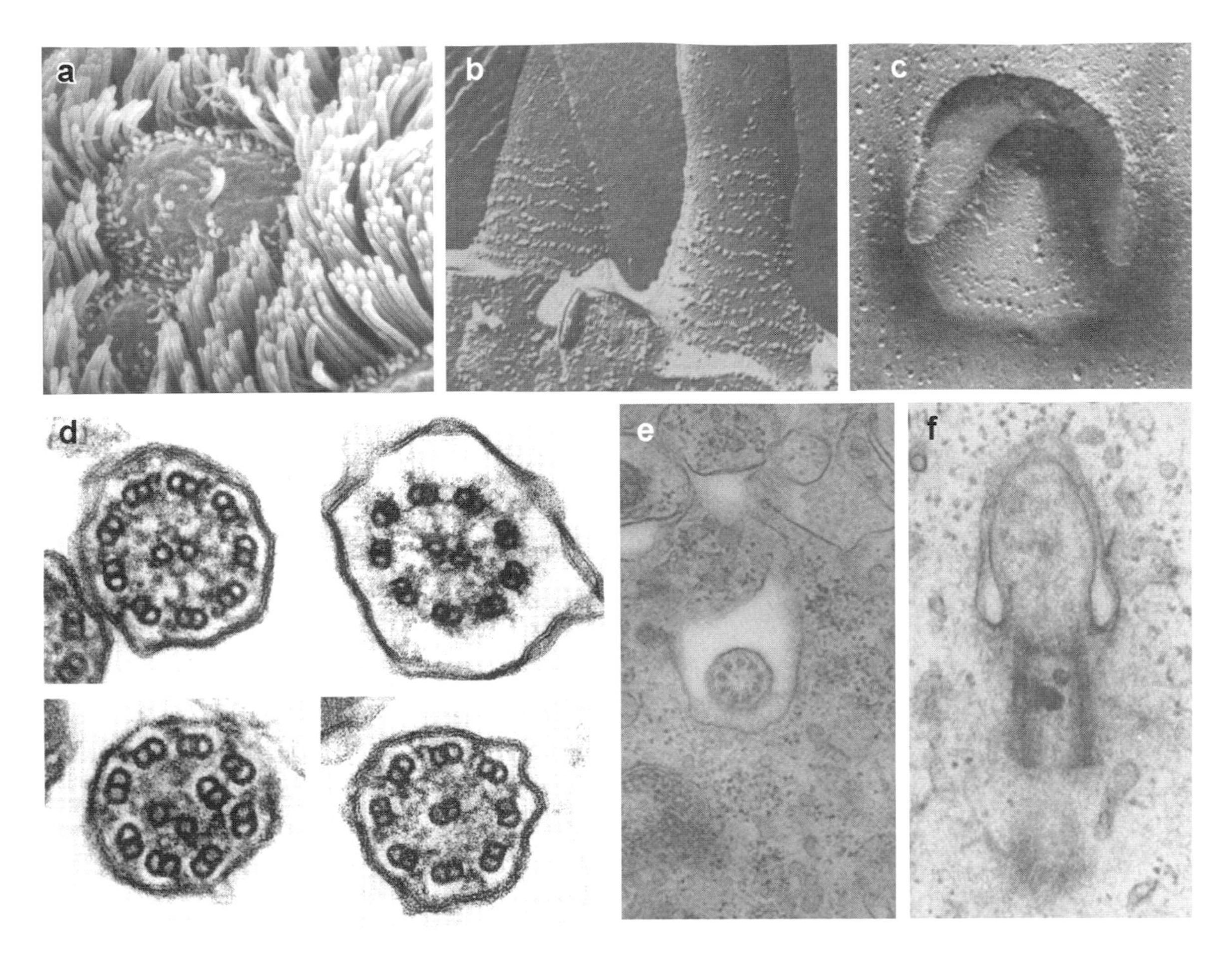

Figure 1

Images of cilia. (*a*) Scanning electron microscopy (SEM) image of oviduct cilia. The central cell shows a primary cilium. Surrounding cells are multiciliated with motile cilia (courtesy of E.R. Dirksen). (*b*) Freeze etch image of motile tracheal cilia showing the multistranded ciliary necklace (from Reference 12, courtesy of Marcel Dekker). (*c*) Freeze etch image of fibroblast primary cilium showing ciliary necklace (from Reference 41, courtesy of *Journal of Cell Biology*). (*d*) Cross sections of human cilia. (*Top left*) Normal motile 9 + 2 axoneme. Other images are of primary ciliary dyskinesia (PCD) cilia. In all images in *d*, ciliary diameter = 0.25 μm. From Reference 3, courtesy of J. Sturgess and *Handbook of Physiology*. (*e*) Cross section of 9 + 0 nonmotile fibroblast cilium (from Reference 64, courtesy of S. Sorokin and *Journal of Cell Biology*). (*f*) Initial step of ciliogenesis of a fibroblast primary cilium (from Reference 64, courtesy of S. Sorokin and *Journal of Cell Biology*).

MOTILE 9 + 2 CILIA

Axonemal Structure

The basic structure of 9 + 2 mammalian cilia was delineated by transmission electron microscopy; the description remains relatively current. The microtubules of the 9 + 2 axoneme polymerize from αβ tubulin heterodimers (9) with the fast polymerizing (+) end at the ciliary tip. Major structures that attach to the microtubules, the outer and inner dynein arms (ODAs and IDAs), the radial spokes, the central-pair projections, and so forth are defined protein complexes, some of whose subunits are AAA-type motors, EF hand proteins, protein kinases, A kinase–anchoring proteins (AKAPs), phosphatases, or proteins with other important

Primary ciliary dyskinesia (PCD): a set of human diseases caused by mutations in ciliary proteins leading to immotile or discoordinated movement, resulting in impaired mucociliary clearance; often coupled to male infertility, hydrocephalus, otitis media, and situs inversus

ODA: outer dynein arm

IDA: inner dynein arm

CBF: ciliary beat frequency

domains for function or assembly. The ODAs and IDAs are force-producing molecular motors that cause the doublet microtubules to slide with respect to one another (10). The doublet sliding is asynchronous with the progression of activity around the axoneme, yielding a helical beat (11). This beat is normally modified by interdoublet links and spoke–central-pair interactions to produce controlled bending, with an effective stroke and a recovery stroke. Maximum beat frequencies range up to approximately 100 Hz, although most reports of mammalian ciliary beat frequency are much lower, perhaps normally 10–20 Hz.

Mechanism of Motion

The ~10–15-μm-long cilium obeys low Reynolds number hydrodynamics, for which viscous forces are paramount. In airway cilia, for example, the effective stroke is vertical, extends into the overlying mucus layer, and propels the mucus toward the pharynx, whereas in the recovery stroke much of the cilium is moving horizontally in the relatively stationary periciliary fluid layer beneath the mucus (12, 13). Freeze etch studies of mammalian sperm axonemes (14) reveal that the structural complexity of the ODAs and IDAs is similar to that in *Chlamydomonas*, except that the ODA of the mammalian sperm axonemes is two (and not three) headed. In general, because the molecular mechanism of motion is conserved from protists to humans, axonemal diameter, which dictates the relationship between doublet sliding and the amount of bending, must also be conserved; axonemal structure is likewise conserved, with little variation.

Recent studies of axonemal structure in PCD and other respiratory diseases reinforce these conclusions. Essentially, genetic defects in respiratory cilia structure that affect beat and transport efficiency lead to PCD (see also Reference 14a). The most prevalent structural defects result in missing ODAs, present in more than 60% of patients with classical PCD clinical profiles (15), sometimes combined with IDA defects. IDA defects are best characterized by computer-assisted image reinforcement (16). In all, arm defects account for approximately 90% of PCD cases when embryonic lethality is not considered. The genetic basis of the ODA defect is most commonly a mutation in *DNAH5*, which codes for one of the ODA heavy chains (17), or in *DNAH1*, coding for a human dynein intermediate chain (18). There is a strong correlation between the number of ODAs (but not of IDAs) present and ciliary beat frequency (CBF); PCD patients have significantly lower CBF (19). These findings are consistent with evidence that ODAs primarily control CBF by increasing or decreasing doublet sliding velocity in the axoneme without greatly affecting beat form, accomplished in many instances by changes in cAMP-dependent ODA light chain phosphorylation (20, 21). Control of CBF via local elevation of Ca^{2+} by mechanostimulation (22) may work partly through this mechanism. The cGMP pathway is also important in this process (see also Reference 22a).

IDAs primarily control parameters related to bend amplitude that affect beat form. This control operates at least in part via signaling kinases and phosphatases that phosphorylate or dephosphorylate radial spoke proteins that act on the velocity of IDA-limited doublet sliding (23). Therefore, although ODA activity controls the overall timing of the stroke cycle, IDA activity of specific doublets controls the amplitudes of the bends originating in the effective and recovery strokes. Radial spoke heads interact with central-pair projections to coordinate doublet sliding activity with bend direction and propagation. With defects in radial spokes, or of central-pair microtubules, IDA activity is not properly regulated, and ciliary beat is paralyzed or abnormal. Some PCD patients have axonemes with the normal complement and composition of dynein arms but with defects in the radial spokes or central-pair structures (**Figure 1**). Because many ciliary proteins that affect normal beat or beat direction may not affect the known major axonemal structures, PCD may occur without obvious axonemal abnormalities.

Aspects of the central pair–radial spoke–dynein arm coordination mechanism have been uncovered by examination of central-pair orientation in *Chlamydomonas*, in which doublet activity is related to central-pair orientation (24) such that orientation is a passive response to bend formation and apparent rotation of the central pair is due to repositioning of the twisted central-pair structure by bend propagation (25). Through central-pair repositioning, different bends propagate along the axoneme at the same rate. This complexity probably is associated with the great potential for changes in the effective stroke direction known for this cell. In mammalian and many other metazoan cilia, the orientation of the effective stroke is fixed, and the orientation of the central pair is roughly perpendicular to that direction. Because all the ODAs, and probably all the IDAs, are unidirectional vectorial force producers, minus-end motors, the axoneme is effectively divided into two operational halves: One half, when active, is responsible for the generation of the effective stroke, whereas the other half generates the recovery stroke. Cilia and basal bodies have only a single enantiomorphic form, with doublet numbers running clockwise when the organelle is viewed from the base toward the tip, so that the two halves define the left and right sides of the cilium and perhaps eventually those of the body (26; see also below). Bend generation in the effective stroke is thought to involve dynein activity on doublets 9, 1, 2, 3, and 4, with doublets 5–6 at the leading edge of the stroke, whereas in the recovery stroke, with doublet 1 in the lead, activity switches to dyneins on doublets 5–6, 7, 8 (27). Ciliary arrest can occur at the switch points, and further beating, for example in respiratory cilia, often requires mechanical stimulation or a burst of cAMP synthesis. When groups of cilia become activated, they stimulate adjacent cilia and initiate metachronal waves that travel across the epithelium (27).

PCD is instructive, not only in demonstrating that the mechanisms of human ciliary motility and its genomic and proteomic control are consistent with what is known from model organisms, but also in delineating the physiological role that motile cilia play in the body. Clinical features of the disease are indications of processes in which ciliary motility is essential (28). Strong phenotypic markers of PCD are chronic rhinitis/sinusitis, otitis media and, as indicated above, male infertility. Female fertility is probably decreased. There is an increased incidence of hydrocephalus, as loss of motile cilia in brain ventricles alters choriod plexus epithelium function and reduces the flow of cerebrospinal fluid through the ventricles (29, 30). There is no correlation with smoking. Cystic fibrosis patients have motile cilia. The function of nodal cilia, the special category of motile 9 + 0 cilia discussed further below, is important for the subset of PCD patients who exhibit reversal of left-right (LR) body asymmetry (situs inversus totalis), which, together with chronic rhinitis and infertility, was identified as Kartageners syndrome.

Left-right (LR) asymmetry: the condition of unpaired organs, such as the heart in the vertebrate body, whereby the organ position on one side of the body is determined; often reversed in animals with PCD

THE CILIARY MEMBRANE

Both 9 + 2 and 9 + 0 ciliary axonemes are surrounded by a ciliary membrane, which extends from and is continuous with the cell membrane but is selectively different from the cell membrane in overall composition. Aside from general descriptions of structure, especially by freeze fracture techniques, and information on lectin or cationic particle binding (31, 32), surprisingly little was known about the ciliary membrane until quite recently. Now, through cilia fractionation and proteomics, a picture of the membrane proteins of the cilium is emerging. In unicellular organisms, in which ciliary response pathways are necessary for survival, control of the 9 + 2 motile cilium depends on specific receptor and channel proteins, including cyclic nucleotide

Ciliary necklace: a specialized region at the base of a cilium defined in freeze fracture electron microscopy; the probable assembly region for transport of cargos into the cilium

receptors, Ca^{2+} channels, and receptors involved in growth control pathways, that are localized to the ciliary membrane (33). In *Chlamydomonas*, membrane proteins in the proteome (7) include six ion pumps or channels, including a homolog of human polycystin 2; three predicted plasma membrane Ca^{2+}-ATPases (PMCAs); and four closely related proteins that have 8 to 12 transmembrane helixes and a PAS domain, a sensory motif involved in detecting diverse stimuli ranging from light or oxygen to redox state and small ligands (34). Evolutionary persistence of sensory function in 9 + 2 cilia of metazoans, and therefore of specific receptors and channels of the mammalian ciliary membrane, would be expected. Because the control of CBF in airway epithelial cells depends upon cyclic nucleotides (35), one might anticipate that, as is true for ciliates, both adenylyl and guanylyl cyclase (AC and GC, respectively) might be localized to the ciliary membrane. However, except in olfactory sensory cilia, for which an odorant-sensitive type 3 AC has been reported, and the adjacent respiratory epithelial cilia, for which lower levels of nonodorant-sensitive AC have been reported (36), there is little specific information as to whether these proteins are in ciliary membranes. There are more than 1000 different odorant receptors, each presumably localized to the ciliary membranes of one olfactory neuron.

Although information on airway ciliary membrane receptors remains sparse, the concept that all cilia have significant sensory function, heavily supported by the new work on primary cilia cited below, led Teilmann & Christensen to compare the distribution of certain receptor tyrosine kinases in primary versus motile cilia of the female reproductive tract. The angiopoietin receptors Tie-1 and Tie-2 localize to motile cilia of the infundibulum (**Figure 2**) and the ampulla of the oviduct (37). The ion channel TRPV4 and polycystins 1 and 2 also localize to ciliary membranes of 9 + 2 motile oviduct cilia (38). The ciliary level of polycystins increases upon ovulation, implying that the ciliary activity of these proteins is associated with the detection of physiochemical changes that establish the environment for oocyte transport, for priming of the ampulla for reception of the oocyte, and for the fertilization and transport of the fertilized oocyte to the uterus.

As we discuss below, polycystins are now commonly known to be present in ciliary membranes of 9 + 0 primary cilia. We predict that similar types of channels and receptors are present in the membranes of all mammalian motile cilia and that they play a role in epithelial homeostasis. Not all important specific channels are localized to the ciliary membrane, of course; for example, CFTR localizes to the apical region of ciliated airway cells but not to the cilia themselves (39).

THE CILIARY NECKLACE

Because specific proteins are localized to or concentrated in the ciliary membrane, as opposed to the rest of the cell membrane, several research groups have postulated that there is a selective barrier at the cilium entrance. This barrier occurs near or more likely in connection with the loading zone for cargo destined for intraciliary transport. The physical manifestation and mode of operation of the barrier are still uncertain, but selection has certain features resembling the passage of material from the cytoplasm into the nucleus, where the barrier is the nuclear envelope (40). One specialized feature of the barrier region that is found on all 9 + 2 and 9 + 0 mammalian and invertebrate cilia that have been studied by freeze fracture electron microscopy, but that is not universally found on sperm, is the ciliary necklace (41). The necklace consists of multiple strands of intramembrane particles that are especially prominent in airway and oviduct cilia (**Figure 1**). The necklace particles line the edges of a cup-like structure with a stem connecting to the center of each basal body doublet just at the transition zone below the origin of the central pair of microtubules and the axoneme. The necklace region

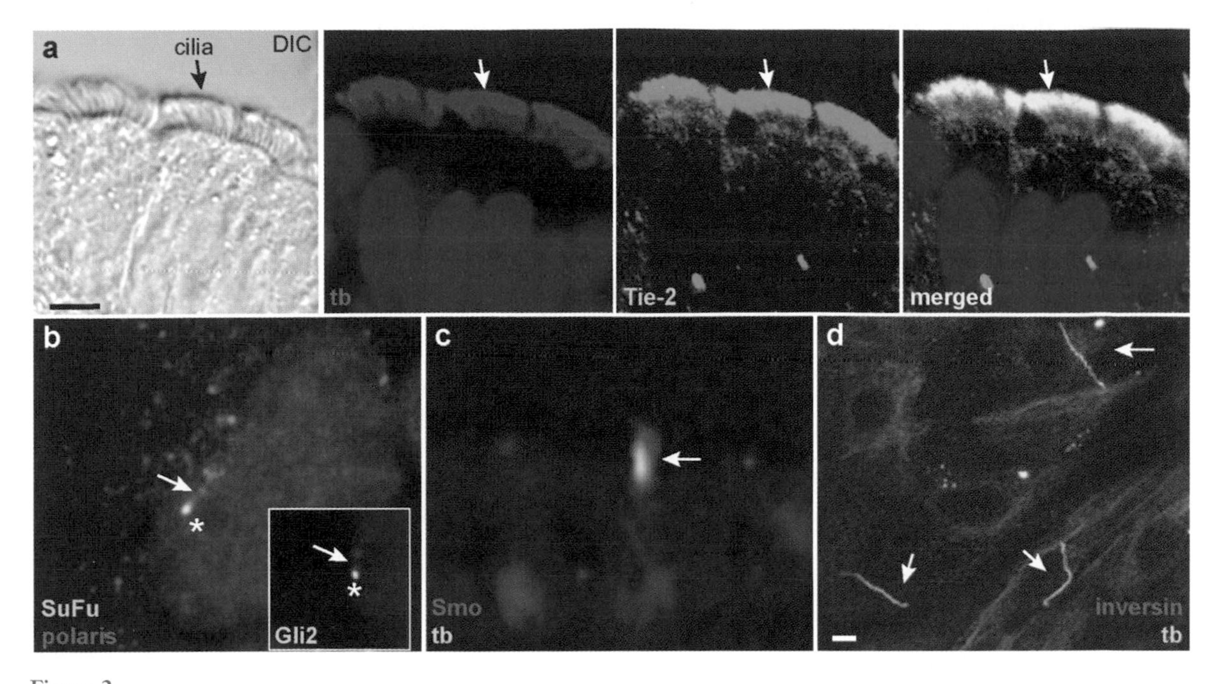

Figure 2

Receptors and signaling molecules localize to cilia (*arrows*). (*a*) Angiopoietin receptor Tie-2 localizes to motile cilia [anti-acetylated α-tubulin (tb)] of the mouse oviduct infundibulum. Nuclei are stained with DAPI (*blue*). Scale bar: 5 μm. From Reference 37 with permission. (*b*) Localization of SuFu (Supressor of Fused) and Gli2 transcription factors to the distal tip of the primary cilium (antipolaris) in mouse primary limb cell cultures (the *asterick* marks the tip of the cilium). From Reference 100 with permission. (*c*) Localization of Smoothened (Smo) to primary cilia (antiacetylated α-tubulin) in cultures of mouse embryonic fibroblasts. From Reference 93 with permission. (*d*) Localization of inversin to primary cilia (tb) of kidney epithelial cells. Similar to Reference 64 (courtesy of L. Eley & J. Goodship).

is readily identified in transmission electron microscopy (TEM) images by this characteristic cross-sectional appearance. Transport proteins have been localized to the necklace region near the repeating intersection of the cup and the membrane (42), which may imply that the cup-like regions are assembly sites for transport of membrane and axonemal cargos.

As befits a barrier, the membrane in the necklace region of airway cilia has a different composition in terms of, for example, lectin binding, anionic charge, and free-cholesterol distribution from the rest of the ciliary and cell membrane (31). Researchers have attempted to isolate and biochemically characterize the necklace region, especially from photoreceptor outer segments (43). A putative guanine nucleotide exchange factor (GEF), retinitis pigmentosa GTPase regulator (RPGR), and its interacting protein are localized to a necklace defining a cross section of the connecting cilium of the photoreceptor (44). RPGR isoforms are also found in the necklace region of motile cilia of the trachea (45). RPGR is mutated in patients with an X-linked phenotype that includes PCD, suggesting that this molecule is required in the development of virtually all mammalian cilia (46). The importance of the necklace in tracheal cilia is reemphasized by its disruption and disappearance upon infection with *Bordetella pertussis* and *Mycoplasma* after attachment of these bacteria to the cilium and prior to cell death. Moreover, when cilia are shed, as in the adult human cochlea, or when deciliation is induced by Ca^{2+} shock, the point of breakage and

GEF: guanine nucleotide exchange factor

Ciliogenesis: the processes of assembly and growth of cilia during cell differentiation

Intraflagellar transport (IFT): the process of transport of materials into and along the cilium for assembly of certain axonemal proteins and placement of receptor and channel proteins in the ciliary membrane

membrane resealing occurs just above the ciliary necklace, and the necklace persists.

If the necklace region is the site of assembly and coupling of membrane protein cargos to the transport machinery, we may reasonably expect to find that membrane proteins targeted to this machinery had ciliary localization signals. Researchers have initiated several searches for localization signals, normally by truncation or site-directed mutagenesis of a known ciliary transmembrane protein, such as polycystin 2 (47). Published information is still fragmentary, but signaling for ciliary localization likely has several different components, depending on the protein species involved and how it is associated with the membrane and membrane scaffold proteins, such as GEFs.

CILIARY MORPHOGENESIS

In multiciliated cells of the mammalian trachea and oviduct, ciliogenesis requires synthesis and assembly of multiple basal bodies and axonemal proteins, accompanied by an enormous increase in membrane area. Human tracheal development has four stages (48). Up to 11 weeks gestation, the trachea is covered by a columnar epithelium with primary cilia. At approximately 12 weeks, ciliogenesis begins with the appearance of fibrogranular masses in the cell cytoplasm. As Dirksen (49, 50) first described, these masses develop central elements, called condensation forms or deuterosomes, around which an explosive development of centrioles, which will become ciliary basal bodies, occurs. The mature centrioles move to the cell membrane, to which they attach, and the ciliary necklace develops as the axoneme begins to elongate (32). The necklace enlarges, and strands are added as the cilium grows. Axonemal dynein is synthesized and found in the cell cytoplasm before ciliogenesis is apparent (51). By 24 weeks, the ciliated border with 200–300 9 + 2 cilia is mature. In the oviduct, centriolar placement and ciliary growth are asynchronous. Cilia first grow at the cell periphery, giving the cell surface a daisy-like appearance: Longer cilia at the periphery encircle shorter cilia at the cell center (52). A distinct tip with an axonemal cap and a unique glycocalyx called the ciliary crown develops on the mature cilium (53), and growth ceases.

At least one transcription factor, hepatocyte nuclear factor 3/forkhead homolog 4 (HFH-4), is known to be involved in regulating the morphogenesis of motile cilia in multiciliated mammalian epithelia (54). In *HFH-4*-null mice, classic 9 + 2 motile cilia are absent in epithelial cells, but 9 + 0 cilia, including nodal cilia, are present, although nodal cilia are probably nonfunctional. Ultrastructural analysis of these animals showed that, in the epithelial cells missing motile cilia, centriole migration and apical docking are abnormal.

The axoneme grows and continuously turns over at its distal tip (55, 56). Axoneme growth and maintenance depend on transport of axonemal precursors from the cell body to the assembling tip. Rosenbaum and collaborators (40), through their work with *Chlamydomonas*, first envisioned and analyzed the molecular basis of the transport process. Historically, the 9 + 2 cilia of this organism were termed flagella, and therefore the transport process building the axoneme was termed intraflagellar transport (IFT). A transport complex consisting of IFT proteins can be visualized moving up and down the cilium, between the membrane and the growing axoneme. Axonemal components such as radial spoke proteins move as cargo, partially assembled, with these complexes and are deposited at the growing tip of the axoneme, where they dock into their proper positions (57). The complexes and their cargos are moved toward the tip by a kinesin-2 molecular motor. After depositing the cargo and picking up axonemal turnover products, the complexes are moved back toward the base by a special isoform of cytoplasmic dynein, dynein 2, also known as dynein 1b (40, 58a). The unloading/loading of cargo proteins and exchange of transport motors at the flagellar tip may occur by a three-step mechanism that may be regulated by flagellar tip proteins (59). IFT protein complexes also participate in the movement of ciliary membrane proteins and signaling molecules (58b, 60). Presumably, vesicles derived from the Golgi apparatus with appropriate scaffold proteins fuse with the cell membrane near the ciliary necklace. This results in the addition of membrane around the elongating cilium and the delivery of peripheral membrane proteins, as well as of transmembrane channels and receptors such as polycystins and TRPV, to the transport apparatus for movement into the growing ciliary membrane. IFT orthologs are prominent in ciliary proteomes from all organisms studied (61), suggesting that this transport mechanism is ancient and that it has been retained for transport in both 9 + 2 and 9 + 0 cilia. IFT has been of great importance for understanding the function of mammalian 9 + 0 primary cilia as sensory organelles, as we discuss below. Hagiwara et al. (62) provide a more detailed review of ciliogenesis in mammalian cilia.

NONMOTILE PRIMARY 9 + 0 CILIA

Primary cilia, which are found on virtually every cell in the body, with some notable exceptions (63) retain the basic axonemal doublet structure as they grow from the basal body. However, they do not assemble the central microtubule complex or IDAs and ODAs, and therefore, like mutants of motile cilia lacking these structures, they do not actively beat. Other aspects of ciliogenesis are normal. There is a ciliary necklace and a ciliary membrane with receptors and channels, comparable with and sometimes identical to those found in motile cilia. Ciliogenesis is via transport mechanisms and IFT proteins completely orthologous to those in *Chlamydomonas*, although certain aspects of growth differ from those of multiciliated cells. In particular, there is no explosive growth of basal bodies from a fibrogranular mass. Instead, the single primary cilium usually originates when a Golgi-derived vesicle encapsulates the distal end of a mother centriole (**Figure 1**). The necklace region develops at the point of encapsulation. Fusion of additional vesicles liberates the growing cilium to the extracellular medium (64).

Although the motile function of 9 + 0 cilia has been lost, the sensory and signal transduction features necessary for survival in single-celled organisms have been retained and perhaps enhanced. There has been a persistent literature on primary cilia from the early 1960s, summarized in part by Wheatley (63, 65). These studies postulated, with some evidence, that the primary cilia could function as chemosensors, mechanosensors, or positional sensors and that their presence was coupled to and perhaps important for the cell cycle and cell differentiation (66). Nevertheless, for more than 30 years after their electron microscopy (EM) characterization, primary cilia were largely dismissed as insignificant or unimportant in leading textbooks of cell biology.

Polycystic kidney disease: a family of genetic diseases whereby the kidney tubule epithelium dedifferentiates and proliferates along the tubule to produce enlarged irregularly spaced expansions leading to loss of function; probably produced by defective signaling in primary cilia

The Revelation of Primary Cilia Biology: Polycystic Kidney Disease

The rapid increase in our understanding of primary cilia biology was primarily due to the work of Pazour et al. (67), which relied on the development of $Tg737^{orpk}$ mouse as a model for human autosomal-recessive polycystic kidney disease. Pazour et al. demonstrated that the mouse homolog of IFT protein 88 in *Chlamydomonas* is the mutated protein of the *Tg737* gene and that with this defect the primary cilia of the mouse kidney are abnormally short or absent.

Long primary cilia are seen by SEM on kidney tubule cells (68), and these can readily be visualized in light microscopy in kidney cell lines such as PtK1 or MDCK cells in confluence in tissue culture (69). These cilia can be bent by fluid flow (70) or mechanically. Praetorius & Spring (71, 72) showed that bending causes intracellular Ca^{2+} to increase and that removal of the cilium abolishes this flow-sensing response. The primary cilia of the kidney tubule act as mechanosensors that dose the cell with periodic increases in Ca^{2+}. Ca^{2+} is a well-known second-messenger molecule whose periodic influx influences specific gene activity. Mechanoregulation of Ca^{2+} is defective in the *Tg737*^{orpk} mouse kidney (73). A clear inference from these studies is that mechanotransduction via the primary cilium is necessary for continued normal function and cell differentiation of the kidney epithelium and that loss of the cilium leads to abnormal function, abnormal cell division, and polycystic kidney disease.

Many mutations that cause polycystic kidney disease are found not in genes that encode IFT proteins or the molecular motors involved in building the cilium. Rather, such mutations are found in genes that encode the polycystins, of which polycystin 1 is a transmembrane receptor and polycystin 2 is a Ca^{2+} channel. If the cilium is normal, but the polycystin is defective, just as if the polycystin is normal and the cilium defective, there will still be polycystic kidney disease; how is this possible? The answer is that the polycystins are proteins of the ciliary membrane (74, 75) and that the signal transduction pathway leading to Ca^{2+} influx must operate through the ciliary membrane to give a normal homeostatic response.

Moreover, polycystin 1 evidently functions as a G protein–coupled receptor that can activate growth control pathways via standard signal transduction cascades (76). Its extracellular domain can interact with carbohydrate or protein moieties. Although its physiological ligand is still unspecified, polycystin 1 may be a chemosensor as well as a mechanosensor. In addition to Ca^{2+} as a signal, receptor phosphorylation cascades and receptor- or mechanically induced regulated intramembrane proteolysis (77) may link ciliary membrane events to the cell cycle.

These features are consistent for all primary cilia studied. Like the motile cilia from which it is evolutionarily derived, the primary cilium incorporates specific receptors and channels into its ciliary membrane. Although some of the specific receptors may be cell type dependent, others, including polycystin 1 and 2, are more widely found. The primary cilium can respond either to mechanical stimuli or to defined ligands. The signal pathway is initiated in the cilium before transmission to the rest of the cell. Normal signaling leads either to homeostasis, that is, maintenance of the differentiated tissue state, or to controlled division and differentiation. In its homeostatic functions, the primary cilium has been compared to a cellular cybernetic probe (78; see also below) or a cellular global positioning device (79). Abnormal signaling can be due to faulty ciliogenesis or to misplacement or mutation of ciliary membrane proteins. This leads to abnormal patterns of tissue growth and cell division and to a variety of human pathologies.

PRIMARY CILIA OF CONNECTIVE TISSUES

In contrast to epithelial cells, primary cilia of connective tissues are embedded in the extracellular matrix. Consequently, these cilia may interact with matrix components and respond to tension on the matrix or to growth factors to control tissue homeostasis and cell cycling. Many of the ideas now applied to epithelial cell primary cilia were originally derived from studies of primary cilia of chondrocytes and fibroblasts.

The Chondrocyte Primary Cilium

Poole and coworkers (78, 80, 81) established that the position, projection, and orientation of the primary cilium in chondrocytes are influenced by the structural organization and mechanical properties of the extracellular matrix. This led to the suggestion that primary cilia in connective tissue cells transduce mechanical, physiochemical, and osmotic stimuli through ciliary receptors and ion channels to control tissue development and to construct a mechanically robust skeletal system. Indeed, *Tg737*^{orpk} mice with defective primary cilia display limb/skull patterning defects, including polydactyly and brachydactyly. Jensen et al. (82) used double-tilt electron tomography to demonstrate that the chondrocyte primary cilium makes physical contact with matrix components such as collagen fibers and proteoglycans, probably via specific extracellular matrix (ECM) receptors, including integrins and NG2 in the chondrocyte cilium (83). This reinforces the idea that mechanical stimuli in cartilage are transmitted through the primary cilium. Integrins potentiating chemosensory, fibronectin-induced Ca^{2+} signaling are also found on primary cilia of MDCK cells (84), although the precise function of this signal pathway is unknown.

The Fibroblast Primary Cilium

Similarly, Albrecht-Buehler (85) used cultures of fibroblasts to illustrate a unique relationship between the orientation of the primary cilium and the direction of cell migration, in which the cilium is aligned with the intended direction of migration. Fibroblast primary cilia could sense the substrate as cells move and relay information from the extracellular milieu to the centrosomal region. In connective tissue, fibroblasts, like chondrocytes, are embedded in a series of protein fibers, some of which may attach and pull on the primary cilium, producing a mechanical stress. Fibroblasts in tissues are most often in a state of growth arrest, but if stimulated by growth factors, they may enter the cell cycle for proliferation and migrate throughout the matrix wherever they are needed, such as in wound healing and tissue reorganization.

As in most cells, the expression of primary cilia in cultured fibroblasts is closely regulated during the cell cycle (86), indicating a role of the cilium in growth control. Primary cilia grow when the cells become confluent and enter G_0. Tucker and coworkers (87, 88) found that stimulation with platelet-derived growth factor (PDGF) and administration of Ca^{2+} ionophores induce calcium fluxes as well as resorption of the primary cilium, which is an early event in the transition from growth arrest to cell proliferation in 3T3 fibroblast cultures. Tucker et al. concluded that signaling produced shortening of the cilia, which was required for normal growth factor–mediated cell cycle entrance.

PDGF: platelet-derived growth factor

PDGF may act directly through receptors in the primary cilium of fibroblasts (89), implying a direct role of the connective tissue primary cilium in communicating both mechanical and physiochemical information that controls migration, cell survival, and growth control in tissue homeostasis. In 1996 Lih et al. (90) demonstrated that PDGF receptor alpha (PDGFRα) is encoded by a growth-arrest-specific gene, such that the receptor is upregulated during growth arrest in cultures of NIH3T3 fibroblasts, which may facilitate the exiting of cells from growth arrest upon mitogenic stimulation by PDGF. The homodimer of PDGFRα, PDGFR$\alpha\alpha$, is specifically activated by PDGF-AA, which regulates several physiological and pathophysiological processes in a variety of tissues. Schneider et al. (89) (**Figure 3**) demonstrated that PDGFRα is targeted to the primary cilium during growth arrest in NIH3T3 cells and primary cultures of mouse embryonic fibroblasts. When PDGF-AA is added, ciliary PDGFR$\alpha\alpha$ is phosphorylated, followed by activation of the mitogenic Mek1/2-Erk1/2 pathway, which also operates in the cilium. Quiescent fibroblasts derived from *Tg737*^{orpk} mutants fail to upregulate PDGFRα, to form normal primary cilia, and to activate Mek1/2-Erk1/2, and these fibroblasts fail to re-enter the cell cycle after stimulation with PDGF-AA. In contrast, PDGF-BB-mediated signaling through PDGFRβ in the plasma membrane is not affected in *Tg737*^{orpk} mutants, indicating a unique function of the primary cilium in balancing the signals required to regulate cell cycle entrance and to maintain tissue homeostasis. Mutations in PDGFRα are known to play a role in the generation of a series of human cancers, supporting the conclusion that perturbation of the growth control pathway from PDGFRα-enriched primary

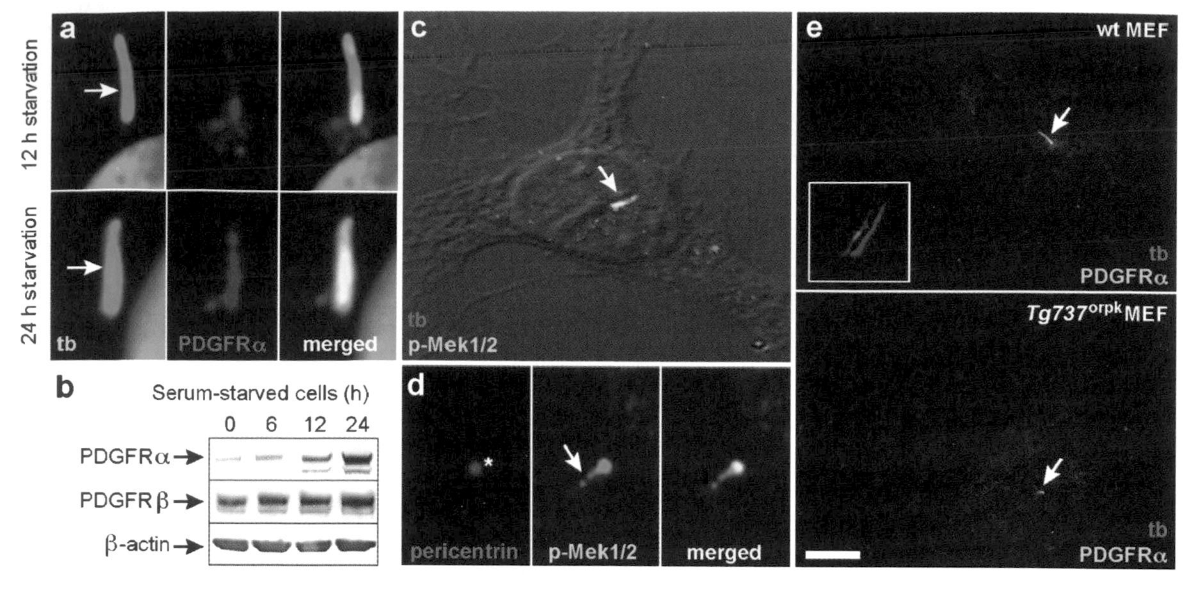

Figure 3

Growth factor–mediated signaling in fibroblast primary cilia (*arrows*). (*a*, *b*) PDGFRα is specifically upregulated and moves into the primary cilium by 24 h of serum starvation of NIH3T3 cells. (*c*, *d*) Mek1/2 is activated by phosphorylation in the cilium and at the ciliary basal body after ligand stimulation of PDGFRαα. (*e*) Comparison between wild-type (wt) and $Tg737^{orpk}$ mouse embryonic fibroblasts shows that primary cilia are deficient in the mutant. Anti-acetylated α-tubulin (tb) identifies the primary cilium and pericentrin spots of the basal body. Scale bar: 10 μm. From Reference 89 with permission.

cilia may be important in the onset or prevention of oncogenesis. Furthermore, ciliary PDGFRα may continuously signal a mechanical stress on the connective tissue, known to activate PDGFRα in cultured aorta smooth muscle cells, which have primary cilia. Wheatley and coworkers (91) demonstrated the absence of primary cilia in early passages of cultured fibroblasts from patients with Werner syndrome and Mulvihill-Smith progeria–like syndrome.

These observations are consistent with the idea that the cilium has a sensory function in developmental processes and early adulthood, and to this end, it is tempting to speculate that there may be other signaling systems in the fibroblast primary cilium, such as inversin and components of the Wnt and Hh signaling pathways (92, 93; see also below), polycystin, or integrin, which may act in concert with PDGFRα to regulate fibroblast function in embryonic patterning and adult tissue homeostasis. The taurine transporter, TauT, which is expressed and localized to primary cilia of NIH3T3 cells (94, 95), may further contribute to this complex of signaling systems in the fibroblast primary cilium. Taurine helps to regulate cell volume control, ion channel activity, and calcium homeostasis in developing and adult mammalian tissues (96) and may thus modulate Ca^{2+}-dependent signaling in the primary cilium. **Figure 4** is primarily based on an extrapolation from those studies that suggest that in the absence of PDGF the fibroblast primary cilium acts essentially as a mechanotransducer, producing a rise in intracellular Ca^{2+}, which acts as in the kidney to maintain the G_0 state of the cell, but that in the presence of ligand, the cilium acts as a chemoreceptor, generating signals that induce normal cell cycling, for example, in wound healing.

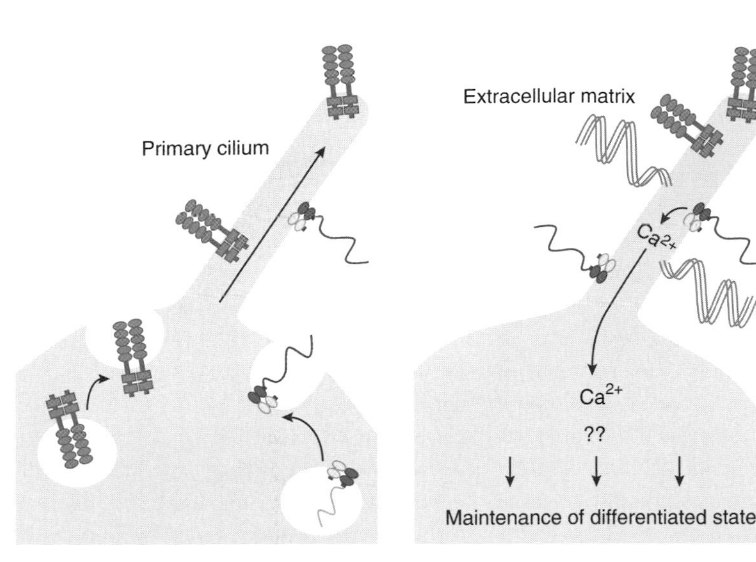

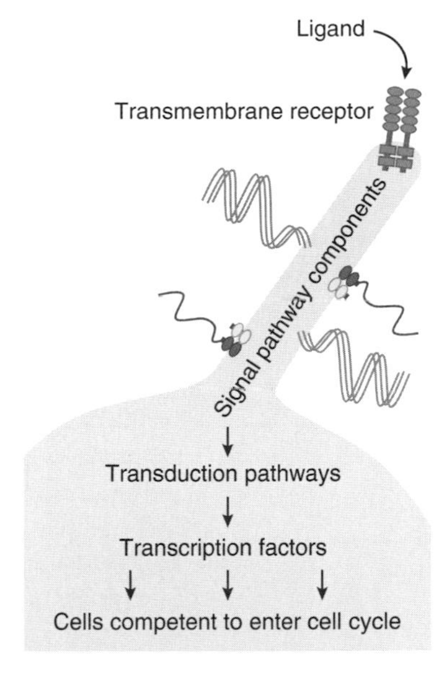

Figure 4

Pathways of membrane differentiation and response in primary cilia. (*Left panel*) In growth arrest (G_0), as the primary cilium grows, specific channels and receptors are concentrated in the ciliary membrane. (*Middle panel*) Channels such as polycystin 2 respond to mechanostimulation by an influx of Ca^{2+} necessary for maintenance of the differentiated cell. (*Right panel*) Receptors such as PDGFRαα respond to specific extracellular ligands by initiating signal transduction pathways that control cell division.

CILIARY HEDGEHOG AND WNT SIGNALING IN DEVELOPMENT

Hedgehog (Hh) and Wnt signaling pathways highly regulate the orchestration of embryonic development and patterning in a range of tissues that direct the growth and spatial plan of the early embryo. Increasing evidence points to a strong relationship between abnormal Hh and/or Wnt signaling and mammalian disorders and pathologies that are caused by defects in cilia, including Bardet Biedl syndrome, Kartageners syndrome, heterotaxia, pulmonary dysfunction, hydrocephalus, holoprosencephaly, infertility, polydactyly, polycystic kidney disease, cancer, and retinal degeneration. Hh operates via primary cilia, and some Wnt signaling pathways require cilia for their function. Protein components participating in either signaling system are coupled to IFT proteins and ciliary assembly and may specifically be regulated in the primary cilia (93, 97–101). These findings open up a whole new realm of ciliary functions in development and tissue homeostasis for exploration.

Cilia and Hh Signaling

Important elements of vertebrate Hh signaling include a family of secreted Hh proteins that, upon binding to the transmembrane patch protein (Ptc), abolish the inhibitory effect of Ptc on the seven-transmembrane receptor smoothened (Smo). This allows Smo to transduce a signal via glioma (Gli) transcription factors to the nucleus for the expression of Hh target genes (102) that control an array of responses at different times and in different cell types (103, 104). In the mouse there are three homologs of Hh proteins: Sonic Hh (Shh), Indian Hh (Ihh), and Desert Hh (Dhh);

these play an essential role in embryonic development, such as in LR asymmetry (105), in limb development, and in neurogenesis. Gli proteins lie downstream in the control of the diverse functions of the Hh pathway. In the mouse Gli1 and Gli2 proteins generally act as activators, and Gli3 as both activator (Gli3A) and repressor (Gli3R) of the Hh pathway, and Smo may control the activation of Gli proteins as well as the proteolytic events that generate the Gli3 repressor (104).

Hh signaling is tightly coupled to the maintenance and function of primary cilia in various mammalian cell types. Three mouse IFT complex B proteins—*Tg737*/polaris as well as the anterograde IFT motor protein subunit Kif3a and the retrograde IFT motor protein subunit Dnchc2—are all necessary for Shh signaling at a step between Smo and Gli proteins, such that loss of IFT causes both neural tube and limb patterning defects at the level of Gli3 processing (100, 104, 106, 107). Haycraft et al. (100) more recently demonstrated that all three full-length Gli transcription factors as well as SuFu (suppressor of fused and repressor of the Hh pathway) colocalize to the distal tip of primary cilia in mouse primary limb bud cells (**Figure 2**). Importantly, loss of polaris in $Tg737^{\Delta2-3\beta-Gal}$ mutants leads to a series of changes in Hh signaling, such as loss of *Ptc* and *Gli1* expression in response to Shh in the limb bud and disruption of Gli2 and full-length Gli3 function in primary limb bud cells; these result in severe polydactyly (108) similar to that following Gli3 loss (109). Haycraft et al. (100) further showed that partial loss of polaris function in $Tg737^{orpk}$ exacerbates the phenotype of *Gli3*-heterozygous mutants. Thus, it is increasingly evident that Hh signaling regulates embryonic development via the primary cilium, in which Gli regulation and processing occur in the cilium, and that loss of cilia results in inefficient Gli3 processing, leading to embryonic and patterning defects. Aberrant activation of Gli oncogenes may also cause cell transformation and tumorigenetic processes by directing target genes in growth control, cell survival, and metastasis (102), suggesting that dysfunctional Hh signaling in primary cilium also plays a major role in cancer.

Hh signaling follows the general rule whereby the primary signaling receptor must be localized to the ciliary membrane to permit normal cell development and signaling begins in responding proteins localized in the cilium itself, but perhaps there is amplification and feedback involved. Corbit et al. (93) demonstrated that whereas Smo localizes to the primary cilium (**Figure 2**), this localization becomes amplified in Shh signaling. In contrast, the Smo antagonist cyclopamine inhibits ciliary localization. Activation of the Hh pathway by Shh markedly upregulates ciliary Smo localization in cultures of MDCK cells, inner medullary collecting duct cells, and mouse embryonic fibroblasts. As discussed above, common motifs for ciliary localization of transmembrane proteins are to be expected, but the exact type of motif may vary, depending on how the transmembrane protein is coupled to the transport machinery at the ciliary necklace. Hh signaling depends upon a specific ciliary localization motif, comprising a conserved hydrophobic and basic residue similar to that of other ciliary seven-transmembrane receptors, including ODR-10 and STR-1 in olfactory cilia in *C. elegans* (110) and somatostatin receptor 3 and serotonin receptor 6 in primary cilia of the mammalian CNS (111, 112). Smo localizes to nodal cilia in early and late headfold stages as well as in the two-, three-, and five-somite stages, supporting a critical role for ciliary Shh signaling in later embryonic development.

Wnt Signaling and the Role of Inversin

The gene *Invs* encodes a protein named inversin that localizes to primary cilia in kidney epithelial cells (113) (**Figure 2**), to nodal cilia, to fibroblast cilia in cell cultures, and to the pituitary gland (92). Inversin is the protein that is mutated in nephronophthisis type 2 (114), and inversin in kidney cells binds anaphase-promoting complex 2 (Apc2), indicating that the primary cilium serves to regulate the cell cycle and that the cilium plays a role in aberrant cell proliferation, which is a hallmark of the cystic process (113). Similarly, inversin contributes to LR determination and is essential for the generation of normal nodal flow (92). A phenotypically similar (iv^-) mouse is defective in nodal cilia LR dynein (115).

Wnt signaling is generally divided into two separate signal transduction pathways known as the canonical and the noncanonical Wnt pathways, which lead to two different end points. The canonical pathway operates through β-catenin, whereas the noncanonical pathway, also known as the planar cell polarity (PCP) pathway, acts through the membrane protein, Van gogh–like 2 (Vangl2). Inversin functions as a molecular switch between the two pathways (97) by targeting the Wnt pathway protein dishevelled for degradation; inversin inhibits the canonical pathway and activates the noncanonical pathway. Fluid flow increases the level of inversin in the cilia. Supporting the conclusion that the canonical Wnt signal transduction pathways are downregulated by the primary cilium, conditional inactivation of KIF3A of kinesin-2 that affects ciliogenesis in kidney tubule cells leads to increased expression of β-catenin and PKD as well as to cell proliferation (116). Similarly, Cano et al. (117) showed that in the *Tg737* ^{orpk} mouse there is an increase in β-catenin expression and localization to dilated ducts in pancreas and increased expression levels of Wnt signaling transcription factors (117). Finally, Ross et al. (118) showed that Vangl2 localizes to cilia and ciliary basal bodies in primary cilia of collecting duct cells as well as in human respiratory epithelial cells. In kidney cells Vangl2 genetically interacts with Bardet Biedl Syndrome (BBS) genes, indicating that cilia are intrinsically involved in PCP processes. PCP proteins in turn feedback into ciliogenesis pathways necessary for Hh signaling (119); in this way the Wnt and Hh pathways interact.

Is polycystin-mediated Ca^{2+} signaling in primary cilia also integrated with Wnt signaling? A Wnt-Ca^{2+} pathway has been implicated as a third Wnt signaling cascade; this pathway is thought to influence both the PCP and canonical Wnt pathways, thereby regulating cell adhesion and cell movements during gastrulation (120). Upon activation the Wnt-Ca^{2+} pathway increases intracellular levels of Ca^{2+}. It is therefore tempting to speculate that this Wnt pathway may interplay with those signaling systems that control Ca^{2+}-mediated processes in mammalian cilia, such as those regulated by polycystins in both primary and motile cilia. Indeed, we may have seen only the tip of the iceberg as to the plethora of ciliary functions associated with Wnt as well as Hh signaling.

NODAL CILIA

The correlation of human ciliary disease with randomization of LR asymmetry manifested as situs inversus led Afzelius (5) to conclude that embryonic ciliary function was necessary for proper LR positioning of organs, such as the heart, within the body. Unexpectedly, such cilia are present at the site of gastrulation, the embryonic node (121). Nodal cilia, like primary cilia, number one per cell with 9 + 0 axonemes, but nodal cilia possess dynein arms with LR dynein (115, 122) and are motile, generating a leftward flow across the node (124). When the mouse intraciliary transport motor KIF3B was disrupted by gene targeting, the node lacked cilia, and LR asymmetry was randomized (123). This led to the hypothesis that nodal flow generated by the cilia was critical for the development of LR asymmetry. Further evidence for the nodal flow hypothesis was provided by examining mutant mice with paralyzed nodal cilia and from experimentally reversing flow, thus reversing LR asymmetry (124).

To produce nodal flow, the cilia must have an asymmetrical component to their movement. Rotation is always clockwise when one looks along the cilium base to tip, suggesting that doublet activity moves around the axoneme clockwise from doublet 9 to 1 to 2 and so on. Just as in 9 + 2 cilia discussed above, for which the effective stroke of the cilium always corresponds to certain doublet activity, the doublets that produce effective flow are a constant half-subset of the axoneme. In this way, the basic asymmetry of the ciliary axoneme determines the basic LR asymmetry of the body. The effective force is enhanced by a posterior tilt of the cilium (11) and by additional mechanisms (125).

Exactly how nodal flow induces LR asymmetry is still uncertain, but it seems likely that, in common with the primary cilia discussed above, induction involves a mechanosensory mechanism involving polycystins, leading to Ca^{2+} influx at the left side of the node, and a chemosensory pathway involving Hh and perhaps additional signaling pathways. McGrath et al. (126) found that LR dynein was present only on a centrally located set of nodal cilia, which are presumably those generating flow, whereas polycystin 2 was found on all nodal cilia. These investigators suggest that the cilia around the periphery of the node are nonmotile primary cilia that sense the flow and respond to initiate asymmetric Ca^{2+} signaling. Such signaling may act further via a Wnt signaling pathway (127). Tanaka et al. (128) propose a second mechanism involving FGF-induced release of membrane-bounded vesicles termed nodal vesicular particles (NVPs) that seem morphologically similar to lung surfactant. NVPs carry Shh and retinoic acid; they are transported leftward to impinge on the primary cilia of the left wall, initiating Shh and other transduction pathways. In support of this hypothesis, Tanaka et al. (128) show that FGF receptor inhibition at the node inhibits the leftward Ca^{2+} influx observed by McGrath et al. (126). The Ca^{2+} signals may be restored by Shh, retinoic acid, and Ihh. Motile 9 + 2 cilia in other tissues may function similarly to form a gradient of signal molecules in the local cellular environment to control tissue homeostasis and function, such as in the oviduct, which is subjected to dramatic changes in the physiochemical milieu during the estrous cycle. Indeed, Sawamoto et al. (129) showed that neuroblast migration from the subventricular zone to the olfactory bulb in the adult brain parallels cerebrospinal fluid flow generated by beating ependymal cilia in which the relocation of cells may be guided by the formation of a gradient of signal molecules along the path of migration.

In summary, nodal cilia are an interesting and important intermediate between 9 + 2 motile cilia and 9 + 0 primary cilia in the body, relying on features of each for their function. Additional perspectives on these features are found in Tabin & Vogan (130), Hirokawa et al. (131), and new reviews on ciliary signaling (132, 133). The interrelationships between signaling pathways in nodal cilia that lead to the determination of LR asymmetry parallel the complexity of signaling in primary cilia of other tissues with different physiological endpoints and probably in motile 9 + 2 cilia as well. Exactly how general the features of ciliary signaling are and whether important variations will prove to be tissue-specific, especially for complex tissues such as respiratory epithelium, remain to be determined.

SUMMARY POINTS

1. Mammalian cilia occur in two major patterns: motile 9 + 2 cilia and usually nonmotile 9 + 0 primary cilia.
2. Both motile and primary cilia have an axoneme surrounded by a ciliary membrane that selectively incorporates specific receptors and ion channels.

3. All cilia have sensory function, in part as mechanotransducers and in part in chemoreception.
4. Motile cilia transduce messengers such as cAMP and Ca^{2+} into increased beat frequency or changes in beat form for efficient mucociliary transport.
5. Primary cilia function as cellular positioning systems with hedgehog and Wnt signaling pathways running through them, as well as in growth control through ciliary PDGFR$\alpha\alpha$ signaling.
6. Mutations in ciliary proteins lead to ciliopathies such as primary ciliary dyskinesia, polycystic kidney disease, retinitis pigmentosa, and hydrocephalus.
7. Motile 9 + 0 nodal cilia create a leftward flow of morphogens that is critical for left-right asymmetry development in the body.

ACKNOWLEDGMENTS

We gratefully acknowledge Charles F. Guerra for help with figures and diagrams and Dr. Lotte B. Pedersen for valuable comments on the IFT section. We apologize to those whose work is not described in this review owing to restricted space. This work was supported in part by The Carlsberg Foundation, Fonden af 1870 and The Danish National Science Research Council grant 57462 (S.T.C.) and in part by NIDDK grants DK41918 and DK41296 (P.S.).

LITERATURE CITED

1. Porter KR. 1957. The submicroscopic morphology of protoplasm. *Harvey Lect.* 51:175–228
2. Satir P. 2005. Tour of organelles through the electron microscope: a reprinting of Keith Porter's classic Harvey Lecture with a new introduction. *Anat. Rec. A* 287:1184–85
3. Satir P, Dirksen ER. 1985. Function-structure correlations in cilia from mammalian respiratory tract. In *Handbook of Physiology-Respiratory System*, ed. AP Fishman, AB Fisher, 1:473–94. Bethesda, MD: Am. Physiol. Soc.
4. Satir P, Sleigh MA. 1990. The physiology of cilia and mucociliary interactions. *Annu. Rev. Physiol.* 52:137–55
5. Afzelius BA. 1976. A human syndrome caused by immotile cilia. *Science* 193:317–19
6. Smith JC, Northey JG, Garg J, Pearlman RE, Siu KW. 2005. Robust method for proteome analysis by MS/MS using an entire translated genome: demonstration on the ciliome of *Tetrahymena thermophila*. *J. Proteome Res.* 4:909–19
7. Pazour GJ, Agrin N, Leszyk J, Witman GB. 2005. Proteomic analysis of a eukaryotic cilium. *J. Cell Biol.* 170:103–13
8. Ostrowski LE, Blackburn K, Radde KM, Moyer MB, Schlatzer DM, et al. A proteomic analysis of human cilia: identification of novel components. *Mol. Cell Proteomics* 1:451–65
9. Nogales E, Whittaker M, Milligan RA, Downing KH. 1999. High-resolution model of the microtubule. *Cell* 96:79–88
10. Satir P. 1997. Cilia and related microtubular arrays in the eukaryotic cell. In *Handbook of Physiology*, ed. JF Hoffman, JD Jamieson, pp. 787–817. New York: Oxford Univ. Press

11. Okada Y, Takeda S, Tanaka Y, Belmonte JC, Hirokawa N. 2005. Mechanism of nodal flow: a conserved symmetry breaking event in left-right axis determination. *Cell* 121:633–44

12. A useful review of ultrastructural aspects of mammalian respiratory tract cilia.

12. Sanderson MJ, Dirksen ER, Satir P. 1990. Electron microscopy of respiratory tract cilia. In *Electron Microscopy of the Lung*, ed. DE Schaufnagel, pp 47–69. New York: Marcel Dekker

13. Boucher RC. 2004. New concepts of pathogenesis of cystic fibrosis lung disease. *Eur. Respir. J.* 23:146–58

14. Vernon GG, Woolley DM. 2002. Microtubule displacements at the tips of living flagella. *Cell Motil. Cytoskel.* 52:151–60

14a. Zariwala MA, Knowles MR, Omran H. 2007. Genetic defects in ciliary structure and function. *Annu. Rev. Physiol.* 69:423–50

15. Noone PG, Leigh MW, Sannuti A, Minnix SL, Carson JL, et al. 2004. Primary ciliary dyskinesia: diagnostic and phenotypic features. *Am. J. Respir. Crit. Care Med.* 169:459–67

16. Escudier E, Couprie M, Duriez B, Roudot-Thoraval F, Millepied MC, et al. 2002. Computer-assisted analysis helps detect inner dynein arm abnormalities. *Am. J. Respir. Crit. Care Med.* 166:1257–62

17. Olbrich H, Haffner K, Kispert A, Volkel A, Volz A, et al. 2002. Mutations in DNAH5 cause primary ciliary dyskinesia and randomization of left-right asymmetry. *Nat. Genet.* 30:143–44

18. Pennarum G, Escudier E, Chapelin C, Bridoux AM, Cacheux V, et al. 1999. Loss-of-function mutations in a human gene related to *Chlamydomonas* IC78 result in primary ciliary dyskinesia. *Am. J. Hum. Genet.* 65:1508–19

19. de Iongh RU, Rutland J. 1995. Ciliary defects in healthy subjects, bronchiectasis, and primary ciliary dyskinesia. *Am. J. Respir. Crit. Care Med.* 151:1559–67

20. Hamasaki T, Nielson JH, Satir P. 1998. Regulation of outer arm dynein activity via light chain phosphorylation. In *Cilia, Mucus and Mucociliary Interactions*, ed. GL Baum, Z Priel, Y Roth, N Liron, E Ostfeld, pp. 21–25. New York: Marcel Dekker

21. Christensen ST, Guerra C, Wada Y, Valentin T, Angeletti RH, et al. 2001. A regulatory light chain of ciliary outer arm dynein in *Tetrahymena thermophila*. *J. Biol. Chem.* 276:20048–54

22. Sanderson MJ, Lansley AB, Evans JH. 2001. The regulation of airway ciliary beat frequency by intracellular calcium. In *Cilia and Mucus*, ed. M Salathe, pp. 39–57. New York: Marcel Dekker

22a. Salathe M. 2007. Regulation of mammalian ciliary beating. *Annu. Rev. Physiol.* 69:401–22

23. Habermacher G, Sale WS. 1997. Regulation of flagellar dynein by phosphorylation of a 138-kD inner arm dynein intermediate chain. *J. Cell Biol.* 136:167–76

24. Wargo MJ, Smith EF. 2003. Asymmetry of the central apparatus defines the location of active microtubule sliding in *Chlamydomonas* flagella. *Proc. Natl. Acad. Sci. USA* 100:137–42

25. Mitchell DR, Nakatsugawa M. 2004. Bend propagation drives central pair rotation in *Chlamydomonas reinhardtii* flagella. *J. Cell Biol.* 166:709–15

26. Afzelius BA. 1999. Asymmetry of cilia and of mice and men. *Int. J. Dev. Biol.* 43:283–86

27. Sanderson MJ, Sleigh MA. 1981. Ciliary activity of cultured rabbit tracheal epithelium: beat pattern and metachrony. *J. Cell Sci.* 47:331–47

28. Afzelius BA. 2004. Cilia-related diseases. *J. Pathol.* 204:470–77

29. Banizs B, Pike MM, Millican CL, Ferguson WB, Komlosi P, et al. 2005. Dysfunctional cilia lead to altered ependyma and choroid plexus function, and result in the formation of hydrocephalus. *Development* 132:5329–39

30. Ibanez-Tallon I, Pagenstecher A, Fliegauf M, Olbrich H, Kispert A, et al. 2004. Dysfunction of axonemal dynein heavy chain Mdnah5 inhibits ependymal flow and reveals a novel mechanism for hydrocephalus formation. *Hum. Mol. Genet.* 13:2133–41
31. **Tuomanen E. 1990. The surface of mammalian respiratory cilia. In *Ciliary and Flagellar Membranes*, ed. RA Bloodgood, pp. 363–88. New York: Plenum**
32. Chailley B, Boisvieux-Ulrich E, Sandoz D. 1990. Structure and assembly of the oviduct ciliary membrane. In *Ciliary and Flagellar Membranes*, ed. RA Bloodgood, pp. 337–62. New York: Plenum
33. Satir P, Guerra C. 2003. Control of ciliary motility: a unifying hypothesis. *Europ. J. Protistol.* 39:410–15
34. Taylor BL, Zhulin IB. 1999. PAS domains: internal sensors of oxygen, redox potential, and light. *Microbiol. Mol. Biol. Rev.* 63:479–506
35. Wyatt TA, Forget MA, Adams JM, Sisson JH. 2005. Both cAMP and cGMP are required for maximal ciliary beat stimulation in a cell-free model of bovine ciliary axonemes. *Am. J. Physiol. Lung Cell Mol. Physiol.* 288:L546–51
36. Lazard D, Barak Y, Lancet D. 1989. Bovine olfactory cilia preparation: thiol-modulated odorant-sensitive adenylyl cyclase. *Biochim. Biophys. Acta* 1013:68–72
37. Teilmann SC, Christensen ST. 2005. Localization of the angiopoietin receptors Tie-1 and Tie-2 on the primary cilia in the female reproductive organs. *Cell Biol. Int.* 29:340–46
38. **Teilmann SC, Byskov AG, Pedersen PA, Wheatley DN, Pazour GJ, Christensen ST. 2005. Localization of transient receptor potential ion channels in primary and motile cilia of the female murine reproductive organs. *Mol. Reprod. Dev.* 71:444–52**
39. Kreda SM, Mall M, Mengos A, Rochelle L, Yankaskas J, et al. 2005. Characterization of wild-type and deltaF508 cystic fibrosis transmembrane regulator in human respiratory epithelia. *Mol. Biol. Cell* 16:2154–67
40. Rosenbaum JL, Witman GB. 2002. Intraflagellar transport. *Nat. Rev. Mol. Cell Biol.* 3:813–25
41. Gilula NB, Satir P. 1972. The ciliary necklace. A ciliary membrane specialization. *J. Cell Biol.* 53:494–509
42. Deane JA, Cole DG, Seeley ES, Diener DR, Rosenbaum JL. 2001. Localization of intraflagellar transport protein IFT52 identifies basal body transitional fibers as the docking site for IFT particles. *Curr. Biol.* 11:1586–90
43. Horst CJ, Johnson LV, Besharse JC. 1990. Transmembrane assemblage of the photoreceptor connecting cilium and motile cilium transition zone contain a common immunologic epitope. *Cell Motil. Cytoskel.* 17:329–44
44. Hong DH, Yue G, Adamian M, Li T. 2001. Retinitis pigmentosa GTPase regulator (RPGRr)-interacting protein is stably associated with the photoreceptor ciliary axoneme and anchors RPGR to the connecting cilium. *J. Biol. Chem.* 276:12091–99
45. Hong DH, Pawlyk B, Sokolov M, Strissel KJ, Yang J, et al. 2003. RPGR isoforms in photoreceptor connecting cilia and the transitional zone of motile cilia. *Invest. Ophthalmol. Vis. Sci.* 44:2413–21
46. Moore A, Escudier E, Roger G, Tamalet A, Pelosse B, et al. 2006. RPGR is mutated in patients with a complex X-linked phenotype combining primary ciliary dyskinesia and retinitis pigmentosa. *J. Med. Genet.* 43:326–33
47. Geng L, Okuhara D, Yu Z, Tian X, Cai Y, et al. 2006. Polycystin-2 traffics to cilia independently of polycystin-1 by using an N-terminal RVxP motif. *J. Cell Sci.* 119:1383–95
48. Gaillard DA, Lallement AV, Petit AF, Puchelle ES. 1989. In vivo ciliogenesis in human fetal tracheal epithelium. *Am. J. Anat.* 185:415–28

31. A unique treatment of respiratory cilary membrane properties, including bacterial attachment.

38. The first images comparing ciliary membrane TRP channel distribution on functional mammalian motile cilia and related primary cilia.

49. Dirksen ER, Crocker TT. 1965. Centriole replication in differentiating ciliated cells of mammalian respiratory epithelium: an electron microscopic study. *J. Microsc.* 5:629–56
50. Dirksen ER. 1991. Centriole and basal body formation during ciliogenesis revisited. *Biol. Cell* 72:31–38
51. Carson JL, Reed W, Lucier T, Brighton L, Gambling TM, et al. 2002. Axonemal dynein expression in human fetal tracheal epithelium. *Am. J. Physiol. Lung Cell Mol. Physiol.* 282:L421–30
52. Dirksen ER. 1974. Ciliogenesis in the mouse oviduct. A scanning electron microscope study. *J. Cell Biol.* 62:899–904
53. Portman RW, LeCluyse EL, Dentler WL. 1987. Development of microtubule capping structures in ciliated epithelial cells. *J. Cell Sci.* 87:85–94
54. Brody SL, Yan XH, Wuerffel MK, Song SK, Shapiro SD. 2000. Ciliogenesis and left-right axis defects in forkhead factor HFH-4-null mice. *Am. J. Respir. Cell Mol. Biol.* 23:45–51
55. Johnson KA, Rosenbaum JL. 1992. Polarity of flagellar assembly in Chlamydomonas. *J. Cell Biol.* 119:1605–11
56. Marshall WF, Rosenbaum JL. 2001. Intraflagellar transport balances continuous turnover of outer doublet microtubules: implications for flagellar length control. *J. Cell Biol.* 155:405–14
57. Qin H, Diener DR, Geimer S, Cole DG, Rosenbaum JL. 2004. Intraflagellar transport (IFT) cargo: IFT transports flagellar precursors to the tip and turnover products to the cell body. *J. Cell Biol.* 19:255–66
58a. Pan J, Snell WJ. 2005. *Chlamydomonas* shortens its flagella by activating axonemal disassembly, stimulating IFT particle trafficking, and blocking anterograde cargo loading. *Dev. Cell* 9:431–38
58b. Qin H, Burnette D, Bae YK, Forscher P, Barr MM, Rosenbaum JL. 2005. Intraflagellar transport is required for the vectorial movement of TRPV channels in the ciliary membrane. *Curr. Biol.* 15:1695–99
59. Pedersen LB, Geimer S, Rosenbaum JL. 2006. Dissecting the molecular mechanisms of intraflagellar transport in *Chlamydomonas*. *Curr. Biol.* 16:450–59
60. Pan J, Snell WJ. 2002. Kinesin-II is required for flagellar sensory transduction during fertilization in *Chlamydomonas*. *Mol. Biol. Cell* 13:1417–26
61. Fliegauf M, Omran H. 2006. Novel tools to unravel molecular mechanisms in cilia-related disorders. *Trends Genet.* 22:241–45
62. Hagiwara H, Ohwada N, Takata K. 2004. Cell biology of normal and abnormal ciliogenesis in the ciliated epithelium. *Int. Rev. Cytol.* 234:101–41
63. Wheatley DN. 1982. *The Centriole: A Central Enigma of Cell Biology*. Amsterdam: Elsevier. 232 pp.
64. Sorokin S. 1962. Centrioles and the formation of rudimentary cilia by fibroblasts and smooth muscle cells. *J. Cell Biol.* 15:363–77
65. Wheatley DN. 2005. Landmarks in the first hundred years of primary (9 + 0) cilium research. *Cell Biol. Int.* 29:333–39
66. Fonte VG, Searls RL, Hilfer SR. 1971. The relationship of cilia with cell division and differentiation. *J. Cell Biol.* 49:226–29
67. Pazour GJ, Dickert BL, Vucica Y, Seeley ES, Rosenbaum JL, et al. 2000. *Chlamydomonas IFT88* and its mouse homologue, polycystic kidney disease gene *tg737*, are required for assembly of cilia and flagella. *J. Cell Biol.* 151:709–18
68. Andrews PM. 1975. Scanning electron microscopy of human and rhesus monkey kidneys. *Lab. Invest.* 32:510–18

67. The key paper first demonstrating the role of ciliary proteins in polycystic kidney disease.

69. Roth KE, Rieder CL, Bowser SS. 1988. Flexible-substratum technique for viewing cells from the side: some in vivo properties of primary (9 + 0) cilia in cultured kidney epithelia. *J. Cell Sci.* 89:457–66
70. Schwartz EA, Leonard ML, Bizios R, Bowser SS. 1997. Analysis and modeling of the primary cilium bending response to fluid shear. *Am. J. Physiol.* 272:F132–38
71. Praetorius HR, Spring KA. 2001. Bending the MDCK cell primary cilium increases intracellular calcium. *J. Membr. Biol.* 184:71–79
72. Praetorius HR, Spring KA. 2003. Removal of the MDCK cell primary cilium abolishes flow sensing. *J. Membr. Biol.* 191:69–76
73. Liu W, Murcia NS, Duan Y, Weinbaum S, Yoder BK, et al. 2005. Mechanoregulation of intracellular Ca^{2+} concentration is attenuated in collecting duct of monocilium-impaired *orpk* mice. *Am. J. Physiol. Renal Physiol.* 289:F978–88
74. Pazour GJ, San Agustin JT, Follit JA, Rosenbaum JL, Witman GB. 2002. Polycystin-2 localizes to kidney cilia and the ciliary level is elevated in *orpk* mice with polycystic kidney disease. *Curr. Biol.* 12:R378–80
75. Yoder BK, Hou X, Guay-Woodford LM. 2002. The polycystic kidney disease proteins, polycystin-1, polycystin-2, polaris, and cystin, are colocalized in renal cilia. *J. Am. Soc. Nephrol.* 13:2508–16
76. Parnell SC, Magenheimer BS, Maser RL, Zien CA, Frischauf AM, Calvet JP. 2002. Polycystin-1 activation of c-Jun N-terminal kinase and AP-1 is mediated by heterotrimeric G proteins. *J. Biol. Chem.* 277:19566–72
77. Chauvet V, Tian X, Husson H, Grimm DH, Wang T, et al. 2004. Mechanical stimuli induce cleavage and nuclear translocation of the polycystin-1 C terminus. *J. Clin. Invest.* 114:1433–43
78. Poole CA, Flint CA, Beaumont BW. 1985. Analysis of the morphology and function of primary cilia in connective tissues: a cellular cybernetic probe? *Cell Motil.* 5:175–93

78. One of the first papers to introduce the enormous potential of the primary cilium as a chemical and physical sensory device.

79. Benzing T, Walz G. 2006. Cilium-generated signaling: a cellular GPS? *Curr. Opin. Nephrol. Hypertens.* 15:245–49
80. Poole CA, Jensen CG, Snyder JA, Gray CG, Hermanutz VL, Wheatley DN. 1997. Confocal analysis of primary cilia structure and colocalization with the Golgi apparatus in chondrocytes and aortic smooth muscle cells. *Cell Biol. Int.* 21:483–94
81. Poole CA, Zhang ZJ, Ross JM. 2001. The differential distribution of acetylated and detyrosinated alpha-tubulin in the microtubular cytoskeleton and primary cilia of hyaline cartilage chondrocytes. *J. Anat.* 199:393–405
82. Jensen CG, Poole CA, McGlashan SR, Marko M, Issa ZI, et al. 2004. Ultrastructural, tomographic and confocal imaging of the chondrocyte primary cilium in situ. *Cell Biol. Int.* 28:101–10
83. McGlashan SR, Jensen CG, Poole CA. 2006. Localization of extracellular matrix receptors on the chondrocyte primary cilium. *J. Histochem. Cytochem.* 54(9):1005–14
84. Praetorius HA, Praetorius J, Nielsen S, Frokiaer J, Spring KR. 2004. β1-integrins in the primary cilium of MDCK cells potentiate fibronectin-induced Ca^{2+} signaling. *Am. J. Physiol. Renal Physiol.* 287:F969–78
85. Albrecht-Buehler G. 1977. Phagokinetic tracks of 3T3 cells: parallels between the orientation of track segments and of cellular structures which contain actin or tubulin. *Cell* 12:333–39
86. Wheatley DN. 1971. Cilia in cell-cultured fibroblasts. 3. Relationship between mitotic activity and cilium frequency in mouse 3T6 fibroblasts. *J. Anat.* 110:367–82

87. Tucker RW, Pardee AB, Fujiwara K. 1979. Centriole ciliation is related to quiescence and DNA synthesis in 3T3 cells. *Cell* 17:527–35
88. Tucker RW, Scher CD, Stiles CD. 1979. Centriole deciliation association with the early response of 3T3 cells to growth factors but not to SV40. *Cell* 18:1065–72
89. **Schneider L, Clement CA, Teilmann SC, Pazour GJ, Hoffmann EK, et al. 2005. PDGFR$\alpha\alpha$ signaling is regulated through the primary cilium in fibroblasts. *Curr. Biol.* 15:1861–66**

89. Strong evidence that the receptor tyrosine kinase PDGFR$\alpha\alpha$ needs to be localized to primary cilia to function in growth control.

90. Lih CJ, Cohen SN, Wang C, Lin-Chao S. 1996. The platelet-derived growth factor α-receptor is encoded by a growth-arrest-specific (gas) gene. *Proc. Natl. Acad. Sci. USA* 93:4617–22
91. de Silva DC, Wheatley DN, Herriot R, Brown T, Stevenson DA, et al. 1997. Mulvihill-Smith progeria-like syndrome: a further report with delineation of phenotype, immunologic deficits, and novel observation of fibroblast abnormalities. *Am. J. Med. Genet.* 69:56–64
92. Watanabe D, Saijoh Y, Nonaka S, Sasaki G, Ikawa Y, et al. 2003. The left-right determinant Inversin is a component of node monocilia and other 9+0 cilia. *Development* 130:1725–34
93. **Corbit KC, Aanstad P, Singla V, Norman AR, Stainier DY, Reiter JF. 2005. Vertebrate Smoothened functions at the primary cilium. *Nature* 437:1018–21**

93. The first evidence that Smo needs to be localized to the primary cilium for regulation of the Hh pathway and correct morphogenesis.

94. Voss JW, Pedersen SF, Christensen ST, Lambert IH. 2004. Regulation of the expression and subcellular localization of the taurine transporter TauT in mouse NIH3T3 fibroblasts. *Eur. J. Biochem.* 271:4646–58
95. Christensen ST, Voss JW, Teilmann SC, Lambert IH. 2005. High expression of the taurine transporter TauT in primary cilia of NIH3T3 fibroblasts. *Cell Biol. Int.* 29:347–51
96. Lambert IH. 2004. Regulation of the cellular content of the organic osmolyte taurine in mammalian cells. *Neurochem. Res.* 29:27–63
97. **Simons M, Gloy J, Ganner A, Bullerkotte A, Bashkurov M, et al. 2005. Inversin, the gene product mutated in nephronophthisis type II, functions as a molecular switch between Wnt signaling pathways. *Nat. Genet.* 37:537–43**

97. The first evidence that ciliary inversin acts as a flow-regulated molecular switch between canonical and noncanonical Wnt pathways in developmental processes.

98. Germino GG. 2005. Linking cilia to Wnts. *Nat. Genet.* 37:455–57
99. Pan J, Wang Q, Snell WJ. 2005. Cilium-generated signaling and cilia-related disorders. *Lab. Invest.* 85:452–63
100. **Haycraft CJ, Banizs B, Aydin-Son Y, Zhang Q, Michaud EJ, Yoder BK. 2005. Gli2 and Gli3 localize to cilia and require the intraflagellar transport protein polaris for processing and function. *PLoS Genet.* 1:e53**

100. Strong evidence for a direct role of the primary cilium in Gli processing and Shh signal transduction in limb bud development.

101. Huangfu D, Anderson KV. 2006. Signaling from Smo to Ci/Gli: conservation and divergence of Hedgehog pathways from *Drosophila* to vertebrates. *Development* 133:3–14
102. Kasper M, Regl G, Frischauf AM, Aberger F. 2006. GLI transcription factors: mediators of oncogenic Hedgehog signaling. *Eur. J. Canc.* 42:437–45
103. Kalderon D. 2005. The mechanism of hedgehog signal transduction. *Biochem. Soc. Trans.* 33:1509–12
104. Huangfu D, Anderson KV. 2005. Cilia and Hedgehog responsiveness in the mouse. *Proc. Natl. Acad. Sci. USA* 102:11325–30
105. Zhang XM, Ramalho-Santos M, McMahon AP. 2001. Smoothened mutants reveal redundant roles for Shh and Ihh signaling including regulation of L/R symmetry by the mouse node. *Cell* 106:781–92

106. Huangfu D, Liu A, Rakeman AS, Murcia NS, Niswander L, Anderson KV. 2003. Hedgehog signaling in the mouse requires intraflagellar transport proteins. *Nature* 426:83–87

107. Liu A, Wang B, Niswander LA. 2005. Mouse intraflagellar transport proteins regulate both the activator and repressor functions of Gli transcription factors. *Development* 132:3103–11

108. Zhang Q, Murcia NS, Chittenden LR, Richards WG, Michaud EJ, et al. 2003. Loss of the Tg737 protein results in skeletal patterning defects. *Dev. Dyn.* 227:78–90

109. Mo R, Freer AM, Zinyk DL, Crackower MA, Michaud J, et al. 1997. Specific and redundant functions of Gli2 and Gli3 zinc finger genes in skeletal patterning and development. *Development* 124:113–23

110. Dwyer ND, Adler CE, Crump JG, L'Etoile ND, Bargmann CI. 2001. Polarized dendritic transport and the AP-1 mu1 clathrin adaptor UNC-101 localize odorant receptors to olfactory cilia. *Neuron* 31:277–87

111. Handel M, Schulz S, Stanarius A, Schreff M, Erdtmann-Vourliotis M, et al. 1999. Selective targeting of somatostatin receptor 3 to neuronal cilia. *Neuroscience* 89:909–26

112. Brailov I, Bancila M, Brisorgueil MJ, Miquel MC, Hamon M, Verge D. 2000. Localization of 5-HT(6) receptors at the plasma membrane of neuronal cilia in the rat brain. *Neuron* 31:277–87

113. Morgan D, Eley L, Sayer J, Strachan T, Yates LM, et al. 2002. Expression analyses and interaction with the anaphase promoting complex protein Apc2 suggest a role for inversin in primary cilia and involvement in the cell cycle. *Hum. Mol. Genet.* 11:3345–50

114. Otto EA, Schermer B, Obara T, O'Toole JF, Hiller KS, et al. 2003. Mutations in INVS encoding inversin cause nephronophthisis type 2, linking renal cystic disease to the function of primary cilia and left-right axis determination. *Nat. Genet.* 34:413–20

115. Supp DM, Witte DP, Potter SS, Brueckner M. 1997. Mutation of an axonemal dynein affects left-right asymmetry in inversus viscerum mice. *Nature* 389:963–96

116. Lin F, Hiesberger T, Cordes K, Sinclair AM, Goldstein LS, et al. 2003. Kidney-specific inactivation of the KIF3A subunit of kinesin-II inhibits renal ciliogenesis and produces polycystic kidney disease. *Proc. Natl. Acad. Sci. USA* 100:5286–91

117. Cano DA, Murcia NS, Pazour GJ, Hebrok M. 2004. *orpk* mouse model of polycystic kidney disease reveals essential role of primary cilia in pancreatic tissue organization. *Development* 131:3457–67

118. Ross AJ, May-Simera H, Eichers ER, Kai M, Hill J, et al. 2005. Disruption of Bardet-Biedl syndrome ciliary proteins perturbs planar cell polarity in vertebrates. *Nat. Genet.* 37:1135–40

119. Park TJ, Haigo SL, Wallingford JB. 2006. Ciliogenesis defects in embryos lacking inturned or fuzzy function are associated with failure of planar cell polarity and Hedgehog signaling. *Nat. Genet.* 38:303–11

120. Habas R, Dawid IB. 2005. Dishevelled and Wnt signaling: is the nucleus the final frontier? *J. Biol.* 4:2

121. Sulik K, Dehart DB, Iangaki T, Carson JL, Vrablic T, et al. 1994. Morphogenesis of the murine node and notochordal plate. *Dev. Dyn.* 201:260–78

122. Supp DM, Brueckner M, Kuehn MR, Witte DP, Lowe LA, et al. 1999. Targeted deletion of the ATP binding domain of left-right dynein confirms its role in specifying development of left-right asymmetries. *Development* 126:5495–504

123. Nonaka S, Tanaka Y, Okada Y, Takeda S, Harada A, et al. 1998. Randomization of left-right asymmetry due to loss of nodal cilia generating leftward flow of extraembryonic fluid in mice lacking KIF3B motor protein. *Cell* 95:829–37

106. The first paper to show that the IFT machinery has an essential and vertebrate-specific role in Hh signal transduction.

107. Strong evidence that IFT proteins are required for proteolytic processing and transcriptional activities of Gli proteins as well as for Hh ligand–induced signaling cascade.

117. Evidence for a role of primary cilia in Wnt signaling.

118. Evidence that primary cilia are required for PCP processes.

119. Strong evidence for a central role for PCP signaling in development of cilia and Hh signaling.

124. Nonaka S, Shiratori H, Saijoh Y, Hamada H. 2002. Determination of left-right patterning of the mouse embryo by artificial nodal flow. *Nature* 418:96–99

125. Buceta J, Ibanes M, Rasskin-Gutman D, Okada Y, Hirokawa N, Izpisua-Belmonte JC. 2005. Nodal cilia dynamics and the specification of the left/right axis in early vertebrate embryo development. *Biophys. J.* 89:2199–209

126. McGrath J, Somlo S, Makova S, Tian X, Brueckner M. 2003. Two populations of node monocilia initiate left-right asymmetry in the mouse. *Cell* 114:61–73

127. Nakaya MA, Biris K, Tsukiyama T, Jaime S, Rawls JA, Yamaguchi TP. 2005. *Wnt3a* links left-right determination with segmentation and anteroposterior axis elongation. *Development* 132:5425–36

128. Tanaka Y, Okada Y, Hirokawa N. 2005. FGF-induced vesicular release of Sonic hedgehog and retinoic acid in leftward nodal flow is critical for left-right determination. *Nature* 435:172–77

129. Sawamoto K, Wichterle H, Gonzalez-Perez O, Cholfin JA, Yamada M, et al. 2006. New neurons follow the flow of cerebrospinal fluid in the adult brain. *Science* 311:629–32

130. Tabin CJ, Vogan KJ. 2003. A two-cilia model for vertebrate left-right axis specification. *Genes Dev.* 17:1–6

131. Hirokawa N, Tanaka Y, Okada Y, Takeda S. 2006. Nodal flow and the generation of left-right asymmetry. *Cell* 125:33–45

132. Davis EE, Brueckner M, Katsanis N. 2006. The emerging complexity of the vertebrate cilium: new functional roles for an ancient organelle. *Devel. Cell* 11:9–19

133. Michaud EJ, Yoder BK. 2006. The primary cilium in cell signalling and cancer. *Cancer Res.* 66:6463–67

126. Evidence for the role of primary cilia as well as motile nodal cilia in left-right asymmetry determination.

128. A new model implicating released morphogens and ciliary chemoreception in left-right determination.

Regulation of Mammalian Ciliary Beating

Matthias Salathe

Division of Pulmonary and Critical Care Medicine, Miller School of Medicine, University of Miami, Miami, Florida 33136; email: msalathe@miami.edu

Annu. Rev. Physiol. 2007. 69:401–22

First published online as a Review in Advance on August 31, 2006

The *Annual Review of Physiology* is online at http://physiol.annualreviews.org

This article's doi:
10.1146/annurev.physiol.69.040705.141253

0066-4278/07/0315-0401$20.00

Key Words

ciliary motility, phosphorylation, cyclic AMP, cyclic GMP, protein kinase A, protein kinase C, calcium, intracellular pH

Abstract

Recent advances in our understanding of the structure-function relationship of motile cilia with the 9 + 2 microtubular arrangement have helped explain some of the mechanisms of ciliary beat regulation by intracellular second messengers. These second messengers include cyclic adenosine monophosphate (cAMP) and cyclic guanosine monophosphate (cGMP) as well as calcium and pH. cAMP activates protein kinase A (PKA), which is localized to the axoneme. The cAMP-dependent phosphorylation of PKA's main target, originally described as p29 in *Paramecium*, seems to increase ciliary beat frequency (CBF) directly. The mechanism by which cGMP increases CBF is less well defined but involves protein kinase G and possibly PKA. Protein kinase C inhibits ciliary beating. The regulation mechanisms of CBF by calcium remain somewhat controversial, favoring an immediate, direct action of calcium on ciliary beating and a second cyclic nucleotide–dependent phase. Finally, intracellular pH likely affects CBF through direct influences on dynein arms.

Axoneme: the detergent-resistant structure of a cilium that is mainly composed of microtubules (9 + 2 or 9 + 0 structures) and the attached proteins including dynein arms

Primary ciliary dyskinesia: an inherited disease that causes cilia to beat inefficiently or be immotile

INTRODUCTION

Cilia have regained the attention of the scientific community over the past few years, in part owing to the increased recognition of the importance of primary cilia (see recent reviews in References 1 and 2). However, motile mammalian cilia remain vital cellular structures that fulfill a variety of functions. Even though electron microscopy has rendered a detailed ultrastructural description of cilia, the full and detailed list of its molecular components, specifically as they relate to the regulation of ciliary beating, is still incomplete. Recent advances in genomics and proteomics have identified and classified both genes necessary for ciliogenesis (e.g., 3) and proteins located inside cilia, mainly in its detergent-resistant structure, the axoneme (4, 5). However, a full catalog of mammalian ciliary proteins does not exist but is required for a complete understanding of mammalian ciliary beat regulation. Even the preliminary catalogs of ciliary proteins indicate that motile cilia may also be important sensors of their environment and thus capable of responding to their changing surroundings (see discussions in Reference 6).

This review focuses on the regulation of mammalian ciliary beating. Even though researchers have made tremendous progress with respect to ciliary beat regulation, many issues remain unresolved. Here, I discuss the main intracellular second messengers that regulate ciliary beating and the known mechanisms by which these second messengers regulate axonemal bending.

STRUCTURE AND FUNCTION

Motile cilia play a crucial role in clearing mucus and debris from the airways under normal conditions, as can be seen in patients with abnormal airway ciliary beating, namely patients with primary ciliary dyskinesia (e.g., 7, 8). Motile cilia also play a role in circulating spinal fluid in the ventricles of the brain, where abnormal ciliary beating has recently been linked to hydrocephalus and other developmental cerebral abnormalities (9, 10). In the fallopian tubes, the role of cilia is less clear, but they seem to contribute to the movement of the ovum from the ovaries to the uterus (e.g., 11). Sperm tails are classified as flagella, but their regulation differs from motile cilia in some aspects. Thus, I do not discuss the regulation of sperm flagella here in any detail. Finally, motile cilia exist in the embryonic node, where they seem critical in controlling left-right asymmetry. Little is known about the regulation of these cilia, however; thus, there is no significant discussion about these structures herein.

To relate mammalian ciliary function to structure, I briefly review the components of a cilium and its detergent-resistant, membrane-devoid structure called the axoneme. Each motile human cilium is approximately 6–7 μm long and approximately 0.2–0.3 μm in diameter. (Cilia in the brain are larger, however; see Reference 12.) The main structural components that compose the axoneme of cilia are microtubules. On cross sections of electron-microscopic images, ciliary microtubules are arranged in the classic 9 + 2 configuration (**Figure 1**). This 9 + 2 configuration has been preserved in motile cilia throughout evolution (13) and actually has been believed to be necessary to allow motion. However, cilia in the embryonic node with a 9 + 0 structure are also motile (14), albeit with a rotational movement rather than the back-and-forth motion of 9 + 2 cilia. Thus, the inner microtubule pair may be required for the normal, but asymmetric, back-and-forth motion of 9 + 2 cilia, and in the pair's absence, the cilium may move rotationally. Chilvers et al. (15) recently confirmed this movement by analyzing beat patterns of airway cilia from patients who lack the central microtubule pair.

More than 200 proteins and more than 200 expressed sequence tags have been identified thus far to make up the mammalian axoneme (4, 13, 16, 17), but this number may be an underestimate. Using purified flagella from *Chlamydomonas* and mass spectrometry, Pazour et al. (5) actually identified 360 ciliary

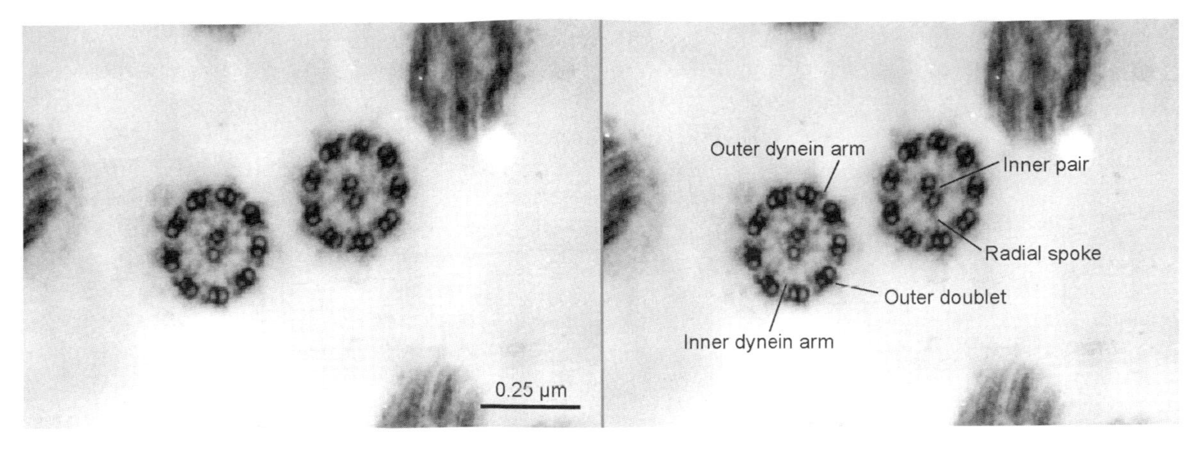

Figure 1

Electron microscopy images of cilia isolated from ovine airway epithelial cells. The cross sections reveal the classic 9 + 2 microtubular structure of motile cilia. The left panel is labeled with the major structures that make up the cilium. Reprinted in modified form from Reference 58, with permission.

proteins. They detected an additional 292 proteins, but there was only moderate confidence in their identification. Because 97 of the 101 previously known flagellar proteins were found through this approach, these mass spectrometric data suggest that more proteins compose the *Chlamydomonas* ciliary structure than previously thought, which is likely also true for mammalian cilia. Notably, the *Chlamydomonas* analysis revealed that a large number of proteins are present in cilia that participate in signaling, thus suggesting that the motile cilium and flagellum participate in important signaling events, a fairly new concept for these motile structures.

The microtubules that compose the 9 + 2 structure are primarily constructed from heterodimers of α- and β-tubulin. The major β-tubulin in mammalian cilia is of the IV isotype (18) and is present in larger quantities in cilia than in the cell body. In addition, α-tubulin can be acetylated, and acetylated tubulin is more enriched in cilia than in the cell body. Researchers have used these two features extensively for the immuno-identification of cilia in cell cultures and tissue sections. Along the microtubules, dynein arms (where dynein heavy chains are situated), radial spokes, and one pair of interdoublet links provide the cilium with its unique electron-microscopic fingerprint. Although these structures are easily seen in a cross section of the axoneme (**Figure 1**), a longitudinal section reveals quickly that the axoneme itself is composed of a repetitive unit with a length of ~96 nm. The repetitive unit consists of four outer dynein arms, three inner dynein arms, one spoke group (three radial spokes), and one pair of interdoublet links (13). The inner arms are more complex than the outer arms, showing three distinct types by electron microscopy. In contrast, the outer arms seem to be all of the same type, at least by electron microscopy, and attached to the doublet microtubules every 24 nm (19). The outer arm dynein from mammalian tracheal cilia is a two-headed bouquet-like molecule with a molecular size of 1–2 million Da (20). Each head contains a heavy chain ATPase of 400,000–500,000 Da. During ciliary beating, these dynein heavy chains interact with adjacent microtubules and move the microtubules relative to each other. The C terminus of β_I, β_{IV}, and β_V tubulins seems critical for this interaction, as antibodies against these structures inhibit at least the movement of isolated bovine axonemes (21).

CBF: ciliary beat frequency

Even though multiple axonemal dynein heavy chains are known—for instance, seven complementary DNAs were cloned from rat airway epithelium (see Reference 22), and a total of ~14 exist—the exact locations and functions of these different molecules remain unknown. To make the issue even more complex, in vitro data of nonmammalian cilia demonstrate that the inner and outer dynein arms are functionally distinct (23). Researchers confirmed this finding in *Chlamydomonas* mutants by demonstrating that outer dynein arms are responsible for adjusting ciliary beat frequency (CBF) and inner dynein arms are responsible for bend formation and beating form (24, 25). Careful analysis of mammalian cilia with respect to the role of outer and inner dynein arms has just gotten underway in patients with primary ciliary dyskinesia, a syndrome caused by a variety of structural ciliary abnormalities, most commonly a failure of outer or inner dynein arm assembly or both (7). Analysis of the ciliary beating patterns in such patients seems to confirm that mammalian cilia regulate their bending form by means of inner dynein arms and their frequency by means of outer dynein arms (15, 26). The dynein variety may explain some of these different roles of inner and outer dynein arms, but confirmation of this hypothesis is lacking.

Given the structure of the cilium, researchers estimate that it must contain more than 4000 inner and outer dynein arms. One wonders how such a complex structure is regulated. If we focus on the outer dynein arms and accept that these structures regulate beating frequency, it is still unclear why more than 2000 such arms are required along a ciliary shaft. One of the first models of flagellar movement, described in 1958 before dynein was discovered (27), required that energy be applied along the flagellar length to keep the beating pattern intact. Thus, this model foresaw that the motor molecules' heavy chains were distributed along the ciliary shaft (27).

Dynein clearly can move microtubules only unidirectionally. The ciliary stroke, however, is a back-and-forth motion, whereby the beating plane is given by the alignment of the central microtubule pair. The cilium, from a resting state (just after the effective stroke), goes through a recovery stroke by swinging almost 180° backward close to the cell surface and then extending almost fully. Next, it goes directly into the effective stroke in a plane perpendicular to the cell surface, reaching a maximum velocity of 1 mm s^{-1} at its tip and describing an arc of approximately 110°. After completing the effective stroke, the cilium rests shortly and then resumes its recovery and effective strokes. The effective stroke is approximately two to three times faster than the recovery stroke (28). When CBF increases, all three phases of the beat cycle are shortened, but the resting period is most markedly affected (28). With dynein moving the microtubules in only one direction, the ciliary motion requires that effective dynein activity alternate between two halves of the axoneme. The switch in dynein activity from one side to the other when moving from the recovery to the effective stroke has been proposed as the switch-point hypothesis (29). The exact mechanism of this switching remains unclear; it may be as simple as a mechanical effect brought about by the bending itself because the bending changes the structural relationships between dynein arms and microtubules. In addition, the necessity of different dynein activities during the beat cycle may explain the need for a significant number of genetically different dyneins to be present in the cilium (as discussed above), assuming that these different dyneins are asymmetrically distributed along the ciliary axis. Unfortunately, these issues have not been resolved.

Using published data on dynein conformational changes, we can calculate the estimated minimum number of sequential dynein arm actions necessary for one single ciliary stroke (30, 31). The two outer doublets separate maximally by ~0.1 μm at the end of a full ciliary bend. (This distance varies, however, from microtubule to microtubule.) A single stroke cycle of a dynein arm can move a microtubule between 4–16 nm. Thus, anywhere between 12 and 50 sequential dynein arm movements are necessary to complete a single ciliary stroke, including its effective and recovery portions. However, during a full stroke, up to 2000 dynein arms can be active; thus, there is an excess of at least 40-fold active dynein arms from the required minimum for any given ciliary beating cycle.

The cilium clearly has to create considerable force during its beat, not only to overcome resistances but also to propel water and/or mucus. Given the known forces that need to be overcome (~20 pN) and the fact that a single outer dynein arm can generate 1–5 pN of force, 4–20 arms (only ~1%) must be active in the cilium all the time (32). When energy dissipation is considered, this number increases to approximately 1000 active dynein arms during a single ciliary beat (32). Not surprisingly, the power generated by the cilium is related to the number of dynein-microtubule interactions (32, 33); thus, any additional activity can increase force generation. In fact, during normal beating, cilia have a reserve for increasing their output force, a feature that has been observed experimentally (34). Recently, Andrade et al. (35) described a possible signaling pathway for force adjustments during ciliary beating via the transient receptor potential vanilloid 4 (TRPV 4) channel that increases intraciliary calcium and thereby CBF (see below). These recent findings are intriguing, as they again suggest that even mammalian motile cilia serve a sensory function and express functional channels and possibly receptors on their membrane. The movement of TRPV channels to the ciliary membrane also was recently shown to be mediated by intraflagellar transport (a feature not discussed here), at least in *Caenorhabditis elegans* sensory cells (36).

MODELING CILIARY BEAT FREQUENCY ADJUSTMENTS

The frequency response of mammalian ciliated cells to increasing calcium concentrations was modeled using the theoretical numbers of dynein arms that must be active during a ciliary beat discussed above (37). I discuss how calcium regulates CBF below, even though this model did not rely on the exact mechanism of calcium-mediated CBF regulation. The model had its basis in three assumptions that are supported in part by data in the literature: (*a*) Ciliary dynein arms move microtubules with either a slow or fast duty cycle; (*b*) a change in $[Ca^{2+}]$ shifts the dynein ATPase activity between fast and slow modes; and (*c*) 25 complete and sequential dynein ATPase cycles (an average of the range of 12–50 mentioned above) are necessary to complete one complete ciliary stroke.

$[Ca^{2+}]_i$: intracellular calcium concentration

Using these assumptions, CBF can be described as follows:

$$CBF = 1/(25 \times T_{dynein\ cycle}),$$

where $T_{dynein\ cycle}$ is the time required (in seconds) for dynein binding, ATP hydrolysis, and the unbinding of P_i and ADP. Adjusting for the lengths of the two duty cycles (fast and slow, not effective and recovery strokes), one can rewrite the equation as

$$CBF = 1/[25(f_{fast} \times T_{fast} + f_{slow} \times T_{slow})],$$

where f_{fast} is the fraction of dynein arms operating in fast mode, f_{slow} is the fraction of the dynein arms operating in slow mode (and equal to $1 - f_{fast}$), T_{fast} is the time (in seconds) required for dynein ATPase cycle in fast mode, and T_{slow} is the time (in seconds) required for dynein ATPase cycle in slow mode. This model was fit to recorded, simultaneous measurements of $[Ca^{2+}]_i$ and CBF from single cells (**Figure 2**) (37).

As seen in **Figure 2**, a theoretical analysis of dynein arm requirements for ciliary beating can be used to analyze and/or predict CBF changes with changing intracellular second-messenger concentrations. The model does not require a specific interaction between the second messenger and the dynein arms, however. It requires only that the ciliary response to Ca^{2+} (or any other second messenger) be so rapid that it is not rate limiting in this situation. That calcium can work quickly to affect ciliary beating, namely within 80 ms, has recently been confirmed experimentally in *Paramecium* cilia (38), thus further supporting the assumptions and output of the model. Interestingly, the output of the fit revealed that the fast duty cycle was approximately half as long as the slow duty cycle, a finding possibly supported by microtubule movement speeds over inner dynein arms of *Chlamydomonas* (39, 40).

REGULATION OF MAMMALIAN CILIARY BEAT FREQUENCY

Mammalian cilia beat to move mucus (e.g., airway mucus) or fluid (e.g., spinal fluid). If cilia fail to beat efficiently, disease ensues (e.g., 7, 9, 10). Although ciliary beat regulation has been studied in many systems, the most information is known from respiratory cilia. Thus, the following discussion mainly focuses on the regulation of airway cilia.

Although many have argued for some time that the engagement of the ciliary tip with the mucus layer in the airway is critical for adequate mucociliary clearance, it remains

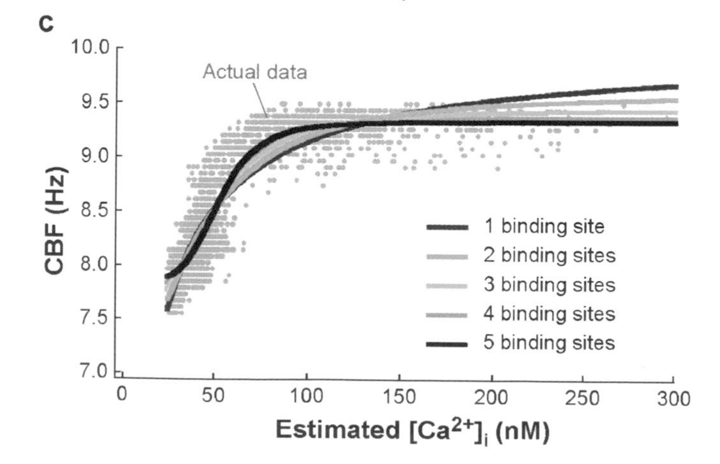

Figure 2

Model of ciliary beat frequency (CBF) regulation by calcium. (*a*) Two outer doublets separate ~0.1 μm at the end of ciliary bending as indicated, and a single stroke of one dynein arm can move a microtubule 4–16 nm. Thus, on average, 25 sequential dynein arm movements are necessary to complete a single ciliary stroke. (*b*) The fraction of fast versus slow dynein arms was computed using the Hill equation and simulated here with the fraction's dependence on $[Ca^{2+}]_i$, using a total K_d of 0.007 μM and one to five cooperative Ca^{2+}-binding sites. For the CBF output in this simulation, we arbitrarily chose a baseline frequency of 7 Hz and a maximal frequency of 12 Hz. (However, this is not the limit of the model in general.) (*c*) The CBF model was fit to actual data, using the slow and fast dynein-motion duration and the total K_d as parameters of the fit (thereby not limiting the model's CBF to any specific range). A model with equal to or more than four cooperative Ca^{2+}-binding sites results in a reasonable fit. The average K_d found by the fit was 0.02 μM; the slow dynein-arm-duty duration was 9 ms; and the fast dynein-arm-duty duration was 4.5 ms. Reprinted with modifications and permission from Reference 37.

unclear whether direct physical action between cilia and the mucus layer is required for mucus transport, as it is obviously not necessary for spinal fluid motion in the brain. Interestingly, the periciliary fluid-level height seems to remain fairly stable at approximately 7 μm if mucus is present, and the addition of fluid to the system seems to swell the mucus layer, with little influence on the periciliary fluid level (40a). Adding fluid to the system and thus the mucus layer, however, improves the efficiency of mucus transport: Patients with the genetic disease pseudohypoaldosteronism have an increased amount of airway surface liquid, and mucociliary clearance in these patients is the highest measured (41). These data therefore suggest that mucociliary clearance in the airways does not operate at full capacity under normal conditions.

Ciliary beating seems coordinated—i.e., cilia beat as part of a metachronal wave. The mechanism of coordination between beating cilia to create such a wave is still speculative. Cilia on a single cell clearly beat together (M. Salathe, unpublished results); however, this is not necessarily true for cilia on different cells in submerged cultures if these ciliated cells are not next to each other, or if there are fewer cilia per apical cell surface than in vivo. This finding implies that, at least in part, the close spatial relationship of cilia is important for some coordinated beating. In fact, cilia are densely packed and close together on all naturally occurring ciliated surfaces (e.g., in the fallopian tubes, the airways, and the ventricles). [Asymmetric beating cilia with a 9 + 0 configuration are not considered here; see Satir & Christensen (41a) for more discussion.] In addition, the environment in which they beat consists at least in part of fluid; thus, significant hydrodynamic forces must exist between beating cilia. These hydrodynamic interactions are believed to be the most important factor for ciliary coordination on epithelial surfaces (42) and may explain why the lengths of metachronal waves are limited (43, 44).

Human cilia beat approximately 12–15 Hz at body temperature when measured in vitro. Thus far, in vivo measurements have been difficult to achieve. The frequency of ciliary beating must be adjusted by the regulation of dynein-microtubule interactions. As indicated above, data in the literature and from the model of CBF calcium regulation may support the movement of dynein arms in two states: a rapid and a slow one. If true, the frequency of ciliary beating would be given by the percentage of dynein arms that beat in the slow or the rapid state (even though this is an oversimplification that does not take into account the different speeds of recovery and effective strokes or differences in the variety of dyneins). The mechanisms of switching the dyneins from slow to fast are not completely understood, but both axonemal phosphorylation and dephosphorylation events (of mainly dynein light chains), as well as possibly influences on the dynein itself or dynein light chains, seem to be responsible. Accumulating evidence shows that mammalian CBF changes in response to changes in the phosphorylation state of ciliary targets, to changes in $[Ca^{2+}]_i$, and to changes in intracellular pH (pH_i). Changes in phosphorylation occur mainly through cyclic adenosine monophosphate (cAMP)-dependent kinase but also possibly through cyclic adenosine monophosphate (cGMP)-dependent kinase and by PKC-mediated phosphorylation events. I discuss these mechanisms in more detail below.

pH_i: intracellular pH

cAMP: cyclic adenosine monophosphate

cGMP: cyclic guanosine monophosphate

In vivo, the changes in airway CBF are mediated mainly through the stimulation of receptors by naturally released ligands. Most effects stem from the activation of G protein–coupled receptors, including P2Y2, adenosine, bradykinin, and possibly muscarinic acetylcholine receptors. However, mechanical influences may also play a role (e.g., 45). In the airway, the most commonly cited paracrine mediator is ATP, released from epithelial cells in response to different stress (e.g., 46, 47); acetylcholine release may also play a role, even though this remains

PKA: protein kinase A or cAMP-dependent protein kinase

somewhat controversial (e.g., 48). We have also found that ciliary beating is regulated specifically by hyaluronan, a glycosaminoglycan present at the apex of the airway epithelium (49, 50). Thus, in addition to ATP and possibly acetylcholine, hyaluronan is another of the few known endogenously produced molecules that regulates human CBF from the luminal side of the airway. The data show that hyaluronan is produced by superficial epithelial cells with an average size of at least 800 kDa, but this large hyaluronan does not affect baseline CBF. Conversely, smaller hyaluronan molecules of 50–200 kDa stimulated CBF through the receptor for hyaluronic acid–mediated motility (RHAMM or CD 168), a receptor we identified at the apex of ciliated airway epithelial cells (**Figure 3**). Because the larger hyaluronan size encountered in normal airways does not affect CBF, we expect hyaluronan to signal via RHAMM only after being degraded in situ (for instance, by reactive oxygen or nitrogen species) and thus play a role in signaling during an airway stress response. Because RHAMM has no transmembrane domain, signaling has been proposed to work through hyaluronan-mediated association of RHAMM with growth-factor receptors, of which only one has been shown to regulate mammalian CBF, namely a member of the hepatic growth-factor receptor family called RON (51). But there is no proof to date that RON is involved in hyaluronan-mediated signaling to increase CBF.

REGULATION OF CILIARY BEAT FREQUENCY BY CYCLIC AMP–DEPENDENT PHOSPHORYLATION

Multiple studies have found that axonemes contain kinases capable of phosphorylating dynein subunits and other ciliary structures. This finding has also been true for mammalian cilia. One of the best-characterized ciliary phosphorylation targets to date remains a 29-kDa outer arm dynein light chain (p29) of *Paramecium* cilia (52). I discuss its discovery and characterization here because p29 has a clear homolog in mammalian cilia. Hamasaki et al. (53) found that permeabilized *Paramecium* pretreated with cAMP and ATP-γ-S (at low Ca^{2+} concentrations) swam approximately 50% faster than controls that were reactivated in the presence of Mg^{2+}-ATP but in the virtual absence of cAMP. Because *Paramecia* in these experiments were permeabilized, these data strongly suggest that cAMP interacted directly with an enzyme linked to the ciliary axoneme. Indeed, a cAMP-dependent protein kinase, also known as protein kinase A (PKA), was found to be attached to *Paramecium* ciliary axonemes (54). The major

Figure 3

Ciliary beat frequency (CBF) responses to hyaluronan. (*a–d*) Staining for hyaluronan of airway sections, using a biotinylated hyaluronan-binding protein and avidin-alkaline phosphatase, shows that hyaluronan is localized to the ciliary border of the epithelium in addition to its known localization in the submucosal interstitium. Incubation with hyaluronidase (37°C overnight) removed specific staining for hyaluronan (*b*), whereas incubation with chondroitinase ABC at pH 7.5 did not change the staining pattern for hyaluronan (*c*). When chondroitinase ABC was used at pH 5.6, at which it has hyaluronidase activity, hyaluronan staining was also removed from the sections (*d*). All bars are 10 μm. (*e–f*) The hyaluronan-induced CBF increase is blocked by anti-receptor for hyaluronic acid–mediated motility (anti-RHAMM) antibodies. CBF of ovine tracheal epithelial cells in primary cultures was measured as described previously (49) in the presence of IgG control antibodies (*e*) or anti-RHAMM antibodies (*f*). Continuous recordings of CBF are shown in response to exogenous 50 μg/ml hyaluronan (each $n \geq 8$). CBF responses to hyaluronan were blocked using functionally blocking anti-RHAMM antibodies. All cells responded to 20 μM ATP with a statistically indistinguishable, transient increase in CBF. The duration of drug application is indicated by labeled horizontal bars (*arrows* indicate continued presence of drug). Reprinted with modifications and permission from Reference 50.

axonemal protein that was phosphorylated in a cAMP-dependent manner was named p29 for its migration pattern on sodium dodecyl sulfate acrylamide gels, and it is a dynein light chain, as evidenced by its co-isolation with the 22S dynein arm (53). The cAMP-dependent phosphorylation of p29 was shown both in permeabilized cells and in isolated

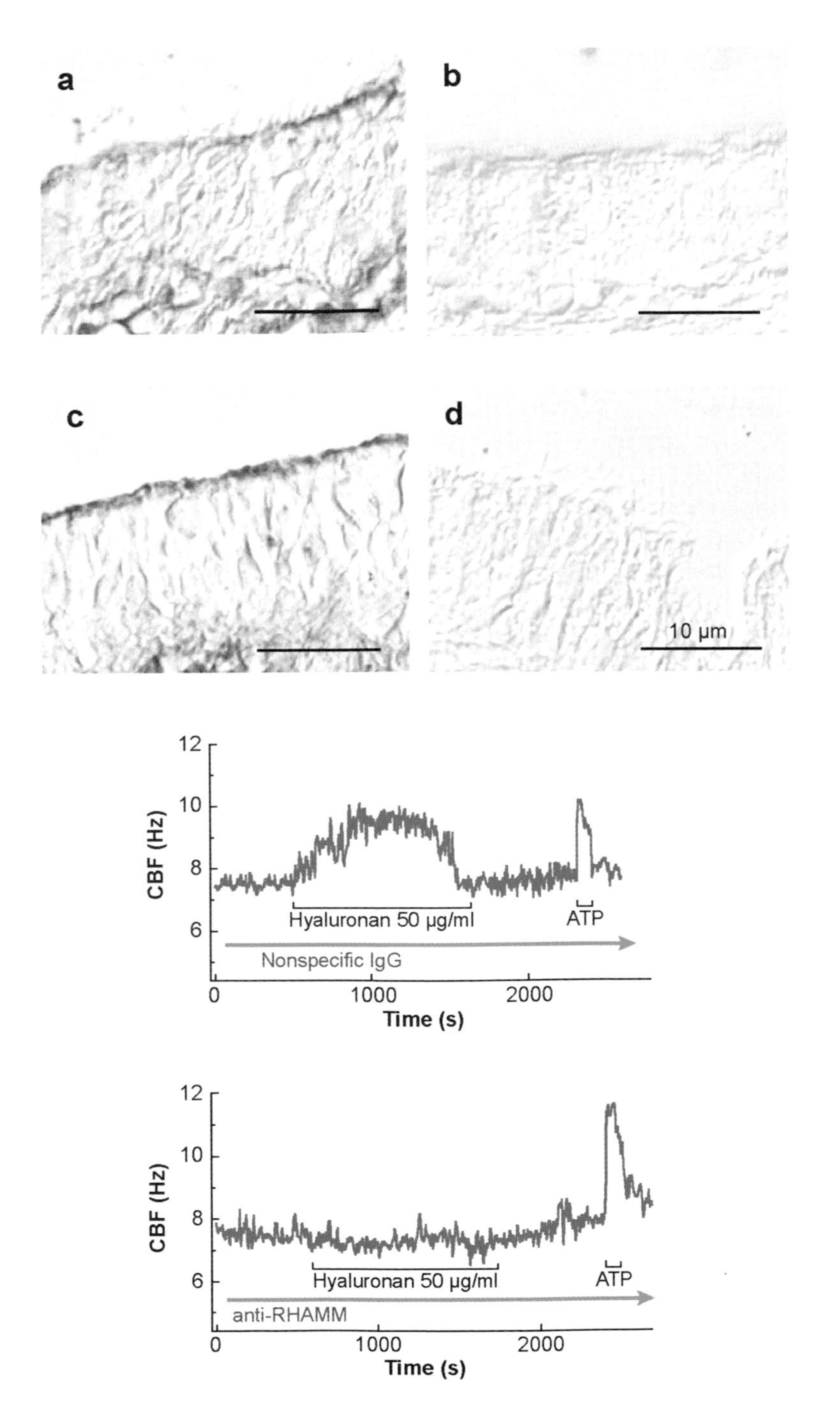

axonemes (52). Interestingly, p29 phosphorylation in vitro was only accomplished by the PKA isolated from *Paramecium* and not by bovine brain PKA (55). Even though these data supported the idea of a specific ciliary PKA isoform (54), this idea did not hold up in mammalian cilia, where any mammalian PKA seems to be able to phosphorylate the mammalian homolog of p29 (see below). Functionally, phosphorylation of p29 increased the velocity of microtubule gliding across outer arm dynein–coated surfaces in vitro and the swimming speed (i.e., beating frequency) of *Paramecium* in vivo (53, 56). In summary, good evidence suggests that PKA-mediated phosphorylation of the *Paramecium* outer arm dynein light chain p29 is directly responsible for an increase in beating frequency that is seen in association with an increase in cytoplasmic cAMP.

Such experiments are more difficult to accomplish with the limited cilia available from mammalian tissue. However, many of the principles described above for *Paramecium* have also been shown to play a crucial role in the regulation of mammalian cilia. Several lines of evidence link PKA and cAMP-dependent, phosphorylation-mediated events to CBF increases in the mammalian cilium. First, an A-kinase anchoring protein was found in human airway cilia (57). In addition, PKA and PKA activity were found in mammalian cilia from multiple species (57–60). Recent data using the recombinant expression of PKA driven by the ciliated-cell-specific Foxj1 promoter in ciliated cells revealed localization of both the regulatory (RII) and catalytic subunit of PKA to cilia (61, 62). Second, axonemal-phosphorylation targets of PKA have been identified by virtue of their cAMP-dependent phosphorylation in ovine respiratory cilia (58), in bovine respiratory cilia (63), and in human airway cilia (57). Third, cAMP-dependent phosphorylation of the axonemal PKA target could be accomplished in isolated, demembranated cilia, showing that the system works without the addition of anything but Mg^{2+}-ATP and cAMP (57, 58). Fourth, isolated bovine axonemes responded with an increase in bending frequency upon exposure to cAMP in vitro (64); however, this finding was less evident in isolated axonemes from human airway epithelial cells (65), possibly pointing to some species differences or additional requirements for cAMP-dependent regulation of ciliary beating. Recent data showing that cAMP regulates CBF in basolaterally permeabilized cells and earlier data showing that the lack of an apical membrane obliterates the response to changing cAMP concentrations (66) may indicate that at least significant parts of the ciliary membrane—as, for instance, shown in nonmammalian cilia (67, 68)—or some level of calcium are required for cAMP to work fully (61). However, these suggestions remain speculative at this time.

To complicate matters, in some biological systems, a PKA-mediated phosphorylation of a single ciliary target may not be sufficient to regulate and/or activate ciliary beating. Supporting this hypothesis, sperm axonemes were not activated by cAMP-dependent phosphorylation alone; the phosphorylation of tyrosine residues seemed to be essential to their activation as well (69). In *Chlamydomonas*, Howard et al. (70) have found that a cAMP-dependent phosphorylation of a regulatory complex between radial spokes and inner arm dynein actually decreases CBF, and some investigators have proposed that PKA activation may actually inhibit mammalian CBF under certain conditions (A. Mehta, personal communication). These issues have not been resolved to date, even though there seems to be a consensus that direct, cAMP-dependent activation of CBF via PKA is the predominate mechanism. In summary, there is good evidence that cAMP regulates mammalian CBF via the activation of axonemal PKA to phosphorylate a dynein light chain, which in turn possibly mediates a switch from the slow to the fast dynein-duty cycle and thus increases CBF.

INTRACELLULAR SOURCES OF CYCLIC AMP FOR CYCLIC AMP–DEPENDENT ACTIVATION OF CILIARY BEATING

For the regulation of airway cilia, cAMP needs to be made available to the axonemes from within the cell. The major, classically considered sources for cAMP are at least nine differentially regulated isoforms of G protein–responsive and Mg^{2+}-sensitive transmembrane adenylyl cyclases (tmACs). At least some of these must be expressed in the apical membrane of airway epithelial cells, but not in the ciliary membrane. These adenylyl cyclases are activated by a variety of receptors that couple to $G_{\alpha}s$. Although researchers have shown that beta-adrenergic agonists increase CBF in a cAMP-dependent manner via β_2-receptors expressed at the apical membrane, a more natural agonist is adenosine, which is produced in a paracrine fashion; i.e., adenosine is made by hydrolysis of apically released ATP (see below) and stimulates the adenosine A2b receptor (71) to increase cAMP and thereby CBF. There is also evidence that increasing intracellular calcium concentration can activate calcium-dependent adenylyl cyclases (e.g., AC1, AC3, AC8) and thereby stimulate CBF in a cAMP-dependent fashion (72).

Interestingly, the production of cAMP in this matter always occurs at the apical membrane and not the ciliary membrane; therefore, cAMP has to diffuse to the axoneme to exert its effects on axonemal PKA and thus on CBF. Diffusion of cAMP is restricted inside cells by phosphodiesterases. Given the overwhelming evidence for this activation pattern of CBF by receptor agonists, however, the diffusion pathway from the apical membrane to the axoneme clearly must be unrestricted to cAMP.

A distinct adenylyl cyclase activity was described in testis approximately 30 years ago (73–75) and was called soluble adenylyl cyclase (sAC). sAC differs enzymatically from tmAC by its preference for Mn^{2+} over Mg^{2+} (76, 77) and its insensitivity to G protein activation or forskolin (77). Chen et al. (78) have shown that mammalian sAC activity is activated directly by HCO_3^- in a pH-independent manner. sAC is also activated by Ca^{2+} if $[HCO_3^-]_i$ is not zero (79). Researchers have found that, besides being expressed in the testis, sAC is expressed in many tissues where it serves important functions (e.g., 80). We have found sAC expression in the axoneme itself (61, 62), and preliminary data suggest that sAC in fact contributes to CBF regulation. The relevance of this enzyme to CBF regulation is currently under investigation.

tmAC: transmembrane adenylyl cyclase

sAC: soluble adenylyl cyclase

PKG: cGMP-dependent kinase

PKC: protein kinase C

CYCLIC GMP REGULATION OF CILIARY BEAT FREQUENCY

cGMP and cGMP-dependent protein kinase (PKG) have also been shown to be involved in CBF regulation, at least in some mammalian species (81–84). There is controversy as to whether a combined signaling by $[Ca^{2+}]_i$ is required (81, 83). However, axonemal beating activity increased with the addition of cGMP to isolated axonemes in the absence of calcium (64). In these experiments, the combined addition of cAMP and cGMP increased axonemal beating beyond the stimulation achieved with either one individually. Although these experiments show that cGMP can influence axonemal and ciliary beating, it remains unclear how cGMP actually mediates these changes. Some experiments suggest that PKG may act through PKA to stimulate CBF (82, 85); however, more recent experiments invoke PKG directly (64, 83). Experiments with isolated axonemes also suggest that PKG activity can be measured in cilia (64), and Western blots showed that PKG was present in axonemes (86). The phosphorylation pattern of axonemal proteins changes on PKG activation; some proteins are phosphorylated, and some dephosphorylated (86). Even though these recent findings question whether data obtained in intact cells can be compared with data from isolated axonemes, how cGMP regulates CBF remains unknown and is not as clearly understood as the cAMP-dependent pathway.

ROLE OF PROTEIN KINASE C

In contrast to PKA, protein kinase C (PKC) has been implicated in slowing ciliary beating (87–90). The finding of decreasing CBF upon PKC activation has been consistent in all mammalian cilia examined. Conversely, cilia from frogs increase CBF upon PKC activation (91), indicating that the frog is not a good model for predicting signaling pathways that regulate CBF in mammals. The mechanisms by which PKC inhibits CBF are not fully understood. A ciliary membrane phosphorylation target for PKC has been found in ovine cilia (92). However, it is not clear whether the phosphorylation of this target really mediates the slowing of CBF. Given other data implicating a role of the ciliary membrane in CBF regulation, this target, however, is a good candidate.

ROLE OF CALCIUM IN CILIARY BEAT REGULATION

It is well established that Ca^{2+} plays a crucial role in CBF regulation. Although cAMP-dependent regulation of CBF seems fairly similar between unicellular organisms and mammals, the CBF regulation by calcium must be different. In *Paramecium*, for instance, the rising $[Ca^{2+}]_i$ slows CBF to the point at which the beat direction is reversed (93). Mammalian cilia, however, never reverse their beating direction; in fact, this reversal would be expected to be fatal, at least if it occurred in the airways. In mammals, elevations of $[Ca^{2+}]_i$ are almost always associated with an increase in CBF (66, 94–100), and decreases in $[Ca^{2+}]_i$ usually cause a decrease in CBF (98). The source of calcium to mediate an increase in CBF can stem from intracellular stores or from calcium influx through the plasma membrane, an event occurring possibly directly (101) or certainly after receptor-mediated calcium release from intracellular stores (e.g., shown by Mn^{2+}-quenching experiments in Reference 102).

Although the mechanisms of CBF regulation by cAMP are fairly well established, the CBF regulation by calcium remains controversial. Initial reports, in which calmodulin inhibitors were used, suggest that calmodulin and Ca^{2+}/calmodulin-dependent kinase (CaM kinase) are involved in the calcium-mediated regulation of CBF, as calmodulin inhibitors blocked calcium-mediated increases in CBF (94, 103). In addition, ciliary targets of CaM kinase were known from *Tetrahymena* (104). However, we were unable to find the CaM kinase targets in mammalian cilia, and we found the calmodulin inhibitors used in these previous experiments to be directly ciliotoxic. In contrast, other calmodulin inhibitors did not inhibit calcium's ability to increase CBF (37). Thus, these data did not support a role for CaM kinase and calmodulin in calcium's regulation of mammalian CBF.

We and others have shown that the CBF regulation by calcium is rapid (within one beat cycle) and occurs over a relatively small change in calcium (37, 105). Following a rapid increase, CBF remains elevated after $[Ca^{2+}]_i$ returns to baseline. A variety of kinase and phosphatase inhibitors—including inhibitors of PKA and PKG, as well as nitric oxide synthase (some shown to work with controls)—proved unable to inhibit the initial increase in CBF due to rising $[Ca^{2+}]_i$ (37, 72, 83). These data strongly support a mechanism that would allow calcium to act directly on a ciliary target, most likely a calcium-binding protein, at least at the early stages of calcium-mediated CBF elevation. Prolonged elevation of CBF, when $[Ca^{2+}]_i$ returns to or reaches baseline, is influenced by other mechanisms (see below). Even though the identification of a calcium-binding protein responsible for this calcium action has remained elusive, this protein likely is an integral part of the axonemal machinery (possibly part of the outer dynein arm). This hypothesis finds support in data showing that the beat frequency of canine cilia, permeabilized with saponin, may be regulated by externally applied Ca^{2+} in concentrations from 0.3 to 10 μM (106). Conversely, these data were challenged in rabbit cilia experiments in which the apical permeabilization of cilia prevented the regulation of ciliary beat by changing calcium concentrations (66). These data point again to a possible role of the ciliary membrane in allowing correct ciliary beat regulation, as discussed above for cAMP.

Another interesting set of data showed that CBF activation in patch-clamped cells requires the presence of cyclic nucleotides for calcium to be effective (107). These findings are intriguing, as they suggest that a low level of cAMP or cGMP must be available for calcium-mediated beat regulation. The source of cyclic nucleotides under baseline conditions remains unclear. We and Zhang (72, 83, 108) have shown that the initial elevation of CBF in response to calcium increases is not dependent on PKA or PKG. Thus, an increasing calcium concentration initially does not likely provide the necessary cyclic nucleotides through the stimulation of tmACs or nitric oxide synthase. However, sAC possibly provides a baseline level of cAMP, as this enzyme is active at normal bicarbonate levels in the cell before it can be stimulated further by increasing bicarbonate or calcium levels. However, this hypothesis remains unexplored.

In contrast to the direct calcium action on ciliary target proteins to mediate the changes in CBF, other authors have hypothesized that calcium always works through kinases or phosphatases. The patch-clamp data from Ma et al. (107) provide the strongest support for this hypothesis. As indicated above, the prolonged increase in CBF seems to depend on additional pathways, either by PKG (108) or PKA (71, 72). These pathways can be activated either through the initial elevated calcium concentration (72, 108) or through extracellular hydrolysis of agonists that cause an increase in calcium into an agonist that stimulates cyclic nucleotide production (71). However, that nitric oxide always has to be activated for the calcium-mediated increase in CBF to occur (86) is likely not accurate, as there are many conflicting reports on the role of nitric oxide in regulating ciliary beating in general and in being involved in specific intracellularly signaling pathways in particular (e.g., 72, 85, 109–112). In addition, data from frog palates and esophagus are not directly applicable to mammalian tissues (see above) (113, 114). Thus, species differences may play a role in the interpretation of the data and may be relevant for different species, even within mammalians.

In summary, elevation of the intracellular calcium concentration clearly increases CBF. The exact mechanism remains uncertain; however, the initial, rapid response of CBF to rising calcium seems to be directly mediated, whereas the prolonged elevation in CBF seems to rely on additional signaling mechanisms, including the ones activated by cAMP and possibly cGMP. Data from patch-clamped ciliated cells show that a baseline level of cyclic nucleotides needs to be present for the initial calcium-mediated increase in CBF to occur. As discussed above, there is an adenylyl cyclase expressed in cilia, namely sAC, that could provide this baseline cAMP. These issues remain under investigation. Interestingly, Delmotte & Sanderson (115) just reported that in small murine airways, CBF is already stimulated to its maximum and does not respond to increasing calcium concentrations anymore. Similar results were found in rat airways in which acetylcholine (mediating its effects via calcium) increased CBF in proximal airways but not in distal airways, although a beta-receptor agonist was able to increase CBF in both proximal and distal airways (116). These data thus confirm that CBF is regulated within tight limits and can reach its maximum by different means, thereby no longer responding to common stimuli.

Interactions between calcium and cyclic nucleotide pathways are multifold and complex. A detailed discussion of these interactions is beyond the scope of this review.

However, the interactions clearly can be triggered intracellularly or extracellularly. Morse et al. (71) provide one example of an extracellular interaction, in which ATP stimulation of P2Y2 receptors was followed by hydrolysis of ATP to adenosine and stimulation of the A2b receptor. The consequence was an initial calcium-mediated increase in CBF followed by a cAMP-mediated increase. A similar, but now intracellular, interaction was shown with short-term stimulation of cells with UTP, in which the intracellular calcium increase triggered a subsequent cAMP-mediated increase in CBF, possibly through the stimulation of calcium-sensitive tmACs (72). Furthermore, increasing calcium concentrations can stimulate PKG-mediated increases in CBF (e.g., 81, 83). Finally, in some mammalian species, beta-agonists cause an immediate cAMP-dependent increase in CBF, followed by a cAMP-dependent increase in $[Ca^{2+}]_i$, which in turn increases CBF (e.g., 117).

REGULATION OF CILIARY BEAT FREQUENCY BY INTRACELLULAR pH

Until recently, little information has been available on the influence of pH_i on airway epithelial CBF. Only one paper investigated the influence of extracellular pH changes (without measuring pH_i) (118). These authors, using non-HCO_3^--based buffers expected to have little effect on pH_i, found that CBF was not significantly modified when extracellular pH varied between 7.5 and 10.5. They observed significantly decreased beat frequencies below pH 7.0 for bronchi and below pH 5.0 for bronchioles. Extreme extracellular pH values (>11.0, <3.0) caused ciliostasis within a few minutes. Kienast et al. (119) found similar results when exposing cell cultures to SO_2, making the bathing solutions extremely acidic.

We have shown that relatively small changes of pH_i, in contrast, have profound effects on CBF and that these changes seem

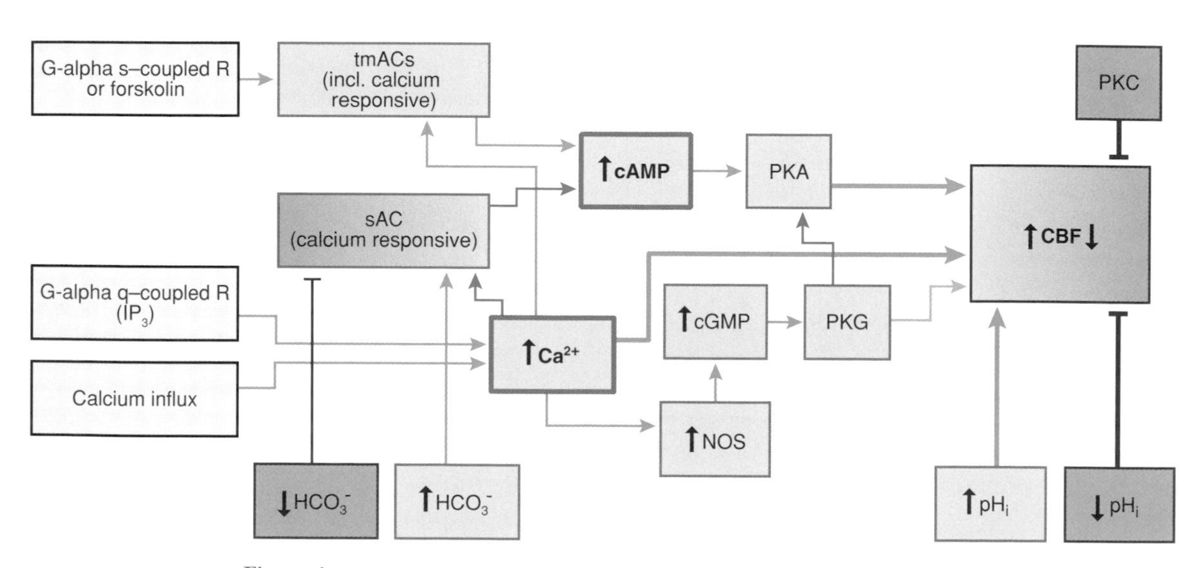

Figure 4

Summary of established and proposed pathways to regulate mammalian ciliary beat frequency (CBF). Green lines and arrows represent known stimulation; red lines and blunt arrows represent known inhibition. The main second messengers cAMP and Ca^{2+} are in bold frames. Blue frames, lines, and arrows represent putative regulation pathways. cAMP, cyclic adenosine monophosphate; CBF, ciliary beat frequency; cGMP, cyclic guanosine monophosphate; NOS, nitric oxide synthase; pH_i, intracellular pH; PKA, protein kinase A; PKC, protein kinase C; PKG, cGMP-dependent protein kinase; sAC, soluble adenylyl cyclase; tmACs, transmembrane adenylyl cyclases.

to influence the axoneme directly (120). Our results show that intracellular alkalization results in faster ciliary beating, whereas intracellular acidification attenuates CBF in human tracheobronchial epithelial cells. Our data did not support a dependence of pH_i-mediated changes in CBF on kinase/phosphatase systems, even though the catalytic efficiency of PKA, for instance, is optimal at near-neutral pH and is inhibited at acidic pH (121); in addition, acidic pH not only inhibits PKA but also activates phosphatase to dephosphorylate PKA targets and vice versa (122).

There is precedence, however, for direct axonemal beat regulation by pH_i. One study showed that human spermatozoa lacking outer dynein arms failed to exhibit higher beat frequency during mild alkalization in contrast to normal spermatozoa, suggesting that outer dynein arms are involved in the response to changing pH independently of HCO_3^- (123). Thus, there is evidence that outer dynein arm activity, the ciliary frequency–determining location (24), may be directly sensitive to pH. Variations in pH_i of airway epithelia may occur in vivo in response to shifting luminal CO_2 concentrations from 5% to 0.02% during a full breathing cycle (124). If a pH_i change occurs in epithelial cells in large airways, it results in faster ciliary beating during inspiration. Furthermore, airway diseases are associated with airway acidification (e.g., asthma, bronchiectasis), and this acidification may contribute to mucociliary dysfunction in these diseases owing to depressed ciliary activity.

CONCLUSION

We have learned a lot about the regulation of mammalian ciliary beat frequency, especially from structure-function relations and from comparative data in unicellular organisms. However, mammals and unicellular organisms clearly differ in CBF regulation. Mammalian CBF is regulated by cAMP (via PKA) and cGMP (via PKG and possibly PKA) as well as calcium and pH. **Figure 4** summarizes the different signaling pathways and some of their possible interactions.

SUMMARY POINTS

1. The regulation of mammalian ciliary beating in some respects mimics and in some respects contrasts the regulation of cilia of unicellular organisms.
2. Structure-function studies have helped identify the ciliary targets that participate in regulating ciliary beating; however, our understanding of such regulation is far from complete.
3. As in unicellular organisms, the outer dynein arm seems to regulate beating frequency, whereas the inner arm regulates bending form. The inner pair of microtubules in cilia with the 9 + 2 configuration is critical for the back-and-forth beating pattern; in its absence, the beating assumes a rotational pattern.
4. The main intracellular second messengers that regulate CBF are cAMP and the intracellular calcium concentration; cGMP also seems to play a role.
5. Frequency adjustments are made via either phosphorylation or dephosphorylation events of axonemal targets (cAMP activates an axonemal PKA to phosphorylate a dynein light chain, increasing CBF); cGMP seems to mediate a phosphorylation or dephosphorylation event to increase CBF; and PKC is involved in slowing CBF.
6. Calcium seems to increase CBF via an initial direct action on the axoneme and subsequently via cross talk to the cAMP or cGMP pathways. Considerable disagreement on the mechanism of the calcium action remains.

7. Additional influences on CBF stem from changes in intracellular pH and possibly intracellular bicarbonate.
8. A significant amount of work is required to understand fully the structural-functional regulation of CBF via almost all signaling pathways.

ACKNOWLEDGMENTS

I greatly acknowledge all my colleagues and collaborators over the years. Owing to space restrictions, I could not include all the work done in this field. Some of the work discussed was supported in part by the NIH (HL-60644 and HL-67206) and a James and Esther King Florida Biomedical Research Program Team Science Project Grant from the State of Florida.

LITERATURE CITED

1. Benzing T, Walz G. 2006. Cilium-generated signaling: a cellular GPS? *Curr. Opin. Nephrol. Hypertens.* 15:245–49
2. Vogel G. 2005. News focus: betting on cilia. *Science* 310:216–18
3. Keller LC, Romijn EP, Zamora I, Yates JR, Marshall WF. 2005. Proteomic analysis of isolated chlamydomonas centrioles reveals orthologs of ciliary-disease genes. *Curr. Biol.* 15:1090–98
4. Ostrowski LE, Blackburn K, Radde KM, Moyer MB, Schlatzer DM, et al. 2002. A proteomic analysis of human cilia: identification of novel components. *Mol. Cell Proteomics* 1:451–65
5. Pazour GJ, Agrin N, Leszyk J, Witman GB. 2005. Proteomic analysis of a eukaryotic cilium. *J. Cell Biol.* 170:103–13
6. Fliegauf M, Omran H. 2006. Novel tools to unravel molecular mechanisms in cilia-related disorders. *Trends Genet.* 22:241–45
7. Mitchison H, Salathe M, Leigh M, Carson J. 2006. Primary ciliary dyskinesia. In *Encyclopedia of Respiratory Medicine*, ed. G Laurent, S Shapiro, pp. 485–89. New York: Academic
8. Moller W, Haussinger K, Ziegler-Heitbrock L, Heyder J. 2006. Mucociliary and long-term particle clearance in airways of patients with immotile cilia. *Respir. Res.* 7:10
9. Banizs B, Pike MM, Millican CL, Ferguson WB, Komlosi P, et al. 2005. Dysfunctional cilia lead to altered ependyma and choroid plexus function, and result in the formation of hydrocephalus. *Development* 132:5329–39
10. Sawamoto K, Wichterle H, Gonzalez-Perez O, Cholfin JA, Yamada M, et al. 2006. New neurons follow the flow of cerebrospinal fluid in the adult brain. *Science* 311:629–32
11. Gieseke C, Talbot P. 2005. Cigarette smoke inhibits hamster oocyte pickup by increasing adhesion between the oocyte cumulus complex and oviductal cilia. *Biol. Reprod.* 73:443–51
12. O'Callaghan C, Sikand K, Rutman A. 1999. Respiratory and brain ependymal ciliary function. *Pediatr. Res.* 46:704–7
13. Satir P, Sleigh MA. 1990. The physiology of cilia and mucociliary interactions. *Annu. Rev. Physiol.* 52:137–55
14. Nonaka S, Tanaka Y, Okada Y, Takeda S, Harada A, et al. 1998. Randomization of left-right asymmetry due to loss of nodal cilia generating leftward flow of extraembryonic fluid in mice lacking KIF3B motor protein. *Cell* 95:829–37

15. Chilvers MA, Rutman A, O'Callaghan C. 2003. Ciliary beat pattern is associated with specific ultrastructural defects in primary ciliary dyskinesia. *J. Allergy Clin. Immunol.* 112:518–24
16. Afzelius B. 1959. Electron microscopy of the sperm tail: results obtained with a new fixative. *J. Biophys. Biochem. Cytol.* 5:269–78
17. Fawcett DW, Porter KW. 1954. A study of the fine structure of ciliated epithelia. *J. Morphol.* 94:221–81
18. Renthal R, Schneider BG, Miller MM, Luduena RF. 1993. βIV is the major β-tubulin isotype in bovine cilia. *Cell Motil. Cytoskelet.* 25:19–29
19. Taylor HC, Satir P, Holwill ME. 1999. Assessment of inner dynein arm structure and possible function in ciliary and flagellar axonemes. *Cell Motil. Cytoskelet.* 43:167–77
20. Hastie AT, Marchese-Ragona SP, Johnson KA, Wall JS. 1988. Structure and mass of mammalian respiratory ciliary outer arm 19S dynein. *Cell Motil. Cytoskelet.* 11:157–66
21. Vent J, Wyatt TA, Smith DD, Banerjee A, Luduena RF, et al. 2005. Direct involvement of the isotype-specific C-terminus of β tubulin in ciliary beating. *J. Cell Sci.* 118:4333–41
22. Andrews KL, Nettesheim P, Asai DJ, Ostrowski LE. 1996. Identification of seven rat axonemal dynein heavy chain genes: expression during ciliated cell differentiation. *Mol. Biol. Cell* 7:71–79
23. Hard R, Blaustein K, Scarcello L. 1992. Reactivation of outer-arm-depleted lung axonemes: evidence for functional differences between inner and outer dynein arms in situ. *Cell Motil. Cytoskelet.* 21:199–209
24. Brokaw CJ, Kamiya R. 1987. Bending patterns of *Chlamydomonas* flagella. IV. Mutants with defects in inner and outer dynein arms indicate differences in dynein arm function. *Cell Motil. Cytoskelet.* 8:68–75
25. Brokaw CJ. 1994. Control of flagellar bending: a new agenda based on dynein diversity. *Cell Motil. Cytoskelet.* 28:199–204
26. de Iongh RU, Rutland J. 1995. Ciliary defects in healthy subjects, bronchiectasis, and primary ciliary dyskinesia. *Am. J. Respir. Crit. Care Med.* 151:1559–67
27. Machin KE. 1958. Wave propagation along flagella. *J. Exp. Biol.* 35:796–806
28. Sanderson MJ, Dirksen ER. 1985. A versatile and quantitative computer-assisted photoelectronic technique used for the analysis of ciliary beat cycles. *Cell Motil.* 5:267–92
29. Satir P, Matsuoka T. 1989. Splitting the ciliary axoneme: implications for a "switch-point" model of dynein arm activity in ciliary motion. *Cell Motil. Cytoskelet.* 14:345–58
30. Holwill MEJ, Foster GF, Hamasaki T, Satir P. 1995. Biophysical aspects and modelling of ciliary motility. *Cell Motil. Cytoskelet.* 32:114–20
31. Holwill ME, Satir P. 1994. Physical model of axonemal splitting. *Cell Motil. Cytoskelet.* 27:287–98
32. Holwill ME. 2001. Dynein motor activity during ciliary beating. In *Cilia and Mucus: From Development to Respiratory Disease*, ed. M Salathe, pp. 19–25. New York: Marcel Dekker
33. Gibbons BH, Gibbons IR. 1972. Flagellar movement and adenosine triphosphatase activity in sea urchin sperm extracted with Triton X-100. *J. Cell Biol.* 54:75–97
34. Johnson NT, Villalon M, Royce FH, Hard R, Verdugo P. 1991. Autoregulation of beat frequency in respiratory ciliated cells: demonstration by viscous loading. *Am. Rev. Respir. Dis.* 144:1091–94
35. Andrade YN, Fernandes J, Vazquez E, Fernandez-Fernandez JM, Arniges M, et al. 2005. TRPV4 channel is involved in the coupling of fluid viscosity changes to epithelial ciliary activity. *J. Cell Biol.* 168:869–74

36. Qin H, Burnette DT, Bae YK, Forscher P, Barr MM, Rosenbaum JL. 2005. Intraflagellar transport is required for the vectorial movement of TRPV channels in the ciliary membrane. *Curr. Biol.* 15:1695–99
37. Salathe M, Bookman RJ. 1999. Mode of Ca^{2+} action on ciliary beat frequency in single ovine airway epithelial cells. *J. Physiol.* 520:851–65
38. Plattner H, Diehl S, Husser MR, Hentschel J. 2006. Sub-second calcium coupling between outside medium and subplasmalemmal stores during overstimulation/depolarisation-induced ciliary beat reversal in *Paramecium* cells. *Cell Calcium* 39:509–16
39. Habermacher G, Sale WS. 1996. Regulation of flagellar dynein by an axonemal type-1 phosphatase in *Chlamydomonas*. *J. Cell Sci.* 109:1899–907
40. Habermacher G, Sale WS. 1997. Regulation of flagellar dynein by phosphorylation of a 138-Kd inner arm dynein intermediate chain. *J. Cell Biol.* 136:167–76
40a. Tarran R, Grubb BR, Gatzy JT, Davis CW, Boucher RC. 2001. The relative roles of passive surface forces and active ion transport in the modulation of airway surface liquid volume and composition. *J. Gen. Physiol.* 118:223–36
41. Kerem E, Bistritzer T, Hanukoglu A, Hofmann T, Zhou Z, et al. 1999. Pulmonary epithelial sodium-channel dysfunction and excess airway liquid in pseudohypoaldosteronism. *N. Engl. J. Med.* 341:156–62
41a. Satir P, Christensen ST. 2007. Overview of structure and function of mammalian cilia. *Annu. Rev. Physiol.* 69:377–400
42. Gheber L, Korngreen A, Priel Z. 1998. Effect of viscosity on metachrony in mucus propelling cilia. *Cell Motil. Cytoskelet.* 39:9–20
43. Gheber L, Priel Z. 1989. Synchronization between beating cilia. *Biophys. J.* 55:183–91
44. Sanderson MJ, Sleigh MA. 1981. Ciliary activity of cultured rabbit tracheal epithelium: beat pattern and metachrony. *J. Cell Sci.* 47:331–47
45. Sanderson MJ, Charles AC, Boitano S, Dirksen ER. 1994. Mechanisms and function of intercellular calcium signaling. *Mol. Cell. Endocrinol.* 98:173–87
46. Homolya L, Steinberg TH, Boucher RC. 2000. Cell to cell communication in response to mechanical stress via bilateral release of ATP and UTP in polarized epithelia. *J. Cell Biol.* 150:1349–60
47. Kawakami M, Nagira T, Hayashi T, Shimamoto C, Kubota T, et al. 2004. Hypo-osmotic potentiation of acetylcholine-stimulated ciliary beat frequency through ATP release in rat tracheal ciliary cells. *Exp. Physiol.* 89:739–51
48. Racke K, Matthiesen S. 2004. The airway cholinergic system: physiology and pharmacology. *Pulm. Pharmacol. Ther.* 17:181–98
49. Lieb T, Forteza R, Salathe M. 2000. Hyaluronic acid in cultured ovine tracheal cells and its effect on ciliary beat frequency. *J. Aerosol Med.* 13:231–37
50. Forteza R, Lieb T, Aoki T, Savani RC, Conner GE, Salathe M. 2001. Hyaluronan serves a novel role in airway mucosal host defense. *FASEB J.* 15:2179–86
51. Sakamoto O, Iwama A, Amitani R, Takehara T, Yamaguchi N, et al. 1997. Role of macrophage-stimulating protein and its receptor, RON tyrosine kinase, in ciliary motility. *J. Clin. Invest.* 99:701–9
52. Hamasaki T, Murtaugh TJ, Satir BH, Satir P. 1989. In vitro phosphorylation of *Paramecium* axonemes and permeabilized cells. *Cell Motil. Cytoskelet.* 12:1–11
53. Hamasaki T, Barkalow K, Richmond J, Satir P. 1991. cAMP-stimulated phosphorylation of an axonemal polypeptide that copurifies with the 22S dynein arm regulates microtubule translocation velocity and swimming speed in *Paramecium*. *Proc. Natl. Acad. Sci. USA* 88:7918–22

54. Hochstrasser M, Carlson GL, Walczak CE, Nelson DL. 1996. *Paramecium* has two regulatory subunits of cyclic AMP-dependent protein kinase, one unique to cilia. *J. Eukaryot. Microbiol.* 43:356–62
55. Walczak CE, Nelson DL. 1993. In vitro phosphorylation of ciliary dyneins by protein kinases from *Paramecium*. *J. Cell Sci.* 106:1369–76
56. Barkalow K, Hamasaki T, Satir P. 1994. Regulation of 22S dynein by a 29-kD light chain. *J. Cell Biol.* 126:727–35
57. Kultgen PL, Byrd SK, Ostrowski LE, Milgram SL. 2002. Characterization of an A-kinase anchoring protein in human ciliary axonemes. *Mol. Biol. Cell* 13:4156–66
58. Salathe M, Pratt MM, Wanner A. 1993. Cyclic AMP-dependent phosphorylation of a 26 kDa axonemal protein in ovine cilia isolated from small tissue pieces. *Am. J. Respir. Cell Mol. Biol.* 9:306–14
59. Wyatt TA, Sisson JH. 2001. Chronic ethanol downregulates PKA activation and ciliary beating in bovine bronchial epithelial cells. *Am. J. Physiol. Lung Cell Mol. Physiol.* 281:L575–81
60. Sisson JH, Mommsen J, Spurzem JR, Wyatt TA. 2000. Localization of PKA and PKG in bovine bronchial epithelial cells and axonemes. *Am. J. Respir. Crit. Care Med.* 161:A449
61. Schmid A, Fregien N, Bai G, Ostrowski L, Conner G, Salathe M. 2006. Simultaneous real-time cAMP and CBF measurements in single airway epithelial cells. *Proc. Am. Thorac. Soc.* 3:A533
62. Schmid A, Sutto Z, Conner GE, Fregien N, Salathe M. 2005. Ciliary beat frequency regulation by soluble adenylyl cyclase in normal and CF airway epithelia. *Pediatr. Pulmonol.* S28:226
63. Wyatt TA, Spurzem JR, Sisson JH. 1997. Acetaldehyde reduces ciliary motility by inhibiting cAMP-mediated phosphorylation. *Am. J. Respir. Crit. Care Med.* 155:A433
64. Wyatt TA, Forget MA, Adams JM, Sisson JH. 2005. Both cAMP and cGMP are required for maximal ciliary beat stimulation in a cell-free model of bovine ciliary axonemes. *Am. J. Physiol. Lung Cell Mol. Physiol.* 288:L546–51
65. Ostrowski L, Wonsetler R, Sears P, Davis C. 2006. Regulation of ciliary beat frequency of isolated cilia from human airway epithelial cells. *Proc. Am. Thorac. Soc.* 3:A421
66. Lansley AB, Sanderson MJ, Dirksen ER. 1992. Control of the beat cycle of respiratory tract cilia by Ca^{2+} and cAMP. *Am. J. Physiol.* 263:L232–42
67. Dentler WL, Pratt MM, Stephens RE. 1980. Microtubule-membrane interactions in cilia. II. Photochemical cross-linking of bridge structures and the identification of a membrane-associated dynein-like ATPase. *J. Cell Biol.* 84:381–403
68. Dentler WL. 1981. Microtubule-membrane interactions in cilia and flagella. *Int. Rev. Cytol.* 72:1–47
69. Dey CS, Brokaw CJ. 1991. Activation of *Ciona* sperm motility: phosphorylation of dynein polypeptides and effects of a tyrosine kinase inhibitor. *J. Cell Sci.* 100:815–24
70. Howard DR, Habermacher G, Glass DB, Smith EF, Sale WS. 1994. Regulation of *Chlamydomonas* flagellar dynein by an axonemal protein kinase. *J. Cell Biol.* 127:1683–92
71. Morse DM, Smullen JL, Davis CW. 2001. Differential effects of UTP, ATP, and adenosine on ciliary activity of human nasal epithelial cells. *Am. J. Physiol.* 280:C1485–97
72. Lieb T, Frei CW, Frohock JI, Bookman RJ, Salathe M. 2002. Prolonged increase in ciliary beat frequency after short-term purinergic stimulation in human airway epithelial cells. *J. Physiol.* 538:633–46
73. Neer EJ. 1978. Physical and functional properties of adenylate cyclase from mature rat testis. *J. Biol. Chem.* 253:5808–12

74. Braun T, Frank H, Dods R, Sepsenwol S. 1977. Mn^{2+}-sensitive, soluble adenylate cyclase in rat testis: differentiation from other testicular nucleotide cyclases. *Biochim. Biophys. Acta* 481:227–35
75. Braun T, Dods RF. 1975. Development of a Mn^{2+}-sensitive, "soluble" adenylate cyclase in rat testis. *Proc. Natl. Acad. Sci. USA* 72:1097–201
76. Jaiswal BS, Conti M. 2001. Identification and functional analysis of splice variants of the germ cell soluble adenylyl cyclase. *J. Biol. Chem.* 276:31698–708
77. Buck J, Sinclair ML, Schapal L, Cann MJ, Levin LR. 1999. Cytosolic adenylyl cyclase defines a unique signaling molecule in mammals. *Proc. Natl. Acad. Sci. USA* 96:79–84
78. Chen Y, Cann MJ, Litvin TN, Iourgenko V, Sinclair ML, et al. 2000. Soluble adenylyl cyclase as an evolutionarily conserved bicarbonate sensor. *Science* 289:625–28
79. Litvin TN, Kamenetsky M, Zarifyan A, Buck J, Levin LR. 2003. Kinetic properties of "soluble" adenylyl cyclase: synergism between calcium and bicarbonate. *J. Biol. Chem.* 278:15922–26
80. Han H, Stessin A, Roberts J, Hess K, Gautam N, et al. 2005. Calcium-sensing soluble adenylyl cyclase mediates TNF signal transduction in human neutrophils. *J. Exp. Med.* 202:353–61
81. Uzlaner N, Priel Z. 1999. Interplay between the NO pathway and elevated $[Ca^{2+}]_i$ enhances ciliary activity in rabbit trachea. *J. Physiol.* 516:179–90
82. Wyatt TA, Spurzem JR, May K, Sisson JH. 1998. Regulation of ciliary beat frequency by both PKA and PKG in bovine airway epithelial cells. *Am. J. Physiol.* 275:L827–35
83. Zhang L, Sanderson MJ. 2003. The role of cGMP in the regulation of rabbit airway ciliary beat frequency. *J. Physiol.* 551:765–76
84. Geary CA, Davis CW, Paradiso AM, Boucher RC. 1995. Role of CNP in human airways: cGMP-mediated stimulation of ciliary beat frequency. *Am. J. Physiol.* 268:L1021–28
85. Sisson JH, May K, Wyatt TA. 1999. Nitric oxide-dependent ethanol stimulation of ciliary motility is linked to cAMP-dependent protein kinase (PKA) activation in bovine bronchial epithelium. *Alcohol. Clin. Exp. Res.* 23:1528–33
86. Gertsberg I, Hellman V, Fainshtein M, Weil S, Silberberg SD, et al. 2004. Intracellular Ca^{2+} regulates the phosphorylation and the dephosphorylation of ciliary proteins via the NO pathway. *J. Gen. Physiol.* 124:527–40
87. Kobayashi K, Salathe M, Pratt MM, Cartagena NJ, Soloni F, et al. 1992. Mechanism of hydrogen peroxide-induced inhibition of sheep airway cilia. *Am. J. Respir. Cell Mol. Biol.* 6:667–73
88. Kobayashi K, Tamaoki J, Sakai N, Chiyotani A, Takizawa T. 1989. Inhibition of ciliary activity by phorbol esters in rabbit tracheal cells. *Lung* 167:277–84
89. Wyatt TA, Schmidt SC, Rennard SI, Tuma DJ, Sisson JH. 2000. Acetaldehyde-stimulated PKC activity in airway epithelial cells treated with smoke extract from normal and smokeless cigarettes. *Proc. Soc. Exp. Biol. Med.* 225:91–97
90. Wong LB, Park CL, Yeates DB. 1998. Neuropeptide Y inhibits ciliary beat frequency in human ciliated cells via nPKC, independently of PKA. *Am. J. Physiol.* 275:C440–48
91. Levin R, Braiman A, Priel Z. 1997. Protein kinase C induced calcium influx and sustained enhancement of ciliary beating by extracellular ATP. *Cell Calcium* 21:103–13
92. Salathe M, Pratt MM, Wanner A. 1993. Protein kinase C-dependent phosphorylation of a ciliary membrane protein and inhibition of ciliary beating. *J. Cell Sci.* 106:1211–20
93. Naitoh Y, Kaneko H. 1972. Reactivated triton-extracted models of paramecium: modification of ciliary movement by calcium ions. *Science* 176:523–24

94. Di Benedetto G, Magnus CJ, Gray PTA, Mehta A. 1991. Calcium regulation of ciliary beat frequency in human respiratory epithelium in vitro. *J. Physiol.* 439:103–13
95. Sanderson MJ, Dirksen ER. 1989. Mechanosensitive and β-adrenergic control of the ciliary beat frequency of mammalian respiratory tract cells in culture. *Am. Rev. Respir. Dis.* 139:432–40
96. Sanderson MJ, Charles AC, Dirksen ER. 1990. Mechanical stimulation and intercellular communication increase intracellular calcium in epithelial cells. *Cell Regul.* 1:585–96
97. Girard PG, Kennedy JR. 1986. Calcium regulation of ciliary activity in rabbit tracheal explants and outgrowth. *Eur. J. Cell Biol.* 40:203–9
98. Salathe M, Bookman RJ. 1995. Coupling of $[Ca^{2+}]_i$ and ciliary beating in cultured tracheal epithelial cells. *J. Cell Sci.* 108(Pt. 2):431–40
99. Verdugo P. 1980. Calcium-dependent hormonal stimulation of ciliary activity. *Nature* 283:764–65
100. Villalon M, Hinds TR, Verdugo P. 1989. Stimulus-response coupling in mammalian ciliated cells: demonstration of two mechanisms of control for cytosolic $[Ca^{2+}]$. *Biophys. J.* 56:1255–58
101. Barrera NP, Morales B, Villalon M. 2004. Plasma and intracellular membrane inositol 1,4,5-trisphosphate receptors mediate the Ca^{2+} increase associated with the ATP-induced increase in ciliary beat frequency. *Am. J. Physiol. Cell Physiol.* 287:C1114–24
102. Salathe M, Lipson E, Ivonnet PI, Bookman RJ. 1997. Muscarinic signal transduction in tracheal epithelial cells: effects of acetylcholine on intracellular Ca^{2+} and ciliary beating. *Am. J. Physiol.* 272:L301–10
103. Verdugo P, Raess BV, Villalon M. 1983. The role of calmodulin in the regulation of ciliary movement in mammalian epithelial cilia. *J. Submicrosc. Cytol.* 15:95–96
104. Hirano-Ohnishi J, Watanabe Y. 1989. Ca^{2+}/calmodulin-dependent phosphorylation of ciliary β-tubulin in *Tetrahymena*. *J. Biochem.* 105:858–60
105. Lansley AB, Sanderson MJ. 1999. Regulation of airway ciliary activity by Ca^{2+}: simultaneous measurement of beat frequency and intracellular Ca^{2+}. *Biophys. J.* 77:629–38
106. Kakuta Y, Kanno T, Sasaki H, Takishima T. 1985. Effect of Ca^{2+} on the ciliary beat frequency of skinned dog tracheal epithelium. *Respir. Physiol.* 60:9–19
107. Ma W, Silberberg SD, Priel Z. 2002. Distinct axonemal processes underlie spontaneous and stimulated airway ciliary activity. *J. Gen. Physiol.* 120:875–85
108. Zhang L, Sanderson MJ. 2003. Oscillations in ciliary beat frequency and intracellular calcium concentration in rabbit tracheal epithelial cells induced by ATP. *J. Physiol.* 546:733–49
109. Salathe M, Lieb T, Bookman RJ. 2000. Lack of nitric oxide involvement in cholinergic modulation of ovine ciliary beat frequency. *J. Aerosol Med.* 13:219–29
110. Tamaoki J, Chiyotani A, Kondo M, Konno K. 1995. Role of NO generation in β-adrenoceptor-mediated stimulation of rabbit airway ciliary motility. *Am. J. Physiol. Cell Physiol.* 37:C1342–47
111. Yang B, Schlosser RJ, McCaffrey TV. 1996. Dual signal transduction mechanisms modulate ciliary beat frequency in upper airway epithelium. *Am. J. Physiol.* 14:L745–51
112. Yang B, Schlosser RJ, McCaffrey TV. 1997. Signal transduction pathways in modulation of ciliary beat frequency by methacholine. *Ann. Otol. Rhinol. Laryngol.* 106:230–36
113. Zagoory O, Braiman A, Gheber L, Priel Z. 2001. Role of calcium and calmodulin in ciliary stimulation induced by acetylcholine. *Am. J. Physiol. Cell Physiol.* 280:C100–9
114. Zagoory O, Braiman A, Priel Z. 2002. The mechanism of ciliary stimulation by acetylcholine: roles of calcium, PKA, and PKG. *J. Gen. Physiol.* 119:329–39

115. Delmotte P, Sanderson MJ. 2006. Ciliary beat frequency is maintained at a maximal rate in the small airways of mice lung slices. *Am. J. Respir. Cell Mol. Biol.* 35:110–17
116. Hayashi T, Kawakami M, Sasaki S, Katsumata T, Mori H, et al. 2005. ATP regulation of ciliary beat frequency in rat tracheal and distal airway epithelium. *Exp. Physiol.* 90:535–44
117. Zhang L, Han D, Sanderson MJ. 2005. Effect of isoproterenol on the regulation of rabbit airway ciliary beat frequency measured with high-speed digital and fluorescence microscopy. *Ann. Otol. Rhinol. Laryngol.* 114:399–403
118. Clary-Meinesz C, Mouroux J, Cosson J, Huitorel P, Blaive B. 1998. Influence of external pH on ciliary beat frequency in human bronchi and bronchioles. *Eur. Respir. J.* 11:330–33
119. Kienast K, Riechelmann H, Knorst M, Schlegel J, Mullerquernheim J, et al. 1994. An experimental model for the exposure of human ciliated cells to sulfur dioxide at different concentrations. *Clin. Investig.* 72:215–19
120. Sutto Z, Conner GE, Salathe M. 2004. Regulation of human airway ciliary beat frequency by intracellular pH. *J. Physiol.* 560:519–32
121. Cox S, Taylor SS. 1995. Kinetic analysis of cAMP-dependent protein kinase: mutations at histidine 87 affect peptide binding and pH dependence. *Biochemistry* 34:16203–9
122. Reddy MM, Kopito RR, Quinton PM. 1998. Cytosolic pH regulates GCl through control of phosphorylation states of CFTR. *Am. J. Physiol.* 275:C1040–47
123. Keskes L, Giroux-Widemann V, Serres C, Pignot-Paintrand I, Jouannet P, Feneux D. 1998. The reactivation of demembranated human spermatozoa lacking outer dynein arms is independent of pH. *Mol. Reprod. Dev.* 49:416–25
124. Willumsen NJ, Boucher RC. 1992. Intracellular pH and its relationship to regulation of ion transport in normal and cystic fibrosis human nasal epithelia. *J. Physiol.* 455:247–69

Genetic Defects in Ciliary Structure and Function

Maimoona A. Zariwala,[1,*] Michael R. Knowles,[1] and Heymut Omran[2]

[1]Department of Medicine, Pathology and Laboratory Medicine, University of North Carolina, Chapel Hill, North Carolina 27599; email: zariwala@med.unc.edu, michael_knowles@med.unc.edu

[2]Department of Pediatrics and Adolescent Medicine, University Hospital, Freiburg, 79106 Germany; email: heymut.omran@uniklinik-freiburg.de

Annu. Rev. Physiol. 2007. 69:423–50

First published online as a Review in Advance on October 20, 2006

The *Annual Review of Physiology* is online at http://physiol.annualreviews.org

This article's doi: 10.1146/annurev.physiol.69.040705.141301

0066-4278/07/0315-0423$20.00

*Corresponding author.

Key Words

primary ciliary dyskinesia (PCD), Kartagener syndrome, situs inversus, dynein

Abstract

Cilia, hair-like structures extending from the cell membrane, perform diverse biological functions. Primary (genetic) defects in the structure and function of sensory and motile cilia result in multiple ciliopathies. The most prominent genetic abnormality involving motile cilia (and the respiratory tract) is primary ciliary dyskinesia (PCD). PCD is a rare, usually autosomal recessive, genetically heterogeneous disorder characterized by sino-pulmonary disease, laterality defects, and male infertility. Ciliary ultrastructural defects are identified in ~90% of PCD patients and involve the outer dynein arms, inner dynein arms, or both. Diagnosing PCD is challenging and requires a compatible clinical phenotype together with tests such as ciliary ultrastructural analysis, immunofluorescent staining, ciliary beat assessment, and/or nasal nitric oxide measurements. Recent mutational analysis demonstrated that 38% of PCD patients carry mutations of the dynein genes *DNAI1* and *DNAH5*. Increased understanding of the pathogenesis will aid in better diagnosis and treatment of PCD.

CILIA AND CILIA-RELATED DISORDERS

Cilia: short hair-like structures on a cell or microorganism that are capable of rhythmic motion. Cilia act in unison to cause movement of the cell or the surrounding medium

Flagella: long, threadlike appendages on certain cells; function as organs of locomotion

Cilia and flagella have a highly ordered basic structure of nine peripheral microtubule doublets arranged in the periphery of the axoneme (1, 2). In most motile cilia, such as respiratory cilia, the nine peripheral doublets surround two central microtubules (9 + 2 axoneme) (**Figure 1**). Each outer doublet is composed of an A and a B tubule. A central pair of microtubules, also structurally and biochemically asymmetric, is present in the center of the ring and extends the axoneme's length. Cilia that contain dynein arms attached to their peripheral microtubule doublets are motile. Such cilia include 9 + 2 respiratory cilia and a subset of cilia that lacks the central pair apparatus, e.g., 9 + 0 nodal monocilia. However, cilia without dynein arm structures, such as 9 + 0 renal monocilia or 9 + 2 vestibular cilia, are immotile. Cilia can also exhibit sensory function, such as mechano- and chemosensation, which has recently attracted much attention.

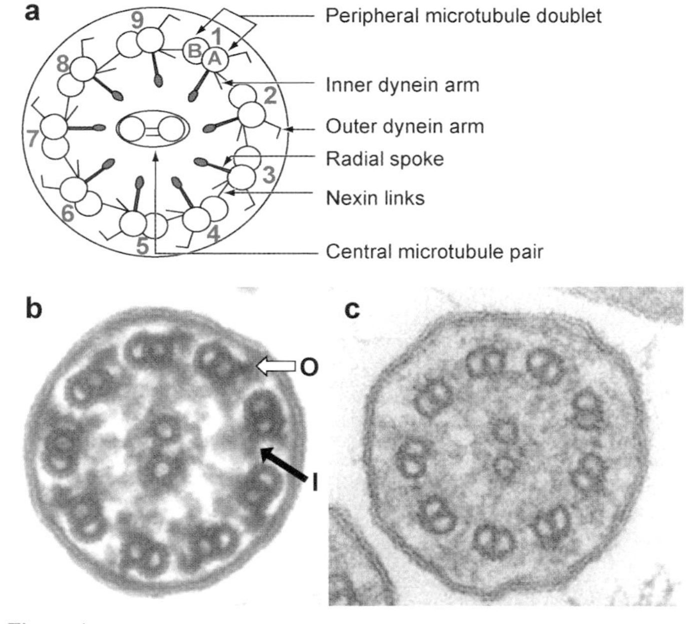

Figure 1

Cross section of cilia. (*a*) Schematic showing cross section of cilia revealing the 9 + 2 arrangement of nine peripheral microtubule doublets surrounding a central microtubule pair. (*b*) Representative electron microscopic image of a normal cilium from the nasal epithelium of a clinically unaffected individual. O, outer dynein arm (*white arrow*); I, inner dynein arm (*black arrow*). The central pair and radial spokes are also visible. (*c*) Representative electron microscopic image of a cilium from the nasal epithelium of an individual with primary ciliary dyskinesia; outer and inner dynein arms are absent. **Figures 1*b*** and ***c*** are reproduced with permission from Johnny L. Carson, **http://pediatrics.med.unc.edu/div/infectdi/pcd/pcd.htm.**

Dysfunction of sensory monocilia has been demonstrated in polycystic kidney disease (3). Proteins mutated in cystic kidney disorders sublocalize to the cilium or ciliary base. This finding suggests that the renal phenotypes observed in diverse disorders such as autosomal dominant and autosomal recessive forms of polycystic kidney disease, nephronophthisis, and Bardet Biedl syndrome result from malfunctional renal monocilia (4). Retinal degeneration is often a hallmark of multisystem disorders in which mutant proteins are localized to the cilium or ciliary base (centrioles). These disorders, also referred to as ciliopathies, include nephronophthisis (also referred to as Senior-Loken syndrome), Bardet Biedl syndrome, and Alstrom syndrome (5). To understand the manifestation of retinal disease, it is important to know that photoreceptors are polarized sensory neurons consisting of a photosensitive outer segment that develops from a nonmotile primary cilium by the movement of membrane and photoreactive proteins, such as opsin, into the cilium (6). The connecting cilium remains in mature photoreceptors as the sole link between the inner and outer segments and is presumably the major transport corridor (6). Recent findings have demonstrated that targeted mutations in mice abolishing intraflagellar transport through the connecting cilium result in severe retinal degeneration, highlighting its prominent role in preserving photoreceptor integrity (7, 8). Sensory cilial dysfunction in other cell types probably leads to other complex disease manifestations, such as developmental anomalies (e.g., polydactyly) and mental retardation, which are observed in Bardet Biedl syndrome. Careful analysis of clinical findings in Bardet Biedl and Alstrom syndromes revealed that 30% and 50%, respectively, of those affected show chronic pulmonary symptoms (9, 10). These observations suggest that some degree of respiratory cilia dysfunction occurs in these disorders as well.

For a more detailed view on the function of sensory cilia, see a recent review by Reference 11 . This current review focuses on genetic defects responsible for dysfunction of motile cilia as observed in primary ciliary dyskinesia.

PRIMARY CILIARY DYSKINESIA

Primary ciliary dyskinesia (PCD) (MIM# 242650) is a rare disorder, usually inherited as an autosomal recessive trait, which affects approximately 1 in 20,000 to 1 in 60,000 individuals (12, 13). Researchers also have elucidated a few variant inheritance patterns. Narayan et al. (14) suggested the role of autosomal dominant or X-linked dominant inheritance in one family in which an affected mother had five affected children born to three different fathers. Furthermore, other researchers recently reported an association between X-linked recessive retinitis pigmentosa, sensory hearing deficits, and PCD (15–20). In addition, Budny et al. (21) described a single family with a novel syndrome that is caused by *OFD1* mutations and characterized by X-linked recessive mental retardation, PCD, and macrocephaly.

Dysfunction of motile cilia and flagella explains the complex PCD phenotype involving various organ systems (22, 23). Multiple motile cilia that line the upper and lower airways mediate mucociliary clearance. These rod-like organelles extend from the airway epithelial cell surface and move extracellular mucus by coordinated beating. In the term newborn, unexplained tachypnoea, chronic rhinitis, and respiratory distress syndrome can be the first signs of PCD. The clinical course of affected individuals is characterized by recurrent respiratory infections caused by defective mucociliary clearance due to immotile or dyskinetic respiratory cilia that result in chronic infection and inflammation of the upper and lower airways (1, 13). The disease typically progresses to bronchiectasis and can cause chronic lung failure in adulthood (**Figure 2**).

Congenital disorders such as hydrocephalus, retinal degeneration, sensory hearing deficits, and mental retardation have been reported in PCD patients (1, 24). Dysfunction of monocilia at the embryonic node is associated with randomization of left-right body asymmetry (25). This explains why half of PCD patients exhibit situs inversus totalis (**Figure 2**). The combination of PCD and situs inversus is also referred to as Kartagener syndrome (MIM# 244400). A current review of 232 PCD patients revealed that 6% had situs ambiguous (heterotaxy), and half of those (3% of the total PCD population) had complex congenital heart disease (26). Thus, genes that cause PCD and situs abnormalities are also candidates for causing congenital heart disease that occurs in association with heterotaxy.

Reduced fertility due to sperm dysmotility and/or reduced sperm count is frequently observed in male PCD patients (27, 28). Ultrastructural defects of sperm tails often resemble those observed in respiratory cilia. The lack of ciliary movement in the Fallopian tubes may contribute to subfertility in affected women. In addition, pulmonary calcium deposition and lithoptysis have been observed in a subset of older patients (29).

Diagnosis of PCD is often delayed despite the presence of typical symptoms early in life and may cause an underestimation of disease prevalence (30). The need for early diagnosis in childhood to delay or even prevent occurrence of bronchiectasis makes molecular research in the PCD field eminently important. Recent advances in the understanding of the disease have led to new diagnostic options and may allow the development of novel therapeutic strategies.

Axoneme: an array of microtubules running longitudinally through the entire length of cilia and flagella; has a characteristic 9 + 2 or 9 + 0 configuration

Dyneins: motor proteins involved in organelle transport toward the cell's central region (minus end of microtubule); responsible for the bending of cilia

Dynein arms: structures that generate forces of ciliary and flagellar movement

Microtubule: protein structures that constitute the majority of the cytoskeletal component and that regulate the shape and control the movements of eukaryotic cells

PCD: primary ciliary dyskinesia

Situs inversus totalis: a congenital disorder with total reversal of all visceral organs

Heterotaxy: abnormal placement of organs of the body due to failure to establish normal left-right (situs solitus) or mirror-image (situs inversus totalis) patterning during embryonic development

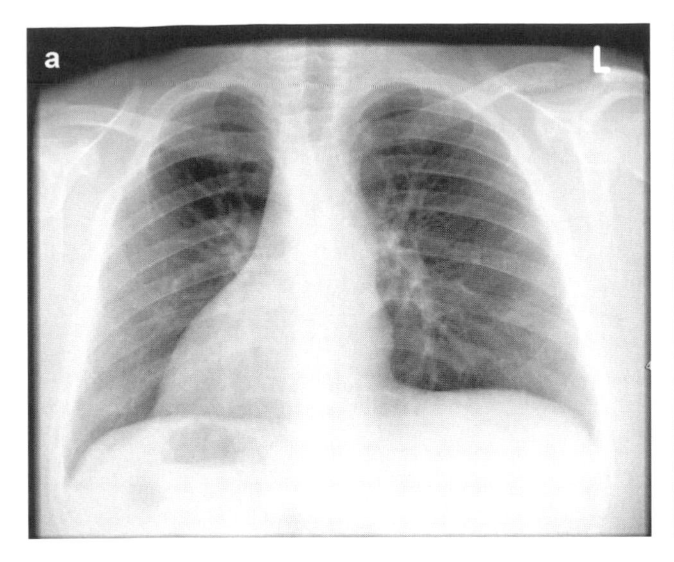

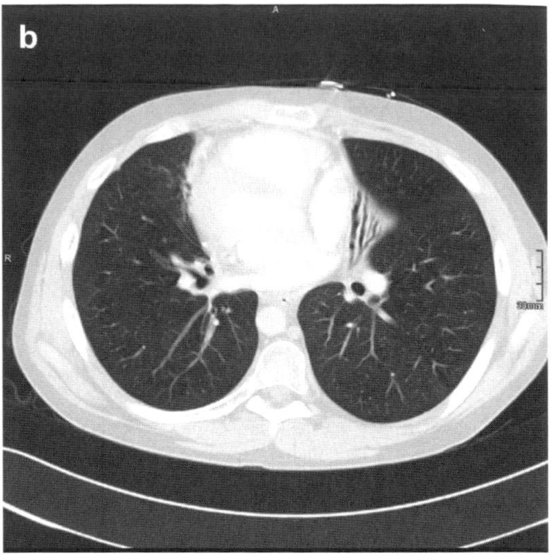

Figure 2

Clinical findings of primary ciliary dyskinesia (PCD). (*a*) Radiograph of the chest and upper abdomen from a child with PCD and situs inversus. Note the dextroposition of the stomach bubble and the right-sided heart. (*b*) CT scan of the thorax of a male PCD patient with situs inversus and atelectasis and bronchiectasis caused by recurrent pulmonary infections of the anatomic middle lobe.

NORMAL ULTRASTRUCTURES OF CILIA AND FLAGELLA

Cilia and flagella are highly complex organelles that are closely related in structure across organisms from humans to protozoa (*Chlamydomonas reinhardtii* and *Trypanozoma brucei*). The cilium is an organelle that emerges as an outgrowth from the basal body (centriole) and is present on almost every human cell.

Chlamydomonas: these biflagellated, unicellular algae, which have well-studied genetics and many known and mapped mutants, are model organisms to study flagellar motility, chloroplast dynamics, biogenesis, and genetics

ODA: outer dynein arms

IDA: inner dynein arms

HC: heavy chain

The motile respiratory cilia as well as sperm tails (flagella) have a 9 + 2 arrangement consisting of nine microtubule doublets surrounding a pair of microtubule singlets (central pair). Radial spokes radiate from the central microtubules toward the peripheral microtubules. Nexin links connect the outer doublets and keep them intact and limit the range of microtubular sliding. Each microtubule A of the doublet has two rows of dynein arms, termed the outer dynein arms (ODA) and inner dynein arms (IDA), and contain motor proteins termed dynein (**Figure 1**) (31). Each cilium contains ~4000 dynein arms in pairs (ODA and IDA). Motile cilia include mucus-propelling cilia of the respiratory tract (~6 microns in length and ~200 per cell), liquid-propelling cilia from ependymal lining of the brain ventricles and oviducts, and sperm flagella. The reviews elsewhere in this volume by Satir & Christensen (31a) and Salathe (31b) discuss details of structure, function, and regulation.

COMPOSITION OF HUMAN OUTER DYNEIN ARM COMPLEXES

Most of our knowledge about ODA composition originates from studies in *Chlamydomonas* (31c). In the *Chlamydomonas* alga, only one distinct type of ODA complex is known. The complex contains three heavy chains (α-, β-, and γ-HCs), two intermediate chains, nine light chains, three docking complex proteins, and at least two associated proteins (32). Like *Chlamydomonas*, *Tetrahymena* also has three distinct HCs, whereas sea urchins, trout, and pigs have ODA complexes that contain only two HCs (33–35). Based on homology searches using the BLAST algorithm (36), Pazour et al. (32) have identified five human ODA HC orthologs of *Chlamydomonas* β- and γ-HCs. Orthologs of the β-HC are *DNAH11* (chromosome 7p21), *DNAH17* (chromosome 17q25), and *DNAH9* (chromosome 17p12). Orthologs of the γ-HC are *DNAH5* (chromosome 5p15) and *DNAH8* (chromosome 6p21). No human ortholog of the *Chlamydomonas* α-HC has been identified. This finding is consistent with the concept that, like other vertebrates, human ODA complexes contain only two ODA HCs. However, because more than one human ODA HC ortholog has been identified, humans may have more than one distinct type of ODA complex, in contrast to *Chlamydomonas*.

The expression pattern of some of the five ODA HCs has been analyzed. Northern blot analyses revealed specific expression of *DNAH5* in the lung, brain, and testis (37). Using the in situ hybridization technique, Kispert et al. (38) found that the cellular expression of the murine ortholog *Mdnah5* (alias *Dnahc5*) is confined in the lung to ciliated respiratory cells of the upper and lower airways. In addition, the ependymal cells lining the brain ventricles, which are also covered by multiple motile cilia, express *Mdnah5* (39). Cells of the embryonic node, which carry one specialized motile monocilium, also express *Mdnah5* (37). Reed et al. (40) studied in detail the expression of DNAH9, also referred to as DNEL1. They demonstrated specific expression in the lung, using Northern blot analysis. Reverse transcriptase-polymerase chain reaction (RT-PCR) analyses revealed RNA messages in cultured respiratory cells of tracheal and nasal origin. In addition, Reed et al. (40) analyzed protein expression, using specific monoclonal anti-DNAH9 antibodies. Immunohistochemistry and Western blot analyses showed that the ODA-HC DNAH9 is present in respiratory cilia and sperm flagella (40). Immunoelectron microscopy demonstrated that DNAH9 localizes to one subfiber of doublet microtubules surrounding the two singlet (central-complex) tubules in respiratory ciliary axonemes. RT-PCR identified *DNAH11* (*Dnahc11*) expression in trachea, testis, and lung (41). The mouse ortholog *lrd* (left-right dynein) is specifically expressed at the embryonic node (42). Detailed expression analyses of the other ODA HCs, DNAH17 and DNAH8, have not yet been performed. In summary, all ODA HCs analyzed so far appear to exhibit specific functions for ciliary and flagellar motility in various cell types (respiratory and ependymal cilia, nodal monocilia, and sperm flagella).

Orthologs: genes that have homologous sequence in different species, are directly related through descent from a common ancestor, and carry out related functions

To investigate the molecular mechanisms involved in human ODA generation and function, Fliegauf et al. (43), using specific antibodies, analyzed the subcellular localization of the axonemal ODA HCs DNAH5 and DNAH9 in respiratory epithelial and sperm cells. Confocal immunofluorecence imaging revealed characteristic patterns of subcellular localization of the analyzed ODA components. In wild-type respiratory cells, the ODA HC DNAH5 is present throughout the entire length of the ciliary axoneme, whereas the ODA HC DNAH9 localizes exclusively to the distal ciliary compartment (**Figure 3**). In contrast, in sperm cells DNAH5 localizes only to the proximal part of the flagellum, whereas DNAH9 antibodies stain the entire length of the sperm tail (**Figure 3**). Thus, human ODA complexes vary in their composition along the respiratory ciliary axoneme and the axoneme of the sperm tail, and this composition also differs between these two cell types.

These data indicate that in respiratory cilia and sperm tails at least two ODA types are present: type 1 (DNAH9 negative and DNAH5 positive) and type 2 (DNAH9 and DNAH5 positive) (**Figure 3**). The spatial diversity of ODA HCs along the axonemes probably contributes to the typical beating characteristics of cilia and sperm flagella and likewise to the various beat modes of other motile cilia types (e.g., ependymal and nodal cilia). Because comparative genome analyses identified at least three other human ODA HCs (*DNAH8*, *DNAH11*, *DNAH17*), an even

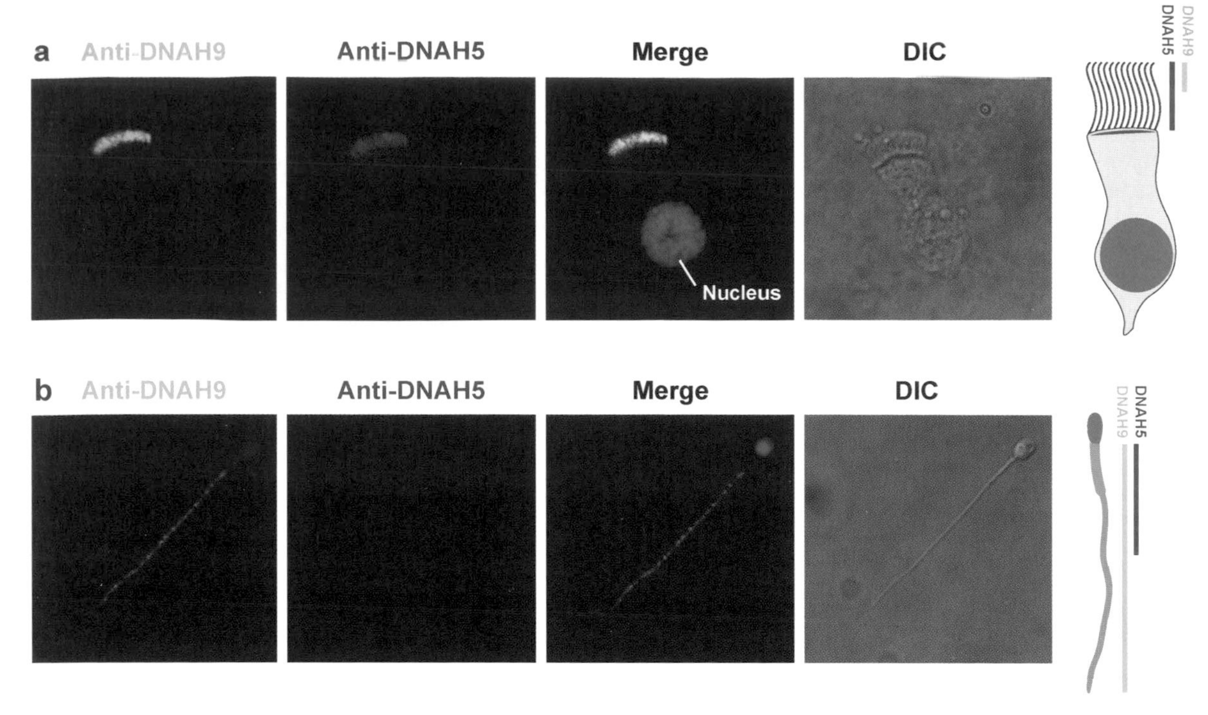

Figure 3

The outer dynein arm (ODA) heavy chains (HCs) DNAH5 and DNAH9 exhibit cell-type-specific distributions along the ciliary/flagellar axoneme. (*a*) DNAH9 (*green*) localizes exclusively to distal parts of respiratory ciliary axonemes, whereas DNAH5 (*red*) localizes along the entire length of the axonemes and within the microtubule-organizing centers at the ciliary base. The nucleus is blue. (*b*) In sperm cells DNAH5 localizes only to the proximal flagellar axoneme, whereas DNAH9 is present throughout the whole sperm flagellum. The head of the sperm is blue. On the right are schematic drawings depicting the distinct sublocalization patterns of the ODA HCs DNAH5 and DNAH9 in respiratory and sperm cells. DIC denotes differential interference contrast.

higher variability of ODA composition in human cilia and sperm flagella should be expected.

COMPOSITION OF HUMAN INNER DYNEIN ARM COMPLEXES

The composition and arrangement of IDA are complex and structurally and functionally diverse. Most of our knowledge regarding IDA composition came from studies in *Chlamydomonas* (reviewed in References 43a,b). *Chlamydomonas* exhibits seven distinct IDA isoforms: one two-headed isoform (I1) and six single-headed isoforms. The isoforms are precisely organized, with a distinct localization pattern along the doublet microtubule of the flagellar axoneme (31c, 43a). The IDA includes several subunits of heavy, intermediate, and light chains that are different from the composition of ODA, indicating different roles for these chains. Similarities exist between *Chlamydomonas* I1 inner arm isoform and ODA and cytoplasmic dyneins (reviewed in References 43a,b). I1 is composed of two HCs, three intermediate chains and three light chains that form a trilobed structure proximal to the first radial spoke involved in the control of flagellar motility. Various human IDA HC genes—*DNAH14*, *DNAH7*, *DNAH1*, *DNAH12*, *DNAH10*, *DNAH2*, and *DNAH6*—have been identified (32). The intermediate-chain genes identified thus far are *IC138* (WDR78) and *IC140* (WDR63). Among the light-chain genes identified are *hp28* and *TCTEL1*.

ULTRASTRUCTURAL DEFECTS IN PRIMARY CILIARY DYSKINESIA

Searching for defects in ciliary ultrastructure via the use of transmission electron microscopy (TEM) remains the current standard for confirming diagnosis of PCD, although this approach has recognized limitations (see below). TEM studies in humans and parallel studies with *Chlamydomonas* have revealed various forms of ultrastructural defects, including absent or shortened ODA or IDA or both, and defects in radial spokes, nexin links, or central pairs.

It is critical to distinguish primary (genetic) defects from acquired defects that result from exposure to environmental and infectious agents. This requires expertise in obtaining the tissue sample, acquiring TEM pictures, and evaluating the ciliary ultrastructure (44). Also, optimal fixation and handling of biopsies are necessary to obtain sufficient ciliated cells for analysis. Furthermore, to distinguish between primary and secondary abnormalities, respiratory epithelial cell cultures are useful, but only a few centers perform these techniques, and a significant amount of cell material is necessary (45).

TEM cannot easily demonstrate all primary ultrastructural defects. For example, IDA defects are difficult to pick up owing to low contrast and organization of IDA. Computer-assisted analysis (46) of TEM cross-section photographs can sometimes improve sensitivity. Significantly, some PCD patients have apparently normal ciliary ultrastructure (47, 48). Representative TEM photographs for a cross section of cilia from a healthy subject and a patient with an ODA defect are shown in **Figure 1 *b*** and ***c***.

Several studies have described the spectrum of ultrastructural defects in PCD. In a recent study (49) that analyzed 50 cross sections of human cilia, the mean numbers of ODA per cilium were 7.5–9.0 for control subjects and <1.6 for subjects with PCD. The mean numbers of IDA were 3–6 for control subjects and <0.6 for subjects with PCD. In addition, of 38 patients with PCD, 45% had defects in ODA and IDA, 24% had isolated ODA defects, 21% had isolated IDA defects, 8% presented with no nexin links, and 3% had ciliary aplasia (49). In another study (50), of 76 PCD patients with known ultrastructural defects, 43% had an isolated ODA defect, 29% had an isolated IDA defect, 24% had a defect in both dynein arms, and 4% had a central apparatus defect. A separate study by Chilvers et al. (51) analyzed 56 PCD patients. Of these patients, 36% had ultrastructural defects of both dynein arms, 29% had ODA defects, 14% had IDA defects, 7% had radial spoke defects, and 14% had transposition defects. Jorissen et al. (52) reported a study in which 42% of subjects had ODA defects, 7% had both IDA and ODA defects, 10% had IDA defects and eccentric central pairs, 7% lacked central pairs, 4% had ciliary aplasia, and 28% showed no anomalies. These studies establish the predominance of ultrastructural defects of dynein arms in PCD patients (the molecular aspects of ODA are covered below).

Evidence from both *Chlamydomonas* and humans suggests that the ODA relates primarily to ciliary beat frequency and force, whereas IDA appears to relate to ciliary bending pattern. Only a small percentage of patients show ciliary aplasia or defects of the radial spoke, nexin link, or central apparatus.

From a genetic standpoint, different mutations lead to different ciliary ultrastructural phenotypes, but the clinical phenotype is similar. In cases in which no ultrastructural defects have been defined, the diagnosis of PCD can be established with a compatible clinical phenotype, sometimes including situs inversus, together with measurements of ciliary beat frequency, measurements of nasal nitric oxide (NO) levels, or dysmotility of cilia or spermatozoa by light microscopy.

TEM: transmission electron microscopy

NO: nitric oxide

GENETIC DEFECTS AFFECTING OUTER DYNEIN ARM FUNCTION

DNAH5 Mutations

The PCD locus on the short arm of chromosome 5 was identified by applying a homozygosity mapping strategy in one large inbred Arab family with PCD and absence of ODA (53). Within the critical genetic region on chromosome 5p15, the human ortholog (*DNAH5*) of the *C. reinhardtii* γ-HC of the ODA was identified. The *DNAH5* genomic region comprises 79 exons and 1 alternative first exon and spans 250 kb (37). Mutations in the *Chlamydomonas* ortholog result in slow-swimming algae with ultrastructural ODA defects (54). Sequencing all 80 *DNAH5* exons in a total of 25 PCD families resulted in the detection, in eight PCD families, of four homozygous mutations and six heterozygous mutations within the coding region of the *DNAH5* gene (37). These mutations were associated with ODA defects and randomization of left-right asymmetry (37). Repeated spermiograms of two brothers with PCD and *DNAH5* mutations revealed low sperm counts (oligozoospermia) and immotile sperm tails in one sibling and the absence of sperm cells (azoospermia) in the other sibling (43).

Hornef et al. (55) analyzed an additional PCD cohort of 109 PCD families for the presence of *DNAH5* mutations. A total of 33 distinct novel mutations and two previously reported mutations were identified. Together with the previously analyzed 25 PCD families (37), the mutational prevalence for *DNAH5* is currently known in a total of 134 PCD families, which can be used to calculate mutation frequency in the PCD population. Affected individuals from 65 of the 134 PCD families exhibited ODA defects. In 38 of these 134 PCD families, *DNAH5* mutations were detected (28%); the mutation detection rate was even higher (49%) when only PCD families with documented ODA defects (32 of 65 PCD families) were considered. Analysis in all analyzed PCD families with inner dynein arm defects detected no *DNAH5* mutations. Thus, *DNAH5* mutations regularly cause ODA defects (55). Interestingly, most of the identified mutations were nonsense mutations, but frameshift, splicing, and missense mutations also were found (**Figure 4**). The most common *DNAH5* mutations clustered within only five exons—34, 50, 63, 76, and 77—which harbor *DNAH5* mutations in 53% of a total of 134 analyzed PCD families. Interestingly, the 10815delT founder mutation is especially prevalent in the Caucasian PCD population in the United States (55).

The ultrastructural analysis, using TEM, of respiratory cilia in three families carrying homozygous *DNAH5* mutations indicated a possible genotype-phenotype correlation (38). Mutations causing premature translational termination of *DNAH5* (5563_5564insA [I1855NfsX6], 8440_8447del [E2814fsX1]) resulted in a complete absence of ODA in respiratory cilia, whereas a splice site mutation at the very 3′ end of the *DNAH5* gene did not cause total absence of ODA. Semiquantitatively assessed cilia of affected subjects with this mutation revealed shortened, stubby ODA in most sections. Furthermore, 54% of the ODA were less than half the average length of ODA in normal subjects, indicating a partial ODA

Figure 4

(*a*) Schematic depicting the genomic localization and gene structure of *DNAH5* on the short arm of chromosome 5. Exons are numbered and indicated by blue boxes. White boxes represent untranslated regions. ATG, start codons; TAA, stop codon. Intron sizes are not drawn to scale. (*b*) Schematic drawing of the putative DNAH5 protein. The six P-loop domains and the microtubule binding domain (MTB) are shown. The positions of all thus far identified *DNAH5* mutations are indicated by red vertical lines. Figure adapted from Reference 55, with permission.

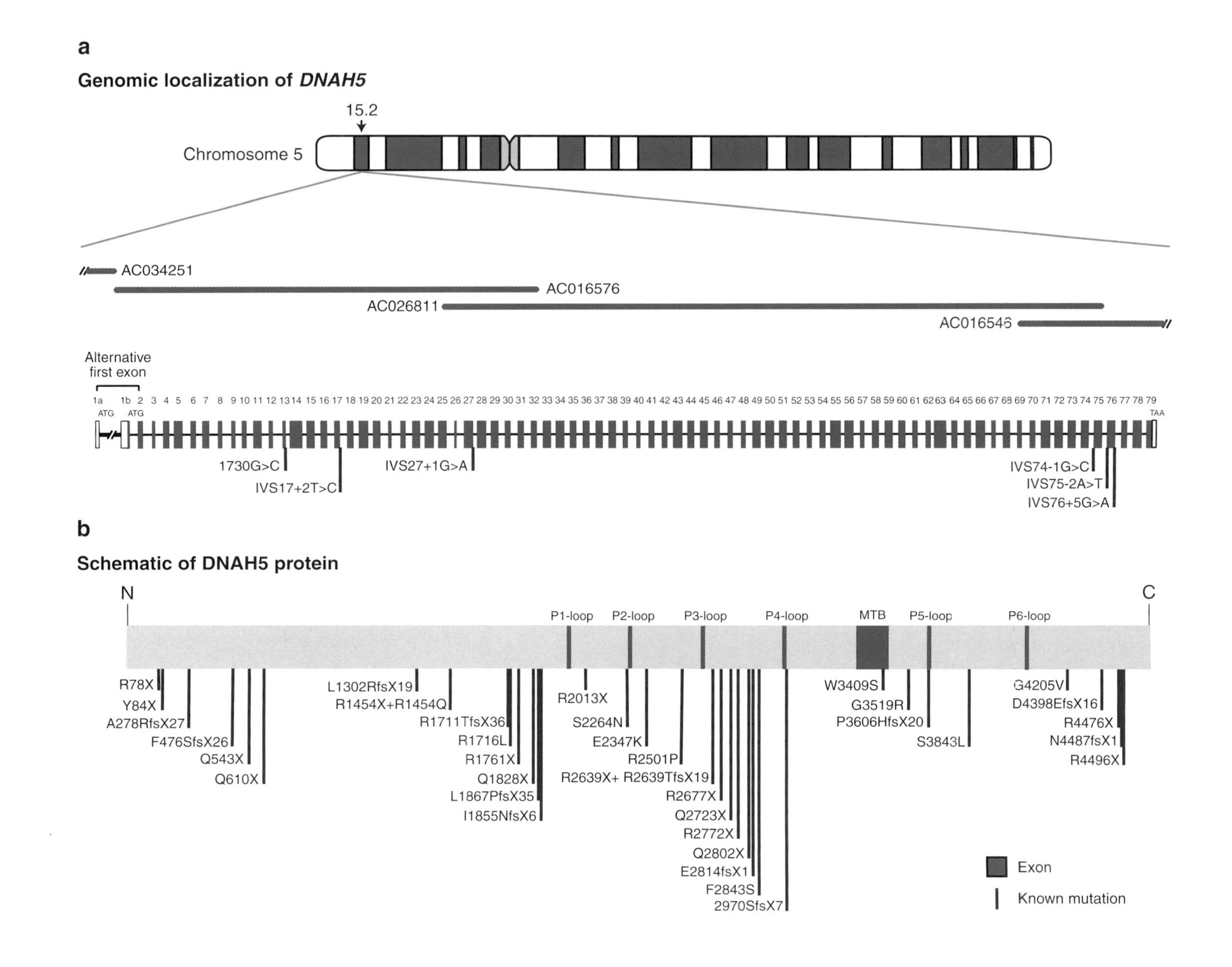

a
Genomic localization of DNAH5
15.2
Chromosome 5
AC034251
AC016576
AC026811
AC016546
Alternative first exon
1a 1b 2 3 4 5 6 7 8 9 10 11 12 13 14 15 16 17 18 19 20 21 22 23 24 25 26 27 28 29 30 31 32 33 34 35 36 37 38 39 40 41 42 43 44 45 46 47 48 49 50 51 52 53 54 55 56 57 58 59 60 61 62 63 64 65 66 67 68 69 70 71 72 73 74 75 76 77 78 79
ATG
ATG
TAA
1730G>C
IVS17+2T>C
IVS27+1G>A
IVS74-1G>C
IVS75-2A>T
IVS76+5G>A
b
Schematic of DNAH5 protein
N
C
P1-loop
P2-loop
P3-loop
P4-loop
MTB
P5-loop
P6-loop
R78X
Y84X
A278RfsX27
F476SfsX26
Q543X
Q610X
L1302RfsX19
R1454X+R1454Q
R1711TfsX36
R1716L
R1761X
Q1828X
L1867PfsX35
I1855NfsX6
R2013X
S2264N
E2347K
R2501P
R2639X+ R2639TfsX19
R2677X
Q2723X
R2772X
Q2802X
E2814fsX1
F2843S
2970SfsX7
W3409S
G3519R
P3606HfsX20
S3843L
G4205V
D4398EfsX16
R4476X
N4487fsX1
R4496X
Exon
Known mutation

deficiency. However, the main problems in interpreting these results are that the composition of ODA complexes differs along the longitudinal length of the ciliary shaft and that TEM does not allow exact allocation of the performed sections.

To gain more insight into the molecular pathology of PCD caused by *DNAH5* mutations, respiratory cells obtained by nasal brushing biopsy from PCD patients were analyzed by high-resolution immunofluorescence image acquisition, through the use of confocal microscopy. To date, 15 PCD patients belonging to 13 PCD families carrying distinct recessive *DNAH5* mutations have been analyzed (43, 45, 55). Interestingly, in these studies regardless of the mutational background, mutant DNAH5 proteins were expressed in respiratory epithelial cells from PCD patients but failed to appropriately localize along the axonemal shaft. In addition, in all 15 analyzed PCD patients with *DNAH5* mutations, we observed mislocalization of mutant DNAH5 with accumulation at the microtubule-organizing centers (**Figure 5**). Because all mutations result in mutant DNAH5 with preservation of the N-terminal third of the protein, this region may contain a domain responsible for correct targeting of DNAH5 to the microtubule-organizing centers at the ciliary base.

All but one PCD patient with *DNAH5* mutations completely lacked mutant DNAH5 in the respiratory ciliary axonemes (**Figure 5**). In these cases *DNAH5* mutations probably affect assembly of both ODA types that contain DNAH5, including type 1 (DNAH9 negative and DNAH5 positive; proximal ciliary axoneme) and type 2 (DNAH9 and DNAH5 positive; distal ciliary axoneme). This interpretation is supported by the complete absence of DNAH9 from the ciliary axoneme. In contrast, in one patient carrying compound heterozygous mutations that are located at the very 3′ end of *DNAH5* (IVS 75 and 76) and that predict aberrant splicing, mutant DNAH5 was absent from the distal ciliary axoneme but preserved within the proximal ciliary shaft (**Figure 5**). Based on this finding one can assume that mutant DNAH5 can be correctly targeted to the site of proximally located type 1 ODA complexes (DNAH9 negative and DNAH5 positive). This suggests that the C-terminal region of DNAH5 is important for assembly of ODA type 2.

To investigate whether similar molecular defects account for ciliary and flagellar (sperm tail) dyskinesia in patients with *DNAH5* mutations, Fliegauf et al. (43) analyzed DNAH5 and DNAH9 localization in sperm and respiratory cells from a PCD patient who carries homozygous frameshift (5563_5564insA [I1855NfsX6]) mutations. In contrast to respiratory epithelial cells from this patient, these investigators identified no differences in the staining pattern in sperm flagella of the patients and the control individual (43). Thus, identical *DNAH5* mutations cause mislocalization of the ODA HCs DNAH5 and DNAH9 in respiratory cells but result in normal distribution of these ODA HCs along the sperm flagella. One possible explanation for these distinct phenotypes is different mechanisms of respiratory cilia and sperm flagella generation. In *Chlamydomonas*, the intraflagellar transport machinery facilitates trafficking of flagella-specific proteins between the two distinct cytoplasmic and flagellar compartments (56). The processes of assembly and maintenance of cilia and flagella by transport particles passing the compartment borders has recently been described as compartmentalized ciliogenesis (57). Human respiratory cilia may use compartmentalized ciliogenesis, whereas, as observed in *Drosophila*, sperm tail generation may resemble cytosolic ciliogenesis, which occurs independently of intraflagellar transport (58).

DNAI1 Mutations

DNAI1 is a human axonemal dynein intermediate chain 1 located on chromosome 9p21-p13 and is an ortholog of the *Chlamydomonas* ODA protein IC78. Mutations in this gene have been observed in PCD patients

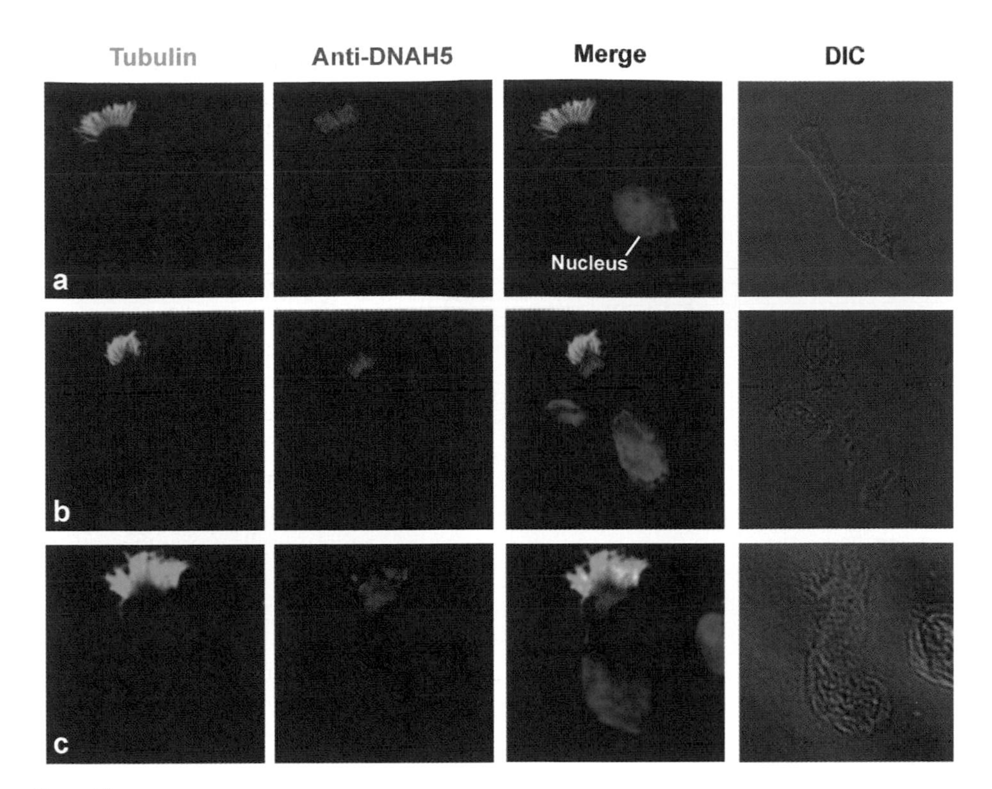

Figure 5

Localization of normal and mutant DNAH5 protein in respiratory epithelial cells from normal subject and PCD patients carrying *DNAH5* mutations (chromosome 5p15.2). Shown is confocal immunofluorescence analysis of human respiratory epithelial cells with anti-DNAH5 antibodies (*red*) and with antibodies against axoneme-specific acetylated α-tubulin (*green*) as control. Nuclei were stained blue. (*a*) In respiratory epithelial cells from a healthy subject, DNAH5 localizes along the entire length of the axonemes and at the ciliary base. (*b*) In contrast, in respiratory cells originating from a PCD patient with compound heterozygous *DNAH5* mutations (R1454Q and 2970SfsX7), mutant DNAH5 is absent from the ciliary axoneme and instead accumulates at the ciliary base. (*c*) In respiratory epithelium of a patient who carries compound heterozygous *DNAH5* mutations affecting splicing at the 3′ end of the gene (IVS75 − 2A>T and IVS76 + 5G>A), mutant DNAH5 is absent from the distal part of the ciliary axonemes but is still detectable within the proximal part. Mutant DNAH5 accumulates at the ciliary base. PCD patients with *DNAI1* mutations show similar patterns of DNAH5 mislocalization. DIC denotes differential interference contrast.

(59–62). This gene was cloned based on the candidate gene approach (59) because *Chlamydomonas* mutants lacking IC78 (oda9 mutants) lost 70% of their motility and were slow swimming (63). Ultrastructural analysis of the cross section of the flagella from the oda9 mutant showed loss of the ODA. IC78 in *Chlamydomonas* is involved in binding to α-tubulin and may be important for axonemal assembly.

Based on the motility defect phenotype and ultrastructural defects of IC78 in the *Chlamydomonas* oda9 mutant, Pennarun et al. (59) cloned and characterized the human ortholog, named *DNAI1*. The *DNAI1* gene has 43% identity with the *Chlamydomonas* gene (32). Human *DNAI1* is a 20-exon gene predicted to encode a 699-amino-acid protein. The C-terminal half contains a region of high similarity that corresponds to sequence elements known as a tryptophan-aspartate (WD) repeat (**Figure 6**). The five human WD repeats share 29–63% identity with the *Chlamydomonas* gene. Initially, sequence

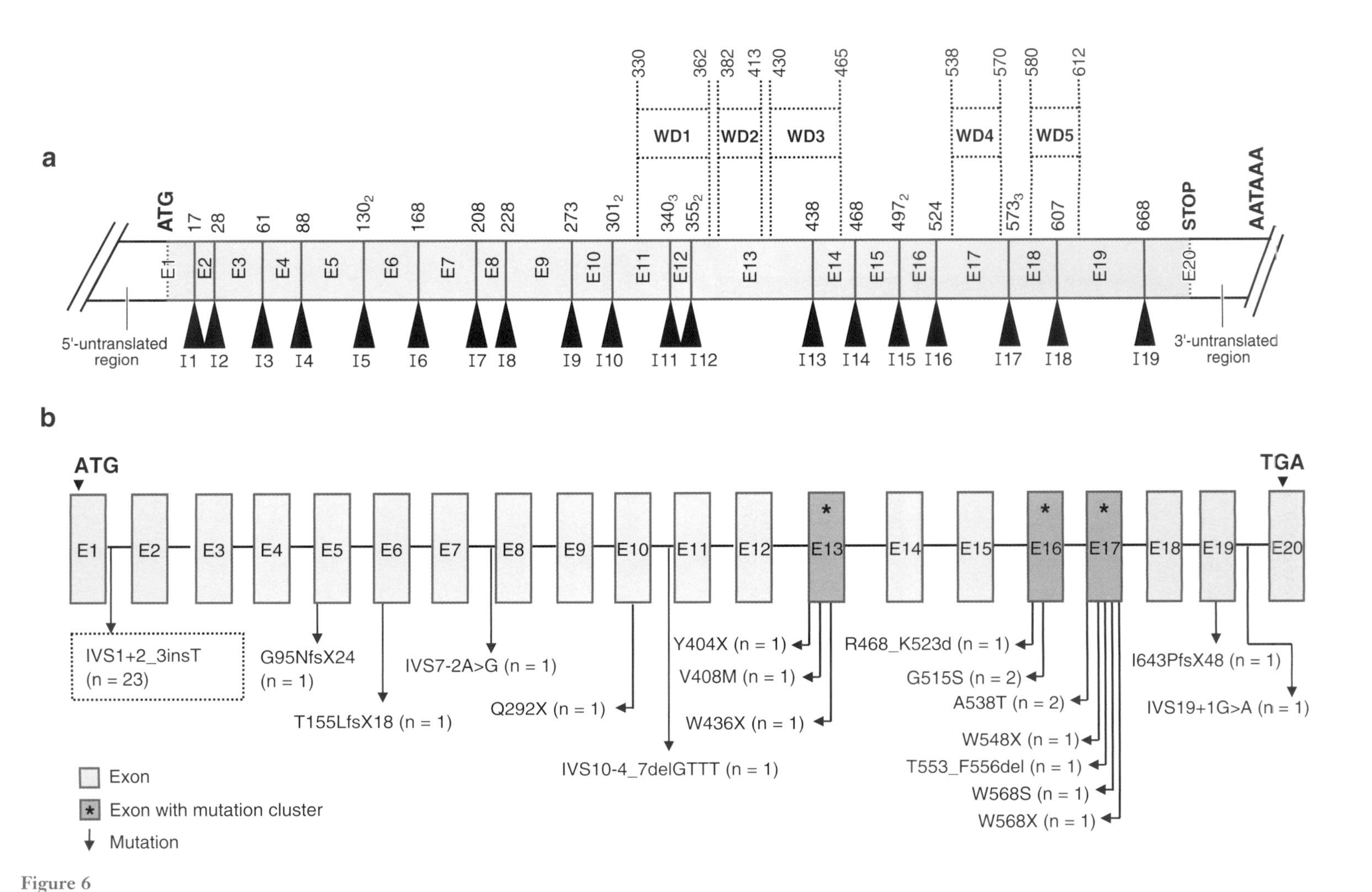

Figure 6

Genomic organization and distribution of mutations in *DNAI1* (chromosome 9p21-p13). (*a*) Genomic organization of the human *DNAI1* as shown previously in Reference 59. The 20 exons are indicated by the boxes containing exon numbers (E#). The locations of the start codon (ATG), stop codon (TGA), untranslated region, and polyadenylation signal (AATAAA) are shown. The number of the first codon in each exon is indicated; exons beginning with the second or third base of a codon are indicated by a subscripted 2 or 3, respectively (130_2, for example). The exons are drawn to scale, and intron-exon boundaries are denoted by blackened triangles. Five conserved WD repeat regions are shown on top of the genomic structure, and the codon number for the start and end of each WD repeat box is indicated. (*b*) Schematic drawing of *DNAI1* as shown previously in References 59 and 62 (reproduced with permission). The positions of all identified mutations are indicated. The number of mutant alleles is indicated in parentheses. A founder mutation (IVS1 + 2_3insT) is denoted in a box on the left side of the figure.

analysis was carried out in six unrelated patients with ODA defects; one PCD patient had two *DNAI1* mutations (IVS1 + 2_3insT and 282_283insAATA), one inherited from each parent (59). Subsequently, Guichard et al. (60) and Zariwala et al. (61) demonstrated three and two unrelated patients harboring biallelic mutations, respectively. All six unrelated patients with mutations from these three studies harbored the IVS1 + 2_3insT mutation on one allele (59–61). Zariwala et al. (62) recently studied 179 unrelated patients from various international geographical locations. They showed that 16 of 179 (9%) PCD patients harbored mutations (14 patients had biallelic mutations). In this study, 12 of 16 unrelated patients presented with a IVS1 + 2_3insT mutation on at least one allele. Combining all four studies (59–62), 226 unrelated patients have been studied thus far for *DNAI1* gene mutations; 22 patients (10%) presented with the mutations, of which 20 patients (9%) harbored biallelic mutations and 2 (1%) harbored monoallelic mutations.

There are 18 different *DNAI1* mutations, of which ~15% were missense mutations and the remaining 85% were nonsense mutations, splice site mutations, small insertions, and deletions (**Figure 6**). Allelic heterogeneity was noted, but certain mutations—the IVS1 + 2_3insT mutation and a mutation cluster in exons 13, 16, and 17—appeared in two or more unrelated families. A total of 29% of the mutant alleles resided in the mutation cluster, whereas ~55% of the mutant alleles showed the IVS1 + 2_3insT mutation. This latter mutation appears to abrogate the splice donor site at the exon–intron 1 junction in the cDNA. This leads to the addition of intron 1 sequences, which is predicted to result in premature translation termination (59).

Based on the high frequency of IVS1 + 2_3insT mutations, Zariwala et al. (62) tested for founder effect, using microsatellite marker from the vicinity of the *DNAI1*. This analysis established that IVS1 + 2_3insT is a founder mutation. The information regarding the founder mutation and mutation cluster is very useful for the development of a clinical molecular genetics test panel for PCD (see below). Additionally, *DNAI1* mutations apparently are more common in Caucasian PCD patients. Although the non-Caucasian data set is small, Zariwala et al. (62) found mutations in only 2% of non-Caucasian alleles (1 of 47) as compared with 11% of the Caucasian alleles. This may be because 55% of mutant alleles were the IVS1 + 2_3insT founder mutation. Mutations in the *DNAI1* gene are always found in association with ODA defects.

DNAH11 Mutations

Mutations of the *DNAH11* gene encoding the axonemal HC dynein 11 may cause PCD (64). The gene comprises 82 exons and spans a genomic region of more than 353 kb on chromosome 7p15.3–21. Mutational analysis by direct sequencing of amplified exons identified a homozygous nonsense mutation (R2852X) in exon 52 in one patient with paternal uniparental isodisomy and cystic fibrosis. The analyzed patient exhibited situs inversus and a severe respiratory phenotype, but electron microscopy revealed normal respiratory cilia axonemes without any obvious ultrastructural dynein arm defects. Thus, the respiratory phenotype in this patient may be related to the recessive *CFTR* (cystic fibrosis transmembrane conductance regulator) (MIM# 602421) mutations responsible for cystic fibrosis. Situs inversus may be explained by the effect of the detected *DNAH11* mutation on nodal monocilial function.

Further sequence analyses in additional PCD patients have not yet confirmed that *DNAH11* mutations cause PCD or result only in isolated randomization of left-right asymmetry that is observed in the iv/iv (inversum viscerum) mouse model (42). Researchers have identified, via the iv/iv mouse model, a homozygous missense mutation of the ortholog mouse gene *Dnah11* (*lrd*) that leads to immotile cilia of the embryonic node. Mutant mice exhibit randomization of the left-right axis but lack any obvious respiratory phenotype. Further analyses are necessary to determine whether *DNAH11* mutations are responsible for a subset of PCD patients or cause only randomization of left-right body asymmetry.

GENETIC DEFECTS AFFECTING NONAXONEMAL PROTEINS

RPGR Mutations (xlRP)

Retinitis pigmentosa (RP) (MIM# 268000) is a group of inherited disorders in which abnormalities of the photoreceptors (rods and cones) on the retinal pigment epithelium lead to progressive loss of vision. The disease incidence is ~1/4000. RP is a genetically heterogeneous disorder that can be inherited as an autosomal dominant, autosomal recessive, or X-linked trait. X-linked RP (xlRP) either (*a*) is recessive and affects only males or (*b*) is dominant and affects both genders, male more severely (65). Patients with xlRP have shown mutations in *RPGR* (RP guanosine triphosphatase regulator) (MIM# 312610). The *RPGR* gene, located on chromosome Xp21.1, is essential for photoreceptor maintenance and viability (66). *RPGR* is expressed in human rods and cones. Disruption of RPGR in mice leads to opsin mislocalization, suggesting that RPGR is involved in protein trafficking across the connecting cilium (67, 68). In addition, expression studies show that *RPGR* is expressed in cochleal, bronchial, and sinus epithelial lining cells (16, 67), indicating a possible functional role of RPGR for the broad phenotype in patients with *RPGR* mutations.

Families in which RP cosegregated with PCD have shown *RPGR* mutations. Two independent families of Polish origin presented with RP and immotile cilia; one case presented with IDA defects (20, 69). The mutation responsible for the disease phenotype has not been described for either of these families. Moore et al. (19) described two probands from a family with RP and symptoms of PCD, including oto-sinu-pulmonary disease, bronchiectasis, and absence of ciliary beat frequency. Ultrastructural analysis showed multiple abnormalities of ciliary structure, including ODA defects with or without IDA defects, and central-complex and nexin link defects. A 57-bp deletion mutation in the *RPGR* gene (631_IVS6 + 9del), predicted to cause a severely truncated protein, is associated with this disease. In addition, in a patient with RP cosegregating with sino-pulmonary disease and with mild hearing loss, Zito et al. (18) found an *RPGR* mutation (845_846delTG), which also is predicted to cause a severely truncated protein. Iannaccone et al. (16) described a patient who presented with RP together with the PCD phenotype, including ear infection and bronchitis. This patient had a missense mutation (Gly173Arg) in *RPGR*.

OFD1 Mutations

Oral-facial-digital type 1 syndrome (OFD1S) (MIM# 311200) is an X-linked dominant human developmental disorder characterized by craniofacial malformations, digital abnormalities, mental retardation, and male lethality (70). Mutations in *OFD1* (formerly known as *CXORF5*), located on chromosome Xp22, are responsible for this disorder. *Ofd1* heterozygous knockout female mice (71) developed kidney cysts that were devoid of cilia. Heterozygous males had severe phenotypes, including embryonic lethality, the absence of cilia in the embryonic node, and midline defects of the cardiac tube, indicating that the Ofd1 protein is important for the left-right axis specification.

Budny et al. (21) reported a large Polish family with a novel X-linked recessive mental retardation (XLMR) syndrome consisting of macrocephaly together with ciliary dysfunction. Nine affected males from this family presented with developmental delay and compatible clinical phenotype of PCD, including chronic respiratory problems and full-term neonatal respiratory distress for one brother. Light microscopic analysis and high-speed video microscopy later confirmed PCD in these brothers, revealing dyskinetic cilia with severely disorganized ciliary beat pattern in an index case. Linkage analysis in the affected families followed by comparative genomic analysis of the refined locus on the X chromosome led to the identification of a frameshift mutation (2122_2125dupAAGA) in the *OFD1* gene (21).

Ofd1 is developmentally regulated in mouse embryo tissue sections and is required for the formation of primary cilia (71). The human OFD1 and mouse Ofd1 proteins are localized at the centrosome and basal body of the primary cilia (71, 72). Additionally, a disorganized beating pattern occurred in the respiratory cilia of the index patient from the X-linked recessive form. This last finding supports the role of *OFD1* in respiratory epithelial ciliary function as well.

STRATEGIES FOR FURTHER GENE IDENTIFICATIONS

Only a few genes (*DNAH5*, *DNAI1*, *DNAH11*, *OFD1*, *RPGR*) have been implicated thus far in the etiology of PCD. Several strategies can identify novel PCD genes. The candidate gene approach utilizes information from genes known to cause specific ultrastructural and functional defects in other species such as *Chlamydomonas*. The human orthologous genes are candidates for human PCD with similar axonemal defects. This approach was responsible for the detection of the first disease-causing gene, *DNAI1*. Several other candidate genes have been tested in various patients, but no mutations have been found to date (see **Table 1** for the complete list of candidate genes).

Comparative computational analyses also can identify candidate genes. Comparative genomics exploits the increasing amount of genetic DNA information provided by sequencing projects from various distinct species. It is assumed that during evolution, genetic information of molecular processes that are no longer exerted is not preserved (73). Based on the hypothesis that the ancestral eukaryote was a ciliated unicellular organism and that cilia were independently lost throughout evolution from several eukaryotic groups, subtractional analyses identified candidate genes necessary for cilia formation and function (57). Similar approaches utilizing information from gene (transcriptomics) and protein (proteomics) expression analyses have generated many candidate genes involved in the function of basal bodies and cilia (31, 74–77). Fliegauf & Omran (73) recently summarized the different "Xomics" studies to unravel molecular mechanisms in cilia-related disorders.

Genome-wide linkage analysis may be exploited to identify candidate loci involved in PCD. Blouin et al. (78) applied this methodology in a large study of 31 multiplex families but have not identified any disease-causing gene. Other studies have described putative PCD loci on chromosomes 19q13.42-q13.43 (79), 15q13.1-q15.1 (80), 16p12.1-p12.2 (80), and 15q24-q25 (81) but have not identified any disease-causing genes. This approach's main limitation is the presence of extensive genetic heterogeneity in PCD. To circumvent the problem of genetic heterogeneity for linkage analysis, one can ascertain a single consanguineous PCD family with multiple affected individuals and search for homozygous segments in affected individuals. Omran et al. (53) successfully applied this homozygosity mapping strategy to localize the *DNAH5* gene. However, large informative families are only rarely available.

MOUSE MODELS FOR PRIMARY CILIARY DYSKINESIA

Mdnah5-Deficient Mouse

The *Mdnah5*-deficient mouse is the only available mouse model for human PCD in which an orthologous PCD gene is mutated (82). The mouse *Mdnah5* gene contains 79 exons, like the human ortholog, *DNAH5*, and is located on murine chromosome 15.

Table 1 Candidate genes analyzed and found negative in PCD patients[a]

Human gene	Human chromosomal location	*Chlamydomonas* ortholog protein	*Chlamydomonas* mutant gene	*Chlamydomonas* mutant axonemal defect	Axonemal component	Number of PCD families tested	Reference(s)
DNAH9	17p12	beta-DHC	*oda4*	Loss of ODA	ODA HC	2	111
DNAH17	17q25	beta-DHC	*oda4*	Loss of ODA	ODA HC	4	112
DNAI2	17q25	IC2	*oda6*	Loss of ODA	ODA IC	16	113, 114
DNAL1	14q24.3	LC1	none	Unknown	ODA LC	86	115
DNAL4	22q13.1	LC6	*oda13*	Assembles ODA[b]	ODA LC	54	116
TCTE3	6q27	LC2	*oda12*	Loss of ODA	ODA LC	36	117
LC8	17q23	LC8	*fla14*	Loss of ODA and other	ODA LC	58	113
DNAH3	16p12	Not known	Not known	Not known	IDA HC	7	118
DNAH7	2q33.1	Not known	Not known	Not known	IDA HC	1	119
HP28	1p35.1	P28	*ida4*	Loss of IDA	IDA LC	61	116, 120
DPCD	10q23	None	None	None	IDA gene	51	87
SPAG6	10p11.2-p12	PF16	*pf16*	Loss of C1 microtubule	CA	54	112
SPAG16	2q35	PF20	*pf20*	Loss of central complex	CA	5	114
FOXJ1/HFH-4	17q22-q25	None	None	None	Expressed[c]	8	121

[a]Abbreviations used: CA, central apparatus; DHC, dynein heavy chain; HC, heavy chain; IC, intermediate chain; IDA, inner dynein arm; LC, light chain; ODA, outer dynein arm; PCD, primary ciliary dyskinesia.
[b]Minor swimming defects.
[c]Expressed in respiratory cilia.

Mdnah5$^{-/-}$ mice exhibit random left-right axis orientation, respiratory distress, chronic inflammatory processes in the paranasal sinuses, and severe inflammatory process (purulent secretions) in the tympanic cavity associted with a failure to respond to noise. Heterozygous animals show no obvious signs of the disease, consistent with a recessive disease trait (82). During early embryonic development, *Mdnah5* (alias *Dnahc5*) is expressed in the node in the respiratory tract and in testis (37, 38). Analyses of respiratory cilia ultrastructure and function revealed an absence of ODA and cilia immotility (82).

In addition to the above, all *Mdnah5*$^{-/-}$ mice developed severe hydrocephalus (82), whose etiology has been linked to ependymal cilia dysfunction (83–85). Detailed studies (39) showed that *Mdnah5* is expressed in the ependymal cells lining the brain ventricles and the aqueduct. Ultrastructural analyses of the ependymal cilia revealed partial ODA deficiency and some residual cilia motility (39). Hydrocephalus has also been reported in patients with PCD, which suggests that cilia dysfunction also contributes to human hydrocephalus formation (39).

Dpcd- and *Poll*-Deficient Mouse

Kobayashi et al. (86) created a DNA polymerase lambda (*poll*) homozygous knockout mice, using homologous recombination. These mice exhibited the classic PCD phenotype, including situs inversus, chronic sinusitis, hydrocephalus, and infertility in male mice. Ultrastructural analysis of the respiratory cilia from these mice revealed defective IDA.

Because the link between *poll* and PCD was not clear, Zariwala et al. (87) examined the targeting construct that was used for creating the mutant mice (86) and noted that, in addition to *poll*, another gene likely was disrupted in these mice. This novel gene was named *Dpcd*, which encodes a 23-kD protein of unknown function. Bertocci et al. (88) generated *poll* knockout mice, using a catalytic domain of the gene in which *Dpcd* was not disrupted. These mice were phenotypically normal and did not have PCD. Thus, *poll* is not a candidate for human PCD, but the phenotype observed by Kobayashi et al. (86) may be due to the loss of *Dpcd*. Zariwala et al. (87) analyzed DNA from 51 unrelated PCD patients (of which 15 had defective IDA) and did not find the mutation in *DPCD*. Therefore, mutations in *DPCD* may be present in only a small subset of PCD patients with IDA defects.

Several other mouse knockout models targeting ciliary genes (*lrd*, *Tektin-t*, *Mdhc7*, *Foxj1/Hfh4*, *Pf20/Spag16*, *Pf16/Spag6*) have been studied, but none of them presented with a respiratory phenotype, which is the hallmark of PCD (42, 64, 84, 89–94).

NOS: nitric oxide synthase

NOVEL DIAGNOSTIC OPTIONS

Nasal Nitric Oxide Measurement

Ciliary beat regulation uses several mechanisms, including intracellular calcium concentration, cAMP, extracellular nucleotides, and NO. Low levels of nasal NO (~1/10 of normal values) have been observed in many PCD patients (50, 61, 95–98), but there is some overlap with NO levels seen in cystic fibrosis patients. NO is formed by the nitric oxide synthase (NOS) enzyme system in the presence of a substrate (L-arginine and oxygen) and cofactor (tetrahydrobiopterin) (99). Three isoforms of NOS (NOS1, NOS2, NOS3) are known. NOS2-mediated NO production is increased in asthma (100). NOS3 localizes close to the base of cilia and may be involved in ciliary beat regulation (101). NOS3 is activated by calmodulin and Ca^{2+} (102).

The mechanism by which low nasal NO occurs in PCD is not clear, but low nasal NO is present, regardless of the type of ciliary ultrastructural defects. Because NO upregulates ciliary beat frequency, low NO in PCD may be related to the lack of motility of cilia. Regardless of the mechanism, the measurement of NO has potential as a screening test for the diagnosis of PCD, and studies are underway to establish the specificity and sensitivity of nasal NO for such an application (103).

High-Speed Videomicroscopy

The diagnosis of PCD is complicated if no specific axonemal alterations can be demonstrated by TEM or high-resolution immunofluorescence imaging. In such cases, an abnormal ciliary beat can establish a diagnosis of PCD (51). In the past, functional assessment of ciliary beat focused mainly on the determination of the ciliary beat frequency. However, an exact evaluation of the beating pattern is more important and can best be achieved by high-speed videomicroscopy. Because functional abnormalities can result from secondary alterations such as inflammatory changes caused by local infections, functional abnormalities should be demonstrated on several occasions or, better yet, after in vitro ciliogenesis.

Chilvers et al. (51) showed that different ultrastructural defects responsible for PCD often result in predictable abnormal ciliary beat patterns. Cilia with transposition defects were characterized by an abnormal circular ciliary beat pattern. Cilia with ODA defects and those with combined ODA and IDA defects often exhibited ciliary immotility, whereas cilia with isolated IDA defects regularly showed ciliary bending defects (reduced amplitude and stiffness). However, Noone et al. (50) reported that two of eight subjects with IDA defects had near-normal ciliary activity that could be demonstrated only by careful high-speed videomicroscopy analysis. This observation again indicates that PCD is a heterogeneous disorder, explaining the absence of a strict correlation between ultrastructural abnormality and a cilia motility defect. This report should alert clinicians not to rule out the diagnosis of PCD in patients with typical clinical findings suggestive of the disease, if these clinicians are unable to properly assess the ciliary beat pattern, which can be achieved only by high-speed videomicroscopy.

Stannard et al. (104) recently identified in three siblings a new distinct PCD variant with a circular ciliary beat pattern, suggestive of a ciliary transposition defect. In these siblings, ultrastructural analysis of the cilia revealed an absence of the central microtubule pair. Fliegauf et al. (43) showed that PCD patients with ODA defects that are of unknown origin or that are caused by *DNAH5* or *DNAI1* mutations have either immotile cilia or cilia with only little flickery cilia movement. This confirms that ODA defects are often associated with cilia immotility.

Mislocalization of Axonemal Components

The characteristic DNAH5 staining pattern of *DNAH5*- or *DNAI1*-mutant respiratory cells suggests that immunofluorescence staining can be applied to detect ODA defects (43). For that purpose, a large cohort of PCD patients and control individuals was analyzed by high-resolution immunofluorescence microscopy in a blind study. All PCD patients previously diagnosed by electron microscopy for ODA defects showed complete or distal absence of DNAH5 staining in the ciliary axonemes plus accumulation of DNAH5 at the microtubule-organizing centers. In contrast, all control individuals, including patients with recurrent respiratory infections of other origin, had normal DNAH5 staining (43).

Immunofluorescence staining has several advantages. First, this method is performed in respiratory epithelial cells obtained by noninvasive transnasal brushings, and this material can be used at the same time for functional analysis of ciliary beat frequency and pattern. Second, samples dried on glass slides can be transported easily to the performing laboratories. Third, secondary ciliary changes do not affect axonemal localization of the ODA component DNAH5 (45). Fourth, the method is able to detect changes along the entire ciliary axoneme. This aspect is of particular significance because TEM localization of the examined cross sections along the ciliary axoneme is not possible.

Interestingly, in a subset of PCD patients the ODA component DNAH5 is not localized to the ciliary axoneme, indicating a defect of two ODA types. However, in another group of patients DNAH5 is still present in the proximal parts of the ciliary compartment, suggestive for a defect of only ODA type 2. These findings may explain some of the variability of ultrastructural defects observed in PCD (50). In addition, immunofluorescence staining results are useful to direct the mutational analyses of PCD patients with ODA defects, as *DNAH5* mutations usually cause a complete absence of DNAH5 from the ciliary axoneme (43, 55). Specifically, only one PCD patient with absence of DNAH5 from the distal ciliary axoneme was identified (55). In contrast, *DNAI1* mutations result in the absence of DNAH5 from only the distal ciliary axoneme (43).

Other ultrastructural defects, e.g., those of the IDA, are also observed in PCD and are probably caused by similar molecular mechanisms. Therefore, subjects with IDA defects should also be amenable to diagnosis with high-resolution immunofluorescence microscopy as robust antibodies become available.

Genetic Testing

PCD is a genetically heterogeneous disorder. Mutations have been identified in *DNAI1* and *DNAH5* in PCD patients and in two rare syndromic PCD variants (*RPGR* and *OFD1*). Mutations in *DNAI1* and *DNAH5* are associated exclusively in patients with ODA defects, with a mutation pickup rate of ~50–60%. Because ODA defects occur in most (~60%) PCD patients, these mutations would be seen in ~38% of the general PCD population. Thus far, 60 different mutant alleles have been identified in *DNAI1* and *DNAH5* (37, 55, 59–62). Despite allelic heterogeneity, some mutations occur commonly either due to the founder effect or as clusters in three exons of *DNAI1* and four exons of *DNAH5* (**Figures 4** and **6**). Using estimations based on published reports (37, 55, 59–62), we have developed a strategy for mutation analysis of *DNAH5* and *DNAI1* that requires sequencing of only 9 exons (out of 100 exons) (55, 62). These 9 exons harbor the most common mutations (IVS1 + 2_3insT mutation and exons 13, 16, and 17 for *DNAI1* and exons 34, 50, 63, 76, and 77 for *DNAH5*). This approach would lead to the identification of at least one mutant allele in approximately 24% of all PCD patients. If biallelic mutations are identified, it is diagnostic for PCD, but if a monoallelic mutation is identified, then full gene sequencing can be conducted for that particular gene to search for the trans-allelic mutation.

Sequence-based assays are time consuming and expensive. Hence, efforts are under way to develop alternative approaches that are more versatile, such as melting curve analysis and microarray chips. These alternative methods would allow the addition of new mutations as they are discovered so that a higher percentage of patients can be detected to improve test specificity further. Genetic testing is available for *DNAI1* and *DNAH5* at the University of North Carolina at Chapel Hill (United States) and the University Hospital Freiburg (Germany).

Novel Therapeutic Options

The primary goal of current therapy for PCD patients is the prevention of progressive lung damage by strict treatment of upper- and lower-airway infections and physical therapy (105). However, therapeutic strategies that correct the inborn error of cilia dysfunction to restore mucociliary clearance are currently not available. Some drugs, such as aminoglycoside antibiotics, can suppress the usage of premature stop codons and permit translation to continue to the normal termination site (106, 107). Pharmacogene therapy with the aminoglycoside antibiotic gentamicin in patients with cystic fibrosis carrying premature nonsense mutations in the *CFTR* gene suppressed premature stop mutations and restored CFTR function (108–110). We found *DNAH5* nonsense mutations in a high proportion of PCD patients and a few in *DNAI1* (55, 62). Because most detected *DNAH5* mutations in PCD patients are nonsense mutations, pharmacogene therapy could correct the primary defect in these PCD patients.

SUMMARY POINTS

1. Cilia and flagella axonemes are complex, evolutionarily conserved structures that perform diverse biological processes. Defective cilia are associated with various ciliopathies, including primary ciliary dyskinesia (PCD).
2. PCD is characterized by oto-sinu-pulmonary disease, randomization of left-right body asymmetry, and male infertility.
3. The diagnosis of PCD is complicated and requires the presence of the characteristic clinical phenotype plus a supportive lab test through the evaluation of ultrastructural defects with electron microscopy, immunofluorescence studies, high-speed videomicroscopy, and/or nasal nitric oxide measurements.
4. Five genes (*DNAH5*, *DNAI1*, *DNAH11*, *OFD1*, and *RPGR*) involved in the pathogenesis of distinct PCD variants have aided our understanding of the molecular processes responsible for PCD.
5. Mutation clusters and founder mutations have been identified in *DNAI1* and *DNAH5* and are helpful for the genetic testing in PCD patients with outer dynein arm defects.
6. Researchers are developing novel diagnostic tools, such as high-resolution immunofluorescence analysis, using antibodies specific for axonemal proteins.
7. Given the high proportion of nonsense mutations in the PCD population, novel therapeutic options, such as pharmacogene therapy, are being explored.

ACKNOWLEDGMENTS

The authors are grateful to all the PCD patients and families. The authors would like to thank Ms. M. Manion, the U.S. PCD foundation, and the German patient support group Kartagener Syndrom und Primaere Ciliaere Dyskinesie e.V. for their continued support. The authors thank Dr. Johnny Carson, Heike Olbrich, and Niki Loges for their help in figure preparation. The authors also thank Drs. Margaret Leigh, Peadar Noone, and Marcus Kennedy. This work was supported by funds from GCRC#00046, MO1 RR00046-42, 1 RO1 HL071798, 5 U54 RR019480, and Deutsche Forschungsgemeinschaft (SFB592, DFG Om6/2, and Om6/4 to H.O.).

LITERATURE CITED

1. Ibanez-Tallon I, Heintz N, Omran H. 2003. To beat or not to beat: roles of cilia in development and disease. *Hum. Mol. Genet.* 12:R27–35
2. Noone PG, Zariwala M, Knowles MR. 2006. Primary ciliary dyskinesia. In *Principles of Molecular Medicine*, ed. MS Runge, C Patterson, pp. 239–50. Totowa, New Jersey: Humana Press

3. Nauli SM, Alenghat FJ, Luo Y, Williams E, Vassilev P, et al. 2003. Polycystins 1 and 2 mediate mechanosensation in the primary cilium of kidney cells. *Nat. Genet.* 33:129–37
4. Hildebrandt F, Otto E. 2005. Cilia and centrosomes: a unifying pathogenic concept for cystic kidney disease? *Nat. Rev. Genet.* 6:928–40
5. **Badano JL, Mitsuma N, Beales PL, Katsanis N. 2006. The ciliopathies: an emerging class of human genetic disorders. *Annu. Rev. Genomics Hum. Genet.* 7:125–48**
6. Besharse JC, Horst CJ. 1990. The photoreceptor connecting cilium. A model for the transition zone. In *Ciliary and Flagellar Membranes*, ed. RA Bloodgood, pp. 389–417. New York: Plenum Publ. Corp.
7. Marszalek JR, Liu X, Roberts EA, Chui D, Marth JD, et al. 2000. Genetic evidence for selective transport of opsin and arrestin by kinesin-II in mammalian photoreceptors. *Cell* 102:175–87
8. Pazour GJ, Baker SA, Deane JA, Cole DG, Dickert BL, Rosenbaum JL, Witman GB, Besharse JC. 2002. The intraflagellar transport protein, IFT88, is essential for vertebrate photoreceptor assembly and maintenance. *J. Cell Biol.* 157:103–13
9. Moore SJ, Green JS, Fan Y, Bhogal AK, Dicks E, et al. 2005. Clinical and genetic epidemiology of Bardet-Biedl syndrome in Newfoundland: a 22-year prospective, population-based, cohort study. *Am. J. Med. Genet. A* 132:352–60
10. Marshall JD, Bronson RT, Collin GB, Nordstrom AD, Maffei P, et al. 2005. New Alstrom syndrome phenotypes based on the evaluation of 182 cases. *Arch. Intern. Med.* 165:675–83
11. Praetorius HA, Spring KR. 2005. A physiological view of the primary cilium. *Annu. Rev. Physiol.* 67:515–29
12. Afzelius BA. 1976. A human syndrome caused by immotile cilia. *Science* 193:317–19
13. van's Gravesande KS, Omran H. 2005. Primary ciliary dyskinesia: clinical presentation, diagnosis and genetics. *Ann. Med.* 37:439–49
14. Narayan D, Krishnan SN, Upender M, Ravikumar TS, Mahoney MJ, et al. 1994. Unusual inheritance of primary ciliary dyskinesia (Kartagener's syndrome). *J. Med. Genet.* 31:493–96
15. Dry KL, Manson FD, Lennon A, Bergen AA, van Dorp DB, Wright AF. 1999. Identification of a 5′ splice site mutation in the RPGR gene in a family with X-linked retinitis pigmentosa (RP3). *Hum. Mutat.* 13:141–45
16. Iannaccone A, Breuer DK, Wang XF, Kuo SF, Normando EM, et al. 2003. Clinical and immunohistochemical evidence for an X-linked retinitis pigmentosa syndrome with recurrent infections and hearing loss in association with an RPGR mutation. *J. Med. Genet.* 40:e118
17. van Dorp DB, Wright AF, Carothers AD, Bleeker-Wagemakers EM. 1992. A family with RP3 type of X-linked retinitis pigmentosa: an association with ciliary abnormalities. *Hum. Genet.* 88:331–34
18. Zito I, Downes SM, Patel RJ, Cheetham ME, Ebenezer ND, et al. 2003. RPGR mutation associated with retinitis pigmentosa, impaired hearing, and sinorespiratory infections. *J. Med. Genet.* 40:609–15
19. Moore A, Escudier E, Roger G, Tamalet A, Pelosse B, et al. 2006. RPGR is mutated in patients with a complex X-linked phenotype combining primary ciliary dyskinesia and retinitis pigmentosa. *J. Med. Genet.* 43:326–33
20. Krawczynski MR, Witt M. 2004. PCD and RP: X-linked inheritance of both disorders? *Pediatr. Pulmonol.* 38:88–89
21. Budny B, Chen W, Omran H, Fliegauf M, Tzschach A, et al. 2006. A novel X-linked recessive mental retardation syndrome comprising macrocephaly and ciliary dysfunction is allelic to oral-facial-digital type I syndrome. *Hum. Genet.* 120:171–78

5. Provides a comprehensive review on various ciliopathies, including primary ciliary dyskinesia.

22. Afzelius BA, Mossberg B, Bergstrom SE. 2001. Immotile cilia syndrome (primary ciliary dyskinesia), including Kartagener syndrome. In *The Metabolic and Molecular Basis of Inherited Disease*, ed. CR Scriver, AL Beaudet, WS Sly, D Valle, B Childs, et al., pp. 4817–27. New York: McGraw-Hill
23. El Zein L, Omran H, Bouvagnet P. 2003. Lateralization defects and ciliary dyskinesia: lessons from algae. *Trends Genet.* 19:162–67
24. Rott HD. 1979. Kartagener's syndrome and the syndrome of immotile cilia. *Hum. Genet.* 46:249–61
25. Nonaka S, Tanaka Y, Okada Y, Takeda S, Harada A, et al. 1998. Randomization of left-right asymmetry due to loss of nodal cilia generating leftward flow of extraembryonic fluid in mice lacking KIF3B motor protein. *Cell* 95:829–37

25. Demonstrated the role of nodal cilia in generating leftward flow of extraembryonic fluid.

26. Kennedy MP, Leigh MW, Dell S, Morgan L, Molina PL, et al. 2006. Primary ciliary dyskinesia and situs ambiguus/heterotaxy: organ laterality defects other than situs inversus totalis. *Proc. Am. Thorac. Soc.* 3:A399 (Abstr.)
27. Afzelius BA, Eliasson R. 1983. Male and female infertility problems in the immotile-cilia syndrome. *Eur. J. Respir. Dis. Suppl.* 127:144–47
28. Munro NC, Currie DC, Lindsay KS, Ryder TA, Rutman A, et al. 1994. Fertility in men with primary ciliary dyskinesia presenting with respiratory infection. *Thorax* 49:684–87
29. Kennedy MP, Noone PG, Carson J, Molina PL, Ghio A, et al. 2006. Calcium stone lithoptysis in primary ciliary dyskinesia. *Respir. Med.* 101:76–83
30. Coren ME, Meeks M, Morrison I, Buchdahl RM, Bush A. 2002. Primary ciliary dyskinesia: age at diagnosis and symptom history. *Acta Paediatr.* 91:667–69
31. Luck DJ. 1984. Genetic and biochemical dissection of the eucaryotic flagellum. *J. Cell Biol.* 98:789–94
31a. Satir P, Christensen ST. 2007. Overview of structure and function of mammalian cilia. *Annu. Rev. Physiol.* 69:377–400
31b. Salathe M. 2007. Regulation of mammalian ciliary function. *Annu. Rev. Physiol* 69:401–22
31c. DiBella LM, King SM. 2001. Dynein motor of the *Chlamydomonas* flagellum. *Int. Rev. Cytol.* 210:227–68
32. Pazour GJ, Agrin N, Walker BL, Witman GB. 2006. Identification of predicted human outer dynein arm genes—candidates for primary ciliary dyskinesia genes. *J. Med. Genet.* 43:62–73

32. Comparative genomics approach to identify human ortholog axonemal dynein proteins.

33. Hastie AT, Dicker DT, Hingley ST, Kueppers F, Higgins ML, Weinbaum G. 1986. Isolation of cilia from porcine tracheal epithelium and extraction of dynein arms. *Cell Motil. Cytoskel.* 6:25–34
34. Witman GB, Wilkerson CG, King SM. 1994. The biochemistry, genetics, and molecular biology of flagellar dynein. In *Microtubules*, ed. J Hyams, C Lloyd, pp. 229–49. New York: Wiley-Liss
35. King SM, Gatti JL, Moss AG, Witman GB. 1990. Outer-arm dynein from trout spermatozoa: substructural organization. *Cell Motil. Cytoskel.* 16:266–78
36. Altschul SF, Gish W, Miller W, Myers EW, Lipman DJ. 1990. Basic local alignment search tool. *J. Mol. Biol.* 215:403–10
37. Olbrich H, Haffner K, Kispert A, Volkel A, Volz A, et al. 2002. Mutations in DNAH5 cause primary ciliary dyskinesia and randomization of left-right asymmetry. *Nat. Genet.* 30:43–44

37. Identified the outer dynein arm heavy-chain gene *DNAH5* as the second disease-causing gene for human PCD.

38. Kispert A, Petry M, Olbrich H, Volz A, Ketelsen UP, et al. 2003. Genotype-phenotype correlations in PCD patients carrying DNAH5 mutations. *Thorax* 58:552–54

39. Ibanez-Tallon I, Pagenstecher A, Fliegauf M, Olbrich H, Kispert A, et al. 2004. Dysfunction of axonemal dynein heavy chain Mdnah5 inhibits ependymal flow and reveals a novel mechanism for hydrocephalus formation. *Hum. Mol. Genet.* 13:2133–41
40. Reed W, Carson JL, Moats-Staats BM, Lucier T, Hu P, et al. 2000. Characterization of an axonemal dynein heavy chain expressed early in airway epithelial ciliogenesis. *Am. J. Respir. Cell Mol. Biol.* 23:734–41
41. Chapelin C, Duriez B, Magnino F, Goossens M, Escudier E, Amselem S. 1997. Isolation of several human axonemal dynein heavy chain genes: genomic structure of the catalytic site, phylogenetic analysis and chromosomal assignment. *FEBS Lett.* 412:325–30
42. Supp DM, Witte DP, Potter SS, Brueckner M. 1997. Mutation of an axonemal dynein affects left-right asymmetry in inversus viscerum mice. *Nature* 389:963–96
43. Fliegauf M, Olbrich H, Horvath J, Wildhaber JH, Zariwala MA, et al. 2005. Mislocalization of DNAH5 and DNAH9 in respiratory cells from patients with primary ciliary dyskinesia. *Am. J. Respir. Crit. Care Med.* 171:1343–49
43a. Porter ME, Sale WS. 2000. The 9 + 2 axoneme anchors multiple inner arm dyneins and a network of kinases and phosphatases that control motility. *J. Cell Biol.* 151:F37–42
43b. Nicastro D, Schwartz C, Pierson J, Gaudette R, Porter ME, McIntosh JR. 2006. The molecular architecture of axonemes revealed by cryoelectron tomography. *Science* 313:944–48
44. Carson JL, Collier AM, Hu S-CS. 1985. Acquired ciliary defects in nasal epithelium of children with acute viral upper respiratory infections. *N. Engl. J. Med.* 312:463–68
45. Olbrich H, Horvath J, Fekete A, Loges NT, van's Gravesande KS, Blum A, Hormann K, Omran H. 2006. Axonemal localization of the dynein component DNAH5 is not altered in secondary ciliary dyskinesia. *Pediatr. Res.* 59:418–22
46. Escudier E, Couprie M, Duriez B, Roudot-Thoraval F, Millepied MC, et al. 2002. Computer-assisted analysis helps detect inner dynein arm abnormalities. *Am. J. Respir. Crit. Care Med.* 166:1257–62
47. Bush A, Ferkol T. 2006. Movement: the emerging genetics of primary ciliary dyskinesia. *Am. J. Respir. Crit. Care Med.* 174:109–10
48. Carda C, Armengot M, Escribano A, Peydro A. 2005. Ultrastructural patterns of primary ciliary dyskinesia syndrome. *Ultrastruct. Pathol.* 29:3–8
49. Carlen B, Stenram U. 2005. Primary ciliary dyskinesia: a review. *Ultrastruct. Pathol.* 29:217–20
50. Noone PG, Leigh MW, Sannuti A, Minnix SL, Carson JL, et al. 2004. Primary ciliary dyskinesia: diagnostic and phenotypic features. *Am. J. Respir. Crit. Care Med.* 169:459–67

50. Comprehensive study describing the clinical phenotype of PCD and diagnostic work up in 78 well-characterized patients.

51. Chilvers MA, Rutman A, O'Callaghan C. 2003. Ciliary beat pattern is associated with specific ultrastructural defects in primary ciliary dyskinesia. *J. Allergy Clin. Immunol.* 112:518–24
52. Jorissen M, Willems T, Van der SB, Verbeken E, De Boeck K. 2000. Ultrastructural expression of primary ciliary dyskinesia after ciliogenesis in culture. *Acta Otorhinolaryngol. Belg.* 54:343–56
53. Omran H, Haffner K, Volkel A, Kuehr J, Ketelsen UP, et al. 2000. Homozygosity mapping of a gene locus for primary ciliary dyskinesia on chromosome 5p and identification of the heavy dynein chain DNAH5 as a candidate gene. *Am. J. Respir. Cell Mol. Biol.* 23:696–702
54. Wilkerson CG, King SM, Witman GB. 1994. Molecular analysis of the gamma heavy chain of *Chlamydomonas* flagellar outer-arm dynein. *J. Cell Sci.* 107(Pt. 3):497–506

55. Large comprehensive study utilizing 134 PCD families identified *DNAH5* mutations in 28% of all PCD patients.

59. Described outer dynein arm intermediate-chain gene *DNAI1* as the first disease-causing gene for human PCD.

62. Large comprehensive study utilizing 226 PCD families reported *DNAI1* mutations in 10% of all PCD patients.

55. **Hornef N, Olbrich H, Horvath J, Zariwala MA, Fliegauf M, et al. 2006. DNAH5 mutations are a common cause of primary ciliary dyskinesia with outer dynein arm defects. *Am. J. Respir. Crit. Care Med.* 174:120–26**
56. Rosenbaum JL, Witman GB. 2002. Intraflagellar transport. *Nat. Rev. Mol. Cell Biol.* 3:813–25
57. Avidor-Reiss T, Maer AM, Koundakjian E, Polyanovsky A, Keil T, et al. 2004. Decoding cilia function: defining specialized genes required for compartmentalized cilia biogenesis. *Cell* 117:527–39
58. Han YG, Kwok BH, Kernan MJ. 2003. Intraflagellar transport is required in *Drosophila* to differentiate sensory cilia but not sperm. *Curr. Biol.* 13:1679–86
59. **Pennarun G, Escudier E, Chapelin C, Bridoux AM, Cacheux V, et al. 1999. Loss-of-function mutations in a human gene related to *Chlamydomonas reinhardtii* dynein IC78 result in primary ciliary dyskinesia. *Am. J. Hum. Genet.* 65:1508–19**
60. Guichard C, Harricane MC, Lafitte JJ, Godard P, Zaegel M, et al. 2001. Axonemal dynein intermediate-chain gene (DNAI1) mutations result in situs inversus and primary ciliary dyskinesia (Kartagener syndrome). *Am. J. Hum. Genet.* 68:1030–35
61. Zariwala M, Noone PG, Sannuti A, Minnix S, Zhou Z, et al. 2001. Germline mutations in an intermediate chain dynein cause primary ciliary dyskinesia. *Am. J. Respir. Cell Mol. Biol.* 25:577–83
62. **Zariwala MA, Leigh MW, Ceppa F, Kennedy MP, Noone PG, et al. 2006. Mutations of *DNAI1* in primary ciliary dyskinesia: evidence of founder effect in a common mutation. *Am. J. Respir. Crit. Care Med.* 174:858–66**
63. Mitchell DR, Kang Y. 1991. Identification of oda6 as a *Chlamydomonas* dynein mutant by rescue with the wild-type gene. *J. Cell Biol.* 113:835–42
64. Bartoloni L, Blouin JL, Pan Y, Gehrig C, Maiti AK, et al. 2002. Mutations in the *DNAH11* (axonemal heavy chain dynein type 11) gene cause one form of situs inversus totalis and most likely primary ciliary dyskinesia. *Proc. Natl. Acad. Sci. USA* 99:10282–86
65. Pagon RA, Daiger SP. 2004. *Retinitis pigmentosa overview.* **http://www.geneclinics.org/**
66. Hong DH, Pawlyk BS, Adamian M, Sandberg MA, Li T. 2005. A single, abbreviated RPGR-ORF15 variant reconstitutes RPGR function in vivo. *Invest. Ophthalmol. Vis. Sci.* 46:435–41
67. Hong DH, Pawlyk B, Sokolov M, Strissel KJ, Yang J, et al. 2003. RPGR isoforms in photoreceptor connecting cilia and the transitional zone of motile cilia. *Invest. Ophthalmol. Vis. Sci.* 44:2413–21
68. Hong DH, Pawlyk BS, Shang J, Sandberg MA, Berson EL, Li T. 2000. A retinitis pigmentosa GTPase regulator (RPGR)-deficient mouse model for X-linked retinitis pigmentosa (RP3). *Proc. Natl. Acad. Sci USA* 97:3649–54
69. Krawczynski MR, Dmenska H, Witt M. 2004. Apparent X-linked primary ciliary dyskinesia associated with retinitis pigmentosa and a hearing loss. *J. Appl. Genet.* 45:107–10
70. Ferrante MI, Giorgio G, Feather SA, Bulfone A, Wright V, et al. 2001. Identification of the gene for oral-facial-digital type I syndrome. *Am. J. Hum. Genet.* 68:569–76
71. Ferrante MI, Zullo A, Barra A, Bimonte S, Messaddeq N, et al. 2006. Oral-facial-digital type I protein is required for primary cilia formation and left-right axis specification. *Nat. Genet.* 38:112–17
72. Romio L, Fry AM, Winyard PJ, Malcolm S, Woolf AS, Feather SA. 2004. OFD1 is a centrosomal/basal body protein expressed during mesenchymal-epithelial transition in human nephrogenesis. *J. Am. Soc. Nephrol.* 15:2556–68
73. Fliegauf M, Omran H. 2006. Novel tools to unravel molecular mechanisms in cilia-related disorders. *Trends Genet.* 22:241–45

74. Li JB, Gerdes JM, Haycraft CJ, Fan Y, Teslovich TM, et al. 2004. Comparative genomics identifies a flagellar and basal body proteome that includes the *BBS5* human disease gene. *Cell* 117:541–52

75. Piperno G, Huang B, Luck DJ. 1977. Two-dimensional analysis of flagellar proteins from wild-type and paralyzed mutants of *Chlamydomonas reinhardtii*. *Proc. Natl. Acad. Sci. USA* 74:1600–4

76. Pazour GJ, Agrin N, Leszyk J, Witman GB. 2005. Proteomic analysis of a eukaryotic cilium. *J. Cell Biol.* 170:103–13

77. Ostrowski LE, Blackburn K, Radde KM, Moyer MB, Schlatzer DM, et al. 2002. A proteomic analysis of human cilia: identification of novel components. *Mol. Cell. Proteomics* 1:451–65

78. Blouin JL, Meeks M, Radhakrishna U, Sainsbury A, Gehring C, et al. 2000. Primary ciliary dyskinesia: A genome-wide linkage analysis reveals extensive locus heterogeneity. *Eur. J. Hum. Genet.* 8:109–18

79. Meeks M, Walne A, Spiden S, Simpson H, Mussaffi-Georgy H, et al. 2000. A locus for primary ciliary dyskinesia maps to chromosome 19q. *J. Med. Genet.* 37:241–44

80. Jeganathan D, Chodhari R, Meeks M, Faeroe O, Smyth D, et al. 2004. Loci for primary ciliary dyskinesia map to chromosome 16p12.1–12.2 and 15q13.1–15.1 in Faroe Islands and Israeli Druze genetic isolates. *J. Med. Genet.* 41:233–40

81. Geremek M, Zietkiewicz E, Diehl SR, Alizadeh BZ, Wijmenga C, Witt M. 2006. Linkage analysis localises a Kartagener syndrome gene to a 3.5 cM region on chromosome 15q24–25. *J. Med. Genet.* 43:e1

82. Ibanez-Tallon I, Gorokhova S, Heintz N. 2002. Loss of function of axonemal dynein *Mdnah5* causes primary ciliary dyskinesia and hydrocephalus. *Hum. Mol. Genet.* 11:715–21

83. Bruni JE, del Bigio MR, Cardoso ER, Persaud TV. 1988. Neuropathology of congenital hydrocephalus in the SUMS/NP mouse. *Acta Neurochir.* 92:118–22

84. Sapiro R, Kostetskii I, Olds-Clarke P, Gerton GL, Radice GL, Strauss III JF. 2002. Male infertility, impaired sperm motility, and hydrocephalus in mice deficient in sperm-associated antigen 6. *Mol. Cell Biol.* 22:6298–305

85. Torikata C, Kijimoto C, Koto M. 1991. Ultrastructure of respiratory cilia of WIC-Hyd male rats. An animal model for human immotile cilia syndrome. *Am. J. Pathol.* 138:341–47

86. Kobayashi Y, Watanabe M, Okada Y, Sawa H, Takai H, et al. 2002. Hydrocephalus, situs inversus, chronic sinusitis, and male infertility in DNA polymerase lambda-deficient mice: possible implication for the pathogenesis of immotile cilia syndrome. *Mol. Cell. Biol.* 22:2769–76

87. Zariwala M, O'Neal WK, Noone PG, Leigh MW, Knowles MR, Ostrowski LE. 2004. Investigation of the possible role of a novel gene, DPCD, in primary ciliary dyskinesia. *Am. J. Respir. Cell Mol. Biol.* 30:428–34

88. Bertocci B, De Smet A, Berek C, Weill JC, Reynaud CA. 2003. Immunoglobulin kappa light chain gene rearrangement is impaired in mice deficient for DNA polymerase mu. *Immunity* 19:203–11

89. Supp DM, Brueckner M, Kuehn MR, Witte DP, Lowe LA, et al. 1999. Targeted deletion of the ATP binding domain of left-right dynein confirms its role in specifying development of left-right asymmetries. *Development* 126:5495–504

90. Tanaka H, Iguchi N, Toyama Y, Kitamura K, Takahashi T, et al. 2004. Mice deficient in the axonemal protein Tektin-t exhibit male infertility and immotile-cilium syndrome due to impaired inner arm dynein function. *Mol. Cell Biol.* 24:7958–64

74. Together with Avidor-Reiss (57), used comparative genomics to define proteins important for ciliary structure and function.

77. First identified more than 200 proteins from isolated human ciliary axonemes, using two-dimensional gel electrophoresis.

91. Neesen J, Kirschner R, Ochs M, Schmiedl A, Habermann B, et al. 2001. Disruption of an inner arm dynein heavy chain gene results in asthenozoospermia and reduced ciliary beat frequency. *Hum. Mol. Genet.* 10:1117–28
92. Chen J, Knowles HJ, Hebert JL, Hackett BP. 1998. Mutation of the mouse hepatocyte nuclear factor/forkhead homologue 4 gene results in an absence of cilia and random left-right asymmetry. *J. Clin. Invest.* 102:1077–82
93. Zhang Z, Sapiro R, Kapfhamer D, Bucan M, Bray J, et al. 2002. A sperm-associated WD repeat protein orthologous to *Chlamydomonas* PF20 associates with Spag6, the mammalian orthologue of *Chlamydomonas* PF16. *Mol. Cell Biol.* 22:7993–8004
94. Zhang Z, Kostetskii I, Tang W, Haig-Ladewig L, Sapiro R, et al. 2006. Deficiency of SPAG16L causes male infertility associated with impaired sperm motility. *Biol. Reprod.* 74:751–59
95. Lundberg JO, Weitzberg E, Nordvall SL, Kuylenstierna R, Lundberg JM, Alving K. 1994. Primarily nasal origin of exhaled nitric oxide and absence in Kartagener's syndrome. *Eur. Respir. J.* 7:1501–4
96. Karadag B, James AJ, Gultekin E, Wilson NM, Bush A. 1999. Nasal and lower airway level of nitric oxide in children with primary ciliary dyskinesia. *Eur. Respir. J.* 13:1402–5
97. Nowinski A, Hawrylkjewicz I, Sulikowska-Rowinska A, Gorecka D. 2002. Primary ciliary dyskinesia: Kartagener's syndrome and fertile male. *Pneumonol. Alergol. Pol.* 70:312–17
98. Narang I, Ersu R, Wilson NM, Bush A. 2002. Nitric oxide in chronic airway inflammation in children: diagnostic use and pathophysiological significance. *Thorax* 57:586–89
99. Gaston B, Drazen JM, Loscalzo J, Stamler JS. 1994. The biology of nitrogen oxides in the airways. *Am. J. Respir. Crit. Care Med.* 149:538–51
100. Yates DH, Kharitonov SA, Thomas PS, Barnes PJ. 1996. Endogenous nitric oxide is decreased in asthmatic patients by an inhibitor of inducible nitric oxide synthase. *Am. J. Respir. Crit. Care Med.* 154:247–50
101. Xue C, Botkin SJ, Johns RA. 1996. Localization of endothelial NOS at the basal microtubule membrane in ciliated epithelium of rat lung. *J. Histochem. Cytochem.* 44:463–71
102. Razani B, Engelman JA, Wang XB, Schubert W, Zhang XL, et al. 2001. Caveolin-1 null mice are viable but show evidence of hyperproliferative and vascular abnormalities. *J. Biol. Chem.* 276:38121–38
103. Brown DE, Hazucha MJ, Minnix SL, Knowles MR, Leigh MW. 2005. Nasal nitric oxide measurement during tidal breathing maneuvers: application for PCD screening. *Proc. Am. Thorac. Soc.* 2:A185 (Abstr.)
104. Stannard W, Rutman A, Wallis C, O'Callaghan C. 2004. Central microtubular agenesis causing primary ciliary dyskinesia. *Am. J. Respir. Crit Care Med.* 169:634–37
105. Bush A. 2000. Primary ciliary dyskinesia. *Acta Otorhinolaryngol. Belg.* 54:317–24
106. Sadek CM, Damdimopoulos AE, Pelto-Huikko M, Gustafsson JA, Spyrou G, Miranda-Vizuete A. 2001. Sptrx-2, a fusion protein composed of one thioredoxin and three tandemly repeated NDP-kinase domains is expressed in human testis germ cells. *Genes Cells* 6:1077–90
107. Martin R, Mogg AE, Heywood LA, Nitschke L, Burke JF. 1989. Aminoglycoside suppression at UAG, UAA and UGA codons in *Escherichia coli* and human tissue culture cells. *Mol. Gen. Genet.* 217:411–18
108. Bedwell DM, Kaenjak A, Benos DJ, Bebok Z, Bubien JK, et al. 1997. Suppression of a CFTR premature stop mutation in a bronchial epithelial cell line. *Nat. Med.* 3:1280–84
109. Howard M, Frizzell RA, Bedwell DM. 1996. Aminoglycoside antibiotics restore CFTR function by overcoming premature stop functions. *Nat. Med.* 2:467–69

110. Wilschanski M, Yahav Y, Yaacov Y, Blau H, Bentur L, et al. 2003. Gentamicin-induced correction of CFTR function in patients with cystic fibrosis and CFTR stop mutations. *N. Engl. J. Med.* 349:1433–41
111. Bartoloni L, Blouin JL, Maiti AK, Sainsbury A, Rossier C, et al. 2001. Axonemal beta heavy chain dynein DNAH9: cDNA sequence, genomic structure, and investigation of its role in primary ciliary dyskinesia. *Genomics* 72:21–33
112. Blouin JL, Albrecht C, Gehrig C, Duriaux-Sail G, Strauss JF III, et al. 2003. Primary ciliary dyskinesia/Kartagener syndrome: searching for genes in a highly heterogeneous disorder. *Am. J. Hum. Genet.* 73(Suppl. 5) (Abstr. No. 2440)
113. Bartoloni L, Mitchison H, Pazour GJ, Maiti AK, Meeks M, et al. 2000. No deleterious mutations were found in three genes (*HFH4*, *LC8*, *IC2*) on human chromosome 17q in patients with primary ciliary dyskinesia. *Eur. J. Hum. Genet.* 8:P484 (Abstr.)
114. Pennarun G, Bridoux AM, Escudier E, Dastot-Le Moal F, Cacheux V, et al. 2002. Isolation and expression of the human *hPF20* gene orthologous to *Chlamydomonas* PF20: evaluation as a candidate for axonemal defects of respiratory cilia and sperm flagella. *Am. J. Respir. Cell Mol. Biol.* 26:362–70
115. Horvath J, Fliegauf M, Olbrich H, Kispert A, King SM, et al. 2005. Identification and analysis of axonemal dynein light chain 1 in primary ciliary dyskinesia patients. *Am. J. Respir. Cell Mol. Biol.* 33:41–47
116. Gehrig C, Albrecht C, Duriaus Sail G, Rossier C, Scamuffa N, DeLozier-Blancet C, Antonarakis SE, Blouin JL. 2002. *Primary ciliary dyskinesia: mutation analysis in dynein light chain genes mapping to chromosome 1 (HP28) and 22 (DNAL4)*. Presented at Eur. Hum. Genet. Conf., Strasbourg, P0305 (Abstr.)
117. Neesen J, Drenckhahn JD, Tiede S, Burfeind P, Grzmil M, et al. 2002. Identification of the human ortholog of the t-complex-encoded protein TCTE3 and evaluation as a candidate gene for primary ciliary dyskinesia. *Cytogenet. Genome Res.* 98:38–44
118. Blouin JL, Gehrig C, Armengot M, Rutishauser M, Jorissen M, et al. 2002. *DNAH3: characterization of the sequence and mutation search in patients with Primary Ciliary Dyskinesia*. Presented at Eur. Hum. Genet. Conf., Strasbourg, P0304 (Abstr.)
119. Zhang YJ, O'Neal WK, Randell SH, Blackburn K, Moyer MB, et al. 2002. Identification of dynein heavy chain 7 as an inner arm component of human cilia that is synthesized but not assembled in a case of primary ciliary dyskinesia. *J. Biol. Chem.* 277:17906–15
120. Pennarun G, Bridoux AM, Escudier E, Amselem S, Duriez B. 2001. The human *HP28* and *HFH4* genes: evaluation as candidate genes for primary ciliary dyskinesia. *Am. J. Respir. Crit. Care Med.* 163:A538 (Abstr.)
121. Maiti AK, Bartoloni L, Mitchison HM, Meeks M, Chung E, et al. 2000. No deleterious mutations in the *FOXJ1* (alias *HFH-4*) gene in patients with primary ciliary dyskinesia (PCD). *Cytogenet. Cell Genet.* 90:119–22

RELATED RESOURCES

1. Afzelius BA. 2004. Cilia-related diseases. *J. Pathol.* 204:470–77
2. Berdon WE, Willi U. 2004. Situs inversus, bronchiectasis, and sinusitis and its relation to immotile cilia: history of the diseases and their discoverers—Manes Kartagener and Bjorn Afzelius. *Pediatr. Radiol.* 34:38–42
3. Chodhari R, Mitchison HM, Meeks M. 2004. Cilia, primary ciliary dyskinesia and molecular genetics. *Paediatr. Respir. Rev.* 5:69–76

4. Gerdes JM, Katsanis N. 2005. Microtubule transport defects in neurological and ciliary disease. *Cell Mol. Life Sci.* 62:1556–70
5. Geremek M, Witt M. 2004. Primary ciliary dyskinesia: genes, candidate genes and chromosomal regions. *J. Appl. Genet.* 45:347–61
6. Leigh MW. 2006. Primary ciliary dyskinesia. In *Disorders of the Respiratory Tract of Children*, ed. V Chernick, TF Boat, RW Wilmott, A Bush, pp. 485–90. Philadelphia: Saunders-Elsevier
7. Stannard W, O'Callaghan C. 2006. Ciliary function and the role of cilia in clearance. *J. Aerosol Med.* 19:110–15

Regulation of Receptor Trafficking by GRKs and Arrestins

Catherine A.C. Moore, Shawn K. Milano, and Jeffrey L. Benovic*

Department of Biochemistry and Molecular Biology, Thomas Jefferson University, Philadelphia, Pennsylvania 19107; email: cachen@mail.jci.tju.edu, skm46@cornell.edu, benovic@mail.jci.tju.edu

Annu. Rev. Physiol. 2007. 69:451–82

First published online as a Review in Advance on October 12, 2006

The *Annual Review of Physiology* is online at http://physiol.annualreviews.org

This article's doi: 10.1146/annurev.physiol.69.022405.154712

0066-4278/07/0315-0451$20.00

*Corresponding author.

Key Words

GPCR, endocytosis, internalization, sorting

Abstract

To ensure that extracellular stimuli are translated into intracellular signals of appropriate magnitude and specificity, most signaling cascades are tightly regulated. One of the major mechanisms involved in the regulation of G protein–coupled receptors (GPCRs) involves their endocytic trafficking. GPCR endocytic trafficking entails the targeting of receptors to discrete endocytic sites at the plasma membrane, followed by receptor internalization and intracellular sorting. This regulates the level of cell surface receptors, the sorting of receptors to degradative or recycling pathways, and in some cases the specific signaling pathways. In this chapter we discuss the mechanisms that regulate receptor endocytic trafficking, emphasizing the role of GPCR kinases (GRKs) and arrestins in this process.

RECEPTOR ENDOCYTOSIS

Introduction

G protein–coupled receptor (GPCR): cell surface receptor with predicted seven-transmembrane (7TM) α-helical domains; couples to heterotrimeric G proteins

Heterotrimeric G protein: heterotrimeric protein complex consisting of an α subunit that binds guanine nucleotides, GDP and GTP, and a βγ dimer (subfamilies: G_i, G_s, G_q, $G_{12/13}$)

Monomeric G protein: monomeric protein that binds guanine nucleotides, GDP and GTP (subfamilies: Ras, Rho, ARF, Rab, Ran)

Endocytic trafficking: targeting of GPCRs to discrete endocytic sites at the plasma membrane, followed by internalization and intracellular sorting to recycling or degradation compartments

Internalization: trafficking of GPCRs from distinct plasma membrane sites, such as CCPs, to intracellular compartments

Recycling: trafficking of GPCRs from intracellular compartments back to the plasma membrane

All tissues must translate an extensive array of extracellular stimuli into appropriate intracellular signals and physiological responses. At the cellular level, this process involves a complex series of biochemical events, many of which are carried out by transmembrane receptors at the plasma membrane interface. G protein–coupled receptors (GPCRs) are one class of receptors positioned at this boundary and are known to play a critical role in many aspects of human physiology, ranging from normal homeostasis to disease states. The physiological output of GPCR activation results from the exquisite spatial, temporal, and kinetic regulation of GPCR ligand binding, heterotrimeric G protein coupling, signal propagation (including to monomeric G proteins), desensitization, and endocytosis. This chapter focuses on endocytosis, which for GPCRs serves as a mechanism to reversibly or irreversibly remove receptors from the cell surface, thus significantly contributing to signal desensitization/resensitization or downregulation, respectively, and allowing for the fine-tuning of signal magnitude and duration.

GPCR endocytic trafficking entails the targeting of receptors to discrete endocytic sites at the plasma membrane, followed by receptor internalization and intracellular sorting to recycling or degradative compartments (**Figure 1**). The structural and molecular machinery utilized can serve to define distinct endocytic pathways, including clathrin-dependent (1, 2), caveolae-dependent (3–5), and clathrin/caveolae-independent pathways. Although these pathways are spatially distinct at the cell surface, they utilize a subset of overlapping machinery and in some cases may merge within the cell, suggesting that the various endocytic pathways are highly integrated. To date, the most well-characterized pathway for GPCR endocytosis occurs through clathrin-coated pits (CCPs) containing clathrin and adapter protein-2 (AP-2) complexes. The nonvisual arrestin proteins, arrestin-2 (β-arrestin1) and arrestin-3 (β-arrestin2), play pivotal roles as adapters and scaffolds in this process; we highlight recent advances in their structural and functional complexities as related to GPCR endocytosis.

For the clathrin-dependent endocytic pathway, receptor phosphorylation–mediated binding to arrestin targets receptors to CCPs (6, 7). G protein–coupled receptor kinases (GRKs), a family of protein kinases that specifically phosphorylate serine/threonine residues, phosphorylate agonist-occupied receptors. Receptor phosphorylation leads to high-affinity arrestin binding and activation, which uncouples the receptor from G protein by steric hindrance and targets the receptor to CCPs through the binding of AP-2 and clathrin. Several elegant structure/function studies have recently advanced mechanistic details of arrestin activation and adapter functions, which are detailed below (see Arrestin as an Endocytic Adapter, below). In addition to these conventional functions, arrestin can scaffold and regulate other proteins implicated in endocytosis, which are also discussed below (see Arrestin as an Endocytic Scaffold, below).

Clathrin-Dependent Pathway

Receptors that follow the clathrin-dependent endocytic pathway are internalized in CCPs in a dynamin-dependent fashion and then proceed to tubular-vesicular early endosomes. There, receptors are subsequently sorted to either recycling endosomes, which traffic receptors back to the plasma membrane, or multivesicular late endosomes, which traffic receptors to lysosomes for degradation. These compartments are often identified by microscopy, using the marker proteins dynamin and Eps15 for CCPs, EEA1 and Rab5 for early endosomes, Rab4 or Rab11 for distinct recycling endosomes, and Rab7 and LAMP for late endosomes. However, the restricted compartmentation that this method suggests clearly underestimates the complexity

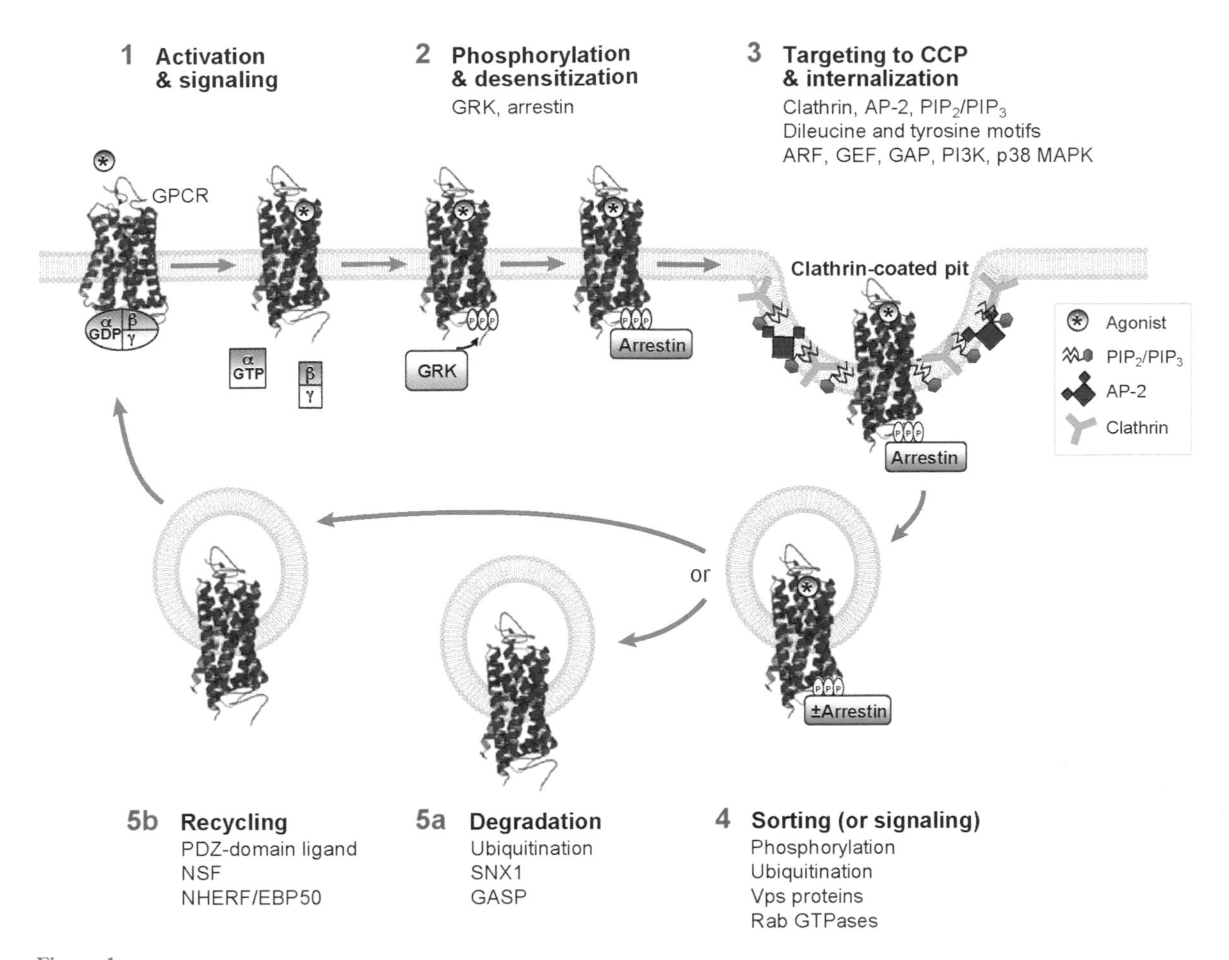

Figure 1

Regulation of G protein–coupled receptor (GPCR) trafficking by G protein–coupled receptor kinases (GRKs) and arrestins. Agonist (*asterisk*) binding to GPCRs leads to receptor activation, G protein coupling, and signal transduction (*step 1*). GRKs then phosphorylate the agonist-activated GPCR on intracellular domains, initiating arrestin recruitment. Arrestin binding to the receptor inhibits G protein coupling and terminates signaling, a process termed desensitization (*step 2*). Receptor/arrestin complexes are then targeted to clathrin-coated pits, where arrestin forms a multicomponent complex with clathrin, adapter protein-2 (AP-2), and phosphoinositides, resulting in receptor internalization (*step 3*). Internalized GPCRs are sorted (*step 4*) to either degradation (*step 5a*) or recycling (*step 5b*) compartments. Topics discussed in this review are listed under each step. Other abbreviations used: ARF, ADP-ribosylation factor; GAP, GTPase-activating protein; GASP, GPCR-associated sorting protein; GEF, guanine nucleotide exchange factor; NHERF/EBP50, Na^+/H^+ exchanger regulatory factor/ezrin-radixin-moesin-binding phosphoprotein of 50 kDa; NSF, *N*-ethylmaleimide-sensitive factor; p38 MAPK, p38 mitogen-activated protein kinase; PDZ, PSD-95/Dlg/ZO-1; PI3K, phosphatidylinositol 3-kinase; PIP_2, phosphatidylinositol 4,5-bisphosphate; PIP_3, phosphatidylinositol 3,4,5-trisphosphate; SNX1, sorting nexin 1; Vps proteins, vacuolar protein–sorting proteins.

of intracellular vesicle transport. Notably, arrestin functions at each of these endocytic compartments. For an abundance of GPCRs, arrestin functions as an adapter to target receptors to CCPs through its AP-2- and clathrin-binding activities. For certain GPCRs, arrestin (*a*) scaffolds extracellular signal–regulated kinase (ERK) signaling complexes on early endosome populations targeted to specific plasma membrane sites; (*b*) stably interacts with receptors on recycling endosome populations, where it dictates the kinetics of receptor recycling; and (*c*) is involved in receptor ubiquitination, which functions on late endosome populations as a degradation signal. These examples highlight the multifunctionality of arrestin at distinct compartments along the endocytic route.

Clathrin-coated pit (CCP): discrete endocytic site at the plasma membrane consisting of a clathrin lattice and AP-2 complexes; forms a cargo-containing budding vesicle

AP-2: adapter protein-2

Nonvisual arrestins: arrestin-2 (β-arrestin1), arrestin-3 (β-arrestin2)

Arrestin: protein originally identified to desensitize rhodopsin; functions as an adapter and scaffold during GPCR endocytosis (subfamilies: visual arrestin-1 and arrestin-4, nonvisual arrestin-2 and arrestin-3)

G protein–coupled receptor kinase (GRK): protein kinase that specifically phosphorylates serine and threonine residues (subfamilies: GRK1/7, GRK2/3, GRK4/5/6)

Degradation: proteolysis of GPCRs in proteosomal or lysosomal compartments

Class A versus Class B Receptors

GPCRs that traffic through the clathrin-dependent endocytic pathway can be divided further into two groups, Class A and Class B receptors, based on the characteristics of agonist-dependent arrestin binding. These two receptor classes were originally assessed by modeling of arrestin translocation profiles obtained by confocal microscopy (8). Class A receptors bind arrestin-3 with greater affinity than arrestin-2 and include the β_2-adrenergic receptor (β_2AR), μ-opioid receptor (MOR), endothelin type A (ET_A), dopamine D1A (D1A), and α_{1b} adrenergic (α_{1b}) receptors. Class B receptors bind arrestin-3 and arrestin-2 with approximately equal affinities and include the angiotensin II type 1A (AT_{1A}), vasopressin V2 (V2), neurotensin 1 (NT1), thyrotropin-releasing hormone (TRH), and substance-P/neurokinin-1 (SP/NK1) receptors. Furthermore, whereas arrestin facilitates the desensitization and internalization of both receptor classes, it either (*a*) dissociates from Class A receptors at or near the plasma membrane or (*b*) maintains a stable association with Class B receptors and colocalizes in endosomes, as previously demonstrated (9).

The prevailing theory suggests that differential receptor-arrestin affinity and trafficking mediate postendocytic receptor sorting to recycling or degradative pathways (for further discussion, see Arrestin as an Endocytic Scaffold, below). For AT_{1A}, V2, NT1, SP/NK1, and oxytocin receptors, stable arrestin association may involve a conserved cluster of serine and threonine residues (10, 11). For the V2 receptor, these residues may function in concert with (11) or independently of (12) arrestin binding to limit receptor recycling. For β_2AR, differential phosphorylation by GRK isoforms may regulate stable receptor-arrestin association (13). GRK2 expression leads to an arrestin plasma membrane translocation pattern, as expected for this Class A receptor. In contrast, GRK5 and GRK6 expression supports arrestin detection in vesicles following β_2AR stimulation, as is normally seen for Class B receptors. For AT_{1A} and V2 receptors, stable arrestin association also requires sustained versus transient arrestin ubiquitination (14, 15). Upon agonist stimulation of AT_{1A} receptors, the deficient endosomal arrestin recruitment displayed by transiently ubiquitinated arrestin-3 lysine mutants is remarkably restored by fusing ubiquitin to arrestin's C-tail, which mimics the stably ubiquitinated form of arrestin (14). Furthermore, expression of this arrestin-ubiquitin chimera is sufficient to impart Class B characteristics to the Class A β_2AR (15). The differential binding and localization of arrestins may function beyond endocytic trafficking, as demonstrated by the role this plays in mitogen-activated protein kinase (MAPK) signaling. For AT_{1A}, V2, and protease-activated receptor 2 (PAR2) receptors, a stable association with arrestin leads to cotrafficking to endosomes and contributes to the sustained cytosolic activity of ERK. Overall, the stability of the receptor-arrestin complex has been shown to influence arrestin's function throughout the endocytic pathway.

Internalization Motifs

GPCRs use arrestin as an adapter for CCP targeting more widely than direct binding to AP-2 via C-tail motifs. However, there are a few notable examples of the latter mechanism. Short linear sequences of amino acids within the intracellular domain of transmembrane proteins often function as determinants for trafficking along the endocytic pathway (16). Internalization motifs that target GPCRs to localized CCP zones (17–19) include the classical dileucine- or tyrosine-based (YXXϕ, where ϕ is a residue with a bulky hydrophobic side chain) motifs described for sorting of non-GPCRs. In general, these motifs mediate direct binding to AP-2 complexes (20) in a lipid-dependent fashion (21) and in turn promote binding to clathrin endocytic components at the plasma membrane.

Two tyrosine-based signals (YXXϕ) have been identified in the C terminus of protease-activated receptor 1 (PAR1). Mutation of the proximal motif $Y_{383}xxL_{386}$ to AxxA impaired agonist-dependent internalization after prolonged exposure while not affecting tonic internalization (22). In contrast, mutation of the distal motif $Y_{420}xxL_{423}L_{424}$ to AxxAA produced receptors that failed to internalize constitutively and thus did not form the intracellular receptor pool that is protected from thrombin cleavage and needed for resensitization independent of de novo receptor synthesis (23). Using an siRNA approach, Paing et al. (23) demonstrated an AP-2 requirement for PAR1 constitutive internalization in both HeLa cells stably expressing Flag-PAR1 and in HUVECs endogenously expressing PAR1. Failure of the $Y_{420}xxL_{423}L_{424}$ mutant to internalize was likely due to disrupted PAR1 and AP-2 association, as suggested by reduced colocalization and in vitro surface plasmon resonance binding of AP-2 with this PAR1 mutant as compared with wild type. A related YXXXϕ motif in the proximal C-tail portion of thromboxane A2 β isoform (TPβ) receptor also functions in tonic internalization, although binding of AP-2 to this region was not addressed (24). Introduction of either this YXXXϕ motif or the classical YXXϕ into the TPα receptor isoform was sufficient to render tonic internalization not normally seen for the wild-type TPα receptor. Interestingly, many GPCRs known to utilize an arrestin-dependent endocytic route also contain an as-yet-uncharacterized canonical YXXϕ motif in their cytoplasmic tails (23). Additional AP-2 binding sites may also exist, as suggested for the α_{1b} receptor, for which AP-2 binding requires a C-tail arginine stretch rather than the YXXϕ motif. This region was implicated in agonist-dependent internalization (25).

Dileucine motifs are present in multiple receptors, including CXCR2, CXCR4, and β_2AR. For CXCR2, mutation of the LLKIL motif to AAKIL and/or LLKAA decreased receptor coimmunoprecipitation of AP-2 but not arrestin, and inhibited receptor endocytosis and IL-8-induced chemotaxis in HEK293 cells (26). For CXCR4 and β_2AR, mutation of the dileucine motif dramatically reduced receptor internalization in response to ligand (27, 28). While dileucine motifs in GPCR C termini may also function as endoplasmic reticulum export signals (29, 30), the mutant expression levels were similar to those of wild type in all three studies.

These receptors also depend on arrestin for efficient internalization, as exemplified by delayed (31) or reduced receptor internalization in mouse embryonic fibroblasts (MEFs) (32) and neutrophils (33) from arrestin-$3^{-/-}$ knockout mice, reduced internalization in HEK293 cells following arrestin-3 siRNA (34), and enhanced internalization in HEK293 cells by arrestin-3/GRK2 coexpression (27, 35). In contrast to unaltered signaling to Ca^{2+} and cyclic AMP (cAMP) upon blocking internalization with CXCR2 and β_2AR dileucine mutants, blocking internalization by arrestin knockout correlates with enhanced signaling to Ca^{2+} and cAMP as well as an increase in CXCR2-mediated neutrophil chemotaxis (32, 33). For CXCR4, however, arrestin-3 knockout decreased CXCR4-mediated lymphocyte chemotaxis (36), and arrestin-3 RNAi reduced signaling to p38, which correlated with decreased CXCR4-mediated HEK293 chemotaxis (37). This is consistent with reduced PAR2-mediated MDA-MB-231 cell migration following arrestin siRNA (38). How these dileucine- and arrestin-based mechanisms are integrated and/or differentially regulated is currently unknown. As for non-GPCRs, this may involve steric and electrostatic properties imparted by receptor phosphorylation, ligand-induced

C-tail conformational changes, and positioning of motifs relative to the receptor transmembrane regions.

Regulated versus Tonic Endocytosis

Tonic endocytosis is the process whereby receptors internalize and traffic in the absence of agonist. Many receptors undergo both agonist-driven and tonic endocytosis but do so via distinct mechanisms. For example, the TPβ receptor utilizes GRK and arrestin for agonist-driven (39) but not for tonic (24) endocytosis, similar to mGluR1a receptors (40). Tonically trafficked TPβ receptors still internalize in CCPs, but this is mediated by the YXXXϕ motif just discussed. As another example, PAR1 that internalizes in response to ligand is degraded in lysosomes, whereas receptors that internalize tonically can recycle (41), which, as mentioned above, may allow for rapid resensitization by replenishing surface receptors independently of de novo receptor synthesis (23). Both pathways appear arrestin independent. Continued efforts to characterize the different mechanisms of agonist-driven versus tonic endocytosis will likely clarify the role of arrestins.

Whereas agonist-dependent endocytosis clearly modulates signals for most GPCRs examined to date, the function of tonic endocytosis, beyond contributing to receptor homeostasis, is much less clear. Indeed, mutationally induced tonic endocytosis can be associated with opposite phenotypes, such as loss of function versus constitutive function (42). For example, a distinct mutation in the V2 receptor is associated with familial nephrogenic diabetes insipidus, in which kidney V2 receptors are unable to promote urine concentration. This mutant receptor (R137H) is constitutively phosphorylated and associated with arrestin and tonically trafficked to intracellular vesicles (43). This results in a loss-of-function phenotype. In contrast, a C5a receptor mutant, which also displays tonic endocytosis and colocalization with arrestin, is constitutively active (44). Engineered mutations in AT_{1A} receptors (R126H) also result in constitutive colocalization with arrestin and tonic endocytosis; however, in this case the receptor signals in response to agonist (45).

Examples of naturally occuring tonic endocytosis may help to shed light on its function. Certain viral-encoded receptors display both constitutive activity and tonic endocytosis, which may allow them to function yet evade the immune system. For example, a human cytomegalovirus–encoded chemokine receptor (US28) is constitutively active but also tonically traffics to perinuclear structures (46). Certain neuronal receptors also display constitutive activity and tonic endocytosis. For example, type 1 cannabinoid (CB1) receptors in cultured hippocampal neurons are constitutively active and display tonic endocytosis in soma and dendrites but not axons, which may regulate the polarized distribution of CB1 receptors to axons versus the somatodendritic compartment (47).

Molecular Machinery

An extensive array of molecular machinery ensures efficient and specific targeting, internalization, and sorting during GPCR endocytosis. The potential for arrestin to function in concert with many of these molecules is becoming clear. The internalization machinery that is discussed below includes, but is not limited to, clathrin, AP-2, dynamin, Eps15, AP180, phosphoinositides, and recently ADP-ribosylation factors (ARFs), phosphatidylinositol 3-kinase (PI3K), and p38 MAPK. The postendocytic sorting machinery discussed below includes the large family of Rab GTPases as well as the ubiquitin ligase system, sorting nexin 1 (SNX1), G protein–coupled receptor–associated sorting protein (GASP), hepatocyte growth factor–regulated tyrosine kinase substrate (HRS), vacuolar protein–sorting (vps) proteins, *N*-ethylmaleimide-sensitive factor (NSF), and Na^+/H^+ exchanger regulatory factor/ezrin-radixin-moesin-binding phosphoprotein of 50 kDa (NHERF/EBP50). There is a growing number of potential cargo-selective adapter proteins, including disabled-2, epsin, Hsc/Hsp interacting protein (HIP), autosomal recessive hypercholesterolemia (ARH), and numb (48–51). If and how pervasively these latter proteins function in GPCR endocytic pathways remain to be determined.

ARRESTIN AS AN ENDOCYTIC ADAPTER

Role of Arrestin and GRK in Receptor Endocytosis

Arrestins bind activated phosphorylated GPCRs and uncouple receptors from heterotrimeric G proteins, thereby terminating G protein signaling, a process termed desensitization. Studies over the past 5–10 years have shown that the biological functions of the nonvisual arrestins, arrestin-2 and arrestin-3, go well beyond this classically established role (52). Nonvisual arrestins are now known to mediate cytosolic and nuclear trafficking and signaling events, including receptor endocytosis, Src and MAPK activation, and gene expression. Arrestins are primarily localized in the cytoplasm and interact with GPCRs following receptor activation and subsequent phosphorylation. Indeed, multiple studies have demonstrated a rapid agonist-dependent translocation of GFP-tagged arrestins to plasma membrane–localized GPCRs (9). In addition, arrestins are also found in the nucleus (54, 55) where arrestin-2 regulates transcription (56). However, relatively few studies have evaluated the localization of endogenous arrestins, and at least one study suggests that a significant amount of endogenous arrestin has a distinct punctate distribution (57). Although the localization of endogenous arrestin is an area that warrants further investigation, we know that arrestins can bind to activated GPCRs and that receptor phosphorylation often enhances binding (58).

GRKs are the primary mediators of agonist-dependent phosphorylation of GPCRs. The seven mammalian GRKs can be divided into three subfamilies: (*a*) GRK1 (rhodopsin kinase) and GRK7; (*b*) GRK2 [β-adrenergic receptor kinase (βARK)] and GRK3 [β-adrenergic receptor kinase 2 (βARK2)]; and (*c*) GRK4, GRK5, and GRK6 (6, 59). A role for receptor phosphorylation in endocytosis was initially provided by the finding that GRK2 overexpression promoted agonist-induced internalization of the M2 muscarinic acetylcholine (M2) receptor, whereas expression of catalytically inactive GRK2 inhibited internalization (60). Indeed, numerous additional studies demonstrated that GRK-mediated phosphorylation promoted GPCR endocytosis (61). Although the primary mechanism for this likely involves the enhanced binding of arrestin to phosphorylated receptors (see below), GRKs may play a more direct role in receptor endocytosis.

GRK2 interacts with several proteins implicated in the endocytic process. GRK2 contains a C-terminal clathrin binding box that interacts directly with clathrin and mediates internalization of the β_1-adrenergic receptor (β_1AR) (62). Additional proteins that interact with GRK2 and appear to regulate receptor internalization include GRK interactor 1 (GIT1), PI3K, and ezrin. GIT1, an ARF GTPase–activating protein that interacts with GRK2, has been implicated in regulating the trafficking of receptors internalized via an arrestin- and dynamin-dependent pathway. Although the functional role of GIT1 interaction with GRK2 has not been elucidated, GIT1 may regulate GPCR trafficking by linking ARF regulation with activated GPCRs.

GRK2 also interacts with PI3K and regulates the recruitment of PI3K to the plasma membrane, where PI3K helps modulate GPCR endocytosis (63). The GRK2/PI3K pathway appears to involve enhanced recruitment of AP-2 to the receptor via the increased production of D-3 phospholipids (64), although the ability of PI3K to phosphorylate the cytoskeletal protein tropomyosin also contributes to βAR endocytosis (65). Recent studies suggest that disruption of the GRK2/PI3K interaction restores βAR signaling and contractile function in heart failure (66). The ability of GRK2 to phosphorylate ezrin, a protein involved in cytoskeletal reorganization, has also been linked to β_2AR endocytosis (67). Taken together, these studies suggest that GRKs orchestrate the formation of protein complexes that regulate receptor signaling and trafficking.

A salient feature of agonist-dependent phosphorylation of GPCRs is the recruitment and binding of arrestins. That arrestins, when overexpressed, promoted β_2AR internalization first established a role for nonvisual arrestins in GPCR trafficking (68, 69). Several integrated steps appear to drive the ability of nonvisual arrestins to mediate receptor endocytosis. The first step in this process involves a conformational change in arrestin, induced by binding to an activated phosphorylated receptor. This conformational change subsequently enhances the interaction of arrestin with proteins critically involved in receptor endocytosis, namely clathrin and AP-2. Finally, arrestins also interact with phosphoinositides such as PIP_2 (phosphatidylinositol 4,5-bisphosphate) and PIP_3 (phosphatidylinositol 3,4,5-trisphosphate); this interaction appears to play an essential role in mediating arrestin-promoted endocytosis of GPCRs. In this section, the structural features of arrestin are discussed and then placed in the context of how nonvisual arrestins function as adapters in the endocytic process.

Structural Basis for Arrestin Function

The X-ray crystal structure of bovine arrestin-1 provided initial insight to arrestin structure (70, 71). More recently, the crystal structures of C-terminally truncated (72) and wild-type (73) bovine arrestin-2 and salamander arrestin-4 (74) have been solved. In general, arrestins have an elongated shape and are almost exclusively made up of β-sheets and connecting loops, with the exception of one short α-helix (**Figure 2*a***). Arrestins are composed of two major domains, termed the N-domain and C-domain, that are held together by a polar core of buried salt bridges. The N- and C-domains are connected by a short hinge. The C-tail is connected by a flexible linker to the C-domain and contains a short β-strand that interacts with a lateral β-strand of the N-domain. Two regions that may undergo substantial conformational rearrangement after receptor binding are residues 282–309 of arrestin-2 and the C-tail (72). The hinge connecting the N- and C-domains may play an important role in this conformational rearrangement because reducing the hinge's length results in reduced activation of arrestin (75). This is consistent with the notion that the N- and C-domains move relative to each other during arrestin's transition into an active conformation.

The arrestin polar core is comprised of charged residues from the N terminus (Asp-29 in arrestin-2), N-domain (Arg-169 and Lys-170), C-domain (Asp-290 and Asp-297), and C-tail (Arg-393), thus bringing different parts of the molecule together to maintain a basal conformation (**Figure 2*a***). The residues involved in formation of the polar core are highly conserved, suggesting that this structural element is conserved in all arrestins and critical for function. Because the buried side chains of the polar core achieve neutrality by an elaborate network of electrostatic interactions, disturbance of the polar core by the introduction of a phosphate group from the receptor may promote structural changes that result in an active conformation of the protein (71). Disruption of polar core interactions can be simulated by charge inversion of an arginine (Arg-175, -169, and -170 in arrestin-1, -2, and -3, respectively) that lies in the center of the polar core. Mutation of this arginine results in an arrestin that binds equally well to phosphorylated and nonphosphorylated GPCRs, suggesting that this residue plays a key role in maintaining the basal conformation of arrestin (71, 76–79).

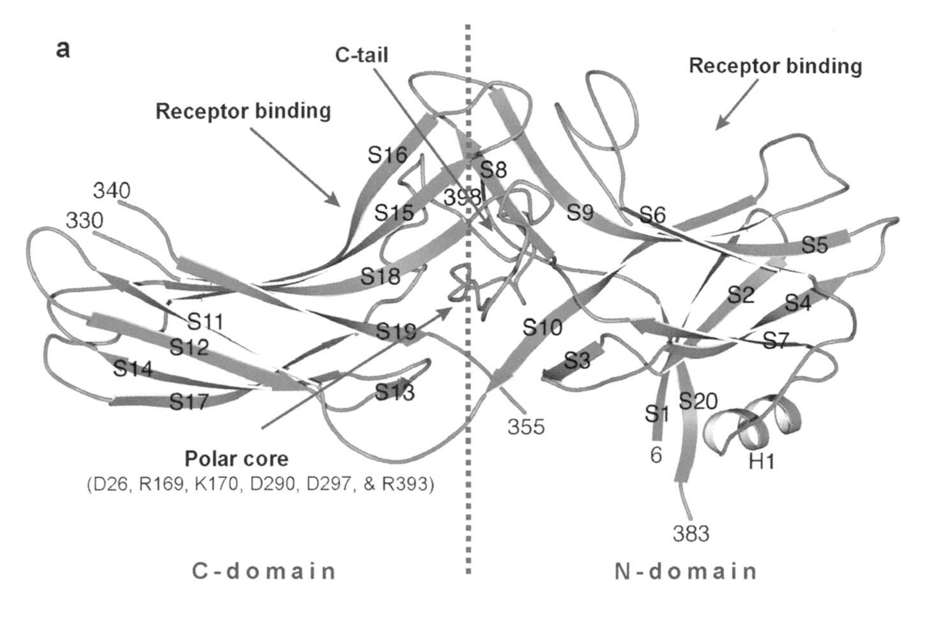

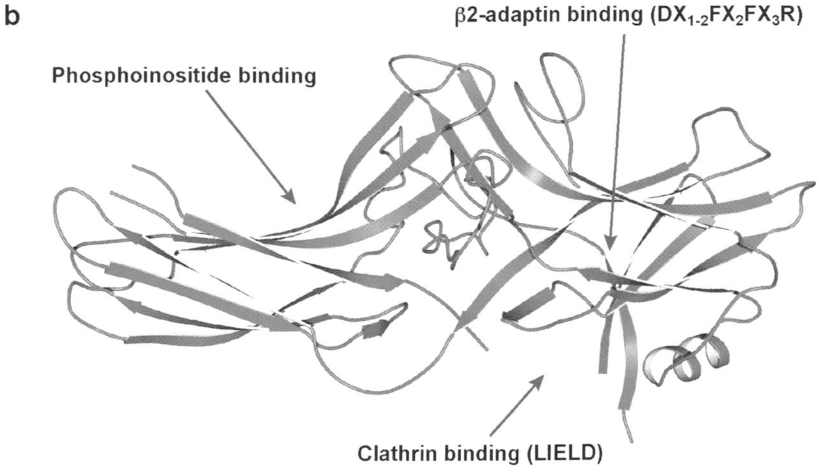

Figure 2

Arrestin-2 secondary structure. (*a*) A ribbon diagram of the arrestin-2 secondary structure, representing the protein's basal conformation, is shown. The N- and C-domains of arrestin-2 are indicated, as is the location of the distinctive polar core, the C-tail, and the primary receptor binding pockets. Residues 331–339, 356–382, and 400–418 are unstructured, whereas residues 398–399 (*red*) are partially obscured. (*b*) A ribbon diagram of arrestin-2 depicting the primary binding sites for phosphoinositides, the β2-adaptin binding domain near the C terminus, and the clathrin binding box in the unstructured region between β-strands 19 and 20.

Indeed, conformational changes induced by incorporation of an R175Q mutation in arrestin-1 have been detected by small-angle X-ray scattering (80). Similarly, conformational changes induced by activating mutations in arrestin-2 (e.g., R169E) have been detected by hydrogen exchange mass spectrometry (81). In addition, bioluminescence resonance energy transfer has detected conformational changes in arrestin-2 induced by receptor binding (82). Taken together, these studies suggest that the activation of arrestin results in significant structural changes that are important in mediating the various protein interactions and cellular functions of arrestin (**Figure 3**).

Arrestin Interaction with GPCRs

An important feature of arrestin interaction with GPCRs is the ability of arrestins to recognize both the activation and phosphorylation states of the receptor. This ability suggests that arrestins contain domains that specifically contact regions of the receptor exposed following receptor activation. Truncated in vitro–translated arrestin-1 (residues 1–191) partially retained the ability

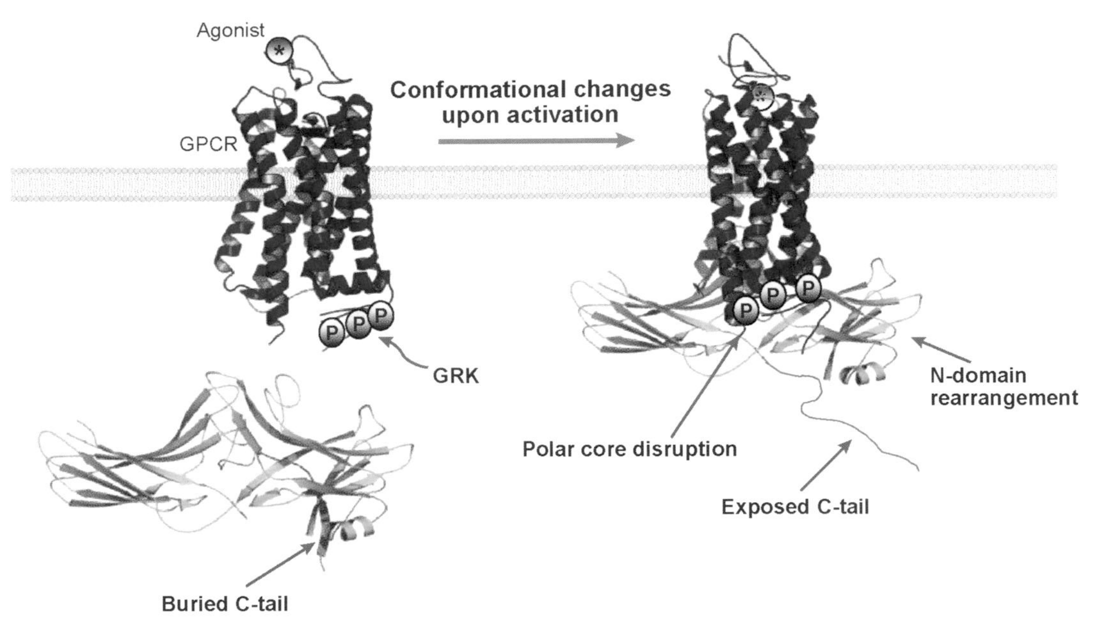

Figure 3

Model depicting conformational changes in arrestin that occur upon receptor binding. Some of the proposed conformational changes that occur upon receptor binding include disruption of the polar core and rearrangements of the N- and C-terminal domains. These changes enhance receptor binding and unmask binding sites for clathrin and β2-adaptin. Abbreviations used: GPCR, G protein–coupled receptor; GRK, G protein–coupled receptor kinase.

to recognize the light-activated state of rhodopsin (83), providing initial localization of such domains. Furthermore, at least three regions within the N-terminal half of arrestin-1 are involved in recognizing light-activated rhodopsin (84). The N-terminal domains of arrestin-2 and -3 also retain the ability to recognize the agonist-activated state of GPCRs, suggesting that the activation recognition region of all arrestins is largely contained within the N-terminal half of the protein (85, 86). This N-terminal receptor-binding region has been localized to β-strands 5 and 6 and adjacent loops, whereas a secondary receptor-binding domain is present within β-strands 15 and 16 in the C-terminal domain (87) (**Figure 2*a***). As expected, swapping these two receptor-binding elements between arrestin-1 and arrestin-2 reverses their receptor specificity (87). Moreover, recent studies have suggested the involvement of additional arrestin surface residues in receptor binding (88).

The importance of receptor phosphorylation in arrestin binding has also been extensively explored. The primary phosphorylation recognition domain within arrestin-1 is localized between residues 163–179 (76, 77), although adjacent positively charged residues within β-strand 1 (Lys-14 and Lys-15) also participate in phosphate binding and may facilitate binding within the primary phosphorylation recognition domain (89). Other positively charged residues, including Arg-18, Lys-20, Lys-55, Arg-56, and Lys-300, may also contribute to phosphate recognition in arrestin-1 (74, 88). Charge inversion of Arg-175 generates an "activated" form of arrestin that demonstrates phosphorylation-independent receptor binding (76). These findings and others provide the basis for a model of GPCR-arrestin interaction that is initiated by the activation- and phosphorylation-dependent binding of arrestin to the receptor, resulting in a conformational change in arrestin that promotes additional interactions with the receptor (76, 77). These studies also identified a phosphorylation switch residue (Arg-175) that is critical in sensing the receptor's phosphorylation state (71, 76, 77). The conserved nature of this mechanism is supported by the observation that mutation of the comparable residue in arrestin-2 (Arg-169) also results in an activated arrestin that binds GPCRs in a phosphorylation-independent manner (78).

On the basis of biophysical and biochemical data, a model for arrestin-receptor interaction has been proposed (58) (**Figure 3**). In this model, arrestin initially probes the functional state of the receptor and binds via its activation and phosphorylation recognition domains. Only when the receptor is activated and phosphorylated can arrestin achieve a high-affinity interaction in which the activation and phosphorylation recognition sites simultaneously engage the receptor. Lysine residues (Lys-14 and Lys-15 and perhaps others) located in the N terminus are suggested to guide receptor-attached phosphates toward the phosphorylation recognition domain and trigger the phosphate-sensitive residue, Arg-175, initiating arrestin activation. The highly charged phosphate groups are proposed to disrupt the buried polar core of arrestin, releasing the constraints imposed by the C-tail on both the N- and C-domains. Phosphate binding to Lys-14 and Lys-15 is also proposed to disrupt intramolecular contacts between β-strand 1, α-helix 1, and the C-tail, further destabilizing the basal conformation of arrestin. The various domains then structurally rearrange and transform arrestin into the activated state, optimizing arrestin-receptor interactions and facilitating arrestin interaction with endocytic proteins.

Arrestin Interaction with Clathrin and AP-2

Multiple studies have revealed that the interaction of arrestin with clathrin and AP-2, the two major proteins in CCPs, is important in nonvisual arrestin–promoted GPCR endocytosis (7). Arrestin-2 and -3 interact specifically, stoichiometrically, and with high affinity with clathrin, and immunofluorescence analyses demonstrate that the activated β_2AR, arrestin-2, and clathrin all colocalize upon agonist addition (69). The primary clathrin-binding determinants in arrestins are localized to hydrophobic and acidic residues in the C-terminal region (residues 373–377 in arrestin-3) (90) (**Figure 2**). These residues constitute a clathrin box motif, LϕXϕD/E (where ϕ is a bulky hydrophobic residue and X represents any polar amino acid), which is a consensus sequence found in clathrin-binding proteins such as AP-2 (91), AP180 (92), amphiphysin (93), and epsin (94). The mutation of this motif in arrestin-3 and its deletion in arrestin-2 disrupt clathrin binding and receptor internalization (90, 95). A similar mutagenesis study on clathrin localized the arrestin binding site to the N-terminal domain of clathrin, specifically residues 89–100 at the extreme N terminus of the heavy chain of clathrin (96). Site-directed mutagenesis identified an invariant Glu-89 and two conserved lysines, Lys-96 and Lys-98, as critical residues in mediating arrestin interactions (96). These hydrophobic and basic residues in clathrin nicely complement the hydrophobic and acidic amino acids in arrestin suggested in mediating arrestin-clathrin interactions. A crystal structure of the clathrin terminal domain (residues 1–363) in complex with an arrestin-3 peptide (residues 369–381) supports the mutagenesis data and the predicted location of the arrestin-clathrin interaction site (97).

Nonvisual arrestins also interact directly with the endocytic adapter protein AP-2 (98, 95). AP-2 is a clathrin adapter protein that promotes the assembly of clathrin lattices and targets receptors to CCPs (99). The domain structure of AP-2 consists of four subunits (α, β2, μ2, and σ2); the β2 subunit (β2-adaptin) interacts with arrestins. Initially, a yeast two-hybrid assay and coimmunoprecipitation experiments demonstrated that β2-adaptin binds to nonvisual arrestins. Furthermore, immunoprecipitation studies in HEK293 cells suggested the formation of a multimeric complex containing the receptor, arrestin-3, and β2-adaptin (98). The β2-adaptin binding region on nonvisual arrestins was localized to the extreme C terminus (95, 100), whereas site-directed mutagenesis revealed that residues Phe-391 and Arg-395 of arrestin-2 were essential in β2-adaptin binding (95). More recent studies have identified an α-helical domain ($[DE]_nX_{1-2}FXX[FL]XXXR$) in nonvisual arrestins and other clathrin adapter proteins, such as ARH and epsin, that contribute to AP-2 binding (101) (**Figure 2*b***). Mutagenesis reveals that Arg-834, Trp-841, Glu-849, Tyr-888, and Glu-902 in the β2-adaptin appendage domain are involved in arrestin binding (95, 102); based on the β2-adaptin/ARH peptide structure, additional residues are predicted to be involved (101). The potential transition of the C-terminal β2-adaptin binding region in arrestin—from a β-sheet in the holoprotein to an α-helix when associated with β2-adaptin—may function as a molecular switch to stabilize the activated conformation of arrestin (101).

Nonvisual arrestins bind to clathrin and AP-2 with high affinity in vitro but appear to be primarily localized in the cytoplasm in cells; this raises questions as to what regulates arrestin association with CCPs in vivo. Previous studies have suggested that the phosphorylation state of arrestin-2 regulates clathrin-mediated endocytosis of the β_2AR (103). Arrestin-2 is constitutively phosphorylated on residue Ser-412 by ERK1/2 in unstimulated HEK293 cells, resulting in reduced interaction with clathrin (103, 104). Upon receptor stimulation, arrestin is recruited to the plasma membrane (103). There, it is dephosphorylated, which is required for targeting the activated receptor-arrestin signaling complex to CCPs (103). A similar mechanism may occur for arrestin-3, for which phosphorylation on Thr-383 in rat arrestin-3 (Thr-382 in bovine) by casein kinase II inhibits arrestin-promoted receptor trafficking (105, 106).

As discussed above, intramolecular contacts of the β2-adaptin binding domain in arrestin may effectively attenuate arrestin-AP-2 association until arrestin is activated (101). Indeed, activation of arrestin-3 by mutation of Arg-170 dramatically enhances the binding of arrestin-3 to β2-adaptin (95). Similarly, the addition of a phosphorylated V2 vasopressin peptide to arrestin-3 enhances the binding of arrestin-3 to clathrin in vitro (107). Thus, the binding of arrestin to a phosphorylated activated GPCR may be the primary switch to promote arrestin dephosphorylation as well as conformational changes that unmask clathrin and AP-2 binding sites.

Arrestin Interaction with Phosphoinositides

Arrestins mediate GPCR regulation by interacting with a diverse array of intracellular proteins. However, like other endocytic machinery such as AP-2 and AP180, arrestins also bind phosphoinositides. Arrestin-2 and -3 contain a high-affinity phosphoinositide binding site located at their C-domain, where three basic amino acids (Lys-233, Arg-237, and Lys-251 in arrestin-3) are implicated in phosphoinositide binding (108). Mutation of these three residues in arrestin-3 (arrestin3-KRK/Q) failed to support β_2AR internalization and receptor recruitment to CCPs. Thus, phosphoinositides may be important in delivering the receptor-arrestin complex to endocytosis sites. The arrestin3-KRK/Q mutant, however, retains both receptor and clathrin binding abilities and is recruited to the plasma membrane upon receptor activation (108). Furthermore, an X-ray crystal structure of arrestin-2 in complex with the inositol polyphosphate hexakisphosphate (IP_6) confirmed the location of phosphoinositide binding (109) (**Figure 2*b***). The major visual arrestin in *Drosophila*, Arr2, also contains a highly electropositive patch of residues located at its C-domain that is involved in phosphoinositide binding (110). Alteration of this site interfered with Arr2 trafficking in photoreceptor cells effecting light adaptation.

The binding affinities of various phosphoinositides for arrestin-2 and -3 have the following relative affinities: IP_6 (~0.08 μM) > PIP_3 (~0.3 μM) > PIP_2 (~1.4 μM) > IP_4 (~4 μM) > IP_3 (~20 μM) (108). Although PIP_2 and PIP_3 are the proposed physiological ligands for arrestins at the plasma membrane (108, 110), it is interesting that soluble IP_6 displays a higher binding affinity for nonvisual arrestins than does either PIP_2 or PIP_3 (108). IP_6 is abundant in cells with concentrations ranging between 15–100 μm and may regulate receptor endocytosis and receptor signaling (111). IP_6 binding to arrestin inhibited both arrestin-2 and -3 binding to light-activated, phosphorylated rhodopsin (108). Moreover, the ability of IP_6 to bind to two distinct sites on nonvisual arrestins appears to mediate homo- and hetero-oligomerization of arrestin-2 and arrestin-3 (109). Mutation of either IP_6 binding site in arrestin-2 disrupted oligomerization, whereas interactions with known binding partners, including clathrin, AP-2, and ERK2, were maintained. Subcellular localization studies showed that arrestin-2 oligomers are primarily cytoplasmic, whereas arrestin-2 monomers display increased nuclear localization (109, 112). This suggests that IP_6 binding to arrestin may regulate arrestin localization and negatively regulate the interaction of arrestin with plasma membrane and nuclear signaling proteins.

ARRESTIN AS AN ENDOCYTIC SCAFFOLD

Role of Arrestin in Novel Endocytic Mechanisms

In addition to their conventional functions, arrestins can scaffold other proteins implicated in endocytosis (**Figure 1**), such as ARF proteins and their guanine nucleotide exchange factors (GEFs) and GTPase-activating proteins (GAPs) as well as potentially p38 MAPK and PI3K. ARF proteins are monomeric guanine nucleotide–binding proteins that comprise a subgroup of the Ras superfamily and are divided into three main classes. Class I includes ARF1, ARF2, and ARF3, which are involved in vesicle coat protein recruitment to specific intracellular compartments. Class II includes ARF4 and ARF5, whose functions are not well defined. Class III consists of ARF6, which resides in distinct endosomal structures as well as at the plasma membrane. ARF6, which regulates phospholipase D, phosphoinositol 4-phosphate-5 kinase (PI_4P-5K), cytoskeletal dynamics, and likely vesicle coat protein recruitment, functions in regulated exocytosis, phagocytosis, cytokinesis, adhesion and migration, and endocytic trafficking (113, 114).

Researchers have recently shown that ARF6 and one of its GEFs (ARNO) and GAPs (GIT) regulate GPCR desensitization, internalization, and trafficking via clathrin-dependent and/or clathrin-independent pathways. Notably, in vitro ARNO and GIT are active on both ARF1 and ARF6 (115–117). Hunzicker-Dunn and colleagues (118, 119) have proposed a model whereby the desensitization of endogenous leutinizing hormone/choriogonadotropin (LH/CG) receptor in isolated ovarian follicular membranes requires activation of ARF6 and ARNO, resulting in the release of arrestin-2 from a plasma membrane docking site and binding to the LH/CG receptor (120–122). This desensitization pathway has been recapitulated for heterologous LH/CG receptors expressed in HEK293 cells (123).

Lefkowitz, Premont, and colleagues (124–127) demonstrated that internalization of heterologous β_2AR in HEK293 cells is inhibited by GIT1 and GIT2 (124–127) and promoted by ARNO (126). They suggested that these findings applied to GPCRs internalized via a clathrin- and arrestin-dependent pathway (β_1AR, A_{2B}, μ-opioid, M1) but not to those following other routes (M2, ET_B) (125). In contrast to these observations with GEFs and

GAPs, direct manipulation of ARF6 by siRNA affected the internalization of GPCRs that traffic through both clathrin-dependent pathways (β_2AR, V2) and clathrin-independent pathways (M2, ET_B) (128). This is consistent with what is observed upon the expression of ARF6 activation mutants (126, 129).

For the β_2AR, the ARF6 pathway is clathrin dependent and may involve an arrestin/ARNO scaffolding complex that associates with ARF6-GDP upon receptor activation and facilitates GDP/GTP exchange (126). For the M2 receptor, the ARF6 pathway has been equated with the clathrin- and arrestin-independent pathway in HeLa cells (129) that Donaldson and colleagues originally described for non-GPCRs such as MHC-I (130). However, for GPCRs, ARF6 dependency does not dictate an arrestin/clathrin- and caveolae-independent pathway, as indicated by ARF6-sensitive endocytosis of the β_2AR and ET_B receptor, which traffic via arrestin/CCP and caveolae pathways, respectively. Furthermore, cell-type variations surely exist, as evidenced by ARF6-dependent transferrin (Tf) receptor trafficking, a marker of the clathrin pathway, in CHO cells but not COS-7 and variably in HeLa and HEK293 cells. How GPCRs are sorted subsequent to ARF6-dependent internalization is beginning to emerge. Immunofluorescence studies suggest that for the M2 receptor, this pathway may merge with the clathrin pathway at the level of early endosomes ($Rab5^+$/TfR^+/$EEA\text{-}1^+$) in HeLa cells (129), consistent with observations for non-GPCRs such as MHC-I (130). The role of ARF proteins and their GEFs and GAPs in GPCR trafficking events is still emerging and may involve arrestin in a receptor-specific fashion. Whether arrestin functions in ARF6-stimulated AP-2 membrane recruitment (131, 132) is not known.

Other endocytic mechanisms that potentially involve arrestin include p38 MAPK–regulated endocytosis (133) and PI3K-regulated endocytosis (65). Upon stimulation of MOR, receptor internalization was blocked in $p38^{-/-}$ MEFs but rescued by exogenous p38 (133). In vitro, p38 phosphorylated the Rab5 effector EEA1 at Thr-1392, and an EEA1 phosphomimic T1392D mutant restored receptor internalization in $p38^{-/-}$ MEFs. Although arrestin's role in p38-dependent MOR endocytosis was not addressed, one study showed that arrestin-2 coimmunoprecipitated and colocalized with p38 signaling complexes upon stimulation of the platelet-activating factor receptor in neutrophils (134). Upon stimulation of the β_2AR, receptor internalization was blocked by PI3K mutants lacking protein kinase activity, and detailed analysis suggested that this was due to decreased phosphorylation of tropomyosin, a PI3K protein substrate (65). Although arrestin's role in PI3K-regulated endocytosis was not directly addressed, one study demonstrated that arrestin-2 complexed with and negatively regulated PI3K upon stimulation of PAR2 (135). As discussed above, GRK2 also interacts with PI3K and may mediate the translocation of PI3K to the β_2AR (63, 64).

Arrestin and Postendocytic Signaling

The scaffolding role of arrestins on endocytic-derived early endosome populations has been well characterized. Another chapter in this volume discusses this in detail (135a), and therefore the discussion here is brief. Following arrestin-dependent internalization of transfected or endogenous PAR2, a multiprotein endosomal arrestin complex consisting of PAR2, arrestin, raf-1, and phospho-ERK was identified by gel filtration, immunoprecipitation, and immunofluorescence (136). This complex was later found to be enriched in partially purified pseudopodia in which ERK activity was prolonged (137), implicated in PAR2-mediated chemotaxis of highly metastatic breast cancer MDA-MB-231 cells (38), and regulated by arrestin-receptor binding stability (138). This arrestin complex may also be enriched in tight junctions, where it may play a role in PAR2-mediated permeability of intestinal epithelial cells following stress and inflammation (139). Similarly, researchers have identified an endosomal arrestin complex consisting of AT_{1A} receptor, arrestin, raf, MEK, and phospho-ERK (140). This complex functions in decreased nuclear ERK (141) but sustained cytosolic ERK activation kinetics (142). Moreover, this complex is regulated by arrestin-receptor binding stability (143) and sustained arrestin ubiquitination (144), as also seen for the V2 receptor (144, 145). While the understanding of arrestin's role in GPCR internalization continues to advance, the function of arrestin in subsequent sorting and signaling compartments, such as endosomes, is emerging.

Arrestin and Postendocytic Recycling

The postendocytic fate of receptors is clearly influenced by C-tail domains and/or binding partners, as exemplified by chimeric studies in which the addition of specific receptor C-tails is sufficient to impart the recycling or degradative properties of the corresponding full-length receptor. Swapping experiments have been carried out between multiple GPCRs. For example, C-tails of the efficiently recycled receptors β_2AR, dopamine D1 (D1), μ-opioid, human LH, and SP/NK1 have been exchanged with, respectively, the C-tails of the inefficiently recycled receptors V2, dopamine D2 (D2), δ-opioid, rat LH, and PAR1. Although arrestin is not currently known to bind all identified motifs and proteins that bestow recycling versus degradation sorting properties, it clearly plays a role at receptor C-tails, as receptor Class A and Class B switching has also been observed upon C-tail swapping. This opens the potential for an integrated mechanism of postendocytic receptor sorting, part of which likely involves arrestin, and thus warrants further discussion.

Recycling sorting motifs that conform to a PDZ (PSD-95, Dlg, ZO-1) domain ligand have been identified at the extreme C terminus of the β_2AR (146) and β_1AR (147) as well as internally in the ET_A receptor (148, 149). Mutagenesis loss-of-function and chimeric gain-of-function studies showed that for βARs such sorting motifs are necessary and sufficient for efficient receptor recycling. Specific point mutations in this region not only disrupt recycling but also inhibit β_2AR binding to the recycling factors NSF (which lacks a PDZ domain) and NHERF/EBP50 (which contains a PDZ domain) (150, 151) and may contribute to altered cardiac myocyte contractility (152). How specific these motifs and factors are to receptor recycling is unclear. Specific point mutations in the PDZ domain–binding motifs also inhibit β_2AR and ET_A receptor internalization (148, 150). NSF and NHERF/EBP50 overexpression also affects β_2AR and TPβ receptor internalization (150, 153). How arrestin fits into this story is also not detailed. Arrestin binds in vitro and coimmunoprecipitates with NSF (154), and mutation of Ser-411 within the PDZ domain ligand of the β_2AR, a GRK5 substrate site in vitro, disrupts NHERF/EBP50 binding and receptor recycling (151).

The generality of recycling motifs and recycling factors is currently being explored. An extensive screen with GST fusions of GPCR C-tails demonstrated NSF binding to multiple receptors, and although of those tested only the β_2AR bound NHERF/EBP50 (155), previous evidence suggests that NHERF/EBP50 also binds other GPCRs (156), including parathyroid hormone (PTH), κ-opioid, and P2Y1 receptors (157–159). Sequence analysis has identified putative internal and distal C-tail PDZ ligand motifs in many GPCRs (148, 155). Additionally, novel recycling signals have been identified by deletion analysis for D1 (160), LH (161), and μ-opioid receptors (162), where grafting of these regions onto the δ-opioid receptor is sufficient to reroute it from a degradative to a recycling pathway. Furthermore, current studies indicate the presence of multiple recycling routes, one of which exhibits slow kinetics ("long" pathway), traverses perinuclear recycling/sorting endosomes, and utilizes Rab11. This has been demonstrated for multiple GPCRs (163), including TPβ (164), SP/NK1 (165), CXCR2 (166), muscarinic M4 (167), and V2 (12) receptors.

Intriguingly, endogenous arrestin has been implicated in GPCR recycling by two different knockdown methods, although a detailed mechanism is currently unknown. How this fits into the regulatory mechanisms imparted by the motifs and proteins just discussed remains open. In HEK293 cells, stably expressed arrestin antisense constructs led to significant defects in A_{2B} adenosine (A_{2B}) receptor recycling, which was restored effectively by either arrestin-2 or arrestin-3 reconstitution but at an increased rate by arrestin-3 (168). Stably expressed *N*-formyl peptide (nFMP) receptors internalize but do not recycle in arrestin-$2/3^{-/-}$ knockout MEFs, and recycling can also be restored by either arrestin-2 or -3 reconstitution (169). Whereas wild-type nFMP receptors require arrestin for recycling, partially phosphorylated receptor mutants do not (170), suggesting an additional level of regulation imparted by phosphorylation status (171). This is reminiscent of the "long" recycling pathway just mentioned and originally described for the V2 receptor, which may also require dephosphorylation (12). Following a common theme, increasing arrestin avidity for bradykinin type 2 receptors results in defective receptor recycling in COS-7 cells (172), suggesting that stable arrestin-receptor complexes negatively regulate receptor recycling, at least in part. As mentioned, this sorting mechanism is reflected in the Class A versus Class B classification system, in which Class B receptors that stably associate with arrestin in endosomes recycle poorly (or slowly) and are commonly targeted for degradation, whereas Class A receptors that dissociate from arrestin at or near the plasma membrane recycle rapidly and resensitize rather than degrade. There are, of course, exceptions, such as for the somatostatin receptor subtype sst_{2A} (173). Continued characterization of the GPCR C-tail and its binding partners will likely help to further clarify arrestin's role in receptor recycling.

Arrestin and Postendocytic Degradation

Receptor C-tail motifs, regulatory binding partners, and covalent modifications mediate postendocytic sorting of receptors to lysosomes. The tyrosine-based signal (YXXϕ), discussed above in the context of internalization, may also function as a lysosomal sorting motif for PAR1 (22). This canonical motif is present in multiple GPCRs (23), although its function has not yet been characterized. Regulatory proteins that bind to receptor C-tails and mediate postendocytic lysosomal sorting include SNX1 and GASP. SNX1, a mammalian homolog of the yeast retromer complex component Vps5p, was originally identified in a yeast two-hybrid screen to bind EGFR (174), associates with PAR1 in GST pulldown and coimmunoprecipitation experiments, and is required for lysosomal sorting and receptor degradation. Ectopic expression of the SNX1 C-terminal domain (175) and siRNA depletion of endogenous SNX1 (176) in HeLa cells markedly inhibited ligand-induced PAR1 degradation. In both cases PAR1 still internalized but subsequently accumulated in $EEA1^{+}$ endosomes, consistent with decreased sorting to a degradative pathway. Although wild-type PAR1 sorts normally in vps26 siRNA–treated cells (176), this mammalian retromer subunit assumes an arrestin fold consisting of two β-sandwich domains stabilized by a polar core (177). This raises the intriguing question of whether arrestin may be part of a SNX1-containing retromer complex and function to sort GPCRs to lysosomes.

The sorting protein GASP associates with the C-tails of δ-opioid receptor (DOR) (178, 179) and the D2 receptor (180), both of which normally traffic to lysosomes and fail to resensitize. In HEK293 cells, dominant negative cGASP rerouted dopamine-induced

D2 receptor trafficking (180) and etorphine-induced DOR trafficking (178), from a degradative to a recycling pathway. In ventral tegmental area rat brain slices, anti-GASP antibody altered D2 receptor responsiveness to quinpirole from nonresensitizing to partially resensitizing (180). SNX1 and GASP interact with other GPCR C-tail GST fusions (155), which for GASP may be regulated by conserved C-tail residues (179). However, whether SNX1 and GASP regulate trafficking of these receptors remains to be determined.

Ubiquitination also functions as a postendocytic lysosomal sorting signal and has been reviewed elsewhere (181). Briefly, nonubiquitinated lysine mutants of the β_2AR (182), CXCR4 (183), PAR2 (184), and NK1 receptor (185), all of which normally undergo activation-dependent ubiquitination, display unaltered internalization but defective degradation upon prolonged exposure to agonist. For the β_2AR, endogenous receptor ubiquitination is also decreased in MEFs from either arrestin-$3^{-/-}$ or arrestin-$2/3^{-/-}$ knockout mice. Additionally, although arrestin-3 is ubiquitinated and binds the E3 ubiquitin ligase MDM2, both proteins of which contribute to β_2AR internalization, the specific ligase that regulates β_2AR ubiquitination and degradation is currently unknown. The E3 ubiquitin ligases AIP4 and c-Cbl mediate receptor ubiquitination for CXCR4 (186) and PAR2 (184), respectively. The role of arrestin in these processes is currently being explored. Additionally, lysosomal trafficking of CXCR4 occurs through an Hrs- and a Vps4-sensitive pathway. Hrs and Vps4 are mammalian homologs of yeast vps molecules that traffic ubiquitinated cargo to the vacuole. For GPCRs, they appear to traffic both ubiquitinated (CXCR4) and nonubiquitinated (DOR) receptors (187). As mentioned above, arrestin's structure is strikingly similar to that of another vps protein, Vps26 (177). For the PAR2 and NK1 receptor, mutation of all predicted intracellular lysine residues not only impeded lysosomal targeting but also unmasked receptor recycling, suggesting that these residues and/or ubiquitination limit wild-type receptor resensitization.

Regulation of Non-GPCRs by Arrestin

Nonvisual arrestins have been implicated in the endocytosis and sorting of cell surface molecules other than classical GPCRs, such as Frizzled (Fz), Smoothened (Smo), Notch, growth factor and lipid receptors, and select ion channels. The interaction of Wnt family glycoproteins with Fz seven-transmembrane (7TM) cell surface receptors leads to Dishevelled (Dsh) recruitment and subsequent canonical signaling to β-catenin/Tcf-mediated transcription, which functions in cell proliferation and developmental cell fate determination. In HEK293 cells, recruitment of arrestin by Dsh was implicated in Fz receptor endocytosis whereby Dsh2 expression promoted arrestin-3 recruitment to Fz4-containing plasma membranes, Dsh2 enhanced arrestin-3 coimmunoprecipitation with Fz4, and arrestin-3 siRNA blocked Wnt5a/PMA-costimulated Fz4 endocytosis (188). The functional outcome of arrestin-3-mediated Fz internalization in mammalian cells is currently unknown. Notably, arrestin-null fly mutants have unaltered Fz receptor levels and localization in both wing and eye imaginal discs (189).

Interaction of Sonic Hedgehog (Shh) glycoproteins with Patched (Ptc) 12-transmembrane (12TM) cell surface receptors leads to relief of Smo 7TM receptor inhibition and subsequent signaling to Gli-mediated transcription, which also functions in cell proliferation and embryonic patterning. In HEK293 cells, the recruitment of arrestin by GRK2-mediated Smo phosphorylation was implicated in Smo endocytosis, whereby GRK2 siRNA decreased Smo phosphorylation, Smo-promoted arrestin-3 recruitment to plasma membrane was blocked by Ptc and enhanced by GRK2 coexpression, and either arrestin-3 or GRK2 siRNA abolished agonist-stimulated Smo endocytosis (190). In zebrafish, arrestin-3 was also implicated in Smo signaling, whereby arrestin-3 knockdown recapitulated Smo mutant phenotypes, which was rescued by arrestin-3 expression or knockdown of Su(fu), an inhibitory component of the Hedgehog pathway (191).

Binding of a juxtaposed cell surface ligand to Notch cell surface receptors leads to proteolytic release of its intracellular domain, which functions as a transcriptional cofactor during cell fate determination. In *Drosophila*, arrestin was implicated as a positive regulator of Deltex (Dx)-dependent Notch ubiquitination and degradation whereby null mutants of Kurtz (Krz), the single *Drosophila* homolog of mammalian nonvisual arrestins, had substantially elevated Notch protein levels and select markers of Notch activity (189). In addition, simultaneous but not single ectopic expression of Krz and Dx, a putative E3 ubiquitin ligase, reduced Notch protein levels, produced phenotypes consistent with Notch downregulation, and greatly enhanced Notch ubiquitination in S2 cells. Krz coimmunoprecipitated Notch only in the presence of Dx, and all three proteins colocalized to intracellular vesicles that are currently unidentified, suggesting that a ternary complex exists.

Arrestin has also been implicated in the regulation of other receptors that signal to ubiquitin ligases, such as the insulin-like growth factor-1 (IGF-1) receptor and Toll-like/interleukin-1 (TL/IL-1) receptors as well as other growth factor and lipid receptors, such as transforming growth factor-β (TGF-β) and low-density lipoprotein (LDL) receptors. IGF-1 receptors function in cell growth and aging. In cell lines P6 (mouse cells overexpressing IGF-1 receptor) and BE (human melanoma cells), arrestin was implicated as a positive regulator of MDM2-mediated IGF-1 receptor ubiquitination and degradation (192, 193). In these cell lines, the addition of arrestin-2 enhanced MDM2-mediated receptor ubiquitination in vitro and reduced receptor expression in cells, whereas the reduction of arrestin-2 by siRNA reduced both basal and ligand-dependent receptor ubiquitination and degradation (193). Conversely, arrestin-2 negatively regulates chronic insulin-induced MDM2-mediated insulin receptor substrate 1 (IRS-1) ubiquitination and degradation (194). Although arrestin coimmunoprecipitates with both IGF-1 and insulin receptors (195, 196), to date only arrestin-promoted IGF-1 receptor internalization has been demonstrated (196). However, insulin-induced arrestin-2 phosphorylation does impair β_2AR internalization (197).

For TL/IL-1 receptors, ligand binding leads to the recruitment and auto-ubiquitination of another E3 ubiquitin ligase, tumor necrosis factor receptor–associated factor 6 (TRAF6), and subsequent signaling to nuclear factor κB (NF-κB)- and AP-1-mediated transcription, which functions in the innate immune response. In cell systems, arrestin was implicated as a negative regulator of TL/IL-1 receptor signaling to TRAF6 ubiquitination and NF-κB activation. Arrestin expression in HEK293 cells reduced TRAF6 oligomerization and ubiquitination, TRAF6 ubiquitination was enhanced in MEFs lacking both arrestin-2 and arrestin-3 and in HeLa cells treated with arrestin siRNA, and arrestin expression reduced AP-1/NF-κB reporter activity and target gene induction (198). Arrestin may act directly on TRAF6, as suggested by time- and agonist-dependent coimmunoprecipitation of endogenous TRAF6 and arrestin. The role played by arrestin in TL/IL-1 receptor internalization is currently unknown.

For TGF-β receptors, ligand binding leads to the recruitment and phosphorylation of Smad transcription factors and subsequent signaling to TGF-β-responsive gene transcription. In HEK293 cells, arrestin was implicated in tonic TGF-β receptor endocytosis. GFP-tagged arrestin-3 redistributed from the cytoplasm to intracellular vesicles upon the expression of type III TGF-β receptors (TβRIII), TβRIII redistributed from the plasma membrane to intracellular vesicles upon GFP-tagged arrestin-3 expression, and arrestin-3 siRNA significantly reduced the internalization of endogenous TGF-β receptors as measured by radioligand binding (199). Interestingly, TGF-β-mediated antiproliferative signaling was slightly enhanced in keratinocytes from arrestin-3$^{-/-}$ knockout mice. Arrestin has also been implicated in the tonic internalization of LDL receptors, whereby arrestin-3 overexpression and knockdown enhanced and inhibited receptor internalization, respectively (200). Notably, in vitro–translated arrestin binds to the GST-LDL receptor.

Arrestin may have a broader role in the trafficking of integral plasma membrane proteins, as suggested by arrestin-3 regulation of the Na^+/H^+ exchanger 5 (NHE5) transporter (201). In vitro, arrestin associated with NHE5 by yeast two-hybrid and GST pull-down assays. In CHO cells, overexpressed arrestin coimmunoprecipitated and colocalized with NHE5 and decreased NHE5 cell surface levels.

FUTURE DIRECTIONS

Selectivity and Specificity of Arrestin Interactions

It is well appreciated that arrestin functions as both an adapter and scaffold for GPCR endocytosis through its binding activities with the receptor, AP-2, clathrin, and a growing list of endocytic machinery. Continued efforts to delineate further the selectivity and specificity of these interactions are needed. Factors that likely dictate regulated pairing or complexing with arrestin include the following. (*1*) Site-specific receptor phosphorylation by distinct GRKs (202–205) may differentially stabilize or recruit arrestin to receptors. (*2*) Region-specific receptor conformations stabilized by select agonists may differentially recruit arrestin to receptors. (*3*) Localization-dependent constraints, such as those imposed by the actin cytoskeleton, membrane curvature, or vesicle coating and uncoating, may influence arrestin binding activities. (*4*) Known arrestin modifications, including phosphorylation and ubiquitination, as well as protein motifs such as SH3 domains, may modulate arrestin binding interfaces. (*5*) Arrestin oligomerization and inositol phosphate binding may regulate select arrestin pools and availability. (*6*) Cell-type-dependent regulation of arrestin expression levels may contribute to binding equilibriums. (*7*) Compensatory mechanisms and functional redundancy may alter arrestin binding activities under select conditions. The relationship between the stability of arrestin interactions and the functional outcomes needs further research. For example, stable receptor-arrestin complexes contribute to sorting and signaling; do transient receptor-arrestin complexes function beyond desensitization and internalization? As another example, if arrestin does not internalize with receptor, does this dictate a "shorter" endocytic pathway, possibly plasma membrane delineated, allowing for rapid resensitization? Or do transient receptor-arrestin interactions function to shuttle specific machinery to or from sites of endocytosis?

Physiology of Arrestin Adapter and Scaffolding Functions

There is a need for the continued development of research paradigms that address the roles of arrestin and GPCR endocytosis in normal physiological processes (206, 207) ranging from ligand delivery to directed cell migration, as well as abnormal physiological processes ranging from pathogen uptake to tumor dissemination (208). An emerging theme is the presence of distinct arrestin pools with varied functional outputs. To date, arrestin is present not only diffusely in the cytosol and membrane but also at discrete sites such as CCPs, in multiple endosomal populations, and in the nucleus. What other plasma membrane or intracellular compartments contain arrestins, and what is the functional readout? Arrestin's role extends well beyond its function in GPCR endocytosis, as other chapters in this volume discuss (135a, 209–211). The use of mass spectrometry, fluorescence resonance energy transfer, MEFs from arrestin knockout mice, and siRNAs are proving extremely useful in dissecting the role of endogenous arrestins. Advances will likely continue as techniques are developed and refined and researchers are intrigued and challenged.

SUMMARY POINTS

1. For an abundance of GPCRs, arrestin functions as an adapter that targets receptors to CCPs through its AP-2 and clathrin binding activities. For a growing number of GPCRs, arrestin functions as a scaffold that complexes with molecules implicated in receptor internalization and sorting to recycling or degradation compartments.
2. The adapter role of arrestin in GPCR trafficking involves several integrated steps. First, a conformational change in arrestin is induced upon binding to activated phosphorylated receptor. This is driven by disruption of the arrestin polar core by phosphoserines and/or phosphothreonines on the receptor. Second, the conformational change in arrestin enhances its interaction with clathrin and AP-2 proteins. This is driven by exposure of clathrin and AP-2 binding determinants in the arrestin C-tail. Third, arrestin interacts with phosphoinositides, which modulates arrestin-promoted receptor trafficking. This is enabled by high-affinity phosphoinositide binding sites in the arrestin C-domain.
3. The scaffolding role of arrestin in GPCR trafficking involves several emerging steps. First, arrestin binds vesicular transport proteins, which function to mediate receptor internalization. Second, arrestin scaffolds signaling complexes, which function to localize signaling to specific compartments. Third, arrestin binds to receptor C-tails, a region that contains recycling and degradation motifs, that binds recycling and sorting factors, and that is modified by ubiquitination and phosphorylation. This may function to integrate different mechanisms of receptor trafficking.
4. Arrestin's role in receptor trafficking extends beyond GPCRs to other cell surface receptors. Continued efforts are needed to delineate the selectivity and specificity of arrestin interactions and to develop research paradigms that address the physiological role of arrestin-mediated receptor trafficking.

LITERATURE CITED

1. Schmid SL. 1997. Clathrin-coated vesicle formation and protein sorting: an integrated process. *Annu. Rev. Biochem.* 66:511–48
2. Kirchhausen T. 2000. Clathrin. *Annu. Rev. Biochem.* 69:699–727
3. Chini B, Parenti M. 2004. G-protein coupled receptors in lipid rafts and caveolae: How, when, and why do they go there? *J. Molec. Endocrin.* 32:325–38
4. Anderson RGW. 1998. The caveolae membrane system. *Annu. Rev. Biochem.* 67:199–225
5. Nichols BJ, Lippencott-Schwartz J. 2001. Endocytosis without clathrin coats. *Trends Cell Biol.* 11:406–12
6. Krupnick JG, Benovic JL. 1998. The role of receptor kinases and arrestins in G protein-coupled receptor regulation. *Annu. Rev. Pharmacol. Toxicol.* 38:289–319

7. Marchese A, Chen CA, Kim YM, Benovic JL. 2003. The ins and outs of G protein-coupled receptor trafficking. *Trends Biochem. Sci.* 28:369–76
8. Oakley RH, Laporte SA, Holt JA, Caron MG, Barak LS. 2000. Differential affinities of visual arrestin, βarrestin1, and βarrestin2 for G protein-coupled receptors delineate two major classes of receptors. *J. Biol. Chem.* 275:17201–10
9. Zhang J, Barak LS, Anborgh PH, Laporte SA, Caron MG, Ferguson SSG. 1999. Cellular trafficking of G protein-coupled receptor/β-arrestin endocytic complexes. *J. Biol. Chem.* 274:10999–11006
10. Oakley RH, Laporte SA, Holt JA, Barak LS, Caron MG. 2001. Molecular determinants underlying the formation of stable intracellular G protein-coupled receptor–β-arrestin complexes after receptor endocytosis. *J. Biol. Chem.* 276:19452–60
11. Oakley RH, Laporte SA, Holt JA, Barak LS, Caron MG. 1999. Association of β-arrestin with G protein-coupled receptors during clathrin-mediated endocytosis dictates the profile of receptor resensitization. *J. Biol. Chem.* 274:32248–57
12. Innamorati G, Le Gouill C, Balamotis M, Birnbaumer M. 2001. The long and the short cycle. Alternative intracellular routes for trafficking of G protein-coupled receptors. *J. Biol. Chem.* 276:13096–103
13. Shenoy SK, Drake MT, Nelson CD, Houtz DA, Xiao K, et al. 2006. β-Arrestin-dependent, G protein-independent ERK1/2 activation by the β2 adrenergic receptor. *J. Biol. Chem.* 281:1261–73
14. Shenoy SK, Lefkowitz RJ. 2005. Receptor-specific ubiquitination of β-arrestin directs assembly and targeting of seven-transmembrane receptor signalosomes. *J. Biol. Chem.* 280:15315–24
15. Shenoy SK, Lefkowitz RJ. 2003. Trafficking patterns of β-arrestin and G protein-coupled receptors determined by the kinetics of β-arrestin deubiquitination. *J. Biol. Chem.* 278:14498–506
16. Bonifacino JS, Traub LM. 2003. Signals for sorting of transmembrane proteins to endosomes and lysosomes. *Annu. Rev. Biochem.* 72:395–447
17. Scott MG, Benmerah A, Mutaner O, Marullo S. 2002. Recruitment of activated G protein-coupled receptors to pre-existing clathrin-coated pits in living cells. *J. Biol. Chem.* 277:3552–59
18. Santini F, Gaidarov I, Keen JH. 2002. G protein-coupled receptor/arrestin3 modulation of the endocytic machinery. *J. Cell Biol.* 156:665–76
19. Blanpied TA, Scott DB, Ehlers MD. 2002. Dynamics and regulation of clathrin coats at specialized endocytic zones of dendrites and spines. *Neuron* 36:435–49
20. Ohno H, Stewart J, Fournier MC, Bosshart H, Rhee I, et al. 1995. Interaction of tyrosine-based sorting signals with clathrin-associated proteins. *Science* 269:1872–75
21. Honig S, Ricotta D, Krauss M, Spate K, Spolaore B, et al. 2005. Phosphatidylinositol-(4,5)-bisphosphate regulates sorting signal recognition by the clathrin-associated adapter complex AP2. *Mol. Cell* 18:519–31
22. Paing MM, Temple BRS, Trejo J. 2004. A tyrosine-based sorting signal regulates intracellular trafficking of protease-activated receptor-1. *J. Biol. Chem.* 279:21938–47
23. Paing MM, Johnston CA, Siderovski DP, Trejo J. 2006. Clathrin adapter AP2 regulates thrombin receptor constitutive internalization and endothelial cell resensitization. *Mol. Cell. Biol.* 26:3231–42
24. Parent JL, Labrecque P, Rochdi MD, Benovic JL. 2001. Role of the differentially spliced carboxyl terminus in thromboxane A_2 receptor trafficking. Identification of a distinct motif for tonic internalization. *J. Biol. Chem.* 276:7079–85

25. Diviani D, Lattion AL, Abuin L, Staub O, Cotecchia S. 2003. The adapter complex 2 directly interacts with the β1b-adrenergic receptor and plays a role in receptor endocytosis. *J. Biol. Chem.* 278:19331–40
26. Fan GH, Yang W, Wang XJ, Qian Q, Richmond A. 2001. Identification of a motif in the carboxyl terminus of CXCR2 that is involved in adaptin 2 binding and receptor internalization. *Biochemistry* 40:791–800
27. Orsini MJ, Parent JL, Mundell SJ, Marchese A, Benovic JL. 2000. Trafficking of the HIV coreceptor CXCR4: role of arrestins and identification of residues in the C-terminal tail that mediate receptor internalization. *J. Biol. Chem.* 274:31076–86
28. Gabilondo AM, Hegler J, Krasel C, Boivin-Jahns V, Hein L, Lohse MJ. 1997. A dileucine motif in the C terminus of the β_2-adrenergic receptor is involved in receptor internalization. *Proc. Natl. Acad. Sci. USA* 94:12285–90
29. Robert J, Clauser E, Patit PX, Ventura MA. 2005. A novel C-terminal motif is necessary for the export of the vasopressin V1b/V3 receptor to the plasma membrane. *J. Biol. Chem.* 280:2300–8
30. Duvernay MT, Zhou F, Wu G. 2004. A conserved motif for the transport of G protein-coupled receptors from the endoplasmic reticulum to the cell surface. *J. Biol. Chem.* 279:30741–50
31. Zhao M, Wimmer A, Trieu K, DiScipio RG, Schraufstatter IU. 2004. Arrestin regulates MAPK activation and prevents NADPH oxidase-dependent death of cells expressing CXCR2. *J. Biol. Chem.* 279:49259–67
32. Kohout TA, Lin FT, Perry SJ, Conner DA, Lefkowitz RJ. 2001. β-Arrestin 1 and 2 differentially regulate heptahelical receptor signaling and trafficking. *Proc. Natl. Acad. Sci. USA* 98:1601–6
33. Su Y, Raghuwanshi SK, Yu Y, Nanney LB, Richardson RM, Richmond A. 2005. Altered CXCR2 signaling in β-arrestin-2-deficient mouse models. *J. Immunol.* 175:5396–402
34. Ahn S, Nelson CD, Garrison TR, Miller WE, Lefkowitz RJ. 2003. Desensitization, internalization, and signaling function of β-arrestins demonstrated by RNA interference. *Proc. Natl. Acad. Sci. USA* 100:1740–44
35. Cheng ZJ, Zhao J, Sun Y, Hu W, Wu YL, et al. 2000. β-Arrestin differentially regulates the chemokine receptor CXCR4-mediated signaling and receptor internalization, and this implicates multiple interaction sites between β-arrestin and CXCR4. *J. Biol. Chem.* 275:2479–85
36. Fong AM, Premont RT, Richardson RM, Yu YRA, Lefkowitz RJ, Patel DD. 2002. Defective lymphocyte chemotaxis in β-arrestin2- and GRK6-deficient mice. *Proc. Natl. Acad. Sci. USA* 99:7478–83
37. Sun Y, Cheng Z, Pei G. 2002. β-Arrestin2 is critically involved in CXCR4-mediated chemotaxis, and this is mediated by its enhancement of p38 MAPK activation. *J. Biol. Chem.* 277:49212–19
38. Ge L, Chenoy SK, Lefkowitz RJ, DeFea K. 2004. Constitutive protease-activated receptor-2-mediated migration of MDA MB-231 breast cancer cells requires both β-arrestin-1 and -2. *J. Biol. Chem.* 279:55419–24
39. Parent JL, Labrecque P, Orsini MJ, Benovic JL. 1999. Internalization of the TXA2 receptor α and β isoforms. *J. Biol. Chem.* 274:8941–48
40. Dale LB, Bhattacharya M, Seachrist JL, Anborgh PH, Ferguson SSG. 2001. Agonist-stimulated and tonic internalization of metabotropic glutamate receptor 1a in human embryonic kidney 293 cells: Agonist-stimulated endocytosis is β-arrestin1 isoform-specific. *Mol. Pharmacol.* 60:1243–53

41. Shapiro MJ, Coughlin SR. 1998. Separate signals for agonist-independent and agonist-triggered trafficking of protease-activated receptor-1. *J. Biol. Chem.* 273:29009–14
42. Parnot C, Miserey-Lenkei S, Bardin S, Corvol P, Clauser E. 2002. Lessons from constitutively active mutants of G protein-coupled receptors. *Trends Endocrinol. Metab.* 13:336–43
43. Barak LS, Oakley RH, Laporte SA, Caron MG. 2001. Constitutive arrestin-mediated desensitization of a human vasopressin receptor mutant associated with nephrogenic diabetes insipidus. *Proc. Natl. Acad. Sci. USA* 98:93–98
44. Whistler JL, Gerber BO, Meng EC, Baranski TJ, Von Zastrow M, Bourne HR. 2002. Constitutive activation and endocytosis of the complement factor 5a receptor: Evidence for multiple activated conformations of a G protein-coupled receptor. *Traffic* 3:866–77
45. Wilbanks AM, Laporte SA, Bohn LM, Barak LS, Caron MG. 2002. Apparent loss-of-function mutant GPCRs revealed as constitutively desensitized receptors. *Biochemistry* 41:11981–89
46. Fraile-Ramos A, Kledal TN, Pelchen-Matthews A, Bowers K, Schwartz TW, Marsh M. 2001. The human cytomegalovirus US28 protein is located in endocytic vesicles and undergoes constitutive endocytosis and recycling. *Mol. Biol. Cell* 12:1737–49
47. Leterrier C, Laine J, Darmon M, Boudin H, Rossier J, Lenkei Z. 2006. Constitutive activation drives compartment-selective endocytosis and axonal targeting of type 1 cannabinoid receptors. *J. Neurosci.* 26:3141–53
48. Mishra SK, Keyel PA, Hawryluk MJ, Agostinelli NR, Watkins SC, Traub LM. 2002. Disabled-2 exhibits the properties of a cargo-selective endocytic clathrin adaptor. *EMBO J.* 21:4915–26
49. Fan GH, Yang W, Sai J, Richmond A. 2002. Hsc/Hsp70 interacting protein (hip) associates with CXCR2 and regulates the receptor signaling and trafficking. *J. Biol. Chem.* 277:6590–97
50. He G, Gupta S, Yi M, Michaely P, Hobbs HH, Cohen JC. 2002. ARH is a modular adaptor protein that interacts with the LDL receptor, clathrin, and AP-2. *J. Biol. Chem.* 277:44044–49
51. Santolini E, Puri C, Salcini AE, Gagliani MC, Pelicci PG, et al. 2000. Numb is an endocytic protein. *J. Cell Biol.* 151:1345–51
52. Reiter E, Lefkowitz RJ. 2006. GRKs and β-arrestins: roles in receptor silencing, trafficking and signaling. *Trends Endocrinol. Metab.* 17:159–65
53. Deleted in proof
54. Scott MG, Le Rouzic E, Perianin A, Pierotti V, Enslen H, et al. 2002. Differential nucleocytoplasmic shuttling of β-arrestins. Characterization of a leucine-rich nuclear export signal in β-arrestin2. *J. Biol. Chem.* 277:37693–701
55. Wang P, Wu Y, Ge X, Ma L, Pei G. 2003. Subcellular localization of β-arrestins is determined by their intact N domain and the nuclear export signal at the C terminus. *J. Biol. Chem.* 278:11648–53
56. Kang J, Shi Y, Xiang B, Qu B, Su W, et al. 2005. A nuclear function of β-arrestin1 in GPCR signaling: regulation of histone acetylation and gene transcription. *Cell* 123:833–47
57. Santini F, Penn RB, Gagnon AW, Benovic JL, Keen JH. 2000. Selective recruitment of arrestin-3 to clathrin coated pits upon stimulation of G protein-coupled receptors. *J. Cell Sci.* 113:2463–70
58. Gurevich VV, Gurevich EV. 2006. The structural basis of arrestin-mediated regulation of G-protein-coupled receptors. *Pharmacol. Ther.* 110:465–502

59. Pitcher JA, Freedman NJ, Lefkowitz RJ. 1998. G protein-coupled receptor kinases. *Annu. Rev. Biochem.* 67:653–92
60. Tsuga H, Kameyama K, Haga T, Kurose H, Nagao T. 1994. Sequestration of muscarinic acetylcholine receptor m2 subtypes. Facilitation by G protein-coupled receptor kinase (GRK2) and attenuation by a dominant-negative mutant of GRK2. *J. Biol. Chem.* 269:32522–27
61. Ferguson SS. 2001. Evolving concepts in G protein-coupled receptor endocytosis: the role in receptor desensitization and signaling. *Pharmacol. Rev.* 53:1–24
62. Shiina T, Arai K, Tanabe S, Yoshida N, Haga T, et al. 2001. Clathrin box in G protein-coupled receptor kinase 2. *J. Biol. Chem.* 276:33019–26
63. Naga Prasad SV, Barak LS, Rapacciuolo A, Caron MG, Rockman HA. 2001. Agonist-dependent recruitment of phosphoinositide 3-kinase to the membrane by β-adrenergic receptor kinase 1. A role in receptor sequestration. *J. Biol. Chem.* 276:18953–59
64. Naga Prasad SV, Laporte SA, Chamberlain D, Caron MG, Barak L, Rockman HA. 2002. Phosphoinositide 3-kinase regulates β2-adrenergic receptor endocytosis by AP-2 recruitment to the receptor/β-arrestin complex. *J. Cell Biol.* 158:563–75
65. Naga Prasad SV, Jayatilleke A, Madamanchi A, Rockman HA. 2005. Protein kinase activity of phosphoinositide 3-kinase regulates β-adrenergic receptor endocytosis. *Nat. Cell Biol.* 7:785–96
66. Perrino C, Naga Prasad SV, Schroder JN, Hata JA, Milano C, Rockman HA. 2005. Restoration of β-adrenergic receptor signaling and contractile function in heart failure by disruption of the βARK1/phosphoinositide 3-kinase complex. *Circulation* 111:2579–87
67. Cant SH, Pitcher JA. 2005. G protein-coupled receptor kinase 2-mediated phosphorylation of ezrin is required for G protein-coupled receptor-dependent reorganization of the actin cytoskeleton. *Mol. Biol. Cell* 16:3088–99
68. Ferguson SS, Downey WE, Colapietro AM, Barak LS, Menard L, Caron MG. 1996. Role of β-arrestin in mediating agonist-promoted G protein-coupled receptor internalization. *Science* 271:363–66
69. Goodman OBJ., Krupnick JG, Santini F, Gurevich VV, Penn RB, et al. 1996. β-arrestin acts as a clathrin adaptor in endocytosis of the β2-adrenergic receptor. *Nature* 383:447–50
70. Granzin J, Wilden U, Choe HW, Labahn J, Krafft B, Buldt G. 1998. X-ray crystal structure of arrestin from bovine rod outer segments. *Nature* 391:918–21
71. Hirsch JA, Schubert C, Gurevich VV, Sigler PB. 1999. The 2.8 Å crystal structure of visual arrestin: a model for arrestin's regulation. *Cell* 97:257–69
72. Han M, Gurevich VV, Vishnivetskiy SA, Sigler PB, Schubert C. 2001. Crystal structure of β-arrestin at 1.9 Å. Possible mechanism of receptor binding and membrane translocation. *Structure* 9:869–80
73. Milano SK, Pace HC, Kim YM, Brenner C, Benovic JL. 2002. Scaffolding functions of arrestin-2 revealed by crystal structure and mutagenesis. *Biochemistry* 41:3321–28
74. Sutton RB, Vishnivetskiy SA, Robert J, Hanson SM, Raman D, et al. 2005. Crystal structure of cone arrestin at 2.3Å: evolution of receptor specificity. *J. Mol. Biol.* 354:1069–80
75. Vishnivetskiy SA, Hirsch JA, Velez MG, Gurevich YV, Gurevich VV. 2002. Arrestin's transition into the active receptor-binding state requires an extended interdomain hinge. *J. Biol. Chem.* 277:43961–67
76. Gurevich VV, Benovic JL. 1995. Visual arrestin binding to rhodopsin: diverse functional roles of positively charged residues within the phosphorylation-recognition region of arrestin. *J. Biol. Chem.* 270:6010–16

77. Gurevich VV, Benovic JL. 1997. Mechanism of phosphorylation-recognition by visual arrestin and the transition of arrestin into a high affinity binding state. *Mol. Pharmacol.* 51:161–69
78. Kovoor A, Celver J, Abdryashitov RI, Chavkin C, Gurevich VV. 1999. Targeted construction of phosphorylation-independent β-arrestin mutants with constitutive activity in cells. *J. Biol. Chem.* 274:6831–34
79. Vishnivetskiy SA, Paz CL, Schubert C, Hirsch JA, Sigler PB, Gurevich VV. 1999. How does arrestin respond to the phosphorylated state of rhodopsin? *J. Biol. Chem.* 274:11451–54
80. Shilton BH, McDowell JH, Smith WC, Hargrave PA. 2002. The solution structure and activation of visual arrestin studied by small-angle X-ray scattering. *Eur. J. Biochem.* 269:3801–9
81. Carter JM, Gurevich VV, Prossnitz ER, Engen JR. 2005. Conformational differences between arrestin2 and preactivated mutants as revealed by hydrogen exchange mass spectrometry. *J. Mol. Biol.* 351:865–78
82. Charest PG, Terrillon S, Bouvier M. 2005. Monitoring agonist-promoted conformational changes of β-arrestin in living cells by intramolecular BRET. *EMBO Rep.* 6:334–40
83. Gurevich VV, Benovic JL. 1992. Cell-free expression of visual arrestin. Truncation mutagenesis identifies multiple domains involved in rhodopsin interaction. *J. Biol. Chem.* 267:21919–23
84. Gurevich VV, Benovic JL. 1993. Visual arrestin interaction with rhodopsin. Sequential multisite binding ensures strict selectivity toward light-activated phosphorylated rhodopsin. *J. Biol. Chem.* 268:11628–38
85. Gurevich VV, Richardson RM, Kim CM, Hosey MM, Benovic JL. 1993. Binding of wild type and chimeric arrestins to the m2 muscarinic cholinergic receptor. *J. Biol. Chem.* 268:16879–82
86. Gurevich VV, Dion SB, Onorato JJ, Ptasienski J, Kim CM, et al. 1995. Arrestin interactions with G protein-coupled receptors. Direct binding studies of wild type and mutant arrestins with rhodopsin, β2-adrenergic, and m2 muscarinic cholinergic receptors. *J. Biol. Chem.* 270:720–31
87. Vishnivetskiy SA, Hosey MM, Benovic JL, Gurevich VV. 2004. Mapping the arrestin-receptor interface. Structural elements responsible for receptor specificity of arrestin proteins. *J. Biol. Chem.* 279:1262–68
88. Hanson SM, Gurevich VV. 2006. The differential engagement of arrestin surface charges by the various functional forms of the receptor. *J. Biol. Chem.* 281:3458–62
89. Vishnivetskiy SA, Schubert C, Climaco GC, Gurevich YV, Velez MG, Gurevich VV. 2000. An additional phosphate-binding element in arrestin molecule. Implications for the mechanism of arrestin activation. *J. Biol. Chem.* 275:41049–57
90. Krupnick JG, Goodman OBJ., Keen JH, Benovic JL. 1997. Arrestin/clathrin interaction. Localization of the clathrin binding domain of nonvisual arrestins to the carboxy terminus. *J. Biol. Chem.* 272:15011–16
91. Shih W, Gallusser A, Kirchhausen T. 1995. A clathrin-binding site in the hinge of the β2 chain of mammalian AP-2 complexes. *J. Biol. Chem.* 270:31083–90
92. Morris SA, Schroder S. Plessmann U, Weber K, Ungewickell E. 1993. Clathrin assembly protein AP180: primary structure, domain organization and identification of a clathrin binding site. *EMBO J.* 12:667–75

93. Ramjaun AR, McPherson PS. 1998. Multiple amphiphysin II splice variants display differential clathrin binding: identification of two distinct clathrin-binding sites. *J. Neurochem.* 70:2369–76
94. Drake MT, Downs MA, Traub LM. 2000. Epsin binds to clathrin by associating directly with the clathrin-terminal domain. Evidence for cooperative binding through two discrete sites. *J. Biol. Chem.* 275:6479–89
95. Kim YM, Benovic JL. 2002. Differential roles of arrestin-2 interaction with clathrin and adaptor protein 2 in G protein-coupled receptor trafficking. *J. Biol. Chem.* 277:30760–68
96. Goodman OBJ, Krupnick JG, Gurevich VV, Benovic JL, Keen JH. 1997. Arrestin/clathrin interaction. Localization of the arrestin binding locus to the clathrin terminal domain. *J. Biol. Chem.* 272:15017–22
97. ter Haar E, Harrison SC, Kirchhausen T. 2000. Peptide-in-groove interactions link target proteins to the β-propeller of clathrin. *Proc. Natl. Acad. Sci. USA* 97:1096–100
98. Laporte SA, Oakley RH, Zhang J, Holt JA, Ferguson SS, et al. 1999. The β2-adrenergic receptor/βarrestin complex recruits the clathrin adaptor AP-2 during endocytosis. *Proc. Natl. Acad. Sci. USA* 96:3712–17
99. Robinson MS, Bonifacino JS. 2001. Adaptor-related proteins. *Curr. Opin. Cell Biol.* 13:444–53
100. Laporte SA, Oakley RH, Holt JA, Barak LS, Caron MG. 2000. The interaction of β-arrestin with the AP-2 adaptor is required for the clustering of β2-adrenergic receptor into clathrin-coated pits. *J. Biol. Chem.* 275:23120–26
101. Edeling MA, Mishra SK, Keyel PA, Steinhauser AL, Collins BM, et al. 2006. Molecular switches involving the AP-2 β2 appendage regulate endocytic cargo selection and clathrin coat assembly. *Dev. Cell* 10:329–42
102. Laporte SA, Miller WE, Kim KM, Caron MG. 2002. β-Arrestin/AP-2 interaction in G protein-coupled receptor internalization: identification of a β-arrestin binding site in β2-adaptin. *J. Biol. Chem.* 277:9247–54
103. Lin FT, Krueger KM, Kendall HE, Daaka Y, Fredericks ZL, et al. 1997. Clathrin-mediated endocytosis of the β-adrenergic receptor is regulated by phosphorylation/dephosphorylation of β-arrestin1. *J. Biol. Chem.* 272:31051–57
104. Lin FT, Miller WE, Luttrell LM, Lefkowitz RJ. 1999. Feedback regulation of β-arrestin1 function by extracellular signal-regulated kinases. *J. Biol. Chem.* 274:15971–74
105. Lin FT, Chen W, Shenoy S, Cong M, Exum ST, Lefkowitz RJ. 2002. Phosphorylation of β-arrestin2 regulates its function in internalization of β_2-adrenergic receptors. *Biochemistry* 41:10692–99
106. Kim YM, Barak LS, Caron MG, Benovic JL. 2002. Regulation of arrestin-3 phosphorylation by casein kinase II. *J. Biol. Chem.* 277:16837–46
107. Xiao K, Shenoy SK, Nobles K, Lefkowitz RJ. 2004. Activation-dependent conformational changes in β-arrestin 2. *J. Biol. Chem.* 279:55744–53
108. Gaidarov I, Krupnick JG, Falck JR, Benovic JL, Keen JH. 1999. Arrestin function in G protein-coupled receptor endocytosis requires phosphoinositide binding. *EMBO J.* 18:871–81
109. Milano SK, Kim YM, Stefano FP, Benovic JL, Brenner C. 2006. Nonvisual arrestin oligomerization and cellular localization are regulated by inositol hexakisphosphate binding. *J. Biol. Chem.* 281:9812–23
110. Lee SJ, Xu H, Kang LW, Amzel LM, Montel C. 2003. Light adaptation through phosphoinositide-regulated translocation of *Drosophila* visual arrestin. *Neuron* 39:121–32
111. Sasakawa N, Sharif M, Hanley MR. 1995. Metabolism and biological activities of inositol pentakisphosphate and inositol hexakisphosphate. *Biochem. Pharmacol.* 50:137–46

112. Storez H, Scott MG, Issafras H, Burtey A, Benmerah A, et al. 2005. Homo- and hetero-oligomerization of β-arrestins in living cells. *J. Biol. Chem.* 280:40210–15
113. D'Souza-Schorey C, Chavrier P. 2006. ARF proteins: roles in membrane traffic and beyond. *Nat. Rev. Mol. Cell Biol.* 7:347–58
114. Donaldson JG. 2003. Multiple roles for Arf6: sorting, structuring, and signaling at the plasma membrane. *J. Biol. Chem.* 278:41573–76
115. Macia E, Chabre M, Franco M. 2001. Specificities for the small G proteins ARF1 and ARF6 of the guanine nucleotide exchange factors ARNO and EFA6. *J. Biol. Chem.* 276:24925–30
116. Frank S, Upender S, Hansen SH, Casanova JE. 1998. ARNO is a guanine nucleotide exchange factor for ADP-ribosylation factor 6. *J. Biol. Chem.* 273:23–27
117. Vitale N, Patton WA, Moss J, Vaughan M, Lefkowitz RJ, Premont RT. 2000. GIT proteins, a novel family of phosphatidylinositol 3,4,5-trisphosphate-stimulated GTPase-activating proteins for ARF6. *J. Biol. Chem.* 275:13901–6
118. Salvador LM, Mukherjee S, Kahn RA, Lamm MLG, Fazleabas AT, et al. 2001. Activation of the luteinizing hormone/choriogonadotropin hormone receptor promotes ADP ribosylation factor 6 activation in porcine ovarian follicular membranes. *J. Biol. Chem.* 276:33773–81
119. Mukherjee S, Casanova JE, Hunzicker-Dunn M. 2001. Desensitization of the luteinizing hormone/choriogonadotropin hormone receptor in ovarian follicular membranes is inhibited by catalytically inactive ARNO. *J. Biol. Chem.* 276:6524–28
120. Mukerjee S, Gurevich VV, Jones JCR, Casanova JE, Frank SR, et al. 2000. The ADP ribosylation factor nucleotide exchange factor ARNO promotes β-arrestin release necessary for luteinizing hormone/choriogonadotropin receptor desensitization. *Proc. Natl. Acad. Sci. USA* 97:5901–6
121. Mukherjee S, Palczewski K, Gurevich VV, Hunzicker-Dunn M. 1999. β-Arrestin-dependent desensitization of luteinizing hormone/choriogonadotropin receptor is prevented by a synthetic peptide corresponding to the third intracellular loop of the receptor. *J. Biol. Chem.* 274:12984–89
122. Mukherjee S, Palczewski K, Gurevich V, Benovic JL, Banga JP, Hunzicker-Dunn M. 1999. A direct role for arrestins in desensitization of the luteinizing hormone/choriogonadotropin receptor in porcine ovarian follicular membranes. *Proc. Natl. Acad. Sci. USA* 96:493–98
123. Mukherjee S, Gurevich VV, Preninger A, Hamm HE, Bader MF, et al. 2002. Aspartic acid 564 in the third cytoplasmic loop of the luteinizing hormone/choriogonadotropin receptor is crucial for phosphorylation-independent interaction with arrestin2. *J. Biol. Chem.* 277:17916–27
124. Premont RT, Claing A, Vitale N, Freeman JLR, Pitcher JA, et al. 1998. β2-Adrenergic receptor regulation by GIT1, a G protein-coupled receptor kinase-associated ADP ribosylation factor GTPase-activating protein. *Proc. Natl. Acad. Sci. USA* 95:14082–87
125. Claing A, Perry SJ, Achiriloaie M, Walker JKL, Albanesi JP, et al. 2000. Multiple endocytic pathways of G protein-coupled receptors delineated by GIT1 sensitivity. *Proc. Natl. Acad. Sci. USA* 97:1119–24
126. Claing A, Chen W, Miller WE, Vitale N, Moss J, et al. 2001. β-Arrestin-mediated ADP-ribosylation factor 6 activation and β2-adrenergic receptor endocytosis. *J. Biol. Chem.* 276:42509–13
127. Premont RT, Claing A, Vitale N, Perry SJ, Lefkowitz RJ. 2000. The GIT family of ADP-ribosylation factor GTPase-activating proteins: functional diversity of GIT2 through alternative splicing. *J. Biol. Chem.* 275:22373–80

128. Houndolo T, Boulay PL, Claing A. 2005. G protein-coupled receptor endocytosis in ADP-ribosylation factor 6-depleted cells. *J. Biol. Chem.* 280:5598–604
129. Delaney KA, Murph MM, Brown LM, Radhakrishna H. 2002. Transfer of M2 muscarinic acetylcholine receptors to clathrin-derived early endosomes following clathrin-independent endocytosis. *J. Biol. Chem.* 277:33439–46
130. Naslavsky N, Weigert R, Donaldson JG. 2003. Convergence of nonclathrin- and clathrin-derived endosomes involves Arf6 inactivation and changes in phosphoinositides. *Mol. Biol. Cell* 14:417–31
131. Paleotti O, Macia E, Luton F, Klein S, Partisani M, et al. 2005. The small G-protein $Arf6_{GTP}$ recruits the AP-2 adaptor complex to membranes. *J. Biol. Chem.* 280:21661–66
132. Krauss M, Kinuta M, Wenk MR, De Camilli P, Takei K, Haucke V. 2003. ARF6 stimulates clathrin/AP-2 recruitment to synaptic membranes by activating phosphatidylinositol phosphate kinase type Iγ. *J. Cell Biol.* 162:113–24
133. Mace G, Miaczynska M, Zerial M, Nebreda AR. 2005. Phosphorylation of EEA1 by p38 MAP kinase regulates μ opioid receptor endocytosis. *EMBO J.* 24:3235–46
134. McLaughlin NJD, Banerjee A, Kelher MR, Gamboni-Robertson F, Hamiel C, et al. 2006. Platelet-activating factor-induced clathrin-mediated endocytosis requires β-arrestin-1 recruitment and activation of the p38 MAPK signalosome at the plasma membrane for actin bundle formation. *J. Immunol.* 176:7039–50
135. Wang P, DeFea KA. 2006. Protease-activated receptor-2 simultaneously directs β-arrestin-1-dependent inhibition and $G\alpha_q$-dependent activation of phosphatidylinositol 3-kinase. *Biochemistry* 45:9374–85
135a. DeWire SM, Ahn S, Lefkowitz RJ, Shenoy SK. 2007. β-Arrestins and cell signaling. *Annu. Rev. Physiol.* 69:483–510
136. DeFea KA, Zalevsky T, Thoma MS, Dery O, Mullins RD, Bunnett NW. 2000. β-Arrestin-dependent endocytosis of proteinase-activated receptor 2 is required for intracellular targeting of activated ERK1/2. *J. Biol. Chem.* 148:1267–81
137. Ge L, Ly Y, Hollenberg M, DeFea K. 2003. A β-arrestin-dependent scaffold is associated with prolonged MAPK activation in pseudopodia during protease-activated receptor-2-induced chemotaxis. *J. Biol. Chem.* 278:34418–26
138. Stalheim L, Ding Y, Gullapalli A, Paing MM, Wolfe BL, et al. 2005. Multiple independent functions of arrestins in the regulation of protease-activated receptor-2 signaling and trafficking. *Mol. Pharmacol.* 67:78–87
139. Jacob C, Yang PC, Darmoul D, Amadesi S, Saito T, et al. 2005. Mast cell tryptase controls paracellular permeability. Role of PAR2 and β-arrestins. *J. Biol. Chem.* 280:31936–48
140. Luttrell LM, Roudabush FL, Choy EW, Miller WE, Field MF, et al. 2001. Activation and targeting of extracellular signal-regulated kinase by β-arrestin scaffolds. *Proc. Natl. Acad. Sci. USA* 98:2449–54
141. Tohgo A, Pierce KL, Choy EW, Lefkowitz RJ, Luttrell LM. 2002. β-Arrestin scaffolding of the ERK cascade enhances cytosolic ERK activity but inhibits ERK-mediated transcription following angiotensin AT_{1a} receptor stimulation. *J. Biol. Chem.* 277:9429–36
142. Ahn S, Shenoy SK, Wei H, Lefkowitz RJ. 2004. Differential kinetic and spatial patterns of β-arrestin and G protein-mediated ERK activation by the angiotensin II receptor. *J. Biol. Chem.* 279:35518–25
143. Wei H, Ahn S, Barnes WG, Lefkowitz RJ. 2004. Stable interaction between β-arrestin 2 and angiotensin type 1A receptor is required for β-arrestin 2-mediated activation of extracellular signal-regulated kinases 1 and 2. *J. Biol. Chem.* 279:48255–61

144. Shenoy SK, Lefkowitz RJ. 2005. Receptor-specific ubiquitination of β-arrestin directs assembly and targeting of seven-transmembrane receptor signalosomes. *J. Biol. Chem.* 280:15315–24
145. Tohgo A, Choy EW, Gesty-Palmer D, Pierce KL, Laporte S, et al. 2003. The stability of the G protein-coupled receptor-β-arrestin interaction determines the mechanism and functional consequences of ERK activation. *J. Biol. Chem.* 278:6258–67
146. Gage RM, Kim KA, Cao TT, von Zastrow M. 2001. A transplantable sorting signal that is sufficient to mediate rapid recycling of G protein-coupled receptors. *J. Biol. Chem.* 276:44712–20
147. Gage RM, Matveeva EA, Whiteheart SW, von Zastrow M. 2005. Type I PDZ ligands are sufficient to promote rapid recycling of G protein-coupled receptors independent of binding to NSF. *J. Biol. Chem.* 280:3305–13
148. Paasche JD, Attramadal T, Kristiansen K, Oksvold MP, Johansen HK, et al. 2005. Subtype-specific sorting of the ET_A endothelin receptor by a novel endocytic recycling signal for G protein-coupled receptors. *Mol. Pharmacol.* 67:1581–90
149. Trejo J. 2005. Internal PDZ ligands: novel endocytic recycling motifs for G protein-coupled receptors. *Mol. Pharmacol.* 67:1388–90
150. Cong M, Perry SJ, Hu LA, Hanson PI, Claing A, Lefkowitz RJ. 2001. Binding of the β2 adrenergic receptor to *N*-ethylmaleimide-sensitive factor regulates receptor recycling. *J. Biol. Chem.* 276:45145–52
151. Cao TT, Deacon HW, Reczek D, Bretscher A, von Zastrow M. 1999. A kinase-regulated PDZ-domain interaction controls endocytic sorting of the β2-adrenergic receptor. *Nature* 401:286–90
152. Xiang Y, Kobilka N. 2003. The PDZ-binding motif of the β2-adrenoceptor is essential for physiologic signaling and trafficking in cardiac myocytes. *Proc. Natl. Acad. Sci. USA* 100:10776–81
153. Rochdi MD, Parent JL. 2003. $G\alpha_q$-coupled receptor internalization specifically induced by $G\alpha_q$ signaling. Regulation by EBP50. *J. Biol. Chem.* 278:17827–37
154. McDonald PH, Cote NL, Lin FT, Premont RT, Pitcher JA, Lefkowitz RJ. 1999. Identification of NSF as a β-arrestin1-binding protein. *J. Biol. Chem.* 274:10677–80
155. Heydorn A, Sondergaard BP, Ersboll B, Holst B, Nielsen FC, et al. 2004. A library of 7TM receptor C-terminal tails. *J. Biol. Chem.* 279:54291–303
156. Weinman EJ, Hall RA, Friedman PA, Liu-Chen LY, Shenolikar S. 2006. The association of NHERF adaptor proteins with G protein-coupled receptors and receptor tyrosine kinases. *Annu. Rev. Physiol.* 68:491–505
157. Sneddon WB, Syme CA, Bisello A, Magyar CE, Rochdi MD, et al. 2003. Activation-independent parathyroid hormone receptor internalization is regulated by NHERF1 (EBP50). *J. Biol. Chem.* 278:43787–96
158. Li JG, Chen C, Liu-Chen LY. 2002. Ezrin-radixin-moesin-binding phosphoprotein-50/Na^+/H^+ exchanger regulatory factor (EBP50/NHERF) blocks U50,488H-induced down-regulation of the human κ opioid receptor by enhancing its recycling rate. *J. Biol. Chem.* 277:27545–52
159. Hall RA, Ostedgaard LS, Premont RT, Blitzer JT, Rahman N, et al. 1998. A C-terminal motif found in the β_2-adrenergic receptor, P2Y1 receptor, and cystic fibrosis transmembrane conductance regulator determines binding to the Na^+/H^+ exchanger regulatory factor family of PDZ proteins. *Proc. Natl. Acad. Sci. USA* 95:8496–501
160. Vargas GA, von Zastrow M. 2004. Identification of a novel endocytic recycling signal in the D1 dopamine receptor. *J. Biol. Chem.* 279:37461–69

161. Galet C, Hirakawa T, Ascoli M. 2004. The postendocytic trafficking of the human lutropin receptor is mediated by a transferable motif consisting of the C-terminal cysteine and an upstream leucine. *Mol. Endocrinol.* 18:434–46

162. Tanowitz M, von Zastrow M. 2003. A novel endocytic recycling signal that distinguishes the membrane trafficking of naturally occurring opioid receptors. *J. Biol. Chem.* 278:45978–86

163. Seachrist JL, Ferguson SS. 2003. Regulation of G protein-coupled receptor endocytosis and trafficking by Rab GTPases. *Life Sci.* 74:225–35

164. Hamelin E, Theriault C, Laroche G, Parent JL. 2005. The intracellular trafficking of the G protein-coupled receptor TPβ depends on a direct interaction with Rab11. *J. Biol. Chem.* 280:36195–205

165. Roosterman D, Cottrell GS, Schmidlin F, Steinhoff M, Bunnett NW. 2004. Recycling and resensitization of the neurokinin 1 receptor. Influence of agonist concentration and Rab GTPases. *J. Biol. Chem.* 279:30670–79

166. Fan GH, Lapierre LA, Goldenring JR, Sai J, Richmond A. 2004. Rab11-family interacting protein 2 and myosin Vb are required for CXCR2 recycling and receptor-mediated chemotaxis. *Mol. Biol. Cell.* 15:2456–69

167. Volpicelli LA, Lah JJ, Fang G, Goldenring JR, Levey AI. 2002. Rab11a and myosin Vb regulate recycling of the M_4 muscarinic acetylcholine receptor. *J. Neurosci.* 22:9776–84

168. Mundell SJ, Matharu AL, Kelly E, Benovic JL. 2000. Arrestin isoforms dictate differential kinetics of A_{2B} adenosine receptor trafficking. *Biochemistry* 39:12828–36

169. Vines CM, Revankar CM, Maestas DC, LaRusch LL, Cimino DF, et al. 2003. *N*-Formyl peptide receptors internalize but do not recycle in the absence of arrestins. *J. Biol. Chem.* 278:41581–84

170. Key TA, Vines CM, Wagener BM, Gurevich VV, Sklar LA, Prossnitz ER. 2005. Inhibition of chemoattractant *N*-formyl peptide receptor trafficking by active arrestins. *Traffic* 6:87–99

171. Boulay F, Rabiet MJ. 2005. The chemoattractant receptors FPR and C5aR: same functions—different fates. *Traffic* 6:83–86

172. Simaan M, Bedard-Goulet S, Fessart D, Gratton JP, Laporte SA. 2005. Dissociation of β-arrestin from internalized bradykinin B2 receptor is necessary for receptor recycling and resensitization. *Cell Signal.* 17:1074–83

173. Tulipano G, Stumm R, Pfeiffer M, Kreienkamp HJ, Hollt V, Schulz S. 2004. Differential β-arrestin trafficking and endosomal sorting of somatostatin receptor subtypes. *J. Biol. Chem.* 279:21374–82

174. Kurten RC, Cadena DL, Gill GN. 1996. Enhanced degradation of EGF receptors by a sorting nexin, SNX1. *Science* 272:1008–10

175. Wang Y, Zhou Y, Szabo K, Haft CR, Trejo J. 2002. Down-regulation of protease-activated receptor-1 is regulated by sorting nexin 1. *Mol. Biol. Cell* 13:1965–76

176. Gullapalli A, Wolfe BL, Griffin CT, Magnuson T, Trejo J. 2006. An essential role of SNX1 in lysosomal sorting of protease-activated receptor-1: evidence for retromer-, hrs-, and tsg101-independent functions of sorting nexins. *Mol. Biol. Cell* 17:1228–38

177. Shi H, Rojas R, Bonifacino JS, Hurley JH. 2006. The retromer subunit vps26 has an arrestin fold and binds vps35 through its C-terminal domain. *Nat. Struct. Mol. Biol.* 13:540–46

178. Whistler JL, Enquist J, Marley A, Fong J, Gladher F, et al. 2002. Modulation of postendocytic sorting of G protein-coupled receptors. *Science* 297:615–20

179. Simonin F, Karcher P, Boeuf JJM, Matifas A, Kieffer BL. 2004. Identification of a novel family of G protein-coupled receptor associated sorting proteins. *J. Neurochem.* 89:766–75
180. Bartlett SE, Enquist J, Hopf FW, Lee JH, Gladher F, et al. 2005. Dopamine responsiveness is regulated by targeted sorting of D2 receptors. *Proc. Natl. Acad. Sci. USA* 102:11521–26
181. Wojcikiewicz RJH. 2004. Regulated ubiquitination of proteins in GPCR-initiated signaling pathways. *Trends Pharmacol. Sci.* 25:35–41
182. Shenoy SK, McDonald PH, Kohout TA, Lefkowitz RJ. 2001. Regulation of receptor fate by ubiquitination of activated β_2-adrenergic receptor and β-arrestin. *Science* 294:1307–13
183. Marchese A, Benovic JL. 2001. Agonist-promoted ubiquitination of the G protein-coupled receptor CXCR4 mediates lysosomal sorting. *J. Biol. Chem.* 276:45509–12
184. Jacob C, Cottrell GS, Gehringer D, Schmidlin F, Grady EF, Bunnett NW. 2005. c-Cbl mediates ubiquitination, degradation, and down-regulation of human protease-activated receptor 2. *J. Biol. Chem.* 280:16076–87
185. Cottrell GS, Padilla B, Pikios S, Roosterman D, Steinhoff M, et al. 2006. Ubiquitin-dependent down-regulation of the neurokinin-1 receptor. *J. Biol. Chem.* 281:27773–83
186. Marchese A, Raiborg C, Santini F, Keen JH, Stenmark H, Benovic JL. 2003. The E3 ubiquitin ligase AIP4 mediates ubiquitination and sorting of the G protein-coupled receptor CXCR4. *Dev. Cell* 5:709–22
187. Hislop JN, Marley A, von Zastrow M. 2004. Role of mammalian vacuolar protein-sorting proteins in endocytic trafficking of a nonubiquitinated G protein-coupled receptor to lysosomes. *J. Biol. Chem.* 279:22522–31
188. Chen W, ten Berge D, Brown J, Ahn S, Hu LA, et al. 2003. Dishevelled 2 recruits β-arrestin 2 to mediate Wnt5A-stimulated endocytosis of frizzled 4. *Science* 301:1391–94
189. Mukherjee A, Veraksa A, Bauer A, Rosse C, Camonis J, Artavanis-Tsakonas S. 2005. Regulation of Notch signaling by nonvisual β-arrestin. *Nat. Cell Biol.* 7:1191–201
190. Chen W, Ren XR, Nelson CD, Barak LS, Chen JK, et al. 2004. Activity-dependent internalization of smoothened mediated by β-arrestin 2 and GRK2. *Science* 306:2257–60
191. Wilbanks AM, Fralish GB, Kirby ML, Barak LS, Li YX, Caron MG. 2004. β-Arrestin 2 regulates zebrafish development through the hedgehog signaling pathway. *Science* 306:2264–67
192. Girnita L, Girnita A, Larsson O. 2003. Mdm2-dependent ubiquitination and degradation of the insulin-like growth factor 1 receptor. *Proc. Natl. Acad. Sci. USA* 100:8247–52
193. Grinita L, Shenoy SK, Sehat B, Vasilcanu R, Girnita A, et al. 2005. β-Arrestin is crucial for ubiquitination and downregulation of the insulin-like growth factor-1 receptor by acting as adapter for the MDM2 E3 ligase. *J. Biol. Chem.* 280:24412–19
194. Usui I, Imamura T, Huang J, Satoh H, Shenoy SK, et al. 2004. β-Arrestin-1 competitively inhibits insulin-induced ubiquitination and degradation of insulin receptor substrate 1. *Mol. Cell. Biol.* 24:8929–37
195. Dalle S, Ricketts W, Imamura T, Vollenweider P, Olefsky JM. 2001. Insulin and insulin-like growth factor I receptors utilize different G protein signaling components. *J. Biol. Chem.* 276:15688–95
196. Lin FT, Daaka Y, Lefkowitz RJ. 1998. β-Arrestins regulate mitogenic signaling and clathrin-mediated endocytosis of the insulin-like growth factor I receptor. *J. Biol. Chem.* 273:31640–43

197. Hupfeld CJ, Resnik JL, Ugi S, Olefsky JM. 2005. Insulin-induced β-arrestin1 ser-412 phosphorylation is a mechanism for desensitization of ERK activation by $G\alpha_i$-coupled receptors. *J. Biol. Chem.* 280:1016–23
198. Wang Y, Tang Y, Teng L, Wu Y, Zhao X, Pei G. 2006. Association of β-arrestin and TRAF6 negatively regulates Toll-like receptor-interleukin 1 receptor signaling. *Nat. Immunol.* 7:139–47
199. Chen W, Kirkbride KC, How T, Nelson CD, Mo J, et al. 2003. β-Arrestin 2 mediates endocytosis of type III TGF-β receptor and down-regulation of its signaling. *Science* 301:1394–97
200. Wu JH, Peppel K, Nelson CD, Lin FT, Kohout TA, et al. 2003. The adapter protein β-arrestin2 enhances endocytosis of the low density lipoprotein receptor. *J. Biol. Chem.* 278:44238–45
201. Szabo EZ, Numata M, Lukashova V, Lannuzzi P, Orlowski J. 2005. β-Arrestins bind and decrease cell-surface abundance of the Na^+/H^+ exchanger NHE5 isoform. *Proc. Natl. Acad. Sci. USA* 102:2790–95
202. Kim J, Ahn S, Ren XR, Whalen EJ, Reiter E, et al. 2005. Functional antagonism of different G protein-coupled receptor kinases for β-arrestin-mediated angiotensin II receptor signaling. *Proc. Natl. Acad. Sci. USA* 102:1442–47
203. Ren XR, Reiter E, Ahn S, Kim J, Chen W, Lefkowitz RJ. 2005. Different G protein-coupled receptor kinases govern G protein and β-arrestin-mediated signaling of V2 vasopressin receptor. *Proc. Natl. Acad. Sci. USA* 102:1448–53
204. Violin JD, Ren XY, Lefkowitz RJ. 2006. G protein-coupled receptor kinase specificity for β-arrestin recruitment to the β2-adrenergic receptor revealed by fluoresence resonance energy transfer. *J. Biol. Chem.* 281:20577–88
205. Potter RM, Maestas DC, Cimino DF, Prossnitz ER. 2006. Regulation of *N*-formyl peptide receptor signaling and trafficking by individual carboxyl-terminal serine and threonine residues. *J. Immunol.* 176:5418–25
206. Palmitessa A, Hess HA, Bany IA, Kim YM, Koelle MR, Benovic JL. 2005. *Caenorhabditis elegans* arrestin regulates neural G protein signaling and olfactory adaptation and recovery. *J. Biol. Chem.* 280:24649–62
207. Scherrer G, Tryoen-Toth P, Filliol D, Matifas A, Laustriat D, et al. 2006. Knockin mice expressing fluorescent delta-opioid receptors uncover G protein-coupled receptor dynamics in vivo. *Proc. Natl. Acad. Sci. USA* 103:9691–96
208. Buchanan FG, Gorden DL, Matta P, Shi Q, Matrisian LM, DuBois RN. 2006. Role of beta-arrestin 1 in the metastatic progression of colorectal cancer. *Proc. Natl. Acad. Sci. USA* 103:1492–97
209. Premont RT, Gainetdinov RR. 2007. Physiological roles of G protein-coupled receptor kinases and arrestins. *Annu. Rev. Physiol.* 69:in press
210. DeFea K. 2007. Stop that cell! β-Arrestin-dependent chemotaxis: a tale of localized actin assembly and receptor desensitization. *Annu. Rev. Physiol.* 69:in press
211. Hupfeld CJ, Olefsky JM. 2007. Regulation of receptor tyrosine kinase by GRKs and β-arrestins. *Annu. Rev. Physiol.* 69:561–77

β-Arrestins and Cell Signaling

Scott M. DeWire,[1] Seungkirl Ahn,[1] Robert J. Lefkowitz,[1,2,3] and Sudha K. Shenoy[2]

[1] Howard Hughes Medical Institute and Departments of [2] Medicine and [3] Biochemistry, Duke University Medical Center, Durham, North Carolina 27710; email: lefko001@receptor-biol.duke.edu

Annu. Rev. Physiol. 2007. 69:483–510

The *Annual Review of Physiology* is online at http://physiol.annualreviews.org

This article's doi: 10.1146/annurev.physiol.69.022405.154749

0066-4278/07/0315-0483$20.00

Key Words

seven-transmembrane receptor, G protein–coupled receptor kinase, scaffold, phosphorylation, ERK, MAPK

Abstract

Upon their discovery, β-arrestins 1 and 2 were named for their capacity to sterically hinder the G protein coupling of agonist-activated seven-transmembrane receptors, ultimately resulting in receptor desensitization. Surprisingly, recent evidence shows that β-arrestins can also function to activate signaling cascades independently of G protein activation. By serving as multiprotein scaffolds, the β-arrestins bring elements of specific signaling pathways into close proximity. β-Arrestin regulation has been demonstrated for an ever-increasing number of signaling molecules, including the mitogen-activated protein kinases ERK, JNK, and p38 as well as Akt, PI3 kinase, and RhoA. In addition, investigators are discovering new roles for β-arrestins in nuclear functions. Here, we review the signaling capacities of these versatile adapter molecules and discuss the possible implications for cellular processes such as chemotaxis and apoptosis.

7TMR: seven transmembrane receptor

GPCR: G protein–coupled receptor

GRK: G protein–coupled receptor kinase

β2AR: β-2 adrenergic receptor

Adapter: a protein that enhances cellular responses by recruiting key proteins into a complex

$AT1_AR$: angiotensin II receptor type 1A

β-ARRESTINS: DISCOVERY, BIOLOGY, AND CLASSICAL FUNCTIONS

The classic paradigm of signal transduction in response to stimulation of seven-transmembrane receptors (7TMRs), also known as G protein–coupled receptors (GPCRs), involves an agonist-induced conformational change that allows the receptor to interact with and dissociate the Gα from the Gβγ subunits of heterotrimeric G proteins (for a historical review, see Reference 1). Distinct subtypes of Gα proteins, such as Gαs, Gαq, and Gαi, signal through discrete pathways via second messenger molecules such as cyclic AMP, inositol triphosphate, diacylglycerol, and calcium. Members of a protein family known as the G protein–coupled receptor kinases (GRKs) initiate the termination of this signaling response (for reviews see References 2–4). GRKs rapidly phosphorylate the receptor, typically on its cytoplasmic tail. β-Arrestins then bind the phosphorylated receptor, which blocks further G protein–initiated signaling through a steric mechanism.

Discovery of β-Arrestins

The discovery of β-arrestins in the late 1980s resulted from the observation that increasingly pure preparations of GRK2 (then referred to as β adrenergic receptor kinase) progressively lost the ability to desensitize G protein activation in a reconstituted β-2 adrenergic receptor (β2AR) system (5). The loss of desensitization could be rescued by addition of high molar excesses of visual arrestin (S-antigen or 48-kDa protein) (6), which researchers had recently discovered to function together with rhodopsin kinase to desensitize rhodopsin signaling. This implied that a homologous protein could exist in nonretinal tissues (5). Molecular cloning confirmed this and revealed two isoforms of the hypothesized protein, termed β-arrestins 1 and 2 (or arrestins 2 and 3) because of their homology with arrestin and their functional regulation of the β2AR (7, 8).

β-Arrestins as Adapters for Internalization

β-Arrestins are expressed ubiquitously in all cells and tissues and function in the desensitization of most 7TMRs except rhodopsin. Although their role in the termination of signaling led to their discovery, later research appreciated that β-arrestins serve a second function in receptor internalization [for a review, see Benovic et al. (8a) in this volume]. By acting as adapters for β(2) adaptin, better known as AP2, and clathrin (9–11), β-arrestins bring activated receptors to clathrin-coated pits for endocytosis, a process critical for receptor recycling and degradation. β-Arrestins also bind to various other proteins implicated in receptor internalization. For example, β-arrestin2 is constitutively bound to the guanine nucleotide exchange factor ARNO (ARF nucleotide binding site opener) and serves as a switch to regulate the activity of the small G protein ARF6 (ADP-ribosylation factor 6), which is bound to β-arrestin2 only upon receptor stimulation (12, 13). When ARNO activates ARF6, the latter is released by β-arrestin2 and assists in the endocytosis of the receptor.

Isoform Differences Between β-Arrestins 1 and 2

The amino acid sequences of the two β-arrestin isoforms are 78% identical; most of the coding differences appear in the C termini. Knockout studies show that mice lacking either β-arrestin1 or -2 are viable (14, 15), whereas the double-knockout phenotype is embryonic lethal (16; R.J. Lefkowitz & F.T. Lin, unpublished data), implying that each β-arrestin functionally substitutes for the other isoform to some degree. However, the molecular studies reviewed here do not support redundant roles for all β-arrestin-mediated functions. For example, internalization of

Table 1 7TMRs and their β-arrestin-dependent properties[a]

Receptor		β-Arrestin recruitment	β-Arrestin-dependent ERK	β-Arrestin isoform regulation of ERK	Cell type	Reference
$AT_{1A}R$	Wild type	Class B	Yes	Reciprocal*	HEK-293	58
	Δ324	Class A	Yes	N.D.	HEK-293	**Figure 4**
β2AR	Wild type	Class A	Yes	Codependent	HEK-293	61
	GRK^-, PKA^-	None	No	N.A.	HEK-293	**Figure 4**
PAR2	Wild type	Class B	Yes	Codependent	HeLa	118
	C-tail deletion	Class A	No	N.A.	HeLa	
V2R		Class B	Yes	Reciprocal*	HEK-293	60
PTH1R		Class B	Yes	Codependent	HEK-293	62
CCR7		Class B	Yes	N.D.	HEK-293	67
PAR 1		N.D.	Yes	Reciprocal**	HEK-293	114

[a]Receptors are designated as recruiting β-arrestins in either a Class A (transient, weak interaction; results in rapid recycling of receptors after internalization) or Class B interaction (strong and long lasting; receptors internalize with β-arrestins, and thus slow recycling occurs). Next, receptors are classified as to whether or not the wild-type or mutant receptor retains the ability to activate ERK via β-arrestins. β-Arrestin isoform specificity for ERK activation is listed in the fourth column: Either both β-arrestin isoforms are required for ERK signaling (termed codependent), or only one particular isoform is required, and the other isoform serves to inhibit ERK signaling (termed reciprocal regulation). One asterick denotes that β-arrestin2 carries the signal to ERK and β-arrestin1 serves to inhibit. Two asterisks denote that β-arrestin1 functions to signal and β-arrestin2 inhibits. N.D., no data; N.A., not applicable.

some 7TMRs is mediated primarily by one isoform, as is the case for the β2AR via β-arrestin2, whereas for others, like the angiotensin II type 1A receptor ($AT1_AR$), both β-arrestin isoforms are equally capable (16, 17). Likewise, either β-arrestin isoform is capable of $AT1_AR$ second messenger desensitization, and the two can functionally substitute for each other, but for the protease-activated receptor 1 (PAR1), only β-arrestin1 can desensitize phosphoinositide turnover (18).

Two Patterns of β-Arrestin Recruitment

Upon 7TMR stimulation and subsequent β-arrestin recruitment to the cytoplasmic membrane, two patterns emerge. For some receptors, there is transient, low-affinity binding characterized by a rapid concentration of β-arrestin at the activated receptor. β-Arrestin is subsequently released after targeting the receptor to clathrin-coated pits. This pattern, termed Class A, is typified by the β2AR. Class A receptors typically undergo rapid recycling to the plasma membrane after their internalization. In contrast, Class B receptors, such as the $AT1_AR$, show a much stronger and more prolonged binding to β-arrestin, such that following recruitment to clathrin-coated pits, the receptor and β-arrestin remain bound together on the surface of endocytic vesicles (17). Because of their prolonged interaction with β-arrestins, Class B receptors recycle to the cell surface much more slowly than Class A receptors. **Table 1** summarizes β-arrestin recruitment patterns for various 7TMRs.

β-Arrestin Regulation by Phosphorylation and Ubiquitination

Mammalian arrestins exist in a constitutively phosphorylated state in the cytosol and are dephosphorylated upon binding activated 7TMRs at the plasma membrane. For β-arrestin1, serine 412 is the site of extracellular signal–regulated kinase (ERK) 1/2—mediated phosphorylation (19). In contrast, for β-arrestin2, which is phosphorylated by casein kinase II, threonine 383 is the primary phosphorylation site, and serine 361 represents a secondary site (20–21). Dephosphorylation of β-arrestins at the plasma membrane is necessary for engaging endocytic partners

ERK: extracellular signal–regulated kinase

Class A interaction: weak and transient interaction of β-arrestins and receptors after agonist binding; receptors undergo rapid recycling back to the membrane after internalization

Class B interaction: strong and long lasting interaction of β-arrestins and receptors after agonist binding; after β-arrestins and receptors internalize together, the receptors undergo slow recycling back to the membrane

Ubiquitination: the covalent attachment of ubiquitin, a 76-amino-acid-residue-containing protein, to lysine residues in the substrate protein

such as clathrin but does not seem to affect receptor binding or desensitization (22). In fact, S412D β-arrestin1, which mimics the phosphorylated form of β-arrestin1, acts as a dominant-negative mutant with respect to receptor internalization (22). Presumably, β-arrestins are rephosphorylated by kinases associated with the internalizing complexes.

β-Arrestins also undergo ubiquitination upon activation of several 7TMRs (23–26). Ubiquitination results from a three-step enzymatic process in which a small protein, ubiquitin, is covalently appended to lysine residues in the substrate proteins (27, 28). Ubiquitination is the result of the sequential action of three enzymes. First, the formation of a high-energy thioester bond by a ubiquitin-activating enzyme, E1, activates the COOH-terminal glycine residue of ubiquitin. Activated ubiquitin is then transferred to an active-site cysteine residue in an E2 (ubiquitin-carrying enzyme). The final step is catalyzed by E3, a ubiquitin protein ligase, which links the COOH terminus of ubiquitin to the ε-amino group of a lysine residue of the substrate protein.

Although originally discovered to function as a protein tag for destruction by the cellular proteasomal machinery, ubiquitin modification is now appreciated for mediating novel outcomes including protein trafficking and signal transduction (29–33). Agonist stimulation of the β2AR leads to β-arrestin ubiquitination, which is mediated by Mdm2, a RING domain containing E3 ubiquitin ligase. β-Arrestin ubiquitination is required for rapid receptor internalization (23). Remarkably, the nature and stability of receptor-arrestin complexes formed upon agonist stimulation reflect the status and longevity of β-arrestin ubiquitination. Stimulation of Class A receptors such as the β2AR leads to transient ubiquitination of β-arrestin (24, 26). In contrast, stimulation of Class B 7TMRs, which recruit β-arrestin to the plasma membrane and subsequently internalize as receptor-arrestin complexes into endosomes, results in sustained β-arrestin ubiquitination (24, 26). The phosphorylation status of the receptor carboxyl tail is one determinant of tight (Class B) versus loose (Class A) binding of β-arrestin thus governing the characteristic endosomal and plasma membrane recruitment patterns. However, these patterns of β-arrestin trafficking are also dictated by the attachment of ubiquitin moieties to specific lysines (e.g., lysines 11 and 12 for the class B $AT1_AR$) within β-arrestin2 (25). Furthermore, when ubiquitin is translationally fused to the C terminus of β-arrestin, endosomal localization occurs even with Class A receptors (24, 25). Thus, β-arrestin ubiquitination status likely plays a key role in determining the stability of the interactions of β-arrestin with 7TMRs and perhaps other elements of the endocytic machinery. These downstream interactions directed by the ubiquitination status of β-arrestin may in turn dictate the trafficking itinerary of endocytosed receptor complexes.

Agonist stimulation of 7TMRs such as the β2AR and the V2R also leads to the ubiquitination of the receptors themselves; surprisingly, β-arrestin2 but not β-arrestin1 is required for this process (23, 34). β-Arrestin2 likely acts as an adapter to bring one or more E3 ubiquitin ligases to the activated receptors. For the above 7TMRs, the specific E3 ligases that collaborate with β-arrestin remain to be elucidated. In contrast, the E3 ligases AIP4 and c-Cbl reportedly mediate the ubiquitination of two other mammalian 7TMRs, CXCR4 and PAR2, respectively (35–37). The exact role, if any, of β-arrestin in these systems remains to be determined.

Researchers recently have discovered a further role for β-arrestins in the mediation of ubiquitination and degradation of single-transmembrane receptors, namely the tyrosine kinase IGF-1R and the *Drosophila* Notch receptor. Although β-arrestin was previously reported to act as an endocytic adapter for the IGF1-R, its function in mediating receptor ubiquitination was only recently appreciated. Girnita et al. (38, 39) unexpectedly found that the β-arrestin1 isoform (unlike β-arrestin2, as indicated above for 7TMRs) acts as an essential adapter to bring the E3 ubiquitin ligase, Mdm2, to the receptor, leading to receptor ubiquitination and degradation. Recently, Mukherjee et al. (40) demonstrated that Kurtz, the unique nonvisual arrestin of *Drosophila*, regulates ubiquitin-dependent downregulation of Notch. Kurtz binds Deltex, a known Notch regulator and a putative E3 ubiquitin ligase for Notch, thus promoting the ubiquitination and degradation of Notch (40). Whether β-arrestin ubiquitination has any role in the signaling mechanisms of these receptors, and whether β-arrestin ubiquitination is even induced by such receptors, remain provocative questions.

Although arrestin homologs have been found only in metazoans, genome sequencing in nonmetazoans has revealed arrestin-like or arrestin domain–containing proteins (41). For *Aspergillus nidulans*, a model fungus, the arrestin-like protein PalF, a component of a fungal ambient pH signaling pathway, becomes phosphorylated and ubiquitinated upon activation of a 7TM receptor, PalH, that responds to alkaline pH. Moreover, PalF phosphorylation and ubiquitination are crucial in linking receptor internalization to the multivesicular body–sorting pathway (42). The pH signaling pathway elicited by the 7TM PalH depends on the arrestin-like protein and its posttranslational modifications but does not require any G protein activity (41). These capabilities of PalF are similar to those of the mammalian β-arrestins, for which phosphorylation and ubiquitination play crucial roles in distinct signaling pathways (see below).

β-ARRESTIN SIGNALING TO THE MITOGEN-ACTIVATED PROTEIN KINASES

In the past seven to eight years, a previously unappreciated function of β-arrestins has come to light: serving as scaffolds for numerous signaling networks. Although we are just beginning to understand some of the signals that β-arrestins regulate, the consequences of these signaling events remain largely a mystery. Below we review current literature relevant to this novel role of β-arrestins as signal transduction scaffolds, with particular attention to the mitogen-activated protein kinases (MAPKs).

MAPK: mitogen-activated protein kinase

JNK: cJun N-terminal kinase

Cellular signal transduction involves highly coordinated cascades of events. The number of possible downstream targets for any given member of a signaling network is vast, and to maintain integrity and specificity of signaling, cells employ molecular scaffolds. These are large chaperone complexes that hold together specific members of a signaling network to give them preferential access to one another, thus ensuring the fidelity of a particular signaling response. Recent considerable data show that, in addition to their classic roles in desensitization and internalization, β-arrestins can also act as signaling scaffolds for many pathways and, in particular, those of the MAPKs.

The MAPKs are a family of serine/threonine kinases that include ERK1/2 (also known as p44/p42MAPK), p38 kinases (isoforms α, β, γ, δ), and the c-Jun N-terminal kinases (JNK1, JNK2, JNK3). The downstream effectors of MAPKs control many cellular functions, including cell cycle progression, transcriptional regulation, and apoptosis. ERK1/2 activation exemplifies the prototypical MAPK signaling module: ERK1/2, a MAPK, is phosphorylated by MEK, a MAPK kinase (MAPKK). MEKs are phosphorylated by a variety of Raf isoforms, which are MAP triple kinases, or MAPKKKs. For each of the MAPK family members, this pattern holds true: A MAPKKK activates a MAPKK, which in turn activates a MAPK.

The First Evidence of β-Arrestin Signaling: c-Src recruitment

The first evidence that β-arrestins could act to facilitate signal transduction from 7TMRs came from studies in receptor internalization–defective systems (43–45). These studies reported that dominant-negative versions of dynamin and β-arrestin1 (45) or chemical blockade of clathrin-mediated internalization (43, 44) diminish receptor signaling to ERK1/2. Although there are a number of ways in which receptors can be internalized, these results suggested that internalization components of at least one of these systems is required for complete MAPK activation under some circumstances.

Shortly thereafter, Luttrell et al. (46) and DeFea et al. (47) discovered that β-arrestin1 can recruit c-Src, a nonreceptor tyrosine kinase family member, to 7TMRs. Src recruitment to the β2AR results in ERK activation, which can be inhibited by expression of a mutant β-arrestin1 defective in receptor binding or Src binding (46, 48). Src phosphorylation is a necessary step in the activation of various mitogenic signaling pathways activated by 7TMRs, and thus β-arrestins were implicated in carrying cellular signals. Likewise, using the neurokinin-1 receptor and its agonist substance P, DeFea et al. (47) showed that Src recruitment by β-arrestin is necessary for the prevention of apoptosis and propagation of mitogenic signals. The above studies were groundbreaking in introducing the idea of a second wave of 7TMR signaling initiated by β-arrestins.

ERK: The Archetype for β-Arrestin Signaling

Soon after the discovery of β-arrestin-dependent Src recruitment, DeFea et al. (49) and Luttrell et al. (50) showed that β-arrestins scaffold specific components of the MAPK cascade. Using the protease-activated receptor 2 (PAR2), DeFea et al. (47) demonstrated that agonist stimulation results in the formation of a complex containing the activated receptor, β-arrestin1, Raf-1, and phosphorylated ERK. In a similar report using the $AT1_AR$, Luttrell et al. (50) described an agonist-induced β-arrestin2, Raf-1, MEK1, and ERK1/2 signaling complex. This study showed that MEK1 indirectly binds β-arrestin2 through contacts with Raf and ERK, whereas the latter components directly bind β-arrestin. These studies clearly demonstrated that β-arrestins can serve as scaffolding molecules that facilitate cell signaling to ERK. Recently, Scott et al. (51) reported that filamin A, an actin-binding protein, directly binds β-arrestin1 and increases its association with ERK. **Figure 1*a*** depicts β-arrestin2 scaffolds for ERK activation, as currently understood.

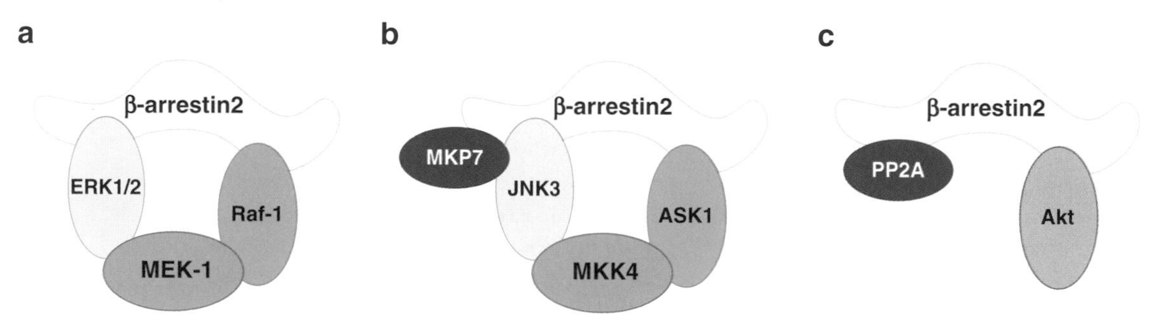

Figure 1

β-Arrestin scaffolds for signaling. (*a*) A depiction of the β-arrestin scaffold for extracellular signal–related kinase (ERK) activation. ERK1/2 and Raf-1 bind β-arrestin2 directly, and the mitogen-activated protein kinase kinase MEK-1 binds indirectly. It is not known which residues of β-arrestin2 bind ERK and Raf-1. (*b*) The β-arrestin scaffold for c-Jun N-terminal kinase (JNK) 3 activation. As does ERK, JNK3 and apoptosis signaling kinase 1 (ASK1) bind β-arrestin2 directly, whereas MAP kinase kinase 4 (MKK4) binds indirectly. JNK3 and the negative regulator MAP kinase phosphatase 7 (MKP7) contact the RRS motif (amino acids 195–202). (*c*) The β-arrestin scaffold for Akt regulation. β-Arrestin2 associates with and is necessary for Akt activation in some systems, although it is not known if Akt and/or PP2A bind directly or indirectly to β-arrestin2.

Once researchers established that β-arrestins were involved in signal transduction to ERK from a variety of 7TMRs, they investigated the functional consequences of these events. One effect of ERK activation is the phosphorylation and activation of Elk-1, an Ets domain–containing transcription factor involved in the transcription of genes that promote cell cycle progression. Using the $AT1_AR$ as a model, Tohgo et al. (52) observed that exogenous β-arrestin1 or -2 expression resulted in decreased agonist-stimulated phosphoinositide hydrolysis, yet increased ERK activation. In other words, β-arrestin expression inhibited G protein signaling (through increased desensitization) but increased ERK phosphorylation through a presumably G protein–independent pathway. However, Elk-1-dependent transcription, which typically results from ERK activation, was decreased when β-arrestins were expressed. (52). Confocal microscopy revealed that this effect is a result of the cytosolic retention of ERK by β-arrestins on endocytic vesicles (50, 53), later termed β-arrestin signalosomes (30). Although these studies clarify that β-arrestin-dependent ERK does not have typical nuclear ERK functions, the precise downstream targets activated by β-arrestin-dependent ERK remain unknown. However, cytoskeletal rearrangement and chemotaxis (54–56) have been linked to β-arrestin signaling. In one report, β-arrestin signalosomes were enriched at the leading edge of chemotactic cells, suggesting that β-arrestin-mediated ERK is responsible (55). In accord with this result, Scott et al. (51) showed that β-arrestins, filamin A, and ERK are all necessary for membrane ruffling in Hep2 cells.

RNA silencing technology has provided a powerful tool for distinguishing β-arrestin-dependent activation of ERK from that initiated by G proteins. Previously, studies of β-arrestin signaling generally relied on overexpression of β-arrestins in cell lines, such as COS-7, with very low endogenous β-arrestin expression (50, 53). Small interfering RNAs (siRNAs) allow for the selective elimination of a particular β-arrestin isoform from an otherwise complete signaling system. Application of siRNA technology to β-arrestin signaling revealed previously unappreciated differences in β-arrestin isoform specificity (57–59). Although siRNA directed toward β-arrestin2 led to a decrease in phosphorylated ERK after stimulation of the $AT1_AR$ in HEK-293 cells, β-arrestin1 siRNA resulted in a surprising increase in ERK activation (59). Ren et al. (60) noted a similar pattern for the V2 vasopressin receptor (V2R). This pattern, termed reciprocal regulation, is still not mechanistically understood, nor does it apply to all 7TMRs. For example, some receptors, such the β2AR and parathyroid hormone receptor, subtype 1 (PTH1R), exhibit a marked decrease in ERK activation following siRNA depletion of either β-arrestin1 or -2 (61, 62). This pattern may be termed codependent regulation, as both β-arrestins are necessary for ERK signaling. **Table 1** lists the patterns for 7TMRs studied to date.

Codependent regulation: when β-arrestin-mediated signaling from a particular receptor relies on both β-arrestin isoforms and the selective elimination of either isoform eliminates β-arrestin-dependent ERK activation

Reciprocal regulation: when β-arrestin-mediated signaling from a particular receptor relies upon only one β-arrestin isoform and the other β-arrestin isoform functions to inhibit that signaling

G protein–mediated and β-arrestin-mediated ERK pathways are both spatially segregated (50, 52, 58) and, according to studies using RNAi, temporally distinct (58) in cells. For the $AT1_AR$ studied in HEK-293 cells, G protein–mediated ERK activity was maximal at 2 min after stimulation, with very little contribution to the cellular pool of phospho-ERK after 10 min (58). β-Arrestin2-mediated ERK activity was miminal until 10 min poststimulation but was responsible for nearly 100% of ERK signaling at times beyond 30 min (58). Blockade of Gαq-dependent signaling by a protein kinase C (PKC) inhibitor yielded an inverse pattern of ERK phosphorylation when compared with that which resulted from β-arrestin2 siRNA. Interestingly, the two pathways to ERK activation were additive over the entire time course. **Figure 2**, a time-course plot of ERK activation by AngII, depicts these findings.

Mutational analysis supports the independence of G protein and β-arrestin signaling pathways for certain receptors. Mutant

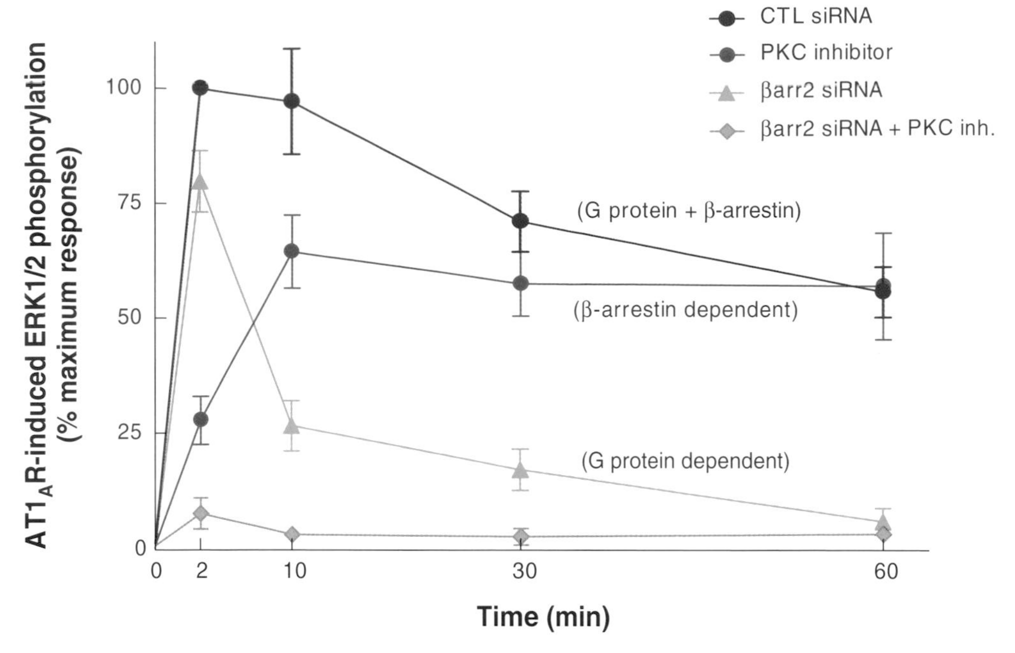

Figure 2

Phospho-ERK activation by the $AT1_AR$ is dependent upon β-arrestin2 expression. A time course of ERK phosphorylation stimulated by the $AT1_AR$ in HEK-293 cells in response to 100-nM angiotensin II. Cells were transfected with control siRNA or β-arrestin2 siRNA for 72 h and treated with either the PKC inhibitor Ro31–8425 or DMSO for 20 min prior to stimulation. Plots are depicted as a percentage of maximal ERK activation at 2 min in the control siRNA, DMSO-treated sample. Results are averages +/− SEM of five experiments. Figure adapted from Reference 58a, with permission.

Biased agonism: when a ligand has the ability to preferentially activate one of a number of possible signaling pathways stemming from a particular receptor

receptors that do not couple to their cognate G proteins but still recruit β-arrestins in response to agonist stimulation provide an excellent model with which to study exclusively β-arrestin-mediated signaling. Thus far, researchers have identified such mutants for two receptors, the β2AR and $AT1_AR$. For the $AT1_AR$, alanine substitution of the first two amino acids of a conserved motif in the second intracellular loop, termed DRY, completely uncouples the receptor from Gαq (63) while retaining the capacity to recruit β-arrestins and activate β-arrestin-dependent ERK (64). The ERK that is phosphorylated in response to AngII stimulation of the $AT1_AR$-DRY/AAY receptor depends completely on β-arrestin2 expression. A similar study employed an evolutionary trace method to create a mutant β2AR ($β2AR^{TYY}$) that is uncoupled from Gαs but retains the ability to recruit and signal through β-arrestins (61). Furthermore, $β2AR^{TYY}$ ERK activation is completely abolished by β-arrestin2 siRNA and is insensitive to protein kinase A (PKA) inhibitors (61).

Biased agonism refers to the preferential activation of one of a number of possible downstream pathways of a receptor by a particular ligand. This concept implies that receptors can exist in an ensemble of distinct conformations in response to ligand binding. For example, in an inactive receptor conformation, G proteins are not dissociated, and β-arrestins are not recruited. Inactive conformations would be expected to predominate in the absence of ligand or upon treatment with an inverse agonist. Conversely, fully active conformations would result from treatment with a full agonist. However, a biased agonist would be expected to elicit a conformational change that can only recruit β-arrestins without coupling to G proteins, or vice versa.

In light of the two possible pathways of signaling elicited by 7TMRs we describe in this review, G protein or β-arrestin mediated, a biased agonist would specifically activate one of these pathways to a greater extent than the other. A prime example of this bias exists for the $AT1_AR$. A mutational approach was successful in identifying a β-arrestin signaling–specific agonist for this receptor (64, 65), termed SI^4I^8 AngII (SII). Although SII cannot activate Gαq signaling, as evidenced by lack of PI hydrolysis (64, 65), calcium mobilization, or diacylglycerol activity, SII can recruit β-arrestin to the $AT1_AR$ in a Class B pattern and stimulates ERK in an entirely β-arrestin2-dependent manner (64). Recently, mutation analysis of the $AT1_AR$ revealed that specific residues in the seventh transmembrane-spanning domain are necessary for SII mediated signaling but not for AngII signaling. This report suggests that SII can induce a conformational change that is distinct from that induced by AngII (66). In vivo, SII mediates different physiological effects than AngII; i.e., whereas AngII stimulates intake of both water and salt in mice, SII increases only salt consumption (67).

Similarly, for the β2AR, the compound ICI118551, a well-established inverse agonist for adenylyl cylase (68), induces ERK phosphorylation that completely depends on β-arrestin2 expression (69). Finally, for the V2R, the inverse agonist SR121463B recruits β-arrestin and stimulates ERK (69). These data further support the independence of G protein and β-arrestin signaling pathways for some 7TMRs.

However, independence of these pathways is not displayed by all receptors. A naturally occurring system of biased agonism exists for the chemokine receptor CCR7: One ligand (termed ELC/CCL19) can activate both G protein and β-arrestin signaling, and another ligand (SLC/CCL21) activates only G protein signaling and does not recruit β-arrestins. In this system, β-arrestin signaling is also dependent upon G protein activity, as pertussis toxin treatment of cells eliminates all activation of ERK through the CCR7 receptor (70). Other Gαi-coupled receptors show a similar pattern. For example, CXCR4 can induce β-arrestin-dependent chemotaxis, which is abolished by pertussis treatment (71). Also, in studies of the $AT1_AR$ and lysophosphatidic acid (LPA) receptor in HEK-293 cells, AngII-induced chemotaxis was slightly reduced, and LPA-induced, β arrestin-dependent chemotaxis was eliminated by pertussis, consistent with the coupling of the LPA receptor to Gαi and the coupling of $AT1_AR$ to both Gαq and Gαi for these cellular responses (56).

Although the downstream effectors of β-arrestin signaling to ERK have not been elucidated, a recent report suggests one physiological outcome with implications for Parkinson's disease. Rotenone, a pesticide, is toxic to dopaminergic (DA) neurons and is used as a model of Parkinson's disease. Stimulation of the group III metabotropic glutamate receptor (mGluRIII) attenuates the toxicity of rotenone on DA neurons. Jiang et. al. (72) found that the mechanism for this stems from L-AP-4 (an mGluRIII agonist) stimulation of ERK that depends on dynamin, β-arrestin2, and c-Src. The β-arrestin-activated ERK stabilizes microtubules, which protects DA neurons from rotenone toxicity (72) and prevents Parkinson's-like symptoms.

β-Arrestin and JNK3 Signaling

Another MAPK activated by β-arrestin signaling pathways is JNK3. Unlike the other two members of this kinase family, JNK1 and JNK2, which are ubiquitously expressed, JNK3 shows tissue specificity, with expression restricted to the brain and to a lesser extent the heart and testis (73). JNK3 was originally identified as a binding partner of β-arrestin2 in a yeast two-hybrid screen, and subsequent investigation revealed that the two upstream kinases, MAP kinase kinase 4 (MKK4) and apoptosis signaling kinase (ASK1), were also present in the complex (74, 75). Thus, β-arrestin2 scaffolds an entire signaling module for JNK3, as was previously shown for ERK1/2. This module responds to signaling from a 7TMR, the $AT1_AR$, in a time-dependent fashion correlating with β-arrestin2 recruitment and receptor binding (74). Interestingly, phosphorylated, active JNK3 can only be detected in association with β-arrestin2 and is excluded from the nucleus. A subsequent study of β-arrestin2 subcellular localization determinants confirmed this finding, as a mutant β-arrestin2 lacking the nuclear export signal showed a redistribution of JNK3 to the nucleus (76). These findings closely parallel the situation with ERK, for which the β-arrestin-activated kinase is found only in the cytoplasm.

Recently, Willoughby et al. (77) showed that a fourth partner in this signaling module, the dual-specificity phosphatase MAP kinase phosphatase 7 (MKP7), dynamically interacted with β-arrestin2. MKP7 seems to function as a "reset switch" in the JNK3 cascade, as it rapidly dissociates from β-arrestin2 upon $AT1_AR$ stimulation and later reassociates to dephosphorylate JNK3 in the complex (77). This is the first report of a negative regulator in a β-arrestin signaling module, which raises the possibility that other β-arrestin scaffolds bind their own phosphatases. **Figure 1*b*** shows a schematic representation of the β-arrestin2 scaffold for JNK3 activation.

p38 Activation via β-Arrestin

The third and final member of the MAPK family, p38, has four isoforms (α, β, γ, δ) and is primarily involved in eliciting a transcriptional response to inflammatory cytokines, growth factors, and cellular stresses. Although researchers have not directly shown that β-arrestins (*a*) scaffold either p38 itself or upstream kinases or (*b*) lead to p38 phosphorylation, three reports have implicated β-arrestins in p38 activation (71, 78, 79). The chemokine receptor CXCR4-induced chemotactic response, which is β-arrestin dependent, is also sensitive to p38 inhibitors (71). These data suggest that at least some cellular functions regulated by β-arrestin signaling involve p38, even though direct β-arrestin-dependent p38 phosphorylation is not documented.

More direct evidence for p38 signaling comes from work with the cytomegalovirus 7TMR US28, which is a constitutively active chemokine receptor homolog (78). US28 is constitutively phosphorylated by GRKs and recruits β-arrestin in the absence of agonist. When the cytoplasmic tail of US28 is truncated, the constitutive recruitment of β-arrestin is lost, and subsequent p38 activation is diminished (78).

Very recently, Bruchas et al. (79) demonstrated that the kappa opioid receptor activates p38 by a mechanism that involves GRK3 phosphorylation and β-arrestin2. In their study, Bruchas et al. used both $GRK3^{-/-}$ MEFs and siRNA directed against β-arrestin2 in striatal astrocytes. They discovered that activated p38 colocalized with β-arrestin2 in features reminiscent of the signalosomes found with β-arrestins and ERK (79).

β-ARRESTIN SIGNALING TO OTHER KINASES

PI3 Kinase and Antiapoptosis

Phosphatidylinositol 3 kinase (PI3K) is a common signaling intermediate of tyrosine kinase receptors such as the insulin-like growth factor-1 receptor (IGF1-R). PI3K catalyzes the phosphorylation of phosphatidylinositol-4,5-bisphosphate (PIP2) to phosphatidylinositol-3,4,5 triphosphate (PIP3). IGF1-R stimulation leads to PI3K activity, which promotes cell growth and inhibits apoptosis. In mouse embryo fibroblasts (MEFs) lacking both β-arrestin1 and -2, IGF-1 was unable to stimulate PI3K activity (80). This effect was rescued by expression of exogenous β-arrestin1 and was independent of the tyrosine kinase activity of the activated IGF1-R (80). This is in contrast to the classical IGF-1-stimulated pathway, whereby agonist binding causes receptor autophosphorylation and subsequent tyrosine phosphorylation of insulin receptor substrate 1 (IRS-1), leading to PI3K and Akt activation.

Akt Activation

One important downstream target of PI3K signaling is Akt, also known as protein kinase B (PKB). Akt is pivotal in promoting cell survival and preventing apoptosis through numerous downstream effectors. In the above-mentioned IGF-1-stimulated PI3K study, Akt was phosphorylated in β-arrestin1-expressing cells but not in β-arrestin1/2 knockout cells (80). This β-arrestin1-dependent Akt signaling led to increased protection from apoptosis (80).

Thrombin, through protease-activated receptors (PARs), can activate Akt by two mechanisms, one Gα dependent and the other β-arrestin1 dependent (81). The function of β-arrestin1-mediated Akt activation is still unclear, as its inhibition does not affect cell cycle progression (81–83).

Recently, β-arrestin2 has also been implicated in Akt signaling. Using the D2 class of dopamine receptors, Beaulieu and colleagues (84) found that dopamine stimulates the formation of a signaling module consisting of β-arrestin2, Akt, and its negative regulator, protein phosphatase 2A (PP2A) (**Figure 1*c***). In mice lacking β-arrestin2, dopamine was unable to regulate Akt signaling, as PP2A and Akt could no longer interact. Furthermore, behavioral effects associated with dopamine administration, such as increased locomotor activity and wall climbing, were lost in these mice, demonstrating the importance of β-arrestin2 in mediating these events (84). This work is another example of a β-arrestin signaling module, much like the JNK3 module, whereby β-arrestin assembles activators and negative regulators of a pathway as a scaffold.

Coordinate Regulation of RhoA by β-Arrestin1 and Gαq

Pertinent to its role in chemotaxis (see Reference 8a), β-arrestin signaling can also mediate rearrangement of cytoskeletal components. In particular, $AT1_AR$ stimulation activates the small GTPase RhoA in a β-arrestin1-dependent fashion, leading to the formation of stress fibers (54). Interestingly, this activation of RhoA depends on both Gαq and β-arrestin1 (54), whereas $AT1_AR$-stimulated, β-arrestin-mediated ERK signaling is independent of G protein activation (64). This implies that the necessity of G protein activity for β-arrestin-mediated functions is specific for a given cellular outcome.

NUCLEAR β-ARRESTIN SIGNALING

β-Arrestin Trafficking and the Nucleus

In addition to the above-mentioned roles that β-arrestins play in cellular signaling in the cytosol, there is new evidence that β-arrestins can also regulate nuclear processes such as transcription of new RNA. Although both β-arrestin1 and -2 contain nuclear localization signals, only β-arrestin2 has a nuclear export signal. As a consequence, β-arrestin1 is present in both the nucleus and cytoplasm, whereas β-arrestin2 is only in the cytoplasm (17). However, upon treatment of cells with leptomycin B, an inhibitor of CRM1-mediated nuclear export, β-arrestin2 accumulates only in the nucleus (76). These data imply functional roles for both β-arrestin1 and -2 in the nucleus.

Transcriptional Regulation by β-Arrestins

Changes in transcriptional responses are a common outcome of signaling via 7TMRs. Although cytoplasmic signaling byβ-arrestins has been extensively described, its effects in the nucleus have not received attention until very recently. One transcription factor family regulated by β-arrestin is NFκB, a nuclear transcription factor that regulates the expression of genes involved in inflammation, the autoimmune response, cell proliferation, and apoptosis (85, 86). Two independent studies identified β-arrestins as binding partners for IκBα, a protein that binds NFκB and inhibits its nuclear translocation and subsequent activity (87, 88). Both β-arrestin isoforms can sequester IκBα-NFκB complexes in the cytosol and attenuate NFκB-dependent nuclear transcription in response to 7TMR or cytokine stimulation. Furthermore, β-arrestin2 can regulate UV-induced NFκB activity via a pathway involving casein kinase II–mediated IkBα phosphorylation (89). In this case, UV irradiation of heterologous cells activates casein kinase II to phosphorylate β-arrestin2. This β-arrestin2 phosphorylation reduces the interaction of β-arrestin with IκBα, causing IκBα degradation and releasing NFκB for nuclear entry and transcriptional activity.

Gesty-Palmer et al. (90), using cDNA microarray technology, recently examined transcriptional regulation attributable to β-arrestin2 signaling in response to LPA. ERK signaling that results from LPA stimulation of MEFs was diminished in β-arrestin2 knockout ($^{-/-}$) MEFs. Most distally, of the seven mRNAs upregulated by LPA treatment in wild-type MEFs, only four persisted in β-arrestin2$^{-/-}$ MEFs. Interestingly, there was an upregulation of several mRNAs in the β-arrestin2$^{-/-}$ MEFs that was not present in wild-type MEFs. Although the precise mechanism was not defined, these data suggest that β-arrestin signaling has nuclear effects that are both positive and negative with regard to gene expression.

β-Arrestins may also control transcription without ever entering the nucleus by activating nuclear receptors such as the retinoic acid receptor (RAR). The RARs control gene expression in response to extracellular stimuli and are regulated by their phosphorylation state. A recent report found that β-arrestin2 signaling regulates all RAR and RXR receptors studied to some extent. In particular, RAR-β2-induced transcription in PC12 cells completely depends on β-arrestin2-mediated ERK activity (91). This β-arrestin-dependent transcriptional regulation inhibits cell growth through expression of nerve growth factor.

Kang et. al. (92) recently observed a much more direct role of β-arrestins in mRNA transcription. This report found that β-arrestin1—which, unlike β-arrestin2, is localized to the nucleus without the addition of leptomycin B—binds directly to the promoter regions of several genes, including *c-fos* and *p27*. Furthermore, β-arrestin1 recruits p300, the histone acetyltransferase, to these promoters via an interaction with its binding partner CREB (92). This recruitment results in an increase in transcription from the promoters bound by β-arrestin1. Both the amount of β-arrestin1 and the level of transcriptional activity at these promoters can be regulated by 7TMR stimulation, in this case the δ-opioid and κ-opioid receptors. However, not all 7TMRs can initiate this signaling, as the β2AR and μ-opioid receptor were unable to elicit such a response (92). In summary, these data provide an excellent mechanism for direct β-arrestin1-mediated control of gene expression upon 7TMR stimulation.

ACTIVATION OF β-ARRESTIN: CONFORMATIONAL CHANGES

Most signal transduction mechanisms are guided by ligand-induced conformational changes in the receptor, which are then translated to downstream effectors, resulting in a specific signal output. Plasma membrane translocation of β-arrestin to activated receptors and its signaling capabilities even in the absence of receptor–G protein interaction strongly suggest the existence of specific receptor conformation(s) for β-arrestin binding (61, 64). Moreover, these findings indicate that β-arrestins must undergo their own activation-dependent conformational changes to facilitate downstream signaling. Further supporting this is evidence that many of the nonreceptor-partner interactions of β-arrestins, such as the formation of MAPK scaffolds, are facilitated upon receptor activation.

Much evidence, including mutagenesis and in vitro biochemical and biophysical characterizations, suggests that visual arrestin undergoes substantial conformational changes as it binds to light-activated, phosphorylated rhodopsin (93–95). The X-ray structures of bovine visual arrestin and β-arrestin1 in the basal inactive state indicate that arrestin is an elongated molecule with two domains (N- and C-domain), connected through a 12-residue linker region (96–99). A notable feature is a hydrogen-bonded network of buried, charged side chains (the polar core) embedded between the N- and C-domains at the fulcrum of the β-arrestin molecule (100). Disruption of the polar core by the phosphate moieties on activated receptors and the rearrangement of the "three-element interaction" (i.e., N and C termini and the single α-helix of β-arrestin) are components of a mechanism by which β-arrestin activation and conformational change are proposed to occur. Xiao et al. (101) recently demonstrated that conformational changes of the nonvisual β-arrestin2 occur in the presence of a phosphorylated, 29-mer peptide that corresponds to the carboxyl tail of the V2R. Addition of the V2R phosphopeptide to β-arrestin2 in vitro led to the exposure of a buried tryptic cleavage site (arginine 394) as well as the release of residues 371 to 379 in the C terminus of β-arrestin2, which contain the sites for clathrin interaction. In this activated conformation induced by the phosphopeptide, β-arrestin binding to clathrin increased at least tenfold.

In another recent study, Bouvier and colleagues (102) monitored the conformational changes in the β-arrestin2 molecule by utilizing bioluminescence resonance energy transfer (BRET) assays. BRET is a distance-dependent nonradiative energy transfer that occurs when an acceptor fluorophore such as the yellow fluorescent protein (YFP) comes within 10–100 Å and captures the energy released by the bioluminescent donor enzyme luciferase upon degradation of the latter's substrate (e.g., coelenterazine). Typically, the donor and acceptor molecules are fused to two independent proteins for testing intermolecular interaction by BRET. However, Bouvier's group adapted this principle to study intramolecular BRET by flanking the β-arrestin molecule with a luciferase moiety at the N terminus and YFP at the C terminus (103). Thus, changes occurring in β-arrestin conformation would cause a rearrangement of the two ends of β-arrestin, producing a BRET signal. Indeed, such intramolecular BRET signals occurred in the luciferase-β-arrestin2-YFP protein upon activation of various 7TMRs. However, the half-life of V2R-stimulated intramolecular BRET in β-arrestin2 (~5 min) was significantly slower than the kinetics of β-arrestin recruitment to the V2R (~0.8 min). Furthermore, quantitatively similar BRET signals were produced in both wild-type β-arrestin and the β-arrestin mutant R169E, which binds nonphosphorylated receptors (104, 105). Accordingly, the authors concluded that the conformational changes in β-arrestin (represented by intramolecular BRET) likely occur due to the binding of β-arrestin-interacting proteins subsequent to the binding of β-arrestin to the receptors and are not due to β-arrestin's interaction with the phosphorylated domains of the 7TMR. However, this conclusion is debatable for at least two reasons. First, the chimeric protein may have constraints that restrict release of the C-terminal region. Hence, this technique may not discern the conformational changes that occur upon binding receptor phosphates. Second, although quantitatively both the wild-type and mutant proteins display similar BRET responses, the wild type may undergo conformational rearrangements qualitatively different than the mutant, wherein the basal constraints in the polar core are absent. The BRET technique cannot distinguish between such qualitative differences.

Thus, receptor binding may activate β-arrestin, causing multiple, parallel, or stepwise conformational changes to induce downstream signaling pathways. Additional work will be necessary to determine whether the β-arrestin conformational changes induced by 7TMR binding are merely a means to facilitate binding of nonreceptor partners, or a process that initiates β-arrestin-mediated signaling pathways such as those leading to MAPK activation, or both.

NEW RECEPTORS FOR β-ARRESTIN

Parallel to the discoveries of the participation of β-arrestin in various signaling pathways has been the tantalizing finding that β-arrestin can bind receptors that are structurally unrelated to 7TMRs. Thus, β-arrestins regulate signaling and/or endocytosis of IGF1R, Frizzled, smoothened, TGFβRIII, LDLR, Na^{+}/H^{+} exchanger NHE5, Toll-like receptor, Interleukin1 receptor, and *Drosophila* Notch (40, 44, 106–111). Accordingly, the biological roles of β-arrestin in signal transduction are likely much broader than we currently appreciate. Because this subject is beyond the scope of this review, these capacities of β-arrestins are not discussed here.

UNANSWERED QUESTIONS AND EMERGING AREAS IN β-ARRESTIN SIGNALING

Dimerization of β-Arrestins

Although homo- and heterodimerization of heptahelical receptors may have important pharmacological and functional implications (112, 113), very few reports have addressed this phenomenon for the nonvisual β-arrestins. Purified inactive full-length β-arrestin1 (418 residues) and β-arrestin1–393, a truncation mutant, exist as monomers, according to gel-filtration chromatography (99). However, another truncation mutant, β-arrestin1–382, exists as a mixture of monomeric and dimeric species. Interestingly, this mutant was previously identified as a β-arrestin species that binds unphosphorylated β2AR reconstituted in liposomes. This phosphorylation-independent mutant was termed constitutively active because it could desensitize receptors in *Xenopus* oocytes even in the absence of GRK-mediated phosphorylation (105). Full-length β-arrestin1 forms only monomeric crystals, whereas the truncated β-arrestin1 (1–382) forms crystals containing exclusively dimers. This scenario is opposite to that observed in visual arrestin, which in solution exists as a tetramer at high concentrations (200 μM) and as mixed monomer-dimer species at lower concentrations. Because arrestin is highly abundant in rod outer segments, it may oligomerize under inactive physiological conditions. In contrast, arrestin-rhodopsin binding occurs at low concentrations of arrestin, suggesting that monomeric forms of arrestin predominating at low concentrations actually can be active.

Researchers have recently implicated phosphoinositides in arrestin oligomerization. Phosphoinositides can function as important modulators of light-dependent trafficking of *Drosophila* visual arrestin (114–116). For mammalian β-arrestins, the soluble inositol hexakisphosphate (IP6) displays tighter binding than the membrane-associated phosphoinositides PIP2 and PIP3 (117, 118). Gaidarov et al. (117) mapped a high-affinity phosphoinositide-binding site located on the C-domain of β-arrestin2. Although mutation of this site did not ablate clathrin binding or membrane translocation of β-arrestin, it prevented localization of β-arrestin in clathrin-coated pits, thus impairing the endocytic properties of β-arrestin. Milano et al. (118) recently cocrystallized β-arrestin1 with IP6 in the basal state and identified an additional low-affinity IP6-binding site. Interestingly, IP6 mediated an interaction between the N-domain of one β-arrestin1 molecule with the C-domain of a second β-arrestin1 molecule, thus forming a head-to-tail symmetry of β-arrestin in the crystal lattice. Mutation of the basic residue patches forming the IP6 contact sites prevented the formation of β-arrestin1 oligomers while increasing the propensity for nuclear localization. These mutants, defective in both IP6 binding and oligomerization, nevertheless bound nonreceptor partners such as clathrin, AP2, and ERK2. Benovic and colleagues (118) hypothesize that the modulation of IP6 levels in the cytosol regulates the oligomerization state of β-arrestin, probably by multiple complex pathways. Cellular IP6 locks β-arrestin in an inactive oligomeric complex, which is a regulatory step that limits plasma membrane and nuclear localization of β-arrestin. Marullo's group (119) showed that homo- and hetero-oligomerization of exogenously expressed β-arrestins occur constitutively in mammalian cells. Cellular expression of a β-arrestin2-FKBP-GFP molecule, containing a FKBP-binding motif, and induction of its oligomerization by the FKBP-dimerizing small molecule AP20187 did not interfere with its trafficking profile with $AT1_A$Rs. Thus, although IP6-induced oligomers form inactive cytosolic pools, oligomerization per se may not interfere with the classical functions of β-arrestin. Whether activated β-arrestins that form signalosomes upon binding 7TMRs are homo- or heterodimers or even monomers currently remains an unsolved question.

Ubiquitination, a Possible Mechanism that Links the Endocytic and Signaling Properties of β-Arrestin

Agonist-stimulated β-arrestin ubiquitination not only regulates receptor internalization and trafficking of receptor-arrestin complexes to the endosomal compartments but also appears important for stabilizing ERK activity on endosomes. A lysine doublet (residues 11, 12) in β-arrestin2 functions as a crucial site for stable ubiquitination in response to angiotensin stimulation and is required for endosomal trafficking as well as for scaffolding phosphorylated ERK in these compartments (25). However, ubiquitination evoked by other Class B receptors, such as the V2R and the neurokinin1 receptor, proceeds at alternate lysine residues, suggesting heterogeneity of the modified receptor-bound β-arrestin. This presents a very provocative scenario in which a specific receptor can provoke ubiquitination at specific site(s), specific to that receptor-β-arrestin pair. These ubiquitination patterns may correspond to particular β-arrestin conformations induced upon receptor binding that expose specific lysine residues for ubiquitination, thus allowing for receptor-specific signaling pathways mediated by β-arrestin. Unlike in phosphorylation, the attachment of ubiquitin adds tertiary structure to the substrate protein, allowing larger conformational changes. Polyubiquitin chains (120), such as those formed on β-arrestin, may specify further conformational complexities and present a foundation for the binding of the many downstream endocytic and/or signaling partners of β-arrestin.

Reciprocal Versus Codependent Regulation of β-Arrestin-Dependent ERK

One of the most intriguing findings regarding β-arrestin signaling is the issue of isoform specificity. Data from siRNA experiments show two distinct patterns for β-arrestin-mediated ERK from 7TMRs. Some receptors, such as the β2AR, depend on both β-arrestins 1 and 2 for G protein–independent ERK, as a loss of either inhibits this branch of signaling (61). In this review, we term this pattern codependent β-arrestin-ERK, as both β-arrestins are necessary and neither one is sufficient by itself. In contrast, β-arrestin-dependent ERK activated by the $AT1_A$R depends exclusively on β-arrestin2 expression. When β-arrestin1 expression is silenced, G protein–independent ERK increases, a phenomenon referred to as reciprocal regulation (59). The opposite is true for PAR1: β-Arrestin1 facilitates signaling to ERK, whereas silencing of β-arrestin2 increases β-arrestin1-dependent ERK (121). **Table 1** classifies various receptors as reciprocal or codependent with regard to β-arrestin signaling.

The molecular mechanisms governing this reciprocal regulation have not been elucidated; however, rescue experiments provide some insight. In **Figure 3**, either control siRNA or human β-arrestin2 siRNA is used in conjunction with expression of a rat β-arrestin2-GFP fusion protein, which is insensitive to silencing by virtue of sequence differences between the species. Expression of rat β-arrestin2 can restore >90% of the ERK activity that is eliminated by β-arrestin2 siRNA (**Figure 3**). Surprisingly, rat β-arrestin1 can also rescue approximately 15% of the ERK activity (S. Ahn & R.J. Lefkowitz, unpublished data). This suggests that β-arrestin1 can function as an ERK scaffold, albeit to a much lesser extent than β-arrestin2. In response to AngII, both β-arrestins normally are recruited to the receptor robustly and equally well. In this experiment, β-arrestin1 siRNA would be predicted to increase β-arrestin2 recruitment through loss of competition and thus increase β-arrestin2-dependent ERK signaling (59). Although this portrayal explains reciprocal regulation, it does not address why receptors such as the β2AR and $AT1_AR$ display such different patterns in the same cell type, HEK-293. One possible explanation for these findings is preferential recruitment of β-arrestin homo- or hetero-oligomers, a subject worthy of future investigation.

Receptor Phosphorylation and β-Arrestin Signaling

Another puzzling question is the extent to which receptor phosphorylation is necessary for, or contributes to, β-arrestin signaling pathways. Here again, the data show two opposite patterns, and the clearest example of each comes from the β2AR and $AT1_AR$. For both these cases, a mutant receptor that is unable to be phosphorylated on the cytoplasmic tail (by either alanine substitution or truncation of all serines and threonines) is available. In **Figure 4**, the β2AR, which lacks both GRK and PKA phosphorylation sites ($\beta 2AR^{GRK-,PKA-}$) and which cannot recruit β-arrestins (61), also cannot mediate β-arrestin-dependent ERK activation. This is expected because GRK phosphorylation typically precedes β-arrestin binding to the receptor. However, in the case of the $AT1_AR$, the corresponding mutant ($AT1_AR^{\Delta 324}$) (122) can elicit the same amount of β-arrestin-dependent ERK as the wild-type receptor (**Figure 4**). Whereas the mutant β2AR cannot recruit β-arrestin at all following stimulation, the mutant $AT1_AR$ can, although it does so in a much weaker Class A pattern than the wild-type Class B receptor (**Table 1**). Taken together, these data suggest that the recruitment of β-arrestin to the plasma membrane, even in a transient Class A pattern, is sufficient to induce β-arrestin-mediated signaling to ERK. A recent study supporting this theory used an artificial system of β-arrestin translocation (103). In the absence of receptor stimulation, β-arrestin recruitment to the membrane was sufficient to trigger ERK activation (103).

However, another report found that mutation of some of the phosphorylation sites in the cytoplasmic tail of Orexin-1 receptor resulted in the loss of sustained ERK activation, yet this mutant still retained some ability to recruit β-arrestins (123). Similarly, cytoplasmic tail truncation mutants of the PAR2 receptor that retained the ability to recruit transiently (Class A), but not stably associate with, β-arrestins lost the ability to activate a prolonged ERK response (124). In another example, using a system of ligand bias for the CCR7 receptor, Kohout et al. (70) found that ELC (CCL19), the ligand capable of inducing receptor phosphorylation, led to an approximate fourfold increase in ERK activation as compared with SLC (CCL21), the ligand that

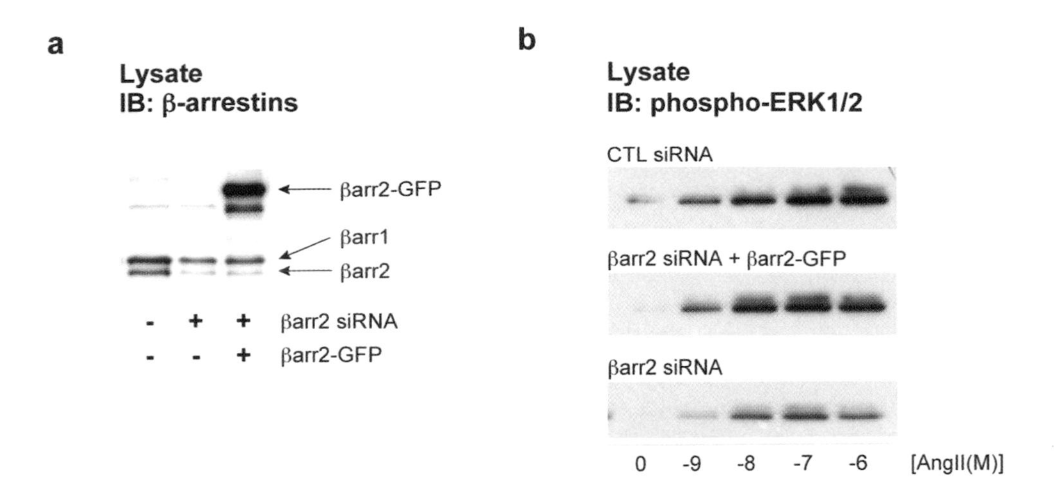

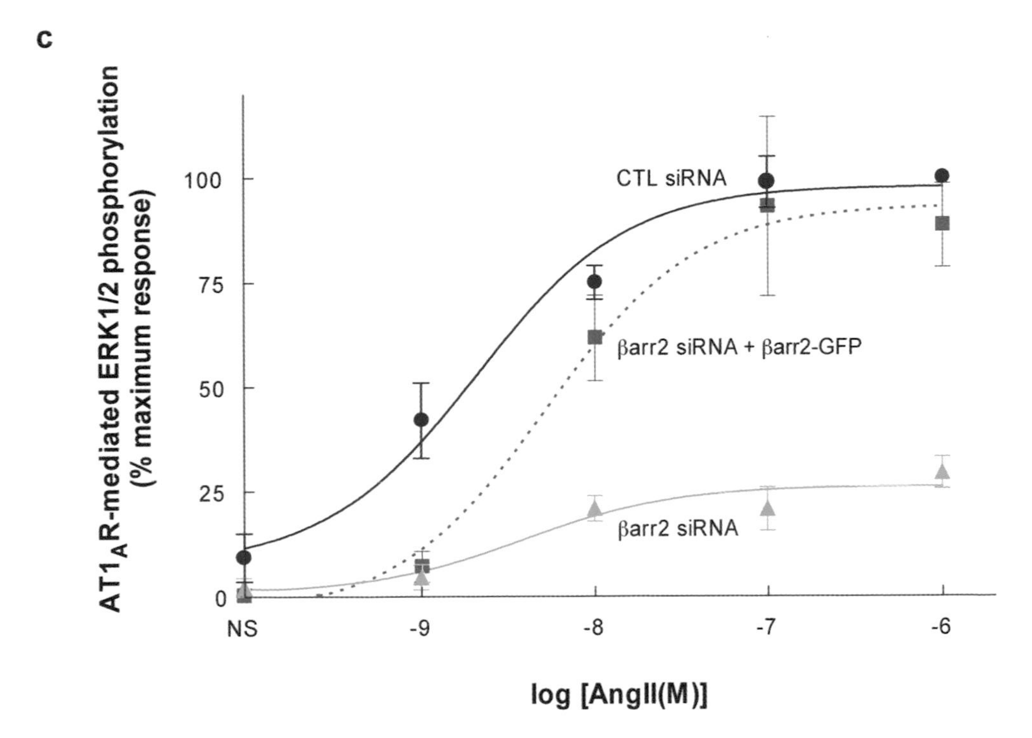

Figure 3

Nonsilenced rat β-arrestin2 (βarr2) expression rescues human β-arrestin2 siRNA effects on ERK activation in HEK-293 cells. Results are averages +/− SEM from five experiments. Cells were treated with either control siRNA (CTL) and the appropriate amount of vector DNA, human β-arrestin2 siRNA and vector DNA, or human β-arrestin2 siRNA and the expression vector for rat β-arrestin2-GFP. Seventy-two hours posttransfection, cells were stimulated with the indicated doses of AngII and analyzed for phospho-ERK. (*a*) Western blot with an antibody specific for β-arrestins. (*b*) Representative phospho-ERK western blots. (*c*) Dose response curve of averaged data for ERK phosphorylation by the $AT1_AR$ taken at 5 min poststimulation in HEK-293 cells. IB denotes immunoblot.

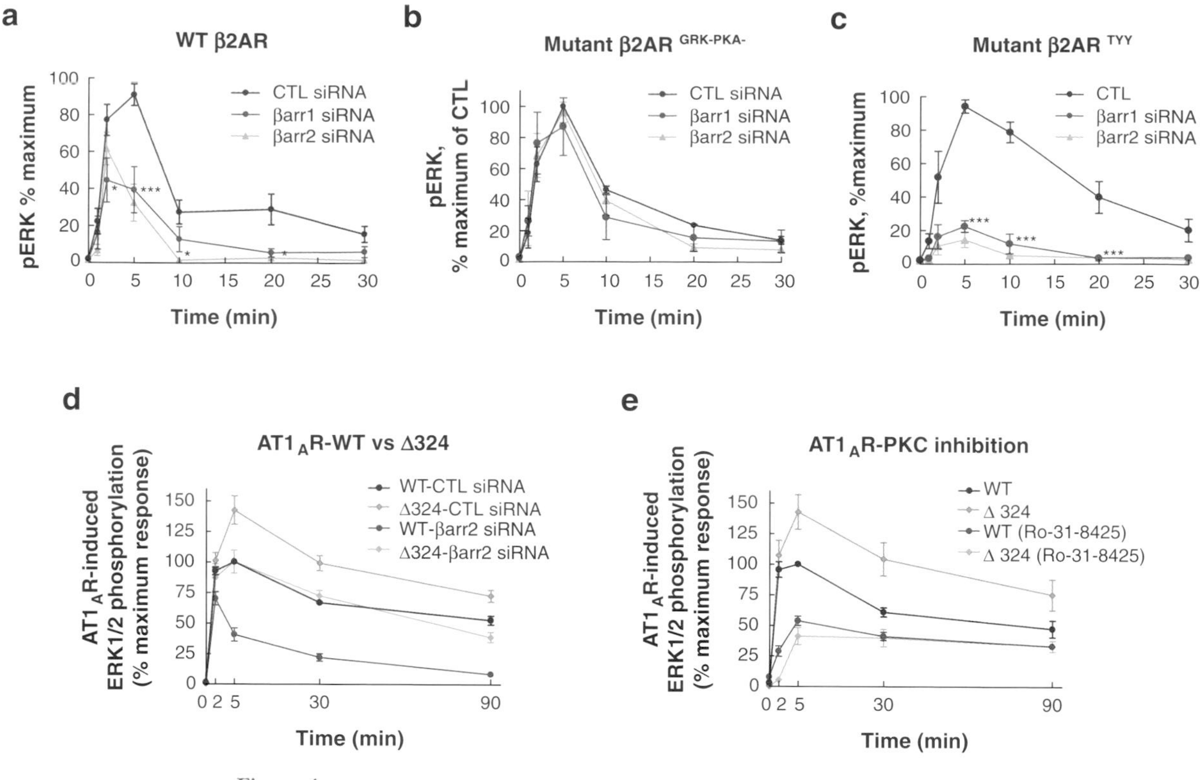

Figure 4

Phosphorylation of receptors is required for β-arrestin signaling for the β-2 adrenergic receptor (β2AR) but not for the angiotensin II receptor type 1A ($AT1_AR$). For *a*, *b*, and *c*, plots represent the average of four to six experiments +/− SEM. *a* and *c* are reproduced from Reference 61, with permission. WT denotes wild type. (*a*) A time course of phospho-ERK activation by the transfected human β2AR in HEK-293 cells in response to 100-nM isoproterenol stimulation. Results were normalized to 100% as control siRNA at 5 min. Cells were transfected with receptor plus control siRNA (CTL siRNA), β-arrestin1 siRNA (βarr1 siRNA) or β-arrestin2 siRNA (βarr2 siRNA). (*b*) The same experiment as in *a* but performed with the GRK^-, PKA^- mutant human β2AR, which has 14 serines and threonines mutated to alanine and thus cannot be phosphorylated by PKA or GRKs. (*c*) The same experiment as in *a* but performed with the TYY mutant human β2AR (which can only signal through β-arrestins and cannot couple to G proteins). Adapted from Reference 61, with permission. (*d*) Time course of ERK activation with 100-nM AngII as previously described for the $AT1_AR$ or cytoplasmic tail truncation mutant, $AT1_AR$-Δ324 (which cannot be phosphorylated by PKC or GRKs), with either control siRNA (CTL siRNA) or β-arrestin2 siRNA (βarr2 siRNA). Results depicted are averages +/− SEM for nine experiments. (*e*) The effect of PKC inhibition of the $AT1_AR$ and truncation mutant, $AT1_AR$-Δ324. The PKC inhibitor, Ro31-8425, at 1 μM or equivalent volume of DMSO was used with 20-min pretreatment. Results depicted are averages +/− SEM for nine experiments.

cannot recruit β-arrestins. Mutants of CCR7 that could not be phosphorylated maximally also did not activate ERK maximally (70). These results of Kohout et al. show a strong correlation between receptor phosphorylation and β-arrestin-mediated ERK activation. However, in this ELC and SLC system, differences between β-arrestin recruitment and receptor phosphorylation cannot be addressed because SLC (CCL21) did not recruit β-arrestins, nor were receptor phosphorylation-site mutants tested for β-arrestin recruitment.

In summary, for some receptors such as the β2AR, Orexin-1, PAR2, and possibly CCR7, receptor phosphorylation seems necessary for β-arrestin-mediated ERK activation. However, for the $AT1_AR$, phosphorylation on the cytoplasmic tail is dispensable for this signaling: The mechanism we propose is that β-arrestin recruitment, even in a transient Class A pattern, is sufficient for β-arrestin-dependent ERK activation. This mechanism is consistent with all the above data except those for the Orexin-1 and PAR-2 receptors, both of which can still recruit β-arrestins in the absence of phosphorylation yet cannot activate β-arrestin-dependent ERK.

CONCLUDING REMARKS

Researchers have identified numerous substrates, including ERK, JNK3, p38, PI3K, Akt, and RhoA, for β-arrestin-dependent signaling in the cytoplasm. Although the specific events leading to the agonist-induced activation of these signaling pathways have been vastly explored, their downstream targets and physiological outcomes are still largely a mystery. In vivo studies have revealed some expected roles of β-arrestins in receptor desensitization (15, 125, 126), but more recently, some manifestations of the signaling capacities of β-arrestins have also been appreciated (84, 127). These findings reveal the consequences of β-arrestin signaling in complex behavioral phenotypes, thus underscoring the importance of this novel signaling mechanism.

Nuclear β-arrestin signaling is an area of great potential for future research. Although β-arrestin-dependent signaling to ERK does not lead to transcriptional responses through ERK-dependent factors such as Elk-1 (52), the interaction of β-arrestins and components of the NFκB pathway can affect transcription negatively (87, 88). Kang et al. (92) recently discovered that β-arrestins can regulate transcriptional changes, and the interaction of β-arrestin1 and p300 provides a promising direct mechanism for activation. Further work is needed to explain β-arrestin1 recruitment to promoters, which may occur through interaction with specific transcription factors.

More tantalizing prospects for the discovery of β-arrestin signaling–dependent physiology include the creation of mice with mutant receptors that perform exclusively β-arrestin signaling. For example, mice expressing the $AT1_AR$-DRY/AAY could be created in the context of the $AT1_AR$ knockout mouse (128) and stimulated with AngII. By measuring the typical AngII-stimulated physiology, one could determine how much of the response is carried out by β-arrestin signaling versus Gαq. Similarly, one also could take advantage of the available biased agonist for the $AT1_AR$, SII.

β-Arrestin signaling represents a new paradigm in cell biology and potentially a new host of therapeutic targets for diseases of 7TMR dysregulation. Although classic pharmacological solutions have focused exclusively on receptor blockade and G protein–mediated responses, β-arrestins, owing to the only very recent appreciation of their signaling capacities, remain an untapped resource. Future research into the intricate β-arrestin-mediated signaling pathways will shed light on how their modulation can impact human health.

SUMMARY POINTS

1. β-Arrestins not only desensitize G protein–dependent signal pathways but independently promote novel pathways of signal transduction.
2. β-Arrestins serve as scaffolds for the activation of a number of signaling pathways, including ERK, JNK, p38, and Akt.
3. β-Arrestins are posttranslationally regulated by phosphorylation and ubiquitination.

4. Receptor binding induces conformational changes in β-arrestin that may facilitate its scaffolding and signaling functions.
5. Investigators have recently discovered nuclear functions of β-arrestins in transcriptional regulation.
6. β-Arrestins can traffic to receptors other than 7TMRs.

FUTURE ISSUES

1. How do conformational changes in β-arrestin contribute to its signaling and endocytic functions?
2. To what extent do homo- and hetero-oligomerization of β-arrestin affect its ability to desensitize, internalize, and carry out cell signaling?
3. Does ubiquitination of β-arrestin connect its trafficking and signaling roles?
4. What are the receptor-specific factors that determine if receptor phosphorylation is necessary for β-arrestin-mediated ERK activation?
5. Why do some receptors, such as the β2AR, require both β-arrestin isoforms for ERK activation, whereas others, such as the $AT1_AR$, only require one?
6. What are the downstream physiological effects of β-arrestin-dependent signaling?

ACKNOWLEDGMENTS

R.J.L. is an investigator with the Howard Hughes Medical Institute (HHMI). S.M.D. and S.A. are supported by HHMI. We acknowledge grant support from the National Institutes of Health (HL 16037 and HL 70631 to R.J.L. and HL080525 to S.K.S.) and the American Heart Association (0530014N to S.K.S.). We thank Donna Addison and Elizabeth Hall for excellent secretarial assistance. We also thank Chris Nelson, Kunhong Xiao, and Kelly Nobles for critical reading of this review.

LITERATURE CITED

1. Lefkowitz RJ. 2004. Historical review: a brief history and personal retrospective of seven-transmembrane receptors. *Trends Pharmacol. Sci.* 25:413–22
2. Pitcher JA, Freedman NJ, Lefkowitz RJ. 1998. G protein-coupled receptor kinases. *Annu. Rev. Biochem.* 67:653–92
3. Pao CS, Benovic JL. 2002. Phosphorylation-independent desensitization of G protein-coupled receptors? *Sci. STKE* 2002:PE42
4. Penela P, Ribas C, Mayor F Jr. 2003. Mechanisms of regulation of the expression and function of G protein-coupled receptor kinases. *Cell Signal.* 15:973–81
5. Benovic JL, Kuhn H, Weyand I, Codina J, Caron MG, Lefkowitz RJ. 1987. Functional desensitization of the isolated beta-adrenergic receptor by the beta-adrenergic receptor kinase: potential role of an analog of the retinal protein arrestin (48-kDa protein). *Proc. Natl. Acad. Sci. USA* 84:8879–82

6. Pfister C, Chabre M, Plouet J, Tuyen VV, De Kozak Y, et al. 1985. Retinal S antigen identified as the 48K protein regulating light-dependent phosphodiesterase in rods. *Science* 228:891–93
7. Lohse MJ, Benovic JL, Codina J, Caron MG, Lefkowitz RJ. 1990. β-Arrestin: a protein that regulates β-adrenergic receptor function. *Science* 248:1547–50
8. Attramadal H, Arriza JL, Aoki C, Dawson TM, Codina J, et al. 1992. Beta-arrestin2, a novel member of the arrestin/beta-arrestin gene family. *J. Biol. Chem.* 267:17882–90

8a. Moore CAC, Milano SK, Benovic J. 2007. Regulation of receptor trafficking by GRKs and arrestins. *Annu. Rev. Physiol.* 69:in press

9. Krupnick JG, Goodman OB, Jr., Keen JH, Benovic JL. 1997. Arrestin/clathrin interaction. Localization of the clathrin binding domain of nonvisual arrestins to the carboxy terminus. *J. Biol. Chem.* 272:15011–16
10. Goodman OB Jr, Krupnick JG, Santini F, Gurevich VV, Penn RB, et al. 1996. Beta-arrestin acts as a clathrin adaptor in endocytosis of the beta2-adrenergic receptor. *Nature* 383:447–50
11. Laporte SA, Miller WE, Kim KM, Caron MG. 2002. β-Arrestin/AP-2 interaction in G protein-coupled receptor internalization: identification of a β-arrestin binding site in β2-adaptin. *J. Biol. Chem.* 277:9247–54
12. Mukherjee S, Casanova JE, Hunzicker-Dunn M. 2001. Desensitization of the luteinizing hormone/choriogonadotropin receptor in ovarian follicular membranes is inhibited by catalytically inactive ARNO$^+$. *J. Biol. Chem.* 276:6524–28
13. Claing A, Chen W, Miller WE, Vitale N, Moss J, et al. 2001. β-Arrestin-mediated ADP-ribosylation factor 6 activation and β2-adrenergic receptor endocytosis. *J. Biol. Chem.* 276:42509–13
14. Conner DA, Mathier MA, Mortensen RM, Christe M, Vatner SF, et al. 1997. β-Arrestin1 knockout mice appear normal but demonstrate altered cardiac responses to β-adrenergic stimulation. *Circ. Res.* 81:1021–26
15. Bohn LM, Lefkowitz RJ, Gainetdinov RR, Peppel K, Caron MG, Lin FT. 1999. Enhanced morphine analgesia in mice lacking beta-arrestin 2. *Science* 286:2495–98
16. Kohout TA, Lin FS, Perry SJ, Conner DA, Lefkowitz RJ. 2001. β-Arrestin 1 and 2 differentially regulate heptahelical receptor signaling and trafficking. *Proc. Natl. Acad. Sci. USA* 98:1601–6
17. Oakley RH, Laporte SA, Holt JA, Caron MG, Barak LS. 2000. Differential affinities of visual arrestin, beta arrestin1, and beta arrestin2 for G protein-coupled receptors delineate two major classes of receptors. *J. Biol. Chem.* 275:17201–10
18. Paing MM, Stutts AB, Kohout TA, Lefkowitz RJ, Trejo J. 2002. β-Arrestins regulate protease-activated receptor-1 desensitization but not internalization or down-regulation. *J. Biol. Chem.* 277:1292–300
19. Lin FT, Miller WE, Luttrell LM, Lefkowitz RJ. 1999. Feedback regulation of beta-arrestin1 function by extracellular signal-regulated kinases. *J. Biol. Chem.* 274:15971–74
20. Kim YM, Barak LS, Caron MG, Benovic JL. 2002. Regulation of arrestin-3 phosphorylation by casein kinase II. *J. Biol. Chem.* 277:16837–46
21. Lin FT, Chen W, Shenoy S, Cong M, Exum ST, Lefkowitz RJ. 2002. Phosphorylation of β-arrestin2 regulates its function in internalization of β2-adrenergic receptors. *Biochemistry* 41:10692–99
22. Lin FT, Krueger KM, Kendall HE, Daaka Y, Fredericks ZL, et al. 1997. Clathrin-mediated endocytosis of the beta-adrenergic receptor is regulated by phosphorylation/dephosphorylation of beta-arrestin1. *J. Biol. Chem.* 272:31051–57

23. Shenoy SK, McDonald PH, Kohout TA, Lefkowitz RJ. 2001. Regulation of receptor fate by ubiquitination of activated beta 2-adrenergic receptor and beta-arrestin. *Science* 294:1307–13
24. Shenoy SK, Lefkowitz RJ. 2003. Trafficking patterns of beta-arrestin and G protein-coupled receptors determined by the kinetics of beta-arrestin deubiquitination. *J. Biol. Chem.* 278:14498–506
25. Shenoy SK, Lefkowitz RJ. 2005. Receptor-specific ubiquitination of beta-arrestin directs assembly and targeting of seven-transmembrane receptor signalosomes. *J. Biol. Chem.* 280:15315–24
26. Perroy J, Pontier S, Charest PG, Aubry M, Bouvier M. 2004. Real-time monitoring of ubiquitination in living cells by BRET. *Nat. Methods* 1:203–8
27. Hershko A, Ciechanover A, Heller H, Haas AL, Rose IA. 1980. Proposed role of ATP in protein breakdown: conjugation of protein with multiple chains of the polypeptide of ATP-dependent proteolysis. *Proc. Natl. Acad. Sci. USA* 77:1783–86
28. Haas AL, Warms JV, Hershko A, Rose IA. 1982. Ubiquitin-activating enzyme. Mechanism and role in protein-ubiquitin conjugation. *J. Biol. Chem.* 257:2543–48
29. Hicke L, Dunn R. 2003. Regulation of membrane protein transport by ubiquitin and ubiquitin-binding proteins. *Annu. Rev. Cell Dev. Biol.* 19:141–72
30. Shenoy SK, Lefkowitz RJ. 2003. Multifaceted roles of beta-arrestins in the regulation of seven-membrane-spanning receptor trafficking and signalling. *Biochem. J.* 375:503–15
31. Welchman RL, Gordon C, Mayer RJ. 2005. Ubiquitin and ubiquitin-like proteins as multifunctional signals. *Nat. Rev. Mol. Cell Biol.* 6:599–609
32. Wojcikiewicz RJ. 2004. Regulated ubiquitination of proteins in GPCR-initiated signaling pathways. *Trends Pharmacol. Sci.* 25:35–41
33. Chen ZJ. 2005. Ubiquitin signalling in the NF-κB pathway. *Nat. Cell Biol.* 7:758–65
34. Martin NP, Lefkowitz RJ, Shenoy SK. 2003. Regulation of V2 vasopressin receptor degradation by agonist-promoted ubiquitination. *J. Biol. Chem.* 278:45954–59
35. Marchese A, Benovic JL. 2001. Agonist-promoted ubiquitination of the G protein-coupled receptor CXCR4 mediates lysosomal sorting. *J. Biol. Chem.* 276:45509–12
36. Marchese A, Raiborg C, Santini F, Keen JH, Stenmark H, Benovic JL. 2003. The E3 ubiquitin ligase AIP4 mediates ubiquitination and sorting of the G protein-coupled receptor CXCR4. *Dev. Cell* 5:709–22
37. Jacob C, Cottrell GS, Gehringer D, Schmidlin F, Grady EF, Bunnett NW. 2005. c-Cbl mediates ubiquitination, degradation, and down-regulation of human protease-activated receptor 2. *J. Biol. Chem.* 280:16076–87
38. Girnita L, Girnita A, Larsson O. 2003. Mdm2-dependent ubiquitination and degradation of the insulin-like growth factor 1 receptor. *Proc. Natl. Acad. Sci. USA* 100:8247–52
39. Girnita L, Shenoy SK, Sehat B, Vasilcanu R, Girnita A, et al. 2005. β-Arrestin is crucial for ubiquitination and down-regulation of the insulin-like growth factor-1 receptor by acting as adaptor for the MDM2 E3 ligase. *J. Biol. Chem.* 280:24412–19
40. Mukherjee A, Veraksa A, Bauer A, Rosse C, Camonis J, Artavanis-Tsakonas S. 2005. Regulation of Notch signalling by non-visual beta-arrestin. *Nat. Cell Biol.* 7:1191–201
41. Herranz S, Rodriguez JM, Bussink HJ, Sanchez-Ferrero JC, Arst HN Jr, et al. 2005. Arrestin-related proteins mediate pH signaling in fungi. *Proc. Natl. Acad. Sci. USA* 102:12141–46
42. Katzmann DJ, Odorizzi G, Emr SD. 2002. Receptor downregulation and multivesicular-body sorting. *Nat. Rev. Mol. Cell Biol.* 3:893–905

43. Luttrell LM, Daaka Y, Della Rocca GJ, Lefkowitz RJ. 1997. G protein-coupled receptors mediate two functionally distinct pathways of tyrosine phosphorylation in rat 1a fibroblasts. Shc phosphorylation and receptor endocytosis correlate with activation of Erk kinases. *J. Biol. Chem.* 272:31648–56
44. Lin FT, Daaka Y, Lefkowitz RJ. 1998. β-Arrestins regulate mitogenic signaling and clathrin-mediated endocytosis of the insulin-like growth factor I receptor. *J. Biol. Chem.* 273:31640–43
45. Daaka Y, Luttrell LM, Ahn S, Della Rocca GJ, Ferguson SS, et al. 1998. Essential role for G protein-coupled receptor endocytosis in the activation of mitogen-activated protein kinase. *J. Biol. Chem.* 273:685–88
46. Luttrell LM, Ferguson SS, Daaka Y, Miller WE, Maudsley S, et al. 1999. Beta-arrestin-dependent formation of beta2 adrenergic receptor-Src protein kinase complexes. *Science* 283:655–61
47. DeFea KA, Vaughn ZD, O'Bryan EM, Nishijima D, Dery O, Bunnett NW. 2000. The proliferative and antiapoptotic effects of substance P are facilitated by formation of a beta-arrestin-dependent scaffolding complex. *Proc. Natl. Acad. Sci. USA* 97:11086–91
48. Miller WE, Maudsley S, Ahn S, Khan KD, Luttrell LM, Lefkowitz RJ. 2000. Beta-arrestin1 interacts with the catalytic domain of the tyrosine kinase c-SRC. Role of beta-arrestin1-dependent targeting of c-SRC in receptor endocytosis. *J. Biol. Chem.* 275:11312–19
49. DeFea KA, Zalevsky J, Thoma MS, Dery O, Mullins RD, Bunnett NW. 2000. Beta-arrestin-dependent endocytosis of proteinase-activated receptor 2 is required for intracellular targeting of activated ERK1/2. *J Cell Biol* 148:1267–81
50. Luttrell LM, Roudabush FL, Choy EW, Miller WE, Field ME, et al. 2001. Activation and targeting of extracellular signal-regulated kinases by beta-arrestin scaffolds. *Proc. Natl. Acad. Sci. USA* 98:2449–54
51. Scott MG, Pierotti V, Storez H, Lindberg E, Thuret A, et al. 2006. Cooperative regulation of extracellular signal-regulated kinase activation and cell shape change by filamin A and β-arrestins. *Mol. Cell Biol.* 26:3432–45
52. Tohgo A, Pierce KL, Choy EW, Lefkowitz RJ, Luttrell LM. 2002. β-Arrestin scaffolding of the ERK cascade enhances cytosolic ERK activity but inhibits ERK-mediated transcription following angiotensin AT1a receptor stimulation. *J. Biol. Chem.* 277:9429–36
53. Tohgo A, Choy EW, Gesty-Palmer D, Pierce KL, Laporte S, et al. 2003. The stability of the G protein-coupled receptor-beta-arrestin interaction determines the mechanism and functional consequence of ERK activation. *J. Biol. Chem.* 278:6258–67
54. Barnes WG, Reiter E, Violin JD, Ren XR, Milligan G, Lefkowitz RJ. 2005. β-Arrestin 1 and Galphaq/11 coordinately activate RhoA and stress fiber formation following receptor stimulation. *J. Biol. Chem.* 280:8041–50
55. Ge L, Ly Y, Hollenberg M, DeFea K. 2003. A beta-arrestin-dependent scaffold is associated with prolonged MAPK activation in pseudopodia during protease-activated receptor-2-induced chemotaxis. *J. Biol. Chem.* 278:34418–26
56. Hunton DL, Barnes WG, Kim J, Ren XR, Violin JD, et al. 2005. Beta-arrestin 2-dependent angiotensin II type 1A receptor-mediated pathway of chemotaxis. *Mol. Pharmacol.* 67:1229–36
57. Ahn S, Nelson CD, Garrison TR, Miller WE, Lefkowitz RJ. 2003. Desensitization, internalization, and signaling functions of beta-arrestins demonstrated by RNA interference. *Proc. Natl. Acad. Sci. USA* 100:1740–44

58. Ahn S, Shenoy SK, Wei H, Lefkowitz RJ. 2004. Differential kinetic and spatial patterns of beta-arrestin and G protein-mediated ERK activation by the angiotensin II receptor. *J. Biol. Chem.* 279:35518–25
58a. Lefkowitz RJ, Shenoy SK. 2005. Transduction of receptor signals by beta-arrestins. *Science* 308:512–17
59. Ahn S, Wei H, Garrison TR, Lefkowitz RJ. 2004. Reciprocal regulation of angiotensin receptor-activated extracellular signal-regulated kinases by beta-arrestins 1 and 2. *J. Biol. Chem.* 279:7807–11
60. Ren XR, Reiter E, Ahn S, Kim J, Chen W, Lefkowitz RJ. 2005. Different G protein-coupled receptor kinases govern G protein and beta-arrestin-mediated signaling of V2 vasopressin receptor. *Proc. Natl. Acad. Sci. USA* 102:1448–53
61. Shenoy SK, Drake MT, Nelson CD, Houtz DA, Xiao K, et al. 2006. Beta-arrestin-dependent, G protein-independent ERK1/2 activation by the beta2 adrenergic receptor. *J. Biol. Chem.* 281:1261–73
62. Gesty-Palmer D, Chen M, Reiter E, Ahn S, Nelson CD, et al. 2006. Distinct beta-arrestin- and G protein-dependent pathways for parathyroid hormone receptor-stimulated ERK1/2 activation. *J. Biol. Chem.* 281:10856–64
63. Gaborik Z, Jagadeesh G, Zhang M, Spat A, Catt KJ, Hunyady L. 2003. The role of a conserved region of the second intracellular loop in AT1 angiotensin receptor activation and signaling. *Endocrinology* 144:2220–28
64. Wei H, Ahn S, Shenoy SK, Karnik SS, Hunyady L, et al. 2003. Independent beta-arrestin 2 and G protein-mediated pathways for angiotensin II activation of extracellular signal-regulated kinases 1 and 2. *Proc. Natl. Acad. Sci. USA* 100:10782–87
65. Holloway AC, Qian H, Pipolo L, Ziogas J, Miura S, et al. 2002. Side-chain substitutions within angiotensin II reveal different requirements for signaling, internalization, and phosphorylation of type 1A angiotensin receptors. *Mol. Pharmacol.* 61:768–77
66. Yee DK, Suzuki A, Luo L, Fluharty SJ. 2006. Identification of structural determinants for G-protein independent activation of mitogen activated protein kinases in the seventh transmembrane domain of the angiotensin II type 1 receptor. *Mol. Endocrinol.* 20(8):1924–34
67. Daniels D, Yee DK, Faulconbridge LF, Fluharty SJ. 2005. Divergent behavioral roles of angiotensin receptor intracellular signaling cascades. *Endocrinology* 146:5552–60
68. O'Donnell SR, Wanstall JC. 1980. Evidence that ICI 118551 is a potent, highly beta 2-selective adrenoceptor antagonist and can be used to characterize beta-adrenoceptor populations in tissues. *Life Sci.* 27:671–77
69. Azzi M, Charest PG, Angers S, Rousseau G, Kohout T, et al. 2003. Beta-arrestin-mediated activation of MAPK by inverse agonists reveals distinct active conformations for G protein-coupled receptors. *Proc. Natl. Acad. Sci. USA* 100:11406–11
70. Kohout TA, Nicholas SL, Perry SJ, Reinhart G, Junger S, Struthers RS. 2004. Differential desensitization, receptor phosphorylation, beta-arrestin recruitment, and ERK1/2 activation by the two endogenous ligands for the CC chemokine receptor 7. *J. Biol. Chem.* 279:23214–22
71. Sun Y, Cheng Z, Ma L, Pei G. 2002. Beta-arrestin2 is critically involved in CXCR4-mediated chemotaxis, and this is mediated by its enhancement of p38 MAPK activation. *J. Biol. Chem.* 277:49212–19
72. Jiang Q, Yan Z, Feng J. 2006. Activation of group III metabotropic glutamate receptors attenuates rotenone toxicity on dopaminergic neurons through a microtubule-dependent mechanism. *J. Neurosci.* 26:4318–28

73. Mohit AA, Martin JH, Miller CA. 1995. p493F12 kinase: a novel MAP kinase expressed in a subset of neurons in the human nervous system. *Neuron* 14:67–78
74. McDonald PH, Chow CW, Miller WE, Laporte SA, Field ME, et al. 2000. Beta-arrestin 2: a receptor-regulated MAPK scaffold for the activation of JNK3. *Science* 290:1574–77
75. Miller WE, McDonald PH, Cai SF, Field ME, Davis RJ, Lefkowitz RJ. 2001. Identification of a motif in the carboxyl terminus of beta-arrestin2 responsible for activation of JNK3. *J. Biol. Chem.* 276:27770–77
76. Scott MG, Le Rouzic E, Perianin A, Pierotti V, Enslen H, et al. 2002. Differential nucleocytoplasmic shuttling of beta-arrestins. Characterization of a leucine-rich nuclear export signal in beta-arrestin2. *J. Biol. Chem.* 277:37693–701
77. Willoughby EA, Collins MK. 2005. Dynamic interaction between the dual specificity phosphatase MKP7 and the JNK3 scaffold protein beta-arrestin 2. *J. Biol. Chem.* 280:25651–58
78. Miller WE, Houtz DA, Nelson CD, Kolattukudy PE, Lefkowitz RJ. 2003. G-protein-coupled receptor (GPCR) kinase phosphorylation and beta-arrestin recruitment regulate the constitutive signaling activity of the human cytomegalovirus US28 GPCR. *J. Biol. Chem.* 278:21663–71
79. Bruchas MR, Macey TA, Lowe JD, Chavkin C. 2006. Kappa opioid receptor activation of p38 MAPK is GRK3- and arrestin-dependent in neurons and astrocytes. *J. Biol. Chem.* 281:18081–89
80. Povsic TJ, Kohout TA, Lefkowitz RJ. 2003. Beta-arrestin1 mediates insulin-like growth factor 1 (IGF-1) activation of phosphatidylinositol 3-kinase (PI3K) and anti-apoptosis. *J. Biol. Chem.* 278:51334–39
81. Goel R, Phillips-Mason PJ, Raben DM, Baldassare JJ. 2002. α-Thrombin induces rapid and sustained Akt phosphorylation by β-arrestin1-dependent and -independent mechanisms, and only the sustained Akt phosphorylation is essential for G1 phase progression. *J. Biol. Chem.* 277:18640–48
82. Goel R, Baldassare JJ. 2002. β-Arrestin 1 couples thrombin to the rapid activation of the Akt pathway. *Ann. NY Acad. Sci.* 973:138–41
83. Goel R, Phillips-Mason PJ, Gardner A, Raben DM, Baldassare JJ. 2004. α-Thrombin-mediated phosphatidylinositol 3-kinase activation through release of Gβγ dimers from Gαq and Gαi2. *J. Biol. Chem.* 279:6701–10
84. Beaulieu JM, Sotnikova TD, Marion S, Lefkowitz RJ, Gainetdinov RR, Caron MG. 2005. An Akt/beta-arrestin 2/PP2A signaling complex mediates dopaminergic neurotransmission and behavior. *Cell* 122:261–73
85. Sen R, Baltimore D. 1986. Inducibility of κ immunoglobulin enhancer-binding protein NF-κB by a posttranslational mechanism. *Cell* 47:921–28
86. Karin M, Greten FR. 2005. NF-κB: linking inflammation and immunity to cancer development and progression. *Nat. Rev. Immunol.* 5:749–59
87. Witherow DS, Garrison TR, Miller WE, Lefkowitz RJ. 2004. β-Arrestin inhibits NF-κB activity by means of its interaction with the NF-κB inhibitor IκBα. *Proc. Natl. Acad. Sci. USA* 101:8603–7
88. Gao H, Sun Y, Wu Y, Luan B, Wang Y, et al. 2004. Identification of beta-arrestin2 as a G protein-coupled receptor-stimulated regulator of NF-κB pathways. *Mol Cell* 14:303–17
89. Luan B, Zhang Z, Wu Y, Kang J, Pei G. 2005. Beta-arrestin2 functions as a phosphorylation-regulated suppressor of UV-induced NF-κB activation. *EMBO J.* 24:4237–46

90. Gesty-Palmer D, Shewy HE, Kohout TA, Luttrell LM. 2005. β-Arrestin 2 expression determines the transcriptional response to lysophosphatidic acid stimulation in murine embryo fibroblasts. *J. Biol. Chem.* 280:32157–67
91. Piu F, Gauthier NK, Wang F. 2006. Beta-arrestin 2 modulates the activity of nuclear receptor RAR beta2 through activation of ERK2 kinase. *Oncogene* 25:218–29
92. Kang J, Shi Y, Xiang B, Qu B, Su W, et al. 2005. A nuclear function of beta-arrestin1 in GPCR signaling: regulation of histone acetylation and gene transcription. *Cell* 123:833–47
93. Puig J, Arendt A, Tomson FL, Abdulaeva G, Miller R, et al. 1995. Synthetic phosphopeptide from rhodopsin sequence induces retinal arrestin binding to photoactivated unphosphorylated rhodopsin. *FEBS Lett.* 362:185–88
94. McDowell JH, Smith WC, Miller RL, Popp MP, Arendt A, et al. 1999. Sulfhydryl reactivity demonstrates different conformational states for arrestin, arrestin activated by a synthetic phosphopeptide, and constitutively active arrestin. *Biochemistry* 38:6119–25
95. Gurevich VV, Gurevich EV. 2004. The molecular acrobatics of arrestin activation. *Trends Pharmacol. Sci.* 25:105–11
96. Granzin J, Wilden U, Choe HW, Labahn J, Krafft B, Buldt G. 1998. X-ray crystal structure of arrestin from bovine rod outer segments. *Nature* 391:918–21
97. Han M, Gurevich VV, Vishnivetskiy SA, Sigler PB, Schubert C. 2001. Crystal structure of beta-arrestin at 1.9 Å: possible mechanism of receptor binding and membrane translocation. *Structure* 9:869–80
98. Hirsch JA, Schubert C, Gurevich VV, Sigler PB. 1999. The 2.8 Å crystal structure of visual arrestin: a model for arrestin's regulation. *Cell* 97:257–69
99. Milano SK, Pace HC, Kim YM, Brenner C, Benovic JL. 2002. Scaffolding functions of arrestin-2 revealed by crystal structure and mutagenesis. *Biochemistry* 41:3321–28
100. Vishnivetskiy SA, Paz CL, Schubert C, Hirsch JA, Sigler PB, Gurevich VV. 1999. How does arrestin respond to the phosphorylated state of rhodopsin? *J. Biol. Chem.* 274:11451–54
101. Xiao K, Shenoy SK, Nobles K, Lefkowitz RJ. 2004. Activation-dependent conformational changes in beta-arrestin 2. *J. Biol. Chem.* 279:55744–53
102. Charest PG, Terrillon S, Bouvier M. 2005. Monitoring agonist-promoted conformational changes of beta-arrestin in living cells by intramolecular BRET. *EMBO Rep.* 6:334–40
103. Terrillon S, Bouvier M. 2004. Receptor activity-independent recruitment of beta-arrestin2 reveals specific signalling modes. *EMBO J.* 23:3950–61
104. Gurevich VV, Benovic JL. 1995. Visual arrestin binding to rhodopsin. Diverse functional roles of positively charged residues within the phosphorylation-recognition region of arrestin. *J. Biol. Chem.* 270:6010–16
105. Kovoor A, Celver J, Abdryashitov RI, Chavkin C, Gurevich VV. 1999. Targeted construction of phosphorylation-independent beta-arrestin mutants with constitutive activity in cells. *J. Biol. Chem.* 274:6831–34
106. Chen W, Hu LA, Semenov MV, Yanagawa S, Kikuchi A, et al. 2001. β-Arrestin1 modulates lymphoid enhancer factor transcriptional activity through interaction with phosphorylated dishevelled proteins. *Proc. Natl. Acad. Sci. USA* 98:14889–94
107. Chen W, Ren XR, Nelson CD, Barak LS, Chen JK, et al. 2004. Activity-dependent internalization of smoothened mediated by beta-arrestin 2 and GRK2. *Science* 306:2257–60
108. Chen W, Kirkbride KC, How T, Nelson CD, Mo J, et al. 2003. Beta-arrestin 2 mediates endocytosis of type III TGF-beta receptor and down-regulation of its signaling. *Science* 301:1394–97

109. Wu JH, Peppel K, Nelson CD, Lin FT, Kohout TA, et al. 2003. The adaptor protein beta-arrestin2 enhances endocytosis of the low density lipoprotein receptor. *J. Biol. Chem.* 278:44238–45
110. Szabo EZ, Numata M, Lukashova V, Iannuzzi P, Orlowski J. 2005. β-Arrestins bind and decrease cell-surface abundance of the Na^+/H^+ exchanger NHE5 isoform. *Proc. Natl. Acad. Sci. USA* 102:2790–95
111. Wang Y, Tang Y, Teng L, Wu Y, Zhao X, Pei G. 2006. Association of beta-arrestin and TRAF6 negatively regulates Toll-like receptor-interleukin 1 receptor signaling. *Nat. Immunol.* 7:139–47
112. Devi LA. 2001. Heterodimerization of G-protein-coupled receptors: pharmacology, signaling and trafficking. *Trends Pharmacol. Sci.* 22:532–37
113. Terrillon S, Bouvier M. 2004. Roles of G-protein-coupled receptor dimerization. *EMBO Rep.* 5:30–34
114. Lee SJ, Montell C. 2004. Light-dependent translocation of visual arrestin regulated by the NINAC myosin III. *Neuron* 43:95–103
115. Lee SJ, Xu H, Kang LW, Amzel LM, Montell C. 2003. Light adaptation through phosphoinositide-regulated translocation of Drosophila visual arrestin. *Neuron* 39:121–32
116. Strissel KJ, Arshavsky VY. 2004. Myosin III illuminates the mechanism of arrestin translocation. *Neuron* 43:2–4
117. Gaidarov I, Krupnick JG, Falck JR, Benovic JL, Keen JH. 1999. Arrestin function in G protein-coupled receptor endocytosis requires phosphoinositide binding. *EMBO J.* 18:871–81
118. Milano SK, Kim YM, Stefano FP, Benovic JL, Brenner C. 2006. Nonvisual arrestin oligomerization and cellular localization are regulated by inositol hexakisphosphate binding. *J. Biol. Chem.* 281:9812–23
119. Storez H, Scott MG, Issafras H, Burtey A, Benmerah A, et al. 2005. Homo- and hetero-oligomerization of beta-arrestins in living cells. *J. Biol. Chem.* 280:40210–15
120. Varadan R, Assfalg M, Haririnia A, Raasi S, Pickart C, Fushman D. 2004. Solution conformation of Lys63-linked di-ubiquitin chain provides clues to functional diversity of polyubiquitin signaling. *J. Biol. Chem.* 279:7055–63
121. Kuo FT, Lu TL, Fu HW. 2006. Opposing effects of beta-arrestin1 and beta-arrestin2 on activation and degradation of Src induced by protease-activated receptor 1. *Cell Signal.* In press
122. Qian H, Pipolo L, Thomas WG. 1999. Identification of protein kinase C phosphorylation sites in the angiotensin II (AT1A) receptor. *Biochem. J.* 343(Pt. 3):637–44
123. Milasta S, Evans NA, Ormiston L, Wilson S, Lefkowitz RJ, Milligan G. 2005. The sustainability of interactions between the orexin-1 receptor and beta-arrestin-2 is defined by a single C-terminal cluster of hydroxy amino acids and modulates the kinetics of ERK MAPK regulation. *Biochem. J.* 387:573–84
124. Stalheim L, Ding Y, Gullapalli A, Paing MM, Wolfe BL, et al. 2005. Multiple independent functions of arrestins in the regulation of protease-activated receptor-2 signaling and trafficking. *Mol. Pharmacol.* 67:78–87
125. Bohn LM, Gainetdinov RR, Sotnikova TD, Medvedev IO, Lefkowitz RJ, et al. 2003. Enhanced rewarding properties of morphine, but not cocaine, in β-arrestin-2 knock-out mice. *J. Neurosci.* 23:10265–73
126. Bohn LM, Gainetdinov RR, Lin FT, Lefkowitz RJ, Caron MG. 2000. Mu-opioid receptor desensitization by beta-arrestin-2 determines morphine tolerance but not dependence. *Nature* 408:720–23

127. Wang Q, Zhao J, Brady AE, Feng J, Allen PB, et al. 2004. Spinophilin blocks arrestin actions in vitro and in vivo at G protein-coupled receptors. *Science* 304:1940–94
128. Morris M, Li P, Callahan MF, Oliverio MI, Coffman TM, et al. 1999. Neuroendocrine effects of dehydration in mice lacking the angiotensin AT1a receptor. *Hypertension* 33:482–86

RELATED RESOURCES

1. Luttrell LM. 2005. Composition and function of G protein-coupled receptor signalsomes controlling mitogen-activated protein kinase activity. *J. Mol. Neurosci.* 26:253–64
2. Shenoy SK, Lefkowitz RJ. 2003. Multifaceted roles of beta-arrestins in the regulation of seven-membrane-spanning receptor trafficking and signaling. *Biochem. J.* 375(Pt. 3):503–15
3. Gurevich VV, Gurevich EV. 2004. The molecular acrobatics of arrestin activation. *Trends Pharmacol. Sci.* 25(2):105–11
4. Marchese A, Chen C, Kim YM, Benovic JL. 2003. The ins and outs of G protein-coupled receptor trafficking. *Trends Biochem. Sci.* 28:369–76
5. Kenakin T. 2002. Drug efficacy at G protein-coupled receptors. *Annu. Rev. Pharmacol. Toxicol.* 42:349–79

Physiological Roles of G Protein–Coupled Receptor Kinases and Arrestins

Richard T. Premont[1] and Raul R. Gainetdinov[2]

Departments of Medicine[1] and Cell Biology,[2] Duke University Medical Center, Durham, North Carolina 27710; email: richard.premont@duke.edu, r.gainetdinov@cellbio.duke.edu

Annu. Rev. Physiol. 2007. 69:511–34

The *Annual Review of Physiology* is online at http://physiol.annualreviews.org

This article's doi:
10.1146/annurev.physiol.69.022405.154731

0066-4278/07/0315-0511$20.00

Key Words

GPCR, GRK, brain, heart, lung, immunity

Abstract

Heterotrimeric G protein–coupled receptors (GPCRs) are found on the surface of all cells of multicellular organisms and are major mediators of intercellular communication. More than 800 distinct GPCRs are present in the human genome, and individual receptor subtypes respond to hormones, neurotransmitters, chemokines, odorants, or tastants. GPCRs represent the most widely targeted pharmacological protein class. Because drugs that target GPCRs often engage receptor regulatory mechanisms that limit drug effectiveness, particularly in chronic treatment, there is great interest in understanding how GPCRs are regulated, as a basis for designing therapeutic drugs that evade this regulation. The major GPCR regulatory pathway involves phosphorylation of activated receptors by G protein–coupled receptor kinases (GRKs), followed by binding of arrestin proteins, which prevent receptors from activating downstream heterotrimeric G protein pathways while allowing activation of arrestin-dependent signaling pathways. Although the general mechanisms of GRK-arrestin regulation have been well explored in model cell systems and with purified proteins, much less is known about the role of GRK-arrestin regulation of receptors in physiological and pathophysiological settings. This review focuses on the physiological functions and potential pathophysiological roles of GRKs and arrestins in human disorders as well as on recent studies using knockout and transgenic mice to explore the role of GRK-arrestin regulation of GPCRs in vivo.

G protein–coupled receptors (GPCRs): a family of seven-transmembrane proteins that transduce extracellular signals from agonist binding into intracellular signaling pathways

G proteins (guanine nucleotide–binding proteins): a family of proteins that use the binding of guanosine triphosphate (GTP) and its hydrolysis to guanosine diphosphate (GDP) as a molecular switch to transmit biological signals from cell surface receptors to effector enzymes inside the cell

Arrestins: adaptor proteins that bind to GRK-phosphorylated GPCRs to block G protein activation, initiate receptor endocytosis and trafficking, and promote G protein–independent signaling through associated signaling molecules

G protein–coupled receptor kinases (GRKs): protein kinases that recognize and phosphorylate activated GPCRs to promote arrestin binding

MECHANISMS OF GPCR SIGNALING AND DESENSITIZATION

G protein–coupled receptors (GPCRs) are the largest superfamily of cell surface receptor proteins (1). When the appropriate agonist ligand binds to a receptor on the cell surface, the intracellular domain of the receptor changes conformation in ways that are still poorly understood and results in a conformation that acts as a guanine nucleotide exchange factor for heterotrimeric guanine nucleotide–binding proteins (G proteins). Activated receptors facilitate GDP release from inactive G protein heterotrimers as well as GTP binding to activate the G protein. The receptors derive their name from this common mechanism. Nevertheless, it has become clear in recent years that these receptors do not signal exclusively via heterotrimeric G proteins, although that is a major part of their function. GPCRs are also called seven-transmembrane-span (7TM) or heptahelical receptors, referring to the overall structural motif shared by all three mammalian families of these receptors. Receptors within each of these three families share extensive primary sequence similarity, although the families are very distinct from each other.

GPCRs regulate some part of nearly all physiological functions. There are more than 800 known GPCRs in the human genome. Ligands for these receptors include large glycoprotein hormones, a multitude of peptides, bioactive lipids, amino acids and amino acid metabolites such as dopamine and norepinephrine, small molecules such as acetylcholine and sucrose, calcium ions, and even photons. Despite this great diversity of ligand types, receptor function is generally very modular: Receptors couple to a subset of the 16 heterotrimeric G protein subtypes, which are functionally grouped into four broad classes: G_s, G_i, G_q, and G_{12}. These G proteins in turn regulate a relatively small number of intracellular G protein effectors. Within this system, some receptors couple primarily to one G protein subtype and one effector (G_s activation of adenylyl cyclase, for instance), whereas others promiscuously activate members of all four G protein subtypes and couple to correspondingly more intracellular signaling pathways. GPCRs also perform isotype-specific functions through unique interaction partners and can activate signaling pathways independent of (or parallel to) those mediated by heterotrimeric G proteins. Understanding these unique functions has been of great interest in recent years and has been reviewed extensively (2, 3).

All GPCR systems exhibit context-dependent activity (4): That is, receptor sensitivity changes depending on the amount of signaling that a particular receptor has stimulated on a given cell, so that in general receptors adjust their sensitivity to the range of agonist concentrations to which they are exposed. Receptors desensitize to prolonged or repeated exposure to high agonist concentration and resensitize when not exposed to agonist for some time. One important mechanism for regulating GPCR responsiveness is the G protein–coupled receptor kinase (GRK)-arrestin pathway (**Figure 1**) (5–9). Just as G proteins recognize activated receptors, the GRKs also recognize activated GPCRs, which leads to catalytic activation of the protein kinase and results in receptor phosphorylation at specific sites on the intracellular loops and carboxyl-terminal tail. Because of this recognition, GRKs strongly prefer the activated receptor as substrate. Once phosphorylated, receptors become substrates for binding of arrestin proteins, which prevent the receptor from activating additional G proteins. GRK phosphorylation and arrestin binding result in a cessation of G protein signaling, even despite the continued presence of the receptor-activating agonist. The GRK-arrestin pathway also performs other functions, such as facilitating receptor internalization from the cell surface through clathrin-coated pits. By binding to additional signaling proteins, GRKs and arrestins also function as switches, converting receptor activation from

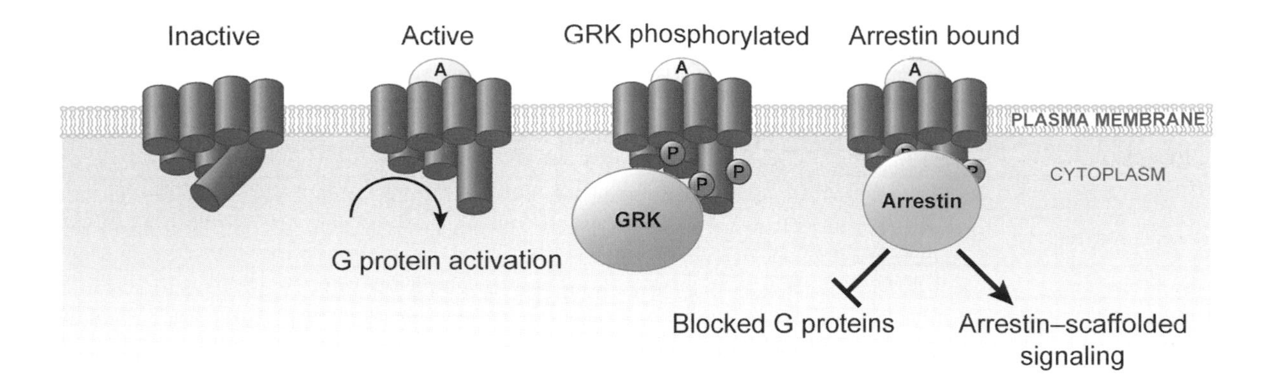

Figure 1

In the absence of an activating agonist ligand (A), the 7TM G protein–coupled receptor (GPCR) on the cell surface is in an inactive conformation and does not activate downstream G proteins or other signaling pathways. When an appropriate agonist binds to the extracellular face of its receptor, the receptor undergoes a conformational change to expose surfaces that act as a guanine nucleotide exchange factor for heterotrimeric G proteins. The activated receptor facilitates release of tightly bound GDP from an inactive G protein bound to the receptor; the receptor then facilitates binding of GTP to the G protein. The GTP-bound G protein undergoes a conformational change that causes its release from the receptor and dissociation into activated α- and βγ-subunits, which each go on to activate downstream effectors (*not shown*). As long as the agonist remains bound to the receptor, this activated receptor can continue to activate G proteins. A G protein–coupled receptor kinase (GRK), which is catalytically activated by this interaction, also recognizes the activated conformation of the receptor. Activated GRKs phosphorylate (P) intracellular domains of the receptor and are then released. The agonist-activated, GRK-phosphorylated receptor binds tightly to an arrestin protein, which interdicts (desensitizes) further G protein activation and couples the receptor to the clathrin-coated-pit internalization pathway and to arrestin-scaffolded (and G protein–independent) signaling pathways.

heterotrimeric G protein signaling pathways to G protein–independent pathways. These pathways have been reviewed extensively (2, 3, 10).

There are seven GRKs in humans, named GRK1 through GRK7, and four arrestin proteins, named arrestins 1 through 4. The GRKs are functionally divided into three classes: GRK1-like, GRK2-like, and GRK4-like. GRK1 (rhodopsin kinase) and the related GRK7 (iodopsin kinase) are primarily found in the retina and regulate the light receptors, the opsins. GRK2 and the related GRK3 are widely expressed, although GRK2 is present at higher levels in tissues, and share a carboxyl-terminal pleckstrin homology domain that mediates PIP_2 and G protein βγ-subunit-mediated translocation of these kinases to the inner leaflet of the plasma membrane near activated receptor substrates. GRK4, GRK5, and GRK6 lack this G protein βγ-subunit binding domain but use direct PIP_2 binding and/or covalent lipid modification with palmitate to reside primarily at the plasma membrane. GRK4 has a limited tissue distribution—it is mainly found in testis—whereas GRK5 and GRK6 are widely expressed (11). Thus, most receptors in the body are potentially regulated by only four of these kinases: GRKs 2, 3, 5, or 6. Arrestins also have visual-specific isoforms, arrestin-1 and arrestin-4, as well as widely expressed somatic isoforms, arrestin-2 (β-arrestin1) and arrestin-3 (β-arrestin2). Likewise, most receptors in the body are subject to regulation only by β-arrestin1 or -2.

PIP_2: phosphatidyl inositol bisphosphate

Most of what has been learned about receptor regulation and signaling via GRK-arrestin pathways has been determined using model cell systems (2, 3, 5–8). Although

Table 1 Major phenotypes in mice deficient in nonvisual GRKs and arrestins

GRK or arrestin	Knockout phenotype
GRK2 [β-adrenergic receptor kinase (βARK), βARK1]	Embryonic lethal, thin myocardium syndrome in embryos (19); enhanced basal and adrenergic responses in cardiac function in adult heterozygotes (18); altered progression of experimental autoimmune encephalomyelitis (48)
GRK3 (βARK2)	Lack of olfactory receptor desensitization (82); altered M2 muscarinic airway regulation (20); blunted kappa-opioid receptor–mediated tolerance in spinal analgesia test (90, 91); disrupted tolerance to the antinociceptive effects of fentanyl but not morphine (89).
GRK4 (IT11)	Normal fertility and sperm function (R.T. Premont, unpublished data). No obvious phenotype.
GRK5	Altered central (96) and lung (51) M2 muscarinic receptor regulation, with normal heart M2 receptor regulation (51).
GRK6	Altered central dopamine receptor regulation (99, 102); deficient lymphocyte chemotaxis (44); increased acute inflammation and neutrophil chemotaxis (45, 46).
β-arrestin1 (arrestin-2)	Altered cardiac responses to beta-adrenergic stimulation (21).
β-arrestin2 (arrestin-3)	Enhanced morphine antinociception (115, 118), reward (125), disrupted morphine tolerance (116), and reduced constipation and respiratory suppression (120); reduced locomotor activity and disrupted dopamine-mediated behaviors (127); deficient lymphocyte chemotaxis (44); altered susceptibility to endotoxic shock and expression of proinflammatory cytokines (49); altered CXCR2-mediated neutrophil chemotaxis (50); altered asthmatic response to allergens (52); decreased bone mass and altered bone architecture (56, 57).

the pathways appear very interesting, it is often unclear whether they actually function in any specific organ in a living organism or have detectable physiological consequences. Studies using knockout and transgenic mice, studies using viral-mediated overexpression or short hairpin RNA (shRNA) knockdown, and genetic studies in human patients recently have begun to probe the physiological roles of GRKs and arrestins. These studies confirm that GRKs and arrestins do regulate GPCR pathways in humans and other animals with a specificity that is only dimly understood and that altering these regulatory pathways can have profound influences on physiological responses and pathophysiology. Owing to the nature of GRK function, involving the regulation of activated receptors, mice lacking GRKs or arrestins often look completely normal until stressed by exposure to a receptor activator. Likewise, because the GRK-arrestin system can have dual roles, those of both suppressing G protein signaling and promoting non–G protein signaling, loss of this regulation can have two opposing effects on physiological systems: (*a*) allowing enhanced or unregulated receptor signaling through loss of desensitization or (*b*) decreasing signaling by preventing the switch from G protein to non–G protein pathways.

shRNA: short hairpin RNA

This chapter reviews the role of nonvisual GRKs and arrestins in a few physiological systems to outline what has been learned so far and how much remains to be explored. **Table 1** summarizes the characterized phenotypes of GRK and β-arrestin knockout mice. Because the large literature on rhodopsin regulation by GRK1/rhodopsin kinase and arrestin regulation in the visual system has been reviewed recently (12–14), we do not discuss this here.

GRKs AND ARRESTINS IN THE CARDIOVASCULAR SYSTEM

GPCRs and their agonists play important roles throughout the cardiovascular system. Blood pressure is controlled by altering heart rate, vascular resistance, and fluid/electrolyte balance, each of which is regulated to a large degree by GPCRs. Heart rate is regulated by adrenergic and muscarinic receptor activation by norepinephrine and acetylcholine, respectively. Vascular tone is regulated by angiotensin II and bradykinin receptor activation, and kidney water and salt reabsorption (fluid balance) is regulated by angiotensin II, vasopressin, and dopamine receptor activation. Because GPCRs play such diverse and fundamental roles in regulating multiple aspects of cardiovascular function, GRKs and arrestins are important in controlling the sensitivity and responses of these organs.

Researchers have studied in detail the GRK regulation of direct cardiac function, using cardiomyocyte overexpression of GRKs and GRK fragments (i.e., GRK2 carboxy terminus as a G protein $\beta\gamma$-subunit scavenger). As this area has been reviewed extensively in recent years (15–17), we do not discuss these transgenic overexpression studies further here. Overall, there is good evidence that GRK2 is important for regulating several cardiac receptor responses and that GRK3 and GRK5 are involved in some receptor responses but not others. Studies in knockout animals also support the important role of GRKs and β-arrestins in cardiac responses. Genetic deletion of GRK2 in mice results in embryonic lethality owing to hypoplasia of the ventricular myocardium, directly demonstrating a critical role of this kinase in cardiac development, whereas GRK2-heterozygous mice exhibit altered cardiac function (18, 19). GRK3 knockout mice also showed altered cardiovascular responses to cholinergic stimulation (20). Deletion of β-arrestin1 in mice does not change resting cardiovascular parameters but results in altered cardiac responses to β-adrenergic stimulation (21).

More recent work has expanded into other aspects of cardiovascular regulation by focusing on overexpression of GRK2 and GRK5 in vascular smooth muscle cells. Hypertension has been associated with altered GRK expression and function. Human patients with hypertension have increased GRK2 expression in lymphocytes (22, 23), and hypertensive rats have increased vascular smooth muscle cell expression of GRK2 (24). Eckhart and colleagues (25) found that mice overexpressing GRK2 from the smooth muscle–specific SM22α promoter had elevated resting blood pressure. These mice also exhibited reduced sensitivity to vascular muscle relaxation by the β-adrenergic agonist isoproterenol and to increased mean arterial blood pressure in response to the vasoconstrictive peptide angiotensin II. SM22α-GRK2-overexpressing mice had a greatly increased smooth muscle cell layer thickness in the aortic medial layer, suggested to result from prolonged exposure to elevated blood pressure. Increased cardiac hypertrophy was also evident, despite normal GRK2 expression in the cardiomyocytes (25). In contrast, although SM22α-GRK5 mice were also hypertensive, they did not exhibit either aortic or cardiac hypertrophy (26). Interestingly, male and female mice overexpressing GRK5 displayed distinctly different mean levels of hypertension; male mice had the highest mean blood pressure. Systemic treatment of the mice with pertussis toxin to inactivate the heterotrimeric G_i proteins normalized the blood pressure, suggesting that G_i-linked receptor pathways are affected by the extra GRK5 (26). Both β-adrenergic and angiotensin pathways are altered, but to differing extents in male and female mice. In total, these studies validate the role of GRKs in regulating vascular tone and responsiveness through the regulation of smooth muscle cell GPCRs. Defining the specificity of vascular receptor–GRK coupling and identifying the precise pathways that are altered to lead to hypertrophy and/or increased blood pressure will require further studies.

Dopamine D1 receptors in the kidney increase sodium (and water) excretion in response to an increased sodium load (27). Essential hypertension in humans has been associated with decreased coupling of dopamine stimulation to sodium excretion (28). Polymorphisms in the human GRK4 gene appear to account for some of this variation. Studies with kidney tubule cells isolated from hypertensive and normotensive humans suggest

eNOS: endothelial nitric oxide synthase

that constitutive activity of some polymorphic variants of the GRK4 protein leads to chronically phosphorylated, inactive D1 receptors (29). This model finds support in (*a*) studies in model cells in which transfected GRK4 variants display marked constitutive kinase activity toward the D1 dopamine receptor (30) as well as in (*b*) a study, in rats, in which GRK4 antisense treatment of kidney tubules in hypertensive rats normalized sodium excretion, urine volume, and blood pressure (31). Various recent studies of hypertensive human populations find a positive correlation between certain GRK4 polymorphisms, or haplotypes, and hypertensive disease (32–34). Thus, hypertension-associated polymorphic variants of GRK4 appear hyperactive, leading to continuous kidney dopamine receptor desensitization and loss of dopamine-stimulated salt and fluid excretion.

GRK2 has a quite unexpected role in the liver, where it is an important regulator of portal blood pressure. Mice with only one active copy of the GRK2 gene are relatively resistant to liver injury–induced portal hypertension (35). Using rats injured with bile duct ligation, Liu et al. (35) found that GRK2 levels increased in injured liver endothelial cells. Previous work had shown that endothelin receptors stimulate the Akt protein kinase to phosphorylate endothelial nitric oxide synthase (eNOS) through a G protein $\beta\gamma$-subunit-activated, phosphatidylinositol 3-kinase–dependent mechanism, increasing nitric oxide production (36). In such a pathway, GRK2 would normally be expected to act to phosphorylate and desensitize the endothelin receptors, leading to reduced signaling to downstream Akt and eNOS and thus to reduced nitric oxide production. However, in endothelial cells from rats injured by bile duct ligation, concomitant with the increased level of GRK2 protein is an increased direct association of GRK2 with Akt (35). GRK2 bound to Akt inhibits the kinase activity of Akt, leading to reduced eNOS phosphorylation and reduced nitric oxide release (35). Adenoviral overexpression of GRK2 leads to increased portal pressure in otherwise normal rats, and the injury-induced deficit in Akt activation can be circumvented by expressing activated Akt in the liver portal system after infection with a recombinant adenovirus (35). Thus, GRK2 acts in an apparently noncatalytic manner to reduce the signaling of Akt. Undoubtedly, this GRK2-mediated inhibition of Akt plays important roles in other tissues as well, and the prevalence of GRK2 expression level changes under pathophysiological conditions indicates that GRK2-mediated inhibition of Akt may be prominent in a variety of disease states.

GRKs AND ARRESTINS IN THE IMMUNE SYSTEM

GPCRs have myriad roles within the immune system. The CC and CXC chemokine peptides (SDF-1, RANTES, and many others) and most arachidonic acid metabolites (prostaglandins, eicosinoids, leukotrienes) activate GPCRs, as do complement (C3a, C5a) and bacterial-derived products (fMLP). As such, G protein pathways are critical to immune cell chemotaxis, homing, activation, and target tissue recruitment of immune cells.

Cells of the immune system have particularly high expression of GRK2 and GRK6 (37, 38). Furthermore, altered expression of GRK2 and GRK6 occurs in immune cells isolated from human patients with a variety of immunological and other diseases as well as in animal models of disease. For example, lymphocytes from humans with rheumatoid arthritis have decreased GRK2 and GRK6 expression, and treating normal lymphocytes with the proinflammatory interferon-γ or interleukin 6 leads to decreases in GRK2 expression (39). Cultured human T lymphocytes exposed to oxidative stress with H_2O_2 or by coculture with activated neutrophils have reduced GRK2 expression (40). Induction of acute adjuvant arthritis in rats results in a reduction in lymphocyte GRK2 and GRK6 levels (41), as does induction of allergic encephalomyelitis (42). Humans with hypertension have increased GRK2 expression in lymphocytes (22, 23). Hypertensive rats have increased lymphocyte GRK2 expression (24). Chronic injection of rats with the β-adrenergic receptor agonist isoproterenol leads to increased expression of GRK2 and GRK5 mRNA in lymphocytes (43). Thus, leukocyte GRK levels are subject to regulation by a wide variety of factors and disease states.

Studies using mice lacking GRK6 or β-arrestin2 have revealed altered regulation of immune cells. In lymphocytes, ablation of GRK6 or β-arrestin2 leads to augmented signaling by SDF-1 (CXCL12) acting through the CXCR4 receptor (44). This increased signaling is associated with a reduced ability of T lymphocytes to chemotax through a filter or through an endothelial cell barrier toward SDF-1 (44). This decreased chemotaxis may be due to either overactive (undesensitized) traditional G protein signaling pathways or decreased β-arrestin-mediated signaling events.

In neutrophils, ablation of GRK6 leads to augmented signaling by leukotriene B4 (LTB4) acting through the BLT1 receptor; however, these receptors do continue to desensitize following repeated LTB4 administration (45). Neutrophils isolated from GRK6 knockout mice chemotax more robustly toward LTB4 and display a higher basal migration activity (45). Arachidonic acid applied to the ear is converted to LTB4 and other active metabolites and stimulates a rapid inflammatory response with both neutrophil influx and edema. GRK6 knockout and -heterozygous mice exhibit an enhanced response characterized by greater tissue swelling and by increased neutrophil infiltration into the tissue, consistent with increased chemotaxis toward LTB4 (45).

Neutrophils also respond to SDF-1 acting through CXCR4 receptors on those cells; this pathway is important in facilitating neutrophil retention in the bone marrow. Neutrophils mobilize to the blood in response to granulocyte colony–stimulating factor (G-CSF), in part by overcoming CXCR4-mediated retention signals. Mice lacking the GRK6 gene are less able to mobilize neutrophils from bone marrow into the blood in response to G-CSF (46). This reduced mobilization appears to result from an inability of G-CSF to overcome SDF-1/CXCR4-mediated retention signals, as neutrophils from GRK6 knockout mice chemotax more robustly to SDF-1 than do wild-type cells, either before or after G-CSF treatment of the mice (46). Calcium influx signaling downstream of CXCR4 fails to desensitize in neutrophils from GRK6 knockout mice, in contrast to what occurs in wild-type cells (46).

Studies using GRK2 knockout mice have also shown that GRK2 is important in immune cells. Because the full two-allele knockout of GRK2 is embryonic lethal (19), in vivo studies have instead focused on heterozygous mice bearing only one active GRK2 gene. In GRK2-heterozygous mice, T lymphocytes display increased chemotaxis toward the CCR5 receptor ligands CCL4 and CCL5 and toward the CCR1 ligand CCL3 (47). The CCR5 receptor exhibits reduced phosphorylation in lymphocytes with reduced GRK2, but enhanced signaling, consistent with a decrease in receptor desensitization (47). GRK2-heterozygous mice also display a marked sensitivity to experimental autoimmune encephalomyelitis, a model for multiple sclerosis, displaying a more rapid onset of lymphoctye infiltration into the brain (48). Despite the more rapid onset, the GRK2-heterozygous mice, unlike wild-type mice, fail to undergo relapse. Thus, decreased GRK2 can have both deleterious and protective effects in this disease model. More complete investigation of the role of GRK2 awaits the availability of mice bearing a conditional GRK2 allele.

Recently, an essential role of β-arrestin2 as a negative regulator of innate immune activation via Toll-like receptor–interleukin 1 receptor signaling was indicated by demonstration of higher expression of proinflammatory cytokines and increased susceptibility to endotoxic shock of β-arrestin2 knockout mice (49). Furthermore, an altered neutrophil

Arrestin-mediated signaling: the use of arrestin proteins as adaptors carrying signaling molecules to phosphorylated receptors to initiate receptor-dependent, G protein–independent signals

LTB4: leukotriene B4

G-CSF: granulocyte colony–stimulating factor

CXCR2 signaling in vivo (50) and defective lymphocyte chemotaxis were also described in mice lacking β-arrestin2 (44).

GPCR desensitization: the loss of cellular responsiveness to repeated or prolonged receptor stimulation; tachyphylaxis

CNS: central nervous system

GRKs AND ARRESTINS IN THE RESPIRATORY SYSTEM

Contraction and relaxation of airway smooth muscles are controlled, at least in part, by multiple GPCRs, and dysfunctional GPCR-mediated signaling has been implicated in asthma. Researchers have investigated the role of GRKs and β-arrestins in airway smooth muscle physiology. Mice lacking GRK3, but not GRK2-heterozygous mice, demonstrate a significant enhancement in the airway response and enhanced sensitivity of the airway smooth muscle response to cholinergic agonist methacholine, indicating that GRK3 may be involved in modulating the cholinergic response of airway smooth muscle (20). GRK5 is also involved in muscarinic regulation of airway responses (51). Although the airway contractile response to a muscarinic receptor agonist was not altered in GRK5-deficient mice, the relaxation component of bilateral vagal stimulation and the airway smooth muscle relaxation resulting from adrenergic stimulation were diminished in GRK5 mutants. Intriguing recent observations also indicate that β-arrestin2 may be involved in the development of allergic asthma (52). Allergen-sensitized mutant mice lacking β-arrestin2 do not accumulate T lymphocytes in their airways and do not demonstrate physiological and inflammatory features of asthma, suggesting that targeting processes controlled by this regulatory protein will provide novel insights in the treatment of asthma.

GRKs AND ARRESTINS IN THE SKELETAL SYSTEM

GRK2 and β-arrestin1 are expressed in osteoblastic cells (53). Transgenic mice expressing the GRK2 carboxyl-terminal inhibitor in mature osteoblasts demonstrated age-dependent enhancement in bone remodeling, as evidenced by an increase in bone density and trabecular bone volume (54, 55). In contrast, mice lacking β-arrestin2 showed decreased bone mass and altered bone architecture (56), revealing an important role of this protein in parathyroid hormone–dependent bone mass acquisition and remodeling (57).

GRKs AND ARRESTINS IN THE NERVOUS SYSTEM

More than 90% of known GPCRs are expressed in the brain (58) and are involved in virtually all vital functions controlled by the nervous system. It is not surprising, therefore, that most of the research on the physiological relevance of GPCR desensitization mechanisms has been performed in analyzing neuronal functions (9). Although sensory GPCRs of the olfactory and gustatory systems potentially are subject to regulation by GRKs and arrestins, only a few such studies have been published to date (58a, 82). As the most extensive evidence for wide variety of physiological functions and distinct tissue distributions of GRKs and arrestins comes from studies of the central nervous system (CNS), here we give a detailed overview of these observations in a more systematic manner by highlighting specific roles of individual neuronal GRKs and arrestins.

Neuronal GRK2

The GRK2 kinase has widespread expression in the brain (59). GRK2 mRNA is distributed in a nearly uniform manner through all cortical layers, the islands of Calleja, several hypothalamic and thalamic nuclei, the hippocampus, the substantia nigra compacta, the ventral tegmental area, the locus coeruleus, and other regions. A lower level of expression was detected in caudate-putamen (60). The expression of GRK2 is developmentally regulated, with a marked increase during the second postpartum week (61).

Alterations in GRK2 expression have been described in several disorders and/or following pharmacological treatments. In patients with major depression, there is an upregulation of GRK2-like immunoreactivity in the prefrontal cortex, whereas long-term antidepressant treatment downregulates GRK2 (62). One interesting hypothesis suggests that the neuronal calcium sensor-1 (NCS-1) contributes to the desensitization of D2 dopamine receptors via interaction with GRK2, thereby attributing abnormalities in NCS-1 expression described in schizophrenia and bipolar disorder to the NCS-1-dependent GRK2 regulation of dopamine receptor signaling (63, 64). Furthermore, GRK2 is increased in caudal caudate and internal globus pallidus in monkeys with experimental parkinsonism, indicating a potential role of this kinase in regulation of dopamine receptors (65). Mice heterozygous for GRK2 deletion were tested for their locomotor responses to the dopamine transporter blocker cocaine or the direct dopamine receptor agonist apomorphine. Cocaine treatment with certain doses resulted in slightly enhanced locomotor responses, whereas no such alterations were found with other doses of cocaine or with climbing responses to the nonselective dopamine agonist apomorphine, suggesting that the impact of partial loss of GRK2 on dopamine receptor–mediated responses is limited (9). Nevertheless, a more pronounced level of GRK2 deficiency may unmask the involvement of this kinase in dopamine receptor regulation. Further studies involving region-selective knockout mice may provide an effective approach with which to examine this possibility.

Several lines of evidence suggest a role for GRK2 in μ-opioid (μOR) regulation. GRK2 levels were increased in the locus coeruleus and cortex of rats chronically treated with morphine (66–68). Similarly, membrane-associated GRK2 levels were increased in brains of human opioid addicts (67). In another investigation, chronic treatment with the opioid antagonist naltrexone resulted in significant upregulation of GRK2 (69). Recent evidence indicates that chronic treatment with etorphine, but not morphine, produces a significant increase in GRK2 protein levels in membranes of the mouse spinal cord (70). In addition, GRK2 is highly expressed in nucleus raphe magnus GABAergic neurons projecting to the spinal cord, where it appears to mediate desensitization of μOR (71). These and other (72, 73) findings suggest that GRK2 contributes to the cellular processes underlying in vivo μOR desensitization and may play an important role in the development of opioid tolerance and dependence. A preliminary investigation, however, found no significant alteration in morphine-induced analgesia in GRK2-heterozygous mice (74). Further studies are necessary to explore in more detail the role of GRK2-mediated processes in μOR regulation and responses to opiates.

NCS-1: neuronal calcium sensor 1 protein

GABA: γ-amino butyric acid

μOR: μ-opioid receptor

The α and β isoforms of synucleins, proteins highly expressed in the brain and linked to the development of Parkinson's and Alzheimer's diseases, can be potently phosphorylated by GRK2 and GRK5 (75). Abnormalities in GRK2/5-like immunoreactivity associated with beta-amyloid accumulation were observed at prodromal and early stages of Alzheimer's disease (76). In neonatal rat, hypoxia/ischemia reduced GRK2 expression in several brain regions, suggesting a role of this kinase in hypoxia-induced brain damage (77). GRK2 is involved in both phosphorylation-dependent and -independent regulation of M1 muscarinic acetylcholine receptors in cultured hippocampal neurons (78). Additionally, GRK2 may have a role in the regulation of corticotropin-releasing factor receptor type 1 in the anterior pituitary gland (79).

Neuronal GRK3

GRK3 is expressed widely in the periphery and the CNS, albeit at lower levels than is GRK2 (59). In the periphery, GRK3 is found in olfactory neurons and dorsal root ganglion neurons, where it may play an important role in the desensitization of odorant receptors and α_2-adrenergic receptors, respectively (80–82).

GRK3 is ubiquitously expressed in the brain in a pattern similar to GRK2 (59, 60).

Substantial in vitro evidence suggests that GRK3 has a role in dopamine receptor regulation (9). Interestingly, the region of chromosome 22q12 containing the GRK3 gene has been identified as a susceptibility locus for bipolar disorder (83). Furthermore, GRK3 expression in the rat frontal cortex can be induced by the dopamine releaser amphetamine (83). Additionally, transmission disequilibrium analyses indicate that two 5′-UTR/promoter polymorphisms may be associated with bipolar disorder (83). These findings were interpreted to suggest that a dysregulation in GRK3 expression alters dopamine receptor desensitization and thereby predisposes affected individuals for this disorder (83, 84). Recent assessment of GRK3 mRNA levels in lymphocytes from bipolar patients, however, did not reveal a major difference between patients and controls (85). Similarly, mutation screening and association study of GRK3 in schizophrenia families provided no evidence for an association between schizophrenia and alleles at polymorphisms in the GRK3 promoter region (86).

5HT: 5-hydroxytryptamine, or serotonin

In mice, GRK3 deficiency does not affect basal locomotor activity, but GRK3 knockout mice demonstrate significantly reduced locomotor or climbing responses to either cocaine or apomorphine (9). Thus, it is unlikely that this kinase is directly involved in the desensitization of dopamine receptors responsible for locomotion. Nevertheless, it is possible that GRK3 is "positively" involved in dopamine receptor signaling leading to locomotor responses or "negatively" regulates other populations of GPCRs, such as, for example, receptors for 5HT (serotonin), that can inhibit dopamine-related behaviors (87). According to another interesting hypothesis, this kinase may be involved in the desensitization of D3 dopamine autoreceptors, thereby affecting processes governing dopamine release (88).

A role for GRK3 in μOR regulation also has been suggested on the basis of altered expression of this kinase following the administration of opiate agonists and antagonists (68, 69). Other researchers, however, failed to observe such changes following chronic morphine (66). In GRK3 knockout mice, morphine-induced analgesia was not changed in a hot-plate analgesia test (74). At the same time, GRK3 deletion influenced opioid analgesic tolerance to a high-efficacy opioid agonist, fentanyl, but did not affect acute antinociceptive responses to either fentanyl or morphine (89). Furthermore, there is evidence suggesting that GRK3 is involved in kappa-opioid receptor regulation. Neuropathic pain–induced activation of the kappa-opioid system in mouse spinal cord and kappa-opioid receptor–dependent tolerance are disrupted in GRK3 knockout mice (90, 91).

Neuronal GRK4

Expression of GRK4 in the brain is limited to cerebellar Purkinje cells (92). GRK4 may regulate the cerebellar metabotropic glutamate 1 receptor and GABA-B receptor (93, 94), suggesting a role of GRK4 in motor coordination and learning. However, mice lacking GRK4 demonstrate no differences in basal level of locomotor activity or motor coordination in a rotorod test (R.T. Premont, unpublished data). GRK4-deficient and control mice also have similar locomotor responses to cocaine (9). Furthermore, no alterations in morphine-induced analgesia have been found in these mutants (74).

Neuronal GRK5

GRK5 is expressed widely in the brain. GRK5 mRNA has been found at high levels in various regions, such as the septum, the cingulate cortex, the septohippocampal nucleus, the anterior thalamic nuclei, the medial habenula, and the locus coeruleus (60). GRK5 expression increases twofold during neural differentiation (95).

Chronic treatment with cocaine resulted in upregulation of GRK5 mRNA in the septum, suggesting a role of this kinase in abnormal plasticity induced by this drug (60). Similarly, an acute administration of morphine, as well as spontaneous and naloxone-induced morphine withdrawal, caused up-regulation of GRK5 mRNA in some brain regions (72). GRK5 deficit may play a role in prodromal and early stages of Alzheimer's disease (76).

Mice lacking GRK5 have an overtly normal phenotype and show only a slight decrease in body temperature (96). Mutant mice were challenged with a number of agonists to identify the specific GPCRs affected by GRK5 deficiency (9, 96). We observed no differences in cocaine-induced stimulation of locomotion or climbing responses following administration of the direct dopamine agonist apomorphine (9, 96). Similarly, morphine-induced analgesia was not altered in a hot-plate test in mutant mice. Furthermore, hypothermic responses induced by serotonin 5-HT1A receptor agonist 8-OH-DPAT did not differ between the genotypes. These observations indicate that responsiveness of dopamine receptors, and of the μOR and 5-HT1A subtype of serotonin receptors relevant for these behavioral and physiological processes, was not affected by deletion of GRK5. However, challenging these mice with the nonselective muscarinic agonist oxotremorine revealed significantly enhanced, centrally mediated muscarinic responses such as hypothermia, tremor, salivation, locomotor suppression, and antinociception (96). Because M2 muscarinic receptors mediate most of these behaviors (97), GRK5 likely regulates M2 muscarinic receptors in vivo. Furthermore, recent studies revealed that M2 receptors in airway are also affected by GRK5 deficiency, whereas regulation of M2 receptors in the heart occurs normally (51). These observations demonstrate that the same GRK can regulate the same GPCR differently in different tissues. Moreover, the hypersalivation response to oxotremorine displayed by the GRK5 mutants suggests that GRK5 also regulates salivary gland M3 receptors in vivo (97). Supersensitivity of central muscarinic receptors may contribute to several brain disorders, including depression, posttraumatic stress disorder, and multiple chemical sensitivities. It would be interesting to explore if abnormalities in GRK5-mediated muscarinic receptor desensitization contribute to these disorders (96).

Neuronal GRK6

GRK6 is ubiquitously expressed in most brain areas studied (98). Interestingly, GRK6 seems to be the most prominent GRK in the caudate putamen, and GRK6 mRNA is also expressed in dopaminergic cell body areas, such as the substantia nigra (60). In striatum, immunohistochemical investigations revealed a high expression of GRK6 protein in the dopamine-receptive striatal GABAergic medium spiny neurons as well as in cholinergic interneurons (99).

Unlike all other GRK-deficient mice, GRK6 knockout mice show significant locomotor supersensitivity to cocaine, amphetamine, and morphine (99). Similar effects were observed with endogenous "trace amine" β-phenylethylamine, which induces psychomotor activation at least partially via indirect activation of the dopamine system (100, 101). Furthermore, researchers observed in mutants (*a*) that an enhanced coupling of striatal D2-like dopamine receptors to G proteins increased the affinity of D2 but not D1 dopamine agonists and (*b*) an enhanced locomotor response to direct dopamine agonists (99, 102). Altogether, these observations indicate that postsynaptic D2-like dopamine receptors in the striatum are physiological targets for GRK6-mediated regulation, suggesting that a pharmacological strategy aimed at modulation of GRK6 expression or activity is beneficial when dopamine signaling is altered, such as in Parkinson's disorder. In fact, GRK6 expression is significantly elevated in several brain regions, including the striatum, in the MPTP-lesioned monkey model of Parkinson's disease (65). Moreover, because dopamine receptor supersensitivity is present in other brain disorders, such as addiction, schizophrenia, and Tourette's syndrome, a potential role of GRK6 in these conditions should be investigated (99).

GRK6 levels were changed in rat brain after chronic treatment with μOR agonists and antagonists (68, 69). Furthermore, decreased immunodensities of GRK6, GRK2, and β-arrestin2 were observed in postmortem brains of opiate addicts (103). Nonetheless, mice lacking GRK6 were indistinguishable from control littermates in the hot-plate morphine analgesia test (74).

Neuronal β-Arrestin1

β-Arrestin1 is expressed ubiquitously in the brain (9, 61, 95, 104–106). Levels of β-arrestin1 mRNA in the brain are estimated to be two- to threefold higher than those of β-arrestin2 mRNA, whereas the ratio of β-arrestin1 to β-arrestin2 protein levels is 10–20-fold (106). Strong immunoreactivity for β-arrestin1 also has been found at postsynaptic densities in the spinal cord (107). β-Arrestin1 greatly increases with neural development, suggesting a specific role for β-arrestin1 in neural differentiation (95).

Chronic systemic morphine treatment resulted in an increase in β-arrestin1/2-like immunoreactivity in rat locus coeruleus (66). Changes in β-arrestin1 mRNA in several brain areas also were detected following acute and systemic treatment with morphine (108). However, direct testing of acute morphine-induced antinociception in mice lacking β-arrestin1 has not revealed significant changes (74).

β-Arrestin1 levels were reduced in leukocytes of patients with depression and elevated by chronic antidepressants imipramine, desipramine, and fluvoxamine in rat cortex and hippocampus. The reduction in β-arrestin1 levels in the lymphocytes of major depression patients was significantly correlated with the severity of depressive symptoms (109). In addition, cerebral hypoxia/ischemia in rat pups increased β-arrestin1 protein expression as well as mRNA levels in the brain (77).

β-Arrestin1 knockout mice are overtly normal, but locomotor-stimulating effects of cocaine and apomorphine-induced climbing are somewhat reduced (9). These observations suggest that, although β-arrestin1 is not likely involved directly in dopamine receptor desensitization, β-arrestin1 may be involved in either G protein–independent dopamine receptor signaling or desensitization/signaling of other GPCRs in neuronal pathways contributing to the locomotor behaviors.

Neuronal β-Arrestin2

β-Arrestin2, much like β-arrestin1, is expressed in virtually all brain regions and in the spinal cord (9, 95, 104, 106, 107). In many brain tissues there is a significant overlap in the expression pattern of β-arrestin2 and β-arrestin1, but each β-arrestin also has a unique distribution in certain brain areas. In particular, β-arrestin2 has relatively higher expression in the medial habenula, in most hypothalamic nuclei, and in the extended amygdala (106).

Various in vivo studies indicate that β-arrestin2 is involved in μOR regulation and effects of opiates. Several groups have shown that acute or chronic treatment with μOR ligands, including morphine, causes significant alterations in the expression of β-arrestin2 in the cortex and striatum (68, 69, 108). The involvement of β-arrestin2 in the modulation of spinal antinociception also is supported by studies in which intrathecal pretreatment with β-arrestin2 antibody potentiated the antinociception induced by μOR agonists in the mouse (110).

Interestingly, earlier in vitro investigations in heterologous cellular systems found that the interaction of μOR with β-arrestin2 depends on agonist efficacy and that the partial agonist morphine produces arrestin-dependent receptor desensitization at a rather slow rate, suggesting that morphine activates μORs without promoting their rapid endocytosis (111–113). However, recent studies convincingly demonstrated that, in striatal and nucleus accumbens neurons, morphine promotes rapid, arrestin-dependent endocytosis of μORs (114). In dissociated primary cultures of rat striatal neurons, morphine promoted a rapid redistribution of both endogenous and recombinant μORs (114). These results strongly support previous in vivo results from β-arrestin2 knockout mice (74, 115, 116) and indicate that morphine is indeed capable of driving rapid endocytosis of μORs in vivo and that β-arrestin2 plays an important role in these effects (117).

β-Arrestin2 knockout mice were instrumental to demonstrating the role of β-arrestin2 in the regulation of μORs and effects of morphine. Mutant mice challenged with morphine showed remarkably enhanced antinociception in hot-plate tests, which is correlated with potentiated μOR–G protein coupling (115, 118). Furthermore, tolerance to morphine's antinociceptive effects in this test was significantly attenuated in β-arrestin2 knockout mice (116). At the same time, manifestations of naloxone-precipitated withdrawal were not altered by β-arrestin2 deletion, suggesting that despite disrupted tolerance mutant mice still can become physically dependent on morphine (116). When morphine effects on spinal cord–mediated antinociception were assessed in the tail-immersion test, the β-arrestin2 knockout mice also showed higher basal nociceptive thresholds as well as significantly enhanced responses to morphine. However, mutants developed tolerance to chronic morphine in this test, albeit with a delayed onset, suggesting the additional involvement of protein kinase C (PKC)-dependent regulatory system in this effect (118). A subsequent study in these mutants documented that relative opiate efficacy is determined by the ability to recruit elements of the GPCR desensitization machinery (119). Opiate agonists that induced robust β-arrestin2 translocation in HEK cells produced similar analgesia in normal and β-arrestin2 knockout mice, whereas morphine and heroin, which did not promote robust β-arrestin2 recruitment under these specific conditions, produced enhanced analgesia in mutant mice. Additionally, well-known side effects of morphine, such as constipation and respiratory suppression, were strikingly reduced in β-arrestin2 knockout mice (120), suggesting a novel approach of modulating GPCR regulatory mechanisms to overcome unwanted effects of the gold-standard analgesic, morphine. A recent study employing rats with overexpression of β-arrestin2 at periaqueductal gray (PAG) further supported a role of this regulatory molecule in μOR regulation and morphine antinociception (121). The demonstration that local β-arrestin2 overexpression by adenovirus results in disruption of the antinociceptive effects of morphine clearly identified PAG as a key area for this regulation.

IPSC: inhibitory postsynaptic current

Recently, a role of β-arrestin2 in delta-opioid receptor function was suggested by the observation that delta-opioid receptor agonists had no effect on miniature inhibitory postsynaptic currents (IPSCs) in β-arrestin2 knockout mice after chronic morphine. Thus, induction of delta-opioid receptor–mediated actions in PAG by chronic morphine may require prolonged μOR stimulation and induction of β-arrestin2 expression (122). Furthermore, when the effects of morphine and fentanyl were analyzed in slices of the locus coeruleus and PAG from β-arrestin2 knockout mice, presynaptic inhibition of evoked IPSCs was enhanced. β-Arrestin2 therefore may attenuate presynaptic inhibition by opioids, thereby revealing a novel mechanism of involvement of β-arrestin2 in opioid effects (123).

Finally, β-arrestin2 knockout mice show significant alterations in locomotor and rewarding properties of morphine. The activation of both μOR and dopamine receptors is required for the locomotor and reinforcing effects of morphine (124, 125). Morphine-induced stimulation of dopamine systems is indirect, originating from a disinhibition of GABAergic cells in dopaminergic cell body regions, leading to increased neuronal firing and increased dopamine release from terminals. In β-arrestin2-deficient mice, morphine, but not cocaine, produces a more pronounced increase in dopamine release and reward in the conditioned place preference test. However, acute morphine induces less locomotor activation in β-arrestin2-mutant mice (125). This paradoxical observation may indicate either (*a*) an important impact of this mutation on neurotransmitter systems other than that of dopamine, perhaps serotonin, that are also involved in morphine effects and exert an inhibitory action on dopamine-dependent hyperactivity (87, 126) or (*b*) a novel positive role of β-arrestin2 in dopaminergic signaling.

In fact, β-arrestin2-deficient mice exhibited somewhat reduced locomotor activation following cocaine administration (125) as well as disrupted climbing response to the direct dopamine agonist apomorphine (9, 127). These in vivo observations are not consistent with an expected role of β-arrestin2 as a controller of desensitization of dopamine receptors but rather indicate that without this regulatory element dopaminergic signaling may be impaired. Recent functional biochemical studies directly demonstrated an important role of β-arrestin2 in D2 dopamine receptor–mediated signaling in vivo (127). D2 dopamine receptor–mediated Akt regulation involves the formation of signaling complexes containing β-arrestin2, protein phosphatase 2A, and Akt. Mice lacking β-arrestin2 show a reduction of amphetamine- and apomorphine-induced behaviors, loss of Akt regulation by dopamine, and disruption of the dopamine-dependent interaction of Akt with PP2A. Furthermore, deletion of β-arrestin2 results in a reduced level of spontaneous locomotor activity in normal as well as hyperdopaminergic dopamine transporter knockout mice (128), directly demonstrating the importance of this regulatory molecule in functional dopaminergic signaling (127). These results for the first time demonstrated in vivo the physiological significance of the dual role of β-arrestin2 in GPCR regulation: In addition to its classical "negative" function in receptor desensitization, β-arrestin2 also acts as a "positive" signaling intermediate as a kinase/phosphatase scaffold. Most importantly, these data indicate that this role of β-arrestin2 is functionally important for the expression of dopamine-mediated behaviors and that the molecular mechanisms involved in the β-arrestin2-dependent Akt/GSK3 (glycogen synthase kinase type 3) signaling cascade may provide novel potential pharmacological targets with which to manage dopamine-related neurological and psychiatric disorders (127, 129).

SUMMARY

GRK and Arrestin Functions

Studies in several systems described above demonstrate that GRKs and arrestins do regulate GPCRs in vivo and that the failure to regulate GPCRs properly can have profound physiological consequences. In the absence of elements of the GRK-arrestin desensitization machinery, GPCRs can remain abnormally supersensitive to agonist stimulation. This manifests as prolonged and exaggerated responses to agonist challenges. In the clearest examples, a receptor subtype appears to require only one particular GRK or arrestin protein to be regulated properly. The D2-like dopamine receptors in striatal medium spiny neurons fail to desensitize in the absence of GRK6, whereas brain M2 muscarinic receptors mediating several responses fail to desensitize in the absence of GRK5. Similarly, β-arrestin2 is required for μ-opioid receptor desensitization, whereas β-arrestin1 appears ineffective at regulating this receptor. Thus, the role of GRKs and arrestins defined in model cell systems and using purified proteins can be validated in the physiological responses of living animals.

In contrast, GRK and arrestin knockout mice are grossly normal (with the one exception of the embryonically lethal GRK2 knockout), which is consistent with the role of the GRK-arrestin system to regulate primarily agonist-activated receptors. In the absence of the stress of a high level of receptor stimulation, the relative inability of these animals to dampen excessive signaling is not immediately evident.

Although GRKs and arrestins have been shown to promote G protein–independent signaling pathways through arrestin-scaffolded signaling proteins in model cell systems, evidence for such pathways in vivo is scant at present. The identification of β-arrestin2 scaffolding–dependent regulation of Akt in the striatum is the clearest example of such function to date. Increased interest in this type of signaling in recent years is sure to focus needed attention on identifying physiological systems in which this role of the GRK-arrestin system can be clearly demonstrated.

GRK and Arrestin Specificity

Because there are several hundred GPCRs but only four widely distributed GRKs and two widely distributed β-arrestin proteins to regulate them, there remains the issue of defining the wiring diagram with which GRKs and arrestins regulate any particular GPCR. Current evidence suggests that this may be quite complex indeed.

Some receptors appear to be regulated by one particular GRK and/or arrestin protein, as with the D2-like receptors in striatum by GRK6 or the μ-opioid receptor by β-arrestin2. In contrast, other receptors do not exhibit a dramatic alteration in signaling with the loss of any one GRK or β-arrestin, suggesting that many GRK subtypes or both β-arrestins all contribute to regulation and can compensate for the loss of any one of their number. The only exception is mice lacking GRK2 (19) or both β-arrestin1 and β-arrestin2 (9); such phenotypes are embryonically lethal. The most curious and informative finding is that a single GPCR subtype may be regulated by distinct GRKs when it is present in different cell types. In the CNS and lung, M2 muscarinic receptors appear to be regulated primarily by GRK5. However, despite the prominent expression of both M2 receptors and GRK5 in the heart, cardiac M2 muscarinic receptors seem regulated quite normally in GRK5 knockout mice. Thus, to understand the regulation of a specific receptor, it is critical to examine that receptor in its native context. Information garnered from model cell systems, or even from examining other organs of an intact animal, may not be informative for the same receptor subtype in a different environment. Given the few receptors and physiological responses examined in detail to date, it remains unclear whether most receptors are regulated by exclusive GRK-arrestin pathways or promiscuously by multiple GRKs and/or both β-arrestins.

Receptors on distinct cell types can also respond differently to loss of a particular GRK or arrestin. The SDF-1-activated CXCR4 receptor appears to be regulated by GRK6 in both T lymphocytes and neutrophils, but loss of GRK6 has opposite effects on chemotaxis in these two cell types: T cells lacking GRK6 chemotax less well toward SDF-1, whereas neutrophils chemotax markedly better toward SDF-1. The basis for this difference remains obscure but may reflect differences in the regulation of downstream pathways leading to chemotaxis in these different cells.

The physiological significance of GRK-arrestin regulation of GPCRs that the studies described here have demonstrated strongly suggests that future development of inhibitors or modulators of these regulatory proteins and their processes is an effective approach by which to fine-tune physiological functions. For example, the ability of some tastants to induce lingering aftertaste may depend on their ability to inhibit GRKs in taste cells that putatively regulate tastant receptors (58a). This and other examples described herein demonstrate the feasibility of pharmacological approaches of modulating GPCR regulation to correct abnormal physiological function in disease.

The lessons learned so far from these in vivo studies only highlight the many further studies that remain to be undertaken and the myriad details of which we remain ignorant. Physiological regulation of nearly all hormone and neurotransmitter receptors by GRK-arrestin pathways remains untested. The widespread regulation of GRK expression levels in various physiological and pathopysiological conditions adds another level of complexity. These GRK dynamics, coupled with examples of substantial physiological changes in GRK6-heterozygous knockout mice in several systems, suggest that GRK-arrestin regulation of a GPCR can differ greatly even in a single organ, depending on the conditions that regulate GRK expression. Furthermore, the finding that GRK4 polymorphisms in humans can alter kidney dopamine receptor responses under apparently basal conditions adds the potential for yet more complexity. Thus, we have a long way to go to define in detail how particular receptors are regulated by these GRK-arrestin pathways.

SUMMARY POINTS

1. The failure to regulate GPCRs properly can have profound physiological consequences.
2. In the absence of elements of the GRK-arrestin desensitization machinery, GPCRs can exhibit prolonged and exaggerated responses to agonist challenge owing to disrupted desensitization.
3. Deficiency in GRK-arrestin mechanisms may also cause a decrease in responses owing to disrupted G protein–independent, arrestin-mediated signaling.
4. Some receptors appear to be regulated by one particular GRK and/or arrestin protein, whereas other receptors appear to be regulated by many GRK subtypes or both β-arrestins.
5. A single GPCR subtype may be regulated by distinct GRKs when it is present in different cell types.
6. Receptors on distinct cell types can also respond differently to loss of a particular GRK or arrestin.
7. The widespread regulation of GRK expression levels in various physiological and pathophysiological conditions suggests that GRK-arrestin regulation of a GPCR can differ greatly even in a single organ, depending on the conditions that regulate GRK expression.

FUTURE ISSUES

1. The complex wiring diagram with which multiple GRKs and arrestins regulate any particular GPCR in a specific cell group needs to be defined.
2. Physiological functions of individual GRKs and arrestins in specific responses in living organisms should be identified.
3. The physiological role of processes governing GPCR desensitization in a specific function requires further understanding.

4. The recent discovery of G protein–independent, arrestin-mediated signaling of GPCRs should spur efforts to identify physiological systems in which this role of the GRK-arrestin system is predominant.
5. GPCR-desensitization-related versus G protein–independent signaling–related functions of GPCR regulation by individual GRKs and arrestins, should be delineated.

ACKNOWLEDGMENTS

R.T.P. is supported by NIH grants DA016347 and GM59989 and grant-in-aid 0655464U from the American Heart Association Mid-Atlantic Affiliate. R.R.G. is supported by the Michael J. Fox Foundation for Parkinson's Research.

LITERATURE CITED

1. Pierce KL, Premont RT, Lefkowitz RJ. 2002. Seven-transmembrane receptors. *Nat. Rev. Mol. Cell Biol.* 3:639–50
2. Luttrell LM. 2005. Composition and function of G protein-coupled receptor signalsomes controlling mitogen-activated protein kinase activity. *J. Mol. Neurosci.* 26:253–64
3. Lefkowitz RJ, Shenoy SK. 2005. Transduction of receptor signals by β-arrestins. *Science* 308:512–17
4. Hausdorff WP, Caron MG, Lefkowitz RJ. 1990. Turning off the signal: desensitization of β-adrenergic receptor function. *FASEB J.* 4:2881–89
5. Premont RT. 2005. Once and future signaling: G protein-coupled receptor kinase control of neuronal sensitivity. *Neuromol. Med.* 7:129–47
6. Penn RB, Pronin AN, Benovic JL. 2000. Regulation of G protein-coupled receptor kinases. *Trends Cardiovasc. Med.* 10:81–89
7. Ferguson SS. 2001. Evolving concepts in G protein-coupled receptor endocytosis: the role in receptor desensitization and signaling. *Pharmacol. Rev.* 53:1–24
8. Penela P, Ribas C, Mayor FJ. 2003. Mechanisms of regulation of the expression and function of G protein-coupled receptor kinases. *Cell Signal* 15:973–81
9. Gainetdinov RR, Premont RT, Bohn LM, Lefkowitz RJ, Caron MG. 2004. Desensitization of G protein-coupled receptors and neuronal functions. *Annu. Rev. Neurosci.* 27:107–44
10. Gurevich VV, Gurevich EV. 2004. The molecular acrobatics of arrestin activation. *Trends Pharmacol. Sci.* 25:105–11
11. Gainetdinov RR, Premont RT, Caron MG, Lefkowitz RJ. 2000. Reply: receptor specificity of G-protein-coupled receptor kinases. *Trends Pharmacol. Sci.* 21:366–7
12. Chen CK. 2002. Recoverin and rhodopsin kinase. *Adv. Exp. Med. Biol.* 514:101–7
13. Arshavsky VY. 2002. Rhodopsin phosphorylation: from terminating single photon responses to photoreceptor dark adaptation. *Trends Neurosci.* 25:124–6
14. Maeda T, Imanishi Y, Palczewski K. 2003. Rhodopsin phosphorylation: 30 years later. *Prog. Retin. Eye Res.* 22:417–34
15. Koch WJ. 2004. Genetic and phenotypic targeting of β-adrenergic signaling in heart failure. *Mol. Cell Biochem.* 263:5–9
16. Hata JA, Williams ML, Koch WJ. 2004. Genetic manipulation of myocardial β-adrenergic receptor activation and desensitization. *J. Mol. Cell Cardiol.* 37:11–21
17. Iaccarino G, Koch WJ. 2003. Transgenic mice targeting the heart unveil G protein-coupled receptor kinases as therapeutic targets. *Assay Drug Dev. Technol.* 1:347–55

18. Rockman HA, Choi DJ, Akhter SA, Jaber M, Giros B, et al. 1998. Control of myocardial contractile function by the level of β-adrenergic receptor kinase 1 in gene-targeted mice. *J. Biol. Chem.* 273:18180–84
19. Jaber M, Koch WJ, Rockman H, Smith B, Bond RA, et al. 1996. Essential role of β-adrenergic receptor kinase 1 in cardiac development and function. *Proc. Natl. Acad. Sci. USA* 93:12974–79
20. Walker JK, Peppel K, Lefkowitz RJ, Caron MG, Fisher JT. 1999. Altered airway and cardiac responses in mice lacking G protein-coupled receptor kinase 3. *Am. J. Physiol.* 276:R1214–21
21. Conner DA, Mathier MA, Mortensen RM, Christe M, Vatner SF, et al. 1997. β-Arrestin1 knockout mice appear normal but demonstrate altered cardiac responses to β-adrenergic stimulation. *Circ. Res.* 81:1021–26
22. Gros R, Tan CM, Chorazyczewski J, Kelvin DJ, Benovic JL, Feldman RD. 1999. G-protein-coupled receptor kinase expression in hypertension. *Clin. Pharmacol. Ther.* 65:545–51
23. Gros R, Benovic JL, Tan CM, Feldman RD. 1997. G-protein-coupled receptor kinase activity is increased in hypertension. *J. Clin. Invest.* 99:2087–93
24. Gros R, Chorazyczewski J, Meek MD, Benovic JL, Ferguson SS, Feldman RD. 2000. G-Protein-coupled receptor kinase activity in hypertension: increased vascular and lymphocyte G-protein receptor kinase-2 protein expression. *Hypertension* 35:38–42
25. Eckhart AD, Ozaki T, Tevaearai H, Rockman HA, Koch WJ. 2002. Vascular-targeted overexpression of G protein-coupled receptor kinase-2 in transgenic mice attenuates β-adrenergic receptor signaling and increases resting blood pressure. *Mol. Pharmacol.* 61:749–58
26. Keys JR, Zhou RH, Harris DM, Druckman CA, Eckhart AD. 2005. Vascular smooth muscle overexpression of G protein-coupled receptor kinase 5 elevates blood pressure, which segregates with sex and is dependent on G_i-mediated signaling. *Circulation* 112:1145–53
27. Zeng C, Sanada H, Watanabe H, Eisner GM, Felder RA, Jose PA. 2004. Functional genomics of the dopaminergic system in hypertension. *Physiol. Genom.* 19:233–46
28. Jose PA, Eisner GM, Felder RA. 1998. Renal dopamine receptors in health and hypertension. *Pharmacol. Ther.* 80:149–82
29. Felder RA, Sanada H, Xu J, Yu PY, Wang Z, et al. 2002. G protein-coupled receptor kinase 4 gene variants in human essential hypertension. *Proc. Natl. Acad. Sci. USA* 99:3872–77

29. Identifies GRK4 gene variants in human essential hypertension.

30. Rankin ML, Marinec PS, Cabrera DM, Wang Z, Jose PA, Sibley DR. 2006. The D1 dopamine receptor is constitutively phosphorylated by G protein-coupled receptor kinase 4. *Mol. Pharmacol.* 69:759–69
31. Sanada H, Yatabe J, Midorikawa S, Katoh T, Hashimoto S, et al. 2006. Amelioration of genetic hypertension by suppression of renal G protein-coupled receptor kinase type 4 expression. *Hypertension* 47:1131–39
32. Sanada H, Yatabe J, Midorikawa S, Hashimoto S, Watanabe T, et al. 2006. Single-nucleotide polymorphisms for diagnosis of salt-sensitive hypertension. *Clin. Chem.* 52:352–60
33. Lohmueller KE, Wong LJ, Mauney MM, Jiang L, Felder RA, et al. 2006. Patterns of genetic variation in the hypertension candidate gene GRK4: ethnic variation and haplotype structure. *Ann. Hum. Genet.* 70:27–41
34. Gu D, Su S, Ge D, Chen S, Huang J, et al. 2006. Association study with 33 single-nucleotide polymorphisms in 11 candidate genes for hypertension in Chinese. *Hypertension* 47:1147–54

35. Liu S, Premont RT, Kontos CD, Zhu S, Rockey DC. 2005. A crucial role for GRK2 in regulation of endothelial cell nitric oxide synthase function in portal hypertension. *Nat. Med.* 11:952–58
36. Liu S, Premont RT, Kontos CD, Huang J, Rockey DC. 2003. Endothelin-1 activates endothelial cell nitric-oxide synthase via heterotrimeric G-protein βγ subunit signaling to protein kinase B/Akt. *J. Biol. Chem.* 278:49929–35
37. De Blasi A, Parruti G, Sallese M. 1995. Regulation of G protein-coupled receptor kinase subtypes in activated T lymphocytes. Selective increase of β-adrenergic receptor kinase 1 and 2. *J. Clin. Invest.* 95:203–10
38. Loudon RP, Perussia B, Benovic JL. 1996. Differentially regulated expression of the G-protein-coupled receptor kinases, βARK and GRK6, during myelomonocytic cell development in vitro. *Blood* 88:4547–57
39. Lombardi MS, Kavelaars A, Schedlowski M, Bijlsma JW, Okihara KL, et al. 1999. Decreased expression and activity of G-protein-coupled receptor kinases in peripheral blood mononuclear cells of patients with rheumatoid arthritis. *FASEB J.* 13:715–25
40. Lombardi MS, Kavelaars A, Penela P, Scholtens EJ, Roccio M, et al. 2002. Oxidative stress decreases G protein-coupled receptor kinase 2 in lymphocytes via a calpain-dependent mechanism. *Mol. Pharmacol.* 62:379–88
41. Lombardi MS, Kavelaars A, Cobelens PM, Schmidt RE, Schedlowski M, Heijnen CJ. 2001. Adjuvant arthritis induces down-regulation of G protein-coupled receptor kinases in the immune system. *J. Immunol.* 166:1635–40
42. Vroon A, Lombardi MS, Kavelaars A, Heijnen CJ. 2003. Changes in the G-protein-coupled receptor desensitization machinery during relapsing-progressive experimental allergic encephalomyelitis. *J. Neuroimmunol.* 137:79–86
43. Oyama N, Urasawa K, Kaneta S, Sakai H, Saito T, et al. 2005. Chronic beta-adrenergic receptor stimulation enhances the expression of G-protein coupled receptor kinases, GRK2 and GRK5, in both the heart and peripheral lymphocytes. *Circ. J.* 69:987–90
44. **Fong AM, Premont RT, Richardson RM, Yu YR, Lefkowitz RJ, Patel DD. 2002. Defective lymphocyte chemotaxis in β-arrestin2- and GRK6-deficient mice. *Proc. Natl. Acad. Sci. USA* 99:7478–83**
45. **Kavelaars A, Vroon A, Raatgever RP, Fong AM, Premont RT, et al. 2003. Increased acute inflammation, leukotriene B4-induced chemotaxis, and signaling in mice deficient for G protein-coupled receptor kinase 6. *J. Immunol.* 171:6128–34**
46. Vroon A, Heijnen CJ, Raatgever R, Touw IP, Ploemacher RE, et al. 2004. GRK6 deficiency is associated with enhanced CXCR4-mediated neutrophil chemotaxis in vitro and impaired responsiveness to G-CSF in vivo. *J. Leukoc. Biol.* 75:698–704
47. Vroon A, Heijnen CJ, Lombardi MS, Cobelens PM, Mayor FJ, et al. 2004. Reduced GRK2 level in T cells potentiates chemotaxis and signaling in response to CCL4. *J. Leukoc. Biol.* 75:901–9
48. Vroon A, Kavelaars A, Limmroth V, Lombardi MS, Goebel MU, et al. 2005. G protein-coupled receptor kinase 2 in multiple sclerosis and experimental autoimmune encephalomyelitis. *J. Immunol.* 174:4400–6
49. Wang Y, Tang Y, Teng L, Wu Y, Zhao X, Pei G. 2006. Association of β-arrestin and TRAF6 negatively regulates Toll-like receptor-interleukin 1 receptor signaling. *Nat. Immunol.* 7:139–47
50. Su Y, Raghuwanshi SK, Yu Y, Nanney LB, Richardson RM, Richmond A. 2005. Altered CXCR2 signaling in β-arrestin-2-deficient mouse models. *J. Immunol.* 175:5396–402

44. Demonstrates the role of β-arrestin2 and GRK6 in lymphocyte chemotaxis.

45. Shows the role of GRK6 in neutrophil chemotaxis and inflammation.

51. Describes M2 muscarinic regulation by GRK5 in the airway, but not in the heart.

52. Uncovers the role of β-arrestin-2 in the development of allergic asthma.

51. **Walker JK, Gainetdinov RR, Feldman DS, McFawn PK, Caron MG, et al. 2004. G protein-coupled receptor kinase 5 regulates airway responses induced by muscarinic receptor activation. *Am. J. Physiol. Lung Cell Mol. Physiol.* 286:L312–19**
52. **Walker JK, Fong AM, Lawson BL, Savov JD, Patel DD, et al. 2003. β-Arrestin-2 regulates the development of allergic asthma. *J. Clin. Invest.* 112:566–74**
53. Bliziotes M, Murtagh J, Wiren K. 1996. β-adrenergic receptor kinase-like activity and β-arrestin are expressed in osteoblastic cells. *J. Bone Miner. Res.* 11:820–26
54. Spurney RF, Flannery PJ, Garner SC, Athirakul K, Liu S, et al. 2002. Anabolic effects of a G protein-coupled receptor kinase inhibitor expressed in osteoblasts. *J. Clin. Invest.* 109:1361–71
55. Wang L, Quarles LD, Spurney RF. 2004. Unmasking the osteoinductive effects of a G-protein-coupled receptor (GPCR) kinase (GRK) inhibitor by treatment with PTH(1–34). *J. Bone Miner. Res.* 19:1661–70
56. Bouxsein ML, Pierroz DD, Glatt V, Goddard DS, Cavat F, et al. 2005. β-Arrestin2 regulates the differential response of cortical and trabecular bone to intermittent PTH in female mice. *J. Bone Miner. Res.* 20:635–43
57. Ferrari SL, Pierroz DD, Glatt V, Goddard DS, Bianchi EN, et al. 2005. Bone response to intermittent parathyroid hormone is altered in mice null for β-arrestin2. *Endocrinology* 146:1854–62
58. Vassilatis DK, Hohmann JG, Zeng H, Li F, Ranchalis JE, et al. 2003. The G protein-coupled receptor repertoires of human and mouse. *Proc. Natl. Acad. Sci. USA* 100:4903–8

58a. Zubare-Samuelov M, Shaul ME, Peri I, Aliluiko A, Tirosh O, Naim M. 2005. Inhibition of signal termination-related kinases by membrane-permeant bitter and sweet tastants: potential role in taste signal termination. *Am. J. Physiol. Cell Physiol.* 289:C483–92

59. Arriza JL, Dawson TM, Simerly RB, Martin LJ, Caron MG, et al. 1992. The G-protein-coupled receptor kinases βARK1 and βARK2 are widely distributed at synapses in rat brain. *J. Neurosci.* 12:4045–55
60. Erdtmann-Vourliotis M, Mayer P, Ammon S, Riechert U, Hollt V. 2001. Distribution of G-protein-coupled receptor kinase (GRK) isoforms 2, 3, 5 and 6 mRNA in the rat brain. *Br. Res. Mol. Br. Res.* 95:129–37
61. Penela P, Alvarez-Dolado M, Munoz A, Mayor FJ. 2000. Expression patterns of the regulatory proteins G protein-coupled receptor kinase 2 and β-arrestin 1 during rat postnatal brain development: effect of hypothyroidism. *Eur. J. Biochem.* 267:4390–96
62. Grange-Midroit M, Garcia-Sevilla JA, Ferrer-Alcon M, La Harpe R, Huguelet P, Guimon J. 2003. Regulation of GRK 2 and 6, β-arrestin-2 and associated proteins in the prefrontal cortex of drug-free and antidepressant drug-treated subjects with major depression. *Br. Res. Mol. Br. Res.* 111:31–41
63. Koh PO, Undie AS, Kabbani N, Levenson R, Goldman-Rakic PS, Lidow MS. 2003. Up-regulation of neuronal calcium sensor-1 (NCS-1) in the prefrontal cortex of schizophrenic and bipolar patients. *Proc. Natl. Acad. Sci. USA* 100:313–17
64. Kabbani N, Negyessy L, Lin R, Goldman-Rakic P, Levenson R. 2002. Interaction with neuronal calcium sensor NCS-1 mediates desensitization of the D2 dopamine receptor. *J. Neurosci.* 22:8476–86
65. Bezard E, Gross CE, Qin L, Gurevich VV, Benovic JL, Gurevich EV. 2005. L-DOPA reverses the MPTP-induced elevation of the arrestin2 and GRK6 expression and enhanced ERK activation in monkey brain. *Neurobiol. Dis.* 18:323–35
66. Terwilliger RZ, Ortiz J, Guitart X, Nestler EJ. 1994. Chronic morphine administration increases β-adrenergic receptor kinase (βARK) levels in the rat locus coeruleus. *J. Neurochem.* 63:1983–86

67. Ozaita A, Escriba PV, Ventayol P, Murga C, Mayor FJ, Garcia-Sevilla JA. 1998. Regulation of G protein-coupled receptor kinase 2 in brains of opiate-treated rats and human opiate addicts. *J. Neurochem.* 70:1249–57
68. Hurle MA. 2001. Changes in the expression of G protein-coupled receptor kinases and β-arrestin 2 in rat brain during opioid tolerance and supersensitivity. *J. Neurochem.* 77:486–92
69. Diaz A, Pazos A, Florez J, Ayesta FJ, Santana V, Hurle MA. 2002. Regulation of μ-opioid receptors, G-protein-coupled receptor kinases and β-arrestin 2 in the rat brain after chronic opioid receptor antagonism. *Neuroscience* 112:345–53
70. Narita M, Suzuki M, Narita M, Niikura K, Nakamura A, et al. 2006. μ-Opioid receptor internalization-dependent and -independent mechanisms of the development of tolerance to μ-opioid receptor agonists: comparison between etorphine and morphine. *Neuroscience* 138:609–19
71. Li AH, Wang HL. 2001. G protein-coupled receptor kinase 2 mediates μ-opioid receptor desensitization in GABAergic neurons of the nucleus raphe magnus. *J. Neurochem.* 77:435–44
72. Fan X, Zhang J, Zhang X, Yue W, Ma L. 2002. Acute and chronic morphine treatments and morphine withdrawal differentially regulate GRK2 and GRK5 gene expression in rat brain. *Neuropharmacology* 43:809–16
73. Ammon-Treiber S, Hollt V. 2005. Morphine-induced changes of gene expression in the brain. *Addict. Biol.* 10:81–89
74. Bohn LM, Gainetdinov RR, Caron MG. 2004. G protein-coupled receptor kinase/beta-arrestin systems and drugs of abuse: psychostimulant and opiate studies in knockout mice. *Neuromol. Med.* 5:41–50
75. Pronin AN, Morris AJ, Surguchov A, Benovic JL. 2000. Synucleins are a novel class of substrates for G protein-coupled receptor kinases. *J. Biol. Chem.* 275:26515–22
76. Suo Z, Wu M, Citron BA, Wong GT, Festoff BW. 2004. Abnormality of G-protein-coupled receptor kinases at prodromal and early stages of Alzheimer's disease: an association with early β-amyloid accumulation. *J. Neurosci.* 24:3444–52
77. Lombardi MS, van den Tweel E, Kavelaars A, Groenendaal F, van Bel F, Heijnen CJ. 2004. Hypoxia/ischemia modulates G protein-coupled receptor kinase 2 and β-arrestin-1 levels in the neonatal rat brain. *Stroke* 35:981–86
78. Willets JM, Nahorski SR, Challiss RA. 2005. Roles of phosphorylation-dependent and -independent mechanisms in the regulation of M1 muscarinic acetylcholine receptors by G protein-coupled receptor kinase 2 in hippocampal neurons. *J. Biol. Chem.* 280:18950–58
79. Kageyama K, Hanada K, Moriyama T, Nigawara T, Sakihara S, Suda T. 2006. G protein-coupled receptor kinase 2 involvement in desensitization of corticotropin-releasing factor (CRF) receptor type 1 by CRF in murine corticotrophs. *Endocrinology* 147:441–50
80. Boekhoff I, Inglese J, Schleicher S, Koch WJ, Lefkowitz RJ, Breer H. 1994. Olfactory desensitization requires membrane targeting of receptor kinase mediated by βγ-subunits of heterotrimeric G proteins. *J. Biol. Chem.* 269:37–40
81. Diverse-Pierluissi M, Inglese J, Stoffel RH, Lefkowitz RJ, Dunlap K. 1996. G protein-coupled receptor kinase mediates desensitization of norepinephrine-induced Ca^{2+} channel inhibition. *Neuron* 16:579–85
82. Peppel K, Boekhoff I, McDonald P, Breer H, Caron MG, Lefkowitz RJ. 1997. G protein-coupled receptor kinase 3 (GRK3) gene disruption leads to loss of odorant receptor desensitization. *J. Biol. Chem.* 272:25425–28

83. Niculescu AB, Segal DS, Kuczenski R, Barrett T, Hauger RL, Kelsoe JR. 2000. Identifying a series of candidate genes for mania and psychosis: a convergent functional genomics approach. *Physiol. Genom.* 4:83–91

84. Barrett TB, Hauger RL, Kennedy JL, Sadovnick AD, Remick RA, et al. 2003. Evidence that a single nucleotide polymorphism in the promoter of the G protein receptor kinase 3 gene is associated with bipolar disorder. *Mol. Psychiat.* 8:546–57

85. Shaltiel G, Shamir A, Levi I, Bersudsky Y, Agam G. 2006. Lymphocyte G-protein receptor kinase (GRK)3 mRNA levels in bipolar disorder. *Int. J. Neuropsychopharmacol.* 9:761–66 doi: 10.1017/S146114570500636X

86. Yu SY, Takahashi S, Arinami T, Ohkubo T, Nemoto Y, et al. 2004. Mutation screening and association study of the β-adrenergic receptor kinase 2 gene in schizophrenia families. *Psychiat. Res.* 125:95–104

87. Gainetdinov RR, Wetsel WC, Jones SR, Levin ED, Jaber M, Caron MG. 1999. Role of serotonin in the paradoxical calming effect of psychostimulants on hyperactivity. *Science* 283:397–401

88. Kim KM, Gainetdinov RR, Laporte SA, Caron MG, Barak LS. 2005. G protein-coupled receptor kinase regulates dopamine D3 receptor signaling by modulating the stability of a receptor-filamin-β-arrestin complex. A case of autoreceptor regulation. *J. Biol. Chem.* 280:12774–80

89. Terman GW, Jin W, Cheong YP, Lowe J, Caron MG, et al. 2004. G-protein receptor kinase 3 (GRK3) influences opioid analgesic tolerance but not opioid withdrawal. *Br. J. Pharmacol.* 141:55–64

90. Xu M, Petraschka M, McLaughlin JP, Westenbroek RE, Caron MG, et al. 2004. Neuropathic pain activates the endogenous κ opioid system in mouse spinal cord and induces opioid receptor tolerance. *J. Neurosci.* 24:4576–84

91. McLaughlin JP, Myers LC, Zarek PE, Caron MG, Lefkowitz RJ, et al. 2004. Prolonged κ opioid receptor phosphorylation mediated by G-protein receptor kinase underlies sustained analgesic tolerance. *J. Biol. Chem.* 279:1810–18

92. Sallese M, Salvatore L, D'Urbano E, Sala G, Storto M, et al. 2000. The G-protein-coupled receptor kinase GRK4 mediates homologous desensitization of metabotropic glutamate receptor 1. *FASEB J.* 14:2569–80

93. Iacovelli L, Salvatore L, Capobianco L, Picascia A, Barletta E, et al. 2003. Role of G protein-coupled receptor kinase 4 and β-arrestin 1 in agonist-stimulated metabotropic glutamate receptor 1 internalization and activation of mitogen-activated protein kinases. *J. Biol. Chem.* 278:12433–42

94. Perroy J, Adam L, Qanbar R, Chenier S, Bouvier M. 2003. Phosphorylation-independent desensitization of GABA(B) receptor by GRK4. *EMBO J.* 22:3816–24

95. Gurevich EV, Benovic JL, Gurevich VV. 2004. Arrestin2 expression selectively increases during neural differentiation. *J. Neurochem.* 91:1404–16

96. Gainetdinov RR, Bohn LM, Walker JK, Laporte SA, Macrae AD, et al. 1999. Muscarinic supersensitivity and impaired receptor desensitization in G protein-coupled receptor kinase 5-deficient mice. *Neuron* 24:1029–36

97. Gomeza J, Shannon H, Kostenis E, Felder C, Zhang L, et al. 1999. Pronounced pharmacologic deficits in M2 muscarinic acetylcholine receptor knockout mice. *Proc. Natl. Acad. Sci. USA* 96:1692–97

98. Fehr C, Fickova M, Hiemke C, Reuss S, Dahmen N. 1997. Molecular cloning of rat G-protein-coupled receptor kinase 6 (GRK6) from brain tissue, and its mRNA expression in different brain regions and peripheral tissues. *Br. Res. Mol. Br. Res.* 49:278–82

84. Presents evidence for a role of GRK3 in bipolar disorder.

96. Describes pronounced behavioral and biochemical muscarinic supersensitivity in GRK5-deficient mice.

99. Gainetdinov RR, Bohn LM, Sotnikova TD, Cyr M, Laakso A, et al. 2003. Dopaminergic supersensitivity in G protein-coupled receptor kinase 6-deficient mice. *Neuron* 38:291–303

100. Sotnikova TD, Budygin EA, Jones SR, Dykstra LA, Caron MG, Gainetdinov RR. 2004. Dopamine transporter-dependent and -independent actions of trace amine β-phenylethylamine. *J. Neurochem.* 91:362–73

101. Premont RT, Gainetdinov RR, Caron MG. 2001. Following the trace of elusive amines. *Proc. Natl. Acad. Sci. USA* 98:9474–75

102. Seeman P, Weinshenker D, Quirion R, Srivastava LK, Bhardwaj SK, et al. 2005. Dopamine supersensitivity correlates with D2 high states, implying many paths to psychosis. *Proc. Natl. Acad. Sci. USA* 102:3513–18

103. Ferrer-Alcon M, La Harpe R, Garcia-Sevilla JA. 2004. Decreased immunodensities of μ-opioid receptors, receptor kinases GRK 2/6 and β-arrestin-2 in postmortem brains of opiate addicts. *Br. Res. Mol. Br. Res.* 121:114–22

104. Attramadal H, Arriza JL, Aoki C, Dawson TM, Codina J, et al. 1992. β-Arrestin2, a novel member of the arrestin/β-arrestin gene family. *J. Biol. Chem.* 267:17882–90

105. Parruti G, Ambrosini G, Sallese M, De Blasi A. 1993. Molecular cloning, functional expression and mRNA analysis of human β-adrenergic receptor kinase 2. *Biochem. Biophys. Res. Commun.* 190:475–81

106. Gurevich EV, Benovic JL, Gurevich VV. 2002. Arrestin2 and arrestin3 are differentially expressed in the rat brain during postnatal development. *Neuroscience* 109:421–36

107. Kittel A, Komori N. 1999. Ultrastructural localization of β-arrestin-1 and -2 in rat lumbar spinal cord. *J. Comp. Neurol.* 412:649–55

108. Fan XL, Zhang JS, Zhang XQ, Yue W, Ma L. 2003. Differential regulation of β-arrestin 1 and β-arrestin 2 gene expression in rat brain by morphine. *Neuroscience* 117:383–89

109. Avissar S, Matuzany-Ruban A, Tzukert K, Schreiber G. 2004. β-Arrestin-1 levels: reduced in leukocytes of patients with depression and elevated by antidepressants in rat brain. *Am. J. Psychiat.* 161:2066–72

110. Ohsawa M, Mizoguchi H, Narita M, Nagase H, Dun NJ, Tseng LF. 2003. Involvement of β-arrestin-2 in modulation of the spinal antinociception induced by μ-opioid receptor agonists in the mouse. *Neurosci. Lett.* 346:13–16

111. Zhang J, Ferguson SS, Barak LS, Bodduluri SR, Laporte SA, et al. 1998. Role for G protein-coupled receptor kinase in agonist-specific regulation of μ-opioid receptor responsiveness. *Proc. Natl. Acad. Sci. USA* 95:7157–62

112. Whistler JL, von Zastrow M. 1998. Morphine-activated opioid receptors elude desensitization by β-arrestin. *Proc. Natl. Acad. Sci. USA* 95:9914–19

113. Kovoor A, Celver JP, Wu A, Chavkin C. 1998. Agonist induced homologous desensitization of μ-opioid receptors mediated by G protein-coupled receptor kinases is dependent on agonist efficacy. *Mol. Pharmacol.* 54:704–11

114. Haberstock-Debic H, Kim KA, Yu YJ, von Zastrow M. 2005. Morphine promotes rapid, arrestin-dependent endocytosis of μ-opioid receptors in striatal neurons. *J. Neurosci.* 25:7847–57

115. Bohn LM, Lefkowitz RJ, Gainetdinov RR, Peppel K, Caron MG, Lin FT. 1999. Enhanced morphine analgesia in mice lacking β-arrestin 2. *Science* 286:2495–98

116. Bohn LM, Gainetdinov RR, Lin FT, Lefkowitz RJ, Caron MG. 2000. μ-opioid receptor desensitization by β-arrestin-2 determines morphine tolerance but not dependence. *Nature* 408:720–23

99. Reports disruption of dopamine receptor desensitization and behavioral abnormalities in GRK6-deficient mice.

115. Found abnormal μ-opioid receptor regulation and enhanced morphine analgesia in mice lacking β-arrestin 2.

117. Beaulieu JM. 2005. Morphine-induced μ-opioid receptor internalization: a paradox solved in neurons. *J. Neurosci.* 25:10061–63
118. Bohn LM, Lefkowitz RJ, Caron MG. 2002. Differential mechanisms of morphine antinociceptive tolerance revealed in βarrestin-2 knock-out mice. *J. Neurosci.* 22:10494–500
119. Bohn LM, Dykstra LA, Lefkowitz RJ, Caron MG, Barak LS. 2004. Relative opioid efficacy is determined by the complements of the G protein-coupled receptor desensitization machinery. *Mol. Pharmacol.* 66:106–12
120. Raehal KM, Walker JK, Bohn LM. 2005. Morphine side effects in β-arrestin 2 knockout mice. *J. Pharmacol. Exp. Ther.* 314:1195–201
121. Jiang B, Shi Y, Li H, Kang L, Ma L. 2006. Decreased morphine analgesia in rat overexpressing β-arrestin 2 at periaqueductal gray. *Neurosci. Lett.* 400:150–53
122. Hack SP, Bagley EE, Chieng BC, Christie MJ. 2005. Induction of δ-opioid receptor function in the midbrain after chronic morphine treatment. *J. Neurosci.* 25:3192–98
123. Bradaia A, Berton F, Ferrari S, Luscher C. 2005. β-Arrestin2, interacting with phosphodiesterase 4, regulates synaptic release probability and presynaptic inhibition by opioids. *Proc. Natl. Acad. Sci. USA* 102:3034–39
124. Elmer GI, Pieper JO, Rubinstein M, Low MJ, Grandy DK, Wise RA. 2002. Failure of intravenous morphine to serve as an effective instrumental reinforcer in dopamine D2 receptor knock-out mice. *J. Neurosci.* 22:RC224
125. Bohn LM, Gainetdinov RR, Sotnikova TD, Medvedev IO, Lefkowitz RJ, et al. 2003. Enhanced rewarding properties of morphine, but not cocaine, in βarrestin-2 knock-out mice. *J. Neurosci.* 23:10265–73
126. Tao R, Auerbach SB. 1995. Involvement of the dorsal raphe but not median raphe nucleus in morphine-induced increases in serotonin release in the rat forebrain. *Neuroscience* 68:553–61
127. Beaulieu JM, Sotnikova TD, Marion S, Lefkowitz RJ, Gainetdinov RR, Caron MG. 2005. An Akt/β-arrestin 2/PP2A signaling complex mediates dopaminergic neurotransmission and behavior. *Cell* 122:261–73

127. Identified role of G protein–independent, β-arrestin 2–mediated signaling in dopaminergic neurotransmission and behavior.

128. Gainetdinov RR, Caron MG. 2003. Monoamine transporters: from genes to behavior. *Annu. Rev. Pharmacol. Toxicol.* 43:261–84
129. Beaulieu JM, Sotnikova TD, Yao WD, Kockeritz L, Woodgett JR, et al. 2004. Lithium antagonizes dopamine-dependent behaviors mediated by an AKT/glycogen synthase kinase 3 signaling cascade. *Proc. Natl. Acad. Sci. USA* 101:5099–104

Stop That Cell! β-Arrestin-Dependent Chemotaxis: A Tale of Localized Actin Assembly and Receptor Desensitization

Kathryn A. DeFea

Division of Biomedical Sciences and Cell, Molecular, and Developmental Biology Program, University of California, Riverside, California 92521; email: katied@ucr.edu

Annu. Rev. Physiol. 2007. 69:535–60

First published online as a Review in Advance on September 13, 2006

The *Annual Review of Physiology* is online at http://physiol.annualreviews.org

This article's doi:
10.1146/annurev.physiol.69.022405.154804

0066-4278/07/0315-0535$20.00

Key Words

cell polarity, scaffolds, nucleation

Abstract

β-arrestins have recently emerged as key regulators of directed cell migration or chemotaxis. Given their traditional role as mediators of receptor desensitization, one theory is that β-arrestins contribute to cell polarity during chemotaxis by quenching the signal at the trailing edge of the cell. A second theory is that they scaffold signaling molecules involved in cytoskeletal reorganization to promote localized actin assembly events leading to the formation of a leading edge. This review addresses both models. It discusses studies demonstrating the involvement of β-arrestins in chemotaxis both in vivo and in vitro as well as recent evidence that β-arrestins directly bind and regulate proteins involved in actin reorganization.

Chemotaxis: directed movement of a cell toward a chemical signal

Chemoattractant: any chemical signal that evokes chemotaxis

Leading edge: the portion of the cell nearest the chemoattractant, usually consisting of a broad lamellipodium, that is a site of active actin assembly

Chemokine receptor (CXCR): a G protein–coupled receptor specific for chemokines

INTRODUCTION

β-Arrestins have recently emerged as essential regulators of chemotaxis, the directed movement of a cell toward or away from a chemical signal, and in the past few years several studies have delineated their role in this process. Chemotaxis requires a chemical ligand (or chemoattractant), interacting with a cell-surface receptor, to coordinate cell polarization and motility in the direction of the chemoattractant (1–3). Chemotactic signals result in gross reorganization of the actin cytoskeleton, leading to the formation of a leading edge, de-adhesion of the trailing edge, and subsequent migration toward the chemoattractant. Because β-arrestins are known as terminators of receptor signaling, clathrin adaptor proteins, and signaling scaffolds, the question arises as to whether their role in chemotaxis reflects receptor desensitization or localized scaffolding of cytoskeletal signaling molecules (**Figure 1**). The former model is consistent with decades-old studies suggesting that desensitization and recycling of chemokine receptors are the major mechanisms by which leukocytes maintain their ability to sense a chemoattractant gradient during inflammation (4–7). In addition to desensitization, recent studies (discussed in detail in this review) suggest that β-arrestins may also contribute to the spatial control over actin assembly proteins. Furthermore, the mechanism by which β-arrestins mediate chemotaxis may differ with cell and receptor type. This review covers the known roles of β-arrestin-1 and -2 in cytoskeletal reorganization and actin assembly, addressing their roles as mediators of receptor desensitization and scaffolding as they pertain to chemotaxis.

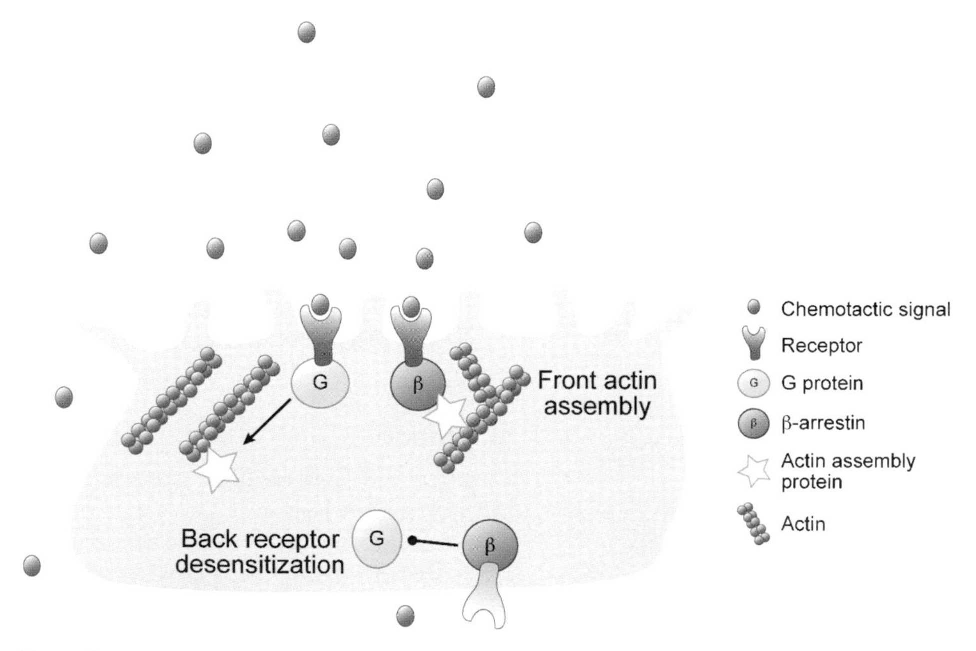

Figure 1

Localized actin assembly versus receptor desensitization. Receptors at the front of the cell are exposed to a higher concentration of chemoattractant than those at the back. β-Arrestin may preferentially uncouple receptors from their cognate G protein in the back of the cell while G protein signaling simultaneously persists in the front, allowing the cell to polarize in the direction of the chemoattractant. β-Arrestin may also scaffold cytoskeletal proteins at the front of the cell, leading to localized actin assembly events and the formation of a leading edge.

β-ARRESTINS AS ESSENTIAL REGULATORS OF LEUKOCYTE CHEMOTAXIS

Chemokine Receptors

Leukocytes are attracted to sites of inflammation, infection, or injury by chemotactic signals released by the injured tissue. Normal immune function requires that leukocytes sense and respond to chemotactic signals, extravasate into the tissue, and control the inflammatory response (8, 9). The first cells recruited are usually granulocytes (e.g., neutrophils, eosinophils, and macrophages); these cells degranulate, releasing a number of bioactive molecules, including additional chemotactic factors (9). Furthermore, the migrating leukocyte may encounter multiple signals, both stimulatory and inhibitory, such that cross-desensitization between chemoattractant receptors is crucial for immune responsiveness (10). Some of the first suggestions that β-arrestins may be involved in chemotaxis emerged from studies on chemokine receptor desensitization in leukocytes (4–7).

Chemokine receptors are G protein–coupled receptors (GPCRs) that mediate chemotaxis in response to inflammatory chemokines released at sites of injury (11). β-Arrestins are required for the desensitization of several chemokine receptors, including CCR2, activated by monocyte chemoattractant protein-1 (4); CXCR1, activated by interleukin (IL)-8 (12); CXCR2, activated by melanoma growth stimulatory activity or IL-8 (12, 13); CXCR4, activated by stromal cell–derived factor 1α or CXCL12 (14, 15); and CCR5, activated by RANTES (regulated upon activation T cell expressed and secreted) (16). A requirement for β-arrestins in chemotaxis was first demonstrated in 2002 with the observation that CXCR4-mediated lymphocyte chemotaxis was defective in β-arrestin-2 and G protein–coupled receptor kinase 6 (GRK6)-knockout mice (17) or after small interfering RNA knockdown of β-arrestin-2 (15). Shortly thereafter, β-arrestins were demonstrated to play a role in CXCR1-, CXCR2-, CXCR3-, and CCR5-mediated chemotaxis as well, but the precise role of β-arrestins in leukocyte chemotaxis has not been fully defined (12, 16, 18–20) (**Table 1**).

Chemokine: a class of inflammatory molecules acting through specific cell-surface receptors that mediate chemotaxis and other inflammatory events

In Vivo Studies on β-Arrestin-Dependent Chemokine-Mediated Migration

Because chemotaxis in vivo is a complex process involving not just the ability of cells to migrate, but to migrate in the correct direction,

Table 1 Comparison of receptors that require β-arrestins for chemotaxis, mitogen-activated protein kinase (MAPK) activation, and desensitization

Receptor	G protein independent?	Cell type	β-Arrestin isoform required for: Chemotaxis	Desensitization/ internalization	MAPK activation
PAR-2	Yes	MDA MB-468, MDA MB-231	1 and 2	1 and 2?	1 and 2 (ERK1/2)
AT1AR	Yes	HEK293	2	1 and 2	2 (Jnk, ERK1/2)
CXCR4	?	Lymphocyte Macrophage HEK293	2	2 (and 1?)	2 (p38MAPK)
CXCR2	?	Lymphocytes Neutrophils (culture only)	2 (and 1?)	1 and 2	?
CXCR1	?	RBL	2 (and 1?)	1 and 2	?
CXCR3	?	HEK293	1	1	?
CCR5	?	HEK293, CHO, RBL	2	1 and 2	Neither

Abbreviations used: AT1AR, angiotensin-II type 1a receptor; CXCR, chemokine receptor; ERK1/2, extracellular signal–regulated kinase 1 and 2.

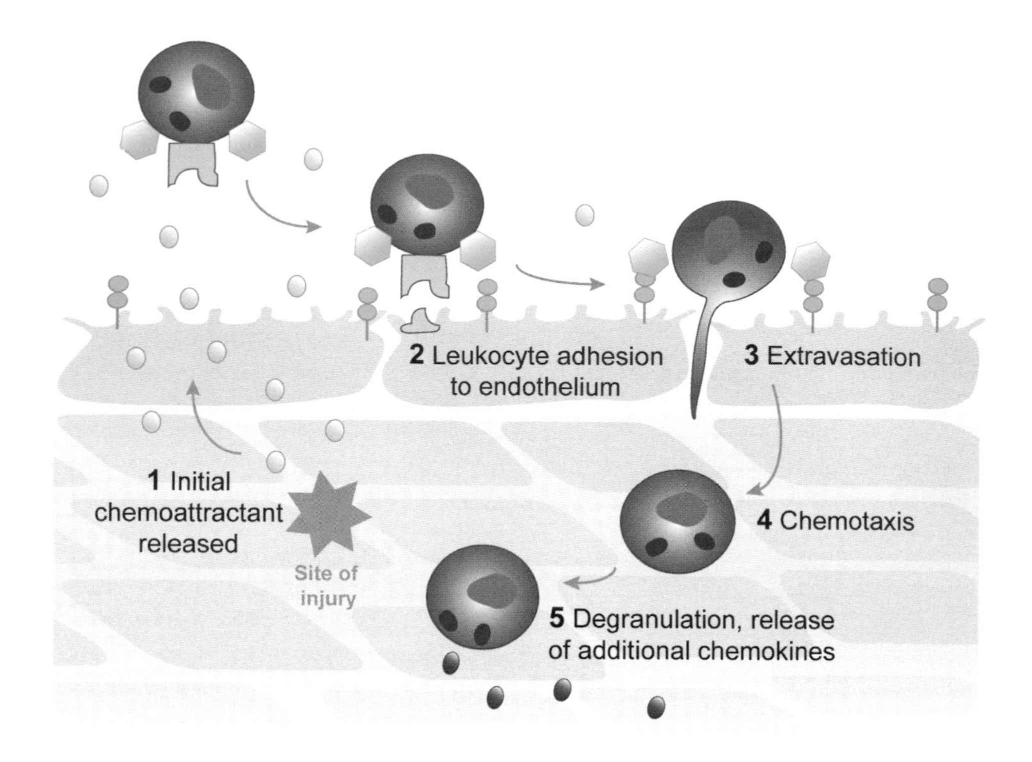

Figure 2

Events involved in leukocyte chemotaxis in vivo. (*1*) In response to injury, infectious pathogens, and inflammation, a chemoattractant agent is released into the blood. (*2*) Leukocytes respond to the chemoattractant by adhering to the vascular endothelium near the site of injury. This process is mediated by a number of cell-surface proteins on both the leukocyte and the vascular endothelial cells. (*3*) The leukocytes undergo cytoskeletal reorganization and extravasate into the tissue, (*4*) where they migrate through the connective tissue to the site of injury. (*5*) At their destination, they release the contents of their granules, which include chemokines, to recruit additional immune cells.

extravasate into the tissue and, in some cases, reattach at a distal site (**Figure 2**), in vitro models of chemotaxis may not always predict the in vivo response correctly. In fact, an in vivo study on the role of β-arrestin-2 in neutrophil chemotaxis found that, contrary to in vitro chemotaxis studies, CXCR2-mediated neutrophil recruitment to sites of inflammation was increased in *β-arrestin-*$2^{-/-}$ mice, although desensitization and internalization were impaired (19). Is this because neutrophils differ from other cell types in their mechanism of chemotaxis, or does this reflect a difference between in vivo and in vitro assays of cell migration? These results contrasted with another in vivo study on allergic asthma in which T cell recruitment to the lung and subsequent cytokine release in response to ovalbumin were impaired in *β-arrestin-*$2^{-/-}$ mice (21). In the same study, neutrophil recruitment in response to lipopolysaccharide (LPS) was unaffected in *β-arrestin-*$2^{-/-}$ mice, suggesting that the role of β-arrestins in chemotaxis does in fact vary depending on the cell type, inflammatory model system, and receptor(s) mediating the response. Furthermore, additional factors that may contribute to neutrophil recruitment in vivo are not reflected in a Transwell filter assay, such as response to other signals present at the site of inflammation and factors involved in tissue extravasation.

Degranulation and Chemokine Release

β-Arrestins have also been implicated in neutrophil degranulation, which is important for leukocyte chemotaxis in vivo, as some of the factors released are chemotactic factors that attract additional immune cells to the site of inflammation. Tyrosine kinases Hck and Fgr are part of a signaling cascade required for IL-8 and zymosan granule release from neutrophils (5, 22, 23). Barlic et al. (24) demonstrated that IL-8 promotes the recruitment of Hck and Fgr into a complex with β-arrestin-1 and that dominant-negative β-arrestin-1 blocks neutrophil degranulation and tyrosine phosphorylation in response to IL-8. In a cultured leukemia cell line (RBL-2H3 cells), GRK3-mediated receptor phosphorylation of the chemokine receptor CC3A (which promotes association of β-arrestins) was essential for expression of the chemokine CCL2, supporting the hypothesis that β-arrestins may contribute to chemotaxis in vivo on multiple levels. Although not all chemokine receptors require β-arrestins for signaling (25), considerable evidence indicates that β-arrestins are important for several steps in a complex inflammatory process involving leukocyte chemotaxis, extravasation, and degranulation (**Figure 2**).

β-ARRESTINS AS GENERAL REGULATORS OF CHEMOTAXIS

Various growth factors, hormones, neuropeptides, peptides, and proteases promote cell migration through the activation of their cognate GPCRs or receptor tyrosine kinases (26). Shortly after the initial demonstration that chemotaxis was impaired in β-arrestin-2-deficient leukocytes, chemotaxis mediated by two different Gαq-coupled receptors—protease-activated receptor-2 (PAR-2) and angiotensin-II type 1a receptor (AT1AR)—was shown to be β-arrestin dependent (27–29) (**Table 1**). In both cases, β-arrestin-dependent chemotaxis could occur independent of G protein coupling, and in the case of PAR-2, the Gαq signaling axis even opposed β-arrestin-dependent migration (27, 30).

Protease-Activated Receptor-2

PAR-2 is activated by various serine proteases, including trypsin, mast cell tryptase, coagulation factors VIIa and Xa, kallekrein, and matripase (31–36), most of which are released at sites of inflammation. It is expressed in a number of motile cells, including neutrophils, mast cells, eosinophils, macrophages, lymphocytes, and tumor cells (29, 33, 37, 38). Several studies suggest that PAR-2-evoked chemotaxis contributes to the metastatic potential of tumor cells, and one study demonstrated that constitutive migration of a highly metastatic tumor cell line was inhibited by knockdown of either β-arrestin-1 or -2 (29, 39, 40). Studies in *PAR-2*$^{-/-}$ mice revealed that ovalbumin-induced eosinophil, neutrophil, and macrophage infiltration to the airway was impaired, whereas nasal administration of PAR-2 agonists I wild-type mice promoted infiltration of leukocytes (38, 41, 42). In contrast, in two other inflammatory models (chemically induced colitis and LPS-induced asthma), PAR-2-activating peptides decreased T cell infiltration into the colonic lamina propria and neutrophil infiltration into the airways. Given recent findings suggesting that the Gαq/Ca^{2+} axis inhibits β-arrestin-dependent signaling downstream of PAR-2 (142a), these contradictory observations may reflect β-arrestin-dependent versus G protein–dependent signaling pathways in different immune cells. Alternatively, PAR-2 is highly expressed in the airway and colonic epithelium, smooth muscle cells, and vascular endothelium (43–46); thus, it may mediate pathways in other cells that negatively regulate immune cell recruitment. The hypothesis that β-arrestin-dependent signaling is important in vivo for PAR-2-stimulated immune cell migration has not been addressed, but such a

PI3K: phosphatidylinositol-3-kinase

hypothesis is consistent with studies demonstrating impaired lymphocyte migration in a model of allergic asthma (21) and impaired PAR-2-stimulated migration in the absence of β-arrestins (28, 29).

Angiotensin IIA Receptor

Angiotensin II (AngII) is a primary mediator of vascular inflammation, hypertension, and atherosclerosis; AngII inhibition is associated with decreased macrophage infiltration into atherosclerotic plaques (47, 48). AngII also stimulates the migration of vascular smooth muscle cells and retinal pericytes, which are important for neovascularization (49, 50). Doctors commonly prescribe angiotensin-converting enzyme inhibitors to treat hypertension. However, that β-arrestin-dependent, AngII-mediated chemotaxis is independent of G protein coupling suggests that chemotactic effects are separable from other vasoconstrictive effects, pointing to the possibility of more specific therapeutic strategies (27, 51–53). In fact, β-arrestin-dependent chemotaxis may represent a G protein–independent signaling arm downstream of a number of GPCRs. Such a hypothesis has major implications for drug development. A number of chemotactic Gαq-coupled receptors have been proposed as therapeutic targets for asthma and inflammatory bowel disease, but Ca^{2+} mobilization and IP3 generation (resulting from Gαq activation) are the most common readouts for receptor activation (11, 54, 55). The emerging role of β-arrestins as mediators of G protein–independent chemotaxis begs the question of whether more specific inhibitors of such GPCRs could be developed.

GPCR cross-activation is involved in platelet-derived growth factor (PDGF)-induced chemotaxis in some cell lines (56), and β-arrestins have in turn been implicated in mediating receptor tyrosine kinase/GPCR cross talk, raising the possibility of an even more general involvement of β-arrestin in cell motility. Still more studies suggest a role for β-arrestins in actin assembly events associated with cell migration, which are critical to the elucidation of the mechanism by which β-arrestins regulate chemotaxis. AT1AR-mediated stress-fiber formation and membrane ruffling, muscarinic M1 receptor, and PAR-2-stimulated pseudopodia formation require β-arrestins (28, 57, 58). Additionally, proteins commonly associated with actin assembly, such as phosphatidylinositol-3-kinase (PI3K), RhoA, and cofilin, can be regulated by β-arrestins (30, 57, 59, 142a). These findings, and how they may pertain to chemotaxis, are discussed in the sections below.

MOLECULAR MECHANISMS OF β-ARRESTIN-DEPENDENT CHEMOTAXIS

Within the scope of this review, there are two hypotheses regarding the requirement for β-arrestins in chemotaxis. The first is that desensitization and recycling of chemotactic receptors are essential for maintaining polarity, and the second is that β-arrestins serve as signaling scaffolds to localize molecules involved in cytoskeletal reorganization. The first model has been proposed for chemotaxis in other organisms as well (60–62), and emerging evidence that β-arrestins specifically bind and regulate actin assembly proteins supports the second model; in reality, both models are probably correct. The next two sections below are devoted to presenting the evidence for each model.

Desensitization and Recycling of Chemotactic Receptors

Researchers first addressed the role of β-arrestins in chemokine signaling, based on the hypothesis that desensitization and recycling of chemokine receptors are essential for chemotaxis (5–7, 62, 63). Thus, a large body of work examines the association and colocalization of β-arrestins with chemokine receptors as well as the effect of dominant-negative β-arrestins on receptor internalization. In one study, β-arrestin-2 associated with CXCR1 but not with a chimeric receptor containing the CXCR2 C-tail, although CXCR2 also bound β-arrestin-2 (12). This same chimera was resistant to internalization but still promoted IL-8-mediated chemotaxis, suggesting that internalization is not essential for chemotaxis. Conversely, dominant-negative β-arrestin-1 inhibited CXCR1-mediated internalization and chemotaxis (20), suggesting that β-arrestin-1 may mediate CXCR1-induced chemotaxis in the absence of β-arrestin-2. Similarly, a mutant CXCR2 lacking the C-terminal phosphorylation sites necessary for β-arrestin binding was unable to internalize or promote chemotaxis (12, 13). However, another study suggested that CXCR2 associated with β-arrestin-1 via its C terminus, but a mutant receptor deficient in β-arrestin-1 binding was still able to undergo ligand-induced endocytosis (64). Still other studies suggest that *N*-formyl-methionyl-leucyl-phenylalanine (fMLP)-mediated chemotaxis does not require desensitization (25). A partial explanation for these seemingly disparate results is that these studies were conducted using different ligands, ligand concentrations, and cell types—all of which may result in slightly different signal transduction pathways (19, 65). For example, CXCR3, similar to many chemokine receptors, is activated by multiple ligands (CCL9, -10, and -11), each of which may elicit distinct outcomes or involve different receptor domains. One study showed that residues in the third intracellular loop were needed for chemotaxis in response to high concentrations of CXCL9 and CXCL11 but not CXCL10, whereas phosphorylation sites in the C-tail were required for both internalization and chemotaxis in response to CCL9 and CCL10 but not CCL11 (65).

Most work investigating the role of desensitization and β-arrestin binding in chemokine receptor chemotaxis relies on the assumption that β-arrestin binding requires GRK phosphorylation of the receptor. However, recent studies suggest that other proteins as well as other amino acids in the cytosolic domains of these receptors are involved in mediating this interaction. In particular, one study suggested that β-arrestin-1 and -2 associated with the dopamine D3 receptor (D3R) under resting conditions, promoting its interaction with the actin-binding protein filamin. GRK phosphorylation reduced the association of D3R with filamin and increased the association of D3R with β-arrestins (66), raising the possibility that β-arrestins may associate with different sites on GPCRs to mediate distinct responses. Two other studies, one regarding CXCR3 and the other regarding CCR5, suggest that the conserved E/DRY motif in the second intracellular loop (common to all family A GPCRs) is an important mediator of β-arrestin-dependent chemotaxis. CCR5 still internalized with an arginine-to-asparagine mutation in the conserved E/DRY motif, but G protein signaling and chemotaxis were inhibited (18, 65). Certainly, these studies suggest that the involvement of β-arrestins in chemotaxis extends beyond their known roles as mediators of desensitization and internalization. One such role is the scaffolding and localization of actin assembly proteins, which is discussed in more detail in the following sections below.

Filamin: an actin-binding protein that bundles actin filaments to promote formation of filipodia and other cellular actin structures

Scaffolding of Actin Assembly

Many of the signaling molecules involved in chemotaxis are common to numerous signaling cascades, raising the age-old question of how specificity of response is achieved. Further complicating matters is that, because chemotaxis requires polarization of the cell such that it is actively moving in one direction, the cell must effectively be sending opposing signals to each pole (1, 3, 67). Activation of a chemotactic receptor is not sufficient for cell migration, nor is simply turning on actin assembly. Instead, signaling molecules involved in cytoskeletal reorganization must be temporally coordinated and spatially controlled. A number of scaffolding proteins, both specific to chemotaxis and ones with more pleiotropic

PIP_2: phosphatidylinositol bisphosphate

functions, have emerged over the past decade as key players in chemotaxis (68–75). In this section, I discuss how β-arrestin may be added to this list. Because actin assembly is the primary target of numerous chemotactic pathways, this next section is devoted to a comprehensive description of essential actin assembly proteins.

Actin polymerization. Actin filaments (F-actin) are polar structures with a fast-growing barbed end that dominates assembly kinetics and a slow-growing pointed end (76). The kinetics of actin polymerization have been determined from in vitro studies (**Figure 3**). The initial rate of de novo polymerization is very slow (lag phase in **Figure 3*a***), but after three or more actin monomers (G-actin) come together to form a stable nucleus, elongation from the exposed barbed ends into a growing filament is a fast first-order reaction (77–79). Spontaneous actin filament assembly in vitro is driven entirely by mass action, and nucleus formation is the rate-limiting step. In vivo this mechanism is suppressed because actin monomers are bound by accessory proteins (i.e., profilin and thymosin-β4) that inhibit spontaneous assembly, and spontaneous assembly is too slow to account for the rapid changes necessary for chemotaxis. For cell migration to occur, the cell must create new barbed ends quickly, which can be accomplished through three general mechanisms (**Figure 3*b***): (*a*) uncapping, (*b*) severing of existing filaments, or (*c*) de novo forming of filaments (76, 80–82). Forward motion involves the formation of various distinct actin-rich structures, which is accomplished in part by the binding of different actin-bundling and cross-linking proteins. For the cell to move along a substrate, it must exert a certain amount of traction, which is accomplished through the formation and disassembly of focal contacts at the leading edge and contractile forces of nonmuscle myosins bound to actin filaments (76, 83) (**Figure 3*c***). Implicit in this model is that the proteins regulating all these events must be tightly controlled. Recent studies suggest that β-arrestins are involved in the spatial regulation of actin assembly and bundling proteins. The possible role of β-arrestin-sequestered kinase activity and other scaffolding functions of β-arrestin are considered below in the context of proteins that affect actin reorganization.

Regulation of actin filament severing. Members of the actin depolymerization factor family, ADF and cofilin, bind specifically to ADP-actin-containing filaments, destabilizing them and promoting dissociation of actin monomers from the pointed end. The end result is the severing and depolymerization of existing filaments, which provides both free barbed ends for elongation and a pool of actin monomers. Cofilin activation is controlled by opposing actions of LIM kinases (LIMKs) (which inactivate it by phosphorylation on Ser 3) and cofilin-specific phosphatases that activate it, such as slingshot homolog 2 and chronophin (84–87). Cofilin activity is further controlled by intracellular pH and phosphatidylinositol bisphosphate (PIP_2) levels, which modulate its actin binding and severing activity (88). During the initial seconds following a chemotactic signal, cofilin severs existing filaments to provide free barbed ends for nucleation and the formation of a leading edge (88–90). Once the leading edge is formed, propagation of migration is thought to involve treadmilling of filaments at the front of the cell, allowing for the recycling of ATP-actin monomers and providing a force for continuous forward motion (90–92). A number of studies dissecting the steps in PAR-2-mediated chemotaxis have begun to uncover a role for β-arrestins as scaffolds that regulate cofilin activity (30). The activation of cofilin-severing activity by PAR-2 requires β-arrestins but is independent of G protein coupling. In fact, G protein–dependent signaling appears to activate the cofilin inhibitory kinase (LIMK), whereas β-arrestins work by antagonizing its activity. Furthermore, the two β-arrestins differ temporally and spatially in their involvement. β-Arrestin-2 is required

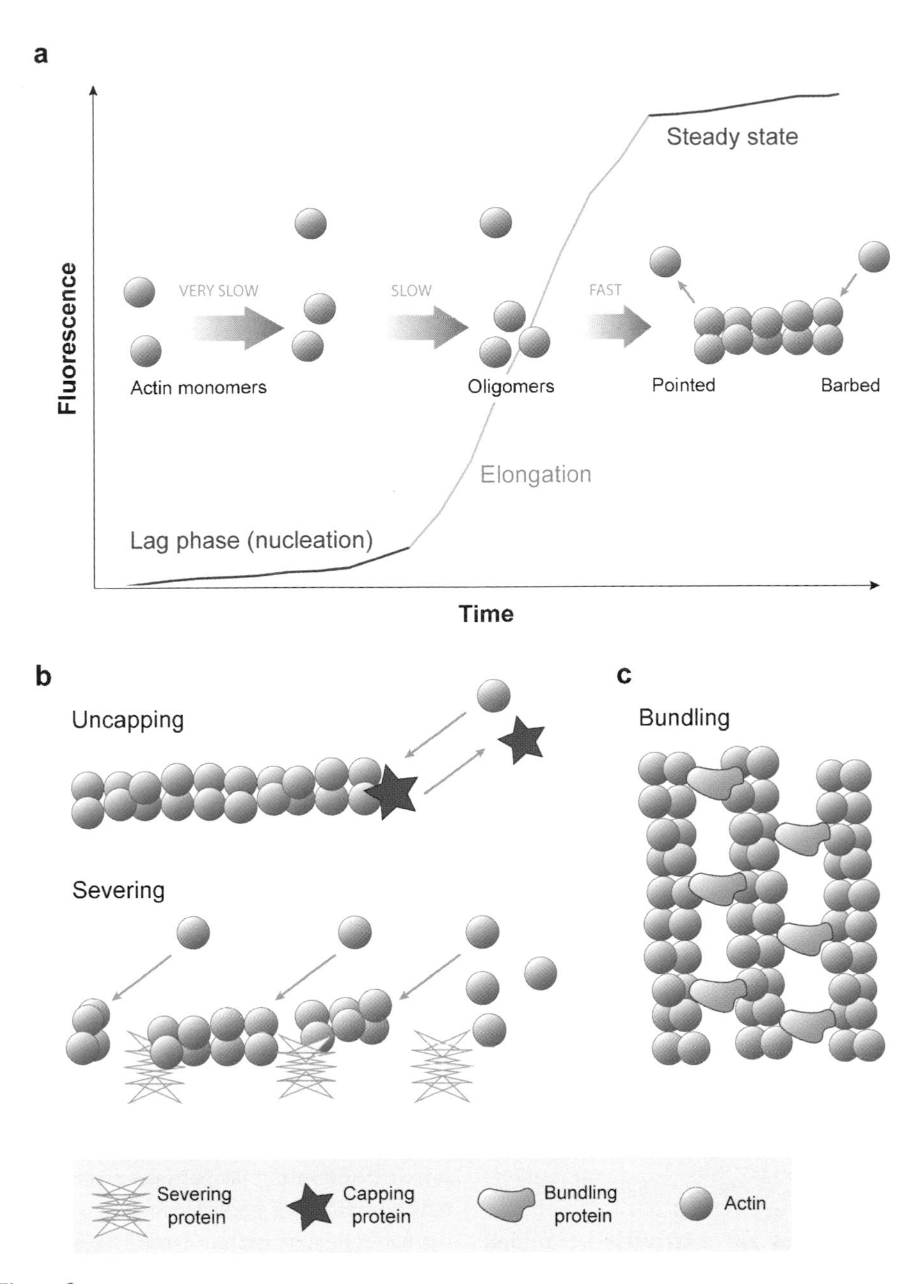

Figure 3

Actin assembly events: nucleation, uncapping, severing, and bundling of filaments. (*a*) Diagram of actin assembly in vitro. The graph depicts typical actin assembly kinetics as measured using pyrene-labeled actin. Pyrene fluorescence increases as it is incorporated into a growing filament and can be used as a readout of actin assembly. The lag phase reflects nucleation; signals that increase the rate of nucleation stimulate actin assembly. The steepest part of the curve reflects filament elongation, and steady state is reached when the addition to the barbed end is equal to disassembly from the pointed end. In the absence of monomer-binding and -capping proteins, steady state is determined by the concentration of free G-actin. (*b*) Free barbed ends can be created by removing capping proteins from barbed ends or severing existing filaments. (*c*) Many cellular actin structures are composed of bundled filaments, achieved by a number of actin-binding proteins.

MAPK: mitogen activated protein kinase

ERK1/2: extracellular signal–regulated kinase 1 and 2

for initial cofilin dephosphorylation and physically associates with cofilin upon PAR-2 stimulation at the back of membrane protrusions. β-Arrestin-1 is important for maintaining a pool of dephosphorylated cofilin and colocalizes with cofilin at the tips of membrane protrusions after prolonged PAR-2 activation (**Figure 4**).

Cofilin and LIMK activities are further regulated by a number of scaffolding proteins that can exert spatial control over filament severing by tipping the scales in favor of either LIMK- or cofilin-phosphatase activities. Members of the 14-3-3 familiy are of particular interest because the binding of 14-3-3 to proteins usually occurs after serine phosphorylation, and because 14-3-3 proteins are binding proteins of GPCRs (93–95). Researchers also found that 14-3-3ζ, which binds to α_2-adrenergic receptor, complexed with phosphorylated (inactive) cofilin and with LIMK and its activating phosphatase (slingshot) (72, 95). The association of 14-3-3 with cofilin protected it from dephosphorylation (75), whereas association with LIMK and slingshot resulted in increased LIMK activity and decreased slingshot association with F-actin. The association of β-arrestin with 14-3-3 has not been reported, but both proteins may bind to similar regions within the third intracellular loop of the α_2-adrenergic receptor (96). Thus, β-arrestin scaffolds may regulate filament-severing activity in part by competing with 14-3-3 proteins, resulting in decreased LIMK activity and increased cofilin dephosphorylation.

Possible roles of β-arrestin-scaffolded MAPKs. Various receptors that require β-arrestins for chemotaxis and actin cytoskeleton reorganization also promote β-arrestin-dependent association with mitogen-activated protein kinases (MAPKs) (27–29, 58, 97–100). In fact, β-arrestins can sequester active extracellular signal–regulated kinase 1 and 2 (ERK1/2) at the leading edge during PAR-2-induced cell migration (28). This may facilitate the specific phosphorylation of proteins involved in chemotaxis, but such targets have not been identified. Various studies suggest potential targets of β-arrestin-dependent MAPK activity; research over the coming years may help elucidate the role of such targets in chemotaxis (**Table 2**).

Regulation of actin filament bundling. The actin-binding protein filamin is important for filipodia formation by various receptors (101–103). Recently, two different studies identified filamin as a β-arrestin-2-binding protein, downstream of D3R, AT1AR, and muscarinic M1 receptor (58, 66). In the case of D3R, β-arrestin and filamin associated with the receptor in the absence of an agonist, and treatment with dopamine resulted in GRK activation and decreased filamin association (66). In the case of AT1AR and muscarinic M1 receptor, the association of filamin with β-arrestin-2 was increased by receptor activation and facilitated the activation and sequestration of ERK1/2 as well as actin cytoskeletal reorganization (58). Various studies have suggested that β-arrestin-dependent ERK1/2 activation is important for actin reorganization and/or chemotaxis, and the role of filamin in filipodia formation is well established. Therefore, the following questions arise from these studies: Does filamin anchor the β-arrestin/ERK1/2 complex to the cytoskeleton via its actin-binding properties so that ERK1/2 can phosphorylate cytosolic/membrane substrates? Does β-arrestin-2-bound ERK1/2 phosphorylate filamin, altering its bundling properties? Filamin does have a number of PXS/TP motifs, a consensus for MAPK phosphorylation. Many seemingly contradictory actin assembly activities are occurring simultaneously in a migrating cell. Some filaments must be bundled for stability, but bundling others would inhibit severing and nucleation. Thus, another question is whether β-arrestin-2 increases or decreases filamin's ability to promote filament bundling (**Figure 5*a***). There is another possible connection between filamin and β-arrestins: through the GTPase RalA. RalA

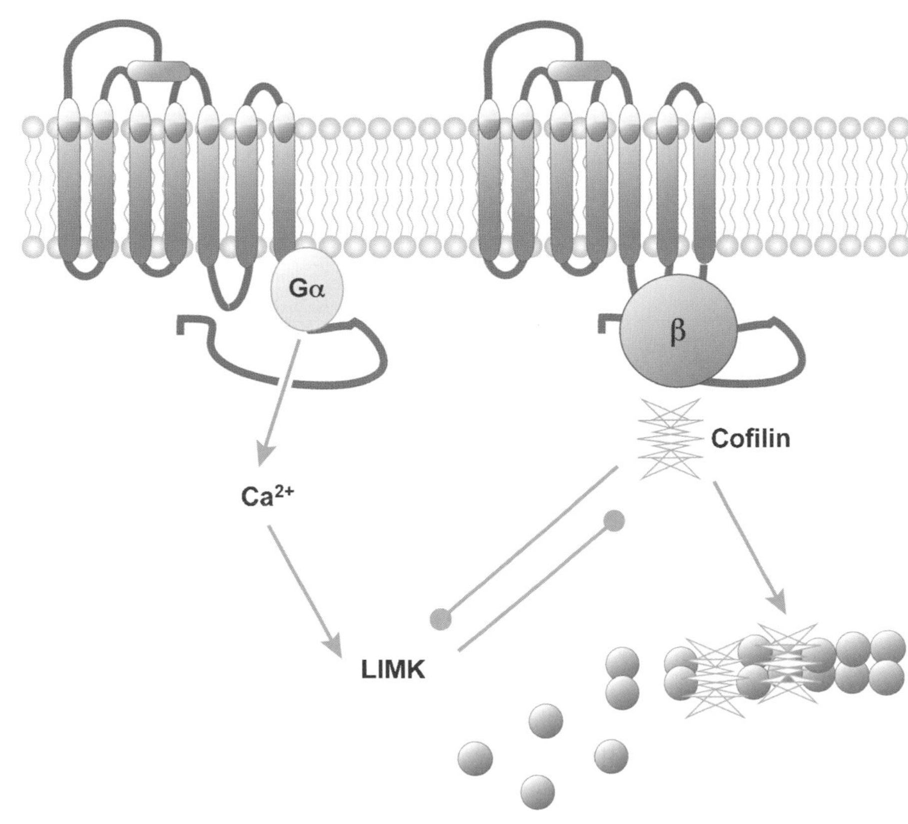

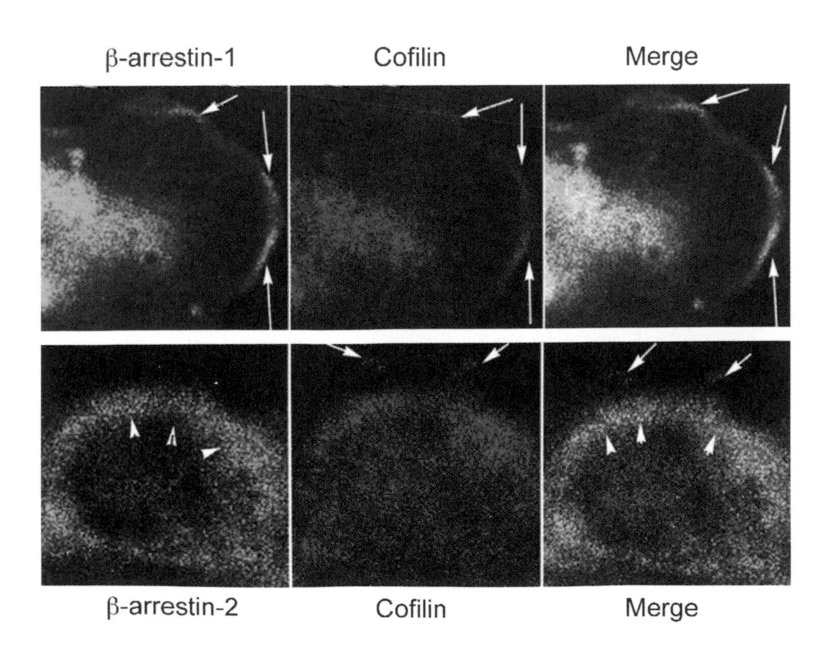

Figure 4

β-arrestin-dependent regulation of cofilin. (*a*) Protease-activated receptor-2 (PAR-2) sends opposing signals to the actin-filament-severing protein, cofilin. One signal is the Gαq/Ca^{2+}-dependent activation of LIM kinase (LIMK), which phosphorylates and inactivates cofilin, and the other signal is the β-arrestin (β)-dependent inhibition of LIMK and the activation of a phosphatase, which results in cofilin activation. (*b*) β-arrestin-1 colocalizes with active cofilin at the tip of lamellipodia, whereas β-arrestin-2 colocalizes at the back. The result is spatially controlled actin filament severing, the creation of barbed ends for elongation, and chemotaxis.

Table 2 Interaction of cytoskeletal proteins with β-arrestins and mitogen-activated protein kinases (MAPKs) and their potential role in β-arrestin-dependent chemotaxis

Cytoskeletal protein	Interacting protein	Receptor	Cell type	Function
Filamin	β-Arrestin-1	AT1AR/muscarinic receptor Dopamine receptor	HEK293	Filament stability, filipodia formation, receptor anchoring MAPK activation?
Cofilin	β-Arrestin-2	PAR-2	Breast-cancer cells	Actin filament severing
p150Spir	Jnk			Increased actin nucleation?
WAVE/Scar	ERK1/2	PDGFR		Increased actin nucleation?
Paxillin	Jnk, ERK1/2	NGF, LPS HGF		Decreased peripheral stress fibers increased FAK association, cell spreading
PI3K	β-Arrestin-1	PAR-2 IGF1R PAR1 Endothelin 1	NIH3T3, Breast-cancer cells HEK293	Inhibition of PI3K activity Localized increase in PIP_2, decreased RhoA-GTPase activity? Activation of PI3K Activation of AkT? Localized decrease in PIP_2 Increased RhoA-GTPase activity?
Akt/PP2A	β-Arrestin-2	Dopamine 2A receptor	In vivo Mouse striatum	Inactivation through scaffolding to PP2A
Ral-GDS	β-Arrestin-1	CXCR4		Ral activation Increased chemokine release? Filamin/actin interaction, filipodia formation?

Abbreviations used: CXCR, chemokine receptor; FAK, focal adhesion kinase; IGF1R, insulin-like growth factor 1 receptor; Jnk, c-Jun N-terminal kinase; PAR, protease-activated receptor; PI3K, phosphatidylinositol-3-kinase; PP2A, protein phosphatase 2A; Ral-GDS, Ral-GDP dissociation stimulator.

promotes filiodia formation in response to the fMLP receptor (a chemokine receptor), tumor necrosis factor α receptor, and IL-1 receptor through interaction with filamin (102, 104). β-Arrestin-1 associates with the Ral-GDP dissociation stimulator (Ral-GDS). Upon activation of the fMLP receptor, β-arrestin binds to the fMLP receptor and then targets Ral-GDS to the membrane, allowing Ral-GDS to bind and activate RalA (104). β-Arrestins may facilitate the filamin-Ral interaction in a similar fashion (**Figure 5*a***).

MAPKs and cell contractility. Stress fibers are F-actin cables that are seen in a cell when it is tightly attached to its substratum. Stress-fiber formation is easily tractable by staining with fluorescent phalloidin (a mushroom toxin that binds tightly to F-actin) and is often used as a readout of actin cytoskeletal reorganization. In an actively migrating cell, stress fibers are usually inhibited in front of the migrating edge and are bound to cross-linking and contractile proteins such as myosin II (105). Various studies suggest that myosin binds actin filaments at the back of migrating cells, stabilizing polarity by inhibiting protrusion formation; while at the leading edge, myosin is important for sensing changes in the substratum. Furthermore, myosin II is a force-generating enzyme that can help to promote traction, through contractile forces exerted on the extracellular matrix, as the cell moves along its substrate (91, 106–108).

Both c-Jun N-terminal kinases (Jnks) and MAPKs have been implicated in the regulation of stress-fiber formation and cell contractility. The expression of constitutively active

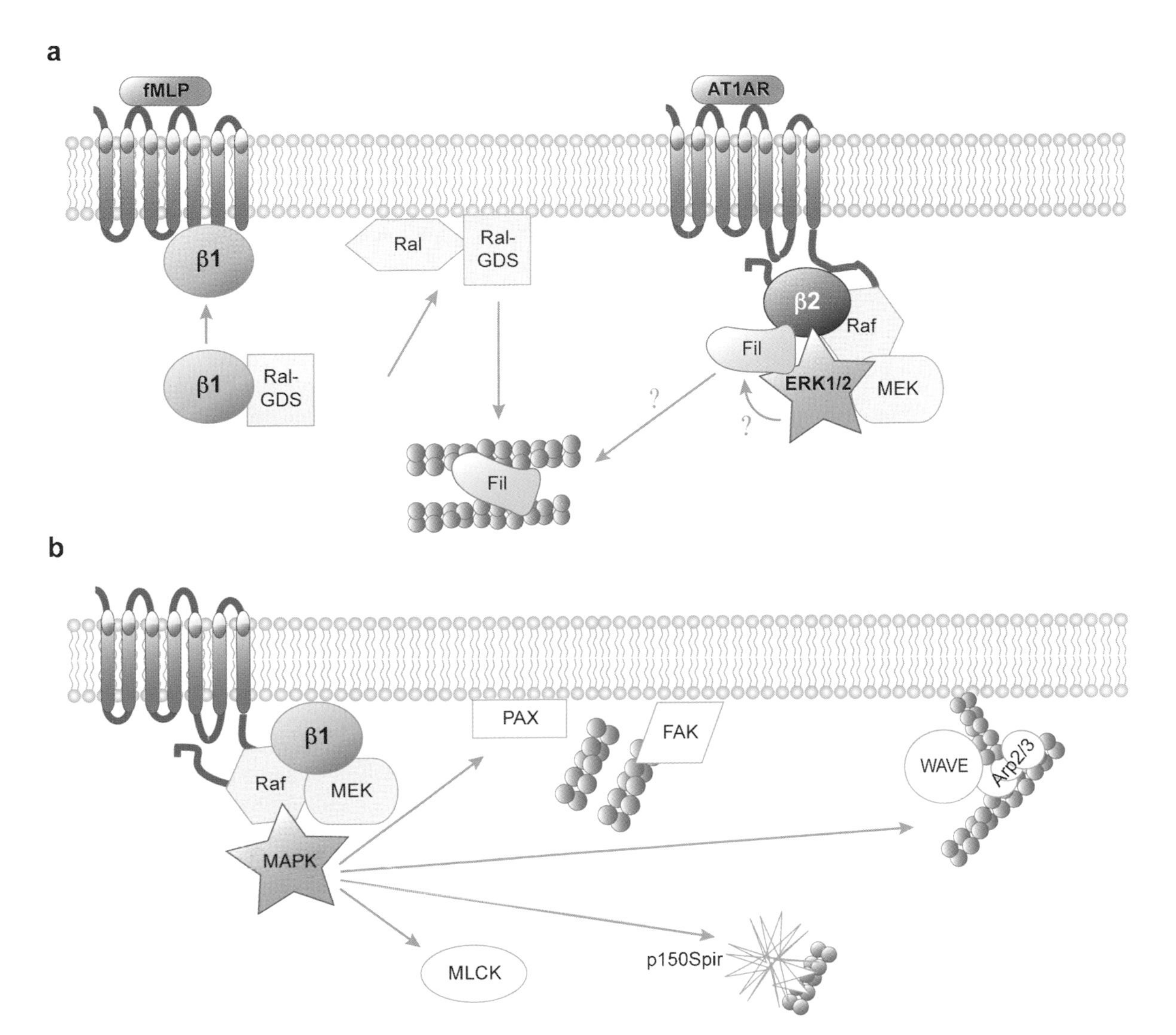

Figure 5

Models of β-arrestin-dependent chemotaxis. (*a*) β-Arrestin-1 associates with Ral-GDP dissociation stimulator (Ral-GDS) in resting cells. Upon activation of the fMLP receptor, they translocate together to the plasma membrane; Ral-GDS is delivered to Ral, and β-arrestin-1 binds to the fMLP receptor. Active Ral promotes F-actin bundling by filamin (Fil) and filipodia formation. AT1AR activation promotes recruitment of β-arrestin-2, ERK1/2, and its upstream activators (Raf and MEK1) into a complex with filamin. β-Arrestin-bound ERK1/2 may phosphorylate filamin to alter its bundling properties, or β-arrestin-2 may cooperate with Ral to promote filamin bundling. (*b*) There are several models for β-arrestin-dependent sequestration of MAPKs in cytoskeletal reorganization: localized phosphorylation of paxillin (PAX), resulting in dissociation from focal adhesion kinase (FAK) and disassembly of focal contacts; phosphorylation of myosin light chain kinase (MLCK), resulting in activation of myosin II and cell contraction; and phosphorylation of WAVE or p150Spir, resulting in increased nucleation.

Actin-related proteins 2 and 3 (Arp2/3): a seven-subunit complex consisting of two actin-related proteins and five accessory proteins that promotes actin nucleation

WAVE/Scar: a family of proteins first identified in patients with Wiscott-Aldrich syndrome that activates Arp2/3 to facilitate actin nucleation

Jnk disrupts stress-fiber formation in fibroblasts (109). In keratinocytes from *MEKK1*$^{-/-}$ (an upstream regulator of Jnk) mice, stress fibers accumulate at the cell periphery in response to chemotactic signals (110). Thus, scaffolding of Jnk activity at the leading edge may play a role in localized disruption of stress-fiber formation at the leading edge. Jnk can phosphorylate the actin-binding protein paxillin on Ser178, and this phosphorylation event is essential for suppressing focal adhesion and stress-fiber formation during cell migration (111). Serine phosphorylation of Jnk may promote paxillin degradation (112). Phosphorylation of S178 also may sterically inhibit constitutive tyrosine phosphorylation at nearby Y181 (113), similar to insulin receptor substrate-1 (IRS-1) (114-116). Tyrosine phosphorylation of paxillin is associated with mature focal complex formation (117); therefore, localized phosphorylation by Jnk may regulate adhesion turnover by inhibiting subsequent phosphorylation by tyrosine kinases (**Figure 5*b***). It is unknown whether β-arrestins are involved in paxillin phosphorylation by Jnk downstream of AT1AR or other receptors, but this hypothesis certainly merits investigation.

ERK1/2 can also phosphorylate paxillin at Ser126 in response to LPS and nerve growth factor. ERK1/2 phosphorylation at Ser126 facilitates subsequent phosphorylation by glycogen synthase kinase on Ser130. Dual phosphorylation on Ser126/130 is required for LPS-induced cell spreading (118). Interestingly, paxillin constitutively associates with MEK1; activation of hepatocyte growth factor receptor promotes the recruitment of Raf and ERK1/2 into a focal adhesion–associated complex and the subsequent phosphorylation of paxillin by ERK1/2 at Ser83 (119, 120). Once again, bearing in mind that adhesion dynamics must be tightly controlled and are turned over during migration, one hypothesis is that β-arrestin competes with paxillin for scaffolding of Raf and ERK1/2, resulting in decreased paxillin/focal adhesion kinase (FAK) association and localized disassembly of focal contacts (**Figure 5*b***). This hypothesis is consistent with studies showing that phosphorylation of myosin light chain kinase (MLCK), which activates myosin II, is essential for focal adhesion turnover at the leading edge and chemotaxis (121–123). One can envision a scenario in which β-arrestin scaffolds Raf, MEK1, and activated ERK1/2 at the leading edge, leading to localized phosphorylation of MLCK and decreased paxillin/FAK association, ultimately promoting focal adhesion turnover (**Figure 5*b***).

MAPKs in the regulation of actin nucleation. The primary cellular mechanism for de novo filament formation requires the activity of the seven-subunit actin-related protein-2 and -3 (Arp2/3) complex. The Arp2/3 complex binds the sides of existing filaments, nucleating actin assembly to form orthogonal arrays of cross-linked filaments, which make up the major structural component of the lamellipodia and migrating cells (124–127). The Arp2/3 complex alone is a poor nucleator of actin assembly. Wiscott-Aldrich syndrome family proteins (WASPs), so named because they were first identified in patients with this rare autoimmune disease, can bind Arp2/3 and accelerate in vitro actin assembly (128, 129). This family contains five members, the immune-system-restricted WASP, its ubiquitously expressed homolog N-WASP, and three homologs of the *Dictyostelium* Scar (suppressor of cAMP receptor), also known as WAVE1–3 (Wasp-Verprolin homology) (128, 130, 131). Various chemotactic signals result in the localization of WASP, N-WASP, and WAVE/Scars to the leading edge, where they appear to direct distinct cytoskeletal structures necessary for cell migration (132, 133). WASPs associate with a number of scaffolding proteins that regulate both their inactive and active conformations (69, 70, 74).

WASPs can be directly activated by PIP_2 and/or RhoA GTPase binding (70, 134), which in turn can be regulated by β-arrestins (57). Furthermore, phosphorylation of the N-WASP C terminus increased its ability to activate Arp2/3, and ERK1/2-dependent phosphorylation of WAVE/Scar was implicated in PDGF-mediated membrane ruffling (135–137). Phosphorylation of cortactin, an activator of N-WASP, by ERK1/2 increased N-WASP activation (138). Little or no investigation into the role of β-arrestins in the regulation of WAVE/Scar proteins has been reported, although PIP_2 levels, RhoA-GTPase activity, and ERK1/2 activation all depend on β-arrestin downstream of a number of chemotaxis-promoting receptors. Moreover, several studies have reported colocalization of WASPs and Arp2/3 with clathrin-coated vesicles (139). This colocalization may reflect a requirement for actin nucleation in endocytosis, but the possibility that it also reflects a role for endosomal scaffolding by β-arrestin in nucleation activity has not been addressed. β-Arrestin scaffolding of MAPKs may also regulate actin nucleation through phosphorylation of consensus sites in the C terminus and polyproline domains of N-WASP or WAVE/Scar. Thus, β-arrestin-dependent sequestration of MAPKs in the cytosol membrane may promote actin nucleation through the phosphorylation of nucleation-promoting factors (**Figure 5*b***).

For some time, Arp2/3 was thought to be the only de novo actin nucleator; however, recently formins and p150Spir were demonstrated to nucleate actin filaments in vitro (81, 140). The MAPK Jnk, which is retained in the cytosol by β-arrestin-2 downstream of AT1AR (98), can phosphorylate and activate p150Spir (141), suggesting that β-arrestin-2 regulates actin nucleation through the spatial regulation of Jnk activity. It is unclear which receptors utilize Spir rather than Arp2/3 for actin nucleation or how Jnk phosphorylation affects its nucleation activity, and the possibility that β-arrestins regulate nucleation through this pathway has not been investigated.

β-Arrestin-dependent scaffolding of other signaling proteins. As discussed above with regard to the actin-filament-severing protein cofilin, β-arrestins can serve as scaffolds for, and/or regulate the activity of, various other proteins that may impact cytoskeletal reorganization and chemotaxis. Among them are PI3K (which phosphorylates the D3 ring of PI or PI4P to create $PI_{34}P2$ and $PI_{345}P3$), Akt (a downstream effector of PI3K), protein phosphatase 2A (PP2A) (which dephosphorylates Akt as well as other kinases), and RhoA-GTPases.

Actin nucleation: association of three actin monomers into a trimer; constitutes the rate-limiting step of actin assembly

β-Arrestins and the PI3K pathway. β-Arrestins have been implicated as both activators and inhibitors of PI3K downstream of insulin-like growth factor receptor, endothelin receptor, and PAR-2 (59, 142, 142a). In the case of PAR-2, β-arrestin-1 directly associates with the p85 regulatory subunit of PI3K in membrane ruffles, and its overexpression inhibits agonist-induced PI3K activity (142a). In the insulin-like growth factor 1–receptor signaling pathway, IRS-1 typically links the receptor to PI3K, but β-arrestin-1 appears to mediate the activation of PI3K and Akt independently of IRS-1. Similarly endothelin-1 receptor and PAR-1 both require β-arrestin-1 for PI3K activation, by a mechanism involving its ability to scaffold Src-family kinases. Similar to ERK1/2, PI3K activity has both proliferative and nonproliferative effects involving nuclear and cytosolic effectors, respectively. Whether β-arrestin-dependent regulation of PI3K activity is important for cell migration downstream of any of these receptors has not been demonstrated.

How does PI3K activity relate to chemotaxis? PI3K is a common target of most chemotactic pathways and is essential for chemotaxis of most cells. PI3K activity directly affects various actin cytoskeletal proteins; once again, each must be spatially and temporally controlled. β-Arrestin-1 may indirectly exert spatial control over filament capping through localized stimulation or inhibition of PI3K, resulting in a subsequent decrease or increase in PIP_2. PIP_2, in turn, regulates numerous processes involved in chemotaxis, such as the activation of WASPs, inactivation of cofilin-filament severing, and disruption of the binding of barbed end–capping proteins to actin filaments. In addition to decreased PIP_2 levels, the PI3K product $PI_{345}P3$ activates a number of PH-domain-containing proteins, including guanine nucleotide exchange factors for RhoA, Cdc42, and Rac-1, which in turn activate LIMK and WASPs (3, 76). Thus, localized regulation of PI3K likely has relatively diverse effects on actin assembly and may depend on other factors present in the same microdomain in which it is activated.

In addition to regulating PI3K, β-arrestin-1 can directly regulate one of its downstream effectors, Akt. Downstream of the dopamine 2A receptor, β-arrestin-2 binds both Akt and its regulatory phosphatase PP2A. PP2A is somewhat promiscuous, but by scaffolding it along with Akt, β-arrestin-2 can promote specific, and possibly localized, inactivation of Akt. Akt is important for various processes, including chemotaxis. Recently, one study demonstrated that spatial regulation of Akt activity is essential for myosin II activation through the activation of p21-activated kinase 1 (PAK1) (143). PAK1 promotes myosin II assembly and cell contraction by phosphorylating MLCK. PAK1 is typically thought of as a Cdc42/Rac effector, but Akt can bypass GTPases to activate PAK1 directly (144). Dopamine is chemotactic for breast-cancer cells in vitro and for immune cells in vivo (145, 146). Thus, β-arrestin-dependent inhibition of Akt by dopamine receptors may reflect yet another mechanism for spatial-regulation actin assembly events important for chemotaxis.

β-Arrestins and RhoA. RhoA-GTPases are small G proteins that are essential for cell motility in most systems. As discussed above, they regulate cofilin-severing activity through the activation of LIMK, actin nucleation through the activation of WASPs, cell contractility through the regulation of MLCK, and many other cellular events (132, 147, 148). The three most commonly studied members of the RhoA-GTPase family are RhoA, Rac-1, and Cdc42, and each is associated with a different actin structure: RhoA typically causes stress-fiber formation; Cdc42 induces filipodia formation; and Rac-1 is important for membrane ruffling and lamellipodia formation (147). The three GTPases work in a temporally and spatially regulated fashion to promote chemotaxis (147, 148). Recently, researchers demonstrated a role for β-arrestin-1 in the regulation of RhoA downstream of AT1AR. Small interfering RNA knockdown of β-arrestin-1 but not β-arrestin-2 obliterated RhoA activation and stress-fiber formation by AT1AR. Interestingly, unlike the β-arrestin-2-dependent regulation of MAPK by AT1AR or cofilin by PAR-2, RhoA activity required cooperation between Gαq-dependent and β-arrestin-dependent pathways (57). Direct interaction of β-arrestin-1 with RhoA was not demonstrated but may exist transiently; the detection of Rho-binding partners can be technically difficult because GTP hydrolysis during the binding assay may reduce the stability of interactions. Thus, β-arrestins may also serve as scaffolds for RhoA, possibly prolonging activity through the inhibition of GTP hydrolysis or facilitating their association with a specific guanine nucleotide exchange factor.

CONCLUDING REMARKS

Chemotaxis is a necessary process for the function of multiple organ systems yet has a downside to it as well. For example, the ability of tumor cells to invade the host's blood and distal organs is associated with increased mortality, and the increased infiltration of immune cells is associated with chronic inflammatory diseases (149). The signals leading to chemotaxis have been a hot area of research for many years, with the number of players on the field growing exponentially. Yet the molecular mechanisms by which the cell maintains the polarity necessary for forward motion have not been completely elucidated. β-Arrestins, traditionally associated with receptor desensitization and more recently with the scaffolding of signaling molecules, need to be added to the growing list of regulators of actin assembly. However, much is to be learned about their roles in chemotaxis.

Unraveling the signal transduction pathways leading to chemotaxis downstream of multiple receptors is a quagmire into which one can sink quickly. Not only do the upstream chemotactic receptors mediate multiple cellular events, such as cell proliferation and metabolism as well as cell migration, but the downstream effectors are promiscuous kinases, GTPases, and multipurpose adaptor proteins. Adding to the murkiness is the complexity imparted by scaffolds that differentially interact with some of the common cytoskeleton-regulating proteins. Thus, not only is the cell charged with the task of spatially restricting actin assembly events in response to a given signal, but it must control the promiscuity of these cytoskeletal signaling proteins to elicit a specific outcome. Dysregulation of β-arrestins inhibits chemotaxis, as is evidenced by decreased infiltration of leukocytes in a mouse model of allergic asthma and by the impaired migratory ability of numerous cell types, but the precise role of β-arrestins and the magnitude of their involvement in chemotaxis appear to differ between receptors and cell types.

When dissecting a signaling pathway in which tight spatial control over various molecules must be maintained, elucidating the role of a protein such as β-arrestin, which can both facilitate and terminate signals, is difficult. In some cases, localized quenching of G protein signaling may contribute to the cell's ability to sense a gradient, as was originally proposed. However, this theory does not take into account the ability of β-arrestin to mediate the regulation of actin assembly proteins and chemotaxis independently of G protein coupling. Certainly, localized signal dampening plays a role in cell polarization and chemotaxis, but increasing evidence suggests that the scaffolding of protein kinases, GTPases, and PI3K is important in the spatial regulation of actin assembly as well. A caveat to addressing this conundrum rigorously is the inability to separate the role of β-arrestin in endocytosis with its role as a signaling scaffold. In fact, if the mechanism by which β-arrestins facilitate subcellular localization of enzymes is by hijacking the endocytotic apparatus, how is one to make this distinction? To make things more complicated, receptors differ in their β-arrestin specificity, and the specific domains that interact with different actin assembly proteins vary. If researchers could identify a specific point mutation that disrupts β-arrestin-dependent chemotaxis for each receptor while signal desensitization and endocytosis are maintained (or vice versa), one could begin to address this question. As it stands now, not only is β-arrestin-dependent signaling to the cytoskeleton important for a wide variety of chemotactic pathways, but it is a potentially proinflammatory/prometastatic pathway that may be enhanced by the inhibition of receptor/G protein coupling.

SUMMARY POINTS

1. Studies on chemokine receptors dating back decades suggest that receptor desensitization and recycling are important for cell polarization during chemotaxis.
2. The requirement of β-arrestins for chemotaxis downstream of multiple receptors suggests that β-arrestins play a key role in mediating cell migration.
3. The involvement of β-arrestins in chemokine release as well as actin reorganization suggests that they play multiple roles in the in vivo chemotactic process.
4. The identification of actin-binding and regulatory proteins as β-arrestin-binding partners suggests that their scaffolding function is important for cell migration.

FUTURE ISSUES

1. β-Arrestins can sequester MAPK in a cytosolic/membrane domain, and numerous actin assembly and bundling proteins are phosphorylated by MAPKs, yet a specific role for β-arrestin-dependent sequestration of MAPKs in chemotaxis has not been shown.
2. The identification of key mutations that eliminate scaffolding without affecting desensitization (or vice versa) is important for distinguishing between the scaffolding and desensitization functions of β-arrestins in different steps of chemotaxis.
3. β-Arrestin can promote chemotaxis in a Transwell filter assay independently of G protein engagement, but other events important for chemotaxis require the synergy of both pathways. More extensive studies that evaluate the speed and distance of migration (as well as more in vivo studies that evaluate immune cell infiltration, tumor cell metastasis, wound healing, and other cellular responses) are necessary to elucidate fully the role of each pathway.

ACKNOWLEDGMENTS

I would like to acknowledge Jonathan Zalevsky (Xencor, Inc.) for careful reading of this review, as well as all members of the DeFea lab for their continued hard work on novel ideas.

LITERATURE CITED

1. Ridley AJ, Schwartz MA, Burridge K, Firtel RA, Ginsberg MH, et al. 2003. Cell migration: integrating signals from front to back. *Science* 302:1704–9
2. Bailly M, Yan L, Whitesides GM, Condeelis JS, Segall JE. 1998. Regulation of protrusion shape and adhesion to the substratum during chemotactic responses of mammalian carcinoma cells. *Exp. Cell Res.* 241:285–99
3. Weiner OD, Servant G, Welch MD, Mitchison TJ, Sedat JW, Bourne HR. 1999. Spatial control of actin polymerization during neutrophil chemotaxis 6. *Nat. Cell Biol.* 1:75–81
4. Aragay AM, Mellado M, Frade JM, Martin AM, Jimenez-Sainz MC, et al. 1998. Monocyte chemoattractant protein-1-induced CCR2B receptor desensitization mediated by the G protein-coupled receptor kinase 2. *Proc. Natl. Acad. Sci.* 95:2985–90
5. Henson PM, Schwartzman NA, Zanolari B. 1981. Intracellular control of human neutrophil secretion. II. Stimulus specificity of desensitization induced by six different soluble and particulate stimuli. *J. Immunol.* 127:754–59
6. Tomhave ED, Richardson RM, Didsbury JR, Menard L, Snyderman R, Ali H. 1994. Cross-desensitization of receptors for peptide chemoattractants: characterization of a new form of leukocyte regulation. *J. Immunol.* 153:3267–75
7. Walensky LD, Roskams AJ, Lefkowitz RJ, Snyder SH, Ronnett GV. 1995. Odorant receptors and desensitization proteins colocalize in mammalian sperm. *Mol. Med.* 1:130–41
8. Stein JV, Nombela-Arrieta C. 2005. Chemokine control of lymphocyte trafficking: a general overview. *Immunology* 116:1–12
9. Moser B, Willimann K. 2004. Chemokines: role in inflammation and immune surveillance. *Ann. Rheum. Dis.* 63(Suppl. 2):84–89

10. Nagasawa T, Hirota S, Tachibana K, Takakura N, Nishikawa S, et al. 1996. Defects of B-cell lymphopoiesis and bone-marrow myelopoiesis in mice lacking the CXC chemokine PBSF/SDF-1. *Nature* 382:635–38
11. Proudfoot AEI. 2002. Chemokine receptors: multifaceted therapeutic targets. *Nat. Rev. Immunol.* 2:106–15
12. Richardson RM, Marjoram RJ, Barak LS, Snyderman R. 2003. Role of the cytoplasmic tails of CXCR1 and CXCR2 in mediating leukocyte migration, activation, and regulation. *J. Immunol.* 170:2904–11
13. Fan GH, Yang W, Wang XJ, Qian Q, Richmond A. 2001. Identification of a motif in the carboxyl terminus of CXCR2 that is involved in adaptin 2 binding and receptor internalization. *Biochemistry* 40:791–800
14. Orsini MJ, Parent JL, Mundell SJ, Benovic JL. 1999. Trafficking of the HIV coreceptor CXCR4: role of arrestins and identification of residues in the C-terminal tail that mediate receptor internalization. *J. Biol. Chem.* 274:31076–86
15. Cheng ZJ, Zhao J, Sun Y, Hu W, Wu YL, et al. 2000. β-Arrestin differentially regulates the chemokine receptor CXCR4-mediated signaling and receptor internalization, and this implicates multiple interaction sites between β-arrestin and CXCR4. *J. Biol. Chem.* 275:2479–85
16. Kraft K, Olbrich H, Majoul I, Mack M, Proudfoot A, Oppermann M. 2001. Characterization of sequence determinants within the carboxyl-terminal domain of chemokine receptor CCR5 that regulate signaling and receptor internalization. *J. Biol. Chem.* 276:34408–18
17. Fong AM, Premont RT, Richardson RM, Yu YR, Lefkowitz RJ, Patel DD. 2002. Defective lymphocyte chemotaxis in β-arrestin2 and GRK6-deficient mice. *Proc. Natl. Acad. Sci. USA* 99:7478–83

17. The first demonstration that β-arrestins were required for chemotaxis downstream of a chemokine receptor.

18. Lagane B, Ballet S, Planchenault T, Balabanian K, Le Poul E, et al. 2005. Mutation of the DRY motif reveals different structural requirements for the CC chemokine receptor 5-mediated signaling and receptor endocytosis. *Mol. Pharmacol.* 67:1966–76
19. Su Y, Raghuwanshi SK, Yu Y, Nanney LB, Richardson RM, Richmond A. 2005. Altered CXCR2 signaling in β-arrestin-2-deficient mouse models. *J. Immunol.* 175:5396–402
20. Barlic J, Khandaker MH, Mahon E, Andrews J, DeVries ME, et al. 1999. β-Arrestins regulate interleukin-8-induced CXCR1 internalization. *J. Biol. Chem.* 274:16287–94
21. Walker JKL, Fong AM, Lawson BL, Savov JD, Patel DD, et al. 2003. β-Arrestin-2 regulates the development of allergic asthma. *J. Clin. Invest.* 112:566–74

21. Suggests a role for β-arrestin-dependent chemotaxis in the pathogenesis of asthma.

22. Welch H, Mauran C, Maridonneau-Parini I. 1996. Nonreceptor protein-tyrosine kinases in neutrophil activation. *Methods* 9:607–18
23. Barlic J, Andrews JD, Kelvin AA, Bosinger SE, DeVries ME, et al. 2000. Regulation of tyrosine kinase activation and granule release through β-arrestin by CXCR1. *Nat. Immunol.* 1:227–33
24. Barlic J, Andrews JD, Kelvin AA, Bosinger SE, DeVries ME, et al. 2000. Regulation of tyrosine kinase activation and granule release through β-arrestin by CXCRI. *Nat. Immunol.* 1:227–33
25. Vines CM, Revankar CM, Maestas DC, LaRusch LL, Cimino DF, et al. 2003. N-formyl peptide receptors internalize but do not recycle in the absence of arrestins. *J. Biol. Chem.* 278:41581–84
26. Affolter M, Weijer CJ. 2005. Signaling to cytoskeletal dynamics during chemotaxis. *Dev. Cell* 9:19–34

27. **The first demonstration that β-arrestin-dependent chemotaxis may represent a G protein–independent event downstream of the AII receptor.**

29. **Suggests a role for β-arrestin-dependent chemotaxis in the pathogenesis of tumor cell metastasis.**

27. **Hunton DL, Barnes WG, Kim J, Ren XR, Violin JD, et al. 2005. β-Arrestin 2-dependent angiotensin II type 1A receptor-mediated pathway of chemotaxis. *Mol. Pharmacol.* 67:1229–36**
28. Ge L, Ly Y, Hollenberg M, DeFea K. 2003. A β-arrestin-dependent scaffold is associated with prolonged MAPK activation in pseudopodia during protease-activated receptor-2-induced chemotaxis. *J. Biol. Chem.* 278:34418–26
29. **Ge L, Shenoy SK, Lefkowitz RJ, DeFea KA. 2004. Constitutive protease-activated-receptor-2 mediated migration of MDA MB-231 breast cancer cells requires both β-arrestin-1 and 2. *J. Biol. Chem.* 279:55419–24**
30. Ge L, Zoudilova M, DeFea K. 2005. *Protease-activated-receptor-2 promotes β-arrestin-dependent, G-protein-independent cofilin activation.* Presented at Annu. Meet. Am. Soc. Biochem. Mol. Biol., San Diego
31. Angelo PF, Lima AR, Alves FM, Blaber SI, Scarisbrick IA, et al. 2006. Substrate specificity of human Kallikrein 6: salt and glycosaminoglycan activation effects. *J. Biol. Chem.* 281:3116–26
32. Cottrell GS, Amadesi S, Grady EF, Bunnett NW. 2004. Trypsin IV: a novel agonist of protease-activated receptors 2 and 4. *J. Biol. Chem.* 279:13532–39
33. Dery O, Corvera CU, Steinhoff M, Bunnett NW. 1998. Proteinase-activated receptors: novel mechanisms of signaling by serine proteases. *Am. J. Physiol.* 274:C1429–52
34. Jiang X, Bailly MA, Panetti TS, Cappello M, Konigsberg WH, Bromberg ME. 2004. Formation of tissue factor-factor VIIa-factor Xa complex promotes cellular signaling and migration of human breast cancer cells. *J. Thromb. Haemost.* 2:93–101
35. Takeuchi T, Harris JL, Huang W, Yan KW, Coughlin SR, Craik CS. 2000. Cellular localization of membrane-type serine protease 1 and identification of protease-activated receptor-2 and single-chain urokinase-type plasminogen activator as substrates. *J. Biol. Chem.* 275:26333–42
36. Corvera CU, Dery O, McConalogue K, Gamp P, Thoma M, et al. 1999. Thrombin and mast cell tryptase regulate guinea-pig myenteric neurons through proteinase-activated receptors-1 and -2. *J. Physiol.* 517(Pt. 3):741–56
37. Howells G, Macey M, Chinni C, Hou L, Fox T, et al. 1997. Proteinase-activated receptor-2: expression by human neutrophils. *J. Cell Sci.* 110:881–87
38. Bolton SJ, McNulty CA, Thomas RJ, Hewitt CRA, Wardlaw AJ. 2003. Expression of and functional responses to protease-activated receptors on human eosinophils. *J. Leukoc. Biol.* 74:60–68
39. Morris DR, Ding Y, Ricks TK, Gullapalli A, Wolfe BL, Trejo J. 2006. Protease-activated receptor-2 is essential for factor VIIa and Xa-induced signaling, migration, and invasion of breast cancer cells. *Cancer Res.* 66:307–14
40. Shi X, Gangadharan B, Brass LF, Ruf W, Mueller BM. 2004. Protease-activated receptors (PAR1 and PAR2) contribute to tumor cell motility and metastasis. *Mol. Cancer Res.* 2:395–402
41. Takizawa T, Tamiya M, Hara T, Matsumoto J, Saito N, et al. 2005. Abrogation of bronchial eosinophilic inflammation and attenuated eotaxin content in protease-activated receptor 2-deficient mice. *J. Pharmacol. Sci.* 98:99–102
42. Schmidlin F, Amadesi S, Dabbagh K, Lewis DE, Knott P, et al. 2002. Protease-activated receptor 2 mediates eosinophil infiltration and hyperreactivity in allergic inflammation of the airway. *J. Immunol.* 169:5315–21
43. Cocks TM, Fong B, Chow JM, Anderson GP, Frauman AG, et al. 1999. A protective role for protease-activated receptors in the airways. *Nature* 398:156–60

44. Kong W, McConalogue K, Khitin LM, Hollenberg MD, Payan DG, et al. 1997. Luminal trypsin may regulate enterocytes through proteinase-activated receptor 2. *Proc. Natl. Acad. Sci. USA* 94:8884–89
45. Chambers LS, Black JL, Poronnik P, Johnson PR. 2001. Functional effects of protease-activated receptor-2 stimulation on human airway smooth muscle. *Am. J. Physiol. Lung Cell Mol. Physiol.* 281:L1369–78
46. Marutsuka K, Hatakeyama K, Sato Y, Yamashita A, Sumiyoshi A, Asada Y. 2002. Protease-activated receptor 2 (PAR2) mediates vascular smooth muscle cell migration induced by tissue factor/factor VIIa complex. *Thromb. Res.* 107:271–76
47. Liu J, Yang F, Yang XP, Jankowski M, Pagano PJ. 2003. NAD(P)H oxidase mediates angiotensin II-induced vascular macrophage infiltration and medial hypertrophy. *Arterioscler. Thromb. Vasc. Biol.* 23:776–82
48. Toko H, Zou Y, Minamino T, Sakamoto M, Sano M, et al. 2004. Angiotensin II type 1a receptor is involved in cell infiltration, cytokine production, and neovascularization in infarcted myocardium. *Arterioscler. Thromb. Vasc. Biol.* 24:664–70
49. Dubey RK, Jackson EK, Luscher TF. 1995. Nitric oxide inhibits angiotensin II-induced migration of rat aortic smooth muscle cell: role of cyclic-nucleotides and angiotensin1 receptors. *J. Clin. Invest.* 96:141–49
50. Nadal JA, Scicli GM, Carbini LA, Scicli AG. 2002. Angiotensin II stimulates migration of retinal microvascular pericytes: involvement of TGF-β and PDGF-BB. *Am. J. Physiol. Heart Circ. Physiol.* 282:H739–48
51. Kurtz TW. 2006. New treatment strategies for patients with hypertension and insulin resistance. *Am. J. Med.* 119:S24–30
52. Shenoy SK, Lefkowitz RJ. 2005. Angiotensin II-stimulated signaling through G proteins and β-arrestin. *Sci. STKE* 2005:cm14
53. Azizi M, Webb R, Nussberger J, Hollenberg NK. 2006. Renin inhibition with aliskiren: Where are we now, and where are we going? *J. Hypertens.* 24:243–56
54. Kelso EB, Lockhart JC, Hembrough T, Dunning L, Plevin R, et al. 2006. Therapeutic promise of proteinase-activated receptor-2 antagonism in joint inflammation. *J. Pharmacol. Exp. Ther.* 316:1017–24
55. Tokuyama H, Ueha S, Kurachi M, Matsushima K, Moriyasu F, et al. 2005. The simultaneous blockade of chemokine receptors CCR2, CCR5 and CXCR3 by a nonpeptide chemokine receptor antagonist protects mice from dextran sodium sulfate-mediated colitis. *Int. Immunol.* 17:1023–34
56. Hobson JP, Rosenfeldt HM, Barak LS, Olivera A, Poulton S, et al. 2001. Role of the sphingosine-1-phosphate receptor EDG-1 in PDGF-induced cell motility. *Science* 291:1800–3
57. Barnes WG, Reiter E, Violin JD, Ren XR, Milligan G, Lefkowitz RJ. 2005. β-Arrestin 1 and $G_{\alpha q/11}$ coordinately activate RhoA and stress fiber formation following receptor stimulation. *J. Biol. Chem.* 280:8041–50
58. **Scott MGH, Pierotti V, Storez H, Lindberg E, Thuret A, et al. 2006. Cooperative regulation of extracellular signal-regulated kinase activation and cell shape change by filamin A and β-arrestins. *Mol. Cell. Biol.* 26:3432–45**
59. Povsic TJ, Kohout TA, Lefkowitz RJ. 2003. β-Arrestin1 mediates insulin-like growth factor 1 (IGF-1) activation of phosphatidylinositol 3-kinase (PI3K) and anti-apoptosis. *J. Biol. Chem.* 278:51334–39
60. Caterina MJ, Hereld D, Devreotes PN. 1995. Occupancy of the *Dictyostelium* cAMP receptor, cAR1, induces a reduction in affinity which depends upon COOH-terminal serine residues. *J. Biol. Chem.* 270:4418–23

58. Identified the actin-bundling protein filamin as a novel β-arrestin-binding partner, suggesting multifaceted roles for β-arrestins in actin assembly and cytoskeletal reorganization.

61. Xiao Z, Zhang N, Murphy DB, Devreotes PN. 1997. Dynamic distribution of chemoattractant receptors in living cells during chemotaxis and persistent stimulation. *J. Cell Biol.* 139:365–74
62. Van Haastert PJ, Wang M, Bominaar AA, Devreotes PN, Schaap P. 1992. cAMP-induced desensitization of surface cAMP receptors in *Dictyostelium*: different second messengers mediate receptor phosphorylation, loss of ligand binding, degradation of receptor, and reduction of receptor mRNA levels. *Mol. Biol. Cell* 3:603–12
63. Hecht I, Cahalon L, Hershkoviz R, Lahat A, Franitza S, Lider O. 2003. Heterologous desensitization of T cell functions by CCR5 and CXCR4 ligands: inhibition of cellular signaling, adhesion and chemotaxis. *Int. Immunol.* 15:29–38
64. Zhao M, Wimmer A, Trieu K, DiScipio RG, Schraufstatter IU. 2004. Arrestin regulates MAPK activation and prevents NADPH oxidase-dependent death of cells expressing CXCR2. *J. Biol. Chem.* 279:49259–67
65. Colvin RA, Campanella GSV, Sun J, Luster AD. 2004. Intracellular domains of CXCR3 that mediate CXCL9, CXCL10, and CXCL11 function. *J. Biol. Chem.* 279:30219–27
66. Kim KM, Gainetdinov RR, Laporte SA, Caron MG, Barak LS. 2005. G protein-coupled receptor kinase regulates dopamine D3 receptor signaling by modulating the stability of a receptor-filamin-β-arrestin complex: a case of autoreceptor regulation. *J. Biol. Chem.* 280:12774–80
67. Bernstein BW, Bamburg JR. 2004. A proposed mechanism for cell polarization with no external cues. *Cell Motil. Cytoskelet.* 58:96–103
68. Miki H, Takenawa T. 2002. WAVE2 serves a functional partner of IRSp53 by regulating its interaction with Rac. *Biochem. Biophys. Res. Commun.* 293:93–99
69. Miki H, Yamaguchi H, Suetsugu S, Takenawa T. 2000. IRSp53 is an essential intermediate between Rac and WAVE in the regulation of membrane ruffling. *Nature* 408:732–35
70. Eden S, Rohatgi R, Podtelejnikov AV, Mann M, Kirschner MW. 2002. Mechanism of regulation of WAVE1-induced actin nucleation by Rac1 and Nck. *Nature* 418:790–93
71. Rohatgi R, Ma L, Miki H, Lopez M, Kirchhausen T, et al. 1999. The interaction between N-WASP and the Arp2/3 complex links Cdc42-dependent signals to actin assembly. *Cell* 97:221–31
72. Soosairajah J, Maiti S, Wiggan O, Sarmiere P, Moussi N, et al. 2005. Interplay between components of a novel LIM kinase-slingshot phosphatase complex regulates cofilin. *EMBO J.* 24:473–86
73. Gumienny TL, Brugnera E, Tosello-Trampont AC, Kinchen JM, Haney LB, et al. 2001. CED-12/ELMO, a novel member of the CrkII/Dock180/Rac pathway, is required for phagocytosis and cell migration. *Cell* 107:27–41
74. Westphal RS, Soderling SH, Alto NM, Langeberg LK, Scott JD. 2000. Scar/WAVE-1, a Wiskott-Aldrich syndrome protein, assembles an actin-associated multi-kinase scaffold. *EMBO J.* 19:4589–600
75. Gohla A, Bokoch GM. 2002. 14-3-3 regulates actin dynamics by stabilizing phosphorylated cofilin. *Curr. Biol.* 12:1704–10
76. Pollard TD, Blanchoin L, Mullins RD. 2000. Molecular mechanisms controlling actin filament dynamics in nonmuscle cells. *Annu. Rev. Biophys. Biomol. Struct.* 29:545–76
77. Cooper JA, Buhle ELJ, Walker SB, Tsong TY, Pollard TD. 1983. Kinetic evidence for a monomer activation step in actin polymerization. *Biochemistry* 22:2193–202
78. Cooper JA, Walker SB, Pollard TD. 1983. Pyrene actin: documentation of the validity of a sensitive assay for actin polymerization. *J. Muscle Res. Cell Motil.* 4:253–62

79. Frieden C, Goddette DW. 1983. Polymerization of actin and actin-like systems: evaluation of the time course of polymerization in relation to the mechanism. *Biochemistry* 22:5836–43
80. Mullins RD, Heuser JA, Pollard TD. 1998. The interaction of Arp2/3 complex with actin: nucleation, high affinity pointed end capping, and formation of branching networks of filaments. *Proc. Natl. Acad. Sci. USA* 95:6181–86
81. Welch MD, Mullins RD. 2002. Cellular control of actin nucleation. *Annu. Rev. Cell Dev. Biol.* 18:247–88
82. Blanchoin L, Pollard TD, Mullins RD. 2000. Interactions of ADF/cofilin, Arp2/3 complex, capping protein and profilin in remodeling of branched actin filament networks. *Curr. Biol.* 10:1273–82
83. Borisy GG, Svitkina TM. 2000. Actin machinery: pushing the envelope. *Curr. Opin. Cell Biol.* 12:104–12
84. Bamburg JR, Wiggan OP. 2002. ADF/cofilin and actin dynamics in disease. *Trends Cell Biol.* 12:598–605
85. Nagata-Ohashi K, Ohta Y, Goto K, Chiba S, Mori R, et al. 2004. A pathway of neuregulin-induced activation of cofilin-phosphatase Slingshot and cofilin in lamellipodia. *J. Cell Biol.* 165:465–71
86. Edwards DC, Sanders LC, Bokoch GM, Gill GN. 1999. Activation of LIM-kinase by Pak1 couples Rac/Cdc42 GTPase signaling to actin cytoskeletal dynamics. *Nat. Cell Biol.* 1:253–59
87. Gohla A, Birkenfeld J, Bokoch GM. 2005. Chronophin, a novel HAD-type serine protein phosphatase, regulates cofilin-dependent actin dynamics. *Nat. Cell Biol.* 7:21–29
88. Bamburg JR. 1999. Proteins of the ADF/cofilin family: essential regulators of actin dynamics. *Annu. Rev. Cell Dev. Biol.* 15:185–230
89. Ghosh M, Song X, Mouneimne G, Sidani M, Lawrence DS, Condeelis JS. 2004. Cofilin promotes actin polymerization and defines the direction of cell motility. *Science* 304:743–46
90. Dawe HR, Minamide LS, Bamburg JR, Cramer LP. 2003. ADF/cofilin controls cell polarity during fibroblast migration. *Curr. Biol.* 13:252–57
91. Cramer LP, Mitchison TJ, Theriot JA. 1994. Actin-dependent motile forces and cell motility. *Curr. Opin. Cell Biol.* 6:82–86
92. Pollard TD, Borisy GG. 2003. Cellular motility driven by assembly and disassembly of actin filaments. *Cell* 112:453–65
93. Couve A, Kittler JT, Uren JM, Calver AR, Pangalos MN, et al. 2001. Association of GABA(B) receptors and members of the 14-3-3 family of signaling proteins. *Mol. Cell Neurosci.* 17:317–28
94. Kagan A, Melman YF, Krumerman A, McDonald TV. 2002. 14-3-3 amplifies and prolongs adrenergic stimulation of HERG K^+ channel activity. *EMBO J.* 21:1889–98
95. Prezeau L, Richman JG, Edwards SW, Limbird LE. 1999. The ζ isoform of 14-3-3 proteins interacts with the third intracellular loop of different α_2-adrenergic receptor subtypes. *J. Biol. Chem.* 274:13462–69
96. Wang Q, Limbird LE. 2002. Regulated interactions of the α_{2A} adrenergic receptor with spinophilin, 14-3-3ζ, and arrestin 3. *J. Biol. Chem.* 277:50589–96
97. Sun Y, Cheng Z, Ma L, Pei G. 2002. β-arrestin2 is critically involved in CXCR4-mediated chemotaxis, and this is mediated by its enhancement of p38 MAPK activation. *J. Biol. Chem.* 277:49212–19
98. McDonald PH, Chow CW, Miller WE, Laporte SA, Field ME, et al. 2000. β-Arrestin 2: a receptor-regulated MAPK scaffold for the activation of JNK3. *Science* 290:1574–77

99. Jacob C, Yang PC, Darmoul D, Amadesi S, Saito T, et al. 2005. Mast cell tryptase controls paracellular permeability of the intestine: role of protease-activated receptor 2 and β-arrestins. *J. Biol. Chem.* 280:31936–48
100. Kohout TA, Nicholas SL, Perry SJ, Reinhart G, Junger S, Struthers RS. 2004. Differential desensitization, receptor phosphorylation, β-arrestin recruitment, and ERK1/2 activation by the two endogenous ligands for the CC chemokine receptor 7. *J. Biol. Chem.* 279:23214–22
101. Dyson JM, Munday AD, Kong AM, Huysmans RD, Matzaris M, et al. 2003. SHIP-2 forms a tetrameric complex with filamin, actin, and GPIb-IX-V: localization of SHIP-2 to the activated platelet actin cytoskeleton. *Blood* 102:940–48
102. Ohta Y, Suzuki N, Nakamura S, Hartwig JH, Stossel TP. 1999. The small GTPase RalA targets filamin to induce filopodia. *Proc. Natl. Acad. Sci. USA* 96:2122–28
103. Jia Z, Barbier L, Stuart H, Amraei M, Pelech S, et al. 2005. Tumor cell pseudopodial protrusions: localized signaling domains coordinating cytoskeleton remodeling, cell adhesion, glycolysis, RNA translocation, and protein translation. *J. Biol. Chem.* 280:30564–73
104. Bhattacharya M, Anborgh PH, Babwah AV, Dale LB, Dobransky T, et al. 2002. β-Arrestins regulate a Ral-GDS-Ral effector pathway that mediates cytoskeletal reorganization. *Nat. Cell Biol.* 4:547–55
105. Ponti A, Machacek M, Gupton SL, Waterman-Storer CM, Danuser G. 2004. Two distinct actin networks drive the protrusion of migrating cells. *Science* 305:1782–86
106. Kolega J. 1998. Cytoplasmic dynamics of myosin IIA and IIB: spatial 'sorting' of isoforms in locomoting cells. *J. Cell Sci.* 111:2085–95
107. Simoes RL, Fierro IM. 2005. Involvement of the rho-kinase/myosin light chain kinase pathway on human monocyte chemotaxis induced by ATL-1, an aspirin-triggered lipoxin A4 synthetic analog. *J. Immunol.* 175:1843–50
108. Verkhovsky AB, Svitkina TM, Borisy GG. 1995. Myosin II filament assemblies in the active lamella of fibroblasts: their morphogenesis and role in the formation of actin filament bundles. *J. Cell Biol.* 131:989–1002
109. Rennefahrt UE, Illert B, Kerkhoff E, Troppmair J, Rapp UR. 2002. Constitutive JNK activation in NIH 3T3 fibroblasts induces a partially transformed phenotype. *J. Biol. Chem.* 277:29510–18
110. Zhang L, Wang W, Hayashi Y, Jester JV, Birk DE, et al. 2003. A role for MEK kinase 1 in TGF-β/activin-induced epithelium movement and embryonic eyelid closure. *EMBO J.* 22:4443–54
111. Huang C, Rajfur Z, Borchers C, Schaller MD, Jacobson K. 2003. JNK phosphorylates paxillin and regulates cell migration. *Nature* 424:219–23
112. Huang C, Jacobson K, Schaller MD. 2004. A role for JNK-paxillin signaling in cell migration. *Cell Cycle* 3:4–6
113. Iwasaki T, Nakata A, Mukai M, Shinkai K, Yano H, et al. 2002. Involvement of phosphorylation of Tyr-31 and Tyr-118 of paxillin in MM1 cancer cell migration. *Int. J. Cancer* 97:330–35
114. Li J, DeFea K, Roth RA. 1999. Modulation of insulin receptor substrate-1 tyrosine phosphorylation by an Akt/phosphatidylinositol 3-kinase pathway. *J. Biol. Chem.* 274:9351–56
115. DeFea K, Roth RA. 1997. Protein kinase C modulation of insulin receptor substrate-1 tyrosine phosphorylation requires serine 612. *Biochemistry* 36:12939–47
116. Aguirre V, Uchida T, Yenush L, Davis R, White MF. 2000. The c-Jun NH_2-terminal kinase promotes insulin resistance during association with insulin receptor substrate-1 and phosphorylation of Ser^{307}. *J. Biol. Chem.* 275:9047–54

117. Ballestrem C, Erez N, Kirchner J, Kam Z, Bershadsky A, Geiger B. 2006. Molecular mapping of tyrosine-phosphorylated proteins in focal adhesions using fluorescence resonance energy transfer. *J. Cell Sci.* 119:866–75
118. Cai X, Li M, Vrana J, Schaller MD. 2006. Glycogen synthase kinase 3- and extracellular signal-regulated kinase-dependent phosphorylation of paxillin regulates cytoskeletal rearrangement. *Mol. Cell. Biol.* 26:2857–68
119. Ishibe S, Joly D, Liu ZX, Cantley LG. 2004. Paxillin serves as an ERK-regulated scaffold for coordinating FAK and Rac activation in epithelial morphogenesis. *Mol. Cell* 16:257–67
120. Liu ZX, Yu CF, Nickel C, Thomas S, Cantley LG. 2002. Hepatocyte growth factor induces ERK-dependent paxillin phosphorylation and regulates paxillin-focal adhesion kinase association. *J. Biol. Chem.* 277:10452–58
121. Klemke RL, Cai S, Giannini AL, Gallagher PJ, de Lanerolle P, Cheresh DA. 1997. Regulation of cell motility by mitogen-activated protein kinase. *J. Cell Biol.* 137:481–92
122. Hong T, Grabel LB. 2006. Migration of F9 parietal endoderm cells is regulated by the ERK pathway. *J. Cell Biochem.* 97:1339–49
123. Totsukawa G, Wu Y, Sasaki Y, Hartshorne DJ, Yamakita Y, et al. 2004. Distinct roles of MLCK and ROCK in the regulation of membrane protrusions and focal adhesion dynamics during cell migration of fibroblasts. *J. Cell Biol.* 164:427–39
124. Svitkina TM, Borisy GG. 1999. Arp2/3 complex and actin depolymerizing factor/cofilin in dendritic organization and treadmilling of actin filament array in lamellipodia. *J. Cell Biol.* 145:1009–26
125. Verkhovsky AB, Chaga OY, Schaub S, Svitkina TM, Meister JJ, Borisy GG. 2003. Orientational order of the lamellipodial actin network as demonstrated in living motile cells. *Mol. Biol. Cell* 14:4667–75
126. Higgs HN, Pollard TD. 2001. Regulation of actin filament network formation through ARP2/3 complex: activation by a diverse array of proteins. *Annu. Rev. Biochem.* 70:649–76
127. Blanchoin L, Amann KJ, Higgs HN, Marchand JB, Kaiser DA, Pollard TD. 2000. Direct observation of dendritic actin filament networks nucleated by Arp2/3 complex and WASP/Scar proteins. *Nature* 404:1007–11
128. Machesky LM, Mullins RD, Higgs HN, Kaiser DA, Blanchoin L, et al. 1999. Scar, a WASp-related protein, activates nucleation of actin filaments by the Arp2/3 complex. *Proc. Natl. Acad. Sci USA* 96:3739–44
129. Miki H, Takenawa T. 1998. Direct binding of the verprolin-homology domain in N-WASP to actin is essential for cytoskeletal reorganization. *Biochem. Biophys. Res. Commun.* 243:73–78
130. Bear JE, Rawls JF, Saxe CL III. 1998. SCAR, a WASP-related protein, isolated as a suppressor of receptor defects in late *Dictyostelium* development. *J. Cell Biol.* 142:1325–35
131. Yarar D, To W, Abo A, Welch MD. 1999. The Wiskott-Aldrich syndrome protein directs actin-based motility by stimulating actin nucleation with the Arp2/3 complex. *Curr. Biol.* 9:555–58
132. Mullins RD. 2000. How WASP-family proteins and the Arp2/3 complex convert intracellular signals into cytoskeletal structures. *Curr. Opin. Cell Biol.* 12:91–96
133. Miki H, Takenawa T. 2003. Regulation of actin dynamics by WASP family proteins. *J. Biochem.* 134:309–13
134. Rohatgi R, Ho HYH, Kirschner MW. 2000. Mechanism of N-WASP activation by CDC42 and phosphatidylinositol 4,5-bisphosphate. *J. Cell Biol.* 150:1299–310

135. Cory GOC, Cramer R, Blanchoin L, Ridley AJ. 2003. Phosphorylation of the WASP-VCA domain increases its affinity for the Arp2/3 complex and enhances actin polymerization by WASP. *Mol. Cell* 11:1229–39
136. Miki H, Fukuda M, Nishida E, Takenawa T. 1999. Phosphorylation of WAVE downstream of mitogen-activated protein kinase signaling. *J. Biol. Chem.* 274:27605–9
137. Suetsugu S, Hattori M, Miki H, Tezuka T, Yamamoto T, et al. 2002. Sustained activation of N-WASP through phosphorylation is essential for neurite extension. *Dev. Cell* 3:645–58
138. Martinez-Quiles N, Ho HY, Kirschner MW, Ramesh N, Geha RS. 2004. Erk/Src phosphorylation of cortactin acts as a switch on-switch off mechanism that controls its ability to activate N-WASP. *Mol. Cell Biol.* 24:5269–80
139. Merrifield CJ, Qualmann B, Kessels MM, Almers W. 2004. Neural Wiskott Aldrich Syndrome Protein (N-WASP) and the Arp2/3 complex are recruited to sites of clathrin-mediated endocytosis in cultured fibroblasts. *Eur. J. Cell Biol.* 83:13–18
140. Quinlan ME, Heuser JE, Kerkhoff E, Dyche Mullins R. 2005. *Drosophila* Spire is an actin nucleation factor. *Nature* 433:382–88
141. Otto IM, Raabe T, Rennefahrt UEE, Bork P, Rapp UR, Kerkhoff E. 2000. The p150-Spir protein provides a link between c-Jun N-terminal kinase function and actin reorganization. *Curr. Biol.* 10:345–48
142. Imamura T, Huang J, Dalle S, Ugi S, Usui I, et al. 2001. β-Arrestin-mediated recruitment of the Src family kinase Yes mediates endothelin-1-stimulated glucose transport. *J. Biol. Chem.* 276:43663–67
142a. Wang P, DeFea K. 2006. Protease-activated receptor-2 simultaneously directs β-arrestin-1-dependent inhibition and Gαq-dependent activation of phosphatidylinositol 3-kinase. *Biochemistry* 45:9374–85
143. Chung CY, Potikyan G, Firtel RA. 2001. Control of cell polarity and chemotaxis by Akt/PKB and PI3 kinase through the regulation of PAKa. *Mol. Cell* 7:937–47
144. Zhou GL, Zhuo Y, King CC, Fryer BH, Bokoch GM, Field J. 2003. Akt phosphorylation of serine 21 on Pak1 modulates Nck binding and cell migration. *Mol. Cell. Biol.* 23:8058–69
145. Watanabe Y, Nakayama T, Nagakubo D, Hieshima K, Jin Z, et al. 2006. Dopamine selectively induces migration and homing of naive $CD8^+$ T cells via dopamine receptor D3. *J. Immunol.* 176:848–56
146. Drell TL IV, Joseph J, Lang K, Niggemann B, Zaenker KS, Entschladen F. 2003. Effects of neurotransmitters on the chemokinesis and chemotaxis of MDA-MB-468 human breast carcinoma cells. *Breast Cancer Res. Treat.* 80:63–70
147. Hall A. 1998. Rho GTPases and the actin cytoskeleton. *Science* 279:509–14
148. Pertz O, Hodgson L, Klemke RL, Hahn KM. 2006. Spatiotemporal dynamics of RhoA activity in migrating cells. *Nature* 440:1069–72
149. Yamaguchi H, Lorenz M, Kempiak S, Sarmiento C, Coniglio S, et al. 2005. Molecular mechanisms of invadopodium formation: the role of the N-WASP-Arp2/3 complex pathway and cofilin. *J. Cell Biol.* 168:441–52

Regulation of Receptor Tyrosine Kinase Signaling by GRKs and β-Arrestins

Christopher J. Hupfeld and Jerrold M. Olefsky

Department of Medicine, Division of Endocrinology and Metabolism, University of California San Diego, La Jolla, California 92093; email: chupfeld@ucsd.edu, jolefsky@ucsd.edu

Annu. Rev. Physiol. 2007. 69:561–77

First published online as a Review in Advance on September 15, 2006

The *Annual Review of Physiology* is online at http://physiol.annualreviews.org

This article's doi:
10.1146/annurev.physiol.69.022405.154626

0066-4278/07/0315-0561$20.00

Key Words

insulin, IGF-1, G proteins

Abstract

Receptor tyrosine kinases (RTKs) are a unique family of cell surface receptors, each containing a common intracellular domain that has tyrosine kinase activity. However, RTKs share many signaling molecules with another unique family of cell surface receptors, the seven-transmembrane receptors (7TMRs), and these receptor families can activate similar signaling cascades. In this review of RTK signaling, we describe the role of cross talk between RTKs and 7TMRs, focusing specifically on the role played in this process by β-arrestins and by G proteins.

INTRODUCTION

RTK: receptor tyrosine kinase

IR: insulin receptor

MAPK: mitogen-activated protein kinase

The Receptor Tyrosine Kinases

The receptor tyrosine kinases (RTKs) are a related family of 59 cell surface receptors with similar structural and functional characteristics (1, 2). Most RTKs are monomers, and their domain structure includes an extracellular ligand-binding domain, a transmembrane domain, and an intracellular domain possessing tyrosine kinase activity. The receptors for insulin and insulin-like growth factor-1 (IGF-1) are more complex, maintaining a disulfide-linked, $\alpha_2\beta_2$ heterotetrameric structure (**Figure 1**) (3).

Ligand binding to RTKs leads to dimerization of monomeric receptors or rearrangement within the quarternary structure of heterotetrameric receptors. These conformational changes lead to activation of the intrinsic receptor tyrosine kinase and to autophosphorylation of tyrosine residues in the intracellular domain (2). Autophosphorylation of the insulin receptor results in a large increase in its kinase activity owing to a further conformational change that gives the catalytic site within the tyrosine kinase domain unlimited access to ATP and to substrates (4).

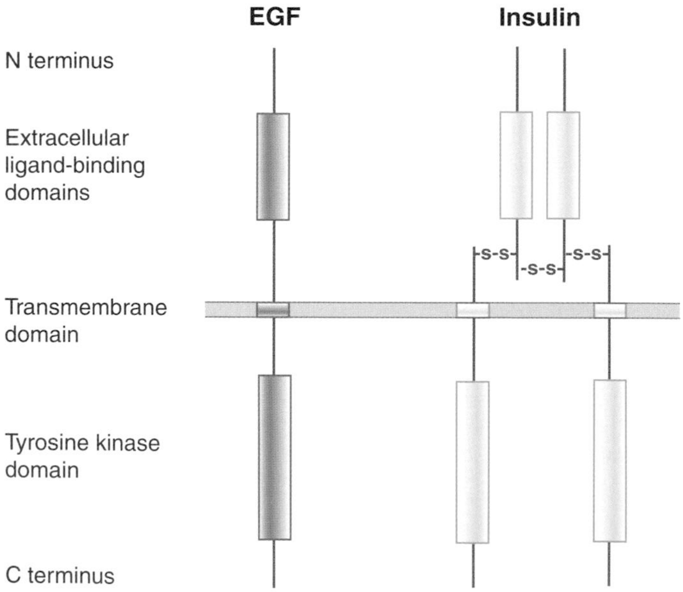

Receptor autophosphorylation on multiple tyrosine residues creates phospho-tyrosine binding motifs recognized by intracellular proteins with phospho-tyrosine binding (PTB) domains or src-homology 2 (SH2) domains (**Figure 2**). For example, insulin receptor substrate 1 (IRS-1) binding to the insulin receptor (IR) occurs through IRS-1 PTB domain recognition of the tyrosine-phosphorylated NPXY motif present in the juxtamembrane region of the IR (5). Shc also interacts with the IR NPXpY motif through its own PTB domain (6, 7). Once bound, Shc and IRS are subsequently phosphorylated on tyrosine residues by the active RTK, creating binding sites for other SH2-domain-containing proteins (8).

Figure 1

Domain structure of a monomeric (e.g., the EGF receptor) and a heterotetrameric (e.g., the insulin receptor) receptor tyrosine kinase. Receptor tyrosine kinases are composed of an extracellular ligand-binding domain, a transmembrane domain, and an intracellular tyrosine kinase domain. Members of the insulin receptor family consist of two identical disulfide-linked dimers, forming an $\alpha_2\beta_2$ heterotetramer.

Downstream Signaling by RTKs

Activation of an RTK leads to a wide array of intracellular responses. The major signaling pathways emanating from RTKs are the p42/44 mitogen-activated protein kinase (MAPK) pathway and the phosphoinositide-3 kinase (PI3-kinase) pathway. RTK-mediated p42/44 MAPK signaling has an important role in the control of the cell cycle and cell migration as well as cell proliferation and differentiation (1). In the case of the IR, p42/44 MAPK signaling is initiated following the binding of Grb2 to tyrosine-phosphorylated Shc or IRS through its SH2 domain. SOS, prebound to Grb2, then catalyzes the exchange of GTP for GDP on the small G protein ras. GTP-bound ras activates the serine kinase raf-1, and a series of downstream serine kinase reactions culminates in the activation of the p42/44 MAPK, also known as the extracellular-regulated kinase (ERK).

Metabolic signaling from the insulin receptor occurs primarily through the

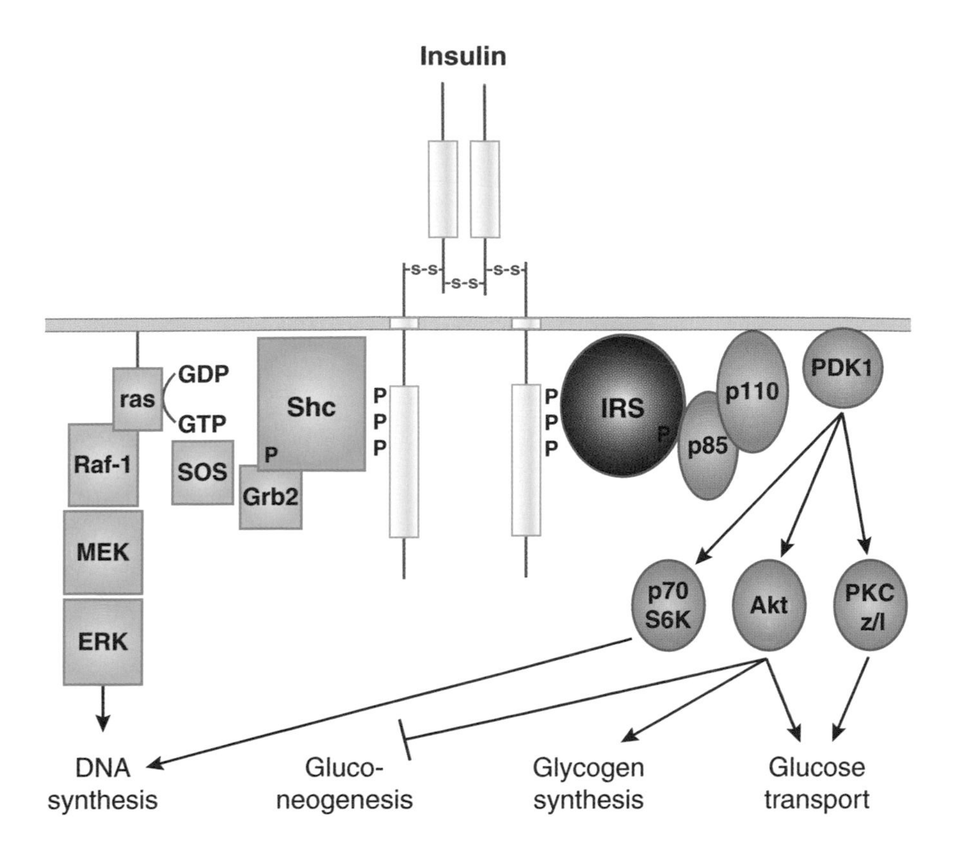

Figure 2

Intracellular signaling by the activated insulin receptor. Scaffold proteins such as Shc and IRS bind to the autophosphorylated insulin receptor and initiate a cascade of intracellular signaling events. This process culminates in the characteristic mitogenic and metabolic responses of a cell to insulin.

IRS:PI3-kinase pathway, although other accessory pathways for metabolic signaling by the insulin receptor clearly exist (9, 10). IRS bound to the insulin receptor is tyrosine phosphorylated by the receptor tyrosine kinase, creating phosphotyrosine binding motifs recognized by SH2-domain-containing proteins, in particular the regulatory p85 subunit of PI3-kinase. Binding of PI3-kinase to IRS, via the former's p85 subunit, induces a conformational change in the kinase that activates the catalytic p110 subunit. The PI3-kinase-IRS association also serves to target the kinase near the plasma membrane and in the vicinity of its membrane lipid substrates, from which 3-phosphoinositides are generated. Other kinases downstream of PI3-kinase, including PDK1, Akt1 and Akt2, and the atypical protein kinase C (PKC) isoforms ζ and λ, regulate many of the enzymes and proteins involved in glucose transport, glycogen synthesis, and gluconeogenesis.

Heterotrimeric G Proteins and RTK Signaling

The term G protein–coupled receptor has previously been used to refer exclusively to cell surface receptors that couple to heterotrimeric G proteins during signaling and possess a classic structure consisting of seven-transmembrane-spanning domains. It has recently become clear, however, that seven-transmembrane receptors (7TMRs) are not the only receptor family that couples to heterotrimeric G proteins.

7TMR: 7 transmembrane receptor

IGF-1R: insulin-like growth factor 1 receptor

βAR: β-adrenergic receptor

In 1995, Luttrell et al. (11) showed that IGF-1R mitogenic signaling was sensitive to pertussis toxin, a specific inhibitor of signaling by heterotrimeric G proteins containing a $G\alpha_i$ subunit. In addition, IGF-1R mitogenesis was inhibited following sequestration of G protein $\beta\gamma$ subunits by the β-ARK peptide, derived from the β-adrenergic receptor (βAR) kinase. These findings suggested that IGF-1-mediated mitogenesis was dependent

EGF: epidermal growth factor

FGF: fibroblast growth factor

upon heterotrimeric G proteins containing the $G\alpha_i$ subunit, and that the mitogenic effect required the presence of $G\beta\gamma$ subunits as well. Consistent with these findings, others have shown that $G\alpha_i$ is present in IGF-1R immunoprecipitates (12, 13) and that IGF-1 treatment leads to GTP loading of $G\alpha_{i2}$ (14).

Other studies have shown that insulin signaling also exhibits a dependence upon heterotrimeric G proteins. In 1999, two labs independently demonstrated the role of $G\alpha_{q/11}$-containing G proteins during insulin signaling in 3T3-L1 adipocytes (10, 15). Imamura et al. (10) found that insulin stimulation caused tyrosine phosphorylation of $G\alpha_{q/11}$ and that injection of 3T3-L1 adipocytes with antibodies against $G\alpha_{q/11}$, or with RGS2 protein (a GAP protein for $G\alpha_q$), inhibited insulin-stimulated translocation of the GLUT4 glucose transporter. Using an adenoviral expression vector, the authors also showed that overexpression of a constitutively active form of $G\alpha_{q/11}$ (Q209L-$G\alpha_q$), even in the absence of insulin, stimulated glucose uptake and GLUT4 translocation to 70% of the maximal insulin-stimulated effect.

In 2003, Usui et al. (16) further defined the signaling mechanisms involved in insulin-stimulated, $G\alpha_{q/11}$-dependent glucose uptake in adipocytes. They investigated the role of cdc42, a Rho GTPase family member previously shown to mediate Gαq-dependent glucose uptake by the bradykinin receptor (17). Microinjection of anti-cdc42 antibody or cdc42 siRNA led to decreased insulin-induced and Q209L-$G\alpha_q$-induced GLUT4 translocation. Insulin treatment caused an increase in cdc42 activity and enhanced the association of cdc42 with the p85 subunit of PI3-kinase, increasing PI3-kinase activity. Furthermore, the effects of insulin, Q209L-$G\alpha_q$, and constitutively active cdc42 on GLUT4 translocation or 2-deoxyglucose uptake were inhibited by microinjection of anti–protein kinase Cλ (PKCλ) antibody or overexpression of a kinase-deficient PKCλ construct. Taken together, these data indicate that cdc42 can mediate insulin signaling to GLUT4 translocation and that it lies downstream of $G\alpha_q$ and upstream of PI3-kinase and PKCλ in this stimulatory pathway.

Heterotrimeric G proteins containing $G\alpha_i$ have also been implicated in insulin-stimulated glucose uptake in adipocytes. It has been reported that the IR tyrosine kinase can phosphorylate $G\alpha_i$ (18), and insulin-stimulated glucose uptake was reduced in isolated rat adipocytes following pertussis toxin treatment (19, 20). However, others have found that glucose uptake and GLUT4 translocation are insensitive to pertussis toxin treatment in 3T3-L1 adipocytes (12). Also, pertussis toxin treatment had no effect on insulin-stimulated glucose oxidation in isolated rat adipocytes (21).

Other RTKs have also been studied in this manner and found to utilize G proteins during signaling. The EGFR can couple to $G\alpha_i$, and EGF (epidermal growth factor) activation of p42/44 MAPK is blocked by pertussis toxin in rat hepatocytes and in pancreatic AR42J cells (22–25). However, others have reported that EGF mitogenic signaling is independent of $G\alpha_i$ in rat1 fibroblasts and in CHO cells, respectively (12, 26). As Waters et al. (27) discuss, these discrepancies are possibly related to inherent differences in receptor expression in these systems. With high endogenous EGFR expression, or in cells in which the EGFR is expressed at a high level from a transfected plasmid, the partial dependence of EGFR signaling on heterotrimeric G proteins may be masked. This has been shown with IGF-1R-mediated p42/44 MAPK activation, where sensitivity to pertussis toxin is lost with increasing levels of IGF-1R expression (11).

Receptor tyrosine kinases such as the PDGFR, the Trk A receptor for NGF (nerve growth factor), the FGFR (fibroblast growth factor receptor), and the VEGFR (vascular endothelial growth factor receptor) have all been shown to interact with G proteins to various degrees, and the reader is directed to the excellent review by Waters et al. (27) for an overview of G protein–mediated signaling by RTKs.

Clearly, RTKs could be considered G protein–coupled receptors from a functional standpoint by virtue of their interaction with heterotrimeric G proteins. In the case of the G protein–coupled 7TMRs, signaling is further modulated by two other intracellular proteins, β-arrestins and G protein–coupled receptor kinases (GRKs). Is this also the case with RTKs?

GRK, β-ARRESTIN, AND RTK SIGNALING

The GRK/β-Arrestin Paradigm

Ligand binding to a 7TMR induces coupling with its cognate heterotrimeric $G\alpha\beta\gamma$ protein. The G protein then rapidly dissociates from the receptor and splits into a separate GTP-bound $G\alpha$ monomer and a $G\beta\gamma$ dimer. Both $G\alpha$-GTP and the $G\beta\gamma$ dimer directly interact with effector proteins to continue the signaling cascade (28).

The $G\beta\gamma$ dimer has received much attention as a modulator of p42/44 MAPK signaling by 7TMRs, primarily from studies demonstrating the critical role of β-arrestins during MAPK signaling and the role of $G\beta\gamma$ as an adapter for GRK/β-arrestin recruitment to activated 7TMRs. The role of $G\beta\gamma$ in this process has been deduced from three main pieces of evidence. First, $G\beta\gamma$ is isoprenylated (29) and thus targeted to the plasma membrane. Second, the highly specific 7TMR serine kinase GRK is recruited to the plasma membrane in proximity to activated 7TMRs via its direct interaction with $G\beta\gamma$ (30, 31). Finally, GRK-mediated serine phosphorylation of C-terminal residues of 7TMRs provides a recognition motif on the receptor for the binding of β-arrestins (32, 33).

The binding of β-arrestin to the activated 7TMR sterically uncouples the receptor from further G protein interaction. Therefore, the activation of some $G\alpha$ protein effectors (i.e., $G\alpha_s$-mediated adenylyl cyclase activation) rapidly desensitizes following β-arrestin binding. However, other G protein–mediated signaling events display a marked dependence upon β-arrestin for normal function, independent of the desensitization process. For example, β-arrestin-bound receptors are targeted to clathrin-coated pits and undergo endocytosis in an arrestin-dependent manner (34). Arrestin-mediated recruitment of the nonreceptor tyrosine kinase Src to activated 7TMRs leads to phosphorylation of Shc (35) as well as to activation of dynamin (36), two important steps during MAPK signaling and receptor endocytosis, respectively. Additionally, p42/44 MAPK activation by some 7TMRs is mediated via an arrestin-dependent ERK activation scaffold containing cRaf-1, MEK1 (MAPK/ERK kinase 1), and ERK2 (37).

GRK: G protein–coupled receptor kinase

β-Arrestin1 function is regulated in part by serine phosphorylation. The 412 serine residue at the C terminus is primarily phosphorylated under basal conditions, and in this state it has a reduced affinity for both clathrin (34) and src (38). Ligand binding and 7TMR activation lead to pronounced dephosphorylation of β-arrestin1 at serine 412 (39). Ser-412 dephosphorylation is likely mediated by the phosphatase PP2A because dephosphorylation is prevented by PP2A inhibition (39). β-Arrestin binding to the activated receptor, however, is unaffected by the phosphorylation status of Ser-412.

WHICH RTKs UTILIZE β-ARRESTIN?

The IGF-1R

β-Arrestin1 (12, 40) and β-arrestin2 (40) both coimmunoprecipitate with the IGF-1R, and IGF-1 treatment rapidly stimulates this association (40). The structural determinates of either the IGF-1R or β-arrestin responsible for this interaction have not been determined, and it is not known if these findings represent a direct receptor-arrestin interaction. If the interaction were direct, it would be reminiscent of other scaffold proteins that bind to the IGF-1R, such as Shc or IRS.

IGF-1R endocytosis is clathrin mediated, a process that is partially dependent upon β-arrestin function. Expression of a dominant-negative mutant of β-arrestin1 (S412D β-arrestin1) in HEK-293 cells impairs IGF-1R internalization, presumably by reducing β-arrestin1-mediated targeting of the IGF-1R to clathrin-coated membrane pits. As further evidence for this, overexpression of wild-type β-arrestin1 or β-arrestin2 increases IGF-1R internalization, as does overexpression of a constituitively active form of β-arrestin1 (S412A β-arrestin1) (40).

As a clathrin adapter, β-arrestin may act as a bridge connecting the IGF-1R to the MAPK pathway, by mediating IGF-1R endocytosis. IGF-1-mediated Shc phosphorylation and p42/44 MAPK activation are dependent upon the endocytosis of the IGF-1R (40, 41). This dependence was demonstrated in IGF-1R-expressing CHO cells (41) by inhibiting receptor endocytosis, using low temperature or with dansylcadaverine (a chemical inhibitor of endocytosis). In HEK-293 cells, coexpression of the IGF-1R either with a dominant-negative dynamin mutant or S412D β-arrestin1 impaired IGF-1R endocytosis and p42/44 MAPK signaling after IGF-1 treatment (40). Finally, in experiments performed in single 3T3-L1 adipocytes, microinjection of an anti-β-arrestin1 antibody inhibited the transcriptional activity of an ERK reporter gene construct (12). Taken together, these data suggest that β-arrestin-mediated, clathrin-dependent IGF-1R internalization is a critical pathway to ERK activation by IGF-1.

In addition to its role as a mitogenic hormone, IGF-1 also leads to activation of the PI3-kinase pathway. Following IGF-1 treatment and IGF-1R tyrosine kinase activation, the SH2-domain-containing p85 subunit of PI3-kinase binds directly to tyrosine-phosphorylated YMXM and YXXM sequences of the IGF-1R and to tyrosine-phosphorylated sequences on IRS-1. In this manner, PI3-kinase is targeted to the plasma membrane, where it can interact with lipid substrates. In 2003, Povsic et al. (42) described a novel PI3-kinase activation pathway emanating from the IGF-1R in mouse embryonic fibroblasts (MEFS). In this study, the authors showed that chemical inhibition of the IGF-1R tyrosine kinase, with subsequent loss of IRS-1 tyrosine phosphorylation, had no effect on IGF-1-mediated PI3-kinase activation and Akt phosphorylation in these cells. IGF-1-mediated PI3-kinase activity was lost, however, in MEFS that were β-arrestin1 −/−. The mechanism of β-arrestin1-dependent PI3-kinase activation in these cells remains unclear. β-arrestin may scaffold PI3-kinase to the IGF-1R, near its lipid substrates at the plasma membrane, bypassing the need for tyrosine-phosphorylated sites on the receptor or on IRS-1. However, an interaction between arrestin and PI3-kinase has not been demonstrated.

Receptor signaling is often modulated by the ubiquitination of either the receptor itself or of downstream signaling proteins. Ubiquitination is a process whereby one or more ubiquitin molecules are covalently attached to a protein in a three-step process involving an activating enzyme (E1), a ubiquitin carrier protein (E2), and a ubiquitin protein ligase (E3). The specificity of the ubiquitin system is due primarily to the many families of E3 ligases, which recognize specific protein substrates through specific recognition sequences (43). Ubiquitination often targets proteins to the 26s proteasome for degradation but also controls a diverse array of cellular functions, including the sorting of plasma membrane proteins to vacuoles or lysosomes, transcription factor and kinase activation, and DNA repair, among others (44).

The E3 ligase MDM2 can ubiquitinate the IGF-1R in vitro, and cells lacking mdm2 had reduced IGF-1R ubiquitination (45). Further studies showed that β-arrestin1 greatly augments mdm2-mediated ubiquitination of the IGF-1R, most likely by acting as a scaffold, recruiting MDM2 to the activated IGF-1R (46).

Co-immunoprecipitation studies showed that the IGF-1R, β-arrestin, and MDM2 all associate upon IGF-1 treatment. Addition of β-arrestin1 to an in vitro cocktail of IGF-1R, E1, E2, and MDM2 greatly enhanced IGF-1R ubiquitination, and in mouse p6 cells overexpressing the IGF-1R, the absence of β-arrestin1 inhibited IGF-1R ubiquitination.

These studies yield a model of IGF-1R signaling that is very similar to that of a typical 7TMR (**Figure 3**). IGF-1 binding to the IGF-1R leads to receptor coupling with a heterotrimeric G protein containing $G\alpha_i$. Downstream signaling dependent on the $G\alpha_i$ and $G\beta\gamma$ subunits, as well as IGF-1R endocytosis mediated by β-arrestin1, are key components

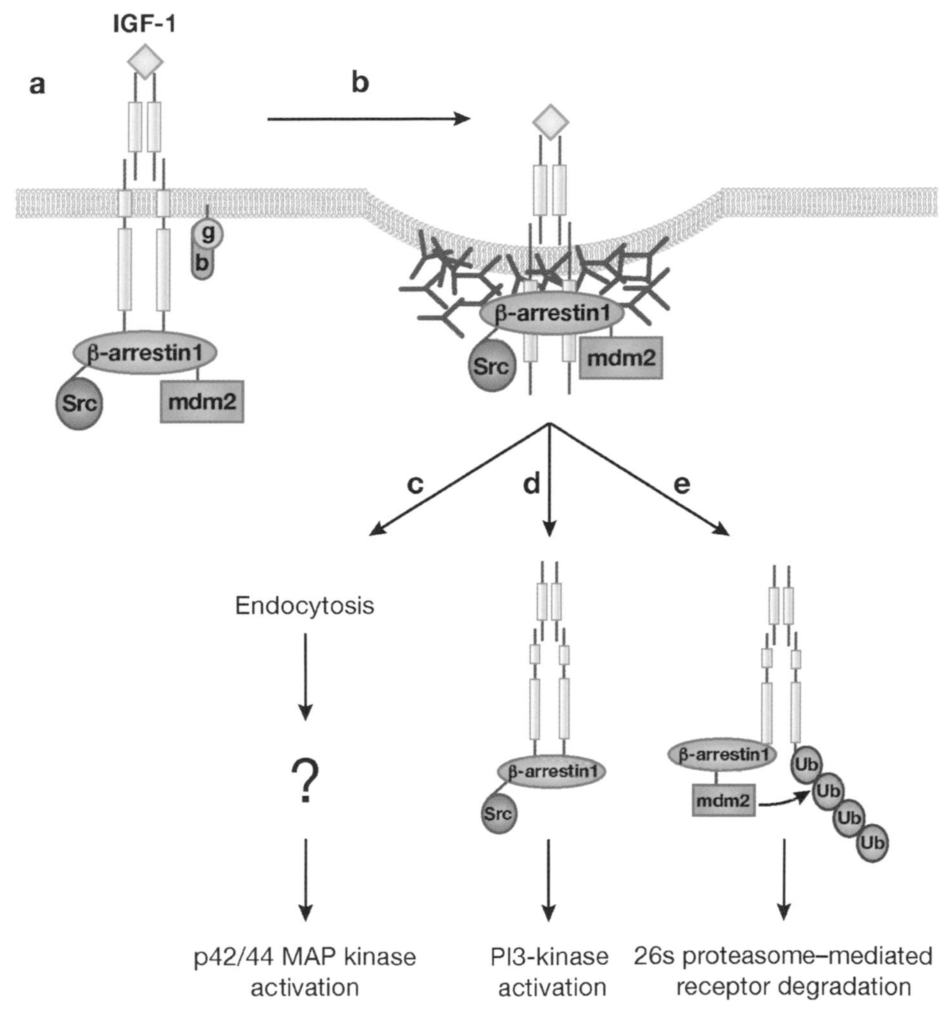

Figure 3

IGF-1R signaling events mimic those of a 7TMR. (*a*) Ligand binding to the IGF-1R initiates heterotrimeric G protein ($G\alpha_i\beta\gamma$) and β-arrestin coupling to the receptor. GTP-for-GDP exchange occurs on $G\alpha_i$. (*b*) β-Arrestin-mediated targeting of the IGF-1R to clathrin-coated plasma membrane pits. (*c*) Clathrin-mediated endocytosis of the IGF-1R. This event is an important part of IGF-1-mediated MAPK signaling, although the mechanistic details of MAPK activation downstream of endocytosis remain unclear. (*d*) β-Arrestin1- and Src-dependent activation of the PI3-kinase/Akt pathway by the IGF-1R. (*e*) β-Arrestin1- and mdm2-dependent ubiquitination of the IGF-1R, leading to proteasomal degradation of the IGF-1R.

of IGF-1R p42/44 MAPK activation. And β-arrestin1-mediated ubiquitination of the IGF-1R promotes receptor degradation, thereby completing the signaling cycle.

The Insulin Receptor

The insulin receptor shares a high degree of structural homology with the IGF-1R. Differences in the extracellular α-subunit sequence confer ligand binding specificity to these related receptors (the IR binds insulin with 100-fold greater affinity than it does IGF-1). However, the intracellular signaling differences between the IR and the IGF-1R are likely due to differences in β-subunit structure. Within the β subunit, the tyrosine kinase domains of the two receptors share 85% homology, whereas the C-terminal domains are less conserved, with 44% homology.

The insulin receptor binds β-arrestin1 in a ligand-dependent manner in rat1 fibroblasts expressing the human IR (HIRcB cells), displaying similar kinetics to IGF-1-mediated IGF-1R/arrestin association (12). Despite this similarity, the IR and IGF-1R display vastly different signaling outcomes in response to various perturbations of β-arrestin1 function. It appears that β-arrestin1 does not function in acute IR-mediated metabolic or mitogenic signaling (discussed below). However, there is evidence that β-arrestin1 does have a role in the desensitization of insulin signaling. As is the case with most receptors, the IR undergoes desensitization upon persistent exposure to ligand. In this manner, insulin treatment eventually leads to insulin resistance. This desensitization process is at least partially mediated by impaired downstream signaling at the level of the IRS proteins. Serine phosphorylation of IRS-1 is a well-recognized event that occurs rapidly upon insulin stimulation, and it leads to impaired IRS association with the IR as well as the ubiquitination and degradation of IRS (47, 48).

The molecular mechanisms of insulin-induced IRS ubiquitination remain unclear, although one study has suggested that elongin B/C may be an E3 ubiquitin ligase involved in this process (48). Usui et al. (49) found that the E3 ligase MDM2 associates with IRS-1 and is required for its ubiquitination and degradation in response to insulin treatment. Because β-arrestin1 interacts with MDM2, these authors proceeded to determine if the β-arrestin1-MDM2 interaction affected insulin-induced IRS degradation. They found that, in cells overexpressing β-arrestin1, insulin-induced IRS-1 ubiquitination was prevented, whereas siRNA-mediated β-arrestin1 knockdown enhanced IRS-1 ubiquitination and degradation, clearly suggesting that β-arrestin1 competes with IRS-1 for MDM2 binding. This finding has important implications for insulin signaling and insulin resistance. If β-arrestin1 protein levels are low, as in adipocytes exposed to chronically high concentrations of insulin (50), insulin-induced, MDM2-mediated IRS-1 ubiquitination and degradation may proceed unimpeded, and the insulin signal is normally, or even rapidly, desensitized.

Other aspects of acute insulin signaling have been examined to determine the role of β-arrestin1 in these processes. Using various endpoints—including p42/44 MAPK phosphorylation, nuclear staining for phosphorylated ERK, bromodeoxyuridine (BrdU) incorporation to detect new DNA synthesis, and a promoter assay for ERK-mediated transcriptional activation—we found no evidence that β-arrestin1 mediates any of these actions in response to insulin treatment (12, 50). In addition, the acute metabolic actions of insulin, including glucose uptake, GLUT4 translocation to the plasma membrane, and Akt activation, are unaffected by impaired β-arrestin1 function (12).

The insulin receptor is internalized by a clathrin-dependent mechanism. Dominant-negative dynamin, which blocks clathrin-induced receptor endocytosis, inhibits IR internalization (51), although the effects of this maneuver on insulin signaling are minimal (51, 52). There are no published reports regarding the role of β-arrestin1 or β-arrestin2 during insulin receptor endocytosis.

It is tempting to speculate that differences in signaling between the IGF-1R and the IR are due to differences in receptor-arrestin interactions, with IGF-1-mediated mitogenic signaling closely linked to β-arrestin1-dependent processes.

The EGFR

There are few data regarding the role of β-arrestin during EGF signaling. β-Arrestin1 is recruited to the EGFR in a ligand-dependent manner (12), and expression of a C-terminal fragment of β-arrestin1(319–418), which binds clathrin but does not direct receptor endocytosis, inhibited EGFR endocytosis following EGF treatment (53). Because EGFR endocytosis is required for full EGF-induced activation of p42/44 MAPK, β-arrestin1 might be expected to influence this signaling pathway. However, other studies have shown no effect of anti-β-arrestin1 antibody injection (12) or reduced cellular β-arrestin1 protein content (50) on EGF-mediated MAPK signaling. Additionally, depletion of cellular β-arrestin2 had no effect on EGF-mediated cell migration, although EGFR internalization and EGF-mediated MAPK signaling were not tested in this study (54).

The PDGFR

Recent data have demonstrated a critical role for β-arrestin during PDGF (platelet-derived growth factor) signaling. Alderton et al. (55) found that β-arrestin1, as well as GRK2 and $G\alpha_i$, were constitutively associated with the PDGFR. This association is dependent upon the formation of a complex between the PDGFR and a 7TMR, the endothelial differentiation gene 1 receptor (EDG-1R), whose ligand is sphingosine 1-phosphate. The PDGFR-EDG1R-arrestin complex is internalized in a clathrin- and src-dependent manner. Similarly, Hobson et al. (56) showed that PDGF treatment of cells transactivates the EDG-1R and induces β-arrestin1 translocation to the plasma membrane.

PDGF: platelet-derived growth factor

LPA: lysophosphatidic acid

The Trk A Receptor

TRK A is the RTK for neurotrophin, also known as NGF. Signaling by this receptor enhances the survival and differentiation of sensory and sympathetic neurons. Ligand binding induces persistent activation of p42/44 MAPK via upstream signaling involving Rap1 and B-Raf (57).

Ligand binding to Trk A induces β-arrestin1 recruitment to the receptor, and P42/44 MAPK signaling by Trk A is pertussis toxin sensitive (58). Pyne and colleagues (27) have shown that Trk A forms a complex with the receptor for lysophosphatidic acid (LPA), an arrestin-dependent 7TMR. Therefore, the Trk A-arrestin association may be indirect as part of a larger signaling complex that includes Trk A and the LPA-R.

The FGFR

There are no published studies describing either an interaction between the FGFR and β-arrestin or a role for β-arrestin during FGFR signaling. However, there is evidence to suggest a role of G proteins (both $G\alpha_s$ and $G\alpha_i$) in FGFR action, and a recent report suggests that the FGFR mitogenic response is dependent on sphingosine 1-phosphate, in a manner similar to the PDGFR (59). It remains to be seen if the FGFR is complexed to a 7TMR and in this manner is dependent upon β-arrestin for full mitogenic signaling.

β-ARRESTINS AS TARGETS FOR RECEPTOR CROSS TALK

Because β-arrestin plays a critical role in modulating signaling pathways emanating from many 7TMRs and RTKs, alterations in β-arrestin function, induced by various forms of receptor cross talk, are likely to have pronounced effects on intracellular signaling.

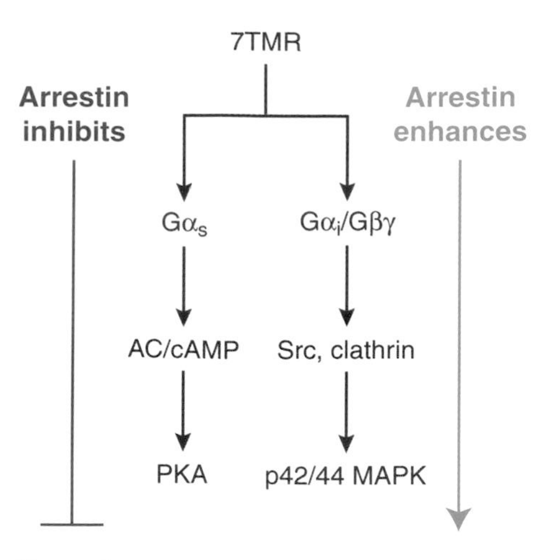

Figure 4

Dual role of arrestin during GPCR signaling. Arrestin (*green*) coordinates p42/44 MAPK signaling by directing receptor endocytosis and by providing a scaffold for key signaling proteins. Arrestin (*red*) leads to homologous desensitization of G protein signaling by physically disrupting the GPCR–G protein interaction. Abbreviations: 7TMR, seven-transmembrane receptor; AC, adenylyl cyclase; PKA, protein kinase A.

Using the simple model in **Figure 4**, one can make predictions regarding changes in cellular signaling following various perturbations of β-arrestin function.

Proteasomal Degradation of β-Arrestin1

Many studies have shown that when β-arrestin protein levels are reduced or absent (siRNA, cell lines derived from arrestin knockout mice), 7TMR signaling that is dependent upon β-arrestin is altered in a predictable manner. $G\alpha_s$-mediated signaling to adenylyl cyclase, normally desensitized owing to β-arrestin binding to the activated Gαs-coupled 7TMR, becomes supersensitized (60–62). Conversely, receptor endocytosis (61–63) and signaling to p42/44 MAPK (62) are impaired in cells lacking β-arrestin. Therefore, for a given cell system, conditions in which β-arrestin degradation is induced would be predicted to have similar consequences.

As mentioned above, chronic treatment with insulin leads to desensitization of insulin signaling, in part owing to serine phosphorylation and the subsequent ubiquitination and degradation of IRS-1. Owing to the similarities in scaffolding function between IRS and β-arrestin, and because β-arrestin1 is present at the insulin receptor basally and increases in response to ligand, we measured β-arrestin1 protein levels following treatment with insulin. We found that chronic insulin treatment was associated with proteasomal-mediated degradation of β-arrestin1, such that after 8 h of insulin treatment, cells retained only 50% of their original β-arrestin1 protein content (50, 63).

Because this represented a possible means of receptor cross talk between the insulin receptor and β-arrestin-dependent 7TMRs, we next measured signaling by 7TMRs in cells depleted of β-arrestin1 by insulin pretreatment. We found that insulin-mediated β-arrestin1 degradation was associated with supersensitization to β-adrenergic $G\alpha_s$ signaling, as measured by enhanced isoproterenol-mediated intracellular cAMP generation in 3T3-L1 adipocytes (63). Reduced β-arrestin1 protein was also associated with reduced β2-AR endocytosis (63).

Reduced β-arrestin1 protein levels following chronic insulin treatment were also associated with impaired p42/44 MAPK signaling by receptors coupled to G proteins and β-arrestin (50). The IGF-1R displayed reduced IGF-1R:β-arrestin1 association, reduced IGF-1-mediated Shc phosphorylation, reduced Grb2 binding to Shc, and reduced p42/44 MAPK phosphorylation. The defect in IGF-1-mediated MAPK phosphorylation was partially rescued by overexpression of wild-type β-arrestin1 protein. 7TMRs dependent upon β-arrestin exhibited similar defects in MAPK signaling. Thus, following stimulation of the β-adrenergic receptor with isoproterenol, cells pretreated with insulin showed reduced β2AR:β-arrestin association, reduced β2AR:src association, and reduced MAPK phosphorylation.

β-Arrestin1 Phosphorylation

β-Arrestin1 function during heterotrimeric G protein–mediated MAPK signaling is also regulated by phosphorylation of the C-terminal Ser-412 residue, providing another possible mechanism for impairing β-arrestin1 function in addition to β-arrestin protein degradation.

Dephosphorylation of Ser-412 is required for the isoproterenol-stimulated association between β-arrestin1 and Src, a nonreceptor tyrosine kinase involved in β1- and β2-adrenergic receptor mitogenic signaling (38). In addition, Ser-412 dephosphorylation is also required for β-arrestin1/clathrin association, which is essential for β2AR endocytosis (34). A mutant β-arrestin1 with a serine-to-aspartic-acid substitution at the 412 residue to mimic phospho-serine displays reduced affinity for Src and clathrin and functions as a dominant-negative inhibitor of G protein–mediated MAPK signaling (34). It has been estimated that ~70% of cellular β-arrestin1 is phosphorylated at Ser-412 under basal unstimulated conditions, consistent with an inhibitory role of this phosphorylated residue during signaling.

Although insulin treatment is associated with a reduction in β-arrestin1 protein content, this reduction (50%) was in some cases less than would be expected to produce a total block of arrestin-dependent signaling events. We therefore asked if insulin-induced Ser-412 phosphorylation of β-arrestin1 could be a mechanism for further insulin-induced inhibition of β-arrestin1-mediated signaling. In 3T3-L1 adipocytes, we found that insulin treatment leads to an increase in Ser-412 phosphorylation, such that the Ser-412 phosphorylation stoichiometry approaches 1.0 after 8 h of insulin. Isoproterenol- and LPA-induced dephosphorylation of Ser-412, required for efficient MAPK signaling by the 7TMRs for these ligands, is also prevented by insulin treatment. Treatment of cells with the MEK inhibitor PD98059 partially prevented insulin-stimulated Ser-412 phosphorylation, suggesting that insulin-activated ERK is partially responsible for this effect. However, Ser-412 phosphorylation may also be regulated through the actions of a phosphatase, and insulin treatment is associated with inhibition of protein phosphatase 2A (PP2A) activity (64). We found that treatment of cells with a chemical inhibitor of protein phosphatase 2A (okadaic acid) mimicked the effect of insulin to increase Ser-412 phosphorylation. In addition, insulin-stimulated Ser-412 phosphorylation was lost in cells expressing the small T antigen, a specific viral-derived inhibitor of PP2A. These findings demonstrate that Ser-412 is maintained in its phosphorylated state through the dual actions of ERK on phosphorylation and PP2A on dephosphorylation (39). In summary, insulin treatment is associated with degradation of β-arrestin1 protein and with phosphorylation of the remaining β-arrestin at Ser-412, changes that lead to impaired β-arrestin1-dependent intracellular signaling events.

GRKS AND RTK SIGNALING

GRK Physiology

Following receptor activation and dissociation of the heterotrimeric G protein into its Gα and Gβγ subunits, a GRK is recruited to the receptor. Several mechanisms underlying membrane targeting of GRKs are known. Free Gβγ subunits, prenylated and thus membrane bound, bind GRK2 and GRK3 via their C-terminal pleckstrin homology domain, whereas GRK5 can associate directly with membrane phospholipids. These interactions target GRKs to the plasma membrane region in proximity to the activated 7TMR, allowing 7TMR serine phosphorylation (65).

GRKs and RTKs

GRKs, the EGFR, and the PDGFR. Like the β-arrestins, GRKs have been implicated as modulators of RTK signaling. In HEK293 cells expressing tagged GRK2, ligand activation of the EGFR led to translocation of GRK2 to the plasma membrane and increased p42/44 MAPK phosphorylation (66). A further study from this group demonstrated that, upon translocation to the plasma membrane, GRK2 colocalized with the EGFR in a $G\beta\gamma$ subunit–dependent manner (67).

Freedman et al. (68) showed that GRK2 can serine phosphorylate both the EGFR and the PDGFR in response to ligand. GRK2-mediated serine phosphorylation of the PDGFR led to reduced PDGFR tyrosine kinase activity. These authors then showed that PDGFR activation leads to tyrosine phosphorylation and activation of GRK2, providing a mechanism for GRK2-mediated feedback inhibition of PDGF signaling (69).

Taken together, these results suggest that $G\beta\gamma$-mediated recruitment of GRK to the plasma membrane is possible following activation of any receptor that couples to a heterotrimeric G protein. GRK may then phosphorylate and desensitize the activated receptor.

GRKs and the insulin receptor. In 3T3-L1 adipocytes, infection with an adenovirus expressing wild-type GRK2 decreased insulin-stimulated GLUT4 translocation and glucose uptake, showing that GRK has an inhibitory role in insulin signaling (70). In addition, adenoviral expression of a kinase-defective GRK2 mutant also decreased these parameters, demonstrating that the inhibitory effect of GRK2 is not dependent upon GRK2 kinase activity. In addition to a kinase domain, GRK2 has an RGS domain, which specifically inhibits the action of GTP-bound $G\alpha_{q/11}$. When an RGS domain–deficient GRK2 was expressed, insulin-stimulated GLUT4 translocation returned to normal. Consistent with this finding, GRK2 expression had no effect on IR tyrosine kinase activity or IRS-1-associated PI3-kinase activity. However, GRK2 expression reduced insulin-stimulated $G\alpha_{q/11}$ tyrosine phosphorylation, cdc42-p85 association, and cdc42-associated PI3-kinase activity, all key components of $G\alpha_q$-mediated signaling. These findings show that GRK2 specifically inhibits the $G\alpha_{q/11}$ pathway of insulin signaling, sequestering $G\alpha_{q/11}$ in a kinase-independent manner.

UNANSWERED QUESTIONS

The precise mechanisms underlying the interaction between the β-arrestins and receptor tyrosine kinases have not yet been determined. For example, which are the RTK domains responsible for this interaction, and does receptor tyrosine kinase activity play a role in this interaction? Does GRK-mediated phosphorylation of an RTK create a recognition site for β-arrestins, analogous to the 7TMRs? RTK-β-arrestin interactions and RTK–heterotrimeric G protein interactions may well be indirect. Ligand-induced formation of large signaling complexes may bring together distinct receptors (both 7TMRs and RTKs), arrestins, G proteins, and other signaling proteins. Two examples of this type of integrative multireceptor signaling have already been described; these involve 7TMR-RTK complexes that form between the β2AR and EGFR (26) and between the EDG1-R and PDGFR (71).

CONCLUSION

Receptor cross talk is of prime importance for a cell that must coordinate multiple extracellular hormonal signals. Each cell surface receptor family possesses unique structural characteristics and leads to specific signaling outcomes in the cell. Despite these specificities, these different receptor families utilize many common intracellular signaling proteins and activate many common signaling pathways. RTK-mediated activation of p42/44 MAPK and PI3-kinase, for example, are in many instances modulated by heterotrimeric G proteins, β-arrestins, and GRKs, molecules previously considered to be associated exclusively with 7TMRs. Because of these similarities, mechanisms of cross talk between RTKs and 7TMRs likely include changes in the levels and function of these common proteins.

Intracellular signaling in cells as diverse as adipocytes, hepatocytes, skeletal and cardiac myocytes, or pancreatic β cells, among others, is affected by cross talk between cell surface receptors. Knowledge of the mechanisms of receptor cross talk will therefore be critical for our understanding of complex intracellular signaling systems.

SUMMARY POINTS

1. Receptor tyrosine kinases utilize components of the seven-transmembrane-receptor signaling machinery, such as heterotrimeric G proteins and β-arrestins, to activate various signaling cascades, including p42/44 MAPK and PI3-kinase.
2. β-Arrestin may be an important site of receptor cross talk between receptor tyrosine kinases and seven-transmembrane receptors.

LITERATURE CITED

1. Schlessinger J. 2000. Cell signaling by receptor tyrosine kinases. *Cell* 103:211–25
2. Hubbard SR, Mohammadi M, Schlessinger J. 1998. Autoregulatory mechanisms in protein-tyrosine kinases. *J. Biol. Chem.* 273:11987–90
3. Lee J, Pilch PF. 1994. The insulin receptor: structure, function, and signaling. *Am. J. Physiol. Cell Physiol.* 266:C319–34
4. Hubbard SR. 1997. Crystal structure of the activated insulin receptor tyrosine kinase in complex with peptide substrate and ATP analog. *EMBO J.* 16:5572–81
5. Eck MJ, Dhe-Paganon S, Trub T, Nolte RT, Shoelson SE. 1996. Structure of the IRS-1 PTB domain bound to the juxtamembrane region of the insulin receptor. *Cell* 85:695–705
6. Gustafson T, He W, Craparo A, Schaub C, O'Neill T. 1995. Phosphotyrosine-dependent interaction of SHC and insulin receptor substrate 1 with the NPEY motif of the insulin receptor via a novel non- SH2 domain. *Mol. Cell Biol.* 15:2500–8
7. Bork P, Margolis B. 1995. A phosphotyrosine interaction domain. *Cell* 80:693–94
8. Virkamaki A, Ueki K, Kahn CR. 1999. Protein-protein interaction in insulin signaling and the molecular mechanisms of insulin resistance. *J. Clin. Invest.* 103:931–43
9. Chiang SH, Baumann CA, Kanzaki M, Thurmond DC, Watson RT, et al. 2001. Insulin-stimulated GLUT4 translocation requires the CAP-dependent activation of TC10. *Nature* 410:944–48
10. Imamura T, Vollenweider P, Egawa K, Clodi M, Ishibashi K, et al. 1999. G_α-q/11 protein plays a key role in insulin-induced glucose transport in 3T3-L1 adipocytes. *Mol. Cell Biol.* 19:6765–74
11. Luttrell LM, van Biesen T, Hawes BE, Koch WJ, Touhara K, Lefkowitz RJ. 1995. Gβγ subunits mediate mitogen-activated protein kinase activation by the tyrosine kinase insulin-like growth factor 1 receptor. *J. Biol. Chem.* 270:16495–98
12. Dalle S, Ricketts W, Imamura T, Vollenweider P, Olefsky JM. 2001. Insulin and insulin-like growth factor I receptors utilize different G protein signaling components. *J. Biol. Chem.* 276:15688–95
13. Hallak H, Seiler AEM, Green JS, Ross BN, Rubin R. 2000. Association of heterotrimeric G_i with the insulin-like growth factor-I receptor. Release of Gβγ subunits upon receptor activation. *J. Biol. Chem.* 275:2255–58

14. Kuemmerle JF, Murthy KS. 2001. Coupling of the insulin-like growth factor-I receptor tyrosine kinase to G_{i2} in human intestinal smooth muscle. $G\beta\gamma$ -dependent mitogen-activated protein kinase activation and growth. *J. Biol. Chem.* 276:7187–94
15. Kanzaki M, Watson RT, Artemyev NO, Pessin JE. 2000. The trimeric GTP-binding protein (G_q/G_{11}) alpha subunit is required for insulin-stimulated GLUT4 translocation in 3T3L1 adipocytes. *J. Biol. Chem.* 275:7167–75
16. Usui I, Imamura T, Huang J, Satoh H, Olefsky JM. 2003. Cdc42 is a Rho GTPase family member that can mediate insulin signaling to glucose transport in 3T3-L1 adipocytes. *J. Biol. Chem.* 278:13765–74
17. Kishi K, Muromoto N, Nakaya Y, Miyata I, Hagi A, et al. 1998. Bradykinin directly triggers GLUT4 translocation via an insulin-independent pathway. *Diabetes* 47:550–58
18. O'Brien RM, Houslay MD, Milligan G, Siddle K. 1987. The insulin receptor tyrosyl kinase phosphorylates holomeric forms of the guanine nucleotide regulatory proteins G_i and G_o. *FEBS Lett.* 212:281–88
19. Ciaraldi TP, Maisel A. 1989. Role of guanine nucleotide regulatory proteins in insulin stimulation of glucose transport in rat adipocytes. Influence of bacterial toxins. *Biochem. J.* 264:389–96
20. Goren HJ, Northup JK, Hollenberg MD. 1985. Action of insulin modulated by pertussis toxin in rat adipocytes. *Can. J. Physiol. Pharmacol.* 63:1017–22
21. Moreno F, Mills I, Garcia-Sainz J, Fain J. 1983. Effects of pertussis toxin treatment on the metabolism of rat adipocytes. *J. Biol. Chem.* 258:10938–43
22. Zhang BH, Ho V, Farrell GC. 2001. Specific involvement of $G_{\alpha i2}$ with epidermal growth factor receptor signaling in rat hepatocytes, and the inhibitory effect of chronic ethanol. *Biochem. Pharmacol.* 61:1021–27
23. Øyvind Melien DS, Ellen JJ, Thoralf C. 2000. Effects of pertussis toxin on extracellular signal-regulated kinase activation in hepatocytes by hormones and receptor-independent agents: evidence suggesting a stimulatory role of G proteins at a level distal to receptor coupling. *J. Cell Physiol.* 184:27–36
24. Øyvind Melien GHT, Dagny S, Eva Ø, Thoralf C. 1998. Activation of p42/p44 mitogen-activated protein kinase by angiotensin II, vasopressin, norepinephrine, and prostaglandin F in hepatocytes is sustained, and like the effect of epidermal growth factor, mediated through pertussis toxin-sensitive mechanisms. *J. Cell Physiol.* 175:348–58
25. Piiper A, Gebhardt R, Kronenberger B, Giannini CD, Elez R, Zeuzem S. 2000. Pertussis toxin inhibits cholecystokinin- and epidermal growth factor-induced mitogen-activated protein kinase activation by disinhibition of the cAMP signaling pathway and inhibition of c-Raf-1. *Mol. Pharmacol.* 58:608–13
26. Maudsley S, Pierce KL, Zamah AM, Miller WE, Ahn S, et al. 2000. The $\beta 2$-adrenergic receptor mediates extracellular signal-regulated kinase activation via assembly of a multi-receptor complex with the epidermal growth factor receptor. *J. Biol. Chem.* 275:9572–80
27. Waters C, Pyne S, Pyne NJ. 2004. The role of G-protein coupled receptors and associated proteins in receptor tyrosine kinase signal transduction. *Semin. Cell Dev. Biol.* 15:309–23
28. Cabrera-Vera TM, Vanhauwe J, Thomas TO, Medkova M, Preininger A, et al. 2003. Insights into G protein structure, function, and regulation. *Endocr. Rev.* 24:765–81
29. Gelb MH, Scholten JD, Sebolt-Leopold JS. 1998. Protein prenylation: from discovery to prospects for cancer treatment. *Curr. Opin. Chem. Biol.* 2:40–48
30. Pitcher JA, Inglese J, Higgins JB, Arriza JL, Casey PJ, et al. 1992. Role of $\beta\gamma$ subunits of G proteins in targeting the β-adrenergic receptor kinase to membrane-bound receptors. *Science* 28:1264–67

31. Koch W, Inglese J, Stone W, Lefkowitz R. 1993. The binding site for the βγ subunits of heterotrimeric G proteins on the β-adrenergic receptor kinase. *J. Biol. Chem.* 268:8256–60
32. Lohse MJ, Benovic JL, Codina J, Caron MG, Lefkowitz RJ. 1990. Beta-arrestin: a protein that regulates β-adrenergic receptor function. *Science* 248:1547–50
33. Violin JD, Ren XR, Lefkowitz RJ. 2006. GRK specificity for β-arrestin recruitment to the β2-adrenergic receptor revealed by fluorescence resonance energy transfer. *J. Biol. Chem.* 281:20577–88
34. Lin FT, Krueger KM, Kendall HE, Daaka Y, Fredericks ZL, et al. 1997. Clathrin-mediated endocytosis of the β-adrenergic receptor is regulated by phosphorylation/dephosphorylation of β-arrestin1. *J. Biol. Chem.* 272:31051–57
35. Luttrell LM, Hawes BE, van Biesen T, Luttrell DK, Lansing TJ, Lefkowitz RJ. 1996. Role of c-Src tyrosine kinase in G protein–coupled receptor and Gβγ subunit-mediated activation of mitogen-activated protein kinases. *J. Biol. Chem.* 271:19443–50
36. Ahn S, Maudsley S, Luttrell LM, Lefkowitz RJ, Daaka Y. 1999. Src-mediated tyrosine phosphorylation of dynamin is required for β2-adrenergic receptor internalization and mitogen-activated protein kinase signaling. *J. Biol. Chem.* 274:1185–8
37. Luttrell LM, Roudabush FL, Choy EW, Miller WE, Field ME, et al. 2001. Activation and targeting of extracellular signal-regulated kinases by β-arrestin scaffolds. *Proc. Natl. Acad. Sci. USA* 98:2449–54
38. Luttrell LM, Ferguson SSG, Daaka Y, Miller WE, Maudsley S, et al. 1999. β-arrestin-dependent formation of 2 adrenergic receptor-Src protein kinase complexes. *Science* 283:655–61
39. Hupfeld CJ, Resnik JL, Ugi S, Olefsky JM. 2005. Insulin-induced β-arrestin1 Ser-412 phosphorylation is a mechanism for desensitization of ERK activation by $G_{\alpha i}$-coupled receptors. *J. Biol. Chem.* 280:1016–23
40. Lin FT, Daaka Y, Lefkowitz RJ. 1998. β-arrestins regulate mitogenic signaling and clathrin-mediated endocytosis of the insulin-like growth factor I receptor. *J. Biol. Chem.* 273:31640–43
41. Chow JC, Condorelli G, Smith RJ. 1998. Insulin-like growth factor-I receptor internalization regulates signaling via the Shc/mitogen-activated protein kinase pathway, but not the insulin receptor substrate-1 pathway. *J. Biol. Chem.* 273:4672–80
42. Povsic TJ, Kohout TA, Lefkowitz RJ. 2003. β-arrestin1 mediates insulin-like growth factor 1 (IGF-1) activation of phosphatidylinositol 3-kinase (PI3K) and anti-apoptosis. *J. Biol. Chem.* 278:51334–39
43. Hershko A, Ciechanover A. 1998. The ubiquitin system. *Annu. Rev. Biochem.* 67:425–79
44. Weissman AM. 2001. Themes and variations on ubiquitylation. *Nat. Rev. Mol. Cell Biol.* 2:169–78
45. Girnita L, Girnita A, Larsson O. 2003. Mdm2-dependent ubiquitination and degradation of the insulin-like growth factor 1 receptor. *Proc. Natl. Acad. Sci. USA* 100:8247–52
46. Girnita L, Shenoy SK, Sehat B, Vasilcanu R, Girnita A, et al. 2005. β-arrestin is crucial for ubiquitination and down-regulation of the insulin-like growth factor-1 receptor by acting as adaptor for the MDM2 E3 ligase. *J. Biol. Chem.* 280:24412–19
47. Zhande R, Mitchell JJ, Wu J, Sun XJ. 2002. Molecular mechanism of insulin-induced degradation of insulin receptor substrate 1. *Mol. Cell Biol.* 22:1016–26
48. Rui L, Yuan M, Frantz D, Shoelson S, White MF. 2002. SOCS-1 and SOCS-3 block insulin signaling by ubiquitin-mediated degradation of IRS1 and IRS2. *J. Biol. Chem.* 277:42394–98

49. Usui I, Imamura T, Huang J, Satoh H, Shenoy SK, et al. 2004. β-arrestin-1 competitively inhibits insulin-induced ubiquitination and degradation of insulin receptor substrate 1. *Mol. Cell Biol.* 24:8929–37
50. Dalle S, Imamura T, Rose DW, Worrall DS, Ugi S, et al. 2002. Insulin induces heterologous desensitization of G protein–coupled receptor and insulin-like growth factor I signaling by downregulating β-arrestin-1. *Mol. Cell Biol.* 22:6272–85
51. Ceresa BP, Kao AW, Santeler SR, Pessin JE. 1998. Inhibition of clathrin-mediated endocytosis selectively attenuates specific insulin receptor signal transduction pathways. *Mol. Cell Biol.* 18:3862–70
52. Kao AW, Ceresa BP, Santeler SR, Pessin JE. 1998. Expression of a dominant interfering dynamin mutant in 3T3L1 adipocytes inhibits GLUT4 endocytosis without affecting insulin signaling. *J. Biol. Chem.* 273:25450–57
53. Kim J, Ahn S, Guo R, Daaka Y. 2003. Regulation of epidermal growth factor receptor internalization by G protein–coupled receptors. *Biochemistry* 42:2887–94
54. Hunton DL, Barnes WG, Kim J, Ren XR, Violin JD, et al. 2005. β-arrestin 2-dependent angiotensin II type 1A receptor-mediated pathway of chemotaxis. *Mol. Pharmacol.* 67:1229–36
55. Alderton F, Rakhit S, Kong KC, Palmer T, Sambi B, et al. 2001. Tethering of the platelet-derived growth factor beta receptor to G-protein-coupled receptors. A novel platform for integrative signaling by these receptor classes in mammalian cells. *J. Biol. Chem.* 276:28578–85
56. Hobson JP, Rosenfeldt HM, Barak LS, Olivera A, Poulton S, et al. 2001. Role of the sphingosine-1-phosphate receptor EDG-1 in PDGF-induced cell motility. *Science* 291:1800–3
57. Huang EJ, Reichardt LF. 2001. Neurotrophins: roles in neuronal development and function. *Annu. Rev. Neurosci.* 24:677–736
58. Rakhit S, Pyne S, Pyne NJ. 2001. Nerve growth factor stimulation of p42/p44 mitogen-activated protein kinase in PC12 cells: role of $G_{i/o}$, G protein–coupled receptor kinase 2, β-arrestin I, and endocytic processing. *Mol. Pharmacol.* 60:63–70
59. Xu CB, Zhang Y, Stenman E, Edvinsson L. 2002. D-Erythro-*N,N*-dimethylsphingosine inhibits bFGF-induced proliferation of cerebral, aortic and coronary smooth muscle cells. *Atherosclerosis* 164:237–43
60. Conner DA, Mathier MA, Mortensen RM, Christe M, Vatner SF, et al. 1997. β-arrestin1 knockout mice appear normal but demonstrate altered cardiac responses to β-adrenergic stimulation. *Circ. Res.* 81:1021–26
61. Kohout TA, Lin FT, Perry SJ, Conner DA, Lefkowitz RJ. 2001. β-Arrestin 1 and 2 differentially regulate heptahelical receptor signaling and trafficking. *Proc. Natl. Acad. Sci. USA* 98:1601–6
62. Ahn S, Nelson CD, Garrison TR, Miller WE, Lefkowitz RJ. 2003. Desensitization, internalization, and signaling functions of β-arrestins demonstrated by RNA interference. *Proc. Natl. Acad. Sci. USA* 100:1740–44
63. Hupfeld CJ, Dalle S, Olefsky JM. 2003. β-arrestin 1 down-regulation after insulin treatment is associated with supersensitization of β2 adrenergic receptor $G_{\alpha s}$ signaling in 3T3-L1 adipocytes. *Proc. Natl. Acad. Sci. USA* 100:161–66
64. Ugi S, Imamura T, Ricketts W, Olefsky JM. 2002. Protein phosphatase 2A forms a molecular complex with Shc and regulates Shc tyrosine phosphorylation and downstream mitogenic signaling. *Mol. Cell Biol.* 22:2375–87
65. Lefkowitz RJ. 1998. G protein–coupled receptors. III. New roles for receptor kinases and β-arrestins in receptor signaling and desensitization. *J. Biol. Chem.* 273:18677–80

66. Gao J, Li J, Ma L. 2005. Regulation of EGF-induced ERK/MAPK activation and EGFR internalization by G protein–coupled receptor kinase 2. *Acta Biochim. Biophys. Sin.* 37:525–31
67. Gao J, Li J, Chen Y, Ma L. 2005. Activation of tyrosine kinase of EGFR induces G βγ-dependent GRK-EGFR complex formation. *FEBS Lett.* 579:122–26
68. Freedman NJ, Kim LK, Murray JP, Exum ST, Brian L, et al. 2002. Phosphorylation of the platelet-derived growth factor receptor-β and epidermal growth factor receptor by G protein–coupled receptor kinase-2. Mechanisms for selectivity of desensitization. *J. Biol. Chem.* 277:48261–69
69. Wu JH, Goswami R, Kim LK, Miller WE, Peppel K, Freedman NJ. 2005. The platelet-derived growth factor receptor-β phosphorylates and activates G protein–coupled receptor kinase-2: a mechanism for feedback inhibition. *J. Biol. Chem.* 280:31027–35
70. Usui I, Imamura T, Satoh H, Huang J, Babendure JL, et al. 2004. GRK2 is an endogenous protein inhibitor of the insulin signaling pathway for glucose transport stimulation. *EMBO J.* 23:2821–29
71. Waters CM, Connell MC, Pyne S, Pyne NJ. 2005. c-Src is involved in regulating signal transmission from PDGF β receptor-GPCR(s) complexes in mammalian cells. *Cell Signal.* 17:263–77

Cumulative Indexes

Contributing Authors, Volumes 65–69

R

S

T

U

V

W

Y

Z

Chapter Titles, Volumes 65–69

Cell Physiology

Comparative Physiology

Endocrinology

Neurophysiology

Respiratory Physiology

Chloride Channels

Functional Imaging in Physiology

Lipid Receptor Processes

Membrane Protein Structure

Proton and Electron Transporters

Trp Channels